Second Supplements to the 2nd Edition of

RODD'S CHEMISTRY OF CARBON COMPOUNDS

ELSEVIER SCIENCE B.V.
Sara Burgerhartstraat 25
P.O. Box 211, 1000 AE Amsterdam, The Netherlands

ISBN: 0-444-82753-6

Second Supplements to the 2nd Edition of

RODD'S CHEMISTRY OF CARBON COMPOUNDS

VOLUME I

ALIPHATIC COMPOUNDS
★

VOLUME II

ALICYCLIC COMPOUNDS
★

VOLUME III

AROMATIC COMPOUNDS
★

VOLUME IV

HETEROCYCLIC COMPOUNDS
★

VOLUME V

MISCELLANEOUS

GENERAL INDEX
★

Second Supplements to the 2nd Edition of

RODD'S CHEMISTRY OF CARBON COMPOUNDS

A modern comprehensive treatise

Edited by
MALCOLM SAINSBURY
School of Chemistry, The University of Bath,
Claverton Down, Bath BA2 7AY, England

Second Supplement to

VOLUME IV HETEROCYCLIC COMPOUNDS

Part E: Six-membered Monoheterocyclic Compounds with
a Hetero-Atom from Groups IV, VI or VII of the Periodic Table

1997
ELSEVIER
Amsterdam – Lausanne – New York – Oxford – Shannon – Tokyo

Contributor to this Volume

R. LIVINGSTONE

Department of Chemical and Biological Sciences,
University of Huddersfield, Queensgate, Huddersfield HD1 3DH, England

Preface to Volume IV E

This entire volume has been written by Professor Livingstone from the University of Huddersfield. It updates the major contribution he made to the First Supplement, published in 1989. The amount of work involved bears dramatic testimony to the intensity of interest in the chemistry of six-membered heterocycles containing an oxygen atom or a sulfur atom.

In this volume the reader will find a comprehensive survey of all that is relevant to the progression of the chemistry of pyrans and thiopyrans and their oxidised and ring-fused analogues in the last decade. It is obvious that these compounds are immensely important and have value in many applications, not just as biologically active substances. These six-membered monoheterocyclic compounds are also intrinsically interesting and many of the more complex representatives present challenges for synthesis.

I would like to thank Professor Livingstone for his dedication to this enormous task. It is difficult to envisage how many hours of work have been invested in presenting the data in such a clear manner and how difficult it will be to match this endeavour in the future.

Malcolm Sainsbury   Bath
                                                              January 1997

Contents
Volume IV E

Heterocyclic Compounds; Six-membered Monoheterocyclic Compounds with a Hetero-Atom from Groups IV, VI, or VII of the Periodic Table

Chapter 20. Six-membered Ring Compounds with one Hetero-Atom: Oxygen
by R. LIVINGSTONE

Chapter 21. Six-membered Ring Compounds with One Hetero-Atom: Sulphur, Selenium, Tellurium, Silicon, Germanium, and Tin
by R. LIVINGSTONE

List of Common Abbreviations and Symbols Used

A	acid
Å	Ångström units
Ac	acetyl
a	axial
as, asymm.	asymmetrical
at.	atmosphere
B	base
Bu	butyl
b.p.	boiling point
c, C	concentration
CD	circular dichroism
conc.	concentrated
crit.	critical
D	Debye unit, 1×10^{-18} e.s.u.
D	dissociation energy
D	dextro-rotatory; dextro configuration
d	density
dec., decomp	with decomposition
deriv.	derivative
E	energy; extinction; electromeric effect
E	entgegen (opposite) configuration
$E1$, $E2$	uni- and bi-molecular elimination mechanisms
E1cB	unimolecular elimination in conjugate base
ESR	electron spin resonance
Et	ethyl
e	nuclear charge; equatorial
f.p.	freezing point
G	free energy
GLC	gas liquid chromatography
g	spectroscopic splitting factor, 2.0023
H	applied magnetic field; heat content
h	Planck's constant
Hz	hertz
I	spin quantum number; intensity; inductive effect
IR	infrared
J	coupling constant in NMR spectra
J	Joule
K	dissociation constant
k	Boltzmann constant; velocity constant
kcal	kilocalories
L	laevorotatory, laevo configuration
M	molecular weight; molar; mesomeric effect
Me	methyl

m	mass; mole; molecule; *meta-*
m.p.	melting point
Ms	mesyl (methanesulphonyl)
[M]	molecular rotation
N	Avogadro number; normal
NMR	nuclear magnetic resonance
NOE	Nuclear Overhauser Effect
n	normal; refractive index; principal quantum number
o	*ortho-*
ORD	optical rotatory dispersion
P	polarisation; probability; orbital state
Pr	propyl
Ph	phenyl
p	*para-*; orbital
PMR	proton magnetic resonance
R	clockwise configuration
S	counterclockwise configuration; entropy; net spin of incompleted electronic shells; orbital state
S_N1, S_N2	uni- and bi-molecular nucleophilic substitution mechanism
S_Ni	internal nucleophilic substitution mechanism
s	symmetrical; orbital
sec	secondary
soln.	solution
symm.	symmetrical
T	absolute temperature
Tosyl	p-toluenesulphonyl
Trityl	triphenylmethyl
t	time
temp.	temperature (in degrees centrigrade)
tert	tertiary
UV	ultraviolet
v	velocity
Z	zusammen (together) configuration
α	optical rotation (in water unless otherwise stated)
$[\alpha]$	specific optical rotation
ϵ	dielectric constant; extinction coefficient
μ	dipole moment; magnetic moment
μ_B	Bohr magneton
μg	microgram
μm	micrometer
λ	wavelength
ν	frequency; wave number
χ, χ_d, χ_μ	magnetic; diamagnetic and paramagnetic susceptibilities
$\sim$	about
(+)	dextrorotatory
(−)	laevorotatory
−	negative charge
+	positive charge

Chapter 20

SIX-MEMBERED RING COMPOUNDS WITH ONE HETERO ATOM: OXYGEN

R. LIVINGSTONE

Over recent years there has been a continuing growth in the chemistry of six-membered heterocycles containing one oxygen atom, especially in the areas relating to tetrahydropyran, tetrahydropyran-2-one, coumarin (2*H*--benzo[*b*]pyran-2-one), and chroman (2*H*-3,4-dihydrobenzo[*b*]pyran). Most of the development has been related to the synthesis of compounds possessing various pharmaceutical properties and others for use as herbicides, fungicides, and insecticides. The number of reports in the area of six-membered heterocycles, containing one oxygen atom and relating to natural products has been very small.

1. Pyran and its derivatives

(a) Pyrans and pyrylium salts

(i) 2H- and 4H-Pyrans (2- and 4-pyrans, α- and γ-pyrans). The acylation of 2-methylbut-1-ene-3-yne, $CH_2 = CMeC \equiv CH$, with RCOCl (R = Pr^i, Bu) in the presence of the catalyst $AlCl_3$ at -78°C yields $RCOCH = C = CMeCH_2Cl$, Z-$RCOCH = CClCMe = CH_2$, and 4-chloro-6-isopropyl(butyl)-3-methyl-2*H*--pyran (1) in ratios of 71:22:7 and 74:18:8, respectively. Similarly, $CH_2 = CMeC \equiv CH$ and RC_6H_4COCl (R = H, 4-Me, 4-MeO, 2-Cl) at -50°C affords Z-$RC_6H_4COCH = CClCMe = CH_2$ (23-32%) and 6-aryl-4-chloro-3--methyl-2*H*-pyran (2) (24-29%), which on photochemical oxidation gives 5--aryl-3-chloro-2-methylfuran (3), resulting from the oxidative extrusion of a CH_2 group (G.G. Melikyan *et al.*, Zh. Org. Khim., 1990, 26, 965).

(1)   (2)   (3)

Treatment of $(EtO)_2CHCHMeCH(OEt)_2$ with $CH_2(CO_2Me)_2$ in Ac_2O containing $ZnCl_2$ yields $EtOCH=CMeCH=C(CO_2Me)(CO_2Me)$ and $EtOCH=C(CO_2Me)_2$. The existence of an equilibrium between the former compound and methyl 2-ethoxy-6-methoxy-2H-pyran-5-carboxylate (4) has been confirmed by NMR spectral data (Zh. A. Krasnaya, T.S. Stytsenko, and V.S. Bogdanov, Izv. Akad. Nauk SSSR, Ser. Khim., 1990, 2561).

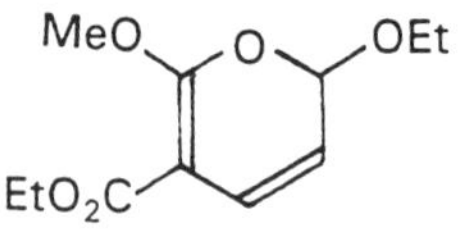

(4) 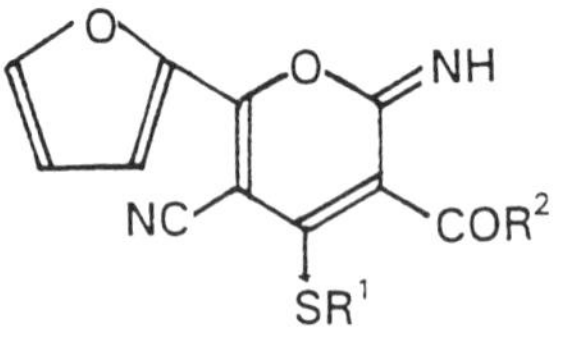　　　　　　　　　　(5)

4-Alkylthio-5-cyano-6-(2-furyl)-2-imino-2H-pyrans 5 (R^1 = alkyl; R^2 = 2-furyl, furfurylamino) are prepared by boiling together for 2 minutes furoyl-2-acetonitrile and $(R^1S)_2C=C(COR^2)CN$ (R^1 = alkyl; R^2 = furyl, furfurylamino) in DMF containing K_2CO_3 [K. Peseke, J. Quincoces Suarez, and M. Almeida Saavedra, Ger. (East) DD 294,942, 1991].

Ethyl diethoxyacrylate reacts with alkynylalkoxycarbene metal complexes, $R^1C\equiv CC(OR^2)=M(CO)_5$ (R^1 = Ph, Bu^t; R^2 = Me, Et; M = Cr, W) to give 6-ethoxy-2H-pyranylidene metal complexes 6 (R = Ph, Bu^t), which on mild oxidation with Me_2SO affords the free ligands 6-ethoxy-4-phenyl- (7; R = Ph) and 6-ethoxy-4-*tert*-butyl-2H-pyran (7; R = Bu^t), respectively (F. Camps *et al.*, Chem. Comm., 1989, 1560).

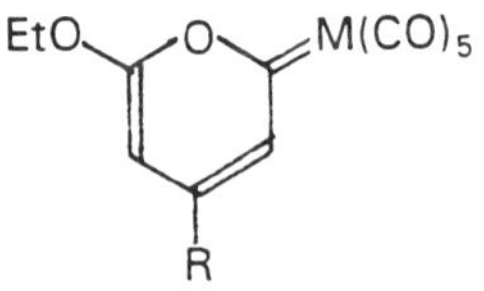

(6) 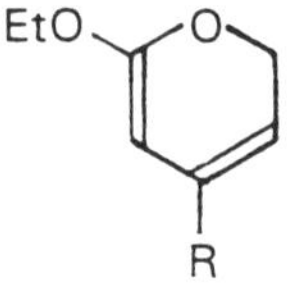　　　　　　　　(7) 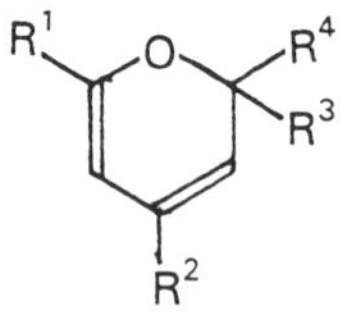　　　　　　　　(8)

2H-Pyrans 8 (R^1-R^4 = Me; R^1 = R^2 = Ph, R^3 = R^4 = Et; R^1 = R^2 = Ph, R^3 = H, R^4 = Me) with two substituents at the 2-position and one at the 4- and 6--positions, which are in equilibrium with R^{1}COCH = CR^{2}CH = CR^{3}R^4 undergo 1,2-addition with organometallic reagents, for example, R$_3^5$ Al, R$_3^5$ Li (R^5 = Me, Et, Bu) and R^{5}MgBr (R^{5}Et, Ph) to give the dienic alcohols, R^{1}R^{5}C(OH)CH = = CR^{2}CH = CR^{3}R^4 as the major products.

It has been reported that the use of cyclopropenes opens a new route to 4H-pyrans (F. Thalhammer, U. Wallfahrer, and J. Sauer, Tetrahedron Letters, 1988, 29, 3231).

The bis(benzoyl)diphenylmethanes 9 (R = Me, MeO, Br) when boiled with P$_2$O$_5$ in xylene under nitrogen are dehydrated to give the 2,6-diaryl-4,4--diphenyl-4H-pyrans 10. These compounds are colourless, but turn blue upon irradiation with sunlight (A. Kurfurst et al., Czech. CS 261,434, 1989).

(9)

(10)

The reaction between 2,6-diphenyl-3-methylpyrylium perchlorate (11) and PhBr and CH$_2$ = CHCH$_2$Cl yields 3-methyl-2,4,6-triphenyl-4H-pyran (12; R = Ph) and 4-allyl-3-methyl-2,6-diphenyl-4H-pyran (12; R = allyl), respectively, which on formylation with DMF-POCl$_3$ afford the corresponding 3-formyl-4H-pyrans 13 (R = Ph, allyl). Other related preparations and properties have been reported (K.F. Suzdalev and A.V. Koblik, Khim. Geterotskl. Soedin., 1990, 603).

ClO_4^-

(11)   (12)   (13)

The reaction of 2,4,4,6-tetraphenyl-4H-pyran (14) with bromine gives 3,5-dibromo-2,4,4,6-tetraphenyl-4H-pyran (15) and with PCl_5 it affords 3,5--dichloro-2,4,4,6-tetraphenyl-4H-pyran (16) (M. Schwarz, P. Sebek, and J. Kuthan, Coll. Czech. Chem. Comm., 1992, 57, 546).

The cyclocondensation of $R^1COCHR^2CHR^3CHR^4COR^5$ with Ac_2O in ether catalyzed by $BF_3 \cdot Et_2O$ yields the penta-substituted 4H-pyrans 17 [$R^1R^2 = (CH_2)_3$, $(CH_2O_4$, $R^1 = Ph$, $R^2 = H$, Me, $R^3 = H$, Me, Pr, Ph, substituted Ph, $R^4R^5 = (CH_2)_3$, $(CH_2)_4$, $R^4 = H$, $R^5 = Ph$, 2-MeC_6H_4] (A.F. Blinokhvatov, O.V. Markovtseva, and M.N. Nikolaeva, Khim. Geterotsikl. Soedin., 1992, 320).

(14)   (15)   R = Br   (17)

(16)   R = Cl

2,5-Di(2,6-diphenyl-4*H*-pyran-4-ylidenemethyl substituted furan, thiophene, and *N*-methylpyrrole 18 (X = O, S, NMe) have been obtained by the Wittig-Horner reaction of the appropriate 2,5-diformyl derivative of the five-membered heterocycle with diethyl pyranylphosphonate. They have been investigated for use as precusors of organic metals and third-order nonlinear optical materials (K. Takimiya *et al.*, Chem. Letters, 1994, 255).

(18)

4-Dicyanomethylene-2-methyl-6-(4-dimethylaminostyryl)-4*H*-pyran (19) has been obtained by heating 4-dicyanomethylene-2,6-dimethyl-4*H*-pyran with 4-Me$_2$NC$_6$H$_4$CHO in ethylene glycol at 90-100°C in the presence of CsF as catalyst (A.P. Lugovskii *et al.*, U.S.S.R. SU 1,456,428, 1989). It has also been prepared by the condensation of 2-methyl-6-(4-dimethylaminostyryl)--4*H*-pyran-4-one with malononitrile in boiling acetic anhydride (T.N. Gerasimova and A.V. Konstantinova, Sib. Khim. Zh., 1993, 41).

(19)

The base-catalyzed Michael addition of $R^1COCH_2COR^2$ ($R^1 = R^2 = Me$, OEt; $R^1 = OEt$, $R^2 = Me$) to $R^3CH = C(CHO)_2$ ($R^3 = 4\text{-}MeOC_6H_4$, $4\text{-}ClC_6H_4$, 2--thienyl) yields $R^1COCH(COR^2)CHR^3CH(CHO)_2$, which on dehydration in acidic medium affords the 2,4-disubstituted 3-acyl-4*H*-pyran-5-carbox-aldehyde 20. ·In the addition of R^4CH_2CN ($R^4 = CN$, CO_2Me) to $R^3CH = C(CHO)_2$ the products are immediately cyclized to afford the 3,4--disubstituted 2-amino-4H-pyran-5-carboxaldehyde 21 (D. Dvorak *et al.*, Coll. Czech. Chem. Comm., 1987, <u>52</u>, 2687). 4-Aryl-2,6-diphenyl-4*H*-pyran-3,5--dicarboxaldehydes 22 (R = Ph, $4\text{-}R^1C_6H_4$; $R^1 = H$, Me, OMe) have been synthesized (M. Venugopal *et al.*, Tetrahedron Letters, 1991, <u>32</u>, 3235).

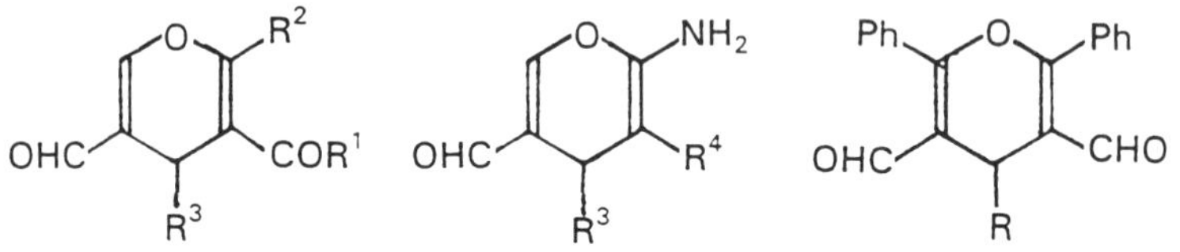

(20)   (21)   (22)

The reaction between $PhCH = C(CN)_2$ and $MeCOCH_2COMe$ in EtOH containing Et_3N yields 5-acetyl-2-amino-3-cyano-6-methyl-4-phenyl-4*H*-pyran (23) along with a small amount of a pyrano[2,3-*b*]pyridine derivative (N. Martin *et al.*, Heterocycles, 1987, <u>26</u>, 2811).

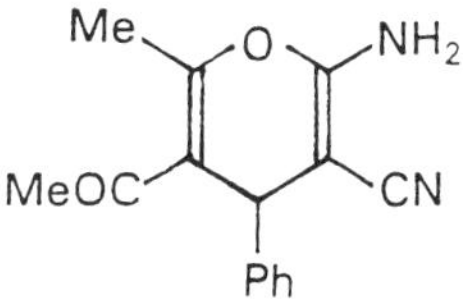

(23) 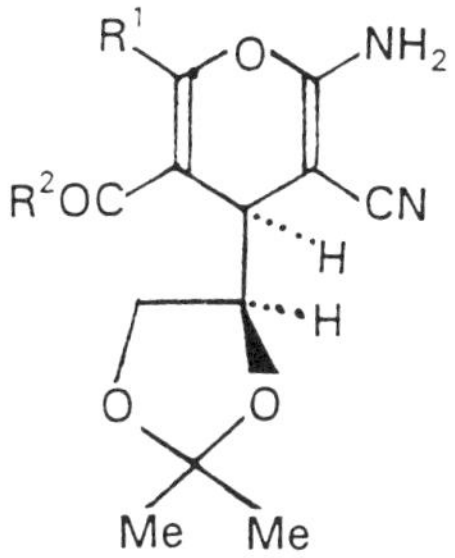   (24)

The first asymmetric synthesis of 4H-pyrans 24 (R^1 = Me, R^2 = OEt, OMe;
R^1 = Ph, R^2 = OEt) starting from (R)-2,3-O-isopropylideneglyceraldehyde has
been developed. The absolute configuration as R at the new stereocentre in
compound 24 has been established (R. Gonzalez *et al.*, Tetrahedron Letters,
1992, *33*, 3809). The asymmetric synthesis of some 3-alkoxycarbonyl-2-
-amino-5-cyano-4,6-diphenyl-4H-pyrans has been reported (N. Martin *et al.*,
Tetrahedron Asymmetry, 1994, *5*, 1435).

5-Acetyl-2-amino-3-cyano-4-(2-furyl)-6-methyl-4H-pyran (25) (S.
Marchalin *et al.*, Coll. Czech. Chem. Comm., 1990, *55*, 718) reacts with
aldehydes, RCHO [R = (un)substituted Ph, 2-furyl] to furnish the substituted
2-methyleneamino derivative 26 (D. Ilavsky *et al.*, *ibid.*, 1994, *59*, 1458).
For the synthesis of the substituted 2-amino-4H-pyrans 27 (R^1 = Me, Et;
R^2 = Ph, substituted Ph; R^3 = Ph, substituted Ph), see N. Martin-Leon *et al.*,
J. Chem. Res., S, 1990, 310; and 3-[2-amino-3,5-di(ethoxycarbonyl)-6-
-methyl-4H-pyran-4-yl]-1-methylpyridinium iodide (28), A.M. Shestopalov *et
al.*, Khim. Geterotsikl. Soedin., 1991, 205.

(25)   R = H
(26)   R₂ = CHR

(27)

(28)

2,2',6,6'-Tetraphenyl-4,4'-bipyranylidene (29) has been prepared for use
in forming charge-transfer complexes by coupling 2,6-diphenylpyrylium
perchlorate using Zn in MeCN (E.M. Gluzman, L.V. Gavriko, and V.A.
Starodub, Elektron. Org. Mater., 1985, 40). The related 2,2',6,6'-tetraaryl-
-4,4'-bipyranylidenes 30 [R = Me, Bu, H(CH₂)₁₀] with 8 flexible chains have
been obtained by electrochemical reduction of the corresponding 2,6-diaryl-
pyrylium salts (C. Amatore *et al.*, Tetrahedron Letters, 1989, 30, 1383).
Several 2,2'-diaryl-3,3'-dihalogeno-6,6'-diphenyl-4,4-bipyranylidenes 31
(Ar = Ph, 4-MeC₆H₄, 4-ClC₆H₄, R = Br, iodo) have been synthesized (M.G.
Marei, Sulphur Letters, 1993, 16, 131).

(29)   Ar = Ph
(30)   Ar = 3,4(RO)₂C₆H₃

(31)

Cyclopenta[*c*]pyran (32; R = H) and 6-*tert*-butylcyclopenta[*c*]pyran (32; R = Bu^t) have been synthesized in a number of steps from BrCH$_2$CH(OMe)$_2$ and 5-(trimethylsilyl)cyclopenta-1,3-diene, and 1-*tert*-butylcyclopenta-1,3--diene, respectively (T. Kaempchen *et al.*, Ann., 1988, 855). Cyclization of diarylidenecyclopentanones and CH$_2$(CN)$_2$ gives the cyclopenta[b]pyran derivatives 33 (R = Ph, 4-tolyl, 4-ClC$_6$H$_4$) (M.A. Sofan *et al.*, *ibid.*, 1989, 935).

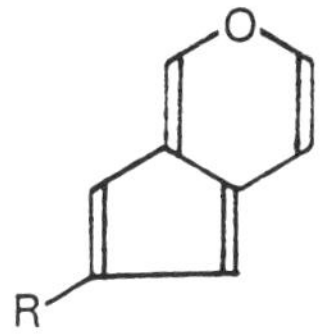

(32) 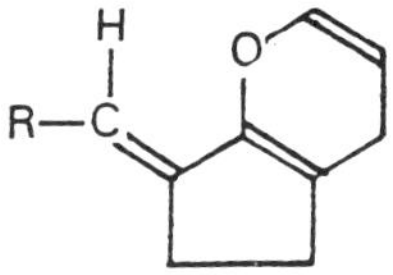   (33) 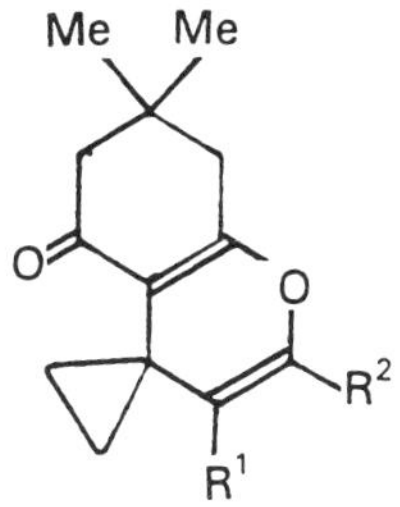   (34)

The unstable and highly reactive intermediate cyclopropylidenedimedone, is trapped by ynamines R^{1}C≡CR^2 (R^1 = Me, R^2 = NEt$_2$; R^1 = Ph, R^2 = NEt$_2$; R^1 = Ph, R^2 = NMePh; R^1 = H, R^2 = OEt; R^1 = H, R^2 = Ph; R^1 = H, R^2 = NMePh) in a Diels-Alder reaction to afford the cycloadducts 34 (E. Vilsmaier and R. Baumheier, Ber., 1989, 122, 1285).

Cyclization of indane-1,3-dione with 4-RC$_6$H$_4$CH = C(CN)R^1 (R = H, NO$_2$; R^1 = CO$_2$Et, CN) yields the indeno[1,2-*b*]pyrans 35 and 36, respectively, of potential biological activity (M. Mashaly and M. Mashaly, Zhonghua Yaoxue Zazhi, 1993, 45, 175).

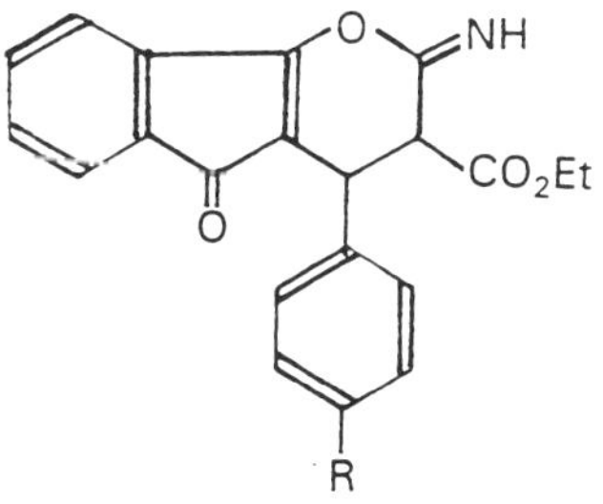

(35) 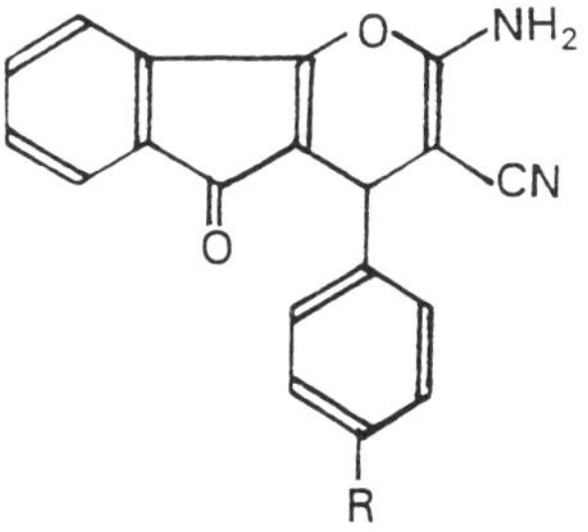   (36)

(ii) Pyrylium salts. 2,6-Dimethyl-4-phenylpyrylium sulphoacetate (1;
$X = HO_2CCH_2SO_3$) had been prepared by treating acetic anhydride with
concentrated sulphuric acid, followed by the addition of $PhCMe = CH_2$ (T.S.
Balaban, A. Dinculescu, A.T. Balaban, Org. Prep. Proced. Int., 1988, 289).
A number of methods have been described for obtaining 2,6-dimethyl-4-
-phenylpyrylium salts 1 [for example, $X = H_2C = C(CH_2SO_3H)CH_2SO_3$,
CF_3CO_2, CCl_3CO_2, Cl_2CHCO_2] with various anions, which influence strongly
their solubility, reactivity and ease of purification (A. Dinculescu *et al.*, Bull.
Soc. Chim. Belg., 1991, <u>100</u>, 665). For the preparation of 4-alkyl-2,6-di-
methylpyrylium salt 2 ($R^1 = H$, $R^2 = Me$, Et, Pr^i), 4-ethyl-2,3,6-trimethyl-
pyrylium salt 2 ($R^1 = R^2 = Me$), 3-alkyl-2,4,6-trimethylpyrylium salt 3 ($R^1 = H$,
$R^2 = Me$, Et, Pr^i), and 2,3,4,5,6-pentamethylpyrylium salt 3 ($R^1 = R^2 = Me$),
see C. Roussel, H.G. Rajoharison, and J. Metzger, Rev. Roum. Chim., 1991,
<u>36</u>, 613.

$$X^- \qquad\qquad CF_3SO_3^- \qquad\qquad CF_3SO_3^-$$

(1)	(2)	(3)

The cyclocondensation of pivaloylacetone and benzoylacetone with
$(EtO)_3CH$ in the presence of perchloric acid affords 4-ethoxy-2-*tert*-
-butylpyrylium perchlorate (4; $R = Bu^t$) and 4-ethoxy-2-phenylpyrylium
perchlorate (4; $R = Ph$), respectively (E.P. Olekhnovich, I.V. Korobka, and
V.V. Mezheritskii, Zh. Org. Khim., 1990, <u>26</u>, 173).

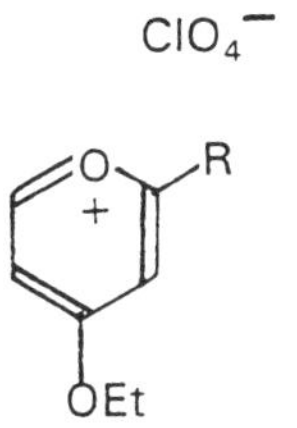

(4)

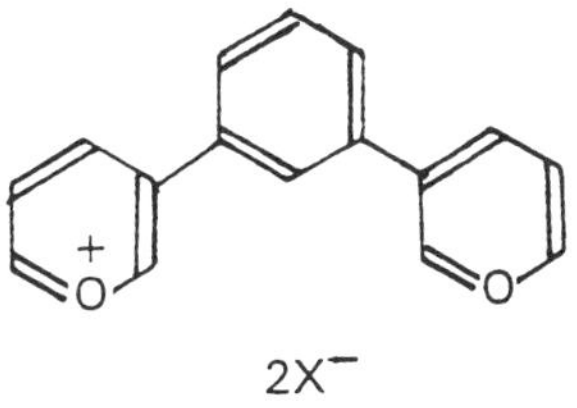

(5)

The acylation of 1,3-[MeC$=$CH$_2$]$_2$C$_6$H$_4$ with Ac$_2$O-H$_2$SO$_4$ yields the bispyrylium salt 5 (X$=$HO$_2$CCH$_2$SO$_3^-$), which has been converted to the related salts 5 (X$=$ClO$_4$, BF$_4$, I) (A.T. Balaban and T.S. Balaban, Rev. Roum. Chim., 1989, <u>34</u>, 41).

The reaction between PhCOMe and PhCOCH$_2$COR (R$=$Me, Et, Pr) in the presence of 70% aqueous HClO$_4$ yields a mixture of 2-alkyl-4,6-diphenyl- and 4-alkyl-2,6-diphenylpyrylium perchlorates 6 and 7. A quantitative analysis of the mixtures has been achieved by the use of UV spectroscopic methods (S.M. Luk'yanov *et al.*, Zh. Org. Khim., 1989, <u>25</u>, 1008). Investigations into the formation of mixtures of symmetrical and unsymmetrical pyrylium salts by the condensations of carbonyl compounds in the presence of acids, for example HClO$_4$ have been reported (*idem, ibid.*, 1991, <u>27</u>, 1588).

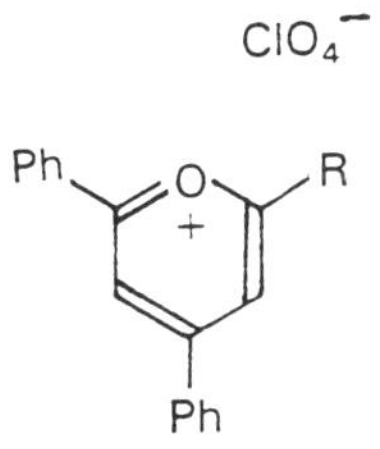

(6)

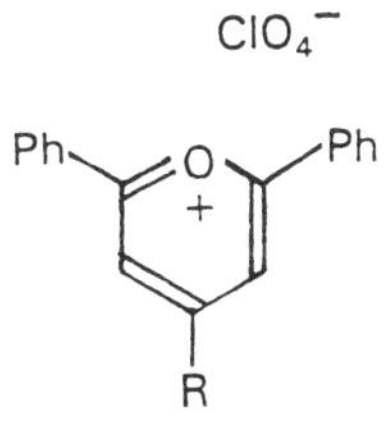

(7)

12

The cyclocondensation of the acetal, $PhC \equiv CCH(OEt)_2$, with ketone, $R^1COCH_2R^2$ ($R^1 = Me$, Ph, Bu^tCH_2CO, $R^2 = H$, Ac, Me, Et, COPh, Bu^tCH_2CO) in the presence of $HClO_4$ affords pyrylium perchlorates 8 (E.P. Olekhnovich *et al.*, *ibid.*, 1990, 26, 2609). The reactions of $Ph_3C^+X^-$ ($X = ClO_4$, F_3CSO_3) with $PhCOCR^1 = CR^2CH = CHAr$ ($Ar = Ph$, $4-MeOC_6H_4$; $R^1 = H$, Me; $R^2 = H$, Ph) to afford pyrylium perchlorates or triflates, along with the mechanism for their formation have been investigated (N. Barry-Kone *et al.*, Bull. Soc. Chim. Fr., 1993, 130, 218). 4-Methyl or 4-phenyl substituted 3,5-dimethyl--2,6-diphenylpyrylium salts are only formed from the appropriate penta-substituted 4H-pyran in the presence of oxygen, unlike the tri- and tetra--substituted 4H-pyrans, which give the respective pyrylium salt in the presence of acids (V.G. Kharchenko, N.I. Kozhevnikova and L.K. Kulikova, Khim.-Farm. Zh., 1990, 24, 36).

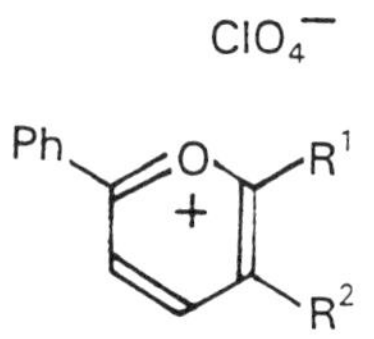

$$ClO_4^-$$

$$X^-$$

(8)

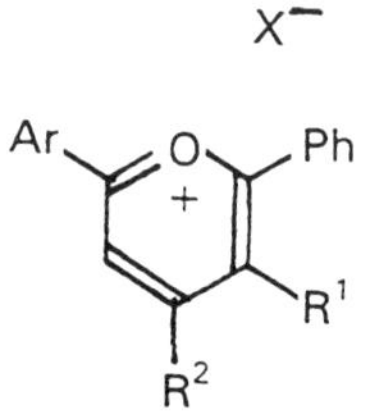

(9)

The 2,4,6-triarylpyrylium tetrafluoroborates 10 ($R^1 = H$, halogeno, alkoxy, Ph, NO_2, Me; $R^2 = H$, MeO, NO_2) have been obtained by the cyclocondensation of $4-R^1C_6H_4CHO$ with $4-R^2C_6H_4COMe$ in toluene containing 54% HBF_4 (T.N. Boiko, S.V. Andreeva, and I.I. Boiko, U.S.S.R. Su 1,505,940, 1989) and triarylpyrylium tetrafluoroborates 10 ($R^1 = R^2 = H$, Me, MeO, Cl, Ph) have been prepared by the reaction between acetophenones, ortho esters, and boron trifluoride etherate (I.I. Boiko, M.E. Dergunova, and T.N. Boiko, Khim., Geterotsikl. Soedin., 1987, 1467). 4-Aryl-2,6-di(4-methoxyphenyl)pyrylium salts of weak organic acids, for example, salts 11 ($R = H$, MeO; $X = HCO_2$, $PhCO_2$, $HOC_6H_4CO_2$) have been prepared by treating the corresponding strong acid salts 11 ($R = H$, MeO; $X = BF_4$, ClO_4, CF_3CO_2) with a methanolic solution of NaOMe in a water-immiscible organic solvent (Et_2O, C_6H_6, CCl_4), with subsequent treatment of the resulting methoxypyran derivative 12 with the required weak acid in the same organic solvent (B.I. Drevko, V.G. Kharchenko, and L.M. Yudovich, *ibid.*, 1,447,824, 1988). Two polymer--bound aryldiphenylpyrylium tetrafluoroborates 13 and 14 ($R = polymer$) have been prepared from Merrifield resin and polystyrene, respectively. They are sensitizers for photochemical induced electron-transfer reactions (J. Mattay, M. Vondenhof, and R. Denig, Ber., 1989, 122, 951).

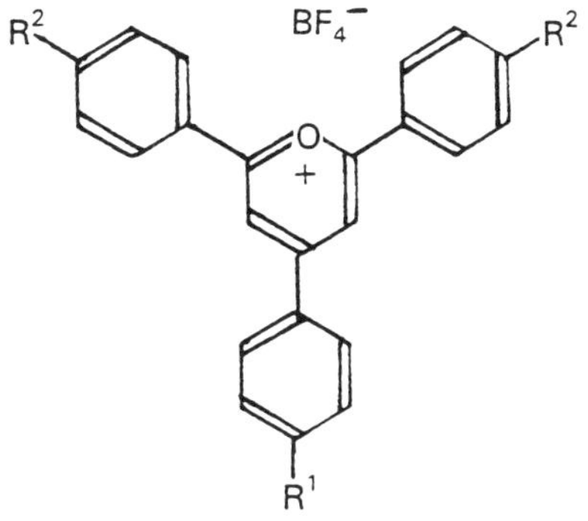

(10)

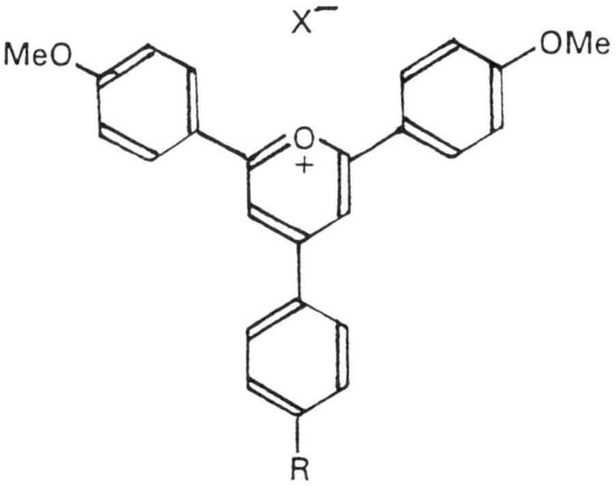

(11)

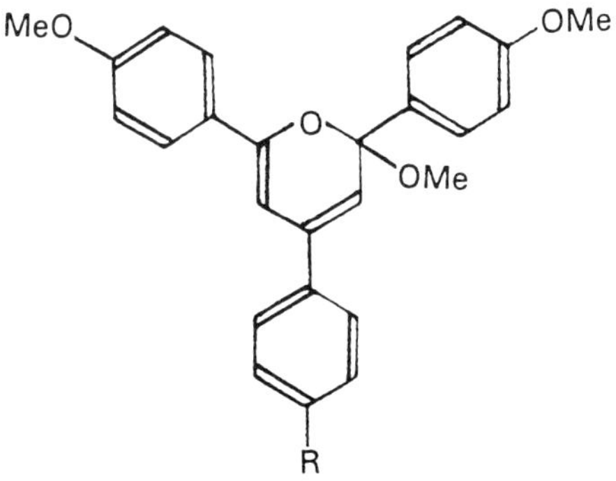

(12)

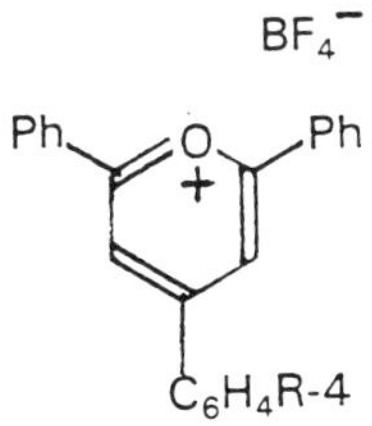

(13)

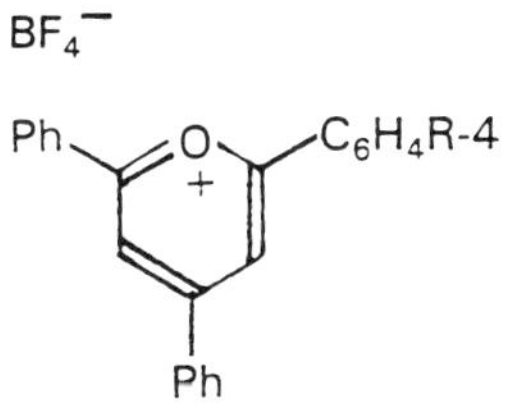

(14)

14

The ring closure of 1,5-di(2-furyl)-3-phenylpenta-1,5-dione (15) with triphenylcarbenium tetrafluoroborate leads to a mixture of 2,6-di(2-furyl)-4-phenylpyrylium tetrafluoroborate (16; R = H) and 2-(2-furyl)-4-phenyl-6-(5-triphenylmethyl-2-furyl)pyrylium tetrafluoroborate (16; R = CPh$_3$) (C. Reichardt and P. Milart, Ann., 1990, 401).

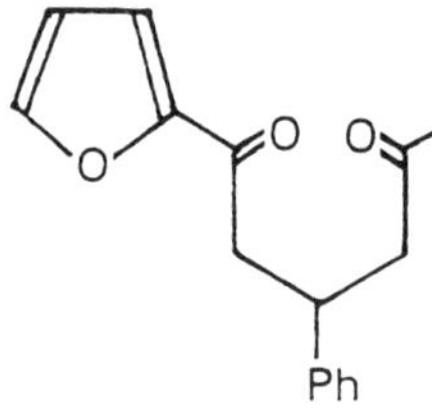

(15) 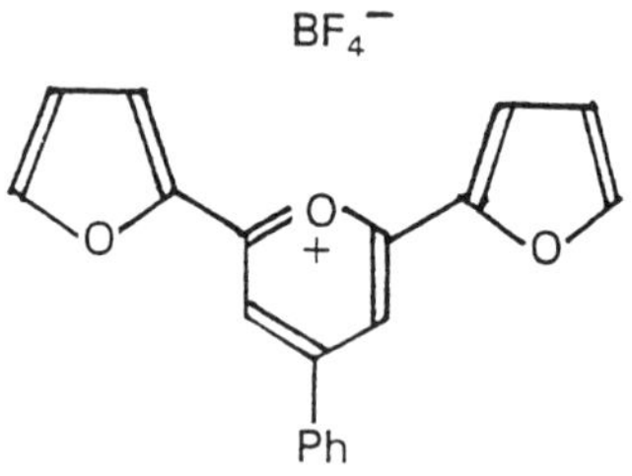   (16)

For the preparation of substituted 2-(4-chlorobutadienyl)pyrylium perchlorates 17 (R^1 = R^2 = aryl; R^3 = R^4 = H; R^{3}R^4 = (CH$_2$)$_n$; R^5,R^6 = alkyl, aryl, or R^5 = H, R^6 = aryl) as dye intermediates and laser dyes, see F. Kropfgans, M. Pulst, and M. Weissenfels, Ger. (East) DD 253,427, 1988; and 18 [R^1 = R^2 = Ph, R^3 = R^4 = H, R^{3}R^4 = (CH$_2$)$_n$, n = 2,3; R^1 = MeOC$_6$H$_4$, R^2 = Ph, R^3 = R^4 = H; R^5 = Ph, 4-MeC$_6$H$_4$, 4-MeOC$_6$H$_4$, 4-Me$_2$NC$_6$H$_4$], *idem, ibid.*, Z. Chem., 1988, <u>28</u>, 286.

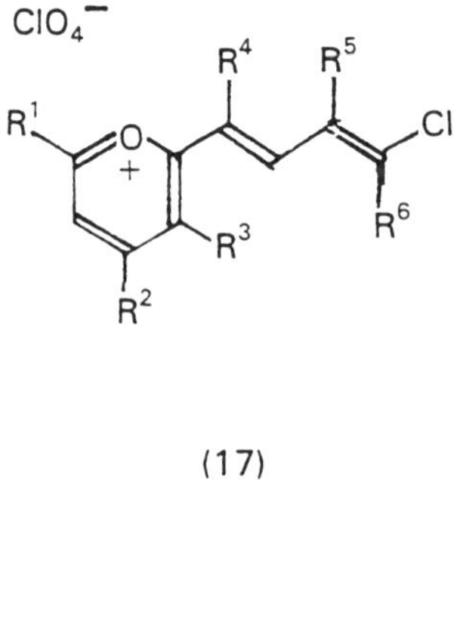

(17) 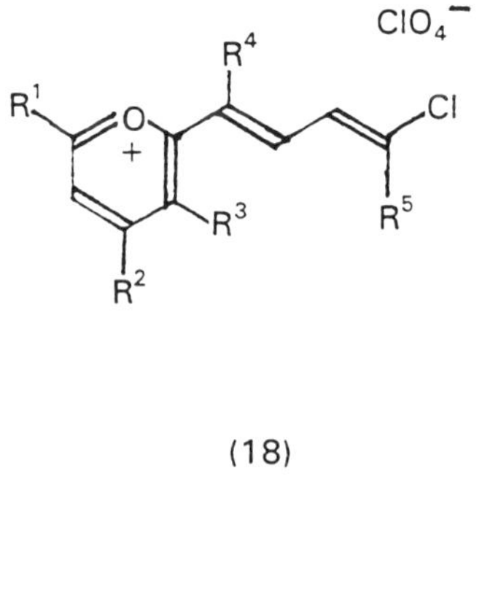   (18)

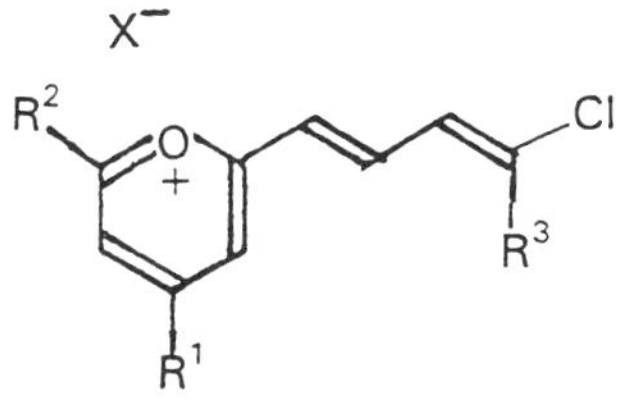

(19)

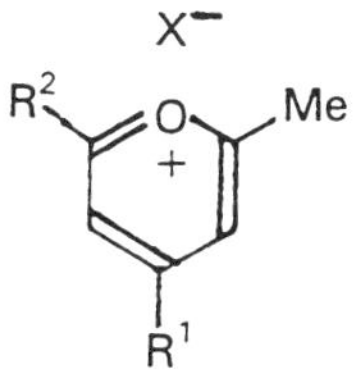

(20)

2-Mono- and 2,6-di(4-chlorobutadienyl)pyrylium salts 19 [R^1,R^2 = (un)-substituted aryl; R^1 = (un)substituted aryl, R^2 = R^3CCl = CHCH = CH; R^3 = (un)substituted aryl; PhCH = CH; X^- = strong acid anion] have been prepared by the condensation of methylpyrylium salts 20 [R^1 = (un)-substituted aryl; R^2 = methyl, (un)substituted aryl] with R^3CCl = CHCH = $N^+R_2^4$ (R^4 = Me, Et; NR_2^4 = morpholino, piperidino) in AcOH containing piperidinium acetate [H. Hartmann, R. Dusi, and A. Maahs, Ger.(East) DD 268,695, 1989].

Alkylpyrylium salts 21 [R^1 = R^2 = H, R^3 = R^4 = Ph; R^1R^2 = $(CH_2)_2$, $(CH_2)_3$, R^3 = R^4 = Ph; R^1 = R^2 = H, R^3 = 4-MeOC$_6$H$_4$, R^4 = Ph; R^1 = R^2 = R^3 = H, R^4 = 4--MeOC$_6$H$_4$] have been condensed with 2-(4-nitrophenyl)-3-(dimethylamino)-acrolein derivatives to afford hemicyanines 22 (R^5 = H, NO_2), which on hydrolysis with Na_2CO_3 in MeCN yielded the (formylallylidene)pyrans 23 (T. Werner *et al.*, Z. Chem., 1990, _30_, 286).

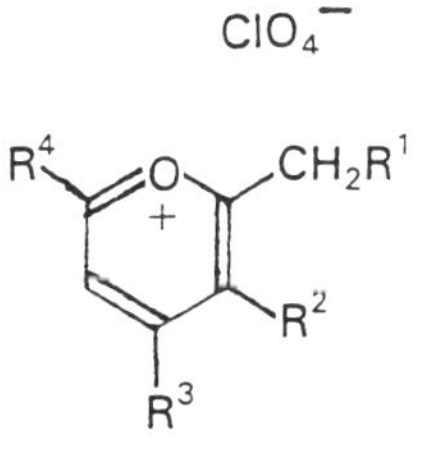

(21)

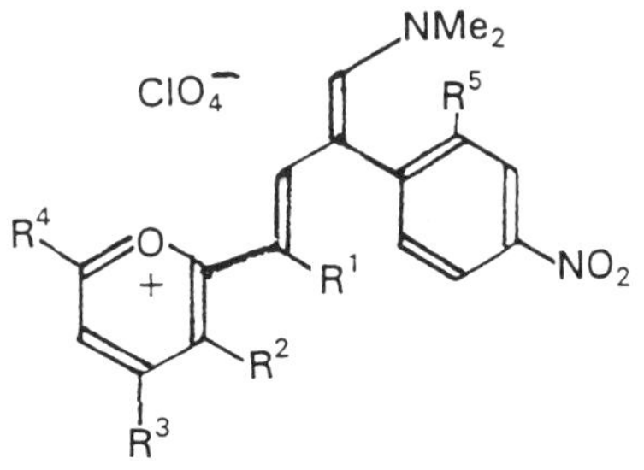

(22)

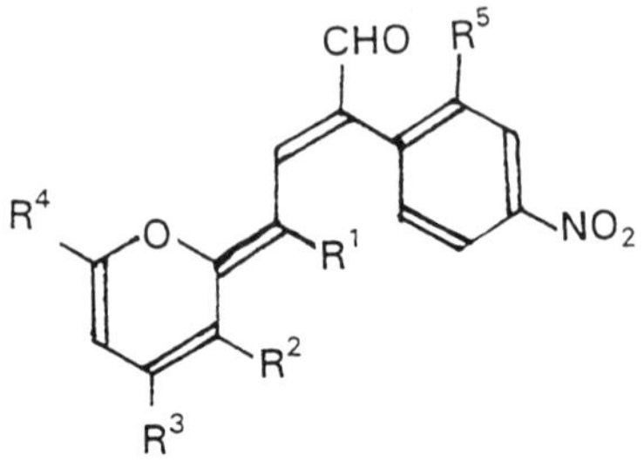

(23)

The use of 3,5-di-*tert*-butyl-1,2-benzoquinone in the synthesis of pyrylium salts has been investigated [V.T. Abaev and O. Yu. Okhlobystin, Karbonil. Soed. v Sinteze Geterotsiklov:Mezhvuz. Nauch. Sb., Saratov, 1989, (Ch. 1), 57. From Zh., Khim., 1990, Abstr. No 3Zh246].

Reaction of 2,4,6-triphenyl-4*H*-pyran with Vilsmeier's reagent, followed by hydrolysis of the product affords 3-formyl-2,4,6-triphenyl-4*H*-pyran (24) (78%) and $PhCOCH = CPhCH_2COPh$ (10%). The former compound 24 on treatment with Ph_3CClO_4 yields 3-formyl-2,4,6-triphenylpyrylium perchlorate (25) (86%) (A.V. Koblik and K.F. Suzdalev, Zh. Org. Khim., 1989, 25, 1342).

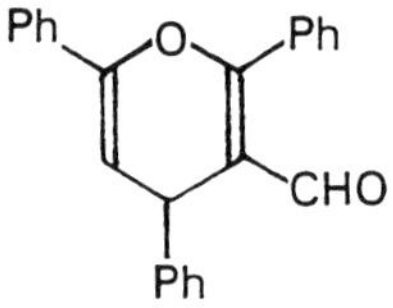

(24) 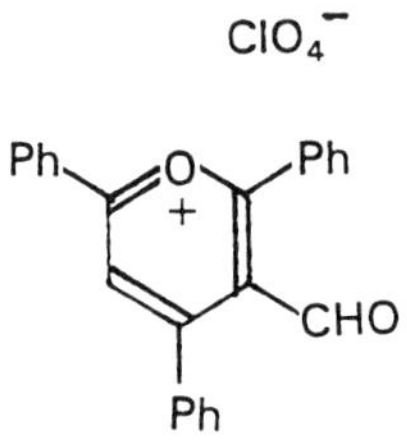   (25)

2-Carboxypyrylium perchlorates 26 (R^1 = Ph, R^2 = Ph, 4-MeOC$_6$H$_4$; R^1 = 4-O$_2$NC$_6$H$_4$, R^2 = Ph) react with amines to afford pyrylium perchlorates 27 along with cyclopentenones 28 (N.V. Kholodova *et al.*, Khim. Geterotsikl. Soedin., 1989, 1564).

(26)   (27)   (28)

2,6-Diaryl-3,5-dichloro-4-phenylpyrylium salts 29 (R = Ph, 4-ClC$_6$H$_4$; X = BF$_4$, ClO$_4$) are prepared by treating the 2,6-diaryl-4-phenylpyrylium salts in CCl$_4$ with a 4-fold excess of ethanolic NaOAc at 65-70°C, followed by chlorine gas at 10-20°, and then boiling with the appropriate acid, HX (X = BF$_4$, ClO$_4$) in acetic acid (N.V. Pchelintseva, S.N. Chalaya, and V.G. Kharchenko, U.S.S.R. SU 1,671,661, 1991; Zh. Org. Khim., 1990, 26, 1854).

(29)   (30)

18

6-Aryl-4-chloro-3,5-dimethyl-2-phenylpyrylium hexachloroantimonates 30
(R = Ph, 4-MeC$_6$H$_4$) and 4-chloro-3,5,6-trimethyl-2-phenylpyrylium hexachloroantimonate (30; R = Me) have been prepared by the acylation of
MeCH = CClCHMeCOPh with RCOCl in MeNO$_2$ containing SbCl$_5$, the reacting
species being RCO$^+$SbCl$_6^-$. The pyrylium salts 30 undergo base-catalyzed
hydrolysis in Et$_2$O containing 10% NaOH to yield PhCOCMe = C = CMeCOR.
The preparation of 4-chloro-2,3,5,6-tetramethylpyrylium hexachloroantimonate has also been reported (O.V. Zubkova, S.V.Borodaev, and S.M.
Kuk'yanov, *ibid*., 1991, <u>27</u>, 2376).

2-Amino-3-aroyl-4,6-diarylpyrylium tetrafluoroborates 31 [R^1-R^3 = (un)-
substituted phenyl], for example, 2-amino-3-benzoyl-4,6-diphenylpyrylium
tetrafluoroborate (31; R^1 = R^2 = R^3 = Ph), have been prepared from
appropriate chalcones and aroylacetonitriles (T. Zimmermann and R.
Radeglia, J. Prakt. Chem./Chem.-Ztg., 1994, <u>336</u>, 623). 2,4,6-Tris-
(dialkylamino)pyrylium perchlorates 32 (R = NMe$_2$ pyrrolidino, piperidino,
morpholino) and 2,6-bis(dialkylamino)pyran-4-ones 33 (R = NMe$_2$, pyrrolidino,
piperidino, morpholino) have been prepared by a multi-step route from
1,1,5,5-tetrachloropenta-1,4-dien-3-one (W. Schroth, R. Spitzner, and E.
Kleinpeter, Tetrahedron Letters, 1988, <u>29</u>, 4695).

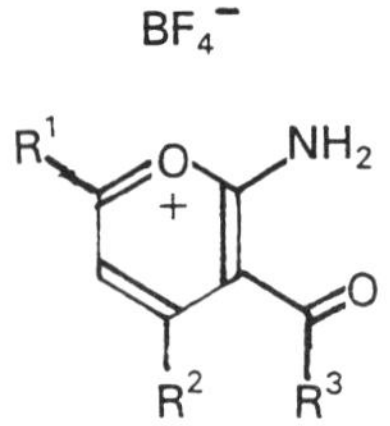

(31) 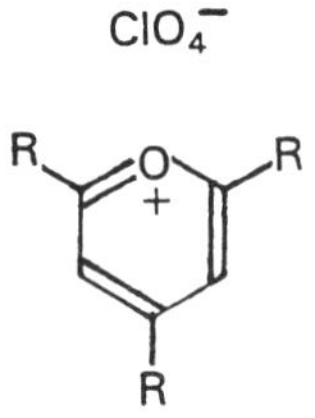   (32) 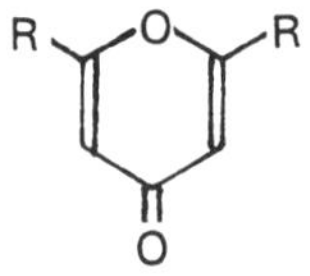   (33)

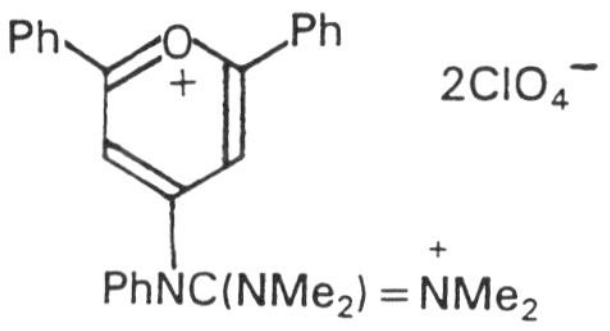

(34)

2-*N*-1,1,3,3-Tetramethylguanidine reacts with 2,6-diphenylpyran-4-one (33; R = Ph) in the presence of phosphorus oxychloride, followed by treatment with perchloric acid to give 1,1,3,3-tetramethyl-2-phenyl-2-(2,6--diphenyl-4-pyrylio)guanidinium diperchlorate (34) (W. Galezowski *et al.*, Canad. J. Chem., 1994, $\underline{72}$, 352).

Sodium borohydride reduction of the 2,4,6-trisubstituted pyrylium salts 35 (R^1-R^3 = Me, Ph, Pr^i, Bu^t) in acetic acid yields mainly the 5,6-dihydro-2*H*--dihydropyrans with *cis*-2- and -6-substituents and the all *cis*-2,4,6-tri-substituted tetrahydropyrans (T.S. Balaban and A.T. Balaban, Tetrahedron Letters, 1987, $\underline{28}$, 1341). Similar reduction of the 2,3,6-substituted 4--phenylpyrylium salts 36 (R^1 = Me, Pr^i, Ph; R^2 = H, Me, Ph; R^3 = Me, Pr^i; X = $HO_2CCH_2SO_3$, ClO_4) affords the 5,6-dihydro-2*H*-pyrans 37 (*idem*, Org. Prep. Proced. Int., 1988, $\underline{20}$, 231).

(35)   (36)   (37)

It has been found that 2,6-di-*tert*-butylpyrylium perchlorate (38) reacted with $NaBH_4$ to give 2,6-di-*tert*-butyl-4H-pyran (39; R = H) and $Bu^tCOCH = CHCH = CHBu^t$, but with NaCN it yielded $Bu^tCOCH = CHCH = -CCNBu^t$. The perchlorate reacted with Bu^tMgBr to give a mixture of 2,6-di--*tert*-butylpyrans 39 (R = H, Bu^t), which on treatment with H_2S and HCl furnished 2,4,6-tri-*tert*-butyl-4H-thiopyran (40), whereas aromatization with chloranil and subsequent treatment with NH_4OAc gave 2,4,6-tri-*tert*-butyl-pyridine (41) (S. Bohm *et al.*, Coll. Czech. Chem. Comm., 1987, $\underline{52}$, 1305).

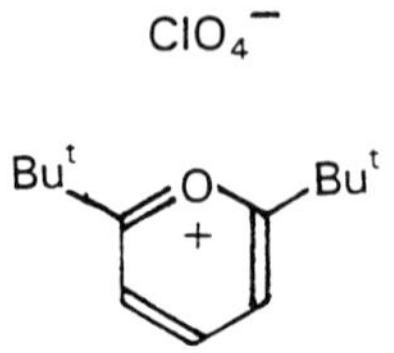

20

ClO_4^-

(38)

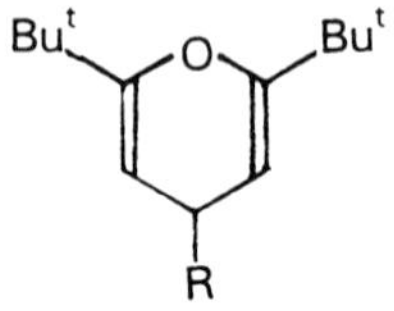

(39)

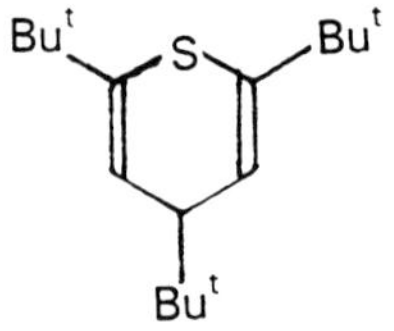

(40)

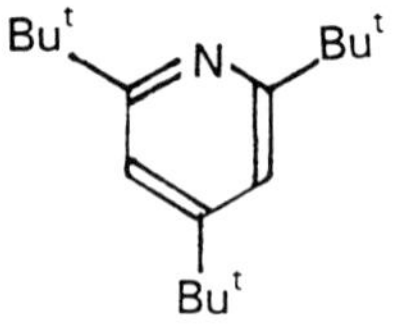

(41)

The reaction between 2,4,6-trimethylpyrylium ion (42) and organocopper reagents, R_2CuM (R = tolyl, anisyl, ClC_6H_4, 1-octyl, $PhCH_2CH_2$; M = Li, MgBr) results in the formation of the 2,4,6-trimethyl-4H-pyrans 43 as the major products along with the 2,4,6-trimethyl-2H-pyrans 44. The reagents 4-Me-C_6H_4Cu and $Me(CH_2)_7Cu$ have also been used (Y. Yamamoto, T. Kume, K. Akiba, Heterocycles, 1987, 26, 1495).

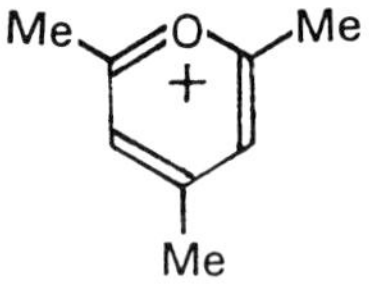

(42)

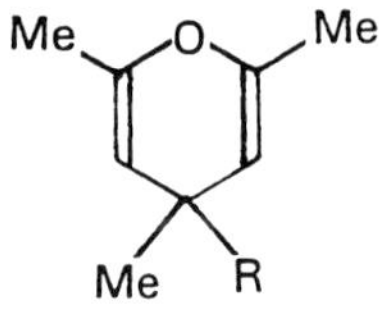

(43)

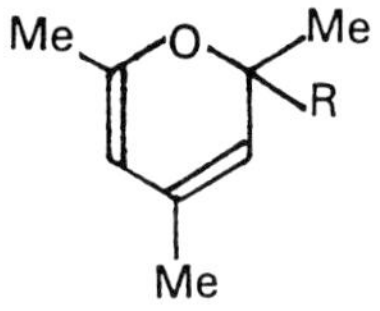

(44)

The regioselective addition reaction between 2-(*tert*-butyldimethylsiloxy)-4,6-dimethylpyrylium triflate (45) and orgonocuprates RCu (R = 4-MeC$_6$H$_4$, CH$_2$=CH, MeCH=CH, CH$_2$=CMe, Me$_2$CHCH$_2$, Bu, Me, Pr^i) affords the 4-substituted 4,6-dimethyl-2-silyloxy-4H-pyrans 46. Treatment of the siloxypyrans 46 (R = 4-MeC$_6$H, CH$_2$=CH, MeCH=CH, CH$_2$=CMe) with N-bromosuccinimide in DMF yields 4-substituted 3-*trans*-bromo-4,6-dimethyl-3,4-dihydro-2H-pyran-2-ones 47. The reaction of 46 (R = 4-MeC$_6$H$_4$) with CH$_2$=N$^+$Et$_2$Cl$^-$ and its oxidation with 3-ClC$_6$H$_4$CO$_2$OH have been reported (T. Kume *et al.*, Tetrahedron Letters, 1987, _28_, 6305).

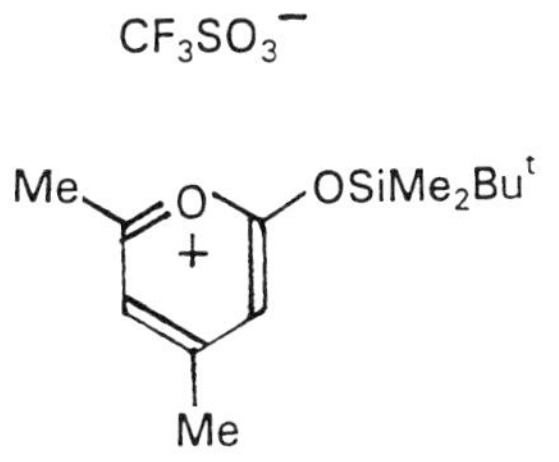

$$CF_3SO_3^-$$

(45)

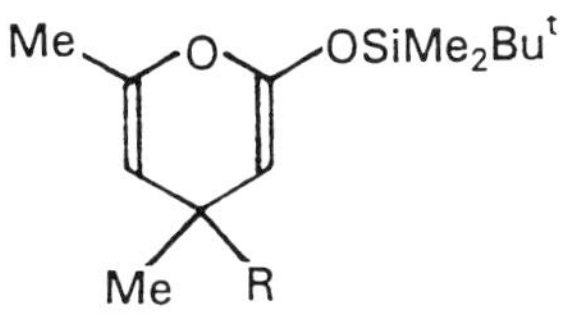

(46)

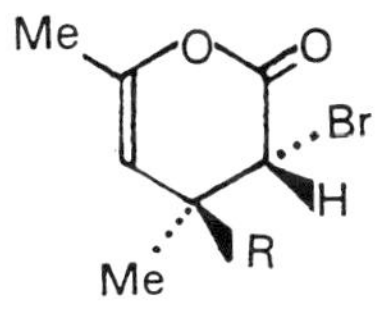

(47)

Lithium diorganocuprates add regioselectively at position 4 of 4,6-dimethyl- and 6-methyl-2-silyloxypyrylium triflates to yield 4-substituted 2-silyloxy-4H-pyrans 48 (R^1 = H, Me; R^2 = allyl, alkenyl, aryl), which react with electrophiles at the 3-position to give the corresponding 3-bromo-, 3-silyloxy-, 3-methylene, and 3-(1-hydroxybutyl)-3,4-dihydro-2H-pyran-2-one derivatives (*idem*, J. Org. Chem., 1989, _54_, 1931).

(48)

(49)

The addition of organolithium reagents RLi [R = Ph, $Me(CH_2)_4C \equiv C$, (*E*)--Bu$_3$SnCH = CH, (*E*)-PhCH = CH, (*E*)-PhCMe = CH, 2-furyl] to pyrylium tetrafluoroborate (49), followed by in situ electrocyclic ring opening of the intermediate 2*H*-pyrans affords *Z,E*-dienals RCH = CHCH = CHCHO with high stereoselectivity (Yu. Yu. Belosludtsev, B.C. Borer, and R.J.K. Taylor, Synthesis, 1991, 320).

The first homolytic substitution of pyrylium salts has been achieved by the methylation of 2,6-disubstituted pyrylium cations 50 with Me radicals generated from Bu^{t}OOH and FeSO$_4$ in acidic aqueous MeCN to give 2,6--disubstituted 4-methylpyrylium cations 51 (G. Doddi and G. Ercolani, Synth. Comm., 1987, <u>17</u>, 817). 2,6-Disubstituted 4-methyl- and 4-ethylpyrylium perchlorates 52 have been obtained by the reaction between 2,6--disubstituted pyrylium perchlorates 52 and alkyl radicals, generated by oxidative cleavage of carboxylic acids R^2CO_2H (R^2 = Me, Et) with $(NH_4)_2S_2O_8$ (G. Doddi, G. Ercolani, and P. laconianni, Gazz., 1989, <u>119</u>, 305). For the conversion of 2,6-diphenylpyrylium and 2,6-di-*tert*-butylpyrylium perchlorates into their 4-alkyl derivatives, by photochemical alkylation with tetraalkyl-stannates in MeCN, see E. Baciocchi *et al.*, Tetrahedron, 1993, <u>49</u>, 3793.

(54)

(50) R^1 = Ph, Bu^t; R^2 = H
(51) R^1 = Ph, Bu^t; R^2 = Me
(52) R^1 = Bu^t, Ph; R^2 = Me, Et; X = ClO$_4$
(53) R^1 = R^2 = Ph; X = BF$_4$

(55)

Tetraalkylsilanes, -germanes, and -stannanes react with 2,4,6-triphenyl-pyrylium tetrafluoroborate (53) *via* photoinduced electron transfer to give 2--alkylated pyrans 54 (R = Bu, Pr^i, Bu^t, PhCH$_2$, Me, CH$_2$=CHCH$_2$) and 4--alkylated pyrans 55 (S. Kyushin, Y. Nakadaira, and M. Ohashi, Chem. Letters, 1990, 2191).

3,5-Dimethyl-2,6-diphenyl-4-phenylethynylpyrylium perchlorate (56) on treatment with an alcohol, ROH (R = Me, Et, Pr^i), yields pyrylium perchlorates 57, which on hydrolysis by CF$_3$CO$_2$H affords the phenacyl derivatives 58 (L.A Murad'yan, G.P. Zolotovskova, and A.V. Koblik, Khim. Geterotsikl. Soedin., 1990, 299). 2,6-Di(4-methoxyphenyl)-4-phenylethynylpyrylium perchlorate (59; R = 4-MeOC$_6$H$_4$) on boiling in H$_2$O, MeOH, or EtOH yields dimer 60 (R = 4-MeOC$_6$H$_4$, X = O$^+$ ClO$_4^-$), which reacts with dry NH$_3$ in C$_6$H$_6$ or with PhNH$_2$ to yield the 2,4-diphenyl-1-(pyran-4-ylidene)-1-(pyridin-4-yl)-but-2-en-4-ones 60 (R = 4-MeOC$_6$H$_4$; X = N, PhN$^+$ ClO$_4^-$, respectively). Analogous reactions are observed for 2,6-diphenyl 4-ethynylpyrylium perchlorate (59; R = Ph) (A.V. Koblik *et al.*, *ibid.*, 1988, 885).

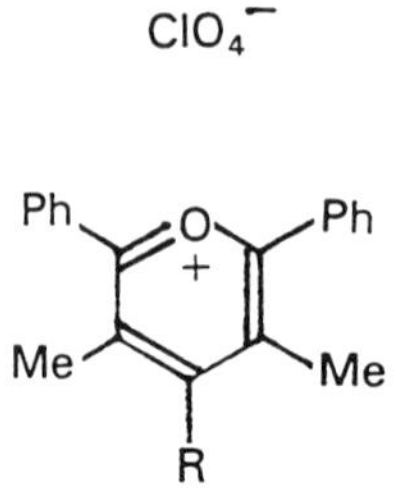

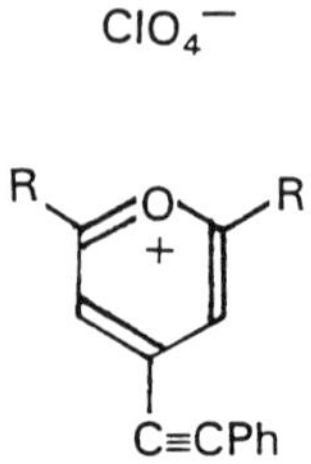

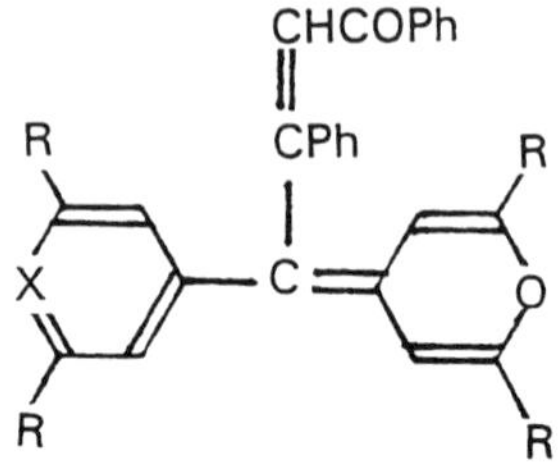

(56)   R = C≡CPh
(57)   R = CH = C(OR)Ph
(58)   R = CH$_2$COPh

(59)

(60)

4,6-Diaryl-2-methylpyrylium salts 61 (R = Ph, 4-MeC$_6$H$_4$) rapidly undergo dimerization reactions resulting in the formation of dimers 62 (4,6-diaryl-2-(2,4-diaryl-6-methylphenyl)pyrylium salts) and the diarylcarbonyl compound 63 (for example, R = Ph) (S.V. Verin *et al.*, Mendeleev Comm., 1993, 47).

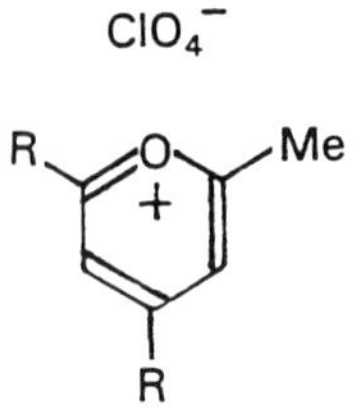

(61)

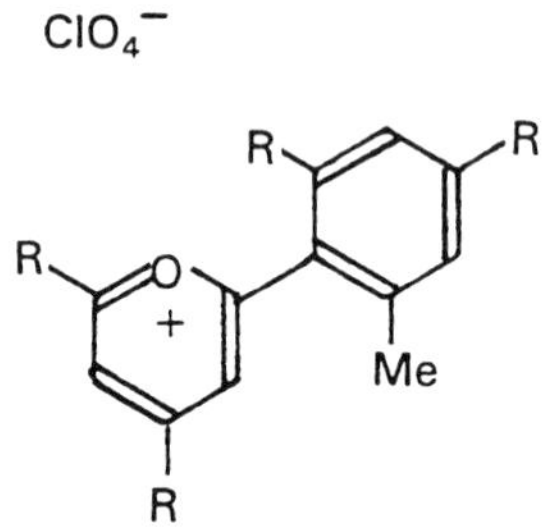

(62)

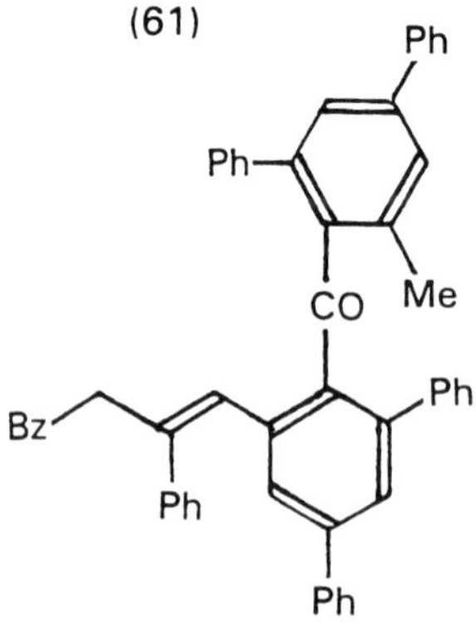

(63)

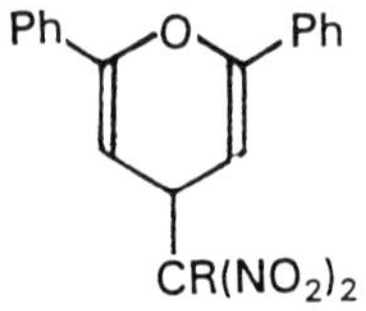

(64)

The reaction of nitro alkane salts with 2,6-diphenylpyrylium salts results in the formation of 4-substituted 2,6-diphenyl-4H-pyrans 64 (R = H, Me, NO$_2$) (A.A. Gakh and A.S. Kiselev, Izv. Akad. Nauk SSSR, Ser. Khim., 1991, 2092).

The condensation of 2,4,6-trisubstituted or 2,3,5,6-tetrasubstituted pyrylium salts possessing methyl groups in the 4- or 2- and 6-positions with variously substituted araldehydes yields intensely coloured styryl-substituted pyrylium salts such as 65 (R = Bu^t, Ph), 66 [R = 3,4,5-(MeO)$_3$C$_6$H$_2$, 4-Me$_2$NC$_6$H$_4$, 3,4-(OCH$_2$O)C$_6$H$_3$, 2-pyrrolyl] and 67 (R = Bu^t, Pr^i, Ph, 2-MeC$_6$H$_4$). These have been characterized by elemental analysis, IR, UV-VIS, and ^{1}H NMR spectroscopy (C.T. Supuran *et al.*, Rev. Roum. Chim., 1992, <u>37</u>, 141; O. Maior *et al., ibid.*, 1991, <u>36</u>, 1315).

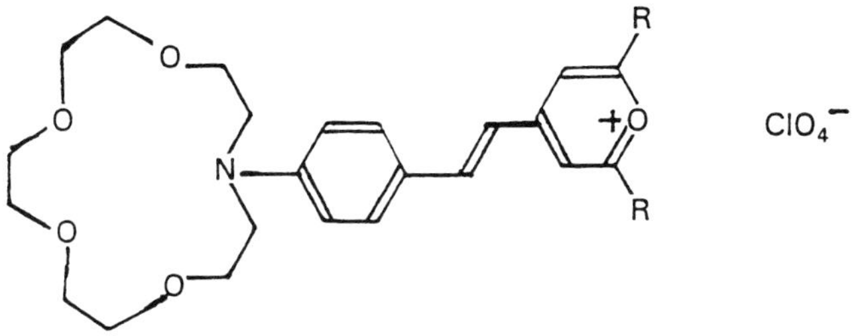

(65)

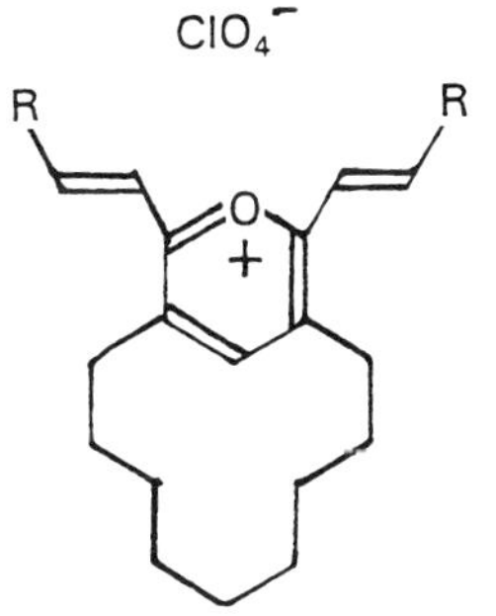

(66)

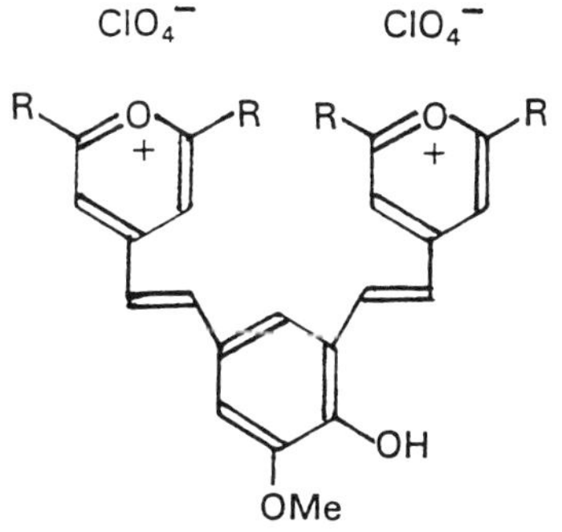

(67)

The reactions between primary amines and non-symmetrically substituted pyrylium salts has been investigated (C. Uncuta *et al., ibid.,* 1989, <u>34</u>, 1425). The crystal structure of 2,6-diphenyl-4-(2-phenylpropan--2-yl)pyrylium perchlorate has been solved and the aromatic character of the pyrylium ring has been described in terms of the HOMA parameter indicating its properties as intermediate between typical aromatic and non-aromatic (M. Gdaniec *et al.*, J. Chem. Soc., Perkin 2, 1989, 613).

2,4,6-Triphenyl(aryl)pyrylium salts react with acyclic 1,2-diketones, CH_3COCOR (R = Me, Ph), in the presence of an appropriate condensing agent (for example, piperidine acetate, Et_3N/AcOH, NaOAc) to give 2,4,5,7a--tetrasubstituted 7a*H*-cyclopenta[b]pyran-7-ones 68 ($R^1 = R^2 = H$), the structure of which has been determined by spectroscopic methods and by a single crystal x-ray analysis of the dibromo derivative 68 ($R^1 = R^2 = Br$) (T. Zimmermann, G.W. Fischer, and L. Kutschabsky, J. Prakt. Chem., 1989, <u>331</u>, 292; Z. Chem., 1988, <u>28</u>, 295). Similarly 2,4,6-triphenyl(aryl)-pyrylium salts react with cycloalkane-1,2-diones to give benzocyclo-alkenones, for example, NaOAc and cyclohexane-1,2-dione lead to the dihydro-2*H*-naphthalenones 69 (n = 1) and cycloheptane-1,2-dione and piperidine acetate, Et_2N, or NaOAc to the benzocycloheptan-5-one 69 (n = 2) (T. Zimmermann *et al.*, J. Prakt. Chem., 1989, <u>331</u>, 306).

(68)   (69)

The ß-ketoaldehyde 70 obtained by treating 9,10-dihydro-9,10-ethano-anthracen-11-one with ethyl formate, exists in two tautomeric forms, one of which has been converted into the pyrylium perchlorate 71 (M. Stanescu, M. Banciu, and A.T. Balaban, Rev. Roum. Chim., 1989, <u>34</u>, 617).

(70)

(71)

(b) 2H-Pyran-2-ones (α or 2-pyrones) and 4H-pyran-4-ones (γ or 4-pyrones)

(i) 2H-Pyrones. The reaction of 2-hydroxy-4-methyltetrahydropyran (1) with a catalyst mixture of $CuO-Cr_2O_3$ containing MnO_2 (0.5%) in dioctyl phthalate at 160°C under nitrogen affords 4-methyl-*2H*-pyrone (2) (57%), useful as monomer and intermediate for drugs, perfumes, and agrochemicals, along with 3-methylpentane-1,5-diol (43%) (Y. Tokito and N. Yoshimura, Jpn. Kokai Tokkyo Koho JP 62 39,535 [87 39,535], 1987).

(1)

(2) R = H
(3) R = Me, Bu, Me_2CHCH_2

28

6-Alkyl-4-methyl-2*H*-pyrones 3 are obtained by a multistep procedure, carried out without isolation and purification of intermediates, from MeCOCH = CRSMe (R.K. Dieter and J.R. Fishpaugh, J. Org. Chem., 1988, 53, 2031). Silylalkenes easily undergo a [4 + 2] cycloaddition reaction with electron-rich 1,3-dienes to yield the corresponding 2*H*-pyrones, for example, 4-methyl-6-(dimethyl*tert*butylsilyl)methyl-2H-pyrone (4) (T. Ito, T. Aoyama, and T. Shioiri, Tetrahedron Letters, 1993, 34, 6583).

(4)   (5)   (6)

4-Aryl- or 4-alkenyl-3-bromo-4,6-dimethyl-3,4-dihydro-2-pyrones 5 (R = 4-MeC$_6$H$_4$, 4-MeOC$_6$H$_4$, Ph, prop-1-en-1-yl, hex-1-en-1-yl, allyl, vinyl) on debromination with AgSbF$_6$ in CH$_2$Cl$_2$ or ClCH$_2$CH$_2$Cl in the presence of 2,6--lutidine, results in the induced migration of the aryl or alkenyl group to yield the 3-substituted 4,6-dimethyl-2H-pyrones 6 (T. Kume and K. Akiba, J. Org. Chem., 1989, 54, 1935). The reaction of 5-monoalkyl derivatives 7 (R^1 = Me, CH$_2$CH$_2$CO$_2$Me, CH$_2$CH$_2$COMe, CH$_2$CH = CHPh, Pr^i, CH$_2$CH$_2$Ph, CHMeCH$_2$COMe, Ph) of Meldrum's acid with α,ß-ynones, R^{2}C≡CCOMe (R^2 = H, Ph), in the presence of K$_2$CO$_3$ and Bu$_4$NCl furnishes 3-alkyl-6-methyl- and 3-alkyl-6-methyl-4-phenyl-2*H*-pyrones 8 (A. Arcadi *et al.*, Synlett., 1993, 741).

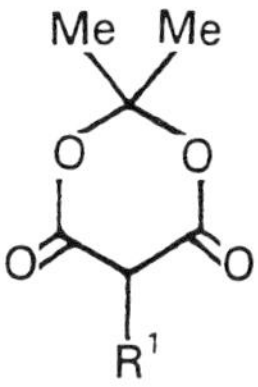

(7)

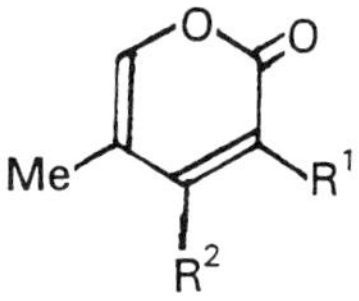

(8)

3,4,6-Trisubstituted-2*H*-pyrones 10 have been obtained by carbonylative cross-coupling-thermolysis of 4-halogenocyclobutenones with alkenyl-, aryl-, and heteroarylstannanes, for example, palladium catalyzed cross-coupling of 4-chloro-2,3-disubstituted-cyclobuten-2-ones 9 ($R^1 = R^2 = Et$, Bu; $R^1 = Me$, $R^2 = Ph$, Me_2CHO) with reagents Bu_3SnR^3 ($R^3 = Ph$, $4\text{-}ClC_6H_4$, $4\text{-}MeOC_6H_4$, 2-furyl, 3-furyl, 2-thienyl, N-methylpyrrol-2-yl, vinyl, 2-ethoxyvinyl, propen-1-yl) and thermolysis yields the 6-substituted 3-alkyl-4-ethyl(phenyl)-2*H*-pyrones 10 (L.S. Liebeskind and J. Wang, Tetrahedron, 1993, <u>49</u>, 5461).

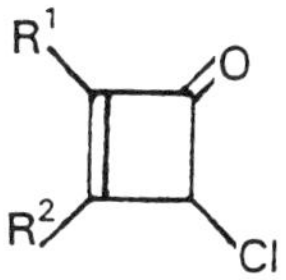

(9)

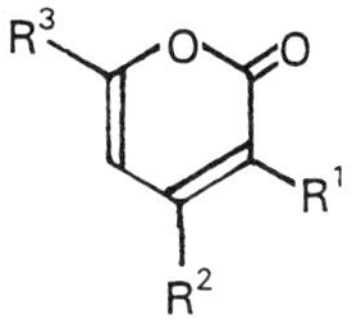

(10)

The process for producing alkyl (lower alkyl group) esters of 4,6-
-dimethyl-2H-pyrone-5-carboxylic acid (4,6-dimethyl-2-oxo-2H-pyran-5-
-carboxylic acid, isodehydroacetic acid) 11 by cyclocondensation of the
corresponding acetoacetic ester $MeCOCH_2CO_2R$ (R = lower alkyl group) in
ROH in the presence of HCl has been investigated (S. Jani, F. Jurak, and M.
Nemes, Hung. Teljes HU 54,670, 1991). The $[Ni(cod)_2]Ph_2P(CH_2)_4PPh_2$-
-catalyzed reaction of hex-3-yne with CO_2 under supercritical conditions
yields 3,4,5,6-tetraethyl-2H-pyrone (12) (M. Reetz, W. Koenen, and T.
Strack, Chimia, 1993, <u>47</u>, 493). Other related preparations of
tetraethylpyrone 12 have been discussed [D. Walther *et al.*, Ger. (East) DD
244,340, 1987]. 3-Alkyl(lower alkyl groups)-5-alkyl(larger alkyl groups)-
-2H-pyrones 13 (R^1 = Me, Et, Pr, MeO; $R^2 = C_{7\text{-}11}$ alkyl) have been prepared
in a number of stages from (±)-pentan-2-ol, for example, 3-methyl-5-(2,4-
-dimethylheptanyl)-2H-pyrone [13; R^1 = Me, $R^2 = Me(CH_2)_2CH(Me)CH_2CH$-
(Me)Me], and studied as cockroach attractants (W.L. Roelofs *et al.*, U.S. US
5,296,220, 1994).

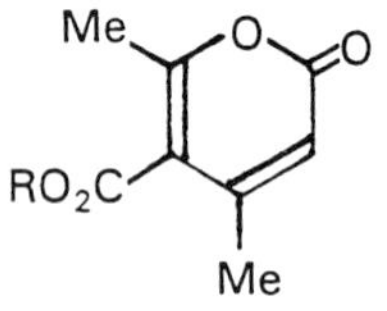

(11) 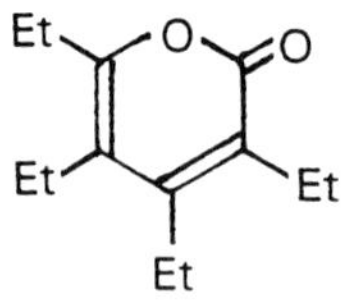   (12) 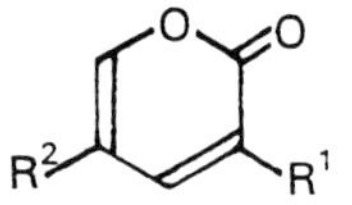   (13)

The condensation reactions of the enaminones (E)-PhCOCH = CHNMe$_2$
and (E,E)-PhCH = CHCOCH = CHNMe$_2$ with ethyl acetoacetate in the
presence of Knoevenagel catalysts afford the 6-substituted-3-acetyl-2H-
-pyrones 14 [R = Ph and (E)-PhCH = CH, respectively]. Similarly with the
dianion of ethyl acetoacetate the ethyl salicylates 15 are obtained (N.
Takeuchi *et al.*, Chem. Pharm. Bull., 1991, <u>39</u>, 1655).

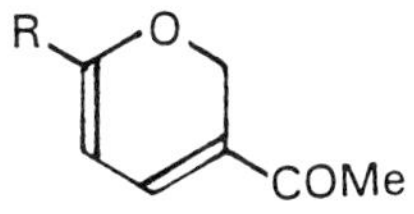

(14) 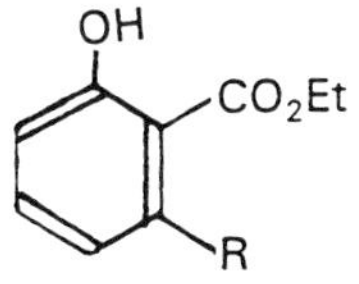 (15)

3-Acetoacetyl-6-aryl-2*H*-pyrones 16 (R = Ph, 4-BrC$_6$H$_4$, 4-ClC$_6$H$_4$, 4-
-O$_2$NC$_6$H$_4$, triazolyl) useful as intermediates for dyes and drugs have been
prepared by boiling together RCCl = CHCH = N$^+$Me$_2$, triacetic lactone, and
Et$_3$N in EtOH, followed by the addition of H$_2$O to the hot solution [B. Hirsch
and N. Hoefgen, Ger. (East) DD 252,604, 1987]. A mixture of 3-chloro-2,3-
-diphenylvinylaldehyde and Et$_3$N on boiling in acetic acid yields 3-aceto-
acetyl-5,6-diphenyl-2*H*-pyrone (17; R = H). Other 3-acetoacetyl-5,6-diaryl-
-2*H*-pyrones 17 (R = Me, MeO, etc.), useful as intermediates for dyes and
pharmaceuticals have been prepared (*idem, ibid.*, 252,188, 1987). A
number of 4,5,6-trisubstituted 3-aryl-2*H*-pyrones have been prepared as
pesticides and herbicides (R. Fischer *et al.*, Eur. Pat. Appl. EP 588,137,
1994).

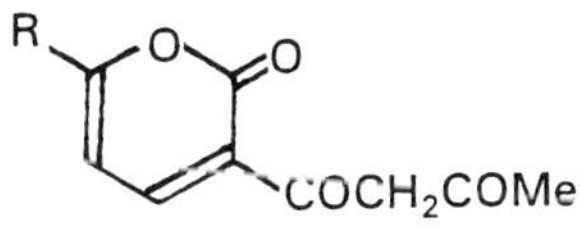

(16) 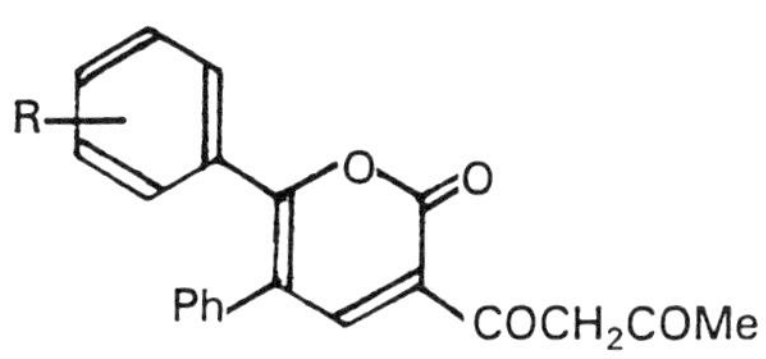 (17)

32

Formylation of 3-substituted diethyl pent-2-enedioates by means of ethyl formate/TiCL$_4$/4-methylmorpholine yields the ethoxymethylene derivatives, EtOCH = C(CO$_2$Et)CR = CHCO$_2$Et (R = H, Cl, F, EtO, SH$_2$Ph, MeSO$_2$, Bu, CO$_2$Et, Ph), which are smoothly cyclized either with formic acid or polyphosphoric acid to afford the corresponding 4-substituted ethyl 2-oxo-2*H*-pyran-5-carboxylates 18 (V. Kvita, Helv., 1990, <u>73</u>, 411). Some 4,5-disubstituted derivatives related to 2*H*-pyrone 18 have been prepared (V. Kvita *et al.*, Eur. Pat. Appl. EP 344,113, 1989).

(18)   (19)   (20)

It has been shown that diphenylcyclopropenethione undergoes nucleophilic attack by pyridinium disubstituted ylides, for instance ylide 19, to produce methyl 6-methoxy-2-thioxo-3,4-diphenyl-2*H*-pyran-5-carboxylate (20). The preparations of methyl 6-methylthio-2-oxo-3,4-diphenyl-2*H*-pyran-5-carboxylate and related derivatives and of analogues of thiopyran-2-one have been reported (Y. Ikemi, J. Heterocyclic Chem., 1990, <u>27</u>, 1597).

Methyl 2-oxo-2*H*-cycloalkano[b]pyran-3-carboxylates 21 (n = 1,2), methyl 6-aryl-2-oxo-2*H*-pyran-3-carboxylates 22 (R = Ph, 4-MeOC$_6$H$_4$, 1-naphthyl, 2-naphthyl), and methyl 5-cyano-6-ethoxy-2-oxo-2*H*-pyran-3-carboxylate (23) are obtained by reacting PhNMeCH = C(CO$_2$Me)$_2$ with cyclopentanone, cyclohexanone, PhCOMe, 4-MeOC$_6$H$_4$COMe, 1- and 2-acetonaphthone, and EtO$_2$CCH$_2$CN, respectively, in THF (A.S. Demir *et al.*, Synth. Comm., 1991, <u>21</u>, 1433). A number of derivatives of esters of 2-oxo-2*H*-pyran-6-carboxylate have been prepared and tested for use as insecticides (M. Mazaki *et al.*, Jpn. Kokai Tokkyo Koho JP 01 151,569 [89 151,569], 1989). Dimethyl acetylenedicarboxylate reacts with acetic acid in the presence of catalyst, Ru$_3$(CO)$_{12}$ to give trimethyl 2-oxo-2*H*-pyran-4,5,6-tricarboxylate (N. Menashe and Y. Shvo, Heterocycles, 1993, <u>35</u>, 611).

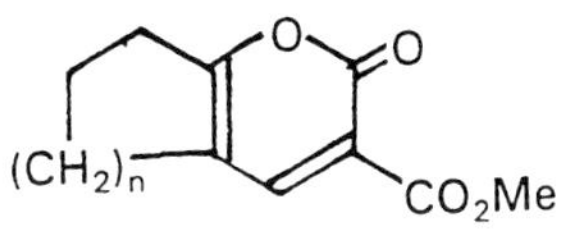

(21) 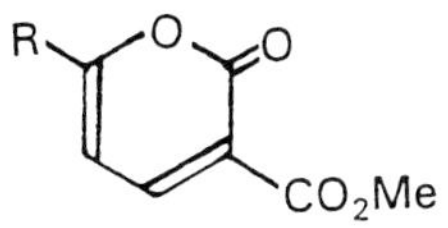   (22) 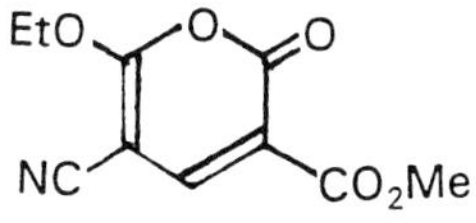   (23)

The Diels-Alder reaction of alkyl 2*H*-pyrone-5-carboxylates 24 (R = Me,
Et, Pr^i) with H_2C = CHOR^1 (R^1 = Et, ClCH$_2$CH$_2$) yields six stable *endo*-8-
-alkoxy-3-oxo-2-oxabicyclo[2.2.2]oct-5-ene-6-carboxylates and six exo
isomers. The formation of a number of adducts with ethenes and their
hydrolysis to give isophthalic acid derivatives or 1,3,5-benzenetricarboxylic
acid derivatives have been reported (T. Shimo, F. Muraoka, and K.
Somekawa, Nippon Kagaku Kaishi, 1989, 1765).

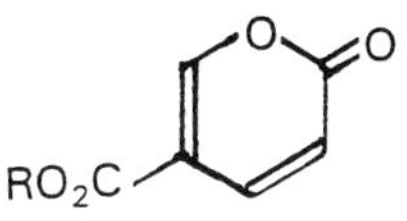

(24) 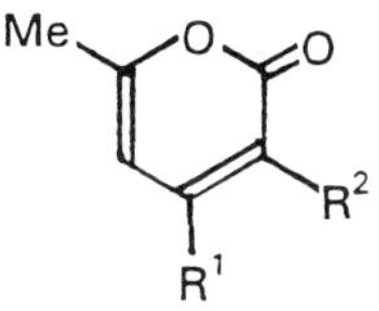   (25) 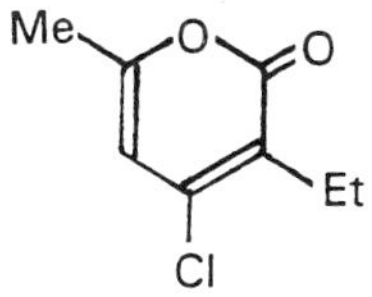   (26)

Dehydroacetic acid reacts with a mixture of PCl_5 and $POCl_3$ to give 4--chloro-3-(1-chlorovinyl)-6-methyl-2*H*-pyrone (25; $R^1 = Cl$, $R^2 = H_2C = CCl$), which on treatment with a base such as Et_3N yields 4-chloro-3-ethynyl-6--methyl-2*H*-pyrone (25; $R^1 = Cl$, $R^2 = HC \equiv C$), catalytically reduced to 4--chloro-3-ethyl-6-methyl-2*H*-pyrone (26) (Y. Azuma *et al.*, Annu. Rep. Tohoku Coll. Pharm., 1988, 221). The reaction of primary and secondary amines with 4-chloro-3-(1-chlorovinyl)-6-methyl-2*H*-pyrone affords the corresponding 4-alkylamino-3-ethynyl-6-methyl-2*H*-pyrone derivatives, which on treatment with dilute HCl are converted to the corresponding 3-acetyl-4--alkylamino-6-methyl-2*H*-pyrones 25 ($R^1 = $ alkylamino, $R^2 = $ acetyl) (*idem, ibid.*, 1989, 219). 4-Chloro-3-(1-chlorovinyl)-6-methyl-2*H*-pyrone and 4--chloro-3-ethynyl-6-methyl-2H-pyrone also react with 4-phenylenediamine to yield the iminopyrone 27 and the aminoethynylpyrone 28, respectively; both of which on treatment with dil HCl afford acetylpyrone 29 (Y. Azuma, A. Sato, and M. Morone, *ibid.*, 1992, <u>39</u>, 251).

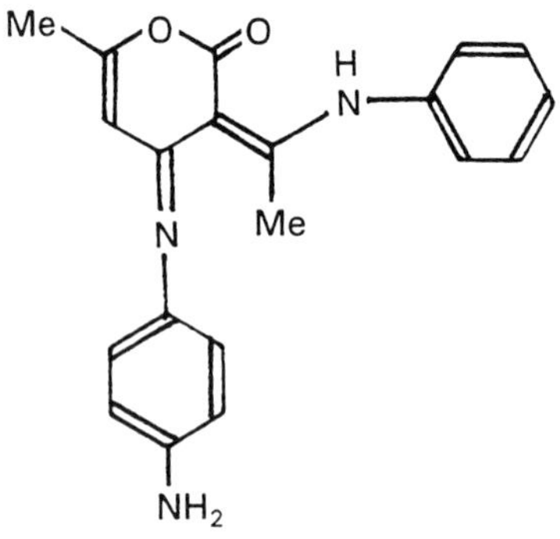

(27)

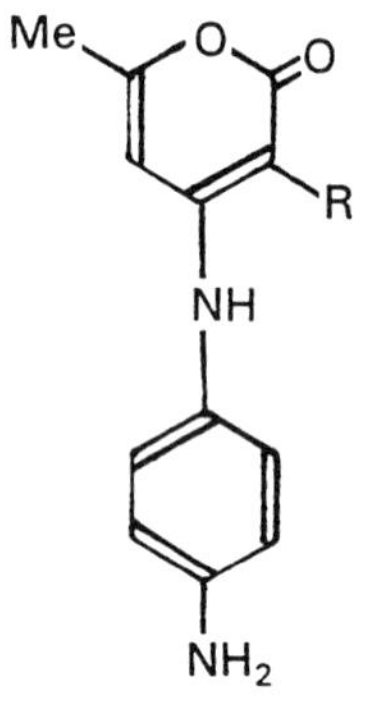

(28) R = C≡CH
(29) R = COMe

The preparation of 6-unsubstituted 2*H*-pyrones, such as 4-chloro-5-(2,2--dichlorovinyl)- and 4,5-bis(trifluoromethyl)-2*H*-pyrone and 2-oxo-2*H*-pyran-6--carbonitrile have been reported (V. Kvita and H. Sauter, Helv., 1990, <u>73</u>, 883).

Treatment of 3-chloroglutaconic acid, $HO_2CCH_2CCl=CHCO_2H$, with PCl_5 yields 4,6-dichloro-2*H*-pyrone (30; $R^1=R^2=Cl$), which on stirring with zinc in acetic acid affords 4-chloro-2*H*-pyrone (30; $R^1=H$, $R^2=Cl$). The related bromo derivatives are obtained in a similar manner and the chlorinated and brominated 2*H*-pyrones are useful for Diels-Alder reactions, for example, with 1,4-anthraquinones in preparing substituted naphthacene-5,12-diones as photoinitiators and photosensitizers (K. Vratislav, Eur. Pat. Appl. EP 369,941, 1990).

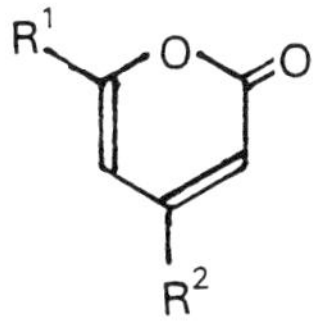

(30)

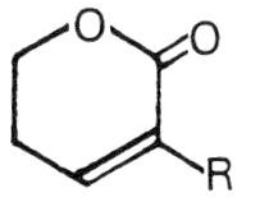

(31)

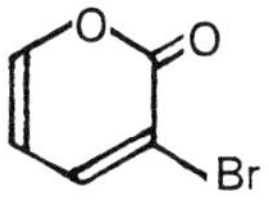

(32)

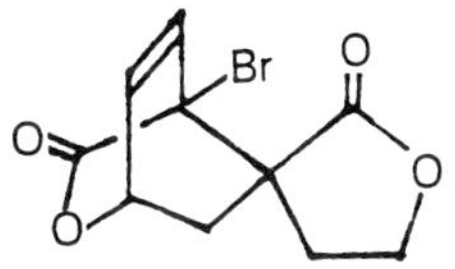

(33)

The reaction of 5,6-dihydro-2*H*-pyran-2-one (31; $R=H$) with Br_2 and EtN(Pr^i)$_2$ gives 3-bromo-5,6-dihydro-2*H*-pyran-2-one (31; $R=Br$), which on bromination with *N*-bromosuccinimide and dehydrobromination with Et$_3$N yields 3-bromo-2*H*-pyrone (32). The cycloaddition reaction between bromopyrone 32 and 2-methylene-4-butanolide affords bromospirofuranone-oxabicyclooctenone 33 (G.H. Posner *et al.*, Tetrahedron Letters, 1991, <u>32</u>, 5295).

4-Amino-6-aryl-2*H*-pyrones 34 (R^1 = Ph, 4-ClC$_6$H$_4$, 4-tolyl, 4-anisyl; R^2_2 N = morpholino, 1-pyrrolidinyl, piperidino, R^2 = Me) are obtained by cyclization of the product formed by the regioselective conjugate 1,4--addition of R^1COCH = C(SMe)NR$^2_2$ with H$_2$C = C(OEt)OZnBr (A. Datta, H. Ila, and H. Junjappa, Synthesis, 1988, 248). For the preparation of 4--substituted 6-amino-3,5-dicyano-2*H*-pyrone 35 (R = 4-MeOC$_6$H$_4$, 4-ClC$_6$H$_4$, 2-thienyl, 2-furyl) see B.Y. Riad and S.M. Hassan, Sulphur Letters, 1989, 10, 1).

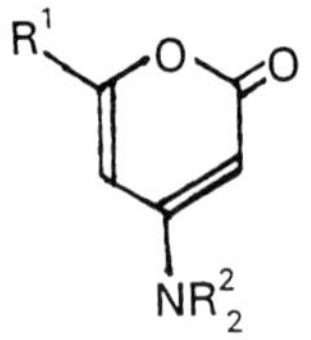

(34) 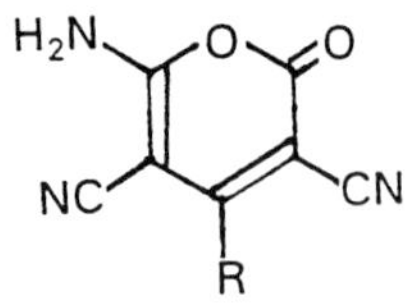 (35) 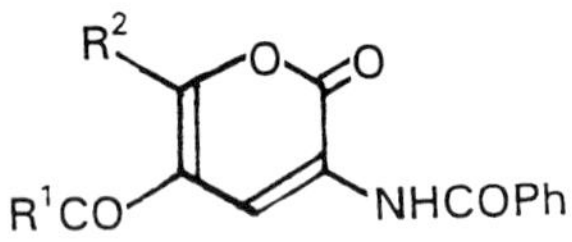 (36)

5,6-Disubstituted 3-benzoylamino-2*H*-pyrones 36 (R^1 = OEt, R^2 = Ph, Me; R^1 = Ph, Me, R^2 = Me) have been obtained either in a one-step reaction from 1,3-dicarbonyl compounds, R^1COCH$_2$COR^2 and methyl 2-benzoylamino-3--dimethylaminopropanate, or from ethyl 2-acyl-3-dimethylaminopropenoates or 2-(dimethylamino)methylene-1,3-diketones and 5-oxo-2-phenyl-1,3--oxazole (J. Svete *et al.*, *ibid.*, 1990, 70). The preparations of dimethyl 3--benzoylamino-2-oxo-2*H*-pyran-5-carboxylate-6-acetate (V. Kepe *et al.*, Heterocycles, 1992, 33, 843), and ethyl 3-benzoylamino-2-oxo-6-triphenyl-phosphoranylidenemethyl-2*H*-pyran-5-carboxylate (M.L. Gelmi and D. Pocar, Synthesis, 1992, 453) have been reported. A number of ethyl or methyl 6--substituted 3-benzoylamino-2-oxo-2*H*-pyran-5-carboxylates and 3-benzoyl-amino-7,8-dihydro-2*H*-benzo[*b*]pyran-2,5(6*H*)-diones have been synthesized and some of them showed local anesthetic, platelet antiaggregating, and other activities (L. Mosti *et al.*, Farmaco, 1994, 49, 45).

The reaction of 4-amino-6-methyl-2*H*-pyrone (37) with RC$_6$H$_4$CHO (R = 4--Cl, 2-NO$_2$, 3-NO$_2$) produces bis(4-amino-6-methyl-2-oxo-2*H*-pyran-3-yl)-arylmethanes 38 (M. Cervera, M. Moreno-Manas, and R. Pleixats, Tetrahedron, 1990, 46, 7885).

(37)

(38)

6-Aryl-5-methyl-4-methylthio-2*H*-pyrones 39 (R^1 = H, Me, Br, NO$_2$;
R^2 = Me, H) have been prepared from 5-[bis(methylthio)methylene] Meldrum's
acid and ketones (X. Huang and G. Wu, Youji Huaxue, 1989, 9, 460). 6-
-Aryl-3-cyano-4-methylthio-2*H*-pyrones 40 (R = 2-benzofuryl, 4-MeOC$_6$H$_4$)
react with nucleophiles (amines and active methylene compounds. For
example, 40 (R = 4-MeOC$_6$H$_4$) and N$_2$H$_4$ yields pyrazolopyranone 41 (V.J.
Ram and M. Verma, Indian J. Chem., 1990, 29B, 624).

(39)

(40)

(41)

Derivatives of 4- and 5-mercapto-2*H*-pyrones 42 (R^1 = R^3-H, R^2 = SH; R^1 = SH, R^2 = OH, R^3 = H; R^1 = H, R^2 = OH, R^3 = SH) have been obtained by the reaction of the corresponding halogeno derivatives 42 (R^1 = R^3 = H, R^2 = Cl; R^1 = Br, R^2 = OH, R^3 = H; R^1 = H, R^2 = OH, R^3 = Br) with potassium ethyl xanthate and/or NaSH (G.S.R. Reddy, K. Sucheta, and N.R. Rao, Indian J. Heterocyclic Chem., 1992, *2*, 95). Several 4-alkylsulphonyl-2*H*-pyrones 43 [R^1 = (substituted) Ph, thienyl, furyl; R^2 = H, (halogeno)alkyl; R^3 = alkyl, Ph, PhCH$_2$; R^4 = H, CO$_2$H, alkoxycarbonyl, halogeno, (alkoxy)alkyl; n = 1,2] have been prepared as agrochemical fungicides. For example, the product from the cyclocondensation of 3-BrC$_6$H$_4$COMe with (MeS)$_2$C = C(CO$_2$Et)$_2$, has been heated at 130°C with polyphosphoric acid to yield, after bromination, 3-bromo-6-(2-bromophenyl)-4-methylthio-2*H*-pyrone (44; n = 0), which has been oxidized to the dioxide 44 (n = 2). The latter gave good to complete control of *Botrytis cinerea* on tomato plants (P. Duvert and J. Huang, Fr. Demande FR 2,665,445, 1992).

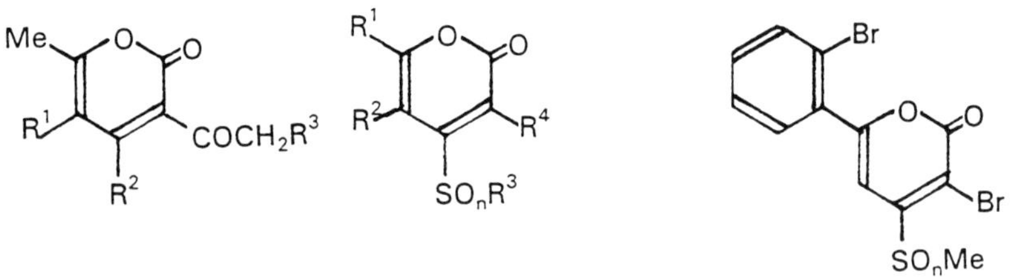

(42)   (43)   (44)

For a simple and general one-pot synthesis of some 2*H*-pyrones, see V. Kepe *et al.*, Tetrahedron, 1990, *46*, 2081; for the preparation of 2*H*--pyrones by cyclooligomerization of alkynes with CO$_2$ in the presence of catalysts comprising Ni-olefin complexes with *N*-donor ligands, R. Kempe *et al.*, Ger. (East) DD 282,459, 1990; for the formation of 2*H*-pyrones by the reaction of 1,3-dicarbonyl compounds with methyl 2-carbomethoxy-3-(*N*--methylanilino)acrylate under acid conditions, C. Tanyeli *et al.*, Heterocycles, 1994, *37*, 1705; and for the preparation of 6-(2-arylvinyl)-4-methyl-2-oxo--2H-pyran-5-carboxylic acids and their ethyl esters, Y. Anghelova and E. Dimitrova, God. Sofii. Univ. "Sv. Kliment Okhridski", Khim. Fak., 1992, *82*, 23.

4,6-Dimethyl- and 4-methoxy-6-methyl-2H-pyrones 45 (R^1 = Me, MeO) undergo benzophenone-sensitized photochemical cyclization with chloroethenes, HCR^2 = CR^3Cl (R^2 = H, R^3 = Cl; R^2 = R^3 = Cl; R^2 = H, R^3 = CN) to yield oxabicyclooctenones 46 (R^1 = Me, MeO; R^2 = H, Cl; R^3 = Cl, CN) and for example 47, along with benzene derivatives (e.g. 48) and/or cyclobutanecarboxylic acids (e.g. 49). Dehydrochlorination of oxabicyclooctenones 46 with Et_3N in EtOH gives 5-ethenyl-2H-pyrones (e.g. 50; R^4 = ClCH = CCl) (T. Shimo *et al.*, Bull. Chem. Soc. Jpn., 1987, <u>60</u>, 621).

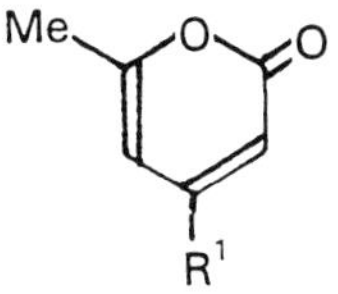

(45) 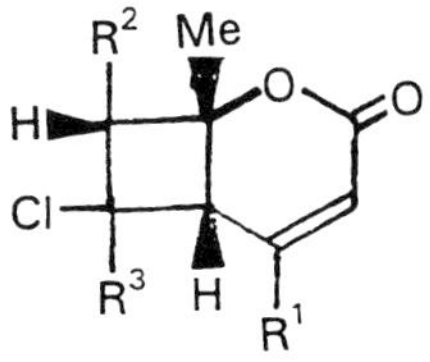   (46) 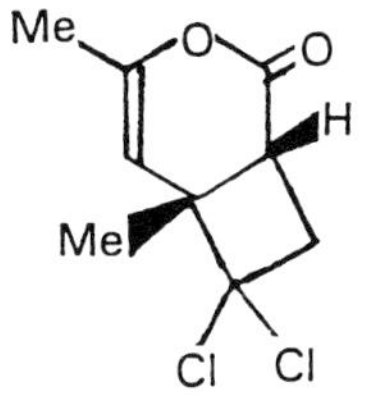   (47)

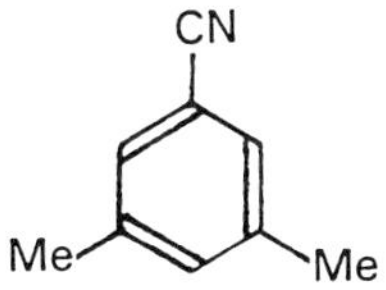

(48) 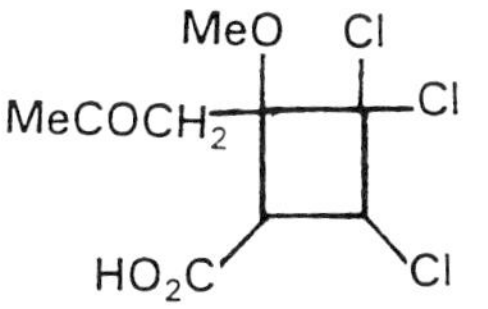   (49) 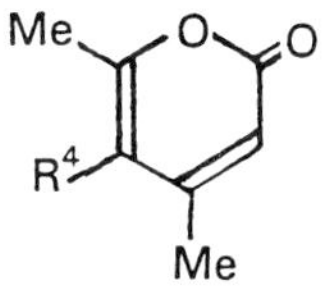   (50)

The structures of the photoproducts of 6-*trans*-styryl-2H-pyrone (51; R = H) and of 6-*trans*-styryl-4-methoxy-2H-pyrone (51; R = MeO), previously characterized as the cis isomers 52, have been revised to cyclobutane dimers 53 (D.G. Mandarino, M. Yoshida, and O.R. Gottlieb, J. Braz. Chem. Soc., 1990, <u>1</u>, 53).

(51) (52) (53)

It has been shown that silica gel and a TADDOL-complexed titanium IV Lewis acid promote mild, practical, asymmetric [4 + 2]-cycloadditions of electron-poor 2*H*-pyrone-3-carboxylates with electron-rich vinyl ethers to form isolable and useful bicyclic lactone adducts (G.H. Posner *et al.*, Tetrahedron Letters, 1994, <u>35</u>, 1321). Bicyclic lactones in high diastereomeric excesses have also been obtained by the cycloaddition reactions between the chiral 2*H*-pyrone derivatives, for example, derivative 54 and various dienophiles, catalyzed by lanthanide shift reagents (I.E. Marko and G.R. Evans, *ibid.*, p.2767). The reactivity of 6-unsubstituted 2*H*-pyrones towards aliphatic and aromatic amines has been studied (V. Kvita, H. Sauter, and G.Ribs, Helv., 1991, <u>74</u>, 407).

Trialkylphosphine ligand 55 effected the Ni(O)-catalyzed cycloaddition of terminally unsubstituted diynes with CO_2 to yield bicyclic 2*H*-pyrones 56 (R^1 = H, X = CH_2, CH_2CH_2; R^1 = Pr^i, X = CH_2CH_2), ordinary trialkylphosphine ligands have been found to be ineffective (T. Tsuda, S. Morikawa, and T. Saegusa, Chem. Comm., 1989, 9). Ni(COD)$_2$-trialkylphosphine catalysts have effected cycloadditions of CO_2 and terminally dialkyl-substituted diynes, for example, $R^1C \equiv CCH_2XCH_2C \equiv CR^1$ (X = O, NPr, CH_2, CH_2CH_2 R^1 = Et; X = CH_2CH_2, R^1 = Me, Bu, EtMeCH) to afford 3,6-dialkyl-4,5-cycloalkano-(hetero)-2*H*-pyrones 57 (T. Tsuda *et al.*, J. Org. Chem., 1988, <u>53</u>, 3140).

(54)   (55)   (56) $R^1 = H$, Pr^i; $R^2 = H$
                                                           (57) $R^1 = R^2 = $ alkyl

Methyl 2-oxo-2*H*-pyran-5-carboxylate undergoes a Diels-Alder reaction with isoprene to give two stereoisomers, methyl 8-methyl-8-vinyl- and 8--isopropenyl-3-oxo-2-oxabicyclo[2.2.2]oct-5-ene-6-carboxylates (58; $R^1 = R^2 = H$, Me), but only one stereoisomer, methyl 8-methoxy-8-vinyl-3-oxo--2-oxabicyclo[2.2.2]oct-5-ene-6-carboxylate (58; $R^1 = OMe$, $R^2 = H$) is obtained in the reaction between methyl 2-oxo-2*H*-pyran-5-carboxylate and $CH_2 = C(OMe)CH = CH_2$) (T. Matsui, S. Kitajima, and M. Nakayama, Bull. Chem. Soc. Jpn., 1988, <u>61</u>, 316).

Methyl 2-oxo-2*H*-pyran-5-carboxylate also reacts with nonconjugated dienes, for example, $H_2C = CHXCH = CH_2$ [$X = CH_2CH_2$, OCH_2, CO_2, CH_2OCH_2, CH_2SCH_2, $(CH_2)_4$], to afford [4 + 2] cycloadducts 59, which lose CO_2 to give intramolecular double Diels-Alder adducts 60 (K. Harano *et al.*, Chem. Pharm. Bull., 1990, <u>38</u>, 1182).

(58)   (59)

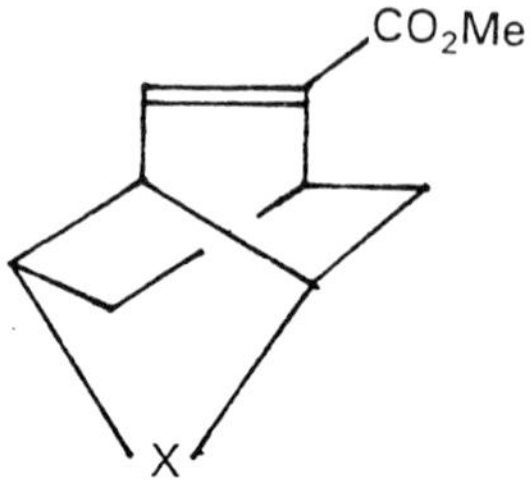

CO$_2$Me

X

(60)

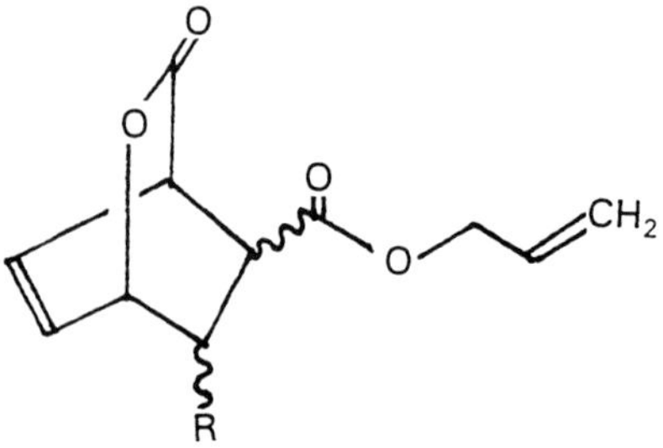

O

R

(61)

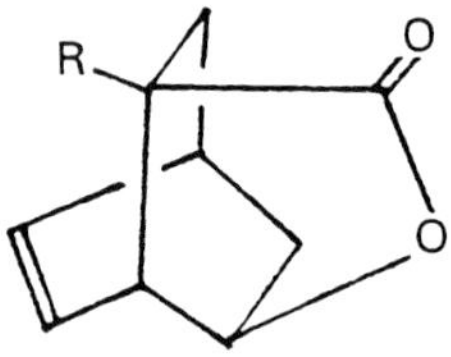

R

O

O

(62)

O

O

CH$_2$

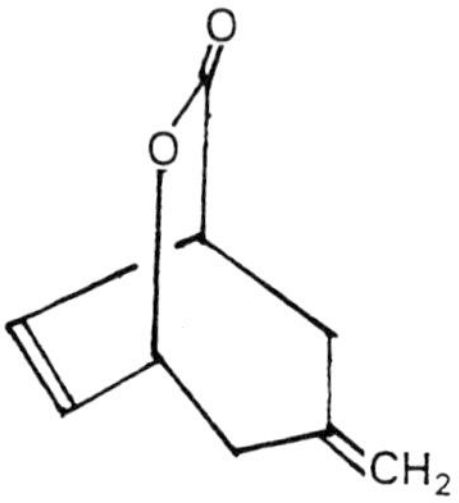

(63)

The reaction of 2*H*-pyrone with α,ω-dienes, for example, H$_2$C = CRCO$_2$CH$_2$CH = CH$_2$ (R = H, Me) yields Diels-Alder adducts 61, which on pyrolysis, lose CO$_2$ to give polycycles 62 (T.M. Swarbrick, I. Marko, and L. Kennard, Tetrahedron Letters, 1991, <u>32</u>, 2549). The Diels-Alder reaction between 2*H*-pyrone and acrylonitrile gives bicyclo[2.2.2]oct-2-en-*endo*--5,*endo*7-dicarbonitrile and the reactions of 2*H*-pyrone with fumaronitrile and maleonitrile yielding monoadducts have been reported (T. Shimo *et al.*, J. Heterocyclic Chem., 1992, <u>29</u>, 811). The [4 + 2]cycloaddition reactions of substituted 2*H*-pyrones with *N*-substituted maleimides have been studied

(*idem*, Nippon Kagaku Kaishi, 1988, 1984) and it has been reported that the photosensitized cycloaddition of 4,6-dimethyl-2*H*-pyrone and methyl 2-oxo-
-2*H*-pyran-5-carboxylate with dienophiles, for example, methacrylonitrile yielded two types of [2 + 2]-cycloadducts (T. Shimo *et al.*, J. Heterocyclic Chem., 1993, <u>30</u>, 419).

2*H*-Pyrone undergoes a [4 + 3] cycloaddition with $AcOCH_2C(=CH_2)$-CH_2SiMe_3 in the presence of $[(Me_2CHO)_3P]_2Pd$ to give the methyleneoxa-bicyclononenone 63 (B.M. Trost and S. Schneider, Angew. Chem., 1989, <u>101</u>, 215).

(ii) 4H-Pyrones. The dirhodium tetraacetate-catalyzed decomposition of diazomalonaldehyde in hot *n*-butyl vinyl ether leads to 2-butoxy-2,3-dihydro-
-4*H*-pyrone (1), which on treatment with acid affords 4*H*-pyrone (2) (E. Wenkert *et al.*, J. Org. Chem., 1990, <u>55</u>, 4975). This is also obtained by the decarboxylation of chelidonic acid (3) in boiling 1,2,3,4-tetrahydro-naphthalene in the presence of a catalytic quantity of copper and 2,2'-
-bipyridyl as the base (C. De Souza, Y. Hajikarimian, and P.W. Sheldrake, Synth. Comm., 1992, <u>22</u>, 755).

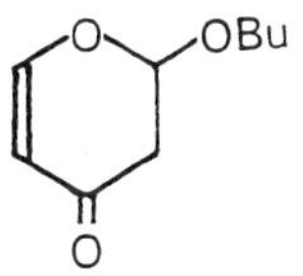

(1) 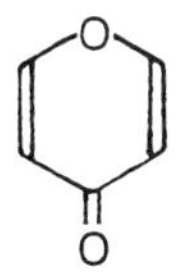 (2) 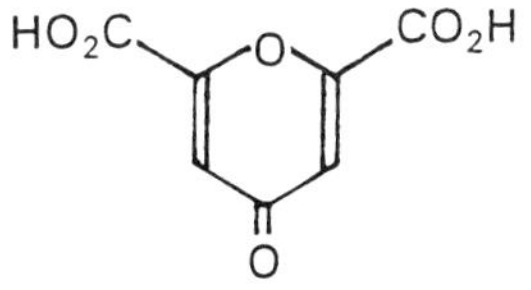 (3)

$MeCH_2COCMe=CHOMe$ on treatment with AcO_2 in AcOH containing BF_3 yields $MeCH_2COCMe=CHOAc$, which on cyclization affords 2,3,5-
-trimethyl-4*H*-pyrone (4) (D. Wirth, D. Gibert, and L. Ferrucio, Eur. Pat. Appl. EP 412,892, 1991). Irradiation of 2-ethyl-3-methyl-4*H*-pyrone (5) in aqueous solution leads to the formation of *trans*-4-ethyl-5-methyl-4,5-
-dihydroxycyclopent-2-enone and *trans*-2-methyl-3-ethyl-4,5-dihydroxycyclo-pent-2-enone. The *trans*-configuration at C-4 and C-5 in the latter case has been confirmed by ^1H NMR spectral data (J.W. Pavlik, S.J. Kirincich, and R.M. Pires, J. Heterocyclic Chem., 1991, <u>28</u>, 537).

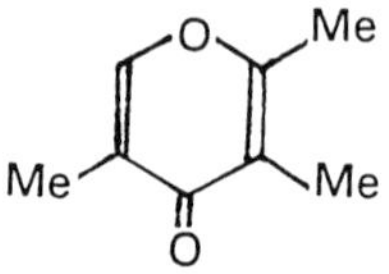

(4)

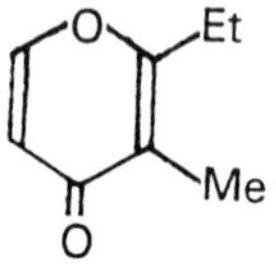

(5)

Benzoylacetone on treatment with acid chlorides, RCOCl (R = Ph, 4-tolyl) and $SbCl_5$ in $MeNO_2$, followed by 30% NaOH solution gives 2-aryl-6-phenyl--4H-pyrones 6 (R = Ph, 4-tolyl) (N.V. Shibaeva *et al.*, Zh. Org. Khim., 1988, <u>24</u>, 2630). A simple synthetic method of obtaining 2,6-diaryl-4*H*-pyrones 8 and/or 4-$R^1C_6H_4COCH_2COCH_2COC_6H_4R^2$-4 with various substituents in the aryl groups, involves the reaction of aroylketenes, generated by pyrolytic decarbonylation of 5-aryl-2,3-dihydrofuran-2,3-diones 7 (R^1 = H, NO_2), with 4-$R^2C_6H_4C(C=CH_2)OSiMe_3$ (R^2 = H, Me, OMe, $OSiMe_3$, Cl, NO_2) (T. Sano, T. Saitoh, and J. Toda, Heterocycles, 1993, <u>36</u>, 2139).

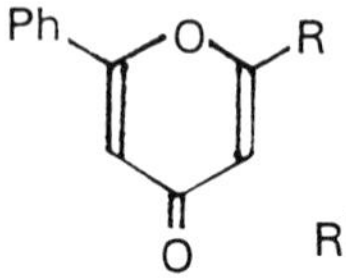

(6) 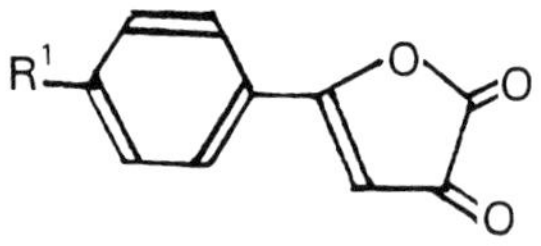   (7) 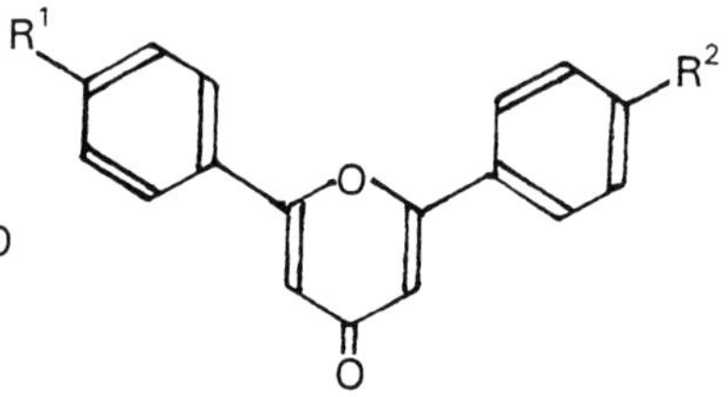   (8)

$R^1C{\equiv}CCH(OH)CHR^3C{\equiv}CR^2$ ($R^1 = n$-Bu, Me, Ph, $R^2 = R^3 = H$; $R^1 = n$-Bu,
$R^2 = Ph$, n-Bu, $R^3 = H$; $R^1 = Ph$, $R^2 = n$-Bu, $R^3 = H$) or $R^1C{\equiv}CCOCR^3 = C = CHR^2$
($R^1 = Me$, $R^2 = R^3 = H$; $R^1 = Ph$, $R^2 = H$, $R^3 = Me$, n-Bu; $R^1 = R^3 = n$-Bu, $R^2 = H$)
on treatment with hot aqueous acid results in the facile formation of di or tri-
-substituted 4H-pyrones 9 (G. Majetich, Y. Zhang, and G. Dreyer,
Tetrahedron Letters, 1993, <u>34</u>, 449). A novel synthesis of 3,5-dimethyl-
-2,6-diphenyl-4H-pyrone (10) from ß-diketone enol esters, involving intramolecular O- to C-acyl shift, followed by cyclization has been discussed
(J.Q. Lu, S.Y. Wang, and P. Zhang, Chin. Chem. Letters, 1992, <u>3</u>, 337).

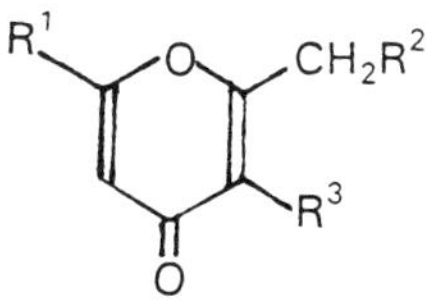

(9)

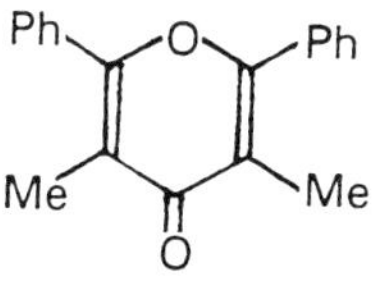

(10)

2-Substituted 4H-pyrones 11 are obtained by the reaction of diazomalon-
aldehyde with $R^2CH = CR^1OSiMe_3$ ($R^1 = Ph$, Me, $R^2 = H$; $R^1 = Me$, $R^2 = Ac$)
(H.-S. Kim and Y.H. Yoon, Bull Korean. Chem. Soc., 1993, <u>14</u>, 159). 3-
-Acetyl-6-aryl-2-methyl-4H-pyrones 12 (R = Ph, 4-BrC_6H_4) are obtained by
boiling the corresponding 6-aryl-2,2-dimethyl-1,3-dioxin-4-ones 13 with
$(MeCO)_2CI I_2$ in dry xylene (Yu.S. Andreichikov, V.L. Gein, and O.V.
Vinokurova, U.S.S.R. SU 1,425,192, 1988). Also the thermolysis of 14
($R^1 = Me$, CI, H, Br) gives aroylketenes, 4-$R^1C_6H_4COCH = C = O$, which
undergoes a [4 + 2] cycloaddition reaction with $R^2C(OH) = CHCOR^3$
($R^2 = R^3 = Me$; $R^2 = CO_2Me$, $R^3 = Me$, Ph, 4-CIC_6H_5) to afford 3-acyl-6-aryl-
-4H-pyrones 15 (*idem*, Khim. Geterotsikl. Soedin., 1989, 161).

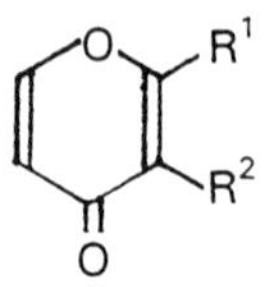

(11)

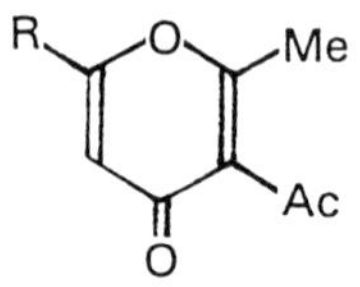

(12)

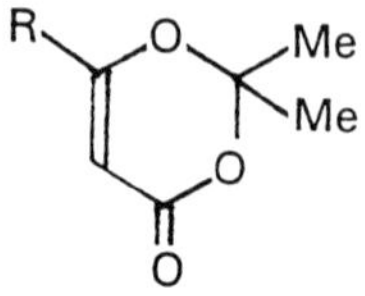

(13)

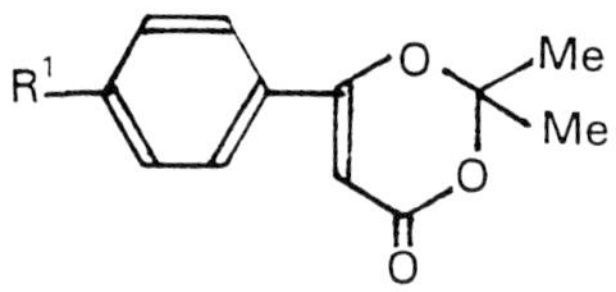

(14)

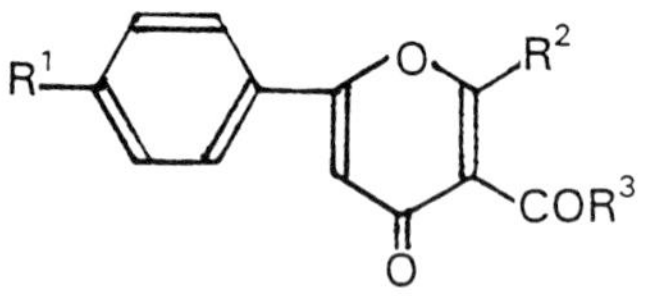

(15)

The reaction of lactone 16 with alcohols, ROH (R = C$_{1-5}$alkyl), in the presence of a base, such as Et$_3$N yields alkyl 4-oxo-4*H*-pyran-2-carboxylates (alkyl comanates) 17 (R = C$_{1-5}$alkyl), useful as synthetic intermediates and antioxidants (M. Tajima, Jpn. Kokai Tokkyo Koho JP 63 218,671 [88 281,671], 1988). The preparations of the methyl esters 18 (R = OMe, SMe) of 2,6-di(trifluoromethyl)-4-oxo-4*H*-pyran-3-carboxylic and thiocarboxylic acid have been reported (S. Babu and M.J. Pozzo, J. Heterocyclic Chem., 1991, *28*, 819). Acyloxypyranones, for example 3-acetoxy-2-methyl-4*H*--pyrone, have been prepared and have activity as activators for inorganic per compounds (for example, sodium perborate, sodium percarbonate) used in textile applications, including bleaching (H. Moeller and B. Kottwitz, Ger. Offen. DE 4,231,465, 1994).

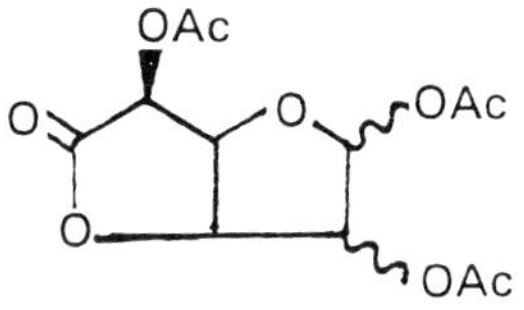

(16)

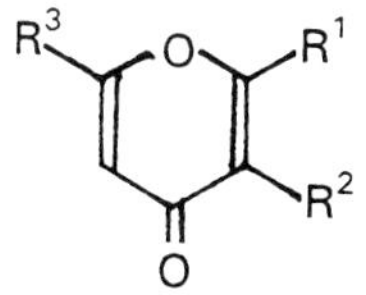

(17) $R^1 = CO_2R$, $R^2 = R^3 = H$
(18) $R^1 = R^3 = CF_3$, $R^2 = COR$
(19) $R^1 = CO_2H$, $R^2 = H$, $R^3 = Ph$

6-Phenyl-2,4,6-trioxohexanoic acid on heating to 100°C in DMSO or AcOH is rapidly converted into 6-phenyl-4-oxo-4*H*-pyran-2-carboxylic acid (19) (M. Stiles and J.P. Selegue, J. Org. Chem., 1991, 56, 4067). A process for preparing 4-oxo-4*H*-pyran-2,6-dicarboxylic acid (chelidonic acid) has been described (B. Harirchian and J.L. Gromley, U.S. US 5,300,657, 1994).

$RCH = CHCOCH = C(SMe)_2$ on oxidation with alkaline H_2O_2 yields epoxides 20 [R = Ph, 4-tolyl, $4\text{-}ClC_6H_4$, 3-anisyl, 4-anisyl, $3,4\text{-}(MeO)_2C_6H_3$, $3,4,5\text{-}(MeO)_3C_6H_2$, 3,4-(methylenedioxy)phenyl], which rearrange in the presence of $Et_2O.BF_3$ to give $HCO(R) = C(OH)CH) = C(SMe)_2$. The latter compound on boiling in AcOH-EtOH cyclizes to yield 5-aryl-2-methylthio-4*H*--pyrones 21 (B. Deb *et al.*, Synthesis, 1987, 893).

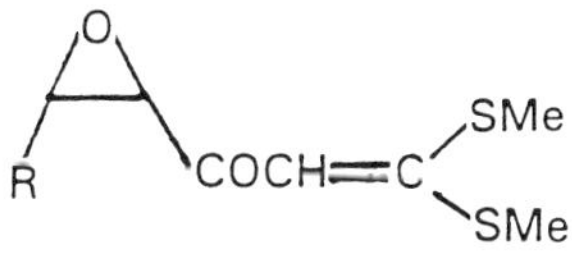

(20) 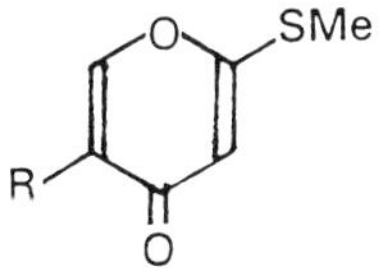   (21) 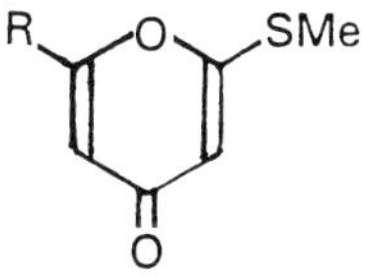   (22)

Condensation of AcCH=C(SMe)$_2$ with RCO$_2$Me (R=Ph, 2-ClC$_6$H$_4$, 4-
-ClC$_6$H$_4$, 3-anisyl, 4-anisyl, 4-tolyl) gives RCOCH=C(OH)CH=C(SMe)$_2$,
which on ring closure afford 6-aryl-2-methylthio-4H-pyrones (22) (L.W.
Singh, H. Ila, and H. Junappa, *ibid.*, p873).

2-Aryl-3-chloro-6-phenyl-4*H*-pyrones (23) are prepared by reacting
PhC≡CCOCHClCOR (R=Ph, ClC$_6$H$_4$) with HCl and 2-(4-bromophenyl)-3-
-chloro-6-phenyl-4*H*-pyrone (23; R=4-BrC$_6$H$_4$) is obtained by the reaction of
BrC$_6$H$_4$COCH$_2$COC≡CPh with *N*-chlorosuccinimide and dibenzoyl peroxide.
3-Iodo-2,6-diphenyl-4*H*-pyrone (24) is obtained from PhC≡CCOCHClCOPh
and ICl (M.G. Marei, M.M. Mishrikey, and I. El S. El-Kholy, J. Chem. Soc.
Pak., 1987, 9, 539).

(23) (24)

PhC≡CCOCH$_2$COR (R=Ph, tolyl, anisyl, halogenophenyl) on nitration
affords 2-aryl-3-nitro-6-phenyl-4*H*-pyrones 25, which on treatment with P$_2$S$_5$
yield the thiones 26. Treatment of thiones 26 with MeNH$_2$ gives the
pyridine derivatives 27 (*idem*, Indian J. Chem., 1988, 27B, 370).

(25) (26) (27)

2-(4-Aminotolyl)-6-phenyl-4*H*-pyrones 28 (R = NH$_2$, NHMe, NMe$_2$, piperidino) are obtained by the bromination of 2-(4-tolyl)-6-phenyl-4H-pyrone by *N*-bromosuccinimide and dibenzoyl peroxide, followed by treatment with the appropriate amine. Reaction of the bromo derivative with NaOEt gives 2--(4-ethoxytolyl)-6-phenyl-4*H*-pyrone (28; R = OEt) (*idem*, Acta Chim. Hung., 1987, <u>124</u>, 733; Indian J. Chem., 1987, <u>26B</u>, 163). The preparation of derivatives of 2-amino-6-phenyl-4*H*-pyrone as antiatherosclerotics and antithombotics (J. Morris, PCT Int. Appl. WO 92 00,290, 1992) and an improved method for the preparation of (5-methoxy-2-oxo-4*H*-pyran-2-yl)-methylammonium bromide (F. Gregan *et al.*, Czech. CS 269,023, 1990) have been reported. 2-Amino-6-aryl-4*H*-pyrones 29 (for example, R = H, 2--MeO, 3-Cl) are obtained by the intramolecular cyclization of the appropriate phenylacetylenic ß-keto amide with methanesulphonic acid (J. Morris and D.G. Wishka, Synthesis, 1994, 43).

(28)   (29)

A number of 4-oxo-4*H*-pyran-3-carboxamide derivatives have been prepared and investigated for use as pharmaceuticals and agrochemicals or their intermediates (H. Yagihara *et al.*, Ger. Offen. DE 3,707,828, 1987; Y. Goto and K. Masamoto, Jpn. Kokai Tokkyo Koho JP 62 187,466 [87 187,466]; 62 212,383 [87 212,383], 1987; 63 10,781 [88 10,781], 1988; 63 66,175 [88 66,175]; 63 91,380 [88 91,380], 1988; Y. Goto and Y. Morishima, *ibid.*, 03 72,474 [91 72,474], 1991). Preparations of 4--oxo-4*H*-pyran-3-carboxanilides as herbicides have been reported (H. Yagihara *et al.*, Jpn. Kokai Tokkyo Koho JP 01 09,983 [89 09,983], 1989).

The reaction between 4*H*-pyran-4-thione (30; R^1 = H) or 2,6-dimethyl--4*H*-pyran-4-thione (30; R^1 = Me) and aryl bromomethyl ketones, R^{2}COCH$_2$Br (R^2 = Ph, 4-BrC$_6$H$_4$, 4-MeOC$_6$H$_4$, 4-MeC$_6$H$_4$, 2-naphthyl, 2-thienyl,

PhCH = CH) yields pyrylium bromides 31, which on rearrangement and desulphurization with DBU afford the vinylogous pyrans 32. Compounds 32 have been shown to exist in the s-cis enone conformation (K. Ohkata, M. Imagawa, and K. Akiba, Heterocycles, 1986, 24, 2817). Several 4-thioxo--4*H*-pyran-3-carboxanilides have been prepared as plant growth inhibitors (H. Yagihara *et al.*, Jpn. Kokai Tokkyo Koho JP 63 154,680 [88 154,680], 1988). For the preparation of alkyl 2-amino-5-cyano-6-(2-furyl)-4-thioxo-4*H*--pyran-3-carboxylate, see K. Peseke, J. Quincoces Suarez, and M. Almeida Saavedra, Ger. (East) DD 294,941, 1991.

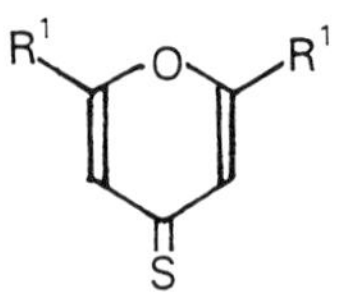

(30) 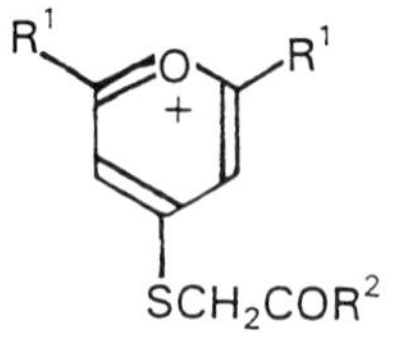   (31) 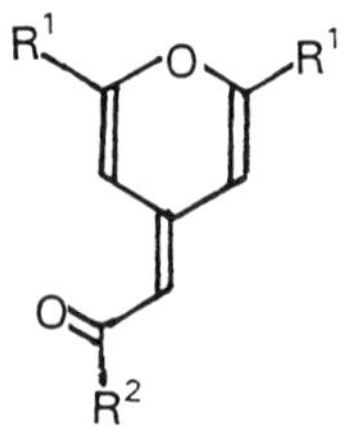   (32)

The synthesis and spectral properties of unsymmetrical heteropolyenes based on 4*H*-pyran (Yu. A. Nesterenko, A.I. Tolmachev, and A.D. Kachkovskii, Ukr. Khim. Zh., 1988, 54, 883) have been investigated. The electrophilic attack of esters to dianions of ß-monosubstituted amino-*a*,ß--unsaturated ketones and the subsequent cyclization of the addition product offers a valuable generalization of the preparation of 4*H*-pyrones and *N*-substituted pyridin-4-ones (G. Bartoli *et al.*, J. Chem. Soc., Perkin 1, 1993, 2081). The syntheses and study of derivatives of 4-pyrones have been reported (T.K. Khanmamedov and O.B. Abdyev, Issled. v Obl. Sinteza Monomer. i Polimer. Produktov, Baku, 1987, 66. From Ref. Zh., Khim., 1988, Abstr. No 5Zh194). The [4 + 3] cycloaddition of photochemically generated 4*H*-pyrones derived oxyallyl zwitterions and furans has been examined as a potential approach to keto-bridged cyclooctenes (F.G. West *et al.*, J. Org. Chem., 1993, 58, 6795).

(iii) Hydroxy-2H-pyran-2-ones (hydroxy-2H-pyrones, hydroxy-2-pyrones, hydroxy-α-pyrones). 3-Hydroxy-2-pyrone (1) useful as additive to foods, cosmetics, and toiletries is manufactured by heating xylonic acid (2, R = H) or certain of its salts 2 (R = alkali metal, alkali earth metal) in the presence of an acid, for example, boiling calcium xylonate with AcOH-H_2O (H. Masuda, T Tsuda, and S. Mihara, Jpn. Kokai Tokkyo Koho JP 04 01,188 [92 01,188], 1992). 3-Hydroxy-2-pyrone (1) is also obtained by heating mucic acid in the presence of P_2O_5 and phosphate salts, for example $NaH_2PO_4.2H_2O$ (A.V. Cheprakov, D.I. Makhon'kov, and I.P. Beletskaya, U.S.S.R. SU 1,558,912, 1990).

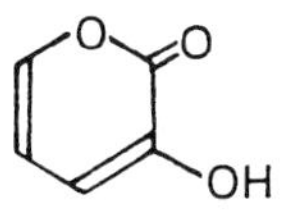

(1)

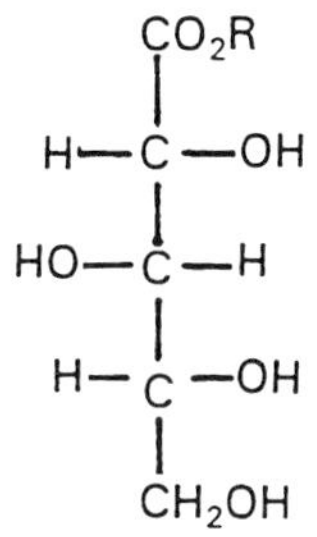

(2)

4-Hydroxy-2-pyrone (3) useful as an intermediate for the preparation of drugs, agrochemicals, and industrial bactericides and other chemicals, is prepared by the decarboxylation of 4-hydroxy-2-oxo-2*H*-pyran-3-carboxylic acid (4), which is obtained by dechlorination of 6-chloro-4-hydroxy-2-oxo-
-2*H*-pyran-3-carboxylic acid, on treatment with hydrogen over Pd/C in THF at 25-30 °C and at ordinary pressure, or by dechlorination and deketonization of pyranodioxinedione 5 (R = lower alkyl, Ph) (S. Kojima *et al.*, Jpn. Kokai Tokkyo Koho JP 02 59,566 [90 59,566], 1990).

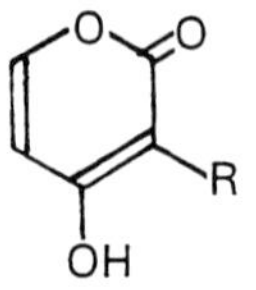

(3) R = H
(4) R = CO$_2$H

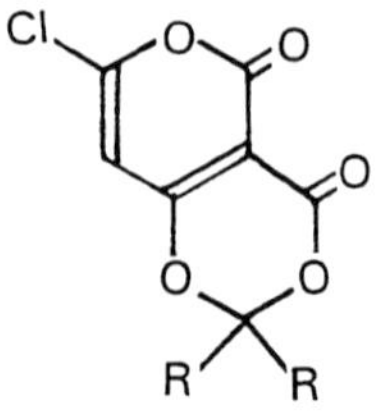

(5)

3-Alkyl-4-hydroxy-6-methyl-2-pyrones 6 (R = Et, Bu, allyl, PhCH$_2$, MeO$_2$CCH$_2$) are prepared by first protecting the diketone moiety of the polyketide, MeCOCH$_2$COCH$_2$CO$_2$Me by Cu(II) complex formation and then by regioselective alkylation to afford MeCOCH$_2$COCHRCO$_2$Me, which on cyclization yield the desired alkylhydroxypyrones (J. Cervello, J. Marquet, and M. Moreno-Manas, Tetrahedron Letters, 1987, 28, 3715).

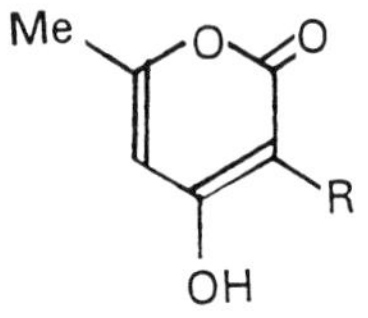

(6)

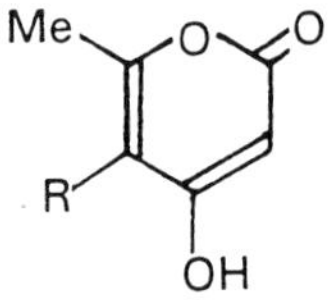

(7)

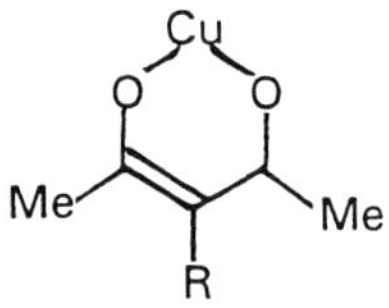

(8)

In the synthesis of 5-alkyl-4-hydroxy-6-methyl-2-pyrones 7 (R = H, Me, Bu, $CH_2CH=CH_2$, $PhCH_2$, PCHMe, Ph_2CH) it has been shown that the yields of $^\downarrow$-carboxylation of MeCOCHRCOMe can be improved by using MeCH(OH):CRCOMe, prepared by mild hydrolysis of the corresponding copper (II) complexes 8 as a starting material. Cyclization of the resulting $MeCOCHRCOCH_2CO_2$ gives the required alkylhydroxypyrone (*idem*, Synth. Comm., 1990, <u>20</u>, 1931).

An improved synthesis of 6-ethyl-4-hydroxy-3,5-dimethyl-2-pyrone (9) has been reported and functionalization at the CH_2 group of the ethyl group described (G. Koester and R.W. Hoffmann, Ann., 1987, 987). Several 3,5,6-trialkyl-2-pyrones 10 (R = alkyl) have been obtained by regioselective cyclotrimerization of aliphatic acyl chlorides, RCH_2COCl (R = Me, Et, Pr, Bu) promoted by $AlCl_3$ (G. Sartori *et al.*, Synthesis, 1993, 851). However, acetyl chloride undergoes cross-condensation with aromatic acid chlorides RC_6H_4COCl (R = H, Me, MeO, NO_2) in the presence of $AlCl_3$ to yield 2-
-acetylindandiones 11 (*idem*, Tetrahedron Letters, 1991, <u>32</u>, 2153).

(9)   (10)   (11)

The reaction of trimethylsilyl enol ethers of aldehydes and acyclic or cyclic ketones, $R^1CH=CR^2-OSiMe_3$ [$R^1 = R^2 = H$, Me; $R^1 = H$, $R^2 = Me$, Et, 4-
-MeC_6H_4; $R^1 = Me$, Et, $R^2 = H$; $R^1R^2 = (CH_2)_3$] with carbon suboxide gives substituted 4-(trimethylsilyloxy)-2-pyrones 12 ($R^3 = SiMe_3$), which on treatment with water yield the corresponding 5,6-disubstituted 4-hydroxy-2-
-pyrones 12 ($R^3 = H$) (L. Bonsignore *et al.*, Heterocycles, 1989, <u>29</u>, 913).

54

(12)

(13)

4-Hydroxy-6-methyl-2-pyrone (triacetic acid lactone) (13; R = H) is
alkylated at its active C-3 atom by thermodynamically controlled Pd-
-catalyzed allylic alkylation, for example the reaction of hydroxypyrone 13
(R = H) with PhCH = CHCH$_2$OAc gives 13 (R = CH$_2$CH = CHPh) which on
reduction yields 4-hydroxy-6-methyl-3-(3-phenylpropyl)-2-pyrone [13;
R = (CH$_2$)$_3$Ph] (M. Moreno-Manas *et al.*, Tetrahedron Letters, 1988, <u>29</u>, 581).
The reaction of hydroxypyrone 13 (R = H) with *cis*-3-acetoxy-5-methylcyclo-
hexene in PhMe containing Pd[CH(COMe)$_2$]$_2$, excess PPh$_3$ and DBU affords
cis- and *trans*-4-hydroxy-6-methyl-3(5-methylcyclohexene-3-yl)-2-pyrone (14
and 15) in a 90.6:9.4 ratio, respectively (M. Moreno-Manas, J. Ribas, and A.
Virgili, J. Org. Chem., 1988, <u>53</u>, 5328).

(14)

(15)

3-(Substituted allyl)-4-hydroxy-6-methyl-2-pyrones 18 [R = (*E*)-PhCH =
-CHCH$_2$, 2-cyclohexen-1-yl, (*E*)-MeCH = CHCHMe, Me$_2$C = CHCHMe,
MeCH = CHCMe$_2$, Me$_2$C = CHCH$_2$, CH$_2$ = CHCMe$_2$] have been obtained by the
following procedure. Regioselective allylation of the model polyketide
system methyl 3,5-dioxohexanoate at the C-2 position is accomplished even
with secondary and tertiary radicals by palladium catalyzed allylation of its
copper (II) complex 16 (M = Cu) by reaction with ROAc, thus yielding the
intermediates 17, which on cyclization afford the 3-allyl-4-hydroxy-6-
-methyl-2-pyrones 18. Initial results have indicated that the cobalt (II)
complex 16 (M = Co) could also be useful (J. Marquet, M. Moreno-Manas,
and M. Prat, Tetrahedron Letters, 1989, <u>30</u>, 3105).

(MeCOCH$_2$COCH$_2$CO$_2$Me)M

(16)

+

ROAc

$\xrightarrow{\text{1.NaH,THF}\quad\text{2.Pd(dba)}_2,\text{ PPh}_3}$

MeCOCH$_2$COCHRCO$_2$Me

(17)

DBU, C$_6$H$_6$ Boil

(18)

3-Acyl-6-alkyl-4-hydroxy- and 3,4-diacyl-6-alkyl-2-pyrones 19 [R^1,R^3 = (un)substituted alkyl optionally bound through O or N atom; R^2 = H, (un)substituted alkyl] have been prepared as elastase inhibitors. Thus $Me(CH_2)_7COCH_2CO_2H$ on cyclization with carbonyldiimidazole in THF gives 4-hydroxy-3-octylcarbonyl-6-octyl-2-pyrone (20) (B. Ternai and M.L. Cook, PCT Int. Appl. WO 88 10,258, 1988).

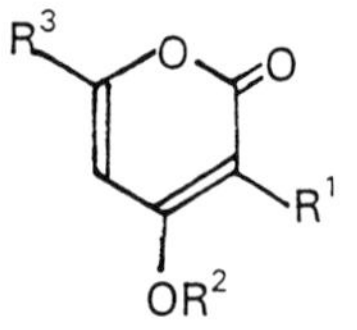

(19) 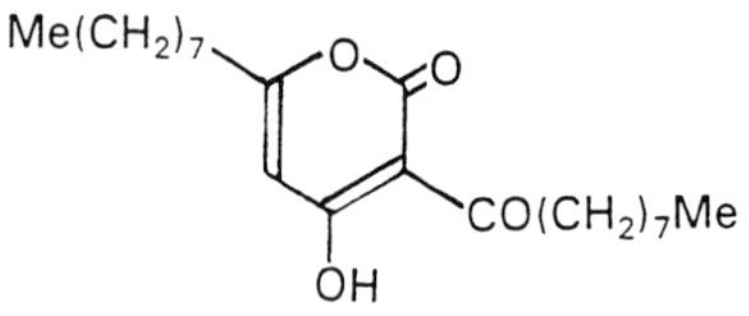   (20)

Photoirradiations of 4-(ω-alkenyloxy)-6-methyl-2-pyrones 21 (X = CH_2 n = 2-4; X = O, n = 3) give site- and regio-specific intramolecular [2 + 2] cycloadducts, the oxatricyclic lactones 22 and/or Dewar-type valence-isomer derivatives, plus cyclobutencarboxylic acids 23 (X = CH_2 n = 4; X = O, n = 3). The reaction path depends upon the alkenyl chain length. The two or three carbon chain gives rise to intramolecular [2 + 2] cycloaddition, while the four carbon chain causes valence-isomerization and cycloaddition (T. Shimo *et al.*, J. Heterocyclic Chem., 1991, 28, 745). The intramolecular photochemical reactions of 4-(ω-alkenyloxy)-6-methyl-2-pyrones possessing an alkoxycarbonyl group on the end carbon of the olefinic carbon chain have been investigated (*idem*, J. Org. Chem., 1991, 56, 7150).

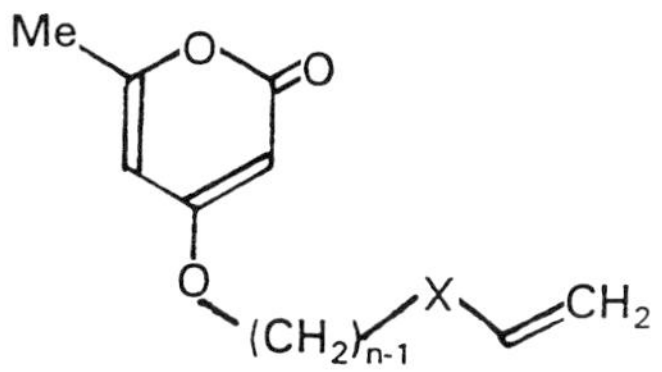

(21)

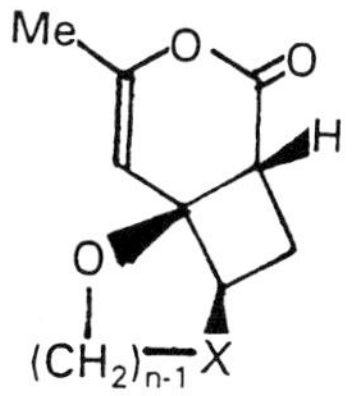

(22)

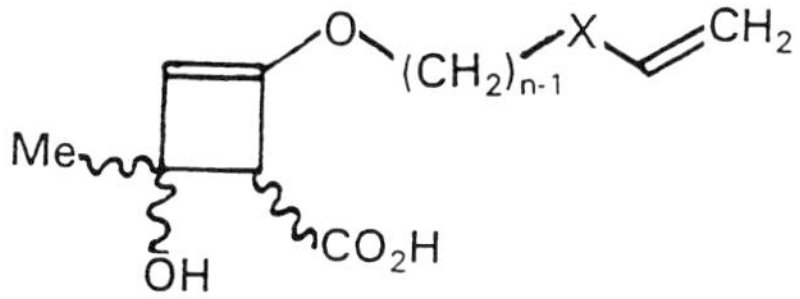

(23)

6-(ω-Alkenyl)-4-methoxy-2-pyrones 24 (n = 1,2,3) and 6-(ω-alkenyl)-3-
-bromo-4-methoxy-2-pyrones 25 (n = 1,2) have been prepared and their
photochemical reactions investigated. Photosensitized reactions of 24
(n = 2,3) yields intramolecular [2 + 2] cycloadducts 26 as tricyclic lactones
site- and regiospecifically. However, 24 (n = 1) gives cyclobutenecarboxylic
acid 27. Cyclobutenecarboxylic acids are also obtained from 25 (n = 1,2) (T.
Shimo, S. Matsuzaki, K. Somekawa, J. Heterocyclic Chem., 1994, $\underline{31}$, 387).

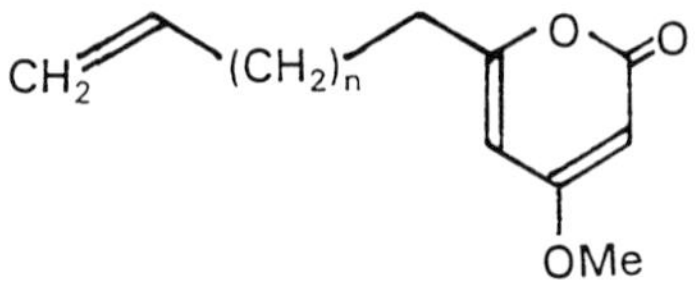

(24)

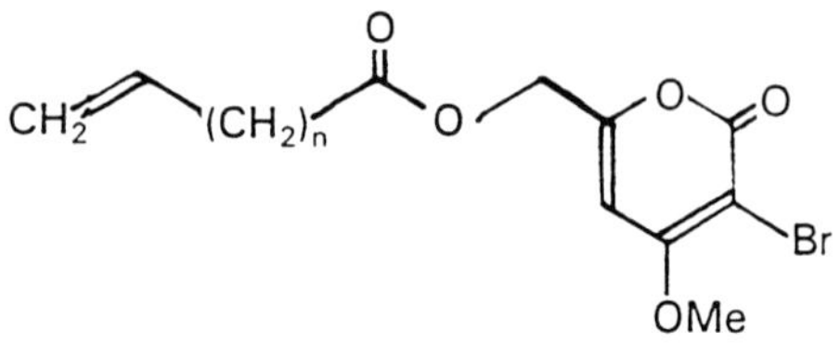

(25)

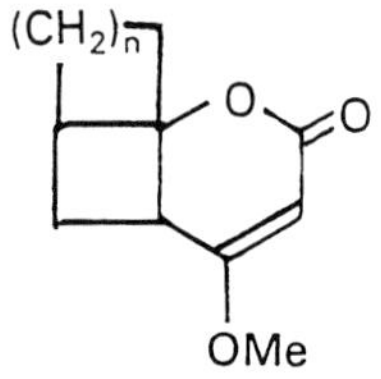

(26)

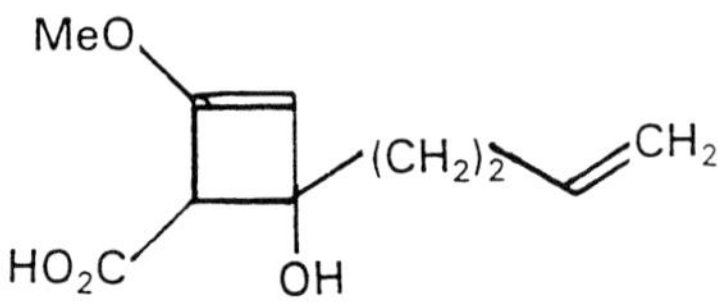

(27)

Arylfurandiones 28 (R=H, Me) react with $CH_2=CHOCHMe_2$ to give 4-
-pyrones 29 and 2-pyrones 30 [3-benzoyl-4-hydroxy-6-phenyl- and 4-
-hydroxy-3-(4-methylbenzoyl)-6-(4-methylphenyl)-2-pyrone, respectively]
(S.N. Shurov *et al.*, Zh. Obshch. Khim., 1992, <u>62</u>, 227). A number of Schiff
bases of 3-aroyl-6-aryl-4-hydroxy-2-pyrones have been synthesized, for
example, by the reaction between the above 2-pyrones and aniline and
substituted 2-phenylenediamines (I. Susnik *et al.*, Monatsh., 1992, <u>123</u>,
817).

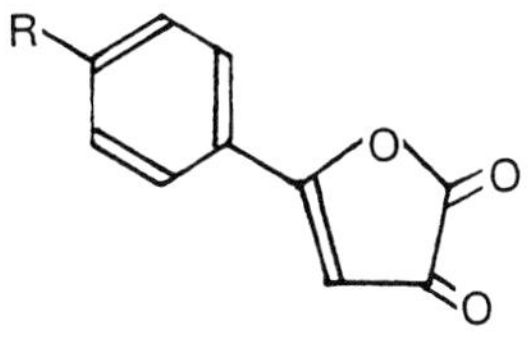

(28)

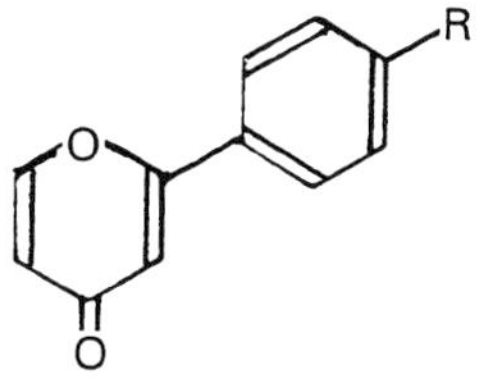

(29)

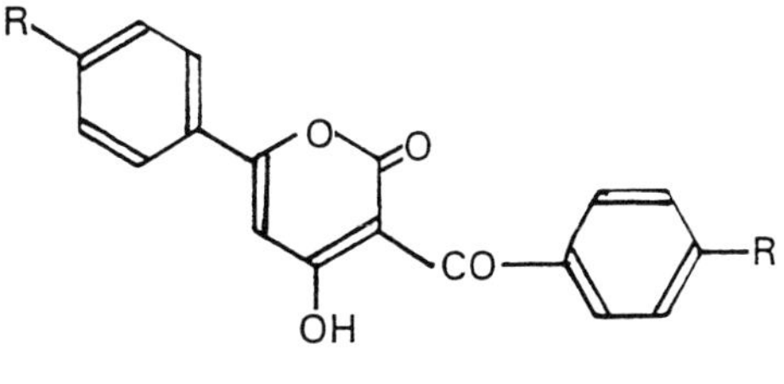

(30)

Treatment of 4-hydroxy-2-pyrone with 4-chloro-2-nitrobenzoyl chloride and Et$_3$N in THF gives 4-(4-chloro-2-nitrobenzoyloxy)-2-pyrone, which on treatment with Et$_3$N, KCN, and 18-crown-6 in MeCN at room temperature gives 3-(4-chloro-2-nitrobenzoyl)-4-hydroxy-2-pyrone (31). A number of related derivatives have been synthesized and investigated for use as herbicides (H. Oishi *et al.*, Jpn. Kokai Tokkyo Koho JP 01 211,577 [89 211,577], 1989).

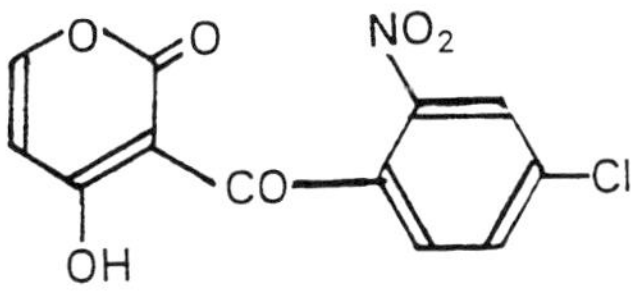

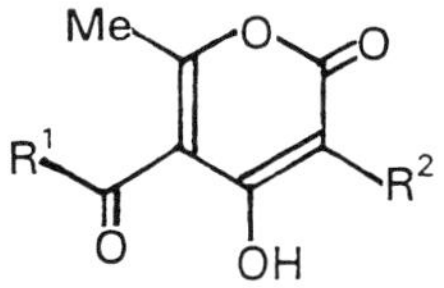

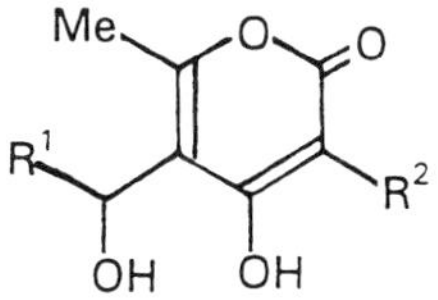

(31)

(32)

(33)

The selective reduction of the 3-substituted 5-carbomethoxy and the 5-
-carbethoxy-4-hydroxy-6-methyl-2-pyrones 32 (R^1 = Me and Et, respectively;
R^2 = H, SPh, $CH_2CH_2CO_2Me$) using 1.1 molecular equivalent of borane-
-methyl sulphide complex yields the corresponding 3-substituted 4-
-hydroxy-5-(1-hydroxyethyl)- and (1-hydroxypropyl)-6-methyl-2-pyrones 33,
respectively (T. Shimizu, S. Hiranuma, and T. Watanabe, Heterocycles,
1993, 36, 2445).

The ester (*E*)-$BrCH_2C(OMe)$ = $CHCO_2Et$ reacts with RCHO [R = Ph, 2-
$MeOC_6H_4$, 4-$MeOC_6H_4$, 3,4-$(MeO)_2C_6H_3$, 1-naphthyl, 2-naphthyl) in the
presence of zinc to give 6-aryl-5,6-dihydro-4-methoxy-2-pyrones 34, which
on radical bromination (2 equivalents NBS) and in situ dehydrobromination
afford 6-aryl-5-bromo-4-methoxy-2-pyrones 35 (M.T. Ayoub, A.M. Dabbagh,
and N.T. Flayyih, J. Iraqi Chem. Soc., 1987, 12, 241).

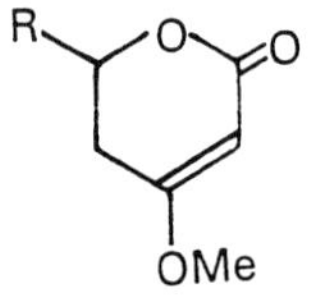

(34) 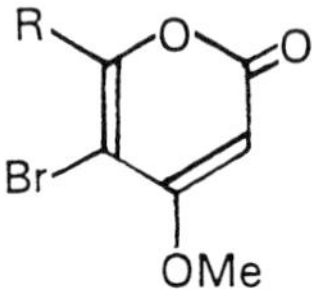   (35)

Treatment of ethoxyethyne, $EtOC \equiv CH$, with CO_2 under 50kg/cm²
pressure at 80°C in THF with a $Ni(COD)_2$-$2PEt_2$ catalyst yields 4,5-diethoxy-
-2-pyrone (36) (T. Tsuda *et al.*, Synth. Comm., 1989, 19, 1575). A
mechanism for the rearrangement of 4,6-dimethoxy-3-methyl-2-pyrone (37)
to 4,6-dimethoxy-5-methyl-2-pyrone (38) has been described (J.K. Cha *et
al.*, Tetrahedron Letters, 1989, 30,3505).

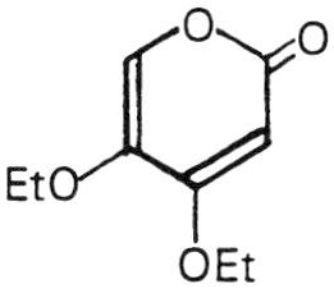

MeO

EtO

OEt

OMe

Me

MeO

Me

OMe

(36) 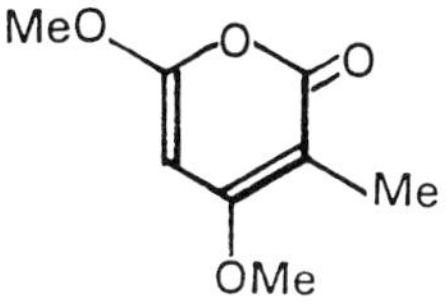   (37) 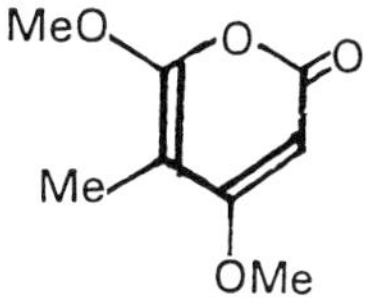   (38)

The main products isolated from the $Rh_2(OAc)_4$-catalyzed reaction of 5-
-bromo-4-methoxy-6-(phenylthio)methyl-2-pyrone with N_2CHCO_2Et and with
$(MeO_2C)_2CN_2$ arise from tandem 1-pot [2,3]-sigmatropic rearrangements of
the corresponding sulphonium ylides followed by [1,3]-allylic Br
rearrangements to afford 2-pyrones 39 and 40, respectively, functionalized
at C-5 and C-6 (P. De March, M. Moreno-Manas, and J.L. Roca, J. Org.
Chem., 1988, <u>53</u>, 5149).

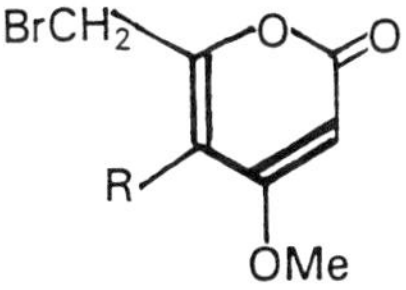

$BrCH_2$

R

OMe

Me

R

OMe

(39) R = $EtO_2CCH(SPh)$ (41) R = H
(40) R = $(MeO_2C)_2C(SPh)$ (42) R = $CH(OH)CMe = CHC_6H_4NO_2$-4

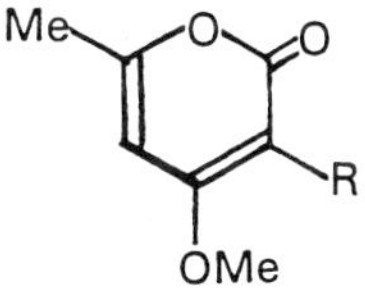

An unusual condensation occurs at C-3 of 4-methoxy-6-methyl-2-pyrone
(41) when it is lithiated at C-3 under kinetically controlled conditions and
then reacted with (*E*)-2-methyl-3-(4-nitrophenyl)propenal to yield the 3-
-alkylated 2-pyrone 42 (M. Moreno-Manas and R. Pleixats, J. Heterocyclic
Chem., 1991, 28, 2041).

Hydrogenation of 3-cinnamoyl-4-hydroxy-6-methyl-2-pyrones 43 (R = Ph,
substituted Ph) over a Pd/C catalyst in EtOAc furnishes the 3-arylpropionyl-
-4-hydroxy-6-methyl-2-pyrones 44 (Y. Rachedi, M. Hamdi, and V. Speziale,
Synth. Comm., 1989, 19, 3437). The reactions of 3-cinnamoyl-4-hydroxy-
-6-methyl-2-pyrone with ammonia and methylamine results in the formation
of 2-(2-phenylvinyl)-6-methyl-4-pyridones 45 (R = H, Me) (M. Iqbal, M.A.
Munawar, and M. Siddiq, J. Chem. Soc. Pak., 1989, 11, 238).

$(43)\,R^1 = COCH = CHR$

$(44)\,R^1 = COCH_2CH_2R$

(45)

(46)

The asymmetric synthesis of 3-(*S*)-α-ethylbenzyl-6-(*R*)-α-ethylphenylethyl-
-4-hydroxy-2-pyrone (46), an HIV Protease inhibitor, *via* a novel α-
-oxoketene/ketene [4 + 2] cyclo-addition reaction has been reported (R.B.
Gammill *et al.*, J. Amer. Chem. Soc., 1994, 116, 12,113).

6-Methyl and 6-phenyl substituted 3-nitro- and 3-phenylazo-4-hydroxy-2-
-pyrones 47 (R^1 = NO_2, N = NPh; R^2 = Me, Ph) on hydrogenation yield the
corresponding 3-amino-4-hydroxy-6-methyl- and -6-phenyl-2-pyrone
derivatives, which on acylation with R^3COCl (R^3 = Ph, 4-tolyl, 4-MeOC$_6$H4,
4-ClC$_6$H$_4$, 3,4-Cl$_2$C$_6$H$_3$) and sulphonylation with R^4SO_2Cl (R^4 = 4-tolyl, 4-
-AcNHC$_6$H$_4$) in pyridine yield derivatives 47 (R^1 = NHCOR^3, NHSO$_2$R^4,
respectively, same R^2-R^4). Bromination of 4-hydroxy-6-methyl-2-pyrone (47;
R^1 = H, R^2 = Me) with Br-AcOH affords 3-bromo-4-hydroxy-6-methyl-2-pyrone
(47; R^1 = Br, R^2 = Me), but all attempts to convert it to 3-amino-4-hydroxy-6-
-methyl-2-pyrone (47; R^1 = NH$_2$, R^2 = Me) failed (S. Radl, V. Houskova,and
V. Zikan, Cesk. Farm., 1989, 38, 219).

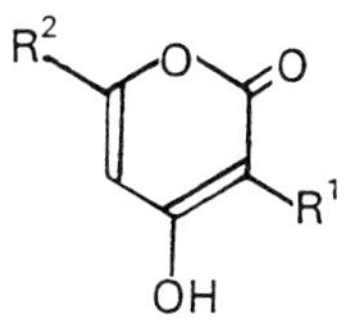

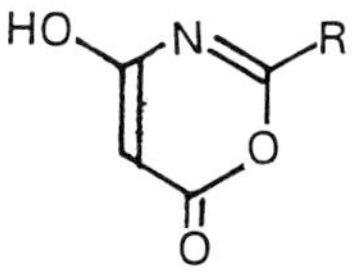

(47) (see text)
(49) $R^1 = CONHCOR$, $R^2 = RCONH$

(48)

4-Aryl-1,3-oxazine-4,6-diones 48 ($R = Ph$, 4-tolyl, 4-MeOC$_6$H$_4$, 4-ClC$_6$H$_4$) with no substituent at the C-5 position of the heterocycle, dimerize in DMSO or in aprotic solvents containing Et$_3$N to yield the 6-benzoylamino-3-benzoyl-aminocarbonyl-4-hydroxy-2-pyrones 49. The structure of 6-(4-methoxy-benzoylamino)-3-(4-methoxybenzoylaminocarbonyl)-4-hydroxy-2-pyrone (49; $R = 4$-MeOC$_6$H$_4$) has been determined by x-ray crystallography (V.E. Zakhs *et al.*, Zh. Org. Khim., 1992, 28, 2611).

The ready conversion of sugar derived 5,6-dihydro-2-pyrones into 3-
-acyloxy- and 3-acylamino-2-pyrones (A. Nin, O. Varela, and R.M. De Lederkremer, Synthesis, 1991, 73) and the preparations of 6-mercapto-methyl-4-methoxy-2-pyrone and some of its derivatives (R. Bacardit *et al.*, J. Heterocyclic Chem., 1989, 26, 1205) have been reported.

Acid-mediated deuteration of bi(acetylpyrandione) 50 ($R^1 = R^2 = H$) affords 50 ($R^1 = H$, $R^2 = D$). Acetylpyranylbutenoates 51 ($R^3 = Me$, Et), precursors of 50, are deuterated in CF$_3$CO$_2$D to yield 50 ($R^1 = R^2 = D$). In both cases chelated acetyl groups were unaffected (L. Crombie and R.V. Dove, Chem. Comm., 1987, 438).

64

(50)

(51)

(iv) Hydroxy-4H-pyran-4-ones (hydroxy-4H-pyrones, hydroxy-4-pyrones, hydroxy-γ-pyrones). The reactions of propionyl and butyryl chlorides R^1CH_2COCl (R^1 = Me, Et), with catalytic amounts of $AlCl_3$ and then with water afford 3,5,6-trisubstituted 2-hydroxy-4-pyrones **1** ($R^1 = R^2$ = Me, Et, R^3 = Et, Pr), the structures of which have been determined by ^1H and ^{13}C NMR, IR, and UV spectroscopy (R.V. Anbrokh *et al.*, Ukr. Khim. Zh. (Russ. Ed.), 1987, **53**, 537).

(1)

(2)

3-Hydroxy-2-methyl-4-pyrone (maltol) (2; R = Me) of 99% purity has been obtained by hydrolysis of streptomycin sulphate with 0.5M NaOH at 90-100°C for 0.25h [T. Constantinescu, S. Budrugeac, and E. Puricel, Rev. Chim. (Bucharest), 1992, 43, 704). Maltol is recovered and purified by heating it in a mixture with a liquid hydrocarbon (for example, *a*-pinene) in which it is too insoluble to form a co-distillate. Thus it is condensed and recovered as pure crystals (A. Fleisher *et al.*, U.S. US 5,221,756, 1993). 3--Hydroxy-2-methyl-4-pyrone (2; R = Me) has been prepared by cyclization on boiling MeCOCH(OBz)COCH = CHNMe$_2$ in AcOH, followed by saponification of the product. Similarly 2-ethyl-3-hydroxy-4-pyrone (2; R = Et) has been prepared from EtCOCH(OBz)COCH = CHNMe^2 (H.J. Wild, Eur. Pat. Appl. EP 339,454, 1989).

(3)

(4)

A general method for preparing 5-aryl-3-hydroxy-2-methyl-4-pyrones, for example, 3-hydroxy-2-methyl-5-phenyl-4-pyrone (3), involves efficient Pd--catalyzed coupling of 3-benzyloxy-5-bromo-2-methyl-4-pyrone (4; R^1 = Br, R^2 = PhCH$_2$) with phenylboronic acid to give efficient intermediate 4 (R^1 = Ph, R^2 = PhCH$_2$), which on hydrolysis with concentrated HCl in AcOH affords 3--hydroxy-2-methyl-5-phenyl-4-pyrone (3) (H. Takao, Y. Endo, and T. Horie, Heterocycles, 1993, 36, 1803).

Treatment of 2-aryl-3-halogeno-6-phenyl-4-pyrones 5 (R^1 = Ph, tolyl, anisyl, halogenophenyl, R^2 = halogeno) with NaOH in EtOH-H$_2$O mixtures affords 2-aryl-3-hydroxy-6-phenyl-4-pyrone 6 (R^1 = Ph, tolyl, anisyl, halogenophenyl). Halogenopyrones 5 on treatment with NaOH in EtOH afford the furanones 7 (M.G. Marei, I. El S. El-Kholy, and M.M. Mishrikey, Indian J. Chem., 1987, 26B, 836).

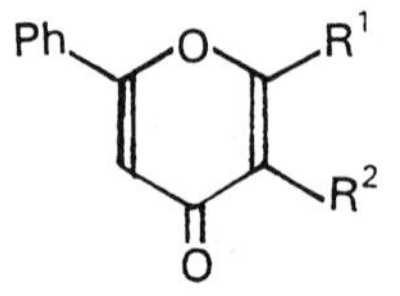

(5)
(6) $R^2 = OH$

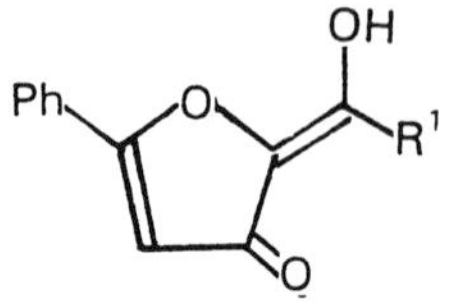

(7)

2-Alkyl-4-oxo-4*H*-pyran-3-yl alkyl/acylsalicylates (salicylic acid maltol esters) and related derivatives 8 ($R^1 = H$, C_{1-5}alkyl, C_{1-5}alkanoyl; $R^2 = H$, C_{1-5}alkyl), useful as antioxidants and antithrombotics, have been obtained by stirring together, for instance, 2-AcOC$_6$H$_4$CO$_2$H, 3-hydroxy-2-methyl-4-pyrone, and DCC in CH$_2$Cl$_2$ at room temperature, followed by removal of *N,N*-dicylohexylureas to give 2-methyl-4-oxo-4*H*-pyran-3-yl acetylsalicylate (8; $R^1 = Ac$, $R^2 = Me$) (B.H. Han, PCT Int. Appl. WO 93 00,339, 1993). For the preparation of 3,5-dihydroxy-4-pyrone, 5-halogeno-3-hydroxy-4-pyrones, and 5-halogeno-4-pyrones, and the syntheses of 3,5-disubstituted 4-pyrones, see H. Takao, Y. Endo, and T. Horie, Heterocycles, 1992, <u>34</u>, 1803).

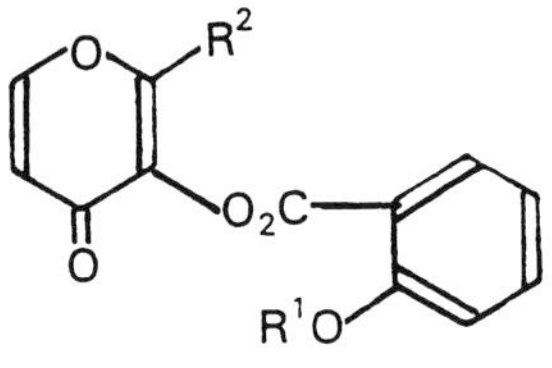

(8)

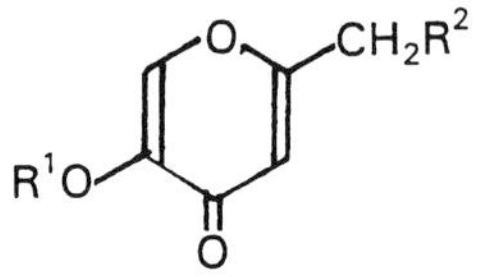

(9)

The preparations of a number of derivatives 9 (R^1 = Me, heptyl, R^2 = OH; R^1 = H, R^2 = O_2CCH_2Cl; R^1 = Me, R^2 = O_2CCH_2Cl, Cl, N_3, $N^+H_3Br^-$, $NHCO_2Et$, $NHCO_2Bu$) of 5-hydroxy-2-hydroxymethyl-4-pyrone (kojic acid) (9; R^1 = H, R^2 = OH) have been reported (R. Cizmaricova, E. Miskikova, and F. Gregan, Pharmazie, 1991, <u>46</u>, 460).

(10)

(11)

Treatment of kojic acid with LiOH, followed by reaction with Et_2SO_4 affords 2-ethoxymethyl-5-hydroxy-4-pyrone (10), useful as a drug for treating pigmentation diseases and as a cosmetic agent for whitening the skin (S. Yamamoto, Eur. Pat. Appl. EP 417,632, 1991). The preparation of kojic acid mono-γ-linolenate (11) a melanin formation inhibitor useful for skin-whitening cosmetics has also been reported (Y. Higa and K. Nakajima, *ibid.*, 347,392, 1989).

Some *S*-substituted 2-thiomethyl-4-oxo-4*H*-pyran-5-yl esters 12 [R^1 = Pr, Pr^i, $(CH_2)_7Me$, 2-methyl-1,3,4-thiadiazol-5-yl, 2-pyridinyl; R^2 = $CHCl_2$, CH_2OR^3, R^3 = substituted Ph] active against bacteria and fungi and stimulants for the growth of plants, have been prepared by replacing bromine of 2--bromomethyl-5-hydroxy-4-pyrone by sulphur containing nucleophiles, followed by acylation of the OH group (M. Veverka, Chem. Pap., 1992, <u>46</u>, 206).

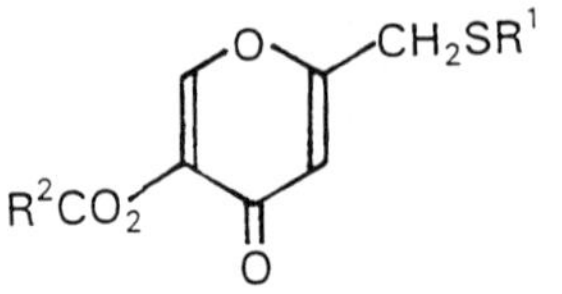

(12)

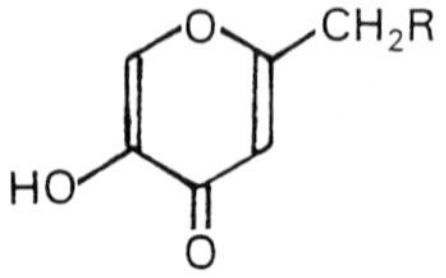

(13)

Kojic acid (9; R^1 = H, R^2 = OH) esters 13 (R = nonsteroidal anti-inflammatory and analgesic carboxylic acid residue), useful as melanin formation inhibitors, analgesics, antithrombotics, and antiinflammatories have been prepared, for example, the reaction of 2-chloromethyl-5-hydroxy--4-pyrone with NaH and ketoprofen affords derivative 13 (R = ketoprofen residue) (A. Nakagawa *et al.*, Jpn. Kokai Tokkyo Koho JP 03 223,275 [91 223,275], 1991). A number of derivatives of kojic acid have been prepared as herbicides, fungicides, agrochemicals, and plant growth regulators, these include some 5-esters of 5-hydroxy-2-hydroxymethyl-4-pyrone 14 [R^1 = (un)substituted Ph, naphthyl; R^2 = H, Me; R^3 = HO, Br, Cl, pyridine residue, 2,4-MeClC$_6$H$_3$OCH$_2$CO$_2$, 2,4-ClC$_6$H$_4$OCH$_2$CO$_2$, azido; n = 0-3] and related derivatives (M. Veverka and E. Kral'ovicova, Czech. CS 269,020, 1990; M. Veverka, *ibid.*, 272,880, 1991; Coll. Czech. Chem. Comm., 1990, 53, 833), and 2-substituted derivatives 15 (R = C$_{1-8}$amino residue; n = 1,2) (H. Kayahara, H. Maeda, and K. Otomo, Jpn. Kokai Tokkyo Koho JP 03 120,267 [91 120,267], 1991).

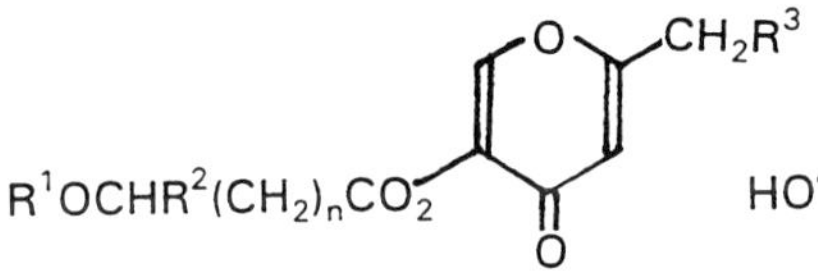

(14)

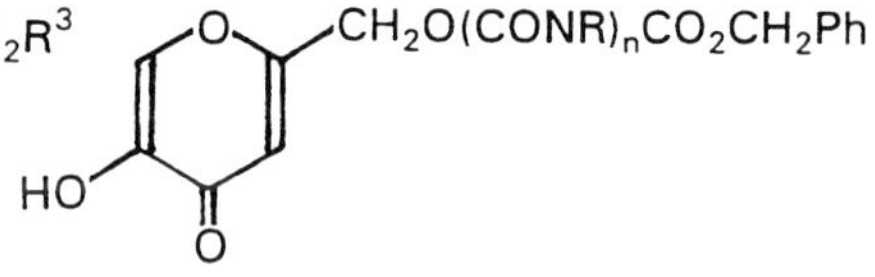

(15)

5-Hydroxy- and 5-methoxy-4-oxo-4*H*-pyranylmethyl 2-alkoxycarbanilates 16 (R^1 =H, R^2 =H, OEt, OPr, OBu, OC_8H_{17}; R^1 = Me, R^2 =H, OC_8H_{17}) have been prepared through the reaction of kojic acid or its 5-methoxy derivative with the appropriate, redistilled 2-$R^2C_6H_4$NCo in boiling xylene optionally containing Et_2N. They have been screened for growth stimulatory activity in chickens and act as plant growth regulators (J. Zamocka, E. Misikova, and J. Durinda, Cesk. Farm., 1992, <u>41</u>, 170; Pharmazie, 1991, <u>4</u>, 610).

(16)   (17)

The preparations of 2-substituted derivatives of 5-hydroxy-4-pyrone of the type 17 (R^1 =H, amino protecting group; R^2 =H, OH protecting group) as intermediates for cephalosporin derivatives have been reported (T. Naito *et al.*, Jpn. Kokai Tokkyo Koho JP 04 91,088 [92 91,088], 1992). Several 5--alkoxy-4-pyrone-3-carboxamides have been prepared and investigated for use as plant growth inhibitors (Y. Goto *et al., ibid.*, 02 108,683 [90 108,683], 1990) and 2-(phosphoryl)sulphenylmethyl-5-hydroxy-4-pyrones and analogues as fungicides (V. Kenecny and M. Uher, Czech. CS 264,978, 1989).

5-Benzyloxy-4-oxo-4*H*-pyran-2-carboxaldehydes (18) is obtained by SeO_2-xylene oxidation of the primary alcohol group in the 2-position of 5--benzyloxy-2-hydroxymethyl-4-pyrone (J. Bransova, M. Uher, and J. Brtko, Chem. Pap., 1993, <u>47</u>, 316).

(18)   (19)

Treatment of kojic acid in water with acetaldehyde and aqueous NaOH at 55°C affords di(5-hydroxy-2-hydroxymethyl-4-oxo-4*H*-pyran-6-yl)methyl-methane (19) as a tyrosinase inhibitor, an antioxidant, and a chelating agent (S. Yamamoto, Jpn. Kokai Tokkyo Koho JP 05 39,284 [93 39,284], 1993).

2. Hydropyrans

(a) Dihydropyrans and derivatives

*(i) 3,4-Dihydro-2*H*-pyrans (2,3-dihydro-4*H*-pyrans).* 3,4-Dihydro-2*H*--pyran (1) is obtained on heating a mixture of pentane-1,5-diol and a cobalt catalyst (prepared by treating cobalt oxalate with hydrogen at 600°C) at 210°C (J. Haji and T. Yamaguchi, Jpn. Kokai Tokkyo Koho JP 01 287,079 [89 287,079], 1989). The rearrangement of tetrahydrofurfuryl alcohol to dihydropyran has been reported (Y.V. Subba Rao *et al.*, J. Org. Chem., 1994, 59, 3998). Treatment of 2,3-dihydro-4*H*-pyran (1) with the benzyltri-ethylammonium borohydride-chlorotrimethylsilane reagent system effects a novel and unusual reaction resulting in the formation of pentane-1,5-diol (S. Baskaran *et al.*, Tetrahedron Letters, 1992, 33, 6371). Treating 3,4-dihydro-2H-pyran (1) with PhIO.BF$_3$ and trioxathiadiodin gives tetra-hydrofurfural in yields of 36 and 80%, respectively (N.S. Zefirov *et al.*, Izv. Akad. Nauk SSSR, Ser. Khim., 1988, 1452).

The OH-protection of alcohols and phenols has been achieved by heating the hydroxylated compound with 3,4-dihydro-2*H*-pyran in the presence of Spanish sepiolite heterogeneous acid clay catalyst (J.M. Campelo *et al.*, Synth. Comm., 1994, 24, 1345). Diols have been monoprotected by selective etherification with dihydropyran in the presence of hexane and silica gel-supported $Al_2(SO_4)_3$ (T. Nishiguchi and K. Kawamine, Chem. Comm., 1990, 1766). 3,4-Dihydro-2*H*-pyran reacts with methacrylic acid in the presence of crosslinked poly(4-vinylpyridine hydrochloride) and phenothiazine to afford tetrahydropyran-2-yl methacrylate (W.R. Hertler, PCT Int. Appl. WO 91 15,453, 1991).

Treatment of *N*-alkyl-*O*-(arylsulphonyl)hydroxylamines 2 (R = Me, Et, Pr, Bu) with 3,4-dihydro-2*H*-pyran (1) results in acid-catalyzed addition to the double bond of 1, followed by cationic rearrangement to give cyclic imidates 3 and 4 from hydride migration (major) and ring expansion (minor), which on lithium aluminium hydride reduction yield amino alcohols 5 and 6, respectively (R.V. Hoffman and J.M. Salvador, J. Chem. Soc., Perkin 1, 1989, 1375).

72

The ß-hydrogen activation *via* loss of methane from 6-[zirconocene-
(methyl)]-3,4-dihydro-2*H*-pyran (7) yields η^2-(3-oxacyclohexyne)zirconocene
(8) which inserts unactivated alkenes and alkynes to give elaborated dihydropyrans, for example, 6-(1-propylpent-1-en-1-yl)-3,4-dihydro-2*H*-pyran (9)
(M.C.J. Harris, R.J. Whitby, and J. Blagg, Synlett, 1993, 705).

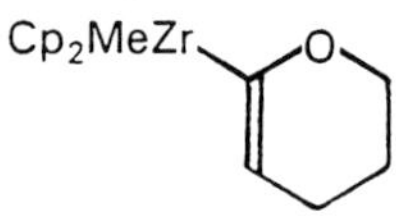

(7) 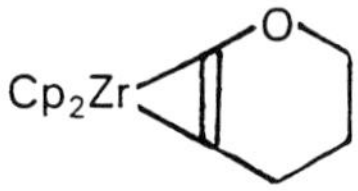   (8) 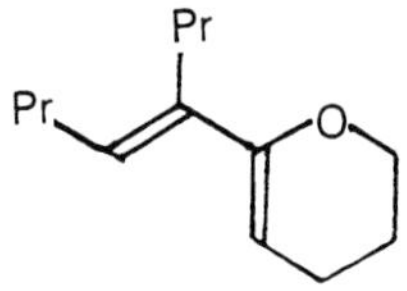   (9)

A procedure for the preparation of 3,4,5-trisubstituted 3,4-dihydro-2*H*-
-pyrans 10 (R^1-R^3 = H, C$_{1-6}$alkyl) from HOCMe$_2$CR^1(OH)CHR^{2}CHR^{3}CH$_2$OH in
the presence of zeolites has been described (W. Hoelderich, R. Fischer, and
W. Mesch, Ger. Offen. DE 3,636,430, 1987). The cyclization of
BuCH = C = CMe(CH$_2$)$_2$CH$_2$COMe in the presence of HgO/4-toluenesulphonic
acid in cyclohexane affords 2-allyl-6-methyl-5-propyl-3,4-dihydro-2*H*-pyran
(11) (*E/Z* = 85:15) (T. Delair, A. Doutheau, and J. Gore, Bull. Soc. Chim. Fr.,
1988, 125).

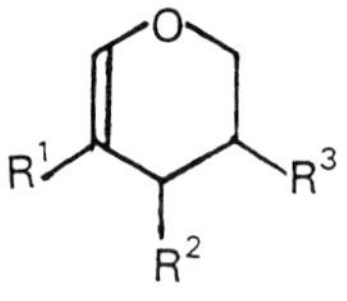

(10) 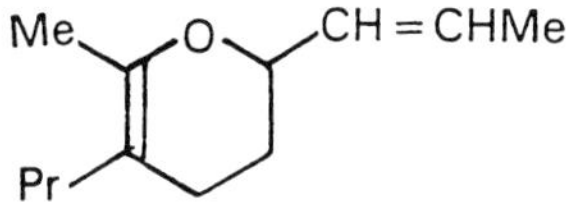   (11)

The ring cleavage of dioxabicyclic compound 12 gives (3*S*)- and (3*R*)-3-
-hydroxymethyl-6-methyl-5-(4-toluenesulphonyl)sulphinyl-3,4-dihydro-2*H*-
-pyran (13) (C. Iwata *et al.*, Tetrahedron Letters, 1987, 28, 3131). For the
preparation of 2-methyl-6-propyl-5-(4-tolylsulphonyl)-3,4-dihydro-2*H*-pyran
(14), see H.K. Jacobs and A.S. Gopalan, J. Org. Chem., 1994, 59, 2014).

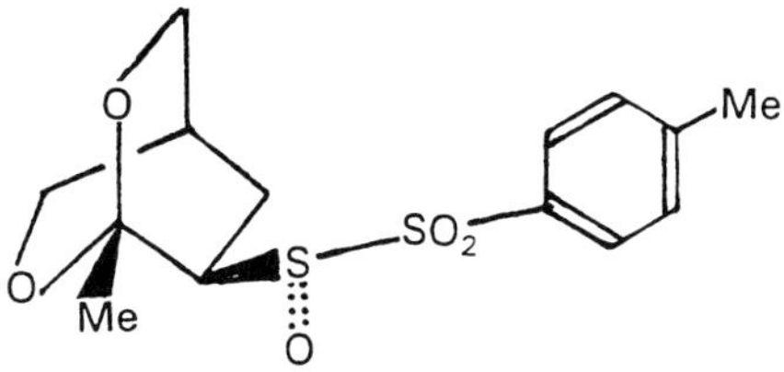

(12)

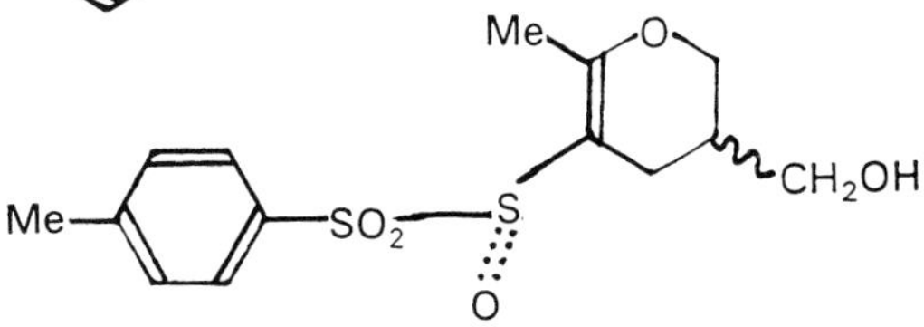

(13)

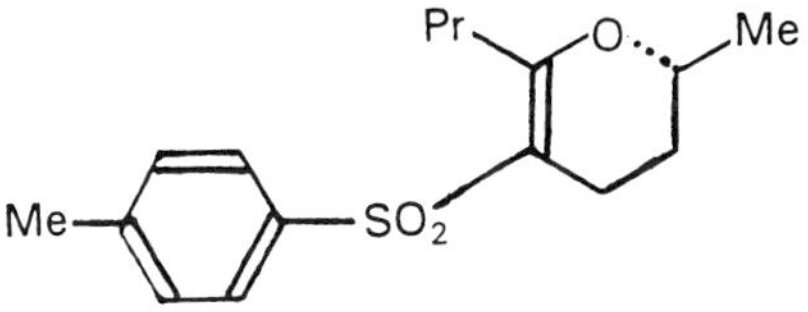

(14)

2-Aryldihydropyrans 15 [the dotted line denotes a double bond at the
4,5- or 5,6-position] have been prepared by the reaction of 3,4-dihydro-2*H*-
-pyran (1) with CF_3SO_3R in the presence of a base and a Pd complex with a
2,2'-bis(diarylphosphino)-1,1'-binaphthyl (BINAP) ligand. They are useful as
intermediates for pharmaceuticals and agrochemicals (T. Hayashi, F. Ozawa,
and A. Kubo, Jpn. Kokai Tokkyo Koho JP 04 149,176 [92 149,176], 1992).

The reactions of 3,4-dihydro-2*H*-pyran (16; R^1 = H) and 2-methoxy-3,4-
-dihydro-2*H*-pyran (16; R^1 = MeO) with hemiacetal vinylogues $HOCHR^2CH=$
$-C(OR^4)CH_2R^3$ (R^2 = H, Me; R^3 = H, Et; R^4 = Me, Et) in the presence of a
catalytic amount of $BF_3.OEt_2$ in $MeNO_2$ yields keto acetals 17. Acidic
hydrolysis of acetals 17 affords different products, depending on substituent
R^1. Hence, when R^1 = H, the hydroxy ketone 18 and heterocycles 19 are
obtained and when R^1 = MeO, keto aldehydes 20 are isolated (R^3 = H, Et)
(J.M. Poirier and G. Dujardin, Tetrahedron Letters, 1987, <u>28</u>, 3337).

(15)

(16) Ar = R^1, double bond
      5,6-position

(17)

(18)

(19)

(20)

2-Ethoxy-3,4-dihydro-2*H*-pyran (21) is obtained by treating acrolein with
ethyl vinyl ether, $H_2C=CHOEt$, and lanthanide chlorides at room temperature
(S. Yu *et al.*, Yingyong Huaxue, 1987, <u>4</u>, 72). The reaction has been carried
out in the presence of a Lewis acid and it was found that conditions could be
changed from about 200°C in an autoclave to ambient temperature and
pressure. The catalytic activities of zinc chloride and iodide were highest
among all catalysts tested (*idem*, Shiyou Huagong, 1989, <u>18</u>, 543). Rare
earth chlorides have been used to catalyze the synthesis of 2-butoxy-3,4-

-dihydro-2*H*-pyran (22) from acrolein and butyl vinyl ether (*idem*, Huaxue Xuebao, 1990, <u>48</u>, 1136; L. Shi and X. Tian, Liaoning Shifan Daxue Xuebao, Ziran Kexueban, 1994, <u>17</u>, 60). The preparation of 2-alkoxy-3,4--dihydro-2*H*-pyrans in a continuous flow reactor instead of an autoclave has been proposed (K.G. Akopyan *et al.*, Prom-st., Stroit. Arkhit. Arm., 1988, 34). The preparation of 2,4,4,6-tetraphenyl-3,4-dihydro-2*H*-pyran-2-ol (P. Sebek *et al.*, Coll. Czech. Chem. Comm., 1992, <u>57</u>, 2383) and the convertion of 6-methyl-2-propionyl-3,4-dihydro-2*H*-pyran into (±)-*exo*- and -*endo*-brevicomin (J. Ishiyama *et al.*, Kenkyu Hokoku Asahi Garasu Zaidan, 1991, <u>58</u>, 153) have been reported.

Heating δ-halogeno acylsilanes, for example, $Ph_2MeSiCO(CH_2)_4Br$, in a polar aprotic solvent (*N*-methylpyrrolidin-2-one) results in the formation of 6--diphenylmethylsilyl-3,4-dihydro-2*H*-pyran (23). The chlorides are less reactive than bromides and require the addition of anhydrous KI to promote the cyclization, which involves nucleophilic displacement of the halide by the carbonyl oxygen of the acylsilane (Y.M. Tsai, H.C. Nieh, and C.D. Cherng, J. Org. Chem., 1992, <u>57</u>, 7010).

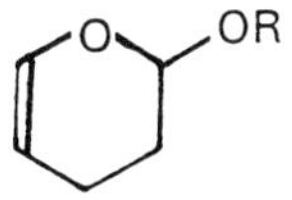

(21) R = Et
(22 R = Bu

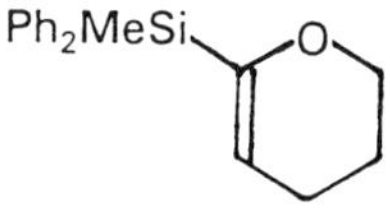

(23)

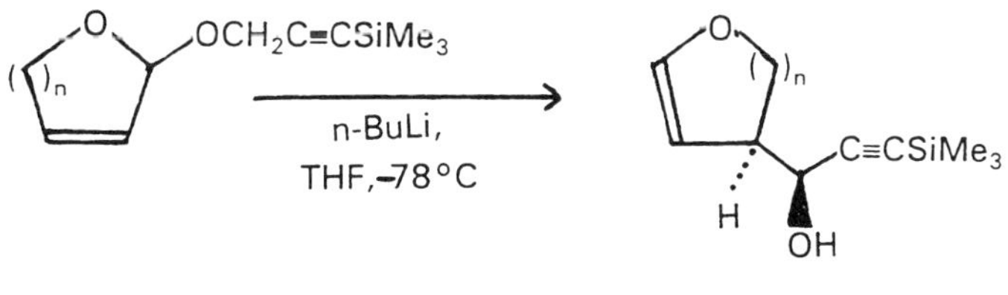

(24)

(25)

A sigmatropic rearrangement of the carbanion from the dihydrofuran 24 (n = 1) proceeds to afford only its [2,3] Wittig product 25 (n = 1). The homologue from 24 (n = 2) gives 25 (n = 2) (K. Tomooka, M. Watanabe and T. Nakai, Tetrahedron Letters, 1990, _31_, 7353).

The addition of dichlorocarbene, generated in situ, to 2-vinyloxy-3,4--dihydro-2_H_-pyrans 26 (R = H, Me) affords 2-(2,2-dichlorocyclopropyloxy)--3,4-dihydro-2_H_-pyran (27; R = H) and 2-(2,2-dichlorocyclopropyloxy)-4--methyl-3,4-dihydro-2_H_-pyran (27; R = Me), respectively (M.G. Voronkov _et al._, Khim. Geterotsikl. Soedin., 1988, 1285).

(26)   (27)

Thermal and catalytic Diels-Alder reactions of divinyl sulphide with acrolein and crotonaldehyde yield 2-vinylthio-3,4-dihydro-2_H_-pyrans 28 (R = H, Me) and bis(3,4-dihydro-2_H_-pyran-2-yl) sulphide (29), respectively. 2--Vinylthio-3,4-dihydro-2_H_-pyran (28; R = H) with hexachlorocyclopentadiene affords hexachloro(dihydropyranylthio)bicycloheptene 30 (L.N. Parshina _et al._, Sulphur Letters, 1987, _6_, 51).

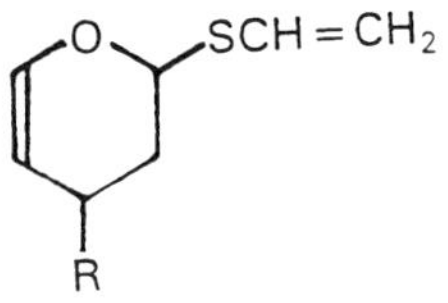

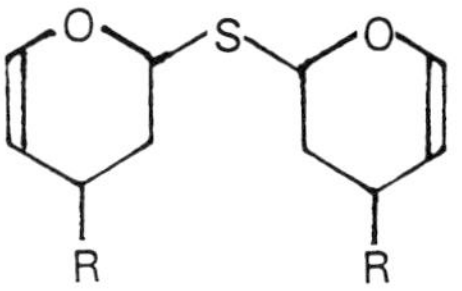

(28)

(29)

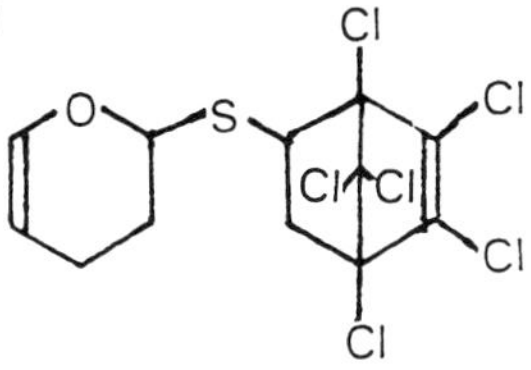

(30)

The cycloaddition of PhCH=CHCOMe with 4-ClC$_6$H$_4$OCH=CH$_2$ in the presence of 4-(HO)$_2$C$_6$H$_4$ and Na$_2$CO$_3$ gives a mixture of *trans-* and *cis*-2-(4-chlorophenoxy)-6-methyl-4-phenyl-3,4-dihydro-2*H*-pyrans (31) (2:1) (W. Himmele *et al.*, Ger. Offen. DE 3,622,599, 1988).

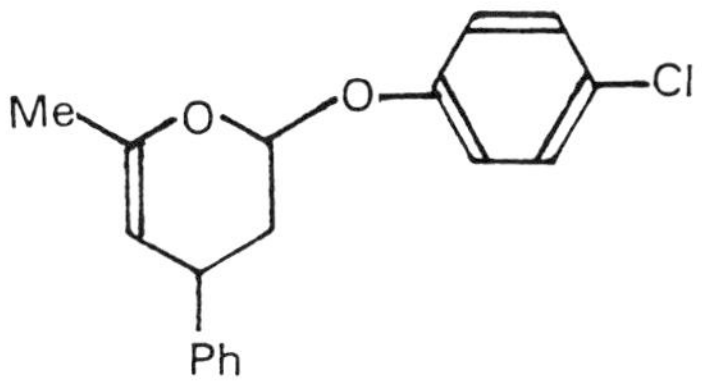

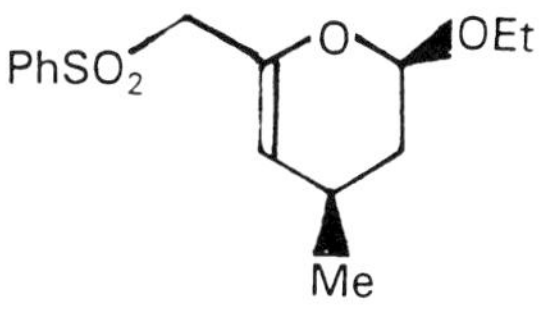

(31)

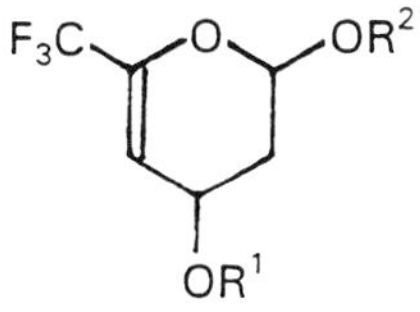

(32)

(33)

Phenyl oxoalkenylsulphinates, for example, E-MeCH = CHCOCH$_2$SO$_2$Ph undergo smooth hetero Diels-Alder reactions with vinyl ethers. For example, with CH$_2$ = CHOEt, in the presence of Eu(fod)$_3$ or TiCl$_2$(Pr^{i}O)$_2$ 2,4-*cis*-2--ethoxy-4-methyl-6-phenylsulphonylmethyl-3,4-dihydro-2H-pyran (32) is formed. The reactions are absolutely endo-selective (E. Wada, H. Yasuoka, and S. Kanemasa, Chem. Letters, 1994, 145).

2,4-Dialkoxy-, 2-alkoxy-4-phenoxy-, and 2,4-diphenoxy-6-(trifluoromethyl)-3,4-dihydro-2H-pyrans, for example, 33 (R^1 = Et, R^2 = Et, CH$_2$CHMe$_2$, CH$_2$CH$_2$Cl; R^1 = Ph, R^2 = Et, Ph) have been prepared by the hetero Diels--Alder reactions of *trans*-ß-(trifluoroacetyl)vinyl ethers, R^{1}OCH = CHCOCF$_3$, with various vinyl ethers (M. Hojo, R. Masuda, and E. Okada, Synthesis, 1989, 215).

The [4 + 2] cycloaddition of RCH = C(CHO)$_2$ (R = cycloalkyl) with CH$_2$ = CHOEt affords 4-cycloalkyl-2-ethoxy-3,4-dihydro-2H-pyran-5-carboxaldehydes 34. The reaction of Me$_2$NCH = C(CHO)$_2$ with BrMg(CH$_2$)$_n$MgBr (m = 5,6) and subsequent cycloaddition of CH$_2$ = CHOEt affords the 3,4--dihydro-2H-pyrans 35 (G.V. Kryshtal *et al.*, Izv. Akad. Nauk SSSR, Ser. Khim., 1990, 2172).

(34)

(35)

ß,ß-Bis(trifluoroacetyl)vinyl ethers undergo hetero-Diels-Alder reactions with electron-rich alkenes such as vinyl ethers, ethyl vinyl sulphide and 1,1--diphenyloxyethene under mild conditions to give 2,4-dialkoxy-, 2-ethoxy-4--phenoxy-, 2-ethylthio-4-alkoxy-, 2,2-diphenoxy-4-ethoxy-, and 2,4-diaryloxy-5-trifluoroacetyl-6-trifluoromethyl-3,4-dihydro-2H-pyrans in excellent yields. For instance, the reaction of Me$_2$CHCH$_2$CH = C(COCF$_3$)$_2$ with CH$_2$ = CHSEt at room temperature furnishes 2-ethylthio-4-(2-methylpropoxy)--5,6-di(trifluoroacetyl)-3,4-dihydro-2H-pyran (36) (M. Hojo, R. Masuda, and E. Okada, Synthesis, 1990, 347).

(36)

(37)

The cycloaddition reactions of 1-alkoxypropa-1,2-dienes, $R^1R^2C=C=CH_2$ ($R^1=EtO$, MeO; $R^2=H$, Me_3Si) with 1-oxa-1,3-dienes, $R^3COCR^4=CHR^5$ (R^3-$R^5=H$, Me), catalyzed by acid-free silica gel give 3-methylene-3,4-dihydro--2H-pyrans 37, which have been transformed into synthetically interesting glutaraldehyde derivatives, for example, $HCOCH_2CH_2C(=CH_2)COSiMe_3$ (M. Conrads and J. Mattay, Ber., 1991, <u>124</u>, 1425). For the preparation of 2-methoxy- and 2-ethoxy-6-methyl-3-methylene-3,4-dihydro-2H-pyran, see *idem, ibid.*, 1989, <u>122</u>, 2207).

The regioselective C-O bond cleavage of the chiral bicyclic ß-arylsulphinyl acetal 38 with $TiCl_4$ in THF at -20°C yielded dihydropyranmethanol 39 as the major product, whereas under basic conditions dihydropyranmethanol 40 was obtained as the major product (C. Iwata *et al.*, Chem. Comm., 1992, 516).

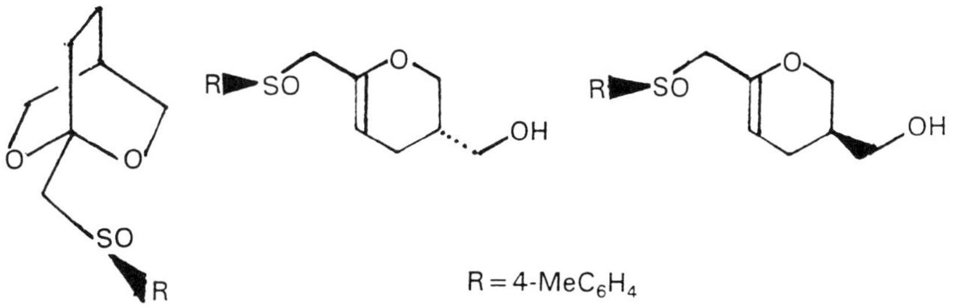

$$R=4\text{-}MeC_6H_4$$

(38)   (39)   (40)

The reactions of 3,4-dihydro-2H-pyran-2-carboxaldehyde (41) with ethyl-metallics based on Li, Ti, and Cu have been examined for stereoselectivity. High selectivity (92:8) has been found for non-chelation controlled addition yielding the *erythro* product 42 when EtLi was used in the presence of $BF_3.OEt_2$. High selectivity (89:11) for chelation controlled addition yielding *threo* product 43 has been observed for ethylcopper reagents in the presence of Mg salts. Ti reagents yielded mixtures rich in products of either chelation or non-chelation controlled addition, depending on the Ti substituents (S.M. Singh and A.C. Oehlschager, Canad. J. Chem., 1988, <u>66</u>, 209). It has been shown that the major product in the hydroformylation of 3,4-dihydro-2H--pyran or 3,6-dihydro-2H-pyran was tetrahydropyran-2-carboxaldehyde

(A. Polo *et al.*, Organometallics, 1992, <u>11</u>, 3525). The oxidation of 2-
-phenylpropene with SeO_2 affords 2-phenylpropenal, which at room
temperature dimerizes to yield 2,5-diphenyl-3,4-dihydro-2*H*-pyran-2-
-carboxaldehyde (44) (T. Laitalainen, P. Kuronen, and A. Hesso, Org. Prep.
Proced. Int., 1993, <u>25</u>, 597). Some appropriate 2,4-disubstituted 3,4-
-dihydro-2*H*-pyran-4-ones have been converted into 2-oxa-3-alkoxy-7-
-hydroxybicyclo[2.2.1]heptanes for use in perfumes (F. Merger *et al.*, Eur.
Pat. Appl. EP 470,469, 1992) and some 5,6-disubstituted 3-benzoyl-4-
-phenyl-3,4-dihydro-2*H*-pyran-4-ones have been prepared (K.G. Pavel *et al.*,
Zh. Org. Khim., 1991, <u>27</u>, 1607).

(41)　　　　　　(42)　　　　　　(43)

(44)　　　　　　(45)

2,5-Dialkylthio-3,4-dihydro-2*H*-pyran-2-carboxaldehydes 45 (R = Me, Et,
Pr, Bu, Me_2CHCH_2, $EtCMe_2$) have been prepared and their antimicrobial
activity determined. Compound 45 (R = Bu) had the highest anti-
staphylococcal activity (L.G. Stepanova *et al.*, Khim.-Farm. Zh., 1988, <u>22</u>,
710).

The self-condensation of $CH_2(CHO)_2$ in water at pH 4.2 affords 4-(1,3-
-dioxopropan-2-yl)-2-hydroxy-3,4-dihydro-2*H*-pyran-5-carboxaldehyde (46)
and (*E*)-2-formylpent-3-ene-1,5-dial, isolated as their K or Na salts (B.T.
Golding, N. Patel, and W.P. Watson, J. Chem. Soc., Perkin 1, 1989, 668).

(46)   (47)   (48)

The treatment of tetrahydropyran-4-one with DMF-POCl$_3$ at 100°C
affords 4-chloro-3,4-dihydro-2*H*-pyran-5-carboxaldehyde (47). Tetrahydro-
thiopyran-4-one reacts in a similar manner (P.R. Giles and C.M. Marson,
Tetrahedron, 1991, <u>47</u>, 1303). The 2,4-di and 2,2,4-trisubstituted 3,4-
-dihydro-2*H*-pyran-5-carboxaldehydes 48 (R^1 = Ph, H, OMe, SMe, R^2 = Ph,
Me, OEt, OMe, SMe, R^3 = 4-ClC$_6$H$_4$, 4-MeOC$_6$H$_4$, 2-, 3-thienyl, 4-O$_2$NC$_6$H$_4$,
4-Me$_2$NC$_6$H$_4$, 2,4,6-Me$_3$C$_6$H$_2$, Ph) have been prepared by the reaction of
R^{3}CH = C(CHO)$_2$ with R^{1}R^{2}C = CH$_2$. Mechanistic and conformational studies
have been reported (D. Dvorak *et al.*, Coll. Czech. Chem. Comm., 1992, <u>57</u>,
2337).

The crystalline potassium salt of a hexameric form of malonaldehyde has
been shown to be racemic potassium hydrogen bis[*trans*-4-(1,3-dioxopropan-
-2-yl)-5-formyl-2-hydroxy-3,4-dihydro-2*H*-pyran (49) (A. Gomez-Sanchez *et
al.*, Tetrahedron Letters, 1992, <u>33</u>, 1361).

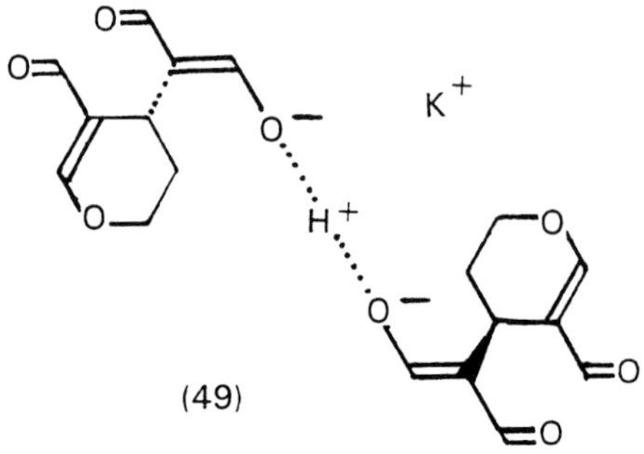

(49)

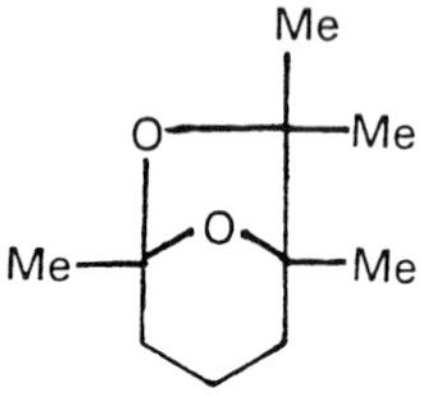

(50)

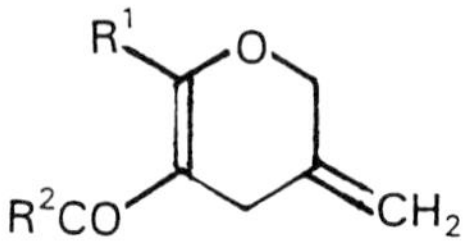

(51)

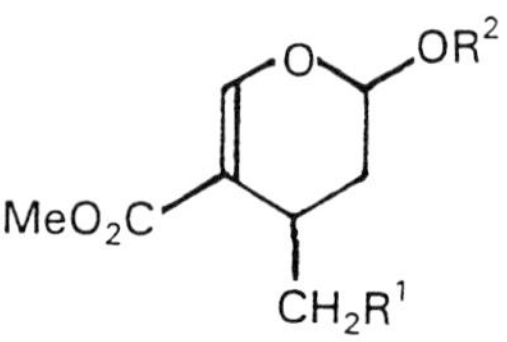

(52)

Methylation of 2-acetyl-2,6-dimethyl-3,4-dihydro-2*H*-pyran, followed by cyclization under acidic conditions furnishes the methylated bicyclic ketal 50 (J.-G. Jun and S.S. Kim, Bull. Korean Chem. Soc., 1994, 15, 612). The reaction of $H_2C=C(CH_2OAc)_2$ with $R^1COCH_2COR^2$ [$R^1=Me$, $R^2=Me$, OMe; $R^1=Ph$, $R^2=Ph$, OMe; $R^1R^2=(CH_2)_3$] in the presence of $Pd(OAc)_2$, PPh_3, and DBU leads to *C*-alkylation and subsequent *O*-alkylation to give 6-substituted 5-acyl-3-methylene-3,4-dihydro-2*H*-pyrans 51 (Y. Huang and X. Lu, Tetrahedron Letters, 1987, 28, 6219). The tandem-Knoevengael hetero--Diels-Alder reaction of $Cl_3CCOCH=CHOEt$ with aldehydes R^1CHCHO ($R^1=Me$, $PhSCH_2$, $PhCH_2OCH_2$) and enol ethers $R^2OCH=CH_2$ ($R^2=Et$, $PhCH_2$, Ph, $ClCH_2CH_2$, $PhSCH_2CH_2$), followed by base-catalyzed methanolysis affords the 2,4-disubstituted methyl 3,4-dihydro-2*H*-pyran-5--carboxylates 52 (L.F. Tietze, H. Meier, and H. Nutt, Ber., 1989, 122, 643).

The Diels-Alder reaction of $H_2C=CHOEt$ with $R^1COC(SPh)=CHOAc$ ($R^1=PhOCH_2$, $4-O_2NC_6H_4OCH_2$, $PhSCH_2$, $Cl_3CCH_2OCH_2$, $BzOCH_2$,

Bu^{t}CO$_2$CH$_2$, Me$_2$NCO$_2$CH$_2$, MeO$_2$C) affords 3,4-dihydro-2*H*-pyrans *endo*-53 (R^2=H, R^3=OEt) and *exo*-53 (R^2=OEt, R^3=H) (M.E. Maier and R.R. Schmidt, Synthesis, 1987, 900).

(53)   (54)   (55)

Ethyl 6-ethyl- and 6-isopropyl-3,4-dihydro-2*H*-pyran-5-carboxylates (54; R^1=H, R^2=Me; R^1=R^2=Me, respectively) are obtained by the cyclo-condensation of 1,3-dibromopropane with R^{1}R^{2}CHCOCH$_2$CO$_2$Et under mildly basic conditions. Photooxygenation of compounds 54 under singlet oxygenation conditions affords dioxetanes 55 and ene products, with polar solvents, favouring the formation of the former (Z. Huang, X. Liang, and Y. Chan, Chin. J. Chem., 1990, 182).

The cycloaddition of enamines with enones yields endo/trans-adducts, for example, ethyl 3-substituted 4-aryl-2-piperidino-6-phenyl-3,4-dihydro-2*H*--pyran-5-carboxylates 56 diastereoselectively and in high yields. Compounds 56 (R^1=H, 3-O$_2$N, R^2=Pr^i; R^1=3-O$_2$N, R^2=Ph) have been transformed to 3--substituted 4-aryl-6-phenyl-4*H*-pyrans 57 [F. Eiden and A. Eckle, Arch. Pharm. (Weinheim, Ger.), 1989, <u>322</u>, 617).

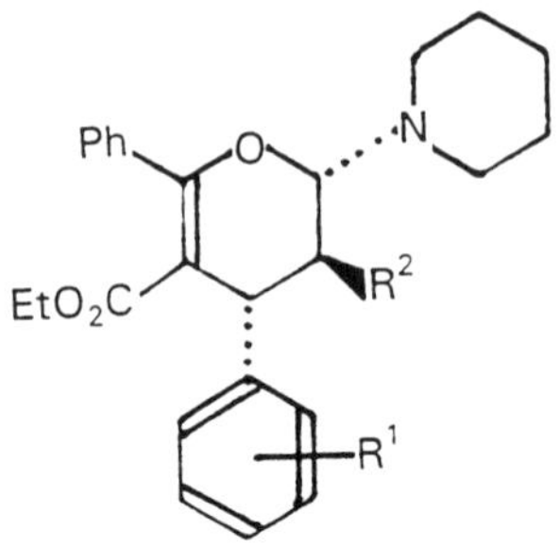

(56) 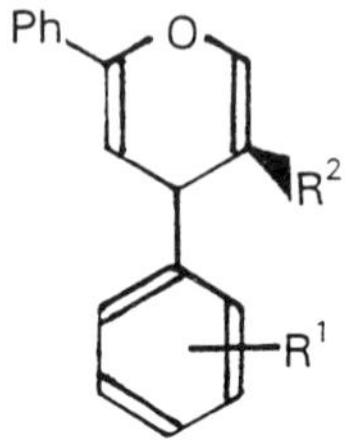   (57)

The inverse-electron-demand Diels-Alder reactions of (E)-
-R^1CH=CHCOCO$_2$Me (R^1=Ph, MeO) with enol ethers [for example,
EtOCH=CH$_2$, (Z)-PhCH$_2$OCH=CHMe] yield diastereomeric 4-substituted
methyl 2-ethoxy-3,4-dihydro-2H-pyran-6-carboxylates 58 (R^1=Ph, MeO;
R^2=α-OEt, ß-OEt) and 4-substituted methyl 2-benzyloxy-3-methyl-3,4-
-dihydro-2H-pyran-6-carboxylates 59 and 60 (R^1=Ph, MeO), respectively
(D.L. Boger and K.D. Robarge, J. Org. Chem., 1988, <u>53</u>, 3373). Similarly
the reaction of methyl 2-oxo-3-alkenoate, PhCH=CHCOCO$_2$Me with
CH$_2$=CH(CH$_2$)$_4$H in CH$_2$Cl$_2$ in the presence of SnCl$_4$ affords methyl
2-butyl-4-phenyl-3,4-dihydro-2H-pyran-6-carboxylate (61) (A. Sera et $al.$,
Chem. Letters, 1990, 2043). The preparation of a number of substituted
methyl 3,4-dihydro-2H-pyran-6-carboxylates from related 4-substituted
methyl 2-oxo-3-butenoates and -3-pentenoates and alkenes have been
reported ($idem$, Bull. Chem. Soc. Jpn., 1994, <u>67</u>, 1912).

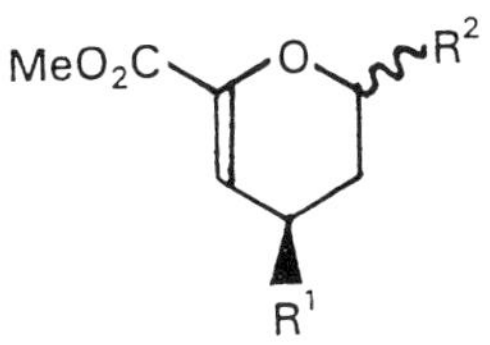

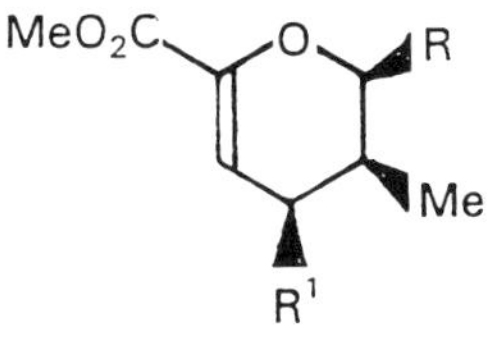

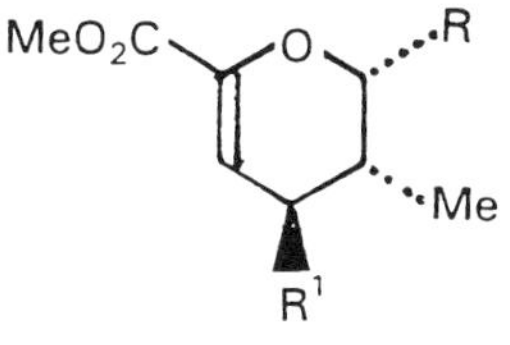

R = OCH$_2$Ph

(58)   (59)   (60)

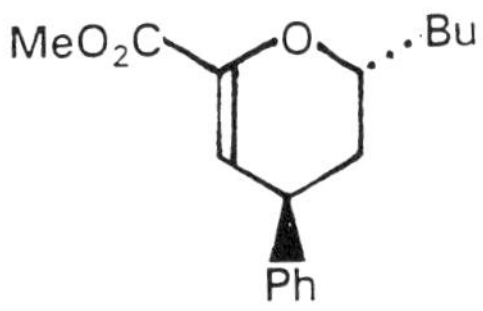

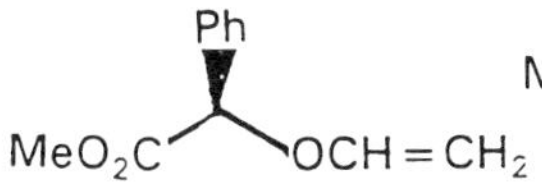

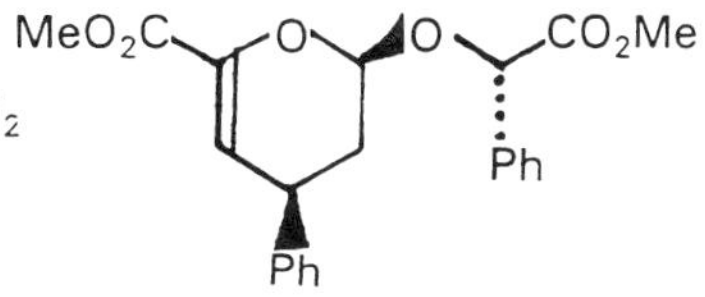

(61)   (62)   (63)

Alkyl vinyl ethers react with methyl (*E*)-benzylidenepyruvate in the presence of a catalytic amounts of Eu(Fod)$_3$ to afford endo-cycloadducts, hence, vinyl ether 62 derived from methyl (*R*)-mandelate yields the 2--substituted methyl 4-phenyl-3,4-dihydro-2*H*-pyran-6-carboxylate 63 (G. Dujardin, S. Molato, and E. Brown, Tetrahedron Asymmetry, 1993, **4**, 193).

Me$_3$SiC≡CSiMe$_3$ reacts with the dinucleophile, MeC(O)CH(CO$_2$Et)CH-(CO$_2$Et)C(O)Me, in the presence of catalysts Pd(dba)$_2$) and dppe to afford diethyl 4-acetyl-6-methyl-3-methylene-3,4-dihydro-2*H*-pyran-4,5-dicarboxylate (64) (40%) and the methylenecyclobutane 65 (30%) (L. Geng and X. Lu, Chin. Chem. Letters, 1991, **2**, 595).

(64)

(65)

Several 2,3-disubstituted 3,4-dihydro-2*H*-pyran-5,6-diacids and related esters 66 (R^1, R^2, R^3 = H, alkyl, alkenyl, aryl, cycloalkyl, aralkyl) have been prepared by acid hydrolysis of acetals 67 (R^4, R^5 = alkyl, alkenyl, cycloalkyl, aralkyl; R^{4}R^5 = alkylene). Thus, diethyl 3-ethyl-2-hydroxy-3,4-dihydro-2*H*-pyran-5,6-dicarboxylate (66; R^1 = R^2 = R^3 = Et) was obtained on treating acetal 67 (R^1 = R^2 = R^3 = Et, R^4 = R^5 = Me) with 1N H$_2$SO$_4$ at room temperature for 5h (K. Komura *et al.*, Jpn. Kokai Tokkyo Koho JP 03 101,674 [91 101,674], 1991).

(66)

(67)

The preparations of some 3,4-dihydro-2*H*-pyran-6-acetic acid esters 68 (R = Me, Et, Bu^t) have been reported (S. Brandaenge and B. Lindqvist, Acta Chem. Scand., 1989, **43**, 807).

(68) (69) (70)

The ozonolysis of ethyl ß-(2,5-dimethyl-3,4-dihydro-2*H*-pyran-2yl)-acrylate (69) gives (*E*)-MeCOCH$_2$CH$_2$CMe(OCHO)CH = CHCO$_2$Et (S. Hillers, A. Niklaus, and O. Reiser, J. Org. Chem., 1993, **58**, 3169). Methyl 6-[(2-phenylethenyl)-3,4-dihydro-2*H*-pyran-5-yl]-methylmethyleneacetate (70) and related derivatives have been prepared and investigated for use as plant protecting agents (B. Mueller *et al.*, Eur. Pat. Appl. EP 534,216, 1993).

3,4-Diacetoxy-2-pentyl-3,4-dihydro-2*H*-pyran (71) has been converted, in four steps *via* stereospecific allylation with Me$_3$SiCH$_2$CH = CH$_2$, into 3-acetoxy-2-pentyl-3,6-dihydro-2*H*-pyran-6-oct-5-enoic acid (72) (A. Tolstikov *et al.*, Zh. Org. Khim., 1991, **27**, 879).

(71) (72)

88

Alkyl 5-alkoxycarbonyl-3-hydroxymethyl-2-methyl-3,4-dihydro-2*H*-pyran-
4-acetates 73 (R^1, $R^2 = C_{1-5}$alkyl) useful as antiviral agents have been
prepared *via* a number of steps from $MeCH(OH)CH(CH_2OR)CH = CHCO_2Me$
($R = Bu^tSiPh_2$) (S. Takano *et al.*, Jpn. Kokai Tokkyo Koho JP 02 11,583 [90
11,583], 1990). Propargylic carbonates react with 2,3-diacylsuccinates in
the presence of Pd(O) and phosphine ligands to yield derivatives of 3-
-methylene-3,4-dihydro-2*H*-pyran, for example, diethyl 3-methylene-6-
-phenyl-4-benzoyl-3,4-dihydro-2*H*-pyran-4,5-dicarboxylate (74) is obtained
from $HC \equiv CCH_2OCO_2Me$ and $EO_2C[CH(COPh)]_2CO_2Et$ (L. Geng and X. Lu,
Tetrahedron Letters, 1990, <u>31</u>, 111).

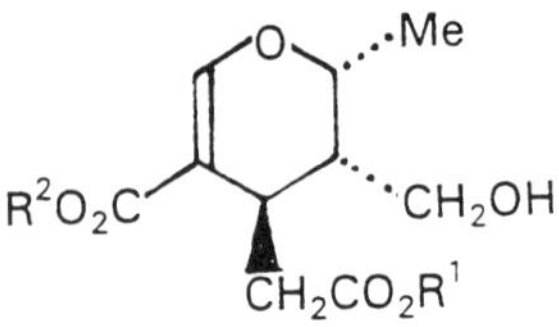

(73)

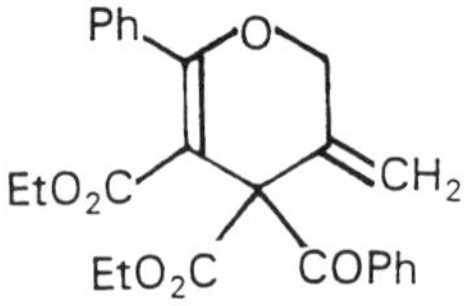

(74)

Alkylation of acetonedicarboxylate with dihalogenoalkanes in the
presence of potassium carbonate in methyl sulphoxide gives either C,C- or
C,O-alkylated products, for example, with dihalomethane 75 is obtained, but
with 1,3-dihalopropanes ethyl 6-ethoxycarbonyl-3,4-dihydro-2*H*-pyran-5-
-acetate (76), is formed (N.S. Zefirov, N.K. Sadovaya, and S.V. Kombarova,
Zh. Org. Khim., 1988, <u>24</u>, 116).

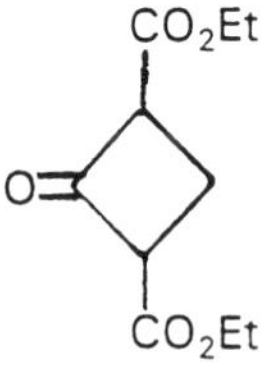

(75)

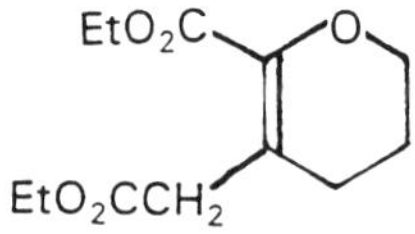

(76)

Acrolein dimer has been resolved by reaction with phosphonates 77 (R = Me, Et, Pr^i, CH$_2$CF$_3$) and it was found that the best yield of Wittig product (*R,E*)-78 was obtained with phosphonate 77 (R = Et) and the best yield of (*S,Z*)-78 with 77 (R = CH$_2$CF$_3$) (T. Rein *et al.*, Angew. Chem. 1994, <u>106</u>, 597).

(77)   (78)

3,4-Dihydro-2*H*-pyran reacts with per- or polyfluoroalkyl iodides, RI [R = Cl(CF$_2$)$_4$, Cl(CF$_2$)$_6$, Cl(CF$_2$)$_8$, F(CF$_2$)$_6$, F(CF$_2$)$_8$], in aqueous MeCN in the presence of Na$_2$S$_2$O$_4$ and NaHCO$_3$ at room temperature to give the 3-(polyfluoro- or perfluoroalkyl)-tetrahydropyran-2-ols 79 (W. Huang, Y. Xie, and L. Lu, Chin. J. Chem., 1991, <u>9</u>, 167). 6-Substituted 3,5-dibromo-2,2-dimethyl-3,4-dihydro-2*H*-pyrans 80 (R = Me, Pr, Pr^i, Bu^t, Ph) have been obtained by a procedure involving a bromoenol-etherification of γ,δ-unsaturated ketones, RCOCH$_2$CH$_2$CH = CMe$_2$ (R. Antonioletti, S. Magnanti, and A. Scettri, Tetrahedron Letters, 1994, <u>35</u>, 2619).

(79)   (80)

The enolate of cyclohexanone condenses with 5-bromo-3,4-dihydro-2*H*-
-pyran (81) during several days in DME at -15°C to give predominantly
alcohol 82 (50% yield) along with compounds 83 and 84, which have been
separated by preparative HPLC (B. Jamart-Gregoire *et al.*, *ibid.*, 1990, <u>31</u>,
7603).

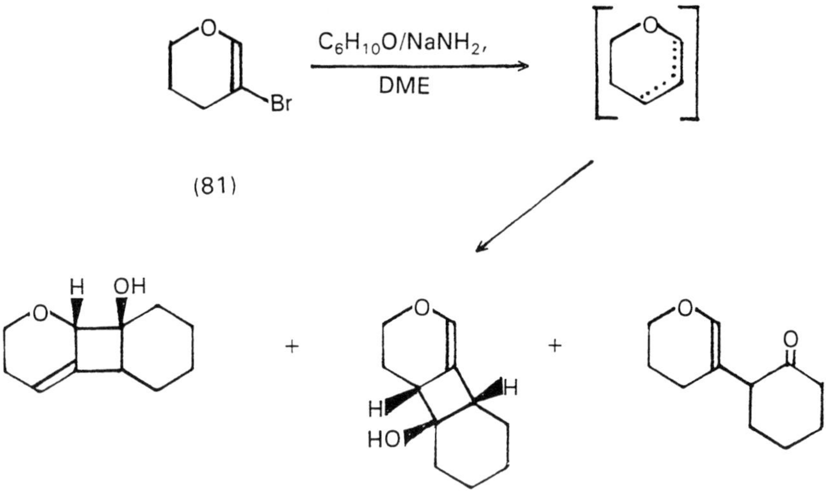

$$C_6H_{10}O/NaNH_2, \quad DME$$

(81)

(82)   (83)   (84)

Treatment of 5-nitro-3,4-dihydro-2*H*-pyran(3-nitro-5,6-dihydro-4*H*-pyran)
(85) with triisobutylaluminium, followed by hydrolysis with 0.2N HCl affords
2-isobutyl-3-nitrotetrahydropyran (86) (70%). When the hydrolysis is carried
out using 3N HCl in order to obtain, through the solvolytic Nef reaction the
corresponding carbonyl compound, surprisingly, only a small amount of 2-
-isobutyltetrahydropyran-3-one (20%) is recovered and the main product of
the reaction is 3-methylbutanal (87) (R. Menicagli, C. Malanga, and V.
Guagnano, *ibid.*, 1992, <u>33</u>, 2867).

AlBu$_3^i$

(85)

0·2N
HCl

3N HCl

Bu^{i}CHO

(86)   (87)

Heating a mixture of BzNHCH = CHCOCO$_2$Me, CH$_2$ = CHOEt, PhMe, and hydroquinone at 135 °C for 12h affords methyl (±)-(2*R*,4*R*)- and (±)--(2*S*,4*R*)-2-ethoxy-4-(benzoylamino)-3,4-dihydro-2*H*-pyran-6-carboxylate (88). A number of derivatives with several and various substituents have been prepared as intermediates for 3-amino-4-deoxy sugars of pharmaceutical interest. (±)-(4*S*)-Benzoylamino-(2*R*)-ethoxy-6-methyl-5--phenylthio-3,4-dihydro-2*H*-pyran on heating with Raney Ni in MeOH at 60 °C yields 4-benzoylamino-2-ethoxy-6-methyl-tetrahydropyran (89) (L.F. Tietze, Ger. Offen. DE 3,632,604, 1988). Methods have been developed for the synthesis of chiral, highly functionalized 6-amino-4-aryl-3,4-dihydro-2*H*--pyrans (J.L. Marco *et al.*, Tetrahedron, 1994, 50, 3509).

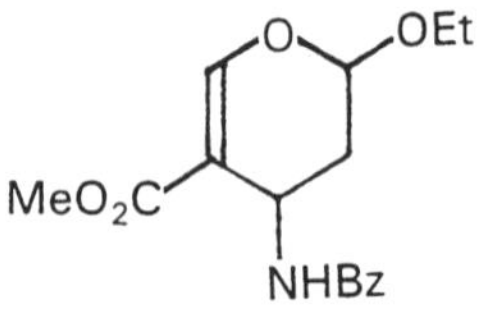

(88)

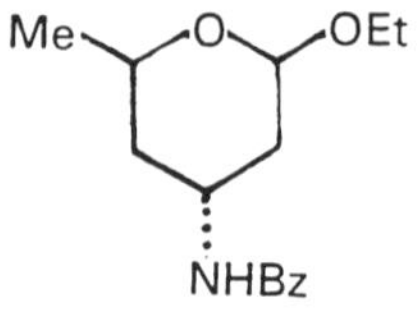

(89)

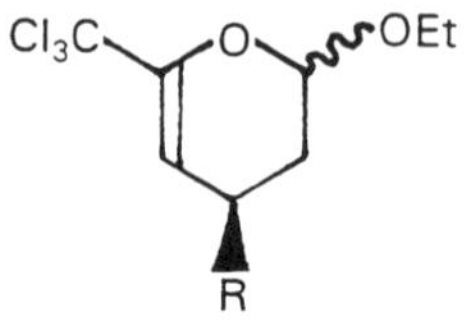

(90)

The Diels-Alder reaction between (*E*)-RCH=CHCOCCl$_3$ (R=phthalimido) and CH$_2$=CHOEt shows increased diastereoselectivity and an increased rate at high pressure. Thus, it yields *cis*- and *trans*-4-phthalimido-6-trichloro-methyl-2-ethoxy-3,4-dihydro-2*H*-pyran (90; R=phthalimido) in a 5:3 ratio at 90°C and 1 bar and a 13.6:1 ratio at 0°C and 6 kbar. Interconversion does not occur under the reaction conditions (L.F. Tietze *et al.*, J. Amer. Chem. Soc., 1988, <u>110</u>, 4065). Methyl 4ß-acetylamino-2α-butoxy-6-methyl-3,4--dihydro-2*H*-pyran-5-carboxylate (91) is obtained on boiling methyl 2-acetyl-aminomethylene-3-oxobutanoate with CH$_2$=CHO in PhMe. The preparation of a number of related derivatives have been reported (R.W. Turner, Eur. Pat. Appl. EP 236,138, 1987). The preparations of a number of 3,4-dihydro-2*H*--pyrans 92 [R^1-R^4=H, alkyl; A=CO$_2$R^5, =CHOR^5 (R^5=alkyl), CH$_2$OR^6 (R^6=alkyl, PhCH$_2$ (un)substituted with alkyl, alkoxy, or halogeno in Ph ring, CHO, alkylcarbonyl, Bz (un)substituted with alkyl, alkoxy, or halogeno); R^3≠alkyl when A=CO$_2$R^5], useful as intermediates for pharmaceuticals or dyes and especially for plant protective agents (W. Speigler *et al.*, Ger. Offen. DE 3,628,576, 1988).

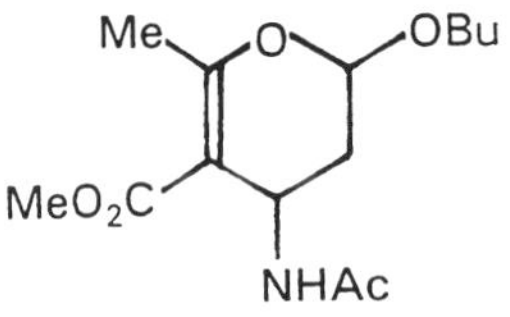

(91)

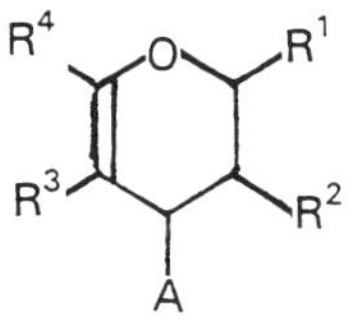

(92)

(ii) 3,4-Dihydro-2H-pyran-2-, 3-, and 4-ones (3,4-dihydro-2-, 3-, and 4-pyrones). HCOCH$_2$CMe$_2$CH$_2$CO$_2$H on slow distillation at 65°C/15mm Hg affords 4,4-dimethyl-3,4-dihydro-2*H*-pyran-2-one (1) (Y. Ding and M. Schen, Faming Zhuanli Shenqing Gongkai Shuomingshu CN 1,037,707, 1989). A preparation of 6-methyl-3,4-dihydro-2*H*-pyran-2-one (2) has been reported (U. Annby, M. Stenkula, and C.M. Andersson, Tetrahedron Letters, 1993, <u>34</u>, 8545).

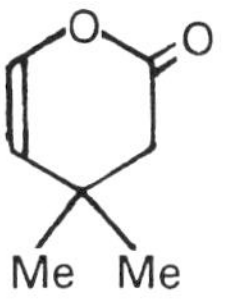

(1) 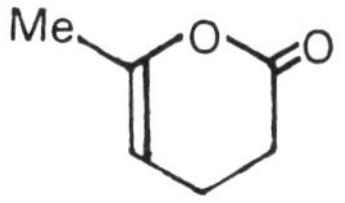   (2) 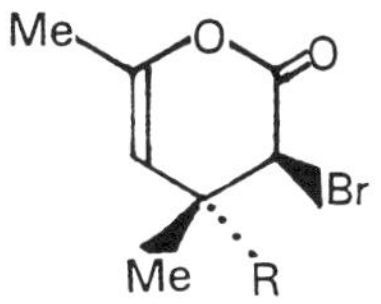   (3) 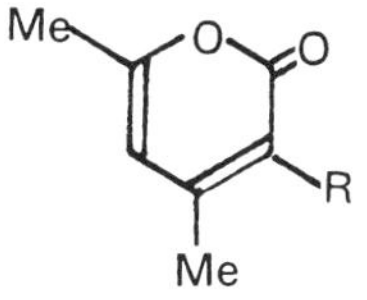   (4)

The debromination of 4-aryl- or 4-alkenyl-3-bromo-4,6-dimethyl-3,4-dihydro-2*H*-pyran-2-ones 3 (R = 4-tolyl, 4-anisyl, Ph, prop-1-enyl, hex-1-ene, vinyl) with AgSbF$_6$ in CH$_2$Cl$_2$ or ClCH$_2$CH$_2$Cl induces rearrangement of the aryl or alkenyl group to give the respective 3-substituted 4,6-dimethyl-2-pyrones 4 (T. Kume *et al.*, *ibid.*, 1988, <u>29</u>, 3825). For the preparation of 6-ethyl-3,4,5-trimethyl-3,4-dihydro-2*H*-pyran-2-one, see Y. Ito *et al.*, *ibid.*, 1987, <u>28</u>, 2247.

5-(2-Ethoxycarbonyl)-6-methyl-3,4-dihydro-2*H*-pyran-2-one (5; R = Me) and 5-(2-ethoxycarbonyl)-6-*iso*butyl-3,4-dihydro-2*H*-pyran-2-one (5; R = Bu^i) have been prepared from 4-acetylheptanedionic acid and 4-isovalerylheptanedionic acid, respectively, in the presence of acetic anhydride. The latter acid when boiled with ferric sulphate as catalyst also yielded pyrone (5; R = Bu^i) (H. Huang *et al.*, Gaodeng Xuexiao Huaxue Xuebao, 1992, <u>13</u>, 630). 6-(4-Bromophenyl-3-[2-(4-bromophenyl)ethenyl]-3,4-dihydro-2*H*-pyran-2-one (6) has been obtained from ethyl 4-carboxy-6-(4-bromophenyl)-2-oxo-3,4-dihydro-2*H*-pyran-3-acetate and 4-BrC$_6$H$_4$CHO (M.M. Mohamed *et al.*, Egypt. J. Chem., 1986, <u>29</u>, 539).

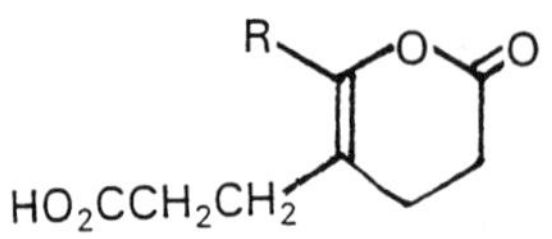

(5)

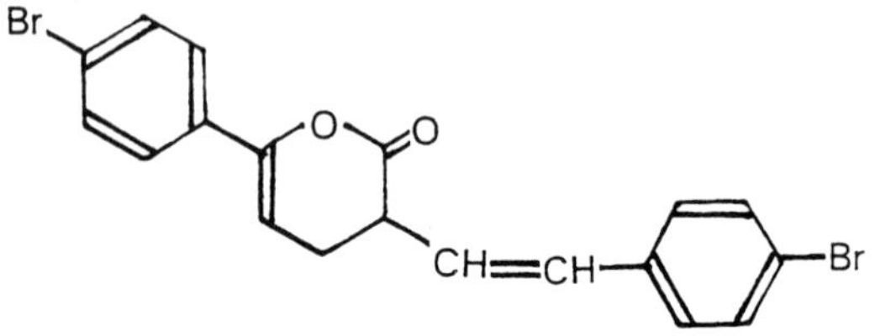

(6)

The reaction between phenylacetic acid and acetic anhydride gives, besides the expected phenylacetic anhydride, 6-benzyl-3,5-diphenyl-2-oxo--3,4-dihydro-2*H*-pyran-4-yl phenylacetate (7) (M. Chakrabarty and S.C. Pakrashi, Indian J. Chem., 1989, <u>28B</u>, 285).

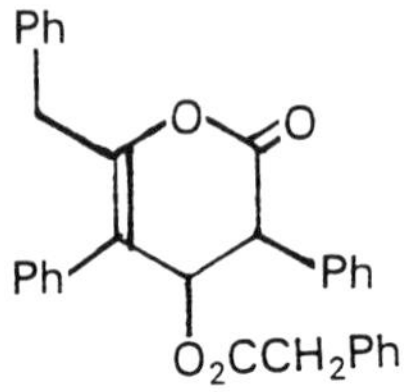

(7)

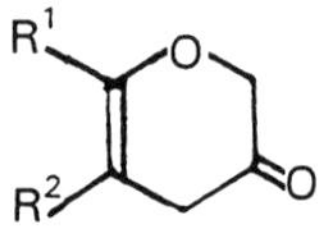

(8)

6-Methyl-5-phenyl-3,4-dihydro-2*H*-pyran-3-one (8; R^1 = Me, R^2 = Ph) and 5-phenyl-3,4-dihydro-2*H*-pyran-3-one (8; R^1 = H, R^2 = Ph) have been obtained by the rhodium (II) perfluorobutyrate-catalyzed decomposition of the appropriate diazoketones, $R^1COCHR^2CH_2COCHN_2$ (D.J. Fairfax *et al.*, J. Chem. Soc., Perkin 1, 1992, 2837).

3-Benzyl-, 3-allyl- and 3-butyl-3,4-dihydro-2*H*-pyran-4-ones (9; $R^1 = PhCH_2$, allyl, Bu) are prepared by the cyclocondensation of $MeCOCHR^1$-CH_2OH with $HC(OEt)_3$ in the presence of $SnCl_4$. Dihydropyrones 9 are transformed into 4,5-disubstituted 5,6-dihydro-2*H*-pyran-2-ones 10 by treatment with R^2Li ($R^2 = Me$, Bu), followed by oxidation with pyridinium chlorochromate (A. Nangia and P.B. Rao, Tetrahedron Letters, 1993, <u>34</u>, 2681).

(9) (10)

6-Alkenyl-3,4-dihydro-2*H*-pyran-4-ones 11 (n = 1,2) undergo intramolecular, photochemical, cycloaddition reactions in high regioselectivity to furnish adducts 12 and 13 (N. Haddad and I. Kusmenkov, *ibid.*, p.6127).

(11) (12)

(13)

96

The reaction of RCHO (R = Ph, cyclopropyl, 2-furyl, cyclohexyl, PhCH$_2$CH$_2$) with H$_2$C = C(OSiMe$_3$)CH = CHOMe in the presence of tryptophan-derived oxaborolidine catalyst 14 yields an aldol product, which on treatment with acid yields the 2-substituted 3,4-dihydro-2*H*-pyran-4-ones 15 (E.J. Corey, C.L. Cywin, and T.D. Roper, *ibid.*, 1992, <u>33</u>, 6907).

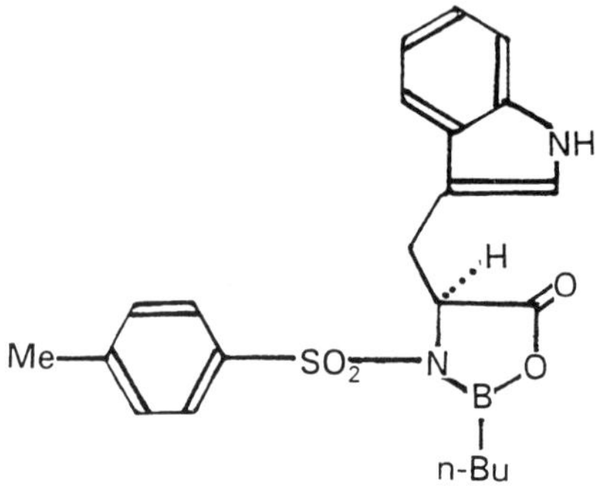

(14) 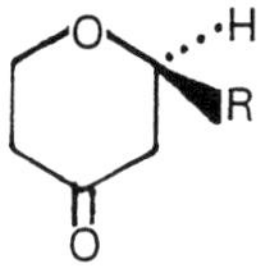   (15)

6-Ethyl-2-methyl-3,4-dihydro-2*H*-pyran-4-one (16) useful as an insect sex pheromone has been obtained by stirring together a mixture of (*S*)-2-tetra-hydropyranyloxyoctane-4,6-dione and 4-MeC$_6$H$_4$SO$_3$H in MeOH at room temperature (K. Mori and H. Kishida, Jpn. Kokai Tokkyo Koho JP 63 33,375 [88 33,375], 1988). Acid treatment of MeCOCH$_2$COCH(OCOMe)Me affords 2-ethyl-6-methyl-3,4-dihydro-2*H*-pyran-4-one (17) (G. Bianchi, M. Grugni, and A. Tava, J. Chem. Res., S, 1987, 302). (+)-(2*R*)- and (-)-(2*S*)--6-ethyl-2-methyl-4-oxo-3,4-dihydro-2*H*-pyran-5-carboxylic acid enantiomers have been prepared in several steps from (-)-(3R)- and (+)-(3S)-3--hydroxybutanoates (P.F. Deschenaux *et al.*, Helv., 1989, <u>72</u>, 731). The Mukaiyama condensation of 2,4-bis(trimethylsiloxy)penta-1,3-dienes with aldehydes and ketones in the presence of excess titanium (IV) chloride provides direct entry to 2-functionalized 3,4-dihydro-2*H*-pyran-4-ones (J.R. Peterson and E.W. Kirchhoff, Synlett, 1990, 394). The preparation of 5--phenyl-3,4-dihydro-2*H*-pyran-4-one spiro compound 18 with 1-methyl-piperidine has been reported (S.R. Schow and S.W. Tam, Bioorg. Med. Chem. Letters, 1993, <u>3</u>, 221).

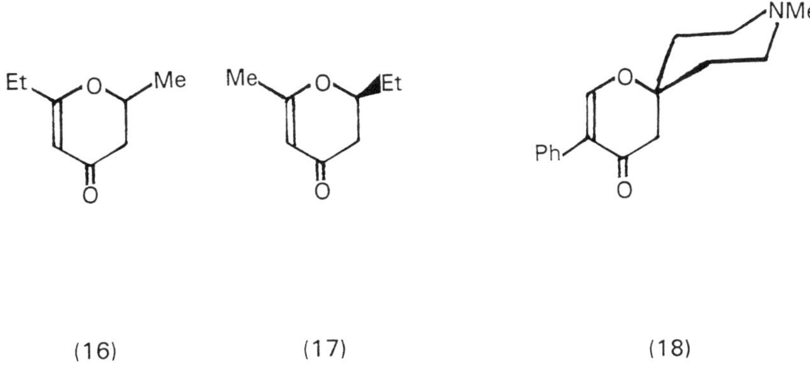

(16)   (17)   (18)

2,2-Dimethyl-3,4-dihydro-2H-pyran-4-one (19; $R^1 = R^2 = Me$) and related compounds 19 [$R^1R^2 = (CH_2)_5$, MeCHCH$_2$NMeCHMeCH$_2$] have been obtained by the cyclization of $R^1R^2C(OH)C \equiv CCH = CHOBu$ by 20% H_2SO_4(T.S. Sadykov, S.S. Kolkhosova, and K.B. Erzhanov, Izv. Akad. Nauk Kaz. SSR, Ser. Khim., 1988, 68). For the preparation of 2,2-dimethyl-6-phenyl-3,4- -dihydro-2H-pyran-4-one, see D. Obrecht, Helv., 1989, _72_, 447; for 5,6- -disubstituted 2,2-dimethyl-3,4-dihydro-2H-pyran-4-ones 20 ($R^1 = Bu^t$, Ph, 2- -MeOC$_6$H$_4$; $R^2 = H$, Br, CO$_2$Et), *idem, ibid.*, 1991, _74_, 27; and for 5,6- -disubstituted 2-methoxy-2-methyl-3,4-dihydro-2H-pyran-4-one 21 ($R^1 = R^2 = H$, Me; $R^1 = H$, $R^2 = Me$); R.S. Coleman and E.B. Grant, Tetrahedron Letters, 1990, _31_, 3677.

(19)   (20)   (21)

Chiral organoaluminium reagents (*R*)-22 and (*S*)-22 (R = Ph, 3,5-xylyl) have been successfully used as chiral Lewis acid catalysts in the asymmeteric hetero-Diels-Alder reaction of various aldehydes with dienes. Thus, PhCHO reacts with MeOCH = CMeC(OSiMe$_3$) = CHMe under the influence of (*R*)-22 (R = Ph) at -20°C to afford, after exposure of the resulting adducts to CF$_3$CO$_2$H, 2,3-*cis*-3,5-dimethyl-2-phenyl-3,4-dihydro-2*H*--pyran-4-one (23) (77%; 95% optically pure) along with 2,3-*trans*-3,5--dimethyl-2-phenyl-3,4-dihydro-2*H*-4-one (24) (7%) (K. Maruoka *et al.*, J. Amer. Chem. Soc., 1988, 110, 310; K. Maruoka and H. Yamamoto *ibid.*, 1989 111, 789). Dimethylphenylpyrone 23 has also been obtained from the same starting materials in the presence of the stable chiral(acyloxy)borane complex 25 (CAB) (R = 2-MeOC$_6$H$_4$). Related complexes 25 [R = Bu, Ph, 2,4,6-Me$_3$C$_6$H$_2$, 2,4,6-(Pr^i)$_3$C$_6$H$_2$, 2-MeOC$_6$H$_4$, 2-Pr^{i}C$_6$H$_4$] have been prepared (Q. Gao *et al.*, J. Org. Chem., 1992, 57, 1951).

(22)

(23)

(24)

(25)

The preparations of (2*S*, 3*R*, 1'*S*)-stegobinone [6-(1-methyl-2-oxobutyl)--2,3,5-trimethyl)-3,4-dihydro-2*H*-pyran-4-one] (26), useful as an insect pheromone inhibitor and its intermediate (2*S*, 3*R*, 1'*R*, 2'*R*)-stegobiol [6-(2--hydroxy-1-methylbutyl)-2,3,5-trimethyl-3,4-dihydro-2*H*-pyran-4-one] (27) have been reported (T. Ebata and K. Mori, Jpn. Kokai Tokkyo Koho JP 03 101,675 [91 101,675], 1991).

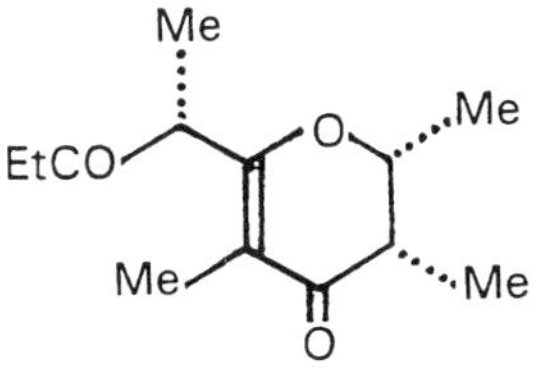

Syn-ß-acyloxyketones 28 (R = H, Et, Ph, Pr^i, PhCH$_2$) and *anti*-ß-acyloxy-
ketone 29 undergo smooth intramolecular enolate/ester condensations on
treatment with TiCl$_4$/EtN(Pr^i)$_2$ to give 6-substituted 2,3-*cis*-2-ethyl-3,5-
-dimethyl-3,4-dihydro-2*H*-pyran-4-ones 30 and 2,3-*trans*-2,6-diethyl-3-
-methyl-3,4-dihydro-2*H*-pyran-4-one (31), respectively (W. Oppolzer and I.
Rodriguez, Helv., 1993, <u>76</u>, 1282).

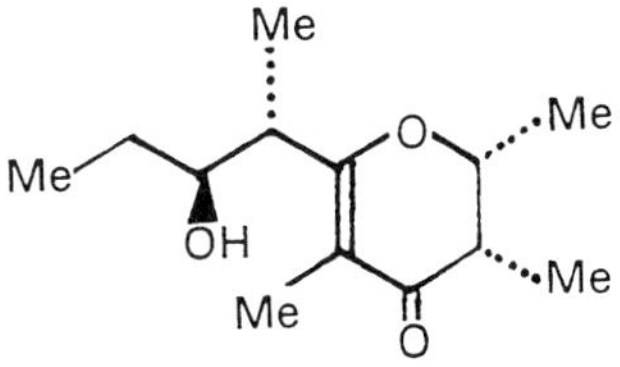

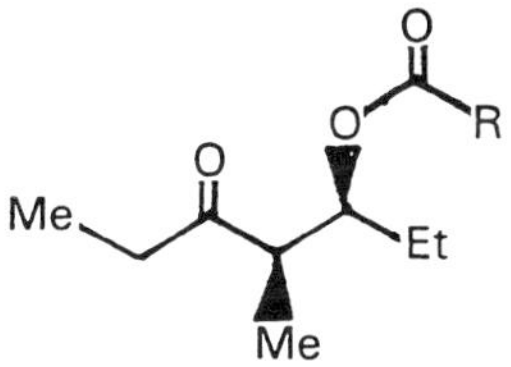

(26)

(27)

(28)

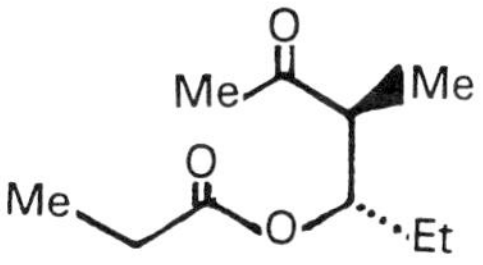

(29)

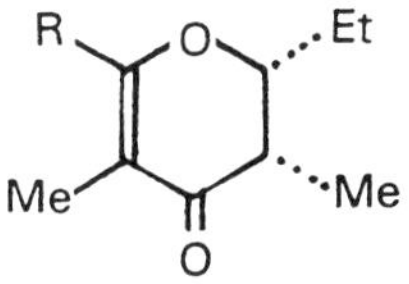

(30)

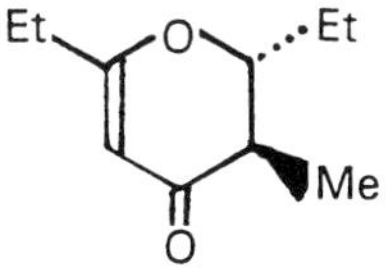

(31)

100

2,3-Disubstituted 3,4-dihydro-2*H*-pyran-4-ones 32 (R^1 = H, Me; R^2 = Pr, MeC = CH$_2$, Ph, PhCH$_2$OCH$_2$, cyclohexyl) have been obtained by cyclization of the *syn*-aldol adducts 33, prepared by the stereoselective aldol condensation of R^{1}CH$_2$COCH = CHX (X = SPh, Cl, iodo) with R^{2}CHO in the presence of borinates, for example, Bu$_2$BO$_3$SCF$_3$ (S. Osborne, Tetrahedron Letters, 1990, <u>31</u>, 2213).

(32)

(33)

2,2,6-Trisubstituted 3,4-dihydro-2*H*-pyran-4-ones 34 (R^1 = Me, CH$_2$Br, CO$_2$Et; R^2 = CO$_2$Et, Ac, CN, CO$_2$Me; R^3 = H, OAc) have been prepared by the Diels-Alder reaction of CH$_2$ = CH(OSiMe$_3$)CR^3 = CHOMe with R^{1}COR^2 in the presence of ZnCl$_2$ (P.C.B. Page *et al.*, Chem. Comm., 1987, 756). The BF$_3$.OEt$_2$-promoted cyclocondensation between electron-rich diene, H$_2$C = C(OSiMe$_3$)CH = CHOMe and racemic thio-substituted aldehydes, for example, R^{2}SCHR^{1}CHO (R^1 = Et, Pr^i, R^2 = Ph, Me), yields the 2-(substituted methyl or phenylthiomethyl)-3,4-dihydro-2*H*-pyran-4-ones 35 (R. Annunziata *et al.*, J. Org. Chem., 1992, <u>57</u>, 3605). Similarly, the reaction of the above diene with halogenated carbonyl compounds, R^{1}COR^2 (R^1 = H, R^2 = CCl$_3$, CH$_2$Cl; R^1 = Ph, R^2 = CF$_3$, CHCl$_2$) in the presence of a Lewis acid yields the 2,2-disubstituted 6-methoxytetrahydropyran-4-ones 36 and the 2,2--disubstituted-3,4-dihydro-2*H*-pyran-4-ones 37. The ^{1}H NMR spectral data of pyranones 36 and 37 have been reported (A. Sera *et al.*, Heterocycles, 1991, <u>32</u>, 273).

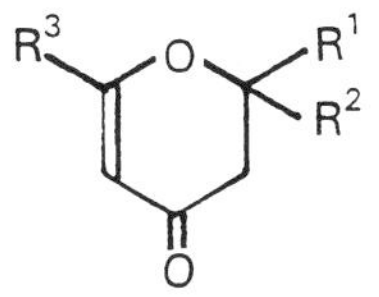

(34)

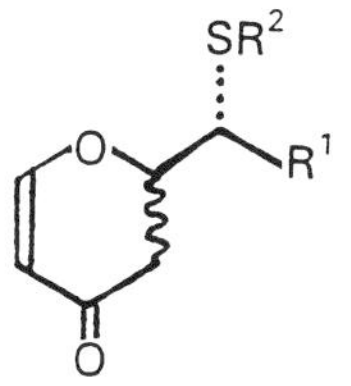

(35)

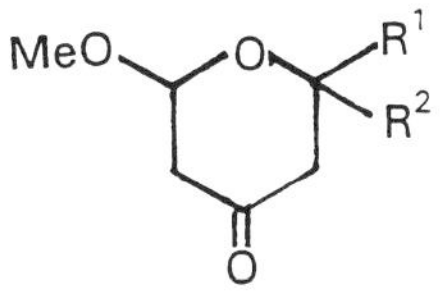

(36)

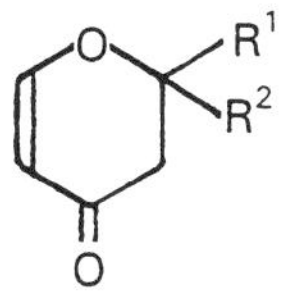

(37)

The direct regioselective synthesis of disubstituted 3,4-dihydro-2*H*-pyran-
-4-ones (hex-1-enopyran-3-3-uloses) 38 (R^1 = OAc, R^2 = H, R^3 = Ac; R^1 = H,
R^2 = OAc, R^3 = Ac; R^1 = OCH$_2$Ph, R^2 = H, R^3 = CH$_2$Ph; R^1 = H, R^2 = OCH$_2$Ph,
R^3 = CH$_2$Ph) and 39 by oxidation of the corresponding per-O-acylated and
per-O-benzylated glycals 40 and 41 has been reported (A. Kirschning, G.
Draeger, and J. Harders, Synlett, 1993, 289).

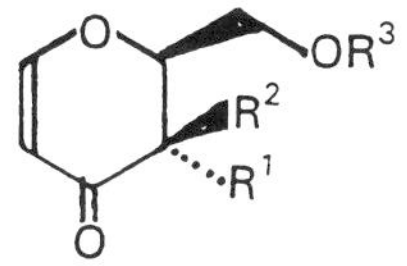

(38)

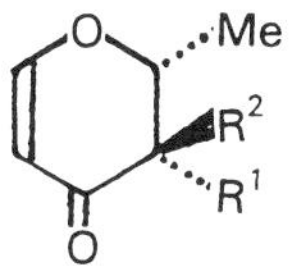

(39)

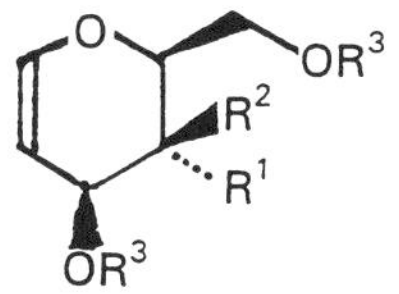

(40)

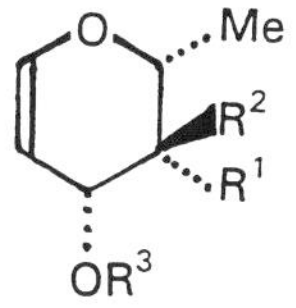

(41)

102

The ZnBr$_2$-catalyzed [4 + 2]cycloaddition of Danishefsky's diene to
R^{1}R^{2}NCHMeCHO (R^1 = H, R^2 = Bu^{t}O$_2$C, PhCH$_2$O$_2$C, 4-MeC$_6$H$_4$SO$_2$) gives *syn*-
-2-(aminoethyl)-3,4-dihydro-2*H*-pyran-4-ones 42, whereas R^{1}R^{2}NCHMeCHO
(R^1 = PhCH$_2$, R^2 = PhCH$_2$, 4-MeC$_6$H$_4$SO$_2$, Bu^{t}O$_2$C) yields the *anti*-pyrones 43
(J. Jurczak, A. Golebiowski, and J. Raczko, J. Org. Chem., 1989, 54,
2495).

Dimethyl chelidonate (dimethyl 4-oxo-4*H*-pyran-2,6-dicarboxylate)
undergoes radical alkylation with RCO$_2$H (R = Bu^t, Pr^i) to afford dimers 44,
which consist of meso and dl stereoisomers at the central bond of the dimer
(M. Tada and N. Mori, Heterocycles, 1991, 32, 749).

(42)   (43)

(44)

N-Bromosuccinimide in the presence of K$_2$CO$_3$ and a catalytic amount of
dibenzoyl peroxide converts allylic acetates and the allylic silyl ethers of
secondary allylic alcohols derived from 3,4-dihydroxy-2-hydroxymethyl-3,4-
-dihydro-2*H*-pyran (D-glucal) (45) into the corresponding 2-substituted 4-oxo-
-3,4-dihydro-2*H*-pyran-3-yl acetates 46 (R = Ac, SiMe$_2$Bu^t, SiPh$_2$Bu^t) (A.
Bouillot *et al.*, Synth. Comm., 1993, 23, 2071). Photocycloaddition of
acetylene to related 3,4-dihydro-2*H*-pyran-4-ones 47 (R^1 = Ac, R^2 = OAc;
R^1 = Ac, H, R^2 = H) proceeds in a stereospecific manner to yield the
corresponding 2-oxabicyclo[4.2.0]oct-7-en-5-ones 48. Some novel
rearrangements of compound 48 have been reported and the nature of the
product strongly depends on the kind of Lewis acid used (T. Naguyen Dinh
et al., Bull. Soc. Chim. Fr., 1993, 130, 287).

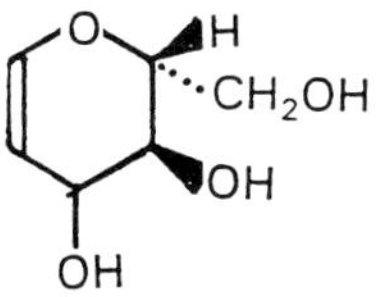

(45)

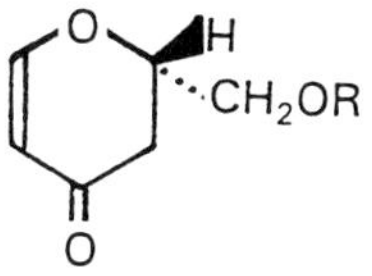

(46)

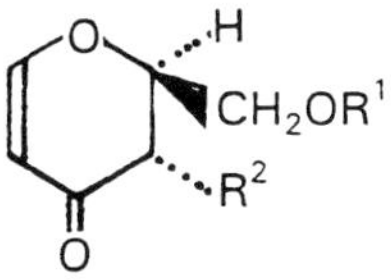

(47)

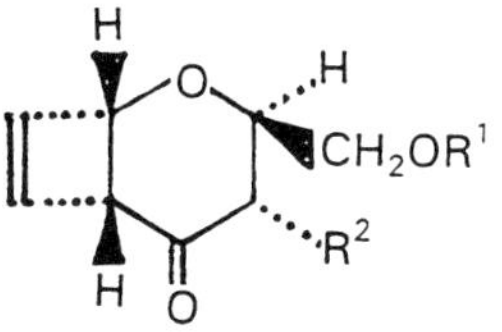

(48)

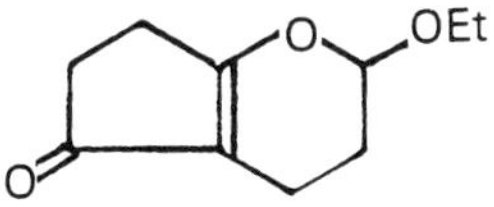

(49)

Treatment of 2-(phenylsulphinyl)methylcyclopenta-1,3-dione with pyridine results in the generation of 2-methylenecyclopenta-1,3-dione, which reacts with $H_2C=CHOEt$ to yield 2-ethoxy-3,4-dihydro-2*H*-cyclopenta[b]-pyran-5-one (49) (W.H. Bunnelle and L.A. Meyer, J. Org. Chem., 1988, <u>53</u>, 4038).

(iii) 3,4-Dihydro-2H-pyran-2,4-diones. 3,4-Dihydro-2*H*-pyran-2,4-dione (1) has been obtained by deketalization, dechlorination, reduction and decarboxylation of the pyranodioxine 2. Thus, compound 2, AcOH, and AcONa have been heated in an autoclave over Pd/C in the presence of hydrogen to give dione 1 (S. Kojima *et al.*, Jpn. Kokai Tokkyo Koho JP 02 101,071, [90 101,071], 1990).

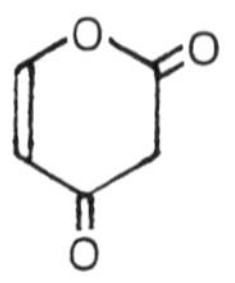

104

(1)

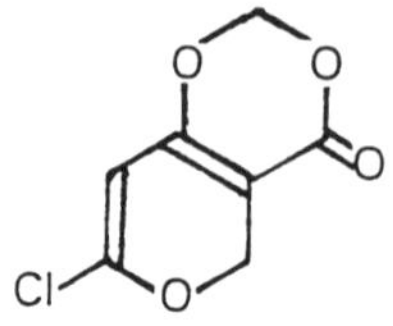

(2)

R$^1_2$C(COCl)$_2$ [R^1 = Me, R^{1}R^1 = (CH$_2$)$_n$, n = 2-4] on cyclization with R^{2}Ac [R^2 = Me(CH$_2$)$_4$, H$_2$C = CHCH$_2$CH$_2$, Me$_2$C = CH, cyclohexyl, 1-cyclohexenyl, PhCH$_2$, 4-tolyl, 4-anisyl, cinnamyl, 2-furyl] gives 6-substituted 3,3-dimethyl- and 3,3,6-trisubstituted 3,4-dihydro-2*H*-pyran-2,4-diones 3, respectively, some of which isomerize to afford 2-pyrones 4 (n = 2,3). The crystal and molecular structure of ethyl 6-ethoxy-3,3-dimethyl-2,4-dioxo-3,4-dihydro-2*H*-pyran-5-carboxylate has been determined (R.W. Saalfrank *et al.*, Ber., 1990, <u>123</u>, 1175). Precursors of prostaglandin and Thomboxane B analogues based on 3-acetyl-6-methyl-3,4-dihydro-2*H*-pyran-3,4-dione (dehydroacetic acid) (5) and its derivatives have been described (F.S. Pashovskii, I.P. Lokot, and F.A. Lakhvich, Vestsi Akad. Navuk Belarusi, Ser. Khim. Navuk, 1993, 81).

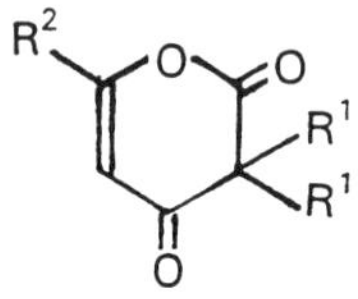

(3)

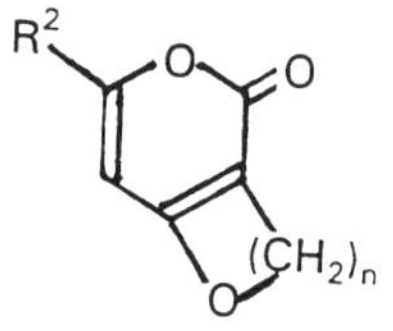

(4)

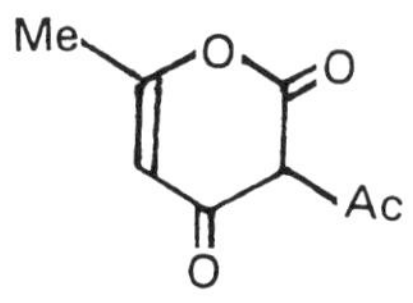

(5)

Disubstituted malonyl chlorides $R^1R^2C(COCl)_2$ [$R^1 = R^2 = $ Me, Et, allyl, CH_2Ph, $R^1R^2 = (CH_2)_3$, 2,2'-biphenylyl] react with $MePhC(OEt) = C = C(OEt)$-NMePh and depending on the bulkiness of the substituents R^1 and R^2, by transallenation yield allene dianilides, $R^1R^2C = C = C(CONPhMe)_2$ and 3,3--disubstituted 5-methylphenylamido-6-methylphenylamino-3,4-dihydro-2*H*--pyran-2,4-diones 6 and 3,3-cyclobutyl-6-dimethylamino-5-methylphenyl-amido-3,4-dihydro-2*H*-pyran-3,4-dione (7), respectively (*idem*, Ber., 1988, <u>121</u>, 1291).

(6)   (7)

(iv) 5,6-Dihydro-2H-pyrans, 3,6-dihydro-2H-pyrans. Both of these parent names are at present used in the literature and the one referred to in the following references will be reported. The major product in the hydroformylation of 3,6-dihydro-2*H*-pyran is tetrahydropyran-2-carboxaldehyde (A. Polo *et al.*, Organometallics, 1992, <u>11</u>, 3525). Insertion reactions of 2--methyl-5,6-dihydro-2*H*-pyran (1) with dihalogenocarbenes (:CX_2; X = Cl, Br), generated in situ from strong bases and CHX_3 yield oxabicycloheptanes 2 and 2-dihalogenomethyl-2-methyl-5,6-dihydro-2*H*-pyrans 3 (U.G. Ibatullin *et al.*, Izv. Akad. Nauk SSSR, Ser. Khim., 1987, 2280).

(1)   (2)   (3)

Heating $PrCH(OCH_2CH_2CMe=CH_2)_2$ at 200°C affords a mixture containing 4-methyl-2-propyl-5,6-dihydro-2*H*-pyran (4), 4-methyl-2-propyl-3,6-dihydro-2*H*-pyran (4-methyl-6-propyl-5,6-dihydro-2*H*-pyran), 4--methylene-2-propyltetrahydropyran and $CH_2=CMeCH_2CH_2OH$ (U.G. Ibatullin and T.F. Petrushina, Khim. Geterotsikl. Soedin., 1993, 134). 3-Methylbut--3-en-1-ol cyclocondenses with R^1R^2CO [$R^1=H$, R^2Et, Bu, Me_2CHCH_2, Ph; $R^1=Me$, $R^2=Me$, Me_2CHCH_2; $R^1R^2=(CH_2)_4$, $(CH_2)_5$] on the surface of silica gel and Al_2O_3 in the absence of solvents and conventional catalysts to afford 6,6- and 2,2-disubstituted 4-methyl-5,6-dihydro-2*H*-pyrans 5 and 6, 2,2--disubstituted 4-methylenetetrahydropyran 7, and 2,2-disubstituted 4--methyltetrahydropyran-4-ol 8 (U.G. Ibatullin, Yu. V. Pavlov, and M.G. Safarov, *ibid*., 1989, 1326). Similarly, pyrans 4, 5 and 6 have been obtained by the cyclocondensation of $CH_2=CMeCH_2CH_2OH$ with R^1R^2CO ($R^1=H$, $R^2=Bu$; $R^1=R^2=Me$) at room temperature on the surface of some metal salts, for example, Na_2SO_4, $MnSO_4$, $FeSO_4$, $CuSO_4$, $CoSO_4$, $Fe_2(SO_4)_3$ $Al_2(SO_4)_3$, CuF_2, $CuCl_2$, CuI_2, $CuMoO_3$ (*idem, ibid*., 1990, 1019). (*R*)-2--Propyl-5,6-dihydro-2*H*-pyran has been obtained by the zirconium-catalyzed kinetic resolution of the racemic pyran (J.P. Morken *et al*., J. Amer. Chem. Soc., 1994, <u>116</u>, 3123).

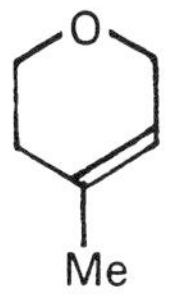

(4) 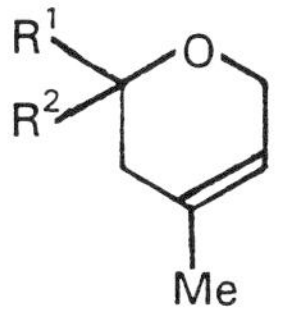   (5) 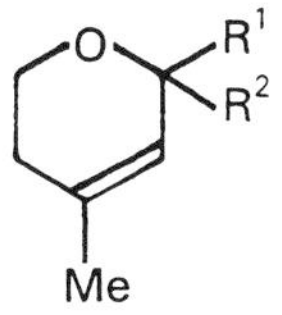   (6)

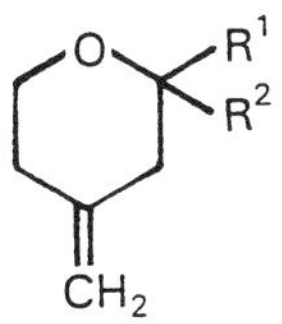

(7) 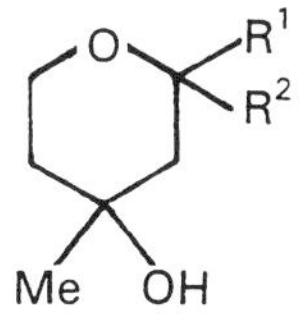   (8)

The cyclization of aldehydes (e.g., HCHO, MeCHO) with an α-olefin dimers, $R^1R^2C=CR^3R^4$ (R^1-R^3 = C_{1-30}alkyl; R^4 = H, C_{1-4}alkyl) (e.g., dec-1-ene dimer) in the presence of a Lewis acid catalyst (e.g., $SnCl_4$, BF_3) affords 4,5,5-trisubstituted and 3,4,5,5-tetrasubstituted 5,6-dihydro-2*H*-pyrans 9, useful as lubricants basestock (K. Senaratne *et al.*, U.S. US 5,196,552, 1992). Several substituted 4-methyl-5,6-dihydro-2*H*-pyrans 10 and 11 and 4-methylenetetrahydropyrans 12 (R^1 = H, Me; R^2-R^6 = H, Me; excluding R^4 = R^6 = Me and R^2 = R^3 = R^5 = H) useful for perfumes, have been prepared (H. Takada *et al.*, Jpn. Kokai Tokkyo Koho JP 63 30,481 [88 30,481], 1988).

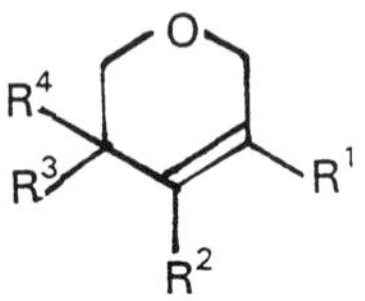

(9)

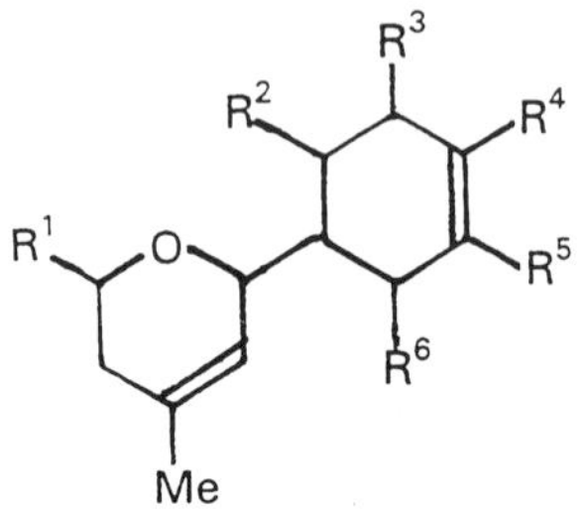

(10)

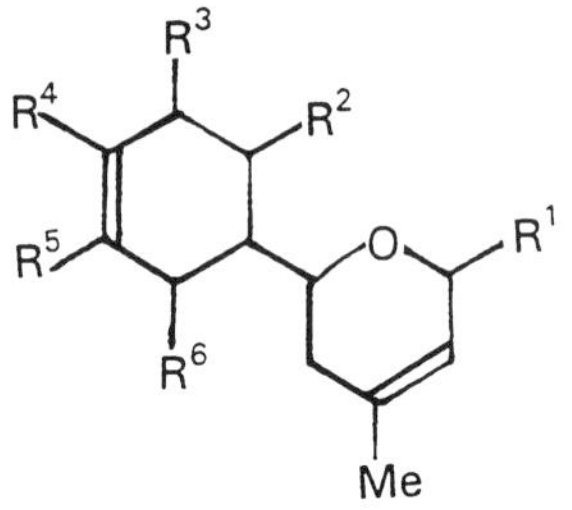

(11)

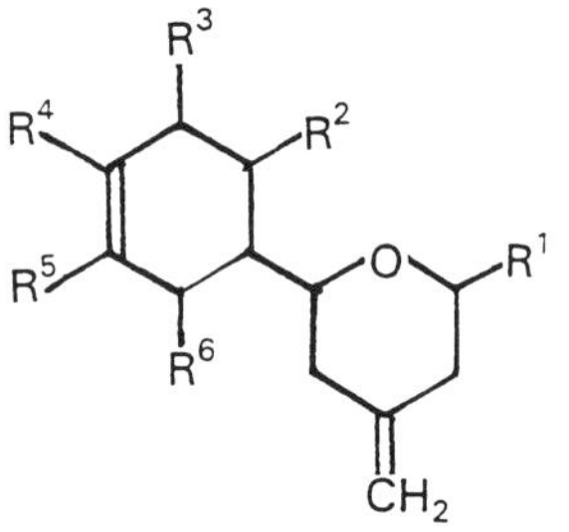

(12)

The reaction of diiodoethyne with lithium aluminium compounds of 6,6-
-dimethyl-5,6-dihydro-2*H*-pyran and -thiopyran affords the 6,6-dimethyl-5,6-
-dihydro-2*H*-pyran-4-yl-triyne 13 (X = O) and the di(6,6-dimethyl-5,6-dihydro-
-2*H*-pyran-4-yl)-tetrayne 14 (X = O) and the corresponding thiopyrans 13 and
14 (X = S) (S.Zh. Zhumagaliev and N.N. Godovikov, Izv. Akad. Nauk Kaz.
SSR, Ser. Khim., 1987, 81).

(13) (14)

The preparations of a number of substituted 5,6-dihydro-2H-pyrans 15
($R^1 = C \leq 15$ alkyl, cycloalkyl, aralkyl, alkaryl, aryl; $R^2, R^3 = H$, C_{1-6}alkyl),
useful as perfumes or their precursors, by the Diels-Alder reaction of
aldehydes R^1CHO with dienes, $R^3CH = CHCR^2 = CH_2$, in the presence of a
Lewis acid catalyst, especially $AlCl_3$ or $SnCl_4$ have been reported (N.L.J.M.
Broekhof and J.J. Hofma, Eur. Pat. Appl. EP 325,000, 1989). Also reported
are the preparations of several oxy-substituted-2- and -6-phenyl-4-methyl-
-5,6-dihydro-2H-pyrans (W.J. Wiegers and A. Van Ouwerkerk, U.S. US
4,914,083, 1990).

3,4-Dihydro-2H-pyran (16) has been converted into 2-(propa-1,2-dienyl-
oxy)-5,6-dihydro-2H-pyran (17) by a sequence of reactions involving
halogenation, reaction with propargyl alcohol, dehydrohalogenation, and
isomerization (J.-P. Dulcère *et al.*, J. Org. Chem., 1993, 58, 5702).

(15)

110

DBU
110°C
5h

Bu^{t}OK, C$_5$H$_{12}$,
35°C, 12h

NBS
HC≡C-CH$_2$OH
-20°C

Bu^{t}OK,
C$_6$H$_6$
60°C, 6h

(16)

Br

(17)

Bu^{t}OK, C$_5$H$_{12}$,
35°C, 12h

Bu^{t}OK, C$_6$H$_6$,
60°C, 6h

Br

The high pressure [4 + 2] cycloaddition of 1-methoxybuta-1,3-diene with D-PhCH$_2$O$_2$CNHCHMeCHO affords a 1:2.5 mixture of 6-substituted 2-methoxy-5,6-dihydro-2H-pyrans 18 and 19 (R^1 = CO$_2$CH$_2$Ph; R^2 = H), respectively. However, 1-methoxybuta-1,3-diene and D-PhCH$_2$O$_2$CN(CO$_2$Bu^t)CHMeCHO yields a 16:1 mixture of 18 and 19 (R^1 = CO$_2$CH$_2$Ph; R^2 = CO$_2$Bu^t). Derivative 18 has been converted by a number of steps into methyl 2,6-di-N-acetyl-a-D-purpurosaminide (20) (J. Jurczak, A. Golebiowski, and J. Raczko, Tetrahedron Letters, 1988, 29, 5975). The preparations of some 6-substituted 2-methoxy-5,6-dihydro-2H-pyrans have also been reported (*idem*, Bull. Pol. Acad. Sci., Chem., 1991, 39, 307).

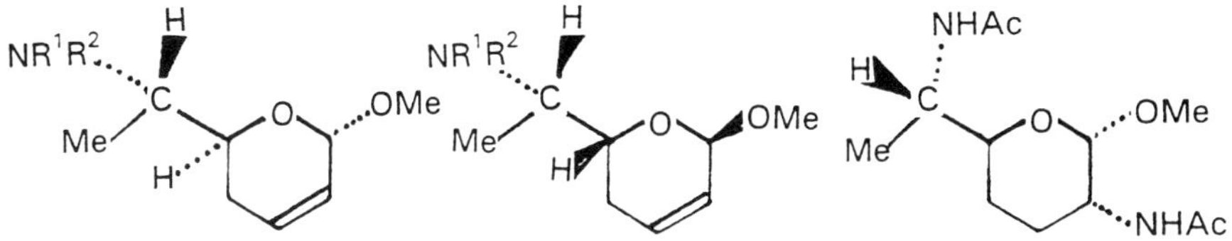

(18)   (19)   (20)

The reaction of 3-aryloxy-4-methyl-3,6-dihydro-2H-pyrans 21 (R = H, 2-Me, 3-Me, 4-Me, 2,6-Me$_2$, 2-Cl, 2,4-Cl$_2$, 4-Br, 2,4-Br$_2$) with peracid gives regioselectively the *trans*-epoxides, 3-aryloxy-4,5-epoxy-4-methyl-3,6-dihydro-2H-pyrans 22 (U.G. Ibatullin *et al.*, Izv. Akad. Nauk SSSR, Ser. Khim., 1988, 623).

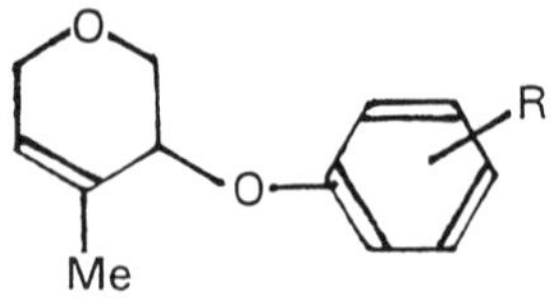

112

(21) 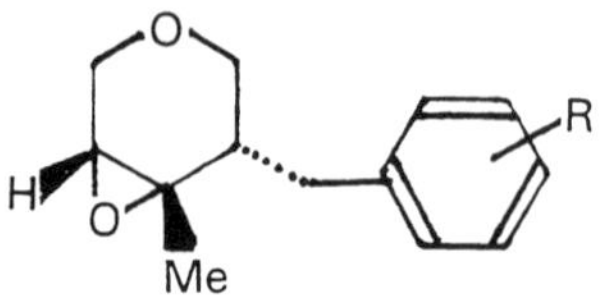 (22)

4-Methyl-5,6-dihydro-2*H*- and 4-methylenetetrahydropyrans react with phenylsulphinyl chloride in Pr^{i}NO$_2$ containing ZnCl$_2$ at -30°C to give 4-methyl-5-phenylsulphinyl- and 4-(phenylsulphinyl)methyl-5,6-dihydro-2*H*--pyran, respectively, which react at *ca*.65° in MeOH containing P(OMe)$_3$ to afford the corresponding 4-methyl-5,6-dihydro-2H-pyran-5-ol (23) and 4--methylenetetrahydropyran-3-ol (24) (Yu. V. Pavlov, V.V. Veselovski, and U.G. Ibatullin, *ibid.*, 1990, 1446).

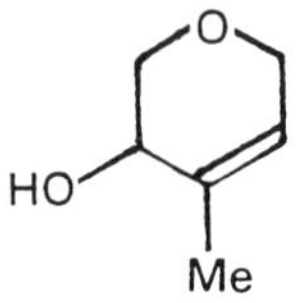

(23) 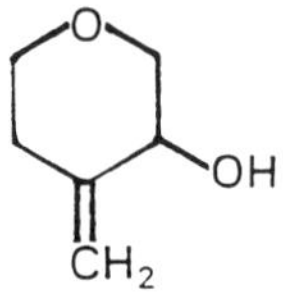 (24)

The 4-methyl-5-phenoxy-5,6-dihydro-2*H*-pyran azomethine derivatives 25 (R = Ph, substituted Ph) have been obtained from 4-methyl-5,6-dihydro-2*H*--pyran *via* bromination, alkylation with 4-HOC$_6$H$_4$CHO and accompanying dehydrobromination, followed by Schiff base formation with RNH$_2$ (M.M. Murza, I.P. Podlipchuk, and M.G. Safarov, Zh. Org. Khim., 1990, <u>26</u>, 384). The preparations of 5,6-dihydro-2*H*-pyran derivatives with mesomorphic properties have been reported, for example, the mesomorphic anilines 26 and 27 (R = H, C$_{1-12}$*n*-alkoxy) have been obtained by condensation of appropriate aminophenoxydihydropyrans with 4-RC$_6$H$_4$CHO (M.M. Murza, A.E. Tarasenko, and M.G. Safarov, Khim. Geterotsikl. Soedin., 1991, 897). The preparations of (2*S*, 6*R*)-, (2*S*, 6*S*)-, (2*R*, 6*S*)-, and (2*R*, 6*R*)-2-acyl-6--methoxy-3,6-dihydro-2*H*-pyrans have been reported (T. Bauer *et al.*, Helv., 1989, <u>72</u>, 482).

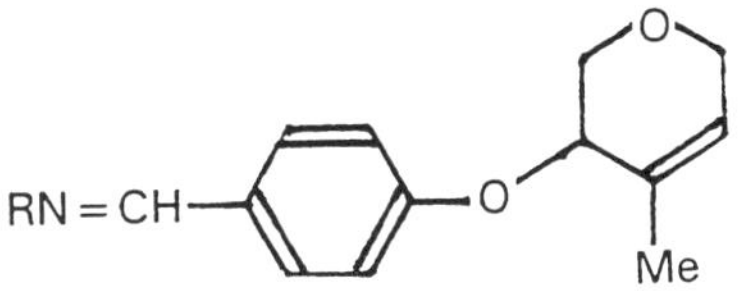

RN=CH— ... —O— ... Me

(25)

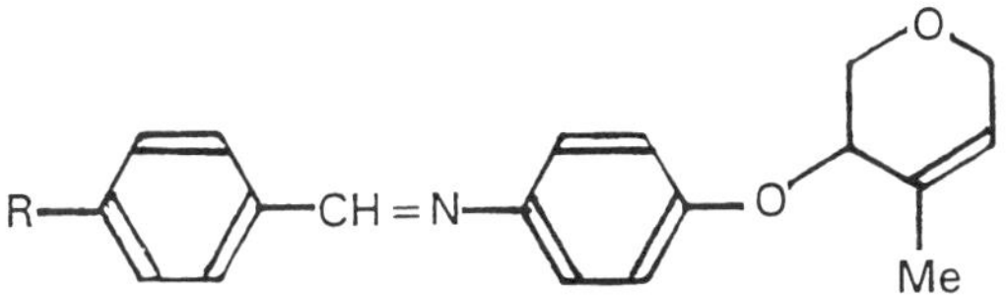

R— ... —CH=N— ... —O— ... Me

(26)

R— ... —CH=N— ... —OCH$_2$

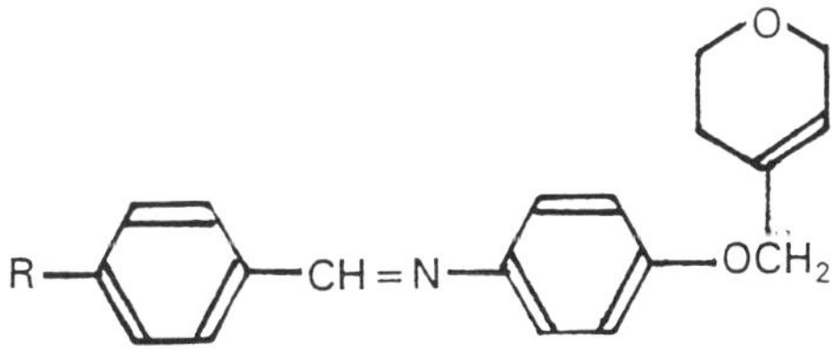

(27)

114

The lithiation of 3-methyl-1-trimethylsilyloxybuta-1,3-diene, $H_2C=CMeCH=CHOSiMe_3$, affords $H_2C=CMeCH=CHOLi$, which on treatment with ß-ionylidene acetaldehyde gives 6-[4-(2,6,6-trimethylcyclo-hex-1-enyl)buta-1,3-dienyl]-4-methyl-5,6-dihydro-2*H*-pyran-2-ol (28). Hydrolysis and ring opening of dihydropyran 28 yields retinal (P. Chabardes *et al.*, Eur. Pat. Appl. EP 430,808, 1989). The cycloaddition reaction of $EtOCH=CHCH=CH_2$ with R^1CHO [$R^1=H(CF_2)_4$, $H(CF_2)_6$] or CF_3COCF_3 yields 2-ethoxy-6-polyfluoroalkyl-5,6-dihydro-2*H*-pyran [29; $R^1=H(CF_2)_4$, $H(CF_2)_6$] or 2-ethoxy-6,6-di(trifluoromethyl)-5,6-dihydro-2*H*-pyran (29; $R^1=R^2=CF_3$), respectively (Yu.G. Shermolovich, S.V. Pazenok, and L.N. Slyusarenko, Zh. Org. Khim., 1990, <u>26</u>, 2213).

(28)

(29)

*Iso*butyraldehyde on treatment with $HC≡CCMe_2OH$ gives $Me_2C=C=CH-CMe_2CHO$, as the major product, and 6-(2-methylbut-3-yn-2-yloxy)-2,2,5,5--tetramethyl-5,6-dihydro-2*H*-pyran (30; $R=CMe_2C≡CH$). Acetalization of $Me_2C=C=CHCMe_2CHO$ by ROH (R = Me, Pr^i, $CMe_2C≡CH$) and 4-MeC_6H_4-SO_3H or $HgSO_4$ affords 6-methoxy-, 6-isopropoxy-, and 6-(2-methylbut-3-yn--2-yloxy)-2,2,5,5-tetramethyl-5,6-dihydro-2*H*-pyran (30), respectively (N.O. Nilsen, L. Skatteboel, and Y. Stenstroem, Acta Chem. Scand., 1987, <u>B41</u>, 459).

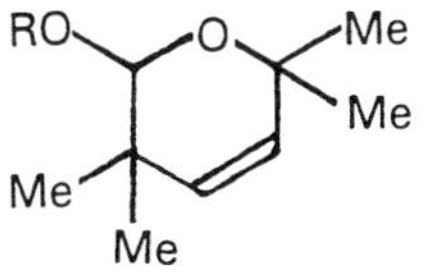

(30)

(31)

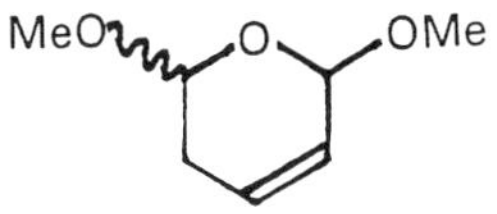

(32)

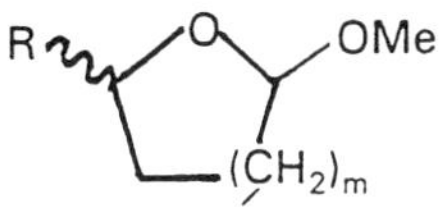

(33)

Monoozonolysis of cyclodienes 31 (n = 1-3) yielded dialdehydes, OHCCH = CH(CH$_2$)$_n$CHO, which in subsequent reactions have been converted into 2,6-dimethoxy-5,6-dihydro-2*H*-pyran (32) and heterocycles 33 (R = CH$_2$CHO; m = 1,2) (K. Griesbaum, I.C. Jung, and H. Mertens, J. Org. Chem., 1990, 55, 6024).

The preparation of some derivatives of 5,6-dihydro-2*H*-pyran-3-carbox-aldehyde (34) and its oxidation product 5,6-dihydro-2*H*-pyran-3-carboxylic acid (35) have been reported and tested in an insecticidal/fungicidal/acaricidal screening programme. It has been found that methyl 5,6-dihydro-2*H*-pyran-3-carboxylate showed a distinct knock down effect against the fruit fly *Drosophila melanogaster* (H. Spreitzer, P. Mueller, and G. Buchbauer, Monatsh., 1990, 121, 963).

Acetoxyalkoxyacetic esters, for example, HC≡CCH$_2$CH$_2$OCH(OAc)CO$_2$Me and (*Z*)-EtCH = CHCH$_2$CH$_2$OCH(OAc)CO$_2$Me cyclize under the influence of SnCl$_4$ in CH$_2$Cl$_2$ to give methyl 4-chloro-5,6-dihydro-2H-pyran-2-carboxylate (36) and methyl 4-chloro-3-ethyl-tetrahydropyran-2-carboxylate (37), respectively. These ring closures proceed through the intermediacy of (methoxycarbonyl)oxonium ions (L.D.M. Lolkema *et al.*, Tetrahedron Letters, 1988, 29, 6365). Methyl and ethyl 2-methyl-5,6-dihydro-2*H*-carboxylate

116

(38; R = Me, Et) have been obtained by the cyclocondensation of $MeCOCH_2CO_2R$ (R = Me, Et) with 1,3-dihalogenopropanes [e.g., $Br(CH_2)_3Cl$] in DMF and gradual addition of NaOH [R. Mueller *et al.*, Ger. (East) DD 266,709, 1989]. Ethyl ester 38 (R = Et) has also been prepared by boiling the above starting materials with K_2CO_3 in aqueous EtOH (R. Frimm *et al.*, Czech. CS 248,932, 1988).

(34) R = CHO
(35) $R = CO_2H$

(36)

(37)

(38)

The main product from the Diels-Alder reaction between *n*-butyl glyoxylate and 2-ethoxybuta-1,3-diene at 60°C is *n*-butyl 4-ethoxy-5,6--dihydro-2H-pyran-6-carboxylate (39). However, the reaction between 2 mols. of *trans*-1-phenylbuta-1,3-diene and *n*-butyl glyoxylate at 80~85°C yields 3-phenyl-4-styrylcyclohexene (40) as the main adduct (J. Huang and X. Chen, Zhongshan Daxue Xuebao, Ziran Kexueban, 1991, 30, 74). The preparation of a number of 2,4-disubstituted 5,6-dihydro-2*H*-pyran-6-carboxylic esters 41 (R^1, R^2 = H, alkyl, OR^3; R^3 = alkyl, dialkylcarbamoyl, provided that $R^1 = R^2 \neq H$; R^4 = alkyl) by stereoselective cycloaddition of butadienes with glyoxylic acid esters in the presence of a binaphthol-Ti complex has been reported and some of them have been found useful as intermediates for antibiotics (K. Mikami *et al.*, Jpn. Kokai Tokkyo Koho JP 04 117,375 [92 117,375], 1992). Also the reactions of methyl glyoxylate and glyoxal monoacetal with 2-methylpenta-1,3-diene to yield methyl 2,4--dimethyl-5,6-dihydro-2*H*-pyran-6-carboxylate and 6-dimethoxymethyl-2,4--dimethyl-5,6-dihydro-2*H*-pyran 42 and 43 [R = CO_2Me, $CH(OMe)_2$, respectively] have been studied under pressure in a microwave oven (A. Stambouli, M. Chastrette, and M. Soufiaoui, Tetrahedron Letters, 1991, 32, 1723).

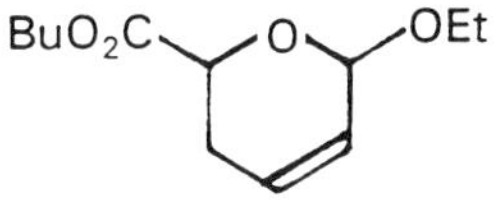

(39)

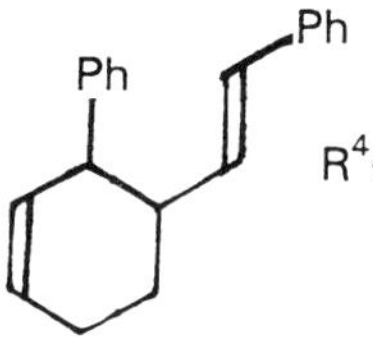

(40)

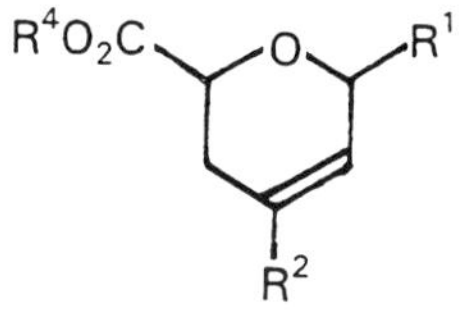

(41)

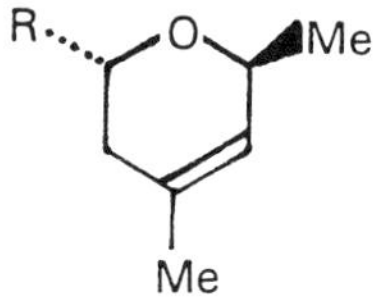

(42)

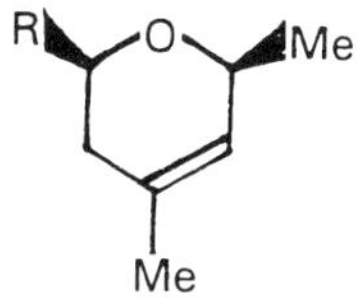

(43)

Methyl trifluoropyruvate reacts with $CH_2=CR^1CR^2=CHR^3$ ($R^1=H$, Me, Cl, $R^2=R^3=H$; $R^1=R^2=H$, $R^3=Me$; $R^1=R^2=Me$, $R^3=HO$) to give the substituted methyl 6-trifluoromethyl-5,6-dihydro-2*H*-pyran-6-carboxylates **44**, which on saponification afford the corresponding acids. Similarly, CF_3COCF_3 with $CH_2=CR^1CR^2=CHR^3$ ($R^1=Me$, $R^2=R^3=H$; $R^1=R^2=H$, $R^3=Me$) yields 4-methyl- and 2-methyl-6,6-di(trifluoromethyl)-5,6-dihydro--2*H*-pyrans (**45** and **46**) (A.S. Golubev *et al.*, Izv. Akad. Nauk SSSR, Ser. Khim., 1991, 141).

118

(44)

(45)

(46)

(47)

The Diels-Alder reaction of *trans*-R¹CH = CHCH = CH₂ (R¹ = Me, MeO)
with (EtO₂C)₂CO or BuO₂CCHO in the presence of a chiral catalyst gives
diethyl 4-methyl- and 4-methoxy-5,6-dihydro-2*H*-pyran-6,6-dicarboxylate
(47; R¹ = Me, MeO, R² = R³ = CO₂Et) and butyl 4-methyl- and 4-methoxy-5,6-
-dihydro-2*H*-pyran-6-carboxylate (47; R¹ = Me, MeO, R² = H, R³ = CO₂Bu).
An enanthiomeric excess of 64% is obtained for butyl *cis*-4-methoxy-5,6-
-dihydro-2*H*-pyran-6-carboxylate (47; R¹ = MeO, R² = H, R³ = CO₂Bu) (M.
Quimpere and K. Jankowski, Chem. Comm., 1987, 676). A process for
producing optically active substituted 5,6-dihydro-2*H*-pyran-6-carboxylic
esters has been reported (K. Mikami *et al.*, Can. Pat. Appl. CA 2,038,243,
1992).

2-Alkyl-4-chloro-6-methyl-5,6-dihydro-2*H*-pyrans 48 (R = alkyl) have been
obtained along with other products by the reaction of silyl ethers of
HC≡CH₂CHMeOH with aldehydes in the presence of TiCl₄ (T.H. Chan and P.
Arya, Tetrahedron Letters, 1989, *30*, 4065). The reaction of a number of 2-
-(cycloprop-1-en-1-yl)ethanol derivatives, for example, 49 (R = Me, H) with
bromine, acid or silver ion leads to ring expansion either to 5,6-dihydro-2*H*-
-pyrans, for example, 4-bromo-2,3-dimethyl-5,6-dihydro-2*H*-pyran (50), or to

2,2-dimethyl-3-methylenetetrahydrofurans 51 (for example R = Br, H) (J.R.
Al-Dulayymi and M.S. Baird, Tetrahedron, 1990, <u>46</u>, 5703). Also reported is
the preparation of 4-bromo-2,2,3,6-tetramethyl-5,6-dihydro-2*H*-pyran (*idem*,
Tetrahedron Letters, 1989, <u>30</u>, 253).

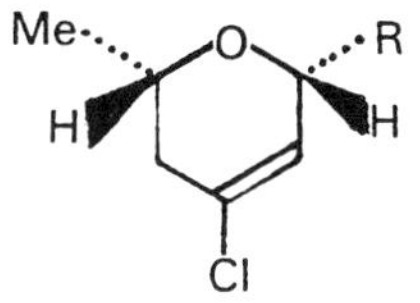

(48)

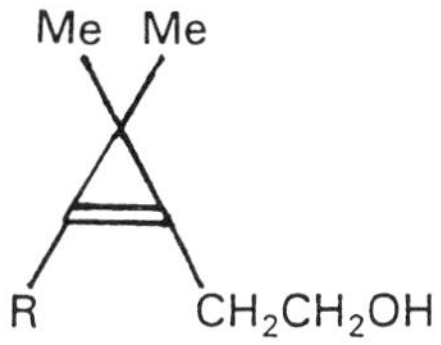

(49)

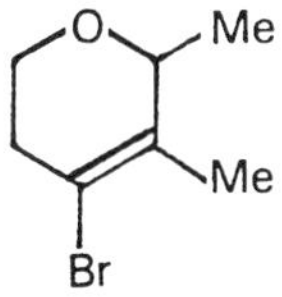

(50)

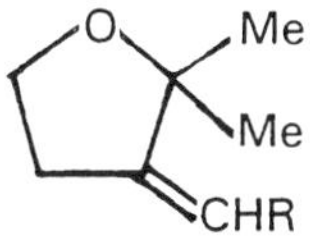

(51)

The intramolecular cyclization of $^\lrcorner$-allenic cyanohydrins,
$R^1R^2C = C = CHCMe_2CH(CN)OH$ [$R^1 = Me$, $R^2 = Me$, Et; $R^1R^2 = (CH_2)_5$], in
MeOH containing $BF_3.2Et_2O$ leads to 6,6-dialkyl-2-cyano-3,3-dimethyl-3,6-
-dihydro-2*H*-pyrans 52 (J. Grimaldi and A. Cormons, *ibid.*, 1987, <u>28</u>, 3487).
The preparations of methyl 3-amino-3,6-dihydro-2*H*-pyran-2-carboxylate (53)
and related carboxylates as medical fungicides have been reported (J.
Mittendorf *et al.*, Eur. Pat. Appl. EP 538,689, 1993).

(52)

(53)

4-(4-Methyl-5,6-dihydro-2*H*-pyran-5-yl)aminobenzylidene(4-alkoxy anilines) 54 (R = Me, Et, Pr, Bu, C_6-C_{10} *n*-alkyl) have been prepared in 3 steps from 4-methyl-5,6-dihydro-2*H*-pyran and they showed smectic and nematic liquid crystal behaviour between 80-105°C (M.M. Murza, T.R. Khlestkina, and M.G. Safarov, Zh. Org. Khim., 1991, 27, 1046).

(54)

The 4-methyl-5,6-dihydro-2*H*-pyran-5-yl thioureas 55 (R = Ph, 4-MeC$_6$H$_4$) with ClCOCH$_2$Cl afford thiazolinones 56 and amide 57 with thiourea yields diaminothiazole 58 (I.N. Bazhitova *et al.*, Khim. Geterotsikl. Soedin., 1992, 323).

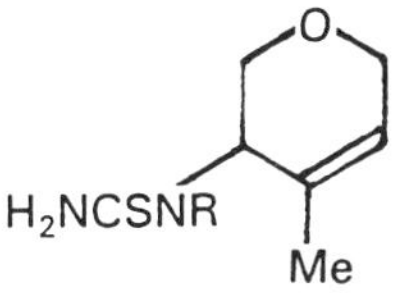

(55)

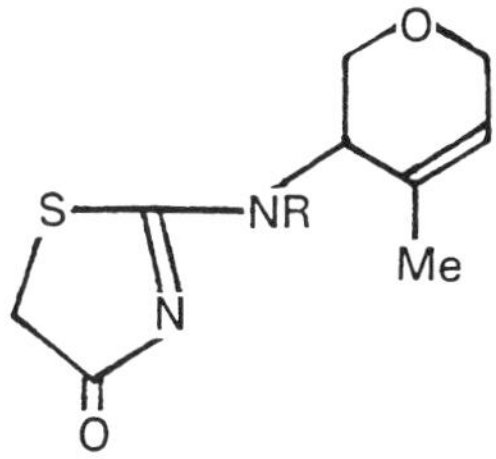

(56)

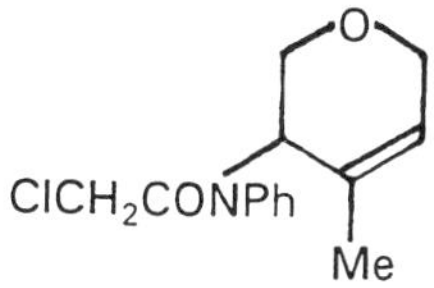

(57)

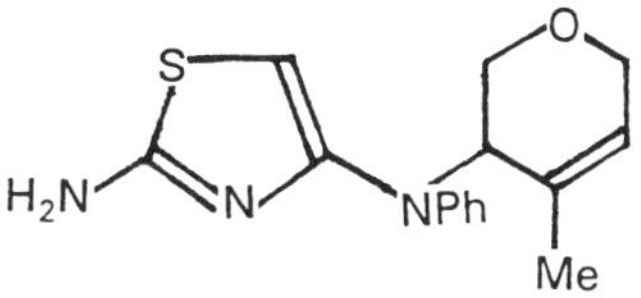

(58)

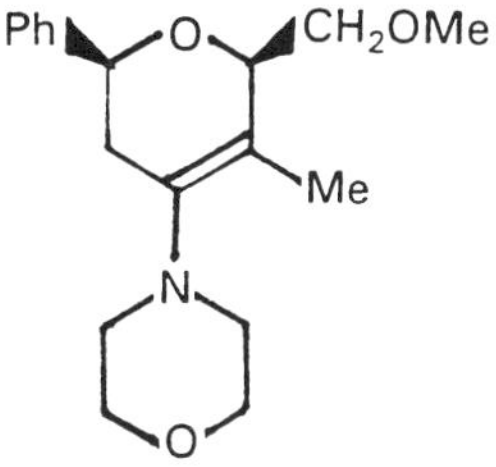

(59)

4-Morpholino-2-(methoxymethyl)-3-methyl-6-phenyl-5,6-dihydro-2*H*-
-pyran (59) has been obtained from the appropriate 2-morpholinopenta-1,3-
-diene and PhCHO, after treatment with AcONa in aqueous AcOH (J.
Barluenga *et al.*, Chem. Comm., 1988, 1247).

Stevens rearrangement of the ammonium salts 60 (R = Et; R_2 = piperidyl,
morpholinyl) and 61 containing a 5,6-dihydro-2*H*-pyranyl group yield
aminoacetic acid derivatives 62 and 63, respectively (A.A. Gevorkyan *et al.*,
Arm. Khim. Zh., 1990, <u>43</u>, 526).

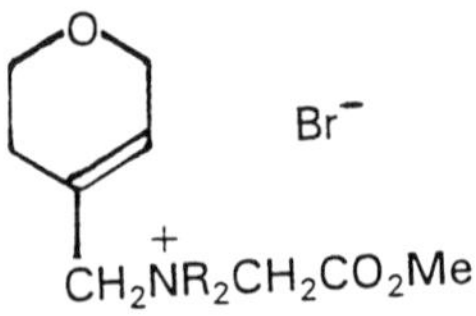

122

Br⁻ — $CH_2\overset{+}{N}R_2CH_2CO_2Me$

(60)

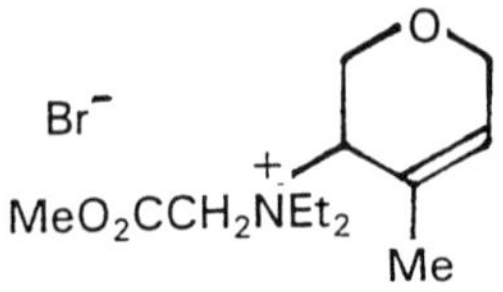

Br⁻ $MeO_2CCH_2\overset{+}{N}Et_2$ Me

(61)

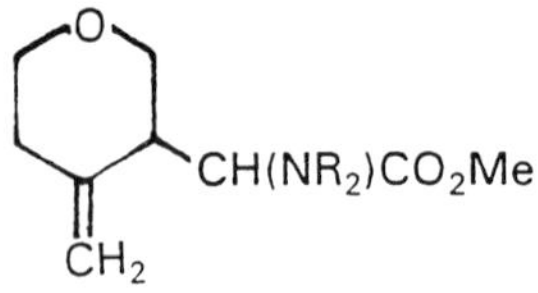

$CH(NR_2)CO_2Me$ CH_2

(62)

$MeO_2C(Et_2N)CH$ Me

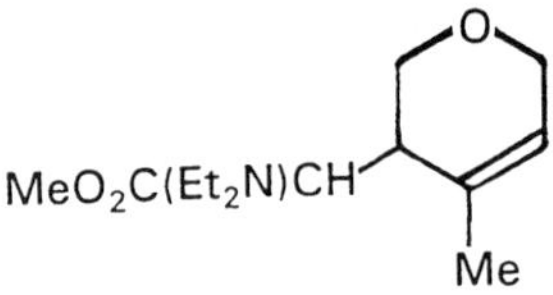

(63)

(Allyloxy)- and (allylthio)benzothiazole derivatives of 5,6-dihydro-2*H*--pyrans, for example 64, easily react with methyl cuprates to yield *C*-Me compounds, for example, the 2-substituted 6-ethoxy-5-methyl-5,6-dihydro--2*H*-pyran 65 (S. Valverde *et al.*, J. Org. Chem., 1990, 55, 2294).

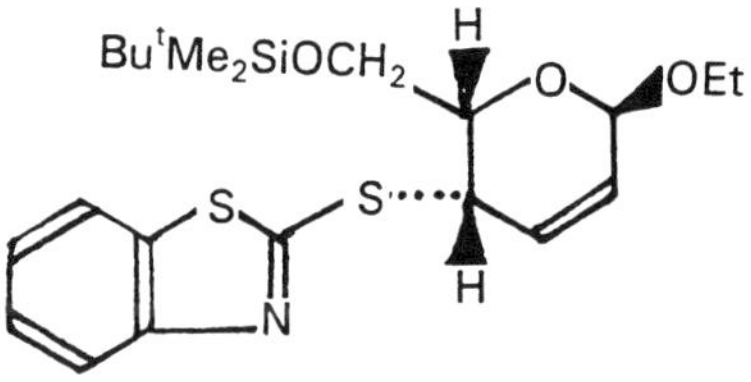

(64)

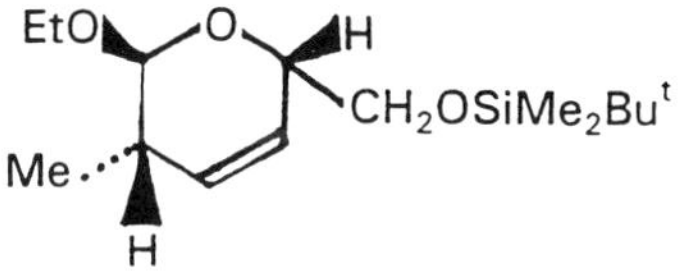

(65)

(v) *5,6-Dihydro-2H-pyran-2-, -5-, and -6-ones (3,6-dihydro-2H-pyran-6-, -3- and -2-ones)*. 3-Methylglutaconic anhydride on LiAlH$_4$ reduction, followed by Jones oxidation yields 4-methyl-5,6-dihydro-2*H*-pyran-2-one (anhydromevalonolactone) (1) (A. Nangia, B.M. Rao, and G. Prasuna, Synth. Comm., 1992, <u>22</u>, 593). The asymmetric hydrogenation of methyl 3,5- -dioxohexanoate, MeCOCH$_2$COCH$_2$CO$_2$Me using Ru$_4$Cl$_2$ [(*S*)-binap](Et$_3$N) as catalyst affords the dihydroxy ester 2, which has been converted into 6- -methyl-5,6-dihydro-2*H*-pyran-2-one (3) (L. Shao *et al.*, Tetrahedron Letters, 1991, <u>32</u>, 7699; Tetrahedron, 1993, <u>49</u>, 1997). The synthesis of enantio- merically pure (*S*) 6-methyl-5,6-dihydro-2*H*-pyran-2-one and (*S*)-6-methyl- tetrahydropyran-2-one from (*S*)-1-(1,3-dithian-2-yl)-2-hydroxypropane has been reported (R. Bernardi and D. Ghiringhelli, Gazz., 1992, <u>122</u>, 395).

(1)   (2)   (3)

6-Substituted 5,6-dihydro-2*H*-pyran-2-ones 4 (R = alkyl, alkenyl, alkynyl) have been prepared by treating the appropriate 2-substituted 6-trimethylsilyl- oxy-3,4-dihydro-2*H*-pyran 5 with allyl methyl carbonate in MeCN in the presence of Pd(OAc)$_2$ at 50-60°C (M. Amaike, M. Iwamoto, and K. Takagi, Jpn. Kokai Tokkyo Koho JP 63 215,676 [88 215,676], 1988). 6-Alkenyl 5,6-dihydro-2*H*-pyran-2-ones 4 (R = alkenyl), useful as flavouring agents have been prepared *via* cyclization of RCH(OH)CH(OH)CH$_2$CN, also Me(CH$_2$)$_4$CH- (OH)CH(OH)CH$_2$CN on heating with concentrated H$_2$SO$_4$ at 80°C affords 6-*n*-pentyl-5,6-dihydro-2*H*-pyran-2-one (4; R = *n*-pentyl) (M. Amaike *et al.*, *ibid.*, 57,583). The preparations of 6-(1-hydroxyhexyl)-5,6-dihydro-2*H*- -pyran-2-one (6) (X. Zhang *et al.*, Chin. Chem. Letters, 1991, <u>2</u>, 919) and 6- -hydroxymethyl-5,6-dihydro-2*H*-pyran-2-one derivatives 7 (R = protecting group) (S. Kono, K. Ogasawara, and Y. Sekiguchi, Jpn. Kokai Tokkyo Koho JP 63 277,669 [88 277,669], 1988) have been reported. For a synthesis of 3-methyl and 3,6-dimethyl-5,6-dihydro-2*H*-pyran-2-ones, see R.G. Hofstraat *et al.*, J. Chem. Soc., Perkin 1, 1988, 2315.

124

(4) (5) (6)

(7) (8) (9)

    (*R*)-1,2-Epoxyoctane has been converted to enantiomerically pure (*R*)-
-hexyl-5,6-dihydro-2*H*-pyran-2-one in 3 steps involving regioselective epoxide
ring cleavage with the carbanion derived from *tert*-butyl propiolate, followed
by partial hydrogenation in presence of Lindlar catalyst, and cyclization under
acidic conditions (U. Goergens and M.P. Schneider, Tetrahedron Asymmetry,
1992, **3**, 831).

    The [3 + n] annulation reactions by means of 3-(trimethylstannyl)-2-
[(trimethylstannyl)methyl]propene, an *iso*butene dianion synthetic equivalent,
have been investigated and it has been shown that the reaction of
$H_2C = C(CH_2SnMe_3)_2$ with RCHO [R = Ph, *n*-hexyl, (*E*)-PhCH = CH], followed
by lithiation with BuLi, carboxylation with CO_2 and acidic hydrolysis yielded
the 6-substituted 4-methyl-5,6-dihydro-2*H*-pyran-2-ones 8 (A. Degl'Innocenti
et al., Synthesis, 1991, 267). Stirring a mixture of MeCN, Ni(COD)$_2$, DPPE,
and allene in an autoclave at 80°C for 2 days under 15 atmospheres CO_2
afforded 4-methyl-5-methylene-5,6-dihydro-2*H*-pyran-2-one (Y. Sasaki, J.
Mol. Catal., 1989, **54**, L9-L12). The preparation of 5-(*S*)-5-[*N*-(*tert*-butoxy-
carbonyl)amino]-5,6-dihydro-2*H*-pyran-2-one has been reported (Y. Ofuna
and K. Shimamoto, Jpn. Kokai Tokkyo Koho JP 02 237,984 [90 237,984],
1990).

    The preparations of some 3-hydroxy-5,6-dihydro-2*H*-pyran-2-ones 9 (R^1-
R^3 = H, alkyl) with 0-3 substituents have been described [G. Frater and U.
Mueller, Patentschrift (Switz.) CH 683,689, 1994]. An intramolecular
Claisen-type condensation of amide 10 or imide 11 led to 6-ethyl-4-
-hydroxy-5-methyl-5,6-dihydro-2*H*-pyran-2-one (12) in racemic or
enantiomerically pure form, respectively. The conversion of pyrone 12 to 6-
-ethyl-5-methyltetrahydropyran-2-one (13), an intermediate of the
pheromone serricornin, has also been reported (S. Brandaenge and H.
Leijonmarck, Tetrahedron Letters, 1992, **33**, 3025).

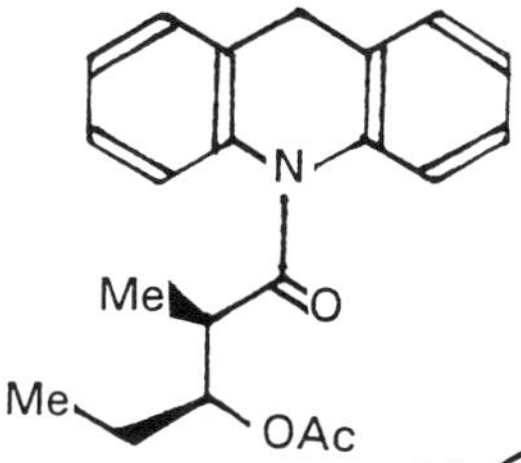

(10)

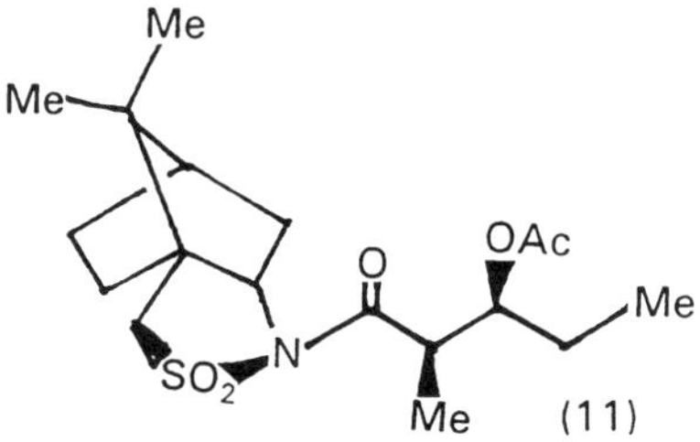

(11)

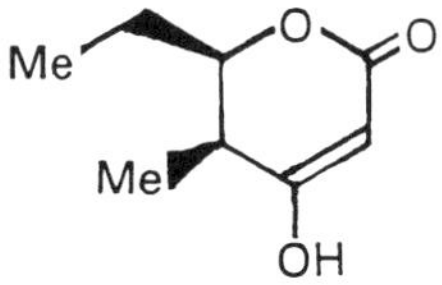

(12)

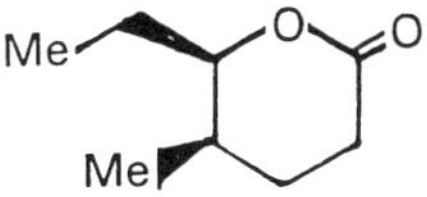

(13)

Hydroxy oxo esters, $R^1CR^2(OH)CH_2COCH_2CO_2Me$ [R^1 = Ph, Me_2CHCH_2, vinyl; R^2 = H; $R^1R^2 = (CH_2)_5$] on treatment with NaOH give 6-substituted and 6-spiro 4-hydroxy-5,6-dihydro-2*H*-pyran-2-ones 14. $R^1CR^2(OH)CH_2COCH_2$-COMe [R^1 = Me, Me_2CHCH_2, Ph; R^2 = Me, H; $R^1R^2 = (CH_2)_4$, $(CH_2)_5$] on treatment with $4\text{-}MeC_6H_4SO_3H$ yields the 2,2-disubstituted 6-methyl-3,4--dihydro-2*H*-pyran-4-ones 15 (J.R. Peterson, T.J. Winter, and C.P. Miller, Synth. Comm., 1988, 18, 949).

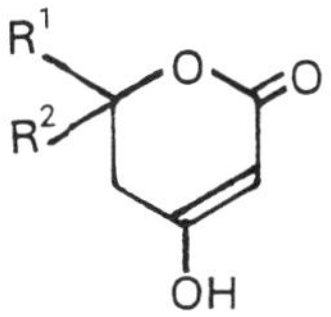

(14)

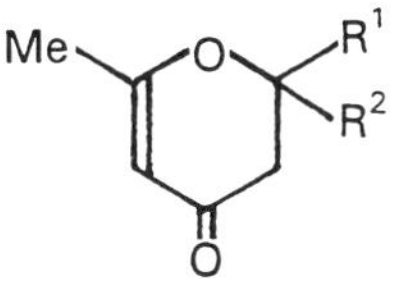

(15)

4-Hydroxy-6-methyl-5,6-dihydro-2*H*-pyran-2-one (16) condenses with RCHO (R = Ph, $4\text{-}ClC_6H_4$, $4\text{-}MeOC_6H_4$, $4\text{-}O_2NC_6H_4$) to furnish bis(oxopyranyl)-arylmethanes 17. Hydroxypyrone 16 also reacts with $2\text{-}HOC_6H_4CHO$ to afford pyranochromenone 18 and with urea and $CH(OEt)_3$ to give ureido-pyrone 19 (Y. Rachedi, M. Hamdi, and V. Speziale, *ibid.*, 1990, 20, 2827). The 6-alkyl- and 3,6-dialkyl-4-hydroxy-5,6-dihydro-2*H*-pyran-2-one tautomers 20 21 (R=H, C_6H_{13}) have been prepared (J.J. Landi, Jr., and K.M. Ramig, Eur. Pat. Appl. EP 524,495, 1993).

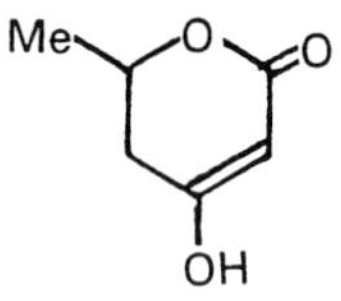

(16)

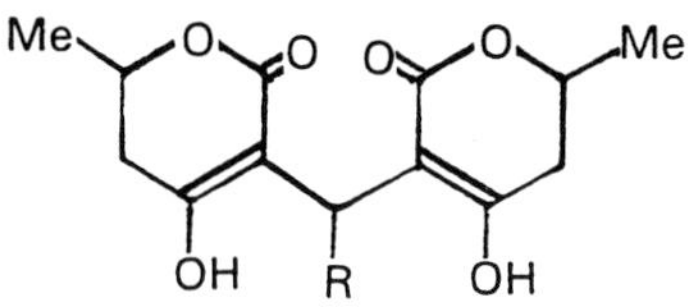

(17)

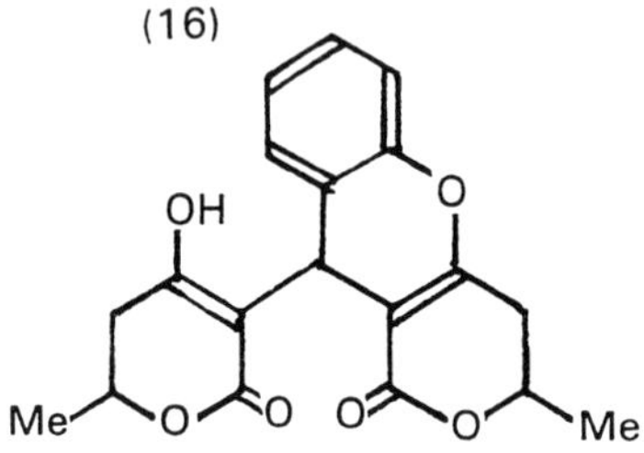

(18)

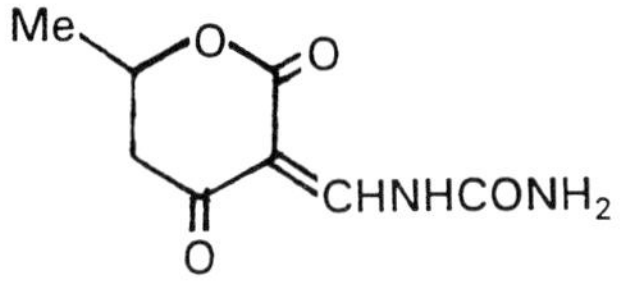

(19)

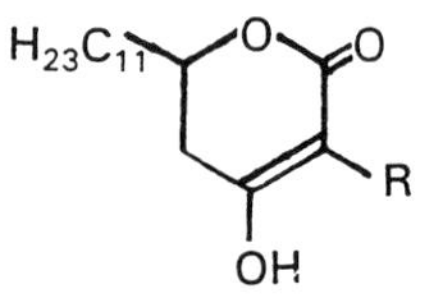

(20)

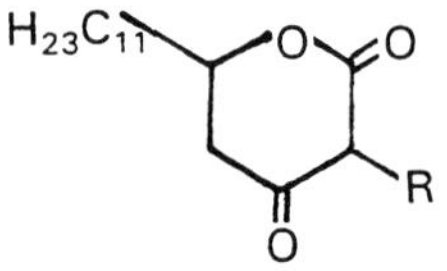

(21)

Conjugate addition of benzenethiol, under mild conditions, to the carbohydrate derived hex-2-enone-δ-lactones, substituted at C-6 or C-7 with an electron-rich insaturation, for example 6-(1-acetoxyprop-2-enyl)-5--methoxy-5,6-dihydro-2*H*-pyran-2-one (22), afforded 3-phenylthio derivatives, that on treatment with tributyltin hydride and AIBN (azoiso-butyronitrile), undergo an efficient radical cyclization to give highly functionalized cyclohexanes and cyclopentanes, for example, 23 and 24 (J.C. Lopez, A.M. Gomez and S. Valverde, Chem. Comm., 1992, 613).

(22)   (23)   (24)

The 6-substituted 5-benzyloxy-5,6-dihydro-2*H*-pyran-2-one 25 undergoes a chemoselective conjugate reduction to the 6-substituted 5-benzyloxytetra-hydropyran-2-one 26, by use of benzenethiol and tributyltin hydride (*idem*, Synlett, 1991, 825).

(25)   (26)

The dienolate produced from 3-methyl-2-phenylthiobut-2-enoic acid and two equivalents of $LiN(Pr^i)_2$, reacts with aldehydes or ketones R^1COR^2 ($R^1 = Ph$, $PhCH_2CH_2$, Pr, Bu, $R^2 = H$; $R^1 = Et$, $R^2 = Me$, Et) to afford 6-mono- or 6,6-disubstituted 4-methyl-3-phenylthio-5,6-dihydro-2*H*-pyran-2-ones (5--alkyl-3-methyl-2-phenylthiopent-2-en-5-olides) 27 (H. Su, K.Shirai, and T. Kumamoto, Chem. Soc. Jpn., 1992, <u>65</u>, 2794). For the preparation of 4,5,6-trimethyl-3-methylthio-5,6-dihydro-2*H*-pyran-2-one (28), see M. Mikolajczyk and W.H. Midura, Synlett, 1991, 245).

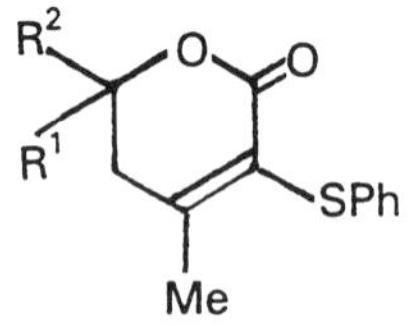

(27) 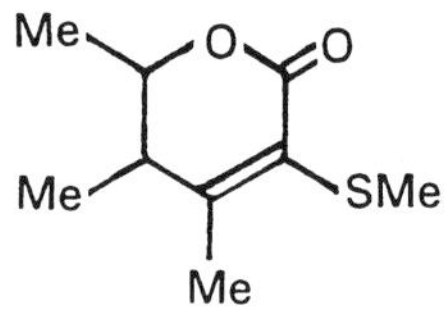   (28)

Dimethyl 3-methylglutaconate on treatment with ketones, NaH, and $ZnCl_2$ yields 6,6-disubstituted methyl 4-methyl-2-oxo-5,6-dihydro-2*H*-pyran--5-carboxylates 29 [R = Me, Et; $R_2 = (CH_2)_4$, $(CH_2)_5$, $(CH_2)_6$, $CH_2CH_2CH(Bu^t)CH_2CH_2$] (J.P. Rocher, M. Ahmar, and J. Paris, J. Heterocyclic Chem., 1988, <u>25</u>, 599).

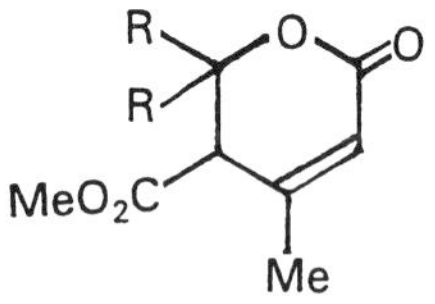

(29) 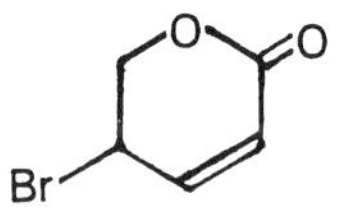   (30)

5-Bromo-5,6-dihydro-2*H*-pyran-2-one (30) has been obtained by treating 5,6-dihydro-2*H*-pyran-2-one with NBS, in the presence of $(PhCO)_2O$ in CCl_4 and then with Et_3N. It has been found to undergo smooth and highly regiocontrolled [2 + 4] cycloadditions between 25-100°C with both electron-poor and electron-rich dienophiles; subsequent radical debrominations produced halogen-free bicyclic lactones, including a precursor to a vitamin D steroid (K. Afarinkia and G.H. Posner, Tetrahedron Letters, 1992, <u>33</u>, 7839).

The cyclocondensation reaction of $MeCOCHMeCH_2OH$ with RCH_2CO_2Et (R = CN, MeCO, EtO_2C) at 60°C in benzene containing 5% aqueous KOH and Katamin AB affords the 3-substituted 4,5-dimethyl-5,6-dihydro-2*H*-pyran-2--ones 31, while $MeCOCH_2CMe_2OH$ and $NCCHRCO_2Et$ (R = C_2-C_5 *n*-alkyl) gives the 3-substituted 3-cyano-4-hydroxy-4,6,6-trimethyltetrahydropyran-2--ones 32 (A.A. Avetisyan, A.A. Kagramanyan, and G.S. Melikyan, Arm. Khim. Zh., 1989, <u>42</u>, 708).

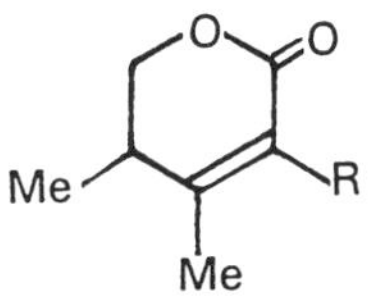

(31)

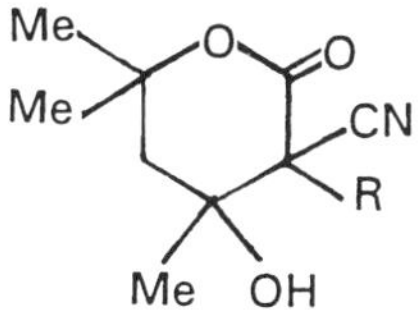

(32)

3-Cyano-4,6,6-trimethyl-5,6-dihydro-2*H*-pyran-2-one (33) condenses with ArCHO (Ar = substituted Ph) to give 5-arylmethylene-3-cyano-4,6,6-tri-methyl-5,6-dihydro-2*H*-pyran-2-one 34 (A.A. Avetisyan *et al.*, *ibid.*, 1990, <u>43</u>, 57). The Mannich reaction of 3-cyano-4,6,6-trimethyl-5,6-dihydro-2*H*--pyran-2-one (33) with HCHO and $Et_2NH.HCl$ gives 3-cyano-5-(2-diethyl-aminoethyl)-4,6,6-trimethyl-5,6-dihydro-2*H*-pyran-2-one hydrochloride (35) (A.A. Avetisyan, A.A. Kagramanyan, and G.S. Melikyan, *ibid.*, 1989, <u>42</u>, 633).

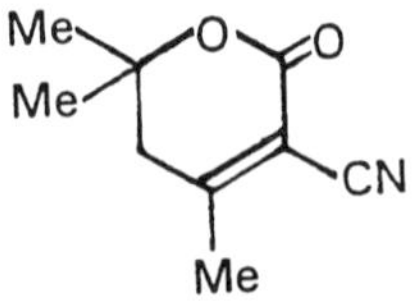

(33) 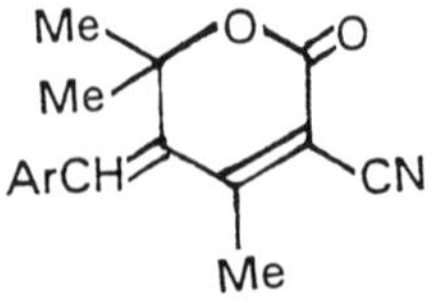   (34) 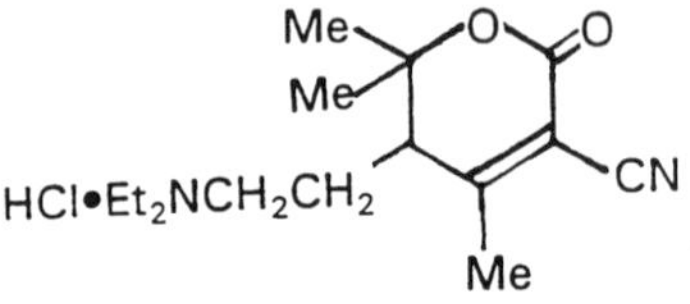   (35)

The Diels-Alder adducts from the reaction of 3-amino-2*H*-1,4-oxazin-2-
-ones 36 (R^1 = NEt_2, NHPr, R^2 = Me; R^1 = NEt_2, R^2 = Ph) with H_2C = CHR^3
(R^3 = CO_2Me, Ph) in boiling toluene, undergo ring transformation furnishing
3,4,6-trisubstituted 6-cyano-5,6-dihydro-2*H*-pyran-2-ones 37 (C.C. Fannes
and G.J. Hoornaert, Tetrahedron Letters, 1992, <u>33</u>, 2049).

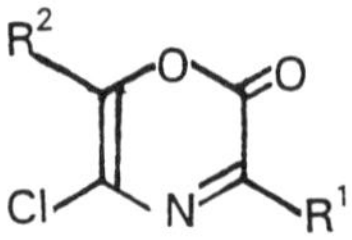

(36) 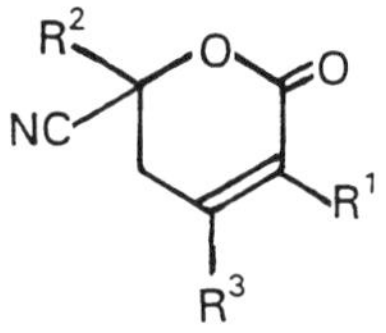   (37)

The preparations of 3,7a-dihydro-1*H*-cyclopenta[c]pyran-3-one 38 (A.
Nangia and P.B. Rao, *ibid.*, p.2375), 4-hydroxy-3,8a-dihydro-1*H*-cyclohexa-
[c]pyran-3-one 39 (G. Frater and U. Mueller, *ibid.*, 1993, <u>34</u>, 2753), and
some (oximinioalkyl)spirodihydropyran-2-ones and related compounds (J.E.
Anderson-McKay and A.J. Liepa, PCT Int. Appl. WO 88 00,945, 1988) have
been reported.

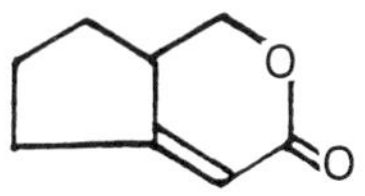

(38)

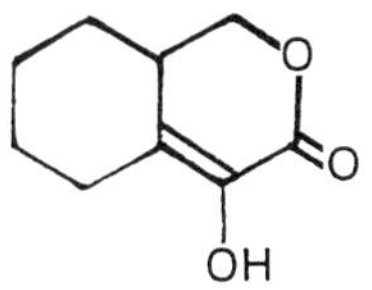

(39)

5,6-Dihydro-2*H*-pyran-5-ones are also reported as 3,6-dihydro-2*H*-pyran-3-ones and 2,3-dihydro-6*H*-pyran-3-ones, therefore, in the following discussion the actual recorded names will be used. An improved procedure for the synthesis of 6-alkoxy-2,3-dihydro-6*H*-pyran-3-ones (2,3-dideoxy-DL-pent-2--enopyranos-4-uloses) involves, for example, treating 6-benzoyloxy-2,3--dihydro-6*H*-pyran-3-one (40) with $Me_3SiCH_2CH_2OH$ in $ClCH_2CH_2Cl$ in the presence of $ZnCl_2.Et_2O$ to yield 6-(2-trimethylsilylethoxy)-2,3-dihydro-6*H*--pyran-3-one (41; R = $CH_2CH_2SiMe_3$), which on treatment with $Pd(OAc)_2$, $NaHCO_3$, and Bu_4NCl in DMF affords the cyclopentenone 42 (B. Mucha and H.M.R. Hoffmann, Tetrahedron Letters, 1989, 30, 4489). The rearrangement of the 6-alkoxypyrones 41 (R = Bu^t, Me, PhCH_2, Cl_3CCH_2, $Me_3SiCH_2CH_2$) to *trans*-4-alkoxy-5-hydroxycyclopent-2-en-1-ones 42 has been optimized (H.C. Kolb and H.M.R. Hoffmann, Tetrahedron, 1990, 46, 5127). For the synthesis of some 2-substituted 6-acyloxy- and 6-alkoxy--2,3-dihydro-6*H*-pyran-3-ones, see V. Constantinou-Kokotou, G. Kokotou, and M.P. Georgiadis, Ann., 1991, 151.

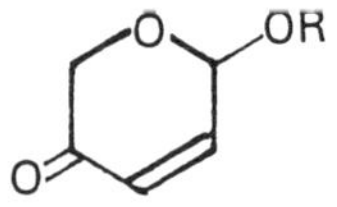

(40) R = Bz
(41) R(see text)

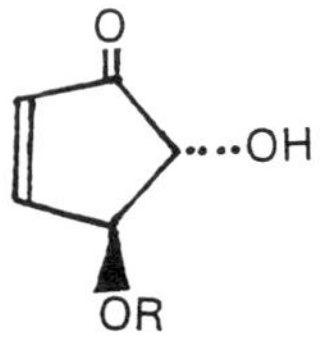

(42)

The preparation of optically pure 2-(4,4-dimethyl-2-oxotetrahydrofur-3-
-yloxy)-5,6-dihydro-2*H*-pyran-5-one (43) using +-pantolactone as a chiral
auxiliary and its successful application in Diels-Alder and Michael type
additions has been described (J. Knol *et al., ibid.*, 1991, 32, 7465).

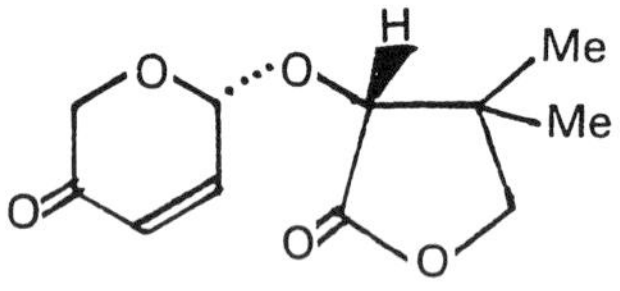

(43)

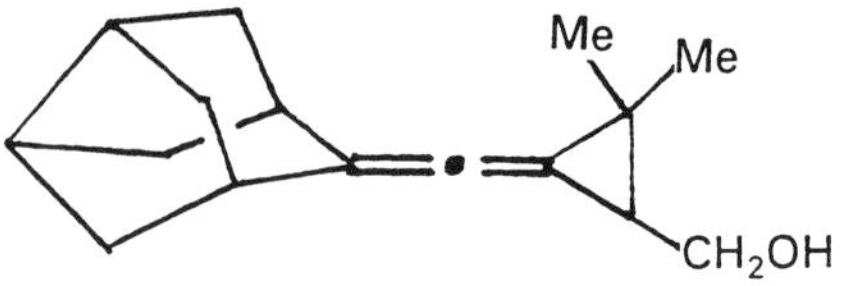

(44)

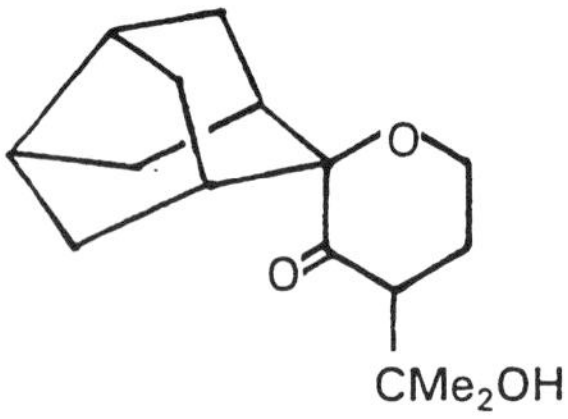

(45)

Photosensitized oxygenation of adamantenylidenecyclopropane 44 yields
the 5,6-dihydro-2*H*-pyran-5-one 45 (T. Akasaka, Y. Misawa, and W. Ando,
ibid., 1990, 31, 1173). Treatment of the 2-(substituted hydroxymethyl)
furans 46 (R^1 = H, Me; R^2 = H, Me, 2,2-dimethyl-1,3-dioxolan-4-yl) with Br in
MeCN-water affords 2,2-disubstituted 6-hydroxy-2,3-dihydro-6*H*-pyran-3-
-ones 47 (S. Pikul, J. Jurczak, and G. Grynkiewicz, Bull. Pol. Acad. Sci.,
Chem., 1987, 35, 293).

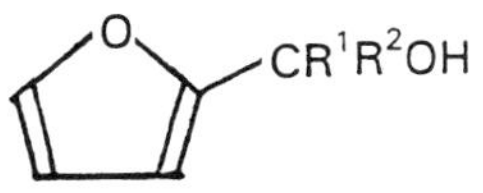

(46)

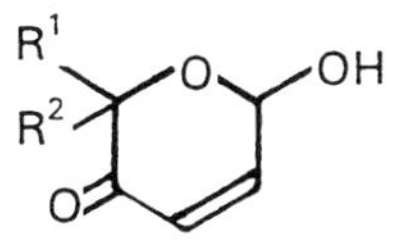

(47)

A number of 4,6-disubstituted 2-alkoxy-5,6-dihydro-2*H*-pyran-5-ones 48 [R^1 = H, alkyl, cycloalkyl, Ph; R^2 = C_{8-20}alkyl; R^3 = H, halogeno, $R^4R^5NCH_2$, Q(49); R^4, R^5 = alkyl, cycloalkyl; A (49) = CH_2, O, N forming an (un)-substituted 5-6-membered heterocycle] useful as agricultural miticides have been prepared (H. Takao, N. Osaki, and N. Yasudomi, U.S. US 4,742,078, 1988). Also reported are the preparations of a number of derivatives of 2-
-hydroxy-5,6-dihydro-2*H*-pyran-5-one (6-hydroxy-3,6-dihydro-2*H*-pyran-3-
-one) (F. Sato, Jpn. Kokai Tokkyo Koho JP 01 199,956 [89 199,956], 1989; T.P. Selby and M.E. Thompson, PCT Int. Appl. WO 92 11,762, 1992) and of 4-(substituted benzoyl)-5-hydroxy-3,6-dihydro-2*H*-pyran-3-one (R.J. Anderson *et al.*, Brit. UK Pat. Appl. GB 2,205,316, 1988).

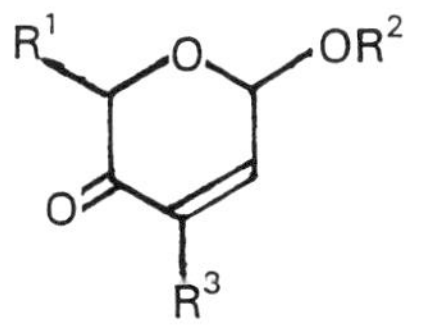

$$Q = H_2CN \quad A$$

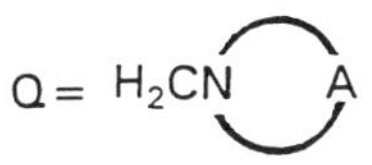

(49)

(48)

134

The unusual reactivity of *tert*-butyl hydroperoxide under acid-catalysis results in the easy conversion of furan derivatives, for example 50 (R=H, OH), into 2-butyl-6-methyl-6-*tert*-butylperoxy-3,6-dihydro-2*H*-pyran-3-one (51) (R. Antonioletti *et al.*, Tetrahedron Letters, 1993, <u>34</u>, 7089).

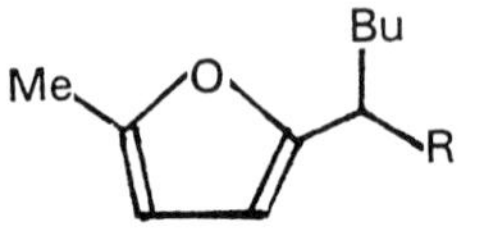

(50) 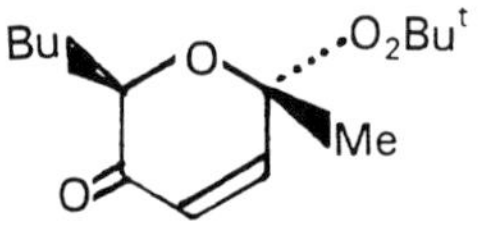   (51)

The acid 4-MeC$_6$HS(O)CH$_2$CH$_2$(=CH$_2$)CO$_2$H when treated with LiN(Pr^i)$_2$ is metallated *a* to the sulphinyl group and the dianion obtained reacts with aldehydes and ketones to give *a*-methylene-*γ*-sulphinyl-*δ*-hydroxy carboxylic acids. Intramolecular cyclization of the appropriate hydroxycarboxylic acid affords, for example, 2,2-disubstituted 5-methylene-5,6-dihydro-2*H*-pyran-6--ones (6,6-disubstituted 3-methylene-3,6-dihydro-2*H*-pyran-2-ones) 52 (R^1=H, R^2=Bu^t, Ph; R^1=Me, R^2=Ph) and 3-methyl-6-*tert*-butyltetrahydro-pyran-2-one (53), respectively, by pyrolytic or hydrogenolytic eliminations of the sulphur residue (P. Bravo, C. De Vita, and G. Resnati, Gazz., 1987, <u>117</u>, 165).

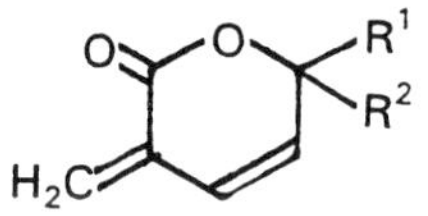

(52) 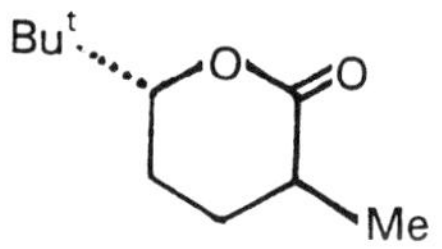   (53)

The following fused ring and spiro compounds contain either a 5,6-
-dihydro-2*H*-pyran-5- or -2-one ring system. Treatment of the bridged
tropone derivative, 2,3-dichloro-8-oxabicyclo[3.2.1]octa-2,6-dien-4-one 54
(R = Cl) with $Me_3SiNR_2^1$ (R_2^1 = Me_2N, piperidino, morpholino, 1-pyrrolidinyl) or
$Me_3SiSMe/AlCl_3$ yields vinylogous amides 54 (R = NR_2^1) and thio derivative 54
(R = MeS), respectively [G. Seitz and R. Van Gemmern, Arch. Pharm.
(Weinheim Ger.), 1987, <u>320</u>, 1138].

(54)

The dichloro- and trichlorocyclobutanopyranones 55 (R^1 = H, Cl) react
with DBU in alcohol to give pyrano[4,3-*b*]pyran-2,5-diones 56, *via* (*Z*)-(2*H*-
-pyran-2-one-3-yl)butenoates 57 (R^2 = Et, Me). The same treatment of the
analogue 58, containing a 3,4-dihydro-2*H*-pyran ring, gives oxooxabicyclo-
octadienecarboxylate 59 (T. Shimo, K. Date, and K. Somekawa, J.
Heterocyclic Chem., 1992, <u>29</u>, 387).

(55)   (56)   (57)

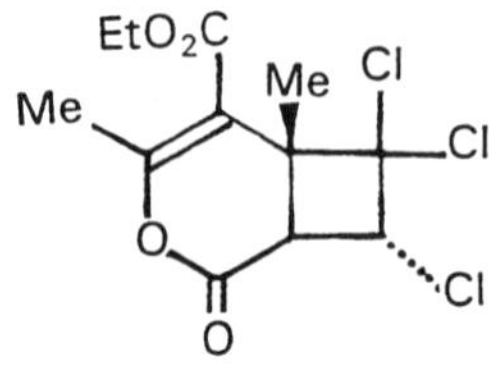

(58)

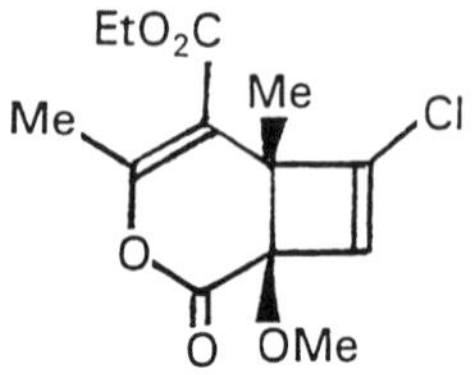

(59)

Compound 3-(PhO)C$_6$H$_4$CH$_2$ONH$_2$ condenses with δ-butyryl-6-
-oxaspiro[4.5]dec-7,9-dione to give 9-hydroxy-8-[1-(3-phenoxybenzylox-
imino)butyl]-6-oxaspiro[4.5]dec-8-en-7-one (60), which gives complete
control of barnyard grass, with no damage to rice. The preparations of a
number of related derivatives have been reported (A.J. Liepa, PCT Int. Appl.
WO 92 14,736, 1992).

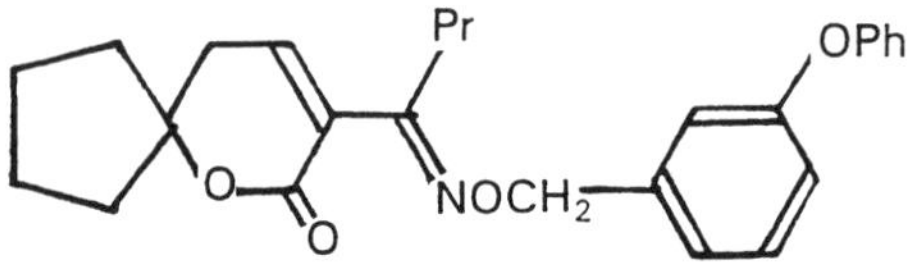

(60)

(vi) 3,6-Dihydro-2H-pyran-2,6-diones, 5,6-dihydro-2H-pyran-2,6-diones.
3,5-(HOCH$_2$CONH)$_2$C$_6$H$_3$NH$_2$ condenses with 3,5-diacetyl-4,6-dihydroxy-2*H*-
-pyran-2-one to give the 3-acetyl-4-hydroxy-5-(substituted methylene)-3,6-
-dihydro-2*H*-pyran-2,6-dione 1 (R = CH$_2$OH) (M.E. Garst, C. Gluchowski, and

L.J. Kaplan, U.S. US 4,725,620, 1988). The (*S*,*S*), (*R*,*R*) and meso-stereo-isomers of dihydropyran 1 (R = HOCH$_2$CHOH) have been synthesized from commonly available (*R*)-(-)- or (*S*)-(+)-2,2-dimethyl-1,3-dioxolane-4--methanol. Also, achiral dihydropyranenamines 1 [R = CH(CH$_2$OH)$_2$, CH$_2$OH] have been synthesized (C. Gluchowski *et al.*, J. Med. Chem., 1991, <u>34</u>, 392).

(1)

(b) Tetrahydropyrans and derivatives

(i) Tetrahydropyran, alkyl and aryl substituted tetrahydropyrans and related derivatives. Tetrahydropyran is obtained on heating pentane-1,5-diol at 190-210°C with butyltin trichloride (G. Tagliavini and D. Marton, Gazz., 1988, <u>118</u>, 483). The reactions of both (*Z*)- and (*E*)-hex-4-en-1-ols with mercuric acetate and mercuric acetate/sodium borohydride in aqueous solution afford 2-methyltetrahydropyran (1) along with 2-ethyltetrahydro-furan (2). However, prolonged reaction times and particularly the addition of acetic acid increases the yield of compound 1 up to 95% (D. Marinkovic and M.L. Mihailovic, J. Serb. Chem. Soc., 1988, <u>53</u>, 295). It has also been found that intramolecular cyclization of alk-4-enyl benzyl ethers with PhSeCl and PhSeBr afforded phenylselenoethers of tetrahydropyran and THF, for example, 2-phenylselenyltetrahydropyran (3) (Z. Bugarcic, S. Konstantinovic, and M.L.J. Mihailovic, *ibid.*, 1990, <u>55</u>, 303). Phenylselenyltetrahydropyrans have been obtained by the electrooxidative phenylselenoetherification of alkenols with PhSeSePh in CH$_2$Cl$_2$ containing Et$_4$NBr (M.L. Mihailovic, S. Konstantinovic, and R. Vukicevic, Tetrahedron Letters, 1987, <u>28</u>, 4343). The reaction of CH$_2$ = CH(CH$_2$)OH with 4-MeOC$_6$H$_4$TeCl$_3$ yields dichloro-telluroether 4, which on reduction gives 2-(4-methoxyphenyltellurylmethyl)-tetrahydropyran (5) (J.V. Comasseto *et al.*, *ibid.*, p.5611).

138

(1) R = Me
(6) R = Bu

(2)

(3)

(4)

(5)

Treatment of tetrahydropyran-2-yl 2,4-dimethoxybenzoate with Grignard-
-derived organocopper reagents, BuCu and CuBr.SMe$_2$ affords 2-butyltetra-
hydropyran (6) (V. Bolitt, C. Mioskowski, and J.R. Falck, *ibid.*, 1989, *30*,
6027). A number of 4-benzyltetrahydropyrans and analogues have been
prepared as lipoxygenase inhibitors (J.F. Dellaria, L.J. Chernesky and D.W.
Brooks, PCT Int. Appl. WO 94 05,281, 1994).

The preparations of a number of tetrahydropyrans containing a complex
substituent have been reported. These include compounds 7 (R^1 = CO$_2$Et,
R^2 = H, R^3 = Me; R^1 = R^3 = H, R^2 = Me) (J.M. Takacs and S.V. Chandramouli,
J. Org. Chem., 1993, *58*, 7315); 8, as a mixture of diastereomers (A.H.
Davidson, N. Eggleton, and I.H. Wallace, Chem. Comm., 1991, 378); and 9
(A. Teniou, L. Toupet, and R. Gree, Synlett, 1991, 195) with the substituent
at the 2-position of tetrahydropyran; 10 a herbicide (U. Misslitz *et al.*, Ger.
Offen. DE 4,227,896, 1994) with the substituent at the 3-position; and 11
and related derivatives as herbicides (J. Kast *et al.*, *ibid.*, 4,018,508, 1991;

Eur. Pat. Appl. EP 456,068, 1991) with the substituent in the 4-position.
Other derivatives similar to compound 11 have been prepared (U. Misslitz *et
al.*, *ibid.*, 456,112; 456,118, 1991).

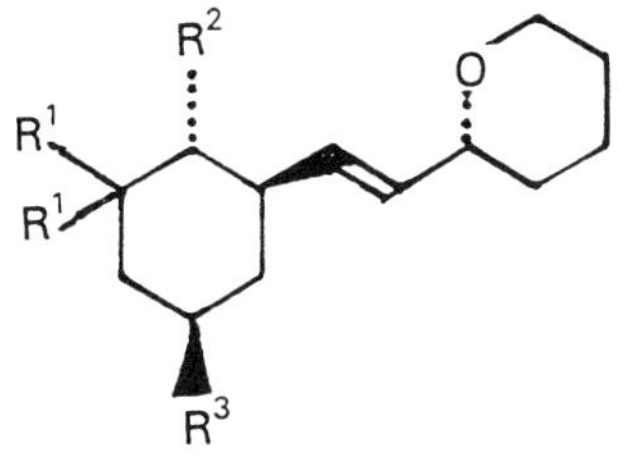

(7)

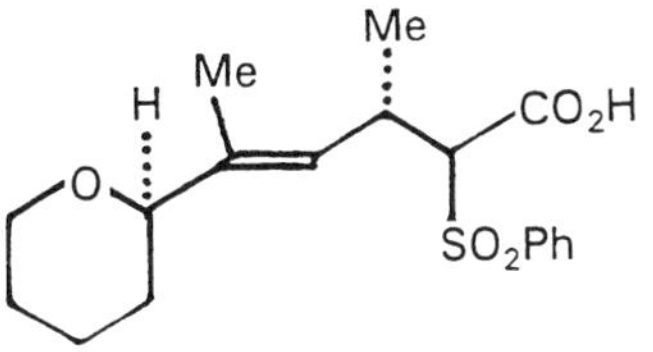

(8)

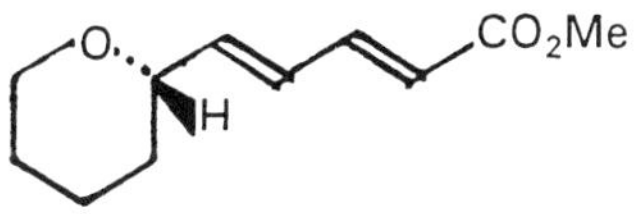

(9)

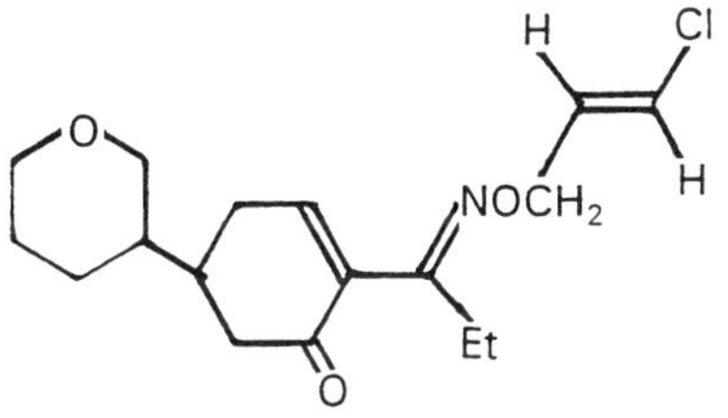

(10)

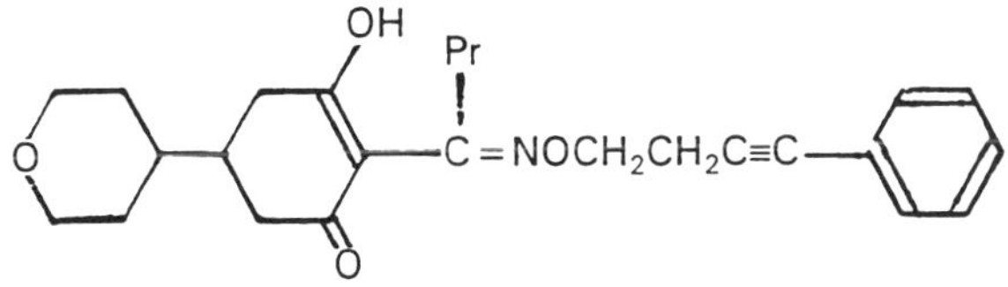

(11)

140

CH$_2$=C(CH$_2$CH$_2$OH)$_2$ on treatment with PhSO$_3$H in aqueous NaOH-
MeOBu^t gives 4-methylenetetrahydropyran (12) (T. Kuekenhoehner *et al.*,
Eur. Pat. Appl. EP 412,361, 1991). Oxidation of 4-methylenetetrahydro-
pyran (12) by HNO$_3$ containing N$_2$O$_3$ or N$_2$O$_4$ gives citric acid and oxalic acid
and similar oxidation of 4-methyl-5,6-dihydro-2*H*-pyran (13) affords oxalic
acid. However, without N$_2$O$_3$ or N$_2$O$_4$ they both give 3-hydroxy-3-methyl-
glutaric acid in addition to the above products (M.S. Sargsyan, S.A.
Mkrtumyan, and A.A. Gevorkyan, Zh. Org. Khim., 1987, <u>23</u>, 2220).

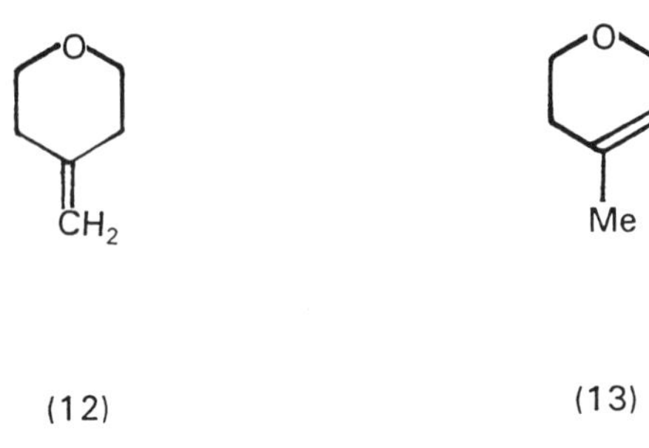

(12)   (13)

The deprotonation of 2-methylenetetrahydropyrans 14 (R=CN, CO$_2$Me)
with Pr$^i_2$NLi and condensation with PrCHO gives endocyclic enol ethers 15,
which on hydrogenation yield exocyclic enol ethers 16 (75-80%) and 17
(4-6%) (A. Takahashi *et al.*, J. Amer. Chem. Soc., 1989, <u>111</u>, 643). The
reaction of racemic 2-(4-tolylsulphinyl)acrolein prepared in situ by the
oxidation of 2-(4-tolylsulphinyl)-3-hydroxyprop-1-ene with 2-methylenetetra-
hydropyran has been investigated (P. Hayes and C. Maignan, Synlett, 1994,
409). The synthesis of 2-*iso*butenyltetrahydropyrans 18 (R=H, Me) based
on 5-methylhexa-1,4-dien-3-one has been reported (I.G. Tishchenko and V.V.
Berezovskii, Vestsi Akad. Navuk BSSR, Ser. Khim. Navuk, 1987, 62).

(14) (15) (16)

(17) (18) (19)

Treatment of 4-methylenetetrahydropyran (12) and Mo ethylhexanoate at 80°C with a 1:1 mixture of Bu^tOH/Bu^tOOH yields 1,6-dioxaspiro[2.5]octane (19) (M. Fischer, Ger. Offen. DE 4,117,370, 1992).

2-Phenylsulphonyltetrahydropyran (20; $R^1 = PhSO_2$, $R^2 = H$) reacts with the organozinc reagent obtained from PhMgBr and $ZnBr_2$ to give 2-phenyl-tetrahydropyran (21) (D.S. Brown and S.V. Ley, Tetrahedron Letters, 1988, 29, 4869). Similarly, compound 20 with Pr^iMgCl, $ZnBr_2$, and $PhC \equiv CH$ in THF affords 2-phenylethynyltetrahydropyran (*idem*, Org. Synth., 1992, 70, 157).

The reaction of benzenesulphones 20 ($R^1 = PhSO_2$, $R^2 = H$, $AcOCH_2$, $HOCH_2$, Ph) with $R^4R^5C = C(R^3)OSiMe_3$ ($R^3 = Ph$, Me, OMe; $R^4 = R^5 = H$, Me; $R^4 = H$, Me, $R^5 = CO_2Me$) in the presence of $AlCl_3$ gives tetrahydropyrans 20 ($R^1 = CR^4R^5COR^3$, R^2 same) (D.S. Brown, S.V. Ley, and M. Bruno, Heterocycles, 1989, 28, 773). The direct substitution of 2-benzene-sulphones 20 ($R^1 = PhSO_2$, $R^2 = H$, Ph, MeO, Bu, $AcOCH_2$) has been studied

142

by using a variety of nucleophiles such as silyl enol ethers, silyl ketene
acetals, alkylsilanes, or Me_3SiCN in the presence of $AlCl_3$ or organozinc
reagents (D.S. Brown *et al.*, Tetrahedron, 1989, **45**, 4293).

2,2-Disubstituted tetrahydropyrans 22 (R = Et, allyl, $PhCH_2$, $PhC \equiv C$) are
obtained by the reaction of $Cl(CH_2)_4COCl$ with RMgBr (1:2 molar ratio),
followed by treatment of the resulting $Cl(CH_2)_4CR_2OH$ with NaH (J.
Barluenga *et al.*, J. Chem. Res., S, 1987, 400). The reaction of 6-
-substituted tetrahydropyran-2-ols 23 ($R^1 = CH_2OSiPh_2Bu^t$, CH_2CH_2Ph) with
organometals in the presence of Lewis acid yields diastereoselectively the
2,6-disubstituted tetrahydropyrans 24 (R^2 = Me, Et, Ph, $CH = CH_2$, allyl) (K.
Tomooka *et al.*, Tetrahedron Letters, 1987, **28**, 6339). It is possible to
distinguish between *cis*- and *trans*-dimethyltetrahydropyrans by the
application of high-resolution electronic spectroscopy (T.J. Cornish and T.
Baer, Anal. Chem., 1990, **62**, 1623).

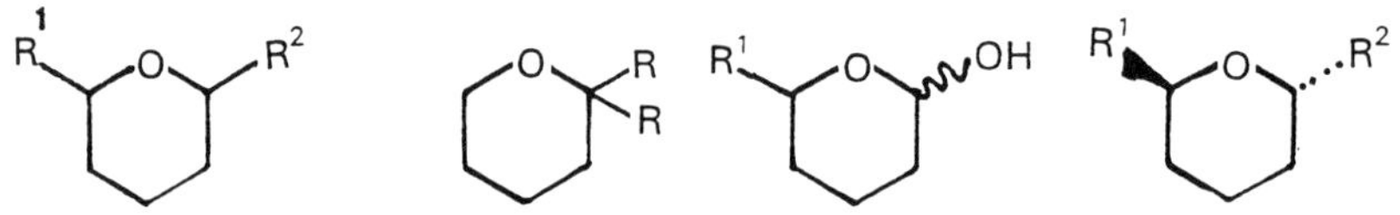

(20)   (22)   (23)   (24)
(21) R^1 = Ph, R^2 = H

Treatment of (E)-$Br(CH_2)_3OCHMeCH_2C(SMe)SO_2C_6H_4Me$-4 with Bu_3SnH-
-AIBN in C_6H_6 affords 3-substituted *trans*-2-methyltetrahydropyran 25 (K.
Ogura *et al.*, *ibid.*, 1993, **34**, 8313). For the preparation of (-)-*cis*-4-methyl-
-2-(2,2-dimethylvinyl)tetrahydropyran [(-)-*cis*-rose oxide] (26) and *cis*-6-
-methyltetrahydropyran-2-acetic acid (27), see K. Homma, H. Takenoshita,
and T. Mukaiyama, Bull. Chem. Soc. Jpn., 1990, **63**, 1898 and for 3-(2-
-hydroxycyclohexyl)-2-methyltetrahydropyran, A. Alexakis *et al.*, Tetrahedron
Letters, 1990, **31**, 1271.

(25)

(26)   (27)

The addition of 2-(chloromagnesiomethyl)-2-propenyl ethers 28
($R^1 = CH_2Ph$, Ph) to epoxides 29 [$R^2 = R^3 = H$, $R^4 = Me$, Ph, $CH(OEt)_2$;
$R^2R^3 = (CH_2)_5$, $R^4 = H$; $R^2R^4 = (CH_2)_3$, $(CH_2)_4$, $(CH_2)_5$, $R^3 = H$] affords the ring
open products 30, which are converted by Pd(O) to 5,6,6-trisubstituted 3-
-methylenetetrahydropyrans 31. Cyclization of the addition products is best
effected by a catalyst system generated in situ from $Pd(OAc)_2$ and $(Pr^iO)_3P$
(J. Van der Louw *et al.*, Tetrahedron Letters, 1989, <u>30</u>, 4863; Tetrahedron,
1992, <u>48</u>, 9901).

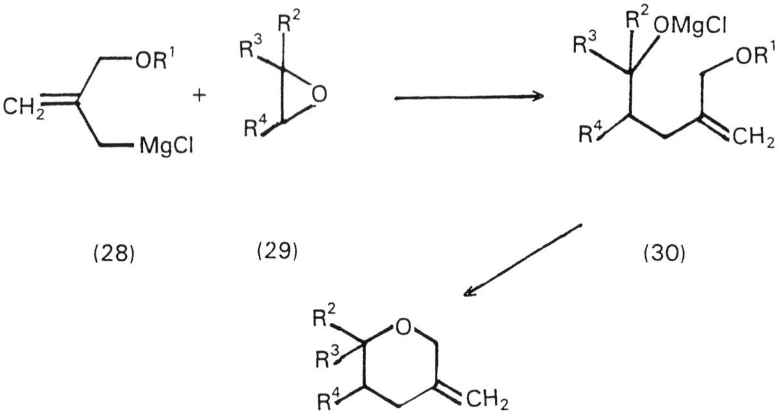

(28) (29) (30)

(31)

The telemerization of butadiene with CH_2O in Pr^iOH in the presence of catalyst $Pd(OAc)_2$ and (+)-DIOP ligand yields 3,6-divinyltetrahydropyrans (32 and 33) in 65% yield with 26% and 18% e.e. respectively (W. Keim *et al.*, Chem. Comm., 1989, 1151). An enantioselective synthesis of four 2--methyl-3,6-divinyltetrahydropyrans 34 by homogeneous Pd-catalyzed telomerization of butadiene with acetaldehyde has also been described (W. Keim and P. Mastrorilli, Gazz., 1993, <u>123</u>, 401).

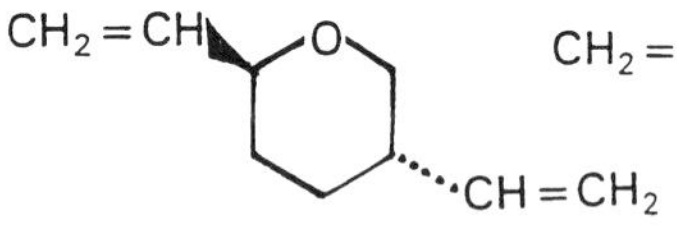

(32) 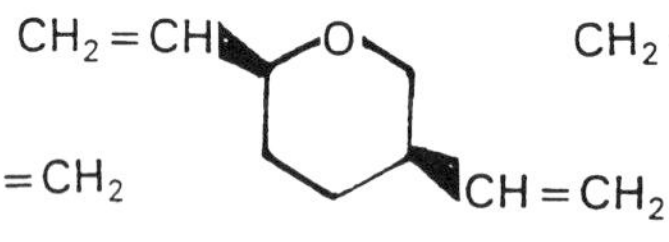 (33) 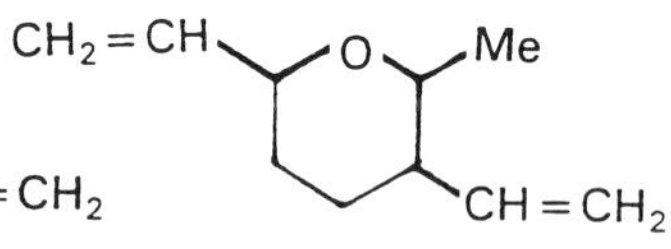 (34)

In order to test a proposal that stereochemistry can be used rationally to control the binding properties of ionophores, a new podant 35 has been prepared and studied (S.D. Erickson and W.C. Still, Tetrahedron Letters, 1990, 31, 4253). Investigations have been carried out on related compounds having binding properties comparable to those of the macrocyclic ethers (S.D. Erickson, M.H.J. Ohlmeyer, and W.C. Still, *ibid.*, 1992, 33, 5925).

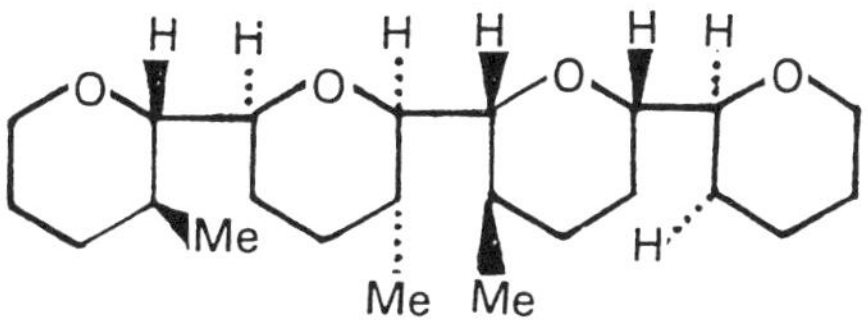

(35)

Oxetane 36, possessing an ether group in the side chain, has been found to rearrange by means of the Lewis acid catalyst, $BF_3.EtO_2$ to give the ring expanded, 4-benzyloxymethyl-4-methyltetrahydropyran (37) (A. Itoh *et al.*, Heterocycles, 1994, 38, 2165).

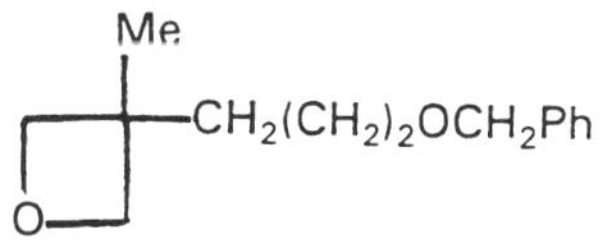

(36) 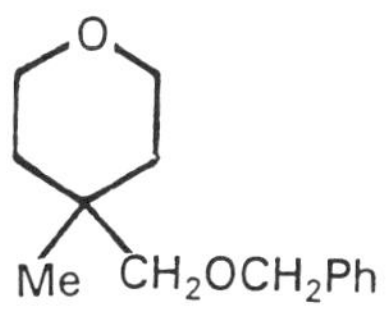   (37) 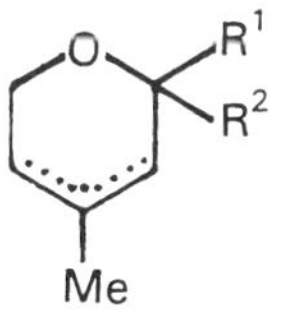   (38)

146

A mixture of 2,2-disubstituted 4-methyl-di and -tetrahydropyrans 38
[R^1 = H, R^2 = Pr, Pr^i, Bu, Me_2CHCH_2; R^1R^2 = $(CH_2)_5$] have been prepared by
the reaction of 1 mol of 3-methylbut-3-en-1-ol with R^1R^2CO in the presence
of 5-20 mol $Fe(SO_4)_3$ or $Al_2(SO_4)_3$ at 20-40°C (U.G. Ibatullin and Yu. V.
Pavlov, U.S.S.R. SU 1,675,299, 1991).

Glycidyl ethers 39 (R^1 = Me, R^2 = Me, Et; R^1 = R^2 = Et) on treatment with
HBr in CH_2Cl_2 yield the hydroxy esters 40, which on hydrogenation over
Raney Ni afford 2,2-dimethyl- and 2-ethyl-2-methyltetrahydropyran-4-
-hydroxyacetates 41 (R^1 = Me, Et, respectively) (N.S. Arutyunyan, L.A.
Akopyan, and S.A. Vartanyan, Arm. Khim. Zh., 1987, $\underline{40}$, 574).

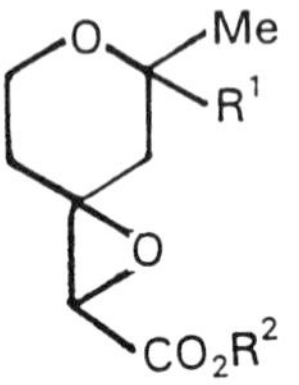

(39) 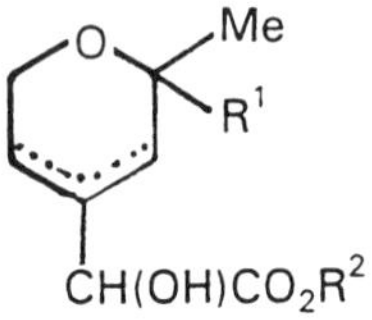   (40) 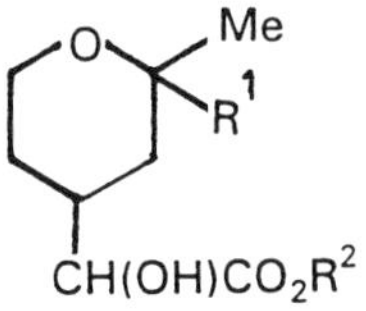   (41)

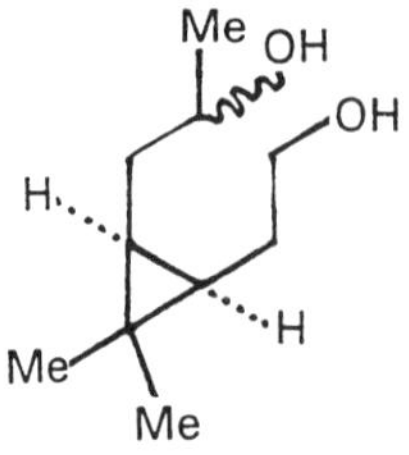

(42) 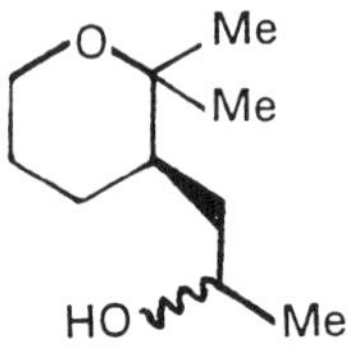   (43) 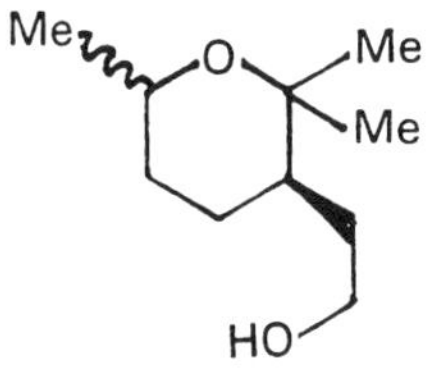   (44)

The sulphuric acid-catalyzed reaction of *cis*-2-[3-(2-hydroxypropyl)-2,2-
-dimethylcyclopropyl]ethanol (42) furnishes a mixture of the 3-substituted
2,2-dimethyl- and 2,2,6-trimethyltetrahydropyrans 43 and 44 and two
related tetrahydrofurans (J. Podlejski and J. Kula, Ann., 1989, 477). The

reaction of 2-(benzyloxymethyl)-3-(bromozincmethyl)buta-1,3-diene with aldehydes, ketones, and imines affords addition products, which undergo Pd(O)-catalyzed cyclization to yield 2-substituted 4,5-dimethylenetetrahydro-pyrans, for example 45, obtained from $H_2C = C(CH_2OCH_2Ph)C(= CH_2)CH_2$-ZnBr and RCHO (R = Ph, cyclopentyl) (J. Van der Louw *et al.*, Tetrahedron Letters, 1989, <u>30</u>, 5497).

Highly regio- and stereoselective reductions of the spiroketals 46 ($R^1 = R^2 = H$; $R^1 = Me$, $R^2 = H$, $CH_2OSiMe_2Bu^t$, CH_2OCH_2Ph) have been achieved by DIBAH and silane-Lewis acid to give tetrahydropyrans 47 as the major product, except where Ph_2SiH_2-$TiCl_4$ was used for 46 ($R^1 = Me$, $R^2 = CH_2OCH_2Ph$) when tetrahydropyran 48 was the major product (H. Oikawa *et al.*, *ibid.*, 1993, <u>34</u>, 5203).

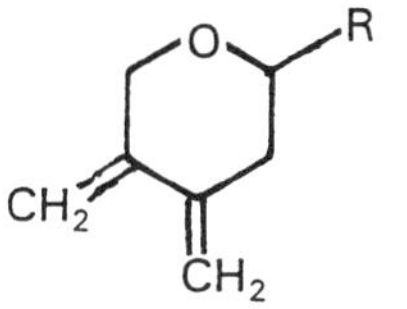

(45)

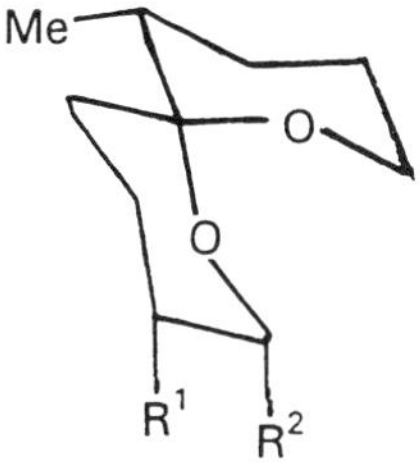

(46)

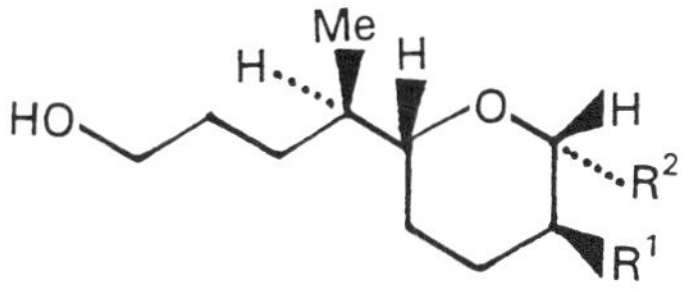

(47)

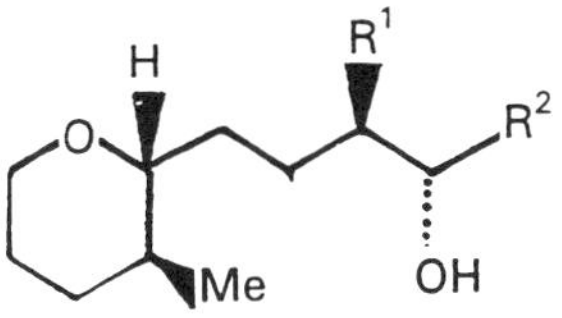

(48)

The *tert*-butyl peracetate-induced decomposition of $H_2C = CHR^1(CH_2)_3$-$C(R^2)_2OOBu^t$ ($R^1 = R^2 = H$; R^1, $R^2 = H$, Me) in the good hydrogen donor solvent CH_2Cl_2 leads to tetrahydropyrans 49 and peroxide adducts

148

$Cl_2CHCH_2CHR^1(CH_2)_3C(R^2)_2OOBu^t$. The substituents on the starting material
influences the ratio of 49 and the peroxide adduct. Thus when ($R^1 = R^2 = H$)
the yield of 49 is 25% and the adduct 9%, whereas when ($R^1 = H$, $R^2 = Me$)
only the adduct (55%) is obtained (E. Montaudon *et al.*, J. Heterocyclic
Chem., 1991, 28, 459).

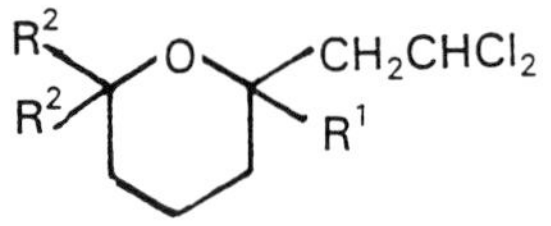

(49)

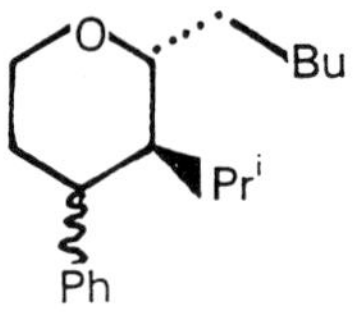

(50)

The unsaturated alcohols derived from a one-pot tandem [2,3]-Wittig-
-anionic oxy-Cope rearrangement of appropriate acyclic bisallylic ethers,
undergo a halogenocyclization reaction with iodine in MeCN to give
substituted tetrahydropyrans, for example, 3-*iso*propyl-2-pentyl-4-phenyl-
tetrahydropyran (50), with a high degree of stereocontrol (N. Greeves and
K.I. Vines, Chem. Comm., 1994, 1469).

Cyclization of $Me_2C = CH(CH_2)_2CHRH$ (R = H, Me) with PhSeCl gives 2,2-
-dimethyl- and 2,2,6-trimethyl-3-phenylselenenyltetrahydropyran (52),
respectively. $MeCH = CH(CH_2)_3OH$ gives a mixture of tetrahydrofuran and
tetrahydropyran derivatives, whereas alk-5-en-1-ols yield only
tetrahydropyran derivatives (Z. Bugarcic, S. Konstantinovic, and M.L.
Mihailovic, J. Serb. Chem. Soc., 1987, 52, 725). The cyclization of
$H_2C = CR^1CHR^2(CH_2)_nCR^3R^4OH$ ($R^1 = H$, Me; $R^2 = H$, Me, OH, Bu^tPh_2SiO;
$R^3,R^4 = H$, Me; n = 1,2) with $PhSeO_3SiCF_3$ to afford tetrahydrofurans and
tetrahydropyrans 53 (n = 1 and 2, respectively) has been reported (S. Murata
and T. Suzuki, Tetrahedron Letters, 1987, 28, 4297).

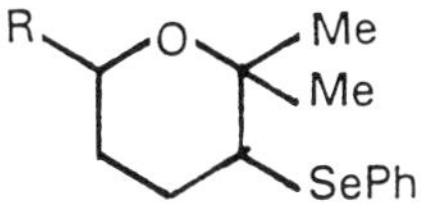

(52)

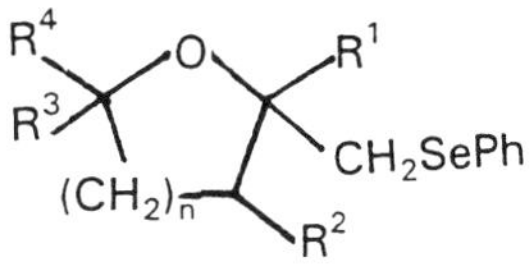

(53)

The Fe(0) bipyridine-catalyzed intramolecular cycloaddition reaction of oxaundecatrienes 54 (R^1, R^2 = H, Me; R^3 = H, Et) with $(CH_2OH)_2$ proceeds regio- and stereoselectively to give tetrahydropyrans 55. The similar reaction of oxaundecatriene 56 with Fe(0) bipyridine affords a 94:6 of tetrahydropyrans 57 and 55 (R^1-R^3 = H), whereas the reaction with Fe(0) pyridine gives a 15:85 ratio of 57 and 55 (R^1-R^3 = H) (J.M. Takacs *et al.*, *ibid.*, 1987, <u>28</u>, 5627).

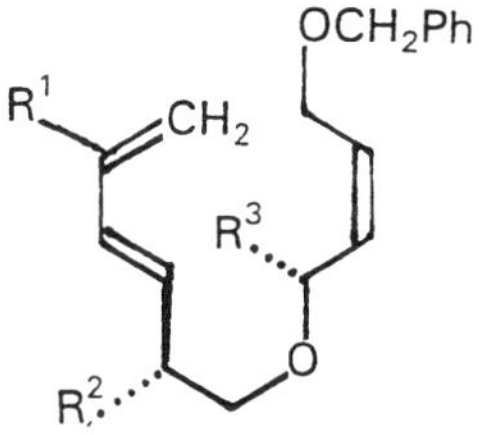

(54)

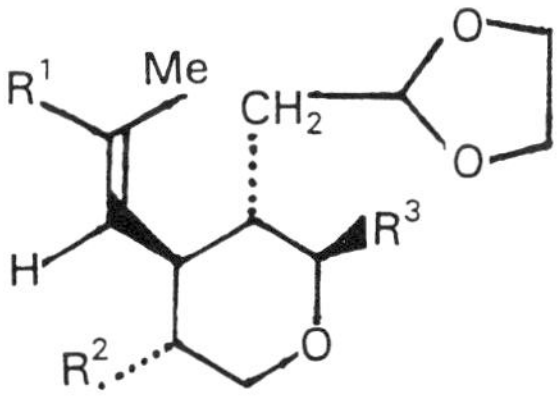

(55)

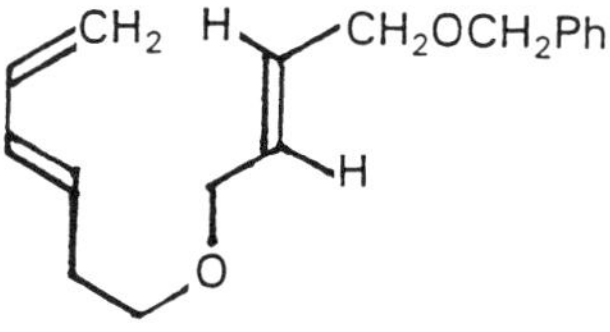

(56)

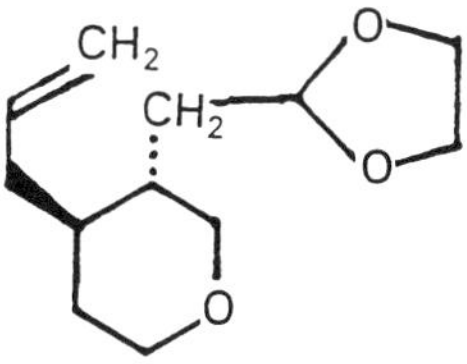

(57)

150

The condensation of 4-(3-amino-1-hydroxyprop-2-yl)-4-benzyl-2,2-di-
methyltetrahydropyran (58) with 3,4-$R^1R^2C_6H_3$CHO, followed by borohydride
reduction of the azomethine products affords 4-benzyl-4-[3-(arylmethyl)-
amino-1-hydroxyprop-2-yl]-2,2-dimethyltetrahydropyran 59 (N.S.
Kharutyunyan, K.M. Garibyan, and R.T. Grigoryan, Arm. Khim. Zh., 1988,
<u>41</u>, 682).

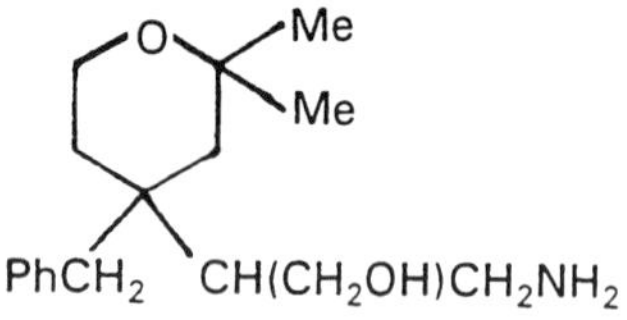

(58)

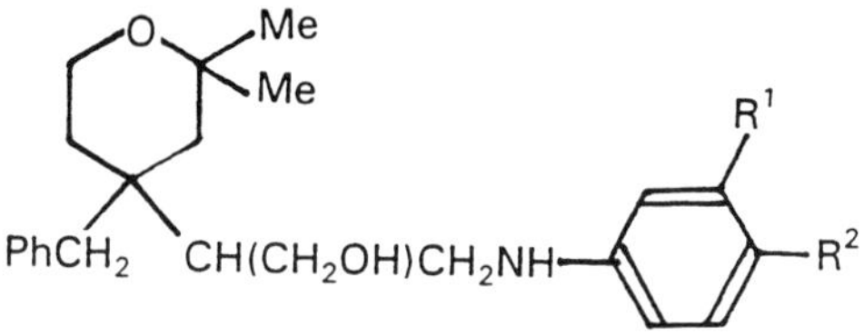

(59)

When *trans, trans*-2-lithio-6-(*trans*-prop-1-enyl)-4-*tert*-butyltetrahydro-
pyrans 60 (R^1 = Bu^t, R^2 = H, R^3 = CH = CHMe) and *trans-cis* 60 (R^1 = Bu^t,
R^2 = CH = CHMe, R^2 = H), generated by aromatic radical-anion induced
reductive lithiation of the corresponding 2-(phenylthio)tetrahydropyrans, are
warmed from -78°C to 0°C, they give, respectively, 3-*tert*-butylocta-4,6-
-*cis,trans*-dien-1-ol and the corresponding *trans, trans*-isomer as the major
products (E.J. Verner and T. Cohen, J. Org. Chem., 1992, <u>57</u>, 1072).

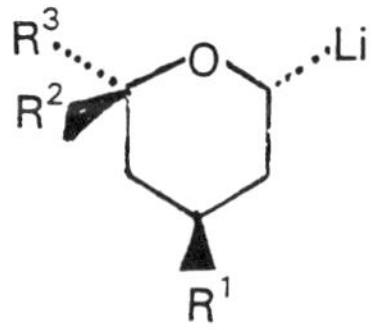

(60)

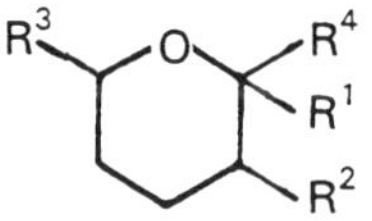

(61)

(62)

2-Aryltetrahydropyrans 61 (R^1 = H, Me, Pr^i; R^2, R^3 = H, Me; R^4 = Ph, 4--MeOC$_6$H$_4$) with up to three additional methyl substituents have been prepared in two steps from tetrahydrofurans 62 (same R^2, R^3) and o-selenoalkyllithiums $R^4CR^1LiSeMe$, *via* 5-aryl-5-methylselenopentan-1-ols (A. Krief, J. Bousbaa, and M. Hobe, Synlett, 1992, 320).

The Diels-Alder reaction of 1,1,2,2,3,3-hexamethyl-4,5-dimethylenecyclopentane (63) with Ph_2C = C = O affords 3-oxa-7,7,8,8,9,9-hexamethyl--4-diphenylmethylenebicyclo[4.3.0]non-1,6-ene (64) and bicyclononenone 65 (H. Mayr and U.V. Heigl, Chem. Comm., 1987, 1804).

(63)   (64)   (65)

The C-O bond in tetrahydropyran is cleaved by $B(SePh)_3$ in the presence of ZnI_2 to give $PhSe(CH_2)_5OH$ (T. Kataoka *et al.*, Heterocycles, 1990, <u>31</u>, 889).

(ii) Tetrahydropyran-ols and derivatives. Stirring 2-methoxytetrahydropyran (1), 2-naphthol, and $BF_3.Et_2O$ in CH_2Cl_2 at 0°C yields 1-(tetrahydropyran-2-yl)-2-naphthol (2) (Y. Fujimori, H. Kondo, and T. Yonetani, Jpn. Kokai Tokkyo Koho JP 02 157,271 [90 157,271], 1990). Treatment of 1--naphthyl tetrahydropyran-2-yl ether in CH_2Cl_2 with $BF_3.Et_2O$ yields 2-(tetrahydropyran-2-yl)-1-naphthol (*idem, ibid.*, 01 68,365 [89 68,365], 1989).

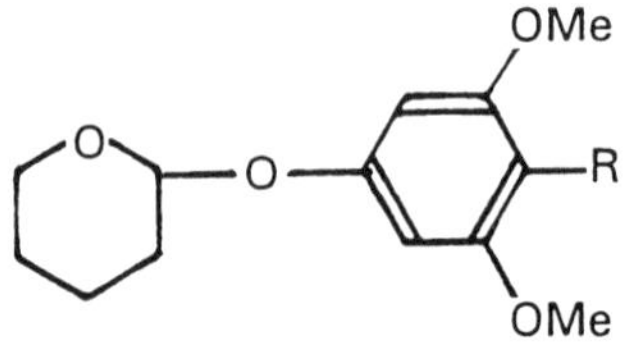

(7)

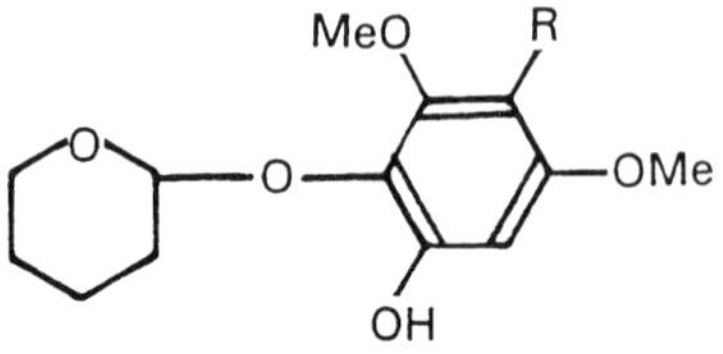

(8)

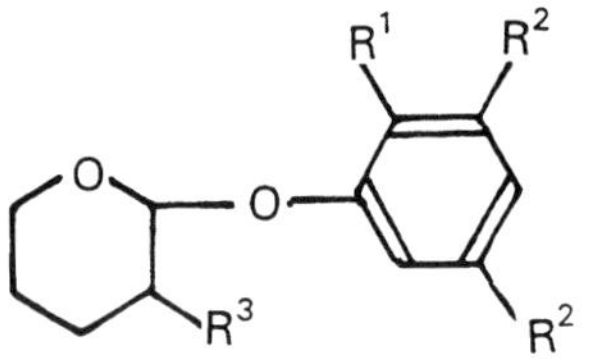

(9)

2-Phenoxytetrahydropyrans 10 (R^1-R^3 = H, OH, lower alkoxy, acyloxy; R^4 = H, lower alkyl, hydroxymethyl, lower alkoxymethyl), useful as drug intermediates, have been prepared from benzenediols 11 and tetrahydropyran derivatives 12 (R^5 = H, lower alkyl) (S. Isayama and H. Tan, Jpn. Kokai Tokkyo Koho JP 62 226,973 [87 226,973], 1987).

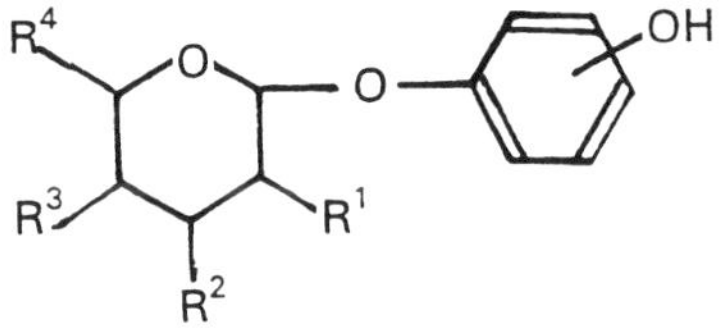

(10)

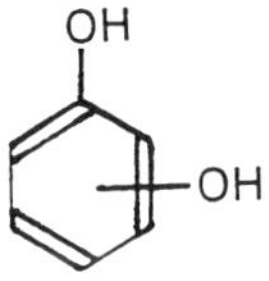

(11)

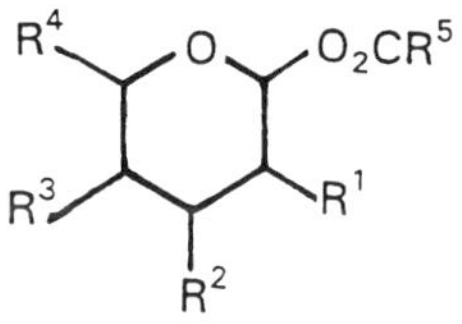

(12)

(13) (14)

Pyrolysis of a mixture of phenol and 3,4-dihydro-2*H*-pyran at 150-180°C affords 2-(tetrahydropyran-2-yl)phenol (13) (37%). An ene-type mechanism has been proposed for the reaction (J.T. Pinhey and Phan Thank Xuan, Austral. J. Chem., 1988, $\underline{41}$, 69). Hydration of 3,4-dihydro-2*H*-pyran yields 5-hydroxypentenal, tetrahydropyran-2-ol (14; R = H), and protected 14 [R = $(CH_2)_4CHO$]. The dependence of formation of these products on the reaction conditions has been investigated (P. Vinczer *et al.*, Acta Chim. Hung., 1990, $\underline{127}$, 581). The preparations of 1-(tetrahydropyran-2-yloxy)-hexadec-9-yne (V.N. Odinokov *et al.*, U.S.S.R. SU 1,505,939, 1989) and 1--(tetrahydropyran-2-yloxy)-11*E*-hexadecen-7-yne (G.A. Tolstikov *et al.*, *ibid.*, 1,708,811, 1992) have been reported.

Tetrahydropyranyl protective groups for alcohols are removed under neutral conditions with Me_3SiCl and NaI in CCl_4 at room temperature, for example, 7-bromoheptyl tetrahydropyranyl ether yields 7-bromoheptanol (C. Li, J. Li, and T.H. Chan, Huaxue Tongbao, 1989, 39). Tetrahydropyranyl ethers are converted to the corresponding iodides and bromides upon reaction with NaI (or LiBr) and $BF_3.Et_2O$ (or $ClSiMe_3$), for example, treatment of benzyl tetrahydropyran-2-yl ether with NaI and $BF_3.Et_2O$ in MeCN gives iodomethylbenzene, $PhCH_2I$ (Y.D. Vankar and K. Shah, Tetrahedron Letters, 1991, $\underline{32}$, 1081). The use of tetrahydropyran in the preparation of octa-2,7--dienyl acetals has been reported (K. Adachi and N. Yoshimura, Jpn. Kokai Tokkyo Koho JP 02 243,681 [90 243,681], 1990).

4-[4-(Tetrahydropyran-2-yloxy)butoxy]phenylalkyl 3-(2,2-dihalogeno-ethenyl)-2,2-dimethylcyclopropane-1-carboxylates 15 ($R^1 = C_{1-4}$n- or branched alkyl; $R^2 = Cl$, Br) have been prepared and studied for use as biologically active substances (J. Kahovcova, Czech. CS 271, 126, 1991).

(15)

Reduction of α- and ß-epoxide diastereomers 16 and 17 with LiAlD$_4$ gives the diastereomerically deuterated diols 18 and 19 (R^1 = R^2 = H), respectively, which on deoxygenation of the thiocarbonate derivatives 18 and 19 (R^{1}R^2 = CS) followed by tetrahydropyranyl ether cleavage and Swern oxidation give aldehydes 20 and 21 (J.A. Marshall and M.W. Andersen, J. Org. Chem., 1992, <u>57</u>, 5851).

(16)

(17)

(18)

(19)

(20)   (21)

The MgBr$_2$.Et$_2$O mediated addition of allyltributyltin to the (3-substituted tetrahydropyran-2-yloxy)acetaldehydes 22 (R = Ac, Bu^{t}CO, Bz, PhCH$_2$) has been shown to be highly diastereoselective. Thus, alcohols 23 and 24 have been prepared with 23 as the predominant product in most cases. The sense and level of induction has been shown to depend on the nature of the protecting group at C-3 of the tetrahydropyran ring (A.B. Charette *et al.*, Synlett, 1993, 81).

(22)   (23)

(24)

158

Diol 25 on treatment with acidic MeOH affords 6-pentyl-2-methoxytetra-
hydropyran-4-ol (26; R = α-OH), which has been epimerized to 26 (R = ß-
-OH). Hydrolysis of the latter followed by oxidation yields (4R, 6R)-4-
-acetoxy-6-pentylvalerolactone (L. Yu and Z. Wang, Chin. Chem. Letters,
1992, 3, 607). For the synthesis of 4-methoxy-2,6-di-*tert*-butylphenyl 2-
-hydroxy-3-methyltetrahydropyran-2-carboxylate (27) see D. Enders *et al.*,
Synlett, 1992, 901).

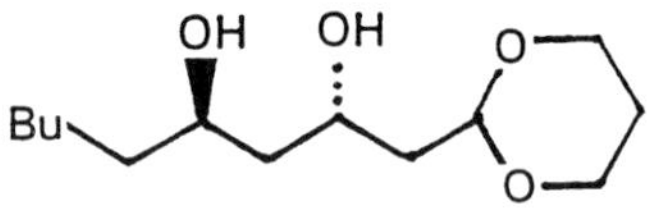

(25)

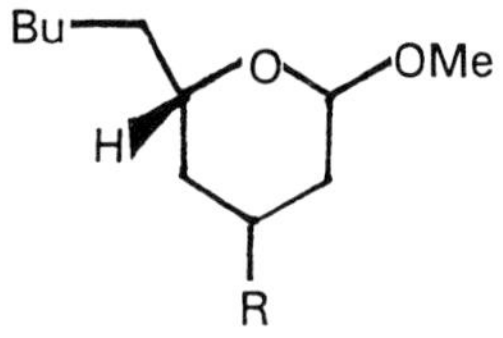

(26)

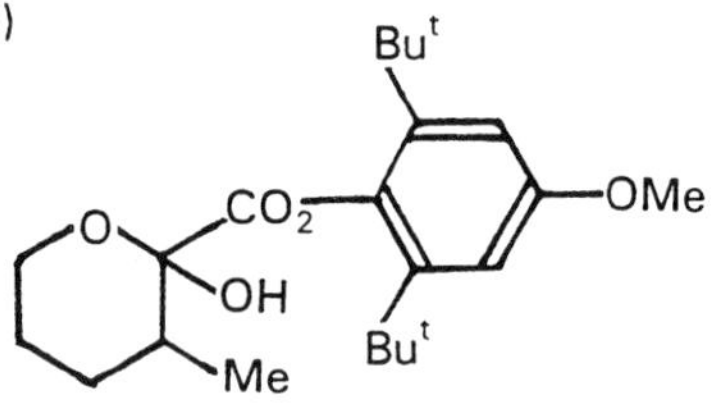

(27)

A number of derivatives of 6-methyltetrahydropyran-2-ol-4,4-dimethyl-
enedithioketal 28 and 29 (R = SiPh$_2$Bu^t, SiMe$_2$Bu^t, CH$_2$OMe, Ac, Bz; X = CO,
CHOH, CHOMe) and 2-substituted 4-methoxy-6-methyltetrahydropyran 30
(R = allyl, CH$_2$CHO, CH$_2$CH$_2$CH$_2$OH) have been prepared as intermediates for
macrolides. Reduction of the 6-methyltetrahydropyran-2-one derivative 31
by (Me$_2$CHCH$_2$)$_2$AlH at -78°C yields 6-methyltetrahydropyran-2-ol-4,4-
-dimethylenedithioketal (28; R = H), which on treatment with imidazole and
Bu^{t}Ph$_2$SiCl at room temperature affords tetrahydropyran 28 (R = SiPh$_2$Bu^t) (T.
Ooishi, T. Nakada, and T. Komatsu, Jpn. Kokai Tokkyo Koho JP 05 17,469
[93 17,469], 1993). The preparation of (2S, 6S)-2-(hydroxymethyl)-6-
-methoxytetrahydropyran, a useful chiral intermediate from D-glucose, has
been reported (K. Jones and W.W. Wood, J. Chem. Soc., Perkin I, 1988,
999).

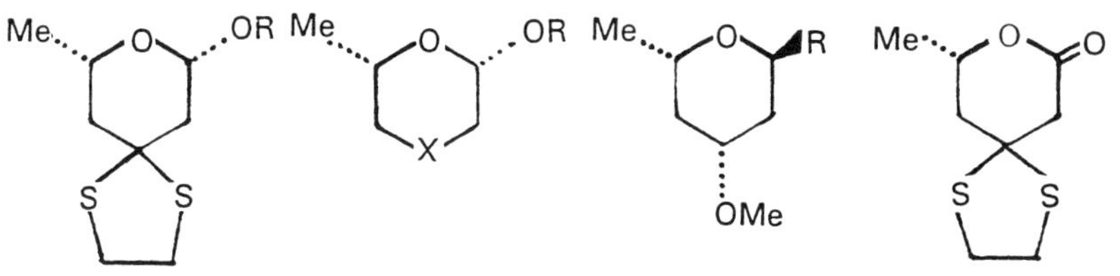

(28)　　　　　(29)　　　　　(30)　　　　　(31)

Tetrahydrofuran-2-ols and tetrahydropyran-2-ols 32 (R^1 = H, alkyl; R^2 = H, alkyl, phenyl; n = 1,2) react with enol acetates and a Lewis acid reagent, derived from N-chlorosuccinimide and Sn(II) chloride, to furnish 2-acetonyl derivatives 33 (R^3, R^4 = H, alkayl) diastereoselectively. A reaction mechanism involving cyclic oxocarbenium ions has been discussed (Y. Masuyama, Y. Kobayashi, and Y. Kurusu, Chem. Comm., 1994, 1123).

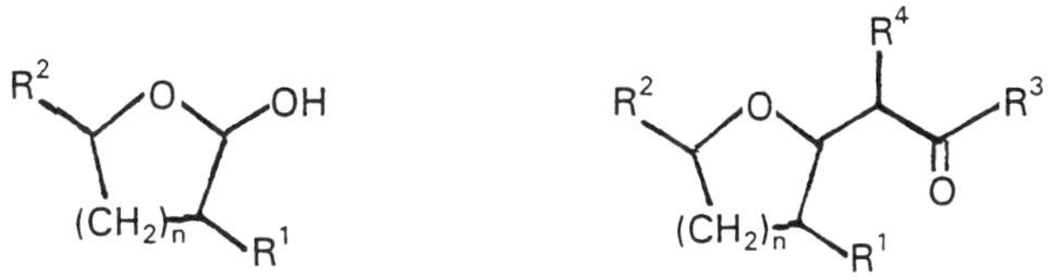

(32)　　　　　　　　　　　(33)

The reaction between diketene and a protected propionaldehyde derivative affords the 5,7-dihydroxy-3-oxoheptanoates 34 (R = alkyl), which on cyclo-condensation yield 2,4-dihydroxytetrahydropyran-2-acetates 35 as a mixture of *cis*- and *trans*-isomers (B. Loubinoux *et al.*, Tetrahedron, 1994, 50, 2047).

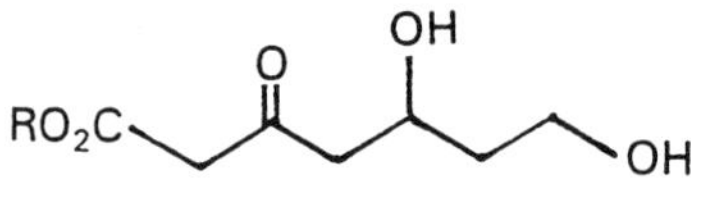

(34)

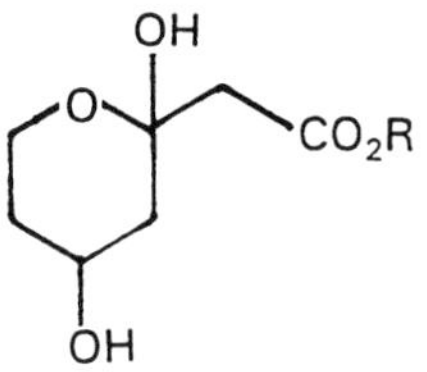

(35)

The reaction between trimethyl orthoformate and 2-methoxy-3,4-di-hydro-2*H*-pyran in the presence of $BF_3.Et_2O$ gives, after neutralization, a mixture containing as a major product, 2,6-dimethoxy-3-dimethoxymethyl-tetrahydropyran (36). The use of 36 and other 2,6-dialkoxy-3-dialkoxy-methyltetrahydropyrans as solubilizing agents for surfactants in liquid alkaline detergents has been claimed (H.T. Leinen and H. Kluschanzoff, Ger. Offen. DE 4,020,270, 1992).

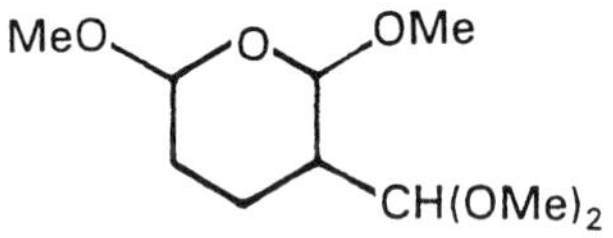

(36)

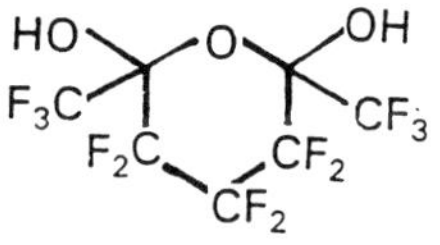

(37)

Perfluoromethylation of FCO(CF$_2$)$_3$COF by Me$_3$SiCF$_3$ in the presence of KF in PhCN affords CF$_3$CO(CF$_2$)$_3$COCF$_3$, which on treatment with water gives the perfluoropyran-2,6-diol 37 (J.D.O. Anderson, D.D. DesMarteau, and W. Navarrini, Eur. Pat. Appl. EP 496,415, 1992; J.D.O. Anderson, W.T. Pennington, and D.D. DesMarteau, Inorg. Chem., 1993, <u>32</u>, 5079).

Optimized conditions for the dimerization of CH$_2$=CMeCH$_2$OH with KH-diglyme give a 5:2 mixture of epimeric 3,5,5-trimethyltetrahydropyran-2-ols (38 and 39). Condensation of CH$_2$=CMeCH$_2$OH and CH$_2$=CMeCHO in the presence of KH yields bicyclononanone 40 (I.A.G. El-Karim *et al.*, Tetrahedron, 1988, <u>44</u>, 1809).

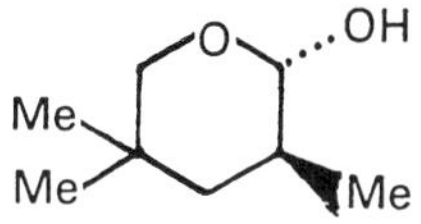

(38)

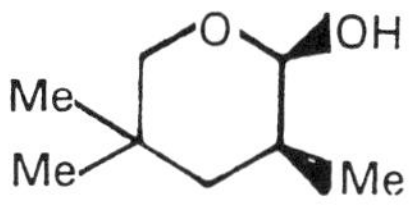

(39)

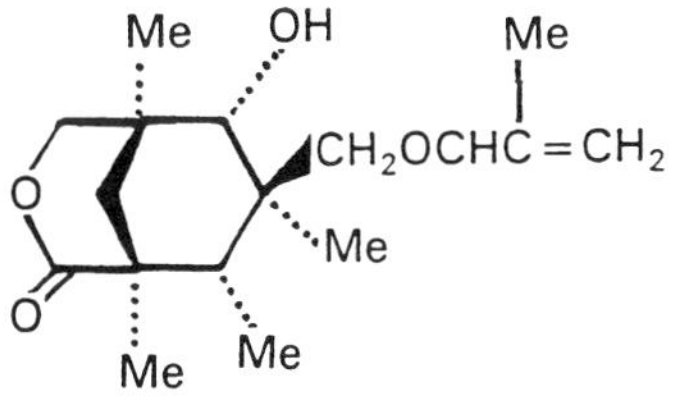

(40)

2-(Dimethylamino)cyclobutanecarboxylic esters 41 (R^1=R^2=Me; R^{1}R^2=H, Et; R^3=Me, Et) in the presence of TiCl$_4$ act as 1,4-zwitterion equivalents and react with ketones, R^{4}COR^5 [R^4=PhCH$_2$CH$_2$, Me(CH$_2$)$_6$, Pr^i, R^5=H; R^4=R^5=Pr] to yield 6-substituted and 6,6-dipropyl substituted 3,3-dimethyltetrahydropyran-2-ol-5-carboxylates 42 and 43 or dihydropyrans, for example, ethyl 5-ethyl-2,2-dipropyl-3,4-dihydro-2*H*-pyran-3-carboxylate (44) (S. Shimada *et al.*, Chem. Letters, 1991, 1149).

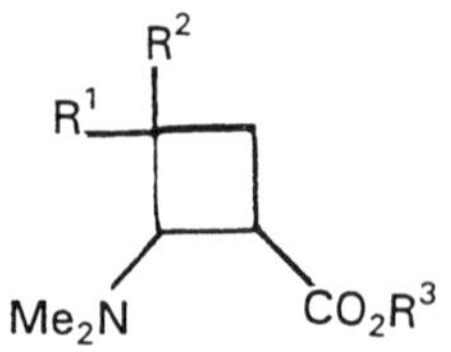

162

(41)

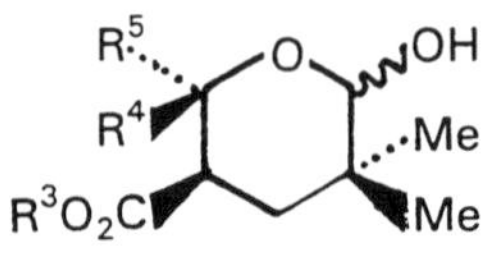

(42)
(43) R⁴, R³O₂C , trans

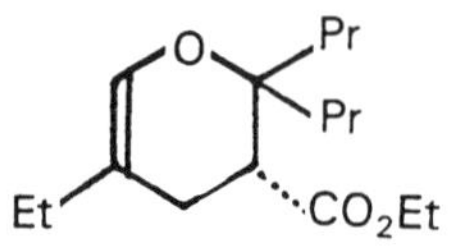

(44)

Allenol ether 45 on treatment with $BF_3.Et_2O$ in CH_2Cl_2 undergoes an intramolecular rearrangement to give 2-(3-oxobut-1-ene-2-yl)tetrahydropyran (46) (A. Ricci *et al.*, Synlett, 1990, 471).

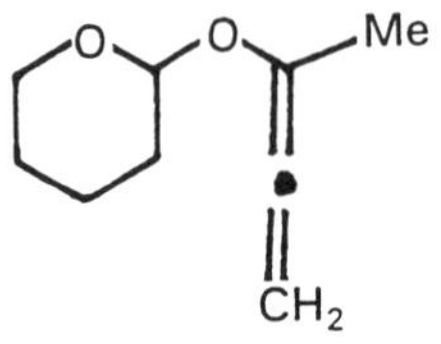

(45)

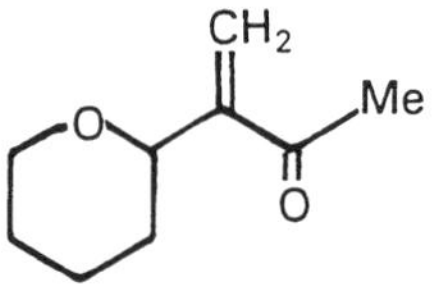

(46)

Acid-catalyzed methanolysis of tetrahydropyranyl acetals 47 (R^1 = Me, R^2 = H, CH_2OH, CH_2OMe; R^1 = Et, R^2 = H) leads to cleavage of both exo- and endocyclic C-O bonds to afford mainly the 3,4-4a,5,10,10a-hexahydro-2*H*-naphtho[2,3-b]pyrans 48 and the 1,2,3,4-tetrahydronaphthalene derivatives 49, formed by intramolecular trapping of both intermediate oxocarbonium ions (R.B. Gupta and R.W. Franck, J. Amer. Chem. Soc., 1987, 109, 6554).

$X = C_6H_3(NO_2)_2\text{-}2,4$

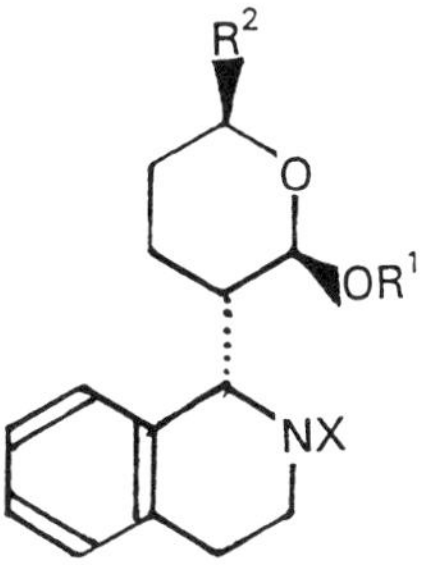

(47)

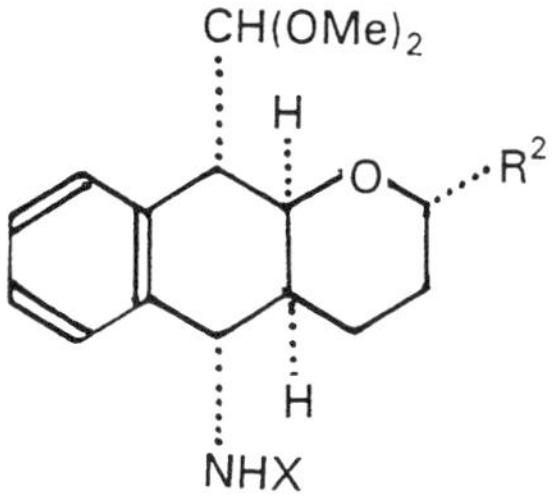

(48)

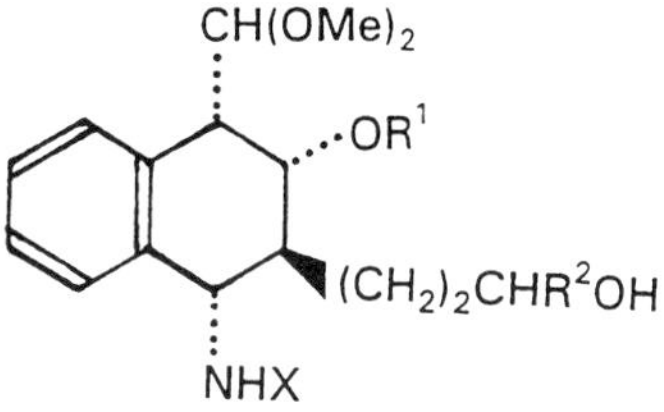

(49)

2-(2-Arylethyl)tetrahydropyran-3-ols 50 (R = H, OMe, NHAc) have been obtained from the epoxides 51 by a disfavoured chemical transformation involving an antibody catalyst (E.M.Coe and C.J. Jones, Polyhedron, 1992, 11, 3123). Similarly, the one-pot stereospecific preparations of the *cis-* and *trans*-2-alkenyltetrahydropyran-3-ols 52 have been achieved by the desilylation-palladium catalyzed cyclization of the silyl ethers of ω-hydroxy-trans and cis-γ,δ-epoxy-α,ß-unsaturated esters 53 (T. Suzuki, O. Sato, and M. Hirama, Tetrahedron Letters, 1990, 31, 4747). The four enantiomerically pure 2,2,6-trimethyl-6-vinyltetrahydropyran-3-ols 54 and 55 have been prepared in three steps from (6S)- or (6R)-6,7-dihydroxygeranyl N-phenyl-carbamate (A. Meou et al., Synthesis, 1991, 681).

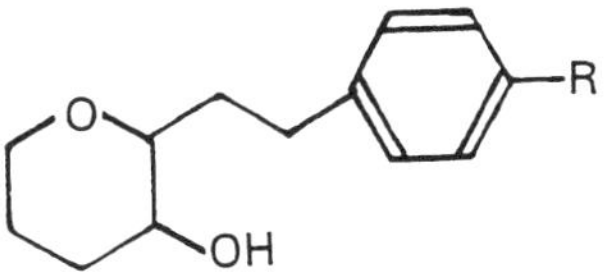

(50)

(51)

164

(52)

(53)

(54)

(55)

The Brønsted acid-catalyzed or Bu_4NF-$TiCl_4$-mediated cyclization of the allylstannane-aldehyde, (Z)-Bu_3SnCH_2CH=$CH(CH_2)_3CHO$, gives *cis*-2-vinyl-tetrahydropyran-3-ol (56), whereas (E)-Bu_3SnCH_2CH=$CH(CH_2)_3CHO$ yields *trans*- 56. To explain this stereochemical outcome, a push-pull mechanism has been proposed (V. Gevorgyan, I. Kadota, and Y. Yamamoto, Tetrahedron Letters, 1993, <u>34</u>, 1313).

(56)

(57) (*cis*)
(58) (*trans*)

Cobalt complexes derived from *trans*-4,5-epoxyhept-6-yn-1-ols and dicobalt octacarbonyl on treatment with a catalytic amount of $BF_3.Et_2O$ in CH_2Cl_2 at -78°C afford *cis*-2-ethynyltetrahydropyran-3-ol (57) derivatives, whereas similar treatment of the *cis*-4,5-epoxides furnishes the corresponding derivatives of *trans*-2-ethynyltetrahydropyran-3-ol (58) (C. Mukai *et al.*, *ibid.*, 1994, <u>35</u>, 2179). The preparations of 2-benzyloxy-methyltetrahydropyran-3-ol and some derivatives and related compounds as platelet activating factor antagonists have been reported (N. Nakamura *et al.*, Eur. Pat. Appl. EP 251,827, 1988).

6-Methoxytetrahydropyran-3-ol (59; R = OH) and 6-substituted 2--methyltetrahydropyran-3-ols 60 (R^1 = OH, R^2, R^3 = OMe, H) on reductive dehydration, followed by treatment with hydrogen peroxide and sodium peroxide afford the respective hydroperoxides 59 (R = OOH) and 60 (R^1 = OOH) (D.A. Casteel and K.E. Jung, J. Chem. Soc., Perkin I, 1991, 2597). (2*S*, 6*S*)-2-Hydroxymethyl-6-methoxytetrahydropyran a useful chiral synthon has been synthesized (K. Jones and W.W. Wood, J. Chem. Soc., Perkin I, 1988, 999).

(59)

(60)

The 2,5-disubstituted tetrahydropyran-3-ol derivatives 61 [R^1 = alkyl, alkenyl, arylalkyl, arylalkenyl, CO_2H, (un)substituted COR^4; R^4 = H, alkyl; R^2 = $CO(CH_2)_nNHR^4$, $CONHNHR^4$; n = 1-4; R^3 = H, Me], which are useful in the treatment of fungal infections, particularly infections of Candida and other yeast and filamentous fungi, have been prepared (P. Ermann *et al.*, U.S. US 5,266,592, 1993).

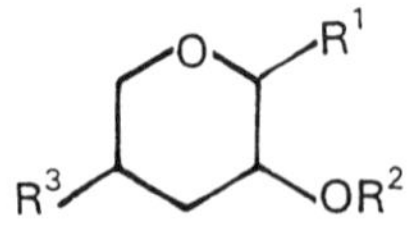

(61)

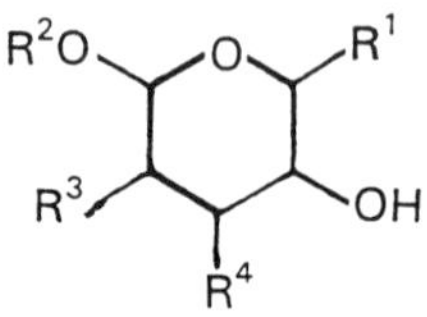

(62)

For the preparations of 3-acetoxy-3-methyltetrahydropyran, see M.L. Mihailovic *et al.*, Ann., 1992, 305; 6-(α-Hydroxy-^{i}propyl)-2,3-dimethyltetra hydropyran-3-ol , S.T. Orszulik, Synth. Comm., 1989, <u>19</u>, 1233; and 2,6--dimethyl-4-methylenetetrahydropyran-3-ol, I.E. Marko and D.J. Bayston, Tetrahedron Letters, 1993, <u>34</u>, 6595; for the fluorination of 2-methoxy--2-methoxycarbonyltetrahydropyran-3-ols J. Venkiah *et al.*, J. Fluorine Chem., 1990, <u>49</u>, 183; and for the preparation of 2-fluoroalkyltetrahydro-pyran-3-ols (62) ($R^1 = C_{1-2}$fluoroalkyl, R^2, R^3, $R^4 = H$, C_{1-15} alkyl, C_{2-15} alkenyl, C_{7-10} aralkyl), useful as bioactive substances and enzyme inhibitors, M. Namekawa *et al.*, Eur. Pat. Appl. EP 548,947, 1993.

2,5-Disubstituted tetrahydropyran-3,4-diols 63 and 64 [$R = CO_2H$, COSH, (un)substituted esters or thioesters, (un)substituted carboamide] and several related compounds, which demonstrate herbicidal activity against a wide variety of plant species, have been prepared (J.E.D. Barton *et al.*, PCT Int. Appl. WO 93 19,599, 1993).

(63)   (64)

4-Alkoxy-4-[(naphthylmethoxy)phenyl]tetrahydropyrans and analogues
(P.N. Edwards and J.M.M.M. Girodeau, Eur. Pat. Appl. EP 375,404, 1990),
4-methylthio-4-[3-(naphth-2-ylmethoxy)phenyl]tetrahydropyran (65) (P.A.R.
Bruneau, R.I. Dowell, and D. Waterson, *ibid.*, 462,830, 1991), 4-[3-(1,3-di-
methyl-2,3-dihydrobenzimidazol-2-one-6-ylthio)phenyl]-4-methoxytetra-
hydropyran (66) (P.A.R. Bruneau, G.C. Crawley, and K. Oldham, *ibid.*,
462,831, 1991), 4-methoxy-4-[3-(naphth-2-ylthio)phenyl]tetrahydropyran
and analogues (P.N. Edwards and T.G.C. Bird, *ibid.*, 409,413, 1991), and 4-
-[5-(heteroarylthio)thien-2-yl]-4-methoxytetrahydropyrans and analogues
(T.G.C. Bird, *ibid.*, 462,812, 1991) have been prepared and their use as
inhibitors of 5-lipoxygenase investigated.

(65)   (66)

A number of derivatives of 4-hydroxy- and 4-methoxytetrahydropyran
with a chain type of substitutent at the C-4 position of the pyran ring have
been prepared, including 67 (R.I. Dowell, P.N. Edwards, and K. Oldham,
ibid., 488,602, 1992), 68 (D. Waterson, *ibid.*, 495,594, 1992), and 69 and
their use as leukotriene inhibitors and ocular inflammation inhibitors and
antiasthmatics has been investigated (R.N. Young *et al.*, *ibid.*, 501,578,
1992), 70 (P. Ple, *ibid.*, 555,067, 1993), 71 (Y. Ducharme *et al.*, PCT Int.
Appl. WO 94 05,658, 1994), 72 (G.C. Crawley *et al.*, J. Med. Chem., 1993,
<u>36</u>, 295), 4-[3-(4-chloromethylphenylthio)-5-fluorophenyl]-4-methoxy-
tetrahydropyran (P.N. Edwards and M.S. Large, Eur. Pat. Appl. EP 570,197,
1993), 4-alkoxypyran-4-yl-substituted arylalkylaryl-, arylalkenylaryl-, and
arylalkynylarylureas (J.F. Dellaria *et al.*, U.S. US 5,346,914, 1994), and (*E*)-
-[5-fluoro-3-((2S,4R)-4-hydroxy-2-methyltetrahydropyran-4-yl)phenylthio]-
-acetophenone oxime (M.S. Large, PCT Int. Appl. WO 94 17,054, 1994).

(67) R^1 = F—⟨phenyl⟩—CH$_2$—⟨phenyl⟩—S— $R^2 = R^3 = H$

(68) R^1 = MeO$_2$SMeN—⟨phenyl⟩—CH$_2$O— $R^2 = R^3 = H$

(69) R^1 = (naphthalene: Ph, NaO—C(=O)—, —O—ethyl) $R^2 = R^3 = H$

(70) R^1 = 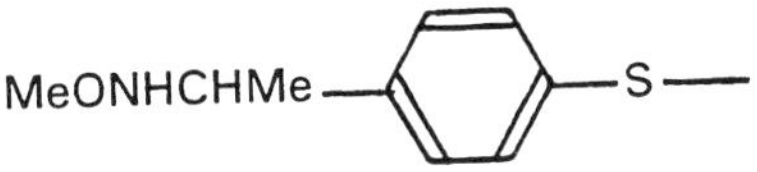

$R^2 = H$, $R^3 = F$

(71) R^2 = 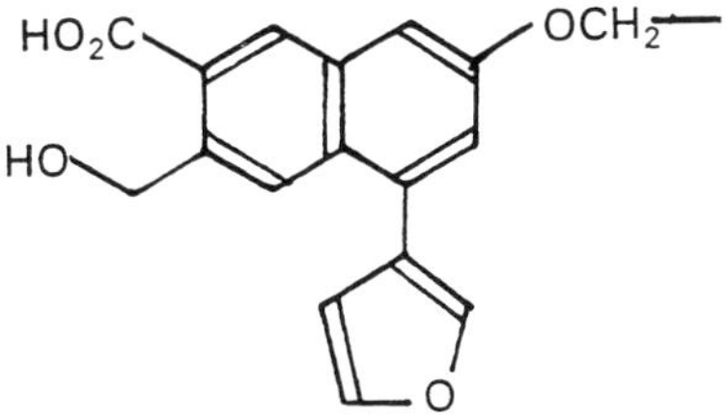

$R^1 = R^3 = H$

(72) R^1 = 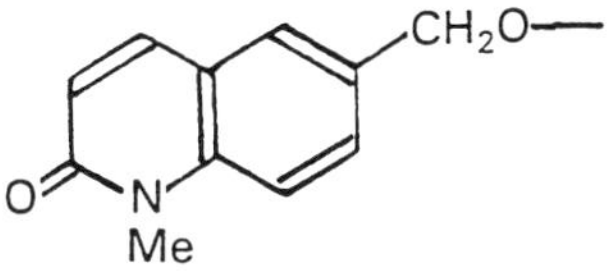

$R^2 = H$, $R^3 = F$

The acid-catalyzed treatment of the 4-hydroxymethyltetrahydropyran-4--ols 73 ($R^1 = R^2 = H$; $R^1 = R^2 = Ac$; $R^1 = H$, $R^2 = Et$) yields the 4-hydroxymethylenetetrahydropyran derivatives 74 ($R^2 = Ac$, Et) and the 4-hydroxymethyl-5,6-dihydro-2*H*-pyran derivatives 75 ($R^2 = H$, Ac, Et) (M.S. Sargsyan *et al.*, Arm. Khim. Zh., 1988, 41, 126).

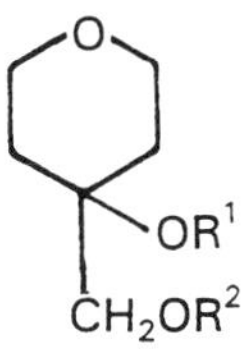

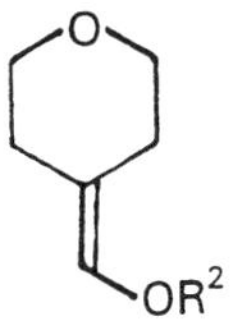

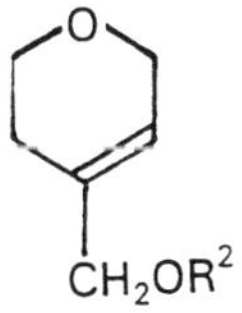

(73) (74) (75)

170

Treatment of 2-ethyl-4-methyltetrahydropyran-4-ol (76) with Ac_2O in $MeSO_3H$ affords 4-acetoxy-2-ethyl-4-methyltetrahydropyran (77; R^1 = Me, R^2 = Et). A number of related esters 77 (R^1 = Me, Et; R^2 = C_{2-4}alkyl, alkenyl), useful as perfume compounds for colognes, cosmetic powders and hair preparations, bleach compounds, and fabric softener compounds, have been prepared (M.A. Sprecker and R.E. Greene, Eur. Pat. Appl. EP 383,446, 1990).

(76)   (77)

Diastereomeric *cis-* and *trans* 4-aryl-3-(dimethylamino)methyltetrahydropyran-4-ols have been obtained from 3-(dimethylamino)methyltetrahydropyran-4-one and their configuration established from IR spectral data. The biological more potent trans isomer has been resolved into its optical antipodes and the absolute stereochemistry of one of the enantiomers 78 determined by x-ray crystallography. Some of the compounds show analgesic activity comparable to codeine (E. Mohacsi *et al.*, J. Heterocyclic Chem., 1990, <u>27</u>, 1623).

(78)   (79)

(*R/S*)-4-Methyl-2-methylenetetrahydropyran-4-ol (79), a vinylidene
analogue of mevalonolactone has been prepared by the iodoetherification of
$HOCH_2CH_2CMe(OH)CH_2CH=CH_2$, followed by DBN-mediated dehydrohalogenation. The preparations of (*S*)-79 and (*S*)-mevalonolactone have been
reported (V. Bolitt *et al.*, J. Org. Chem., 1991, $\underline{56}$, 4238).

The 2-alkoxy-6-(protected hydroxymethyl)tetrahydropyran-4-ols 80
($R^1 = C_{1-4}$alkyl, $R^2 = Ph_3C$) have been prepared by $LiBH_4$ reduction of epoxides
81 ($R^3 =$ leaving group, for example R^4SO_3, R^4 = alkyl or aryl) in the presence
of a titanium derivative TiR_4 ($R =$ alkoxy, acyl) in an organic solvent. Thus,
treatment of epoxide 81 ($R^1 = Me$, $R^2 = Ph_3C$, $R^3 = MeSO_3$) with $LiBH_4$ and
$Ti(OPr^i)_4$ in THF at -10°C and then at 20°C yielded 2-methoxy-6-(triphenyl-
methoxy)methyltetrahydropyran-4-ol (80; $R^1 = Me$, $R^2 = Ph_3C$) (P. Leon, V.
Massonneau, and X. Radisson, PCT Int. Appl. WO 92 08,711, 1992). The
regiochemical control of the ring opening of the epoxy ring in 3,4-epoxytetra-
hydropyran has been investigated (M. Chini *et al.*, Tetrahedron, 1994, $\underline{50}$,
1261).

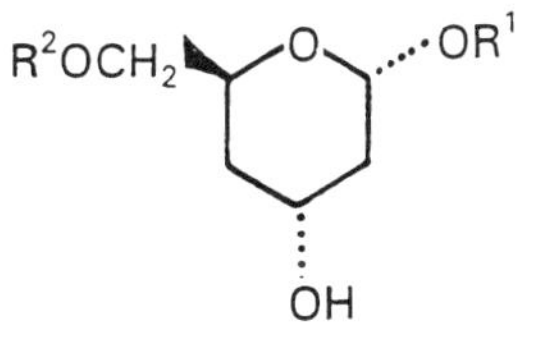

(80)

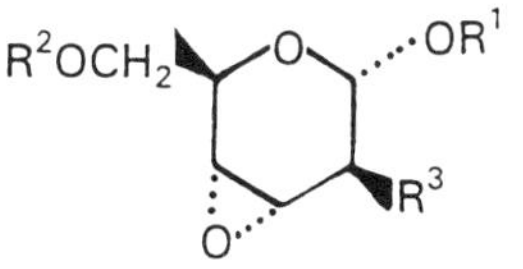

(81)

The preparations of a number of 2,4-disubstituted and 2,3,4-tri-
substituted tetrahydropyranyl 4-ethers useful in perfume compositions (M.A.
Sprecker *et al.*, U.S. US 4,962,090, 1990), 2,2-disubstituted 6-{2-[2-(1-
-alkanoyloxybut-3-enyl)cyclohexyl]ethyl}tetrahydropyran-4-ols and
derivatives as intermediates for HMG CoA reductase inhibitors (M.E. Duggan,
ibid., 4,873,345, 1989), methoxytetrahydropyrans, a new series of selective
and orally potent 5-lipoxygenase inhibitors (G.C. Crawley *et al.*, J. Med.
Chem., 1992, $\underline{35}$, 2600), and elaiophyline and derivatives as anthelmintics
(G. Kretzschmar *et al.*, Eur. Pat. Appl. EP 359,187, 1990) have been
reported.

172

(iii) Hydroxyalkyltetrahydropyrans and related compounds. The oxidation
of $HO(CH_2)_4CH = CH_2$ with $Ph(OAc)_4$ affords mixtures of the 2-hydroxy-
methyltetrahydropyran derivatives 1 [R = Ac, $(CH_2)_4CH = CH_2$] along with
$H_2C = CH(CH_2)_4OAc$ and $H_2C = CH(CH_2)_3CHO$ (M.L. Mihailovic *et al.*, J. Serb.
Chem. Soc., 1985, 50, 327).

(1)

(2)

(3)

(4)

2-Acetoxymethyl-2,3-dihydro-4*H*-pyran has been added to a mixture of
2-nitroimidazole, pyridine, and $AlCl_3$ and boiled to yield 1-(6-acetoxymethyl-
tetrahydropyran-2-yl)-2-nitroimidazole (2; $R^1 = 3$, $R^2 = Ac$). Related
derivatives 2 ($R^1 = 3,4$, etc; $R^2 = H$, acyl) useful as radiosensitizers have been
prepared (S. Sakaguchi, Y. Miyata, and T. Mori, Jpn. Kokai Tokkyo Koho JP
63 170,375 [88 170,375], 1988). For a highly stereoselective synthetic
method to obtain *cis*-6-alkyl-2-hydroxymethyltetrahydropyrans, including
(+)- and (-)-*cis*-(6-methyltetrahydropyran-2-yl)acetic acids, see T. Mandai *et
al.*, Tetrahedron Letters, 1993, 34, 111.

In the presence of Lewis acids or trimethylsilyltriflate, γ,δ-, δ,ϵ-, and ϵ,ζ-
-epoxy ketones, for example 5, and esters undergo cyclization to the
corresponding oxocarbenium ions, which react in situ with a variety of
organosilanes, e.g. Ph_3SiH, and organoaluminium reagents to give the
substituted oxacyclic product, e.g. tetrahydrofuran 6 and tetrahydropyran 7
(C.H. Fotsch and A.R. Chamberlin, J. Org. Chem., 1991, 56, 4141).

(5)   (6)   (7)

In epoxide 8 (R^1 = CH$_2$CH$_2$CO$_2$Me, R^2 = H) the presence of a saturated chain at the remote (from the hydroxy group) secondary epoxide position leads, as expected, under acid condition to the THF system 9 with a substituted hydroxymethyl group at the 2-position, whereas the replacement of an electron-rich double bond at that position leads to the formation of the tetrahydropyran system, with no hydroxy group in the chain at the 2-position. Thus, the hydroxy epoxide 8 [R^1 = (*E*)-CH = CH-CO$_2$Me, R^2 = Me] on treatment with a catalytic amount of camphorsulphonic acid in CH$_2$Cl$_2$ at -40°C to 25°C gives at 66:34 mixture of 2-(2-methoxycarbonylethenyl)-2--methyltetrahydropyran-3-ol (10) and tetrahydrofuran 11 (K.C. Nicolaou *et al.*, J. Amer. Chem. Soc., 1989, <u>111</u>, 5330).

(8)   (9)

174

(10)   (11)

2-(3-Hydroxyprop-2-yl)tetrahydropyran (12) has been prepared by the
addition of 3,4-dihydro-2*H*-pyran in toluene at 10-15°C to (*R*)-1-[(4-nitro-
benzoyl)oxy]propan-2-ol and camphor-10-sulphonic acid in toluene under
nitrogen, followed by the addition of aqueous methanolic KOH (H.
Kumobayashi *et al.*, Eur. Pat. Apl. EP 421,774, 1991).

(12)   (13)   (14)

The reduction of bicyclic acetal 13 with Dibal in CCl_4 affords 2-(2-
hydroxyethyl)-6-methyltetrahydropyran (14; R^1 = Me, R^2 = H)
stereoselectively, but reduction of 13 with Ph_2SiH_2-$TiCl_4$ yields 14 (R^1 = H,
R^2 = Me) (K. Ishihara, A. Mori, and H. Yamamoto, Tetrahedron Letters, 1987,
28, 6613). The cyclization of *anti*-$HOCH_2CH_2CH(OH)CMe(SPh)CH_2CMe_2$-
CH_2CH_2OH affords THF derivative 15 and 2-(3-hydroxy-1-phenylthiopropyl)-
-2,4,4-trimethyltetrahydropyran (16) as a 12:88 mixture of isomers. The syn
isomer of the starting material yields the anti isomer of 16 predominantly (S.
McIntyre, F.H. Sansbury, and S. Warren, *ibid.*, 1991, **32**, 5409). The
preparation of tetrahydropyrans 17 and 18 have also been reported (F.H.
Sansbury and S. Warren, *ibid.*, p.3425).

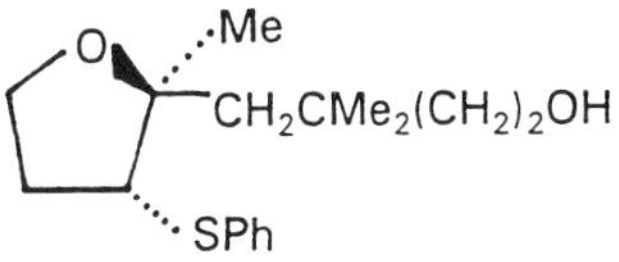

(15)

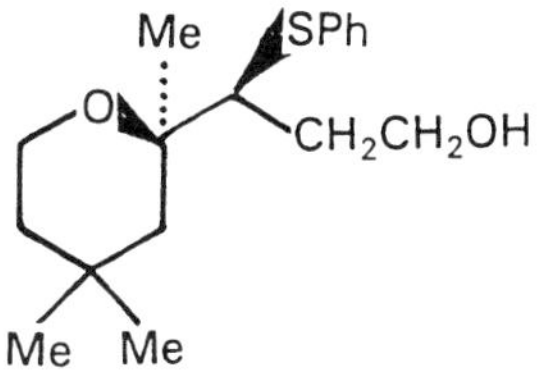

(16)

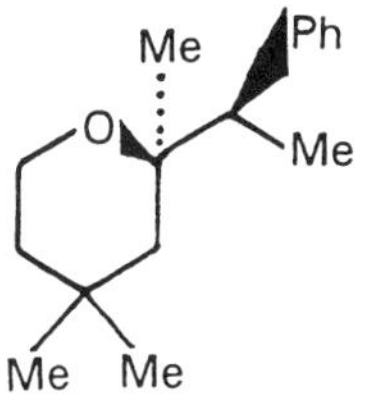

(17)

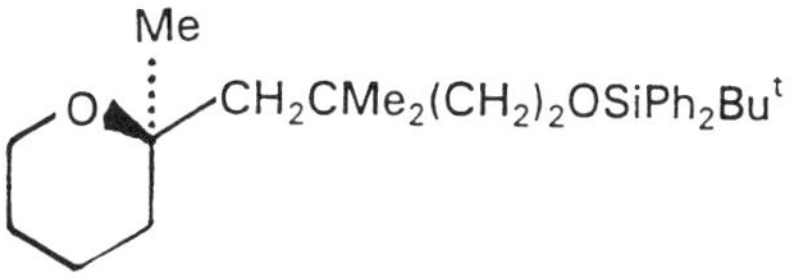

(18)

4-Hydroxymethyltetrahydropyran (19) has been obtained by maintaining dimethyl tetrahydropyran-4,4-dicarboxylate at 200°C with 200 bar hydrogen in MeOH containing a catalyst comprising CoO 64, CuO 18, MnO_2 7, MoO_2 4, and Na_2O 0.2 wt.% (R. Fischer *et al.*, Ger. Offen. DE 4,141,220, 1993). Related compounds 20 [$R^1 = CH_2OH$; R^2-$R^5 = H$, (cyclo)alkyl, aryl, aralkyl] have been prepared by hydrogenation of 20 ($R^1 = CO_2R^6$, $R^6 = H$, alkyl) at 250°C and 260 bar hydrogen in an autoclave containing the above catalyst (J. Henkelmann *et al.*, Eur. Pat., Appl. EP 546,396, 1993). 4-Hydroxymethyltetrahydropyran (19) has also been prepared by heating 2-(tetrahydrofur-3-yl)ethanol at 150°C with FSO_3H (H.J. Weyer *et al.*, Ger. Offen. DE 4,233,430, 1994).

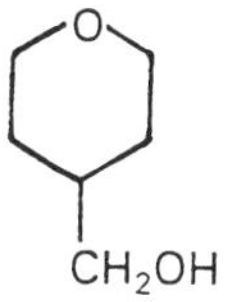

(19)

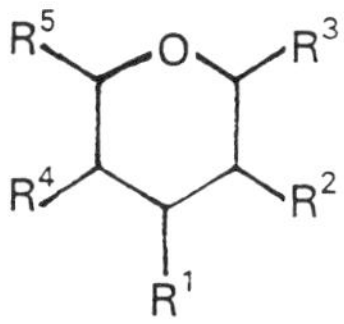

(20)

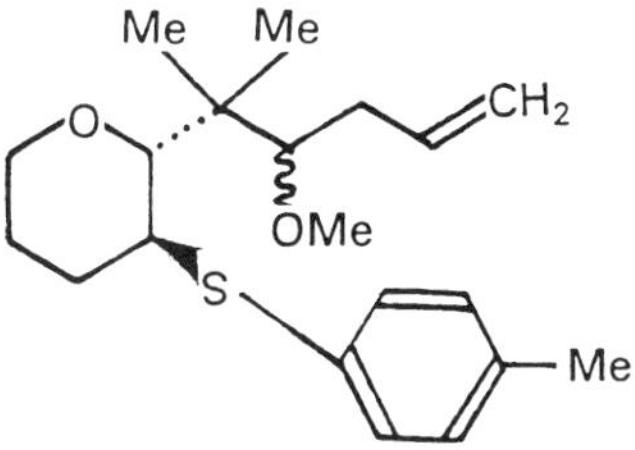

(21)

176

The sequential coupling of 4-MeC$_6$H$_4$SCl with 3,4-dihydro-2H-pyran, Me$_2$C = CH(OMe) and H$_2$C = CHCH$_2$MgCl gives 2-(3-methoxy-2-methylhex-5--en-2-yl)-3-(4-methylphenylthio)tetrahydropyran (21) (I.P. Smolyakova, V.A. Smit, and B. Osinov, Tetrahedron Letters, 1991, _32_, 2601). A number of related derivatives have been prepared (I.P. Smolyakova *et al.*, Izv. Akad. Nauk SSSR, Ser. Khim., 1991, 1100).

(iv) Tetrahydropyran-carboxaldehydes. The preparation of optically active tetrahydropyran-3-carboxaldehyde (3-formyltetrahydropyran) (1) from the racemate using ephedrine (B.S. Moore and F.J. Urban, U.S. US 5,149,821, 1992) and an efficient preparation of tetrahydropyran-(3R)--carboxaldehyde, a key intermediate for the synthesis of a novel 5-lipoxygenase inhibitor (F.J. Urban and B.S. Moore, Tetrahedron Asymmetry, 1992, _3_, 731) have been reported. Acid-catalyzed rearrangement of 5,6--dihydro-4H-1,3-dioxocins 2 (R^1 =H, Me; R^2 = Me, Ph, Pr) gives the 2--substituted and 2,2-disubstituted tetrahydropyran-3-carboxaldehydes 3 (H. Frauenrath and M. Sawicki, Tetrahedron Letters, 1990, _31_, 649).

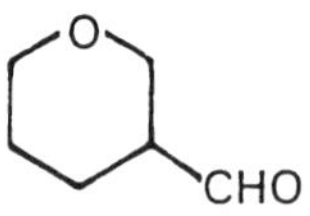

(1) 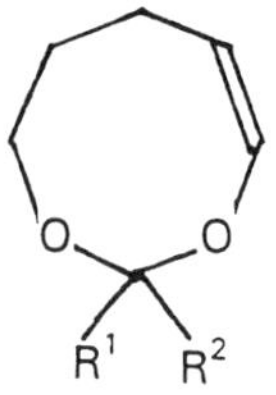   (2) 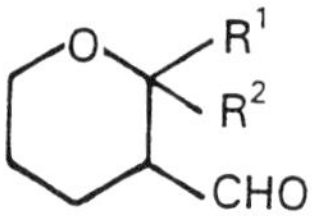   (3)

4-Hydroxymethyltetrahydropyran has been treated with tetramethyl-piperidin-1-yloxyl 4 in a mixture of CH$_2$Cl$_2$ and H$_2$O containing NaOCl, NaHCO$_3$, and NaCl at room temperature to yield tetrahydropyran-4-carbox-aldehyde (5) (J. Suzuki, Jpn. Kokai Tokkyo Koho JP 06 211,827 [94 211,827], 1994), this compound is also obtained by the dehydrogenation of 4-hydroxymethyltetrahydropyran over a Cu, Ag and/or Au catalyst (K. Brenner *et al.*, Ger. Offen. DE 4,039,918, 1992).

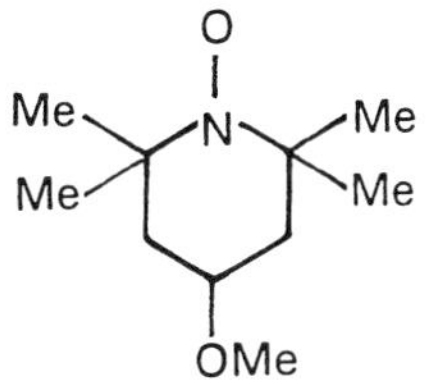

(4)

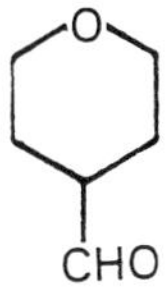

(5)

(v) Tetrahydropyrancarboxylic acids and esters. Optically active tetra-hydropyran-2- and -3-carboxylic acids have been obtained by amidation of racemic 1 and 2 by α-aminoesters, distillation of the diastereomeric amides, and hydrolysis (E. Fritz-Langhals, Eur. Pat. Appl. EP 521,496, 1993; Angew. Chem., Int. Ed., 1993, 105, 753).

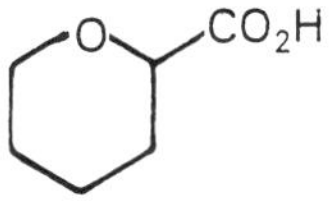

(1)

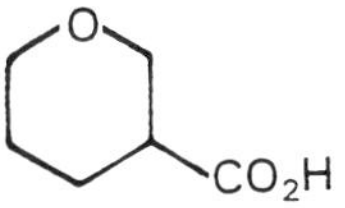

(2)

For the preparation of methyl 4-chlorotetrahydropyran-2-carboxylate, see L.D.M. Lolkema *et al.*, Tetrahedron, 1994, 50, 7115; methyl 4-chloro-6,6--dimethyltetrahydropyran-2-carboxylate, *idem, ibid.*, p.7129; and methyl 4--methoxycarbonylmethylenetetrahydropyran-2-carboxylates, S.P. Munt and E.J. Thomas, Chem. Comm., 1989, 480). The intramolecular alkylation of enolates of *tert*-butyl α-alkoxycarboxylates proceeds stereoselectively to yield 2,2,6-trisubstituted tetrahydropyrans, for example, 2-MeC$_6$H$_4$SO$_3$-(CH$_2$)$_3$CH(CHMe$_2$)OCHMeCO$_2$Bu^t in THF-HMPA affords *tert*-butyl 2-methyl-6--*iso*propyltetrahydropyran-2-carboxylate (3) (T. Fujisawa *et al.*, Bull. Chem. Soc. Jpn., 1992, 65, 3487).

$$\text{Pr}^i \quad \text{O} \quad \text{Me} \quad \text{CO}_2\text{Bu}^t$$

$$\text{CO}_2\text{R}$$

(3)　　　　　　　　　　(4)　　　　　　　　　　(5)

Esters 4 (R = alkyl, phenylalkyl) of tetrahydropyran-4-carboxylic acid (4; R = H) have been prepared from 2,7-dioxaspiro[4.4]nonane-1,6-dione (5) with alcohols in the presence of an acidic catalyst, for example, compound 5 with methanol and Al$_2$O$_3$ gives methyl tetrahydropyran-4-carboxylate (4; R = Me) along with 3-(2-methoxyethyl)-2,3-dihydrofuran-2-one and 5-oxaspiro[2,4]-heptan-4-one (R. Fischer *et al.*, Ger. Offen. DE 4,300,419, 1994). Tetrahydropyran-4-carboxylic acid (4; R = H) and its esters 4(R = C$_{1-6}$alkyl, C$_{5-8}$cycloalkyl, aryl) have been obtained from appropriate 2,3-dihydrofuran-2--ones (D. Borchers *et al.*, Eur. Pat. Appl. EP 584,663, 1994) and by the catalytic ring cleavage-cyclization of α-(2-hydroxyethyl)-γ-butyrolactone and its derivatives (R. Becker *et al.*, *ibid.*, 284,969, 1988).

Tetrahydropyrancarboxylate esters 6 (R^1 = alkyl; R^2 = H, CO$_2$R^3; R^3 = alkyl) have been prepared by the cyclocondensation of R^{5}COCH$_2$CO$_2$R^4 (R^4 = alkyl, R^5 = MeCO, OR^4) with (RCH$_2$CH$_2$)$_2$O (R = Cl, Br, ClCO$_2$, ClSO$_2$) in the presence of an alcoholate and a dipolar aprotic solvent. Thus, methyl tetrahydropyran (6; R^1 = R^2 = H) has been obtained by heating (ClCH$_2$CH$_2$)$_2$O with MeCOCH$_2$CO$_2$Me in MeOH/DMF containing KI and NaOMe (T. Kuekenhoehner *et al.*, *ibid.*, 546,409, 1993). A method for the purification of tetrahydropyran-4-carboxylic esters 7 (R^1 = alkyl; R^2,R^3 = H, alkyl) has been reported (R. Fischer *et al.*, *ibid.*, 546,397, 1993).

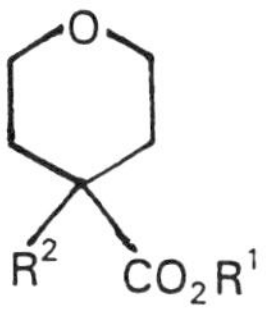

(6) 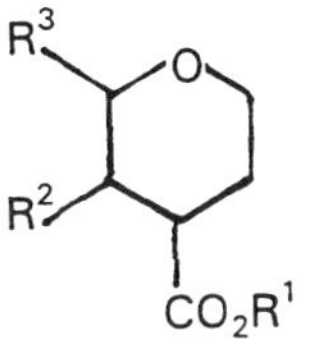   (7) 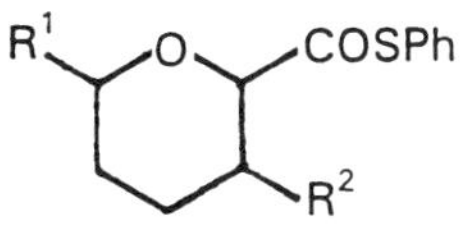   (8)

Bu^{t}SiMe$_2$OCHR^1(CH$_2$)$_2$CHR^{2}CHO (R^1 = H, Me, Ph; R^2 = H, Ph) has been converted into the corresponding (Z)-1-nitro-1-(phenylthio)alkene by condensation with (phenylthio)nitromethane. Subsequent reaction with HF/pyridine (-78 to 0°C) and ozone-potassium *tert*-butoxide in THF (-78°C) afforded 3,6-disubstituted 2-phenylthiocarbonyltetrahydropyrans 8 (A.C.M. Barrett, J.A. Flygare, and C.D. Spilling, J. Org. Chem., 1989, 54, 4723). For the preparation of 2-(methoxy- and *tert*-butoxycarbonyl)methylenetetra-hydropyrans see, S. Brandaenge and B. Lindqvist, Acta Chem. Scand., 1989, 43, 807; and tetrahydropyran-2-yl phenylacetate, A. Steinmann, Eur. Pat. Appl. EP 564,405, 1993).

In the presence of sodium thiophenoxide and thiophenol compound 9 undergoes a reductive cleavage at the S(7)-C(8) bond to give, after acidification, a sulphinic acid, which undergoes methylation to yield (2*S*, 3*S*)--3-methylsulphonyl-2-(methylsulphonylmethyl)tetrahydropyran (10). The preparation of (2*R*, 3*S*)-3-methylsulphonyl-2-[(1*R*)-1-(methylsulphonyl)but-3--enyl]tetrahydropyran (11) has been described (B. Beagley *et al.*, J. Chem. Soc., Perkin I, 1992, 2371).

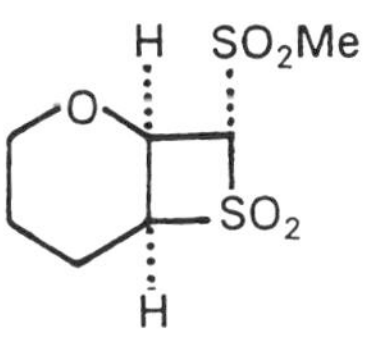

(9) 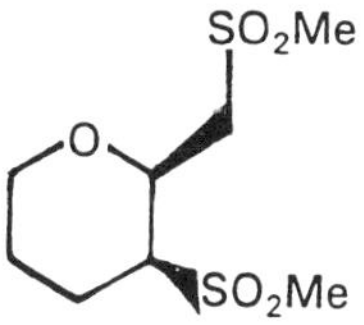   (10) 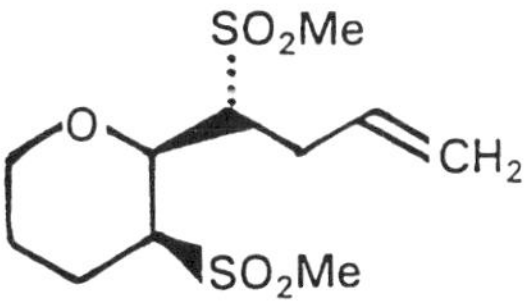   (11)

180

(vi) Halogenotetrahydropyrans. The ring cleavage of either *cis-* or *trans-*
-2-alkyl- or -phenyl-3-chlorotetrahydropyrans 1 (R = alkyl, Ph) by freshly
prepared magnesium gives (E)-$HOCH_2CH_2CH_2CH = CHR$ (X. Ding *et al.*,
Huaxue Xuebao, 1987, <u>45</u>, 1138). The stereoselective synthesis of
enantiomerically enriched endo-substituted halogenotetrahydropyrans, for
example, (+)-methyl 3-chloro- and 3-bromotetrahydropyran-2-acetates 2
(R = Cl, Br), by an intramolecular Michael addition of a suitable chiral γ-
-halogeno-$α$,ß-unsaturated ester has been described (V.S. Martin and J.M.
Palazon, Tetrahedron Letters, 1992, <u>33</u>, 2399).

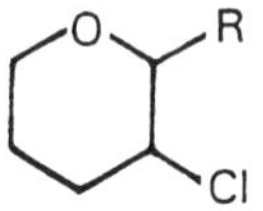

(1) 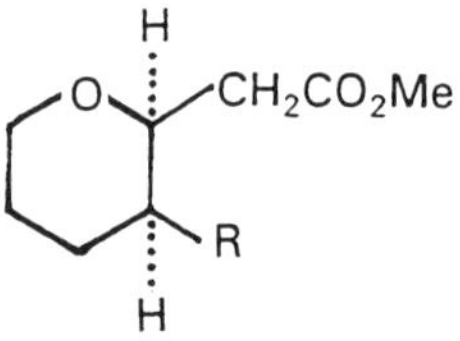   (2)    (3)

Oxidation of 4-substituted 4-chlorotetrahydropyrans 3 (R = CO_2Me, CN)
by 70% HNO_3 and $NaNO_2$ yields 85 and 97% 4-chlorotetrahydropyran-4-
-carboxylic acid (3; R = CO_2H), respectively; $HO_2CCH_2CCl(CO_2H)CH_2CO_2H$
is not formed (M.S. Sargsyan *et al.*, Arm. Khim. Zh., 1987, <u>40</u>, 548).
Oxidation of 4-chloro-4-methyltetrahydropyran (3; R = Me) by HNO_3 affords
3-hydroxy-3-methylglutaric acid and oxalic acid, resulting from the oxidation
of the initial dehydrochlorination or hydroxylation-dechlorination products of
3 (R = Me) (*idem, ibid.*, 1988, <u>41</u>, 222).

Cyclization of $H_2C = CHCH_2CH_2OCHClCO_2$Me catalysed by copper (I)
chloride-6,6'-dimethyl-2,2'-bipyridine complex leads to methyl 4-chlorotetra-
hydropyran-2-carboxylate (4) (J.H. Udding, H. Hiemstra, and W.N.
Speckamp. J. Chem. Soc., Perkin 2, 1992, 1529).

(4)   (5)

Treatment of acetals derived from 2-substituted benzaldehyde-chromium-tricarbonyl complexes and methanol with homoallylic alcohol and $TiCl_4$ furnishes the corresponding (RS, RS, SR, RS)-*cis*-2-aryl-4-chlorotetrahydro-pyran complexes completely diastereoselectively. Thus, by this method, optically pure (R)-(-)-2-anisaldehyde-chromiumtricarbonyl is converted, after decomplexation to optically pure (R, S)-(+)-*cis*-2-(2-anisyl)-4-chlorotetra-hydropyran (5) (S.G. Davies, T.J. Donohoe, and M.A. Lister, Tetrahedron Asymmetry, 1991, *2*, 1085). The asymmetric synthesis of optically pure (R, R, S)- and (S, R, R)-2-(2-anisyl)-3-ethyl-4-chlorotetrahydropyran from homo-chiral 2-anisaldehyde-chromiumtricarbonyl and (Z)- and (E)-hex-3-en-1-ol have been reported (*idem, ibid.*, p.1089).

6-Substituted 4-chloro- and -bromo-2-phenyltetrahydropyrans 6 (R^1 = Cl, Br, R^2 = Me, Pr^i, Ph) have been prepared by stereoselective cyclocondensa-tion reaction of $PhCH(OH)CH_2CH=CH_2$ with R^2CHO in the presence of $AlCl_3$ and $AlBr_3$, respectively (L. Coppi, A. Ricci, and M. Taddei, J. Org. Chem., 1988, *53*, 911). For the preparations of a number of 4-chloro-2,6-diethyl-tetrahydropyrans 7 R^1 = H, R^2 = H, Me; R^1R^2 = $(CH_2)_3$] with substituents at position -3 and positions -3 and 5, see D. Marton *et al.*, J. Organometal. Chem., 1990, *391*, 295.

(6)   (7)

182

(8)

(9) $R^4 = Cl$
(10) $R^4 = OH$

The reaction of $Me_2C(OMe)_2$ with pent-4-en-2-ol in CH_2Cl_2 containing $TiCl_4$ affords 4-chloro-2,2,6-trimethyltetrahydropyran (8) *via* a trans-acetalization-cationic cyclization sequence (N.A. Nikolic *et al.*, J. Org. Chem., 1989, *54*, 2748). Treating the 4-chlorotetrahydropyrans 9 [R^1 = Me, alkoxyethyl, R^2 = H, alkoxymethyl, R^3 = H, Pr, Me_2CHCH_2, $(CH_2)_5$] with 30% H_2O_2 gives the tetrahydropyran-4-ols 10 (A.S. Arakelyan, A.I. Dvoryanchikov, and A.A. Gevorkyan, Zh. Org. Khim., 1987, *23*, 2323).

Treatment of tetrahydropyran with an excess of boiling sulphuryl chloride in the dark affords 2,3,3-trichlorotetrahydropyran (11) (94%) (L. De Buyck, Bull. Soc. Chim. Belg., 1992, *101*, 303). The preparations of 2,3-dichloro-3--methyltetrahydropyran (*trans* and *cis* 10:1) and 2,3,3-trichloro-5-methyltetrahydropyran (*cis* and *trans* 10:1) (L. De Buyck and A. Verhue, *ibid.*, 1993, *102*, 209) and of ω-substituted *a,a*-dichloro aldehydes and derivatives *via* formolysis or acetolysis of 2,3,3-trichlorotetrahydrofuran and 2,2,3-tri-chlorotetrahydropyran (L. De Buyck and P. Vanhulle, *ibid.*, 1992, *101*, 329) have been reported. Treatment of 2-(hydroxymethyl)tetrahydropyran with PBr_3 in toluene containing pyridine gives 2-(bromomethyl)tetrahydropyran. A number of related compounds have been prepared (A. Chene and P. Caruhel, Fr. Demande FR 2,591,593, 1987).

(11) (12) (13)

The Lewis acid-promoted condensation of allylalkoxysilanes with
carbonyl compounds has been investigated, with reference to the preparation
of 4-chloro-2,6-dipropyltetrahydropyran (12) and 6-methyltetrahydropyran-2-
-acetic acid (13), a natural product isolated from the glandular secretion of
the civet cat (Z.Y. Wei *et al.*, J. Org. Chem., 1989, 54, 5768). The use of
$NbCl_5$, $MoCl_5$, $TaCl_5$ and WCl_6 as reagents to effect the replacement of the
hydroxyl group in tetrahydropyran-4-ol and 2-hydroxymethyltetrahydropyran
by chlorine has been investigated (E.M. Coe and C.J. Jones, Polyhedron,
1992, 11, 3123). Tributyl(2-cyclohexyl)stannane on stirring with $BuSnCl_3$
gives cyclohexenyltin chloride derivative 14, which reacts with EtCHO to
give 9-chloro-2,4-diethyl-3-oxabicyclo[3.3.1]nonane (15) (D. Marton, D.
Furlani, and G. Tagliavini, Gazz., 1987, 117, 189).

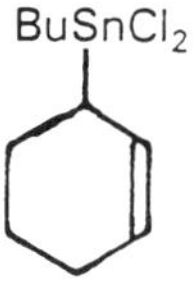

(14) 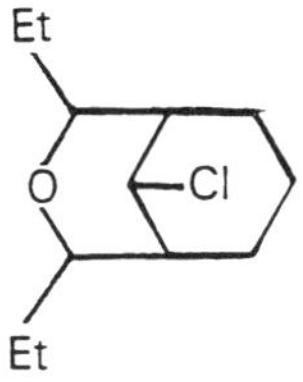　　　　　　　　　　(15)

N-(Phenylseleno)phthalimide cyclization of bromohydrins 16 (R = H, Ac)
yields exclusively the 6-substituted 3-bromo-2,2,6-trimethyltetrahydropyrans
17, which have been elaborated into 3-bromo-6-ethenyl-2,2,6-trimethyltetra-
hydropyrans (18), a possible analogue of the Aplysia pyranoids (G.
Wrensford *et al.*, Tetrahedron Letters, 1990, 31, 4257).

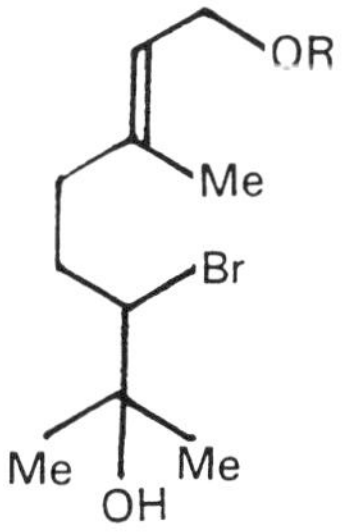

(16) 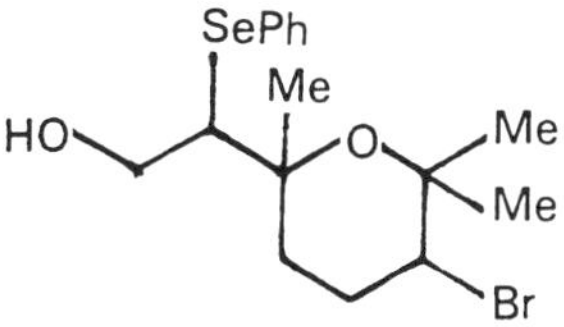　　　　　　　(17) 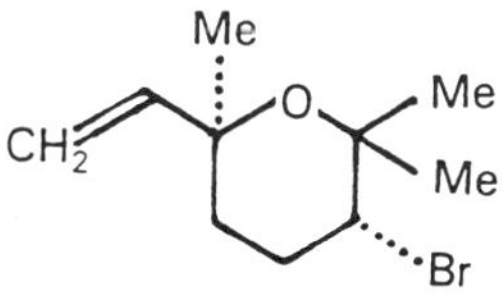　　　　　　　(18)

184

The cyclization of (*E*)- or (*Z*)-HOCHRCH$_2$CH$_2$CH$=$C(OMe)CO$_2$Me (R$=$H, Me) mediated by various electrophilic reagents (3-ClC$_6$H$_4$CO$_2$OH, PhSeCl, NBS, iodine) gives tetrahydropyran derivatives, for example, treatment of (*E*)HOCHMeCH$_2$CH$_2$CH$=$C(OMe)CO$_2$Me with NBS affords methyl 3-bromo-2-methoxy-6-methyltetrahydropyran-2-carboxylate (19). The cyclization of (*Z*) and (*E*) isomers are stereospecific (V. Faivre *et al.*, Tetrahedron, 1989, <u>45</u>, 7765).

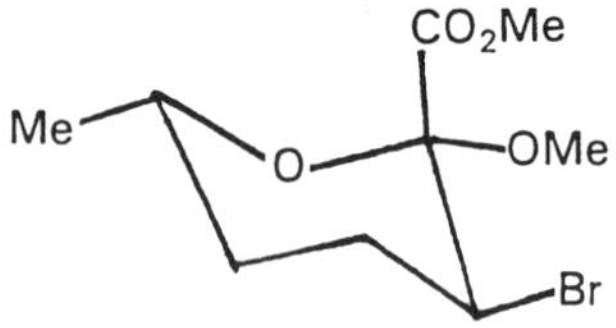

(19)

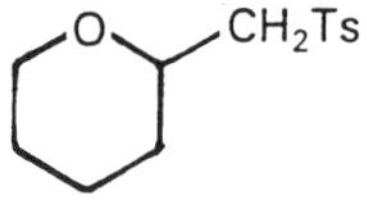

(20)

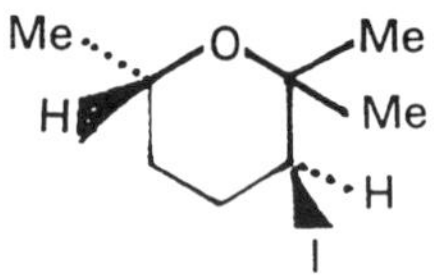

(21)

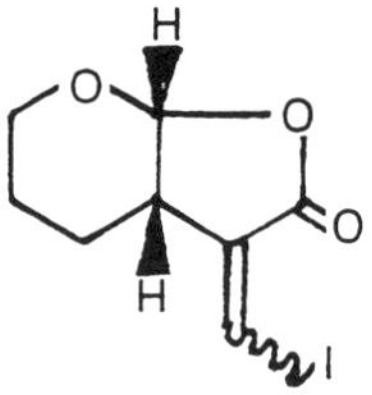

(22)

The radical mediated addition of 4-toluenesulphonyl iodide to alkenol, H$_2$C$=$CH(CH$_2$)$_3$CH$_2$OH occurs readily to give TsCH$_2$CHI(CH$_2$)$_3$CH$_2$OH as a single regioisomer, which on treatment with K$_2$CO$_3$ in MeOH cyclizes to give 2-(4-toluenesulphonylmethyl)tetrahydropyran (20). An anomalous reaction where Me$_2$C$=$CHCH$_2$CH$_2$CHMeOR (R$=$H, SiMe$_2$Bu^t) cyclizes directly to afford 3-iodo-2,2,6-trimethyltetrahydropyran (21) has been discussed (G.L. Edwards and K.A. Walker, Tetrahedron Letters, 1992, <u>33</u>, 1779).

The reaction of 3,4-dihydro-2H-pyran with propiolic acid and *N*-iodo-succinimide gives an iodo ester, which has been cyclized under free-radical conditions to form an (*E*)-iodomethylene lactone 22 (M.J. Lynch, J. Simpson, and R.T. Weavers Austral. J. Chem., 1993, <u>46</u>, 203).

The preparations of 2,5- and 4,5-disubstituted 3,3-difluorotetrahydro-pyrans (T. Morikawa *et al.*, Tetrahedron, 1992, **48**, 8915); 3-fluoro-2--hydroxymethyl-6,6-dimethyltetrahydropyran (23) (A. Arnone *et al.*, J. Chem. Soc., Perkin 1, 1991, 1315); and *cis*-4-fluoro-2,6-bis(methoxyphenyl)-4--methyltetrahydropyran (J.W. Fabler and D.L. Linebarrier, Organometallics, 1990, **9**, 3182) have been reported.

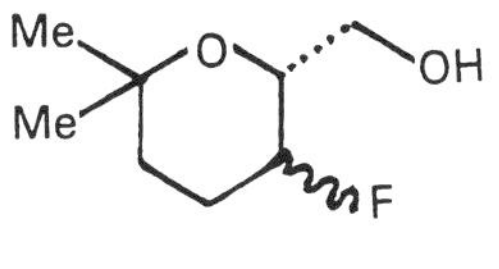

(23) 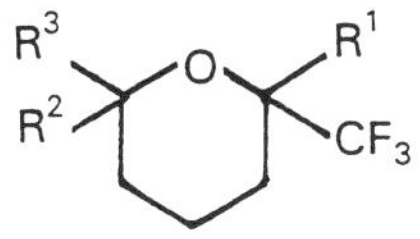   (24)

As an extension of the ene reaction of trifluoromethyl ketones, α-tri-fluoromethylated tetrahydropyrans 24 ($R^1 = CF_3$, Me, $R^2 = Ph$, 4-MeC_6H_4, Bu, $R^3 = Me$; $R^2 = pentyl$, heptyl, $R^3 = H$), including 2-trifluoromethyl-2,6--dimethyl-6-(4-methylphenyl)tetrahydropyran (24; $R^1 = R^3 = Me$, $R^2 = 4$--MeC_6H_5) and 2,2-di(trifluoromethyl)-6-methyl-6-(4-methylphenyl)tetrahydro-pyran (24; $R^1 = CF_3$, $R^2 = 4$-MeC_6H_5, $R^3 = Me$), fluorine analogues of curcumene ether have been prepared from trifluoromethylated homoallyl alcohols, the products of the ene reaction (T. Nagai *et al.*, Heterocycles, 1992, **33**, 51). For the preparation of 2,3,3,4,4,5,5-heptafluorotetrahydro-pyran, see I.T. Bazyl *et al.*, Zh. Org. Khim., 1990, **26**, 2228).

(vii) Cyano-, nitro-, and amino-tetrahydropyrans and related compounds.
2-Cyano-6-methoxytetrahydropyran (1; R = H) has been lithiated with LiN-$(Pr^i)_2$ and then alkylated with alkyl halides to give 2-alkyl-2-cyano-6-methoxy-tetrahydropyrans 1 (R = CH_2R^1; R^1 = Bu, Pr^i, $CH_2CH = CMe_2$, CH_2Ph), which have been converted into the corresponding cyclohexenones 2 (G.C. Laredo and L.A. Maldonado, Heterocycles, 1987, **25**, 179). 2-Cyanomethyltetra-hydropyran (3) derivatives have been prepared from tetrahydropyran-2-ones, *via* di*iso*butylaluminium hydride reduction to the 2-ols, and olefination by Wittig reaction to α,β-unsaturated cyano derivates, which undergo instantaneous cyclization (C.M. Passarotti *et al.*, Boll. Chim. Farm., 1993, **132**, 150).

(1)

(2)

(3)

(4)

Stephen reduction of 4-cyano-4-(3,4-dimethoxyphenyl)tetrahydropyran (4; R = CN) with $SnCl_2$ in EtOAc containing HCl yields 4-(3,4-dimethoxyphenyl)tetrahydropyran-4-carboxaldehyde (4; R = CHO), which has been converted into the 2-(4-aryltetrahydropyran-4-yl)ethyldimethylamine 4 (R = $CH_2CH_2NMe_2$) (A.A. Agekyan *et al.*, Arm. Khim. Zh., 1989, 42, 705).

The preparations of a number of 3-[(tetrahydropyran-2-yloximino)methyl]-phenylureas 5 (R^1,R^2 = H, alkyl, alkoxy, cyclopropyl; R^3 = H, alkyl, alkoxy; R^4 = H, alkyl, F, Cl, Br; R^5 = H, F; X = O, S) as herbicides (K.H.G. Pilgram and G.H. Bozarth, Eur. Pat. Appl. EP 261,710, 1988); and 8-(tetrahydro-pyran-2-yloxy)octanal from 8-nitrooctanol *via* conversion of the alcohol to the tetrahydropyranyl derivative (R. Ballini, E. Marcantoni, and M. Petrini, Synth. Comm., 1992, 22, 641) have been reported.

(5)

(6)

The reaction of N-tetrahydropyran-2-yl protected nitrone 6 with organolithium reagents RLi [R = benzo[b]thien-2-yl, Ph, fur-3-yl, 5-(pyrid-2-yl)thien-2--yl, 5-(pyrid-2-yl)thien-3-yl, 5-methylthien-2-yl], followed by acid hydrolysis yields hydroxylamines, which are isolated as RCHMeN(OH)CONH$_2$ by treatment with Me$_3$SiNCO (A. Basha, J.D. Ratajczyk, and D.W. Brooks, Tetrahedron Letters, 1991, 32, 3783).

2-(N-Substituted 1-amino)ethyl-6-methoxy-3,6-dihydro-2H-pyran 7 (R^1 = CO$_2$CH$_2$Ph, R^2 = CO$_2$Bu^t) has been converted into 5-acetylamino-2-(1--acetylamino)ethyl-6-methoxytetrahydropyran (methyl 2,6-di-N-acetyl-a-D-purpurosaminide B) (8) in 8 steps (J. Jurczak, A. Golebiowski, and J. Raczko, *ibid*., 1988, 29, 5975).

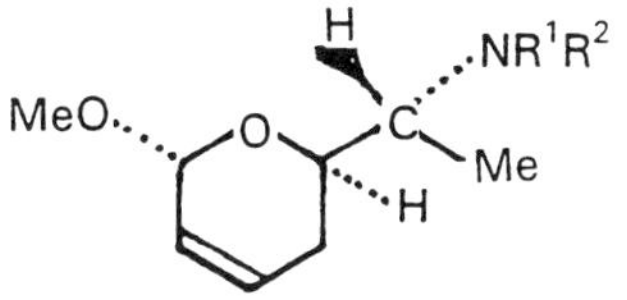

(7)

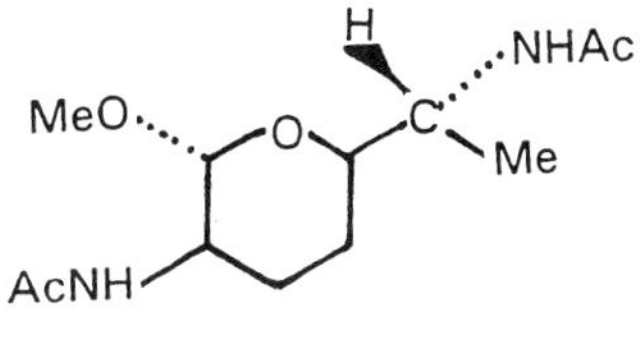

(8)

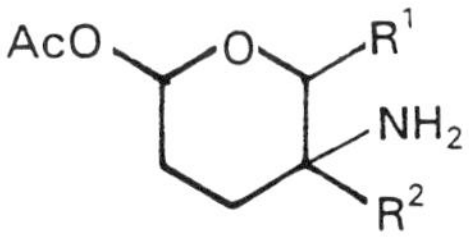

(9)

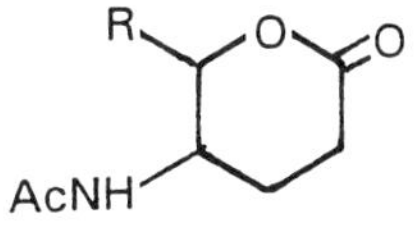

(10)

Novel 2,3-disubstituted 6-acetoxy-3-aminotetrahydropyrans 9 [R^1 = 3,4--methylenedioxyphenyl, 3,4-(MeO)$_2$C$_6$H$_3$, R^2 = H, CH$_2$NH$_2$] and 6-substituted 5-acetylaminotetrahydropyran-2-one 10 (R = 3,4-methylenedioxyphenyl) mimicing compounds with antipsychotic activity (sympathomimetic amines) have been synthesized (M.P. Georgiadis *et al*., J. Heterocyclic Chem., 1991, 28, 697). A number of related derivatives have been synthesized *via* 6--hydroxy-3,6-dihydro-2H-pyran-3-ones and their *cis*-platinum (II) complexes (M.P. Georgiadis, S.A. Haroutounian, and J.C. Bailar, Jr., *ibid*., 1988, 25, 995).

Treatment of 3,4-epoxy-2,4-dimethyl-6-phenyltetrahydropyran (11) with Me$_3$SiNMe$_2$ and AlCl$_3$ affords the tetrahydrofuran derivative 12 (major product) and 4-chloro-2,4-dimethyl-6-phenyl-3-trimethylsiloxytetrahydropyran 13; not the expected 4-amino-2,4-dimethyl-6-phenyl-3-trimethylsiloxytetra-hydropyran (U. Sequin and A. Schneider, Heterocycles, 1987, 26, 2455).

(11) (12) (13)

The reactions of substituted benzaldehydes with 4-amino-2,2-dimethyl-tetrahydropyran affords the corresponding 4-arylmethylimino derivatives, which on subsequent reduction gives the 4-arylmethylamino-2,2-dimethyl-tetrahydropyran, for example, 4-benzylamino-2,2-dimethyltetrahydropyran (14) (K.S. Shaginyan et al., Arm. Khim. Zh., 1990, 43, 723). The diastereomers of 1-[4-(benzyloxy)-3-methoxymethyl)phenyl]-2-[benzyl(2,2-di-methyltetrahydropyran-4-yl)amino]ethanol (15) have been separated (K.M. Garibyan, A.O. Tosunyan, and S.A. Vartanyan, ibid., 1987, 40, 312).

(14) (15)

4-Substituted 4-(2-benzylamino)ethyl-2,2-dimethyltetrahydropyrans 16 (R^1 = Ph, R^2 = H, OMe, R^3 = OMe, OH; R^1 = CH_2Ph, R^2 = R^3 = OMe) and 4- -(substituted benzylamino)-5-benzyl-2,2-dimethyltetrahydropyran 17 (R^4 = H, Me) have been prepared from tetrahydropyranylacetonitriles 18 (R^5 = CN) and 4-amino-5-benzyl-2,2-dimethyltetrahydropyran (19), respectively, *via* condensation of compounds 18 (R^5 = CH_2NH_2) and 19 with aromatic aldehydes and/or ketones and subsequent reduction with $NaBH_4$ (N.S. Arotyunyan *et al.*, *ibid.*, 1989, <u>42</u>, 240).

(16) R^5 = $CH_2NHCH_2C_6H_3R^2R^3$-3,4
(18) R^5 = (see text)

(17) R^5 = $CHR^4C_6H_3R^2R^3$-3,4
(19) R^5 = H

The preparations of 4-amino-3,5-bis(methoxymethoxy)-6-methyl-6- -(1,4,5,8-tetramethoxynaphthalen-2-yl)tetrahydropyran derivatives as intermediates for neoplasm inhibitors (A. Terajima, M. Kawasaki, and F. Matsuda, Jpn. Kokai Tokkyo Koho JP 62 153,282 [87 153,282], 1987); 5- -lipoxygenase-inhibiting ester compounds, for example, naphth-2-ylmethyl (Z)-3-amino-3-(4-methoxytetrahydropyran-4-yl)prop-2-enonate (T.G.C. Bird and A.A.C. Olivier, Eur. Pat. Appl. EP 581,464, 1994); and 5-lipoxygenase- -inhibiting oxime ether derivatives, for example O-{4-[5-fluoro-3-(4-methoxy-tetrahydropyran-4-yl)phenylthio]benzyl}oxime (P.N. Edwards and M.S. Large, *ibid.*, 570,196, 1993) have been reported.

Manganic acetate-mediated annulation of $MeO_2CCH_2CO_2K$ with a 2- -amino-3,4-dihydro-2*H*-pyran derivative yields a fused γ-lactone 20. Carboxylic acids with less active methylene protons fail in this reaction (M. Del Rosario-Chow, J. Ungwitayatorn, and B.L. Currie, Tetrahedron Letters, 1991, <u>32</u>, 1011).

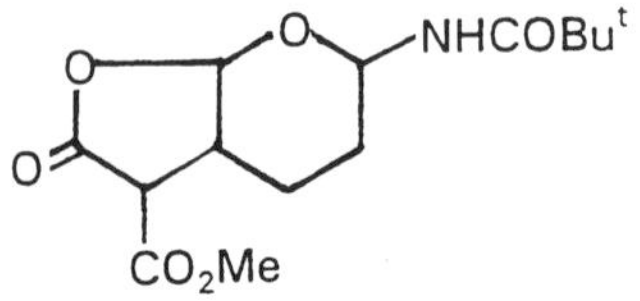

190

(20)

(viii) Tetrahydropyran-2-, -3-, and -4-ones

Tetrahydropyran-2-ones. Tetrahydropyran-2-ones have been prepared from aliphatic saturated 1,5-diols in the presence of an oxidising agent (e.g. NaOCl) and a catalytic amount of a Ru-containing catalyst (e.g. $RuCl_3.H_2O$) (J.S. Plotkin, PCT Int. Appl. WO 90 14,344, 1990). The $IrH_5[(Pr^i)_3P]_2$-catalyzed lactonization of pentane-1,5-diol gives tetrahydropyran-2-one (δ-valerolactone) (1) (Y. Lin, X. Zhu, and Y. Zhou, J. Organomet. Chem., 1992, **429**, 269). Cyclopentanone and PhCHO in benzene reacted with oxygen at 25°C to yield tetrahydropyran-2-one (S. Murahashi and Y. Oda, Eur. Pat. Appl. EP 548,774, 1993). A mixture of 3-methyltetrahydrofuran-2-one and tetrahydropyran-2-one (1) has been obtained by charging an autoclave with anisole, $H_2C=CHCH_2CH_2OH$, $Pd(OAc)_2$, $Ph_2P(CH_2)_4PPh_2$, Ph_3P, and $4\text{-}MeC_6H_4SO_3H$, flushing with CO, pressurizing with CO until a partial pressure of 40 bar was reached and heating to ambient temperature (E. Drent, *ibid.*, 341,773, 1989). Tetrahydropyran-2-ones have been prepared by the alkylation of *N*-benzylpiperi-2,6-dione and subsequent reduction of the resulting carbinolamide intermediate with $NaBH_4$, followed by acid catalyzed cyclization (H. Yoda *et al.*, Chem. Express, 1989, **4**, 515).

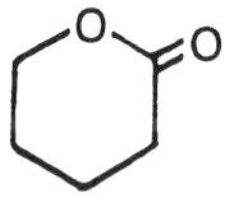

(1)

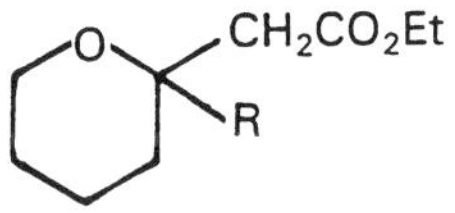

(2)

The steric course of the microbial degradation of racemic unsaturated hydroxy fatty acid to tetrahydrofurans and tetrahydropyrans depends upon subtle substrate structural modifications and the microorganism used [C. Fuganti *et al.*, Prog. Flavour Precursor Stud. Proc. Int. Conf. 1992 (Pub. 1993), 451]. Tetrahydropyran-2-one (1) on treatment with $CH_2 = C(OEt)$-$OSiMe_2Bu^t$ and Et_3SiH, $CH_2 = CHCH_2SiMe_3$, Me_3SiCN and Me_3SiSCH_2Ph and catalyst (trityl hexachloroantimonate in $SbCl_5$-Me_3SiCl-SnI_2 (Ph_3C^+ ClO_4^- and PhC^+ SbF_5^- have also been used) gives 2-substituted ethyl tetrahydropyran--2-acetates 2 (R = H, $CH_2 = CH$-CH_2, CN, SCH_2Ph) [C. Mazal and J. Jonas, Stud. Org. Chem. (Amsterdam), 1988, <u>35</u> (Chem. Heterocyclic Compd.), 412].

The lipase-catalyzed conversion of ($\pm$)-2-methylglutaric anhydride into (*S*)-3-methyl- and (*R*)-5-methyltetrahydropyran-2-one [(*S*)-2-methyl- and (*R*)-4-methyl-δ-valerolactone] *via* a regio- and enantioselective sequential esterification has been reported (R. Ozegowski, A. Kunath, and H. Schick, Ann., 1994, 1019).

3-Alkyltetrahydropyran-2-ones 3 (R = H, Me, Et, Bu) are prepared by hydrolysis of $Cl(CH_2)_3CR(CO_2Et)_2$ with NaOH in 50% EtOH, followed by cyclization and decarboxylation with aqueous HCl (J. Xu, Huaxue Shiji 1987, <u>9</u>, 244). 3-Ethyltetrahydropyran-2-one(ethylpentanolide) (3; R = Et) on treatment with $(Pr^i)_2NLi$ and then $R_2^1NCH = CR^2NO_2$ (R_2^1N = morpholino, R^2 = H) in $MeOCH_2CH_2OMe$ affords (*E*)-3-ethyl-3-(nitrovinyl)tetrahydropyran-2-one (4). The yield of derivative 4 increases to 95% when the zinc enolate of 3 (R = Et) is used. Other nitroolefination reagents $R^1NCH = CR^2NO_2$ ($R_2^1N = Me_2N$, R^2 = H, Me; R_2^1N = morpholino, R^2 = Me) have been used (M. Node *et al.*, Synthesis, 1987, 729). It has also been reported that zinc enolates derived from 3-methyltetrahydropyran-2-one (R = Me) have enhanced reactivity over the corresponding lithium enolates in asymmetric nitroolefination (K. Fuji *et al.*, Chem. Pharm. Bull., 1994, <u>42</u>, 999).

(3)
(6) R = R^2

(4)

192

The reaction of chiral nitro enamines 5 (R^1 = H, Me, Et) with Zn enolates
of 3-alkyltetrahydropyran-2-one 6 (R^2 = Me, Et, allyl) yields the 3-alkyl-3(2-
-alkyl-2-nitrovinyl)tetrahydropyran-2-ones 7 with a high enantiomeric excess
through an addition-elimination process. The best results are obtained with
the reaction of 5 (R^1 = Et) with 6 (R^2 = Et) (*idem*, J. Amer, Chem. Soc., 1989,
<u>111</u>, 7921).

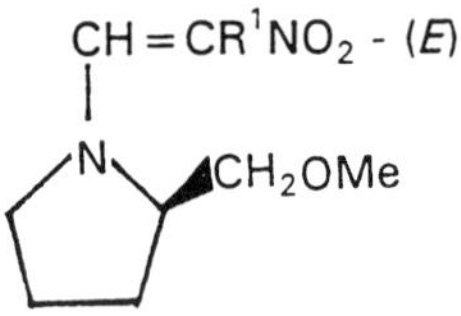

(5) 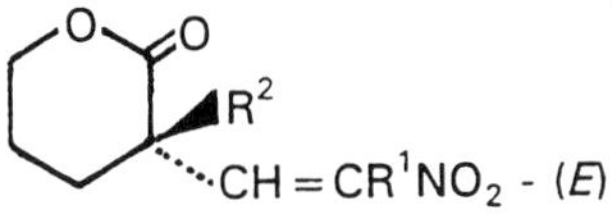   (7)

Ring opening of 3-methylglutaric anhydride with (*S*)-*a*-methyl-1-naphth-
alenemethanol gives the (*S*)-naphthylethyl methyl (*S*)-3-pentanedioate (8),
which on hydrogenolysis, followed by reduction and cyclocondensation
yields (*R*)-4-methyltetrahydropyran-2-one [(*R*)-3-methylvalerolactone] (9)
(P.D. Theisen and C.H. Heathcock, J. Org. Chem., 1993, <u>58</u>, 142). 4-
-Methyltetrahydropyran-2-one and 3-methylpentane-1,5-diol have been
prepared from 4-methyltetrahydropyran-2-ol (Y. Tokito and N. Yoshimura,
Jpn. Kokai Tokkyo Koho JP 63 198,639, 196,531 [88 198,639, 196,531],
1988.

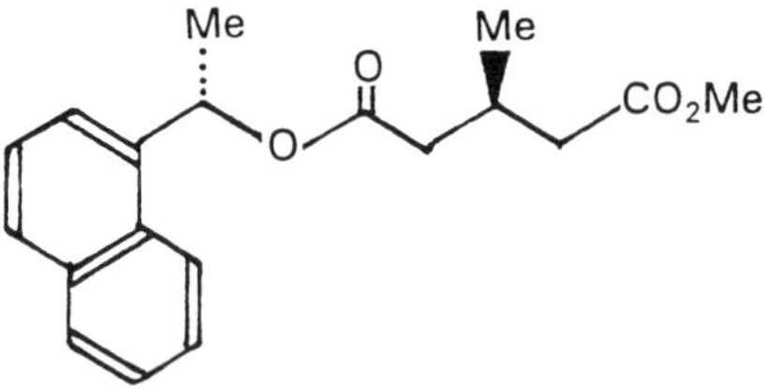

(8) 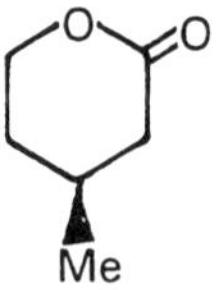   (9)

Under heterogeneous conditions a lipase (Amano P) catalysed the asymmetric ring opening of 3-substituted glutaric anhydrides 10 (R = Me, Et, Pr, Pr^i) with BuOH in Pr$^i_2$O to give (R)-BuO$_2$CCH$_2$CHRCH$_2$CO$_2$H, which have been converted to (R)-4-alkyltetrahydropyran-2-ones 11 having 60-93% enantiomeric excess (Y. Yamamoto et al., Agric. Biol. Chem., 1988, 52, 3087). 4-Spirotetrahydropyran 12 has been obtained by the reaction between the conjugated diene, 1,2-dimethylenecyclohexane and the epoxide, oxirane (M.S. Sell, H. Xiang and R.D. Rieka, Tetrahedron Letters, 1993, 34, 6007).

(10) X = O
(11) X = H$_2$

(12)

(13) R = Me
(14) R (see text)
(15) R = Me(CH$_2$)$_4$

Bakers' yeast reduction of MeCO(CH$_2$)$_2$CH$_2$CN gives MeCH(OH)(CH$_2$)$_2$-CH$_2$CN, which undergoes H$^+$/EtOH catalyzed lactonization to afford 6--methyltetrahydropyran-2-one (13). The reduction product with the S--configuration is obtained in high enantiomeric excess (A. Gopalan et al., Synth. Comm., 1991, 21, 1321). 6-(2-Phenylvinyl)-, 6-(2-phenylethyl)-, and 6-(3-phenylpropyl)tetrahydropyran-2-one (14) [R = CH = CHPh, CH$_2$CH$_2$Ph, CH$_2$(CH$_2$)$_2$Ph, respectively] have been obtained via Bakers' yeast reduction of the corresponding keto acids PhCH = CHCO(CH$_2$)$_3$CO$_2$H and Ph(CH$_2$)$_n$CH$_2$-CO(CH$_2$)$_3$CO$_2$H (n = 1,2) (M. Aquino et al., Tetrahedron, 1991, 47, 7887). Optically active 6-alk(en)yltetrahydropyran-2-ones have been prepared by enantiomer-selective enzymic hydrolysis carried out on the racemates (J.P. Bourdineaud, C. Ehret and M. Petrzilka, PCT Int. Appl. WO 94 07,887, 1994).

6-Pentyltetrahydropyran-2-one (δ-decalactone) (15) has been prepared in 3 steps starting from diethyl hexanedioate by Dieckmann condensation-pentylation, hydrolysis-decarboxylation and Baeyer-Villiger oxidation (Y. Zheng et al., Xiangtan Daxue Ziran Kexue Xuebao, 1988, 10, 45). 6-

-Pentyltetrahydropyran-2-one (15) is also obtained on subjecting 2-pentyl-cyclopentane to the Baeyer-Villiger reaction (S. Chen and Z. Liang, Guangzhou Huagong, 1992, 20, 42). The oxidation of 2-(pent-2-ynyl)cyclo-pentanone by peracetic acid in the presence of acetic anhydride affords 6-(pent-2-ynyl)tetrahydropyran-2-one (94%). In the absence of acetic anhydride the yield is 62% (H. Fujisawa, Jpn. Kokai Tokkyo Koho JP 06 157,503 [94 157,503], 1994). For the preparation of 6-(2-cyclohexenyl-vinyl)tetrahydropyran-2-one and analogues as HMG-CoA reductase inhibitors, see J.R. Regan and K.E. Neuenschwander, U.S. US 5,132,312, 1992).

Amidation of $ClCO(CH_2)_3CH(OAc)(CH_2)_3COCl$ with (+)-(1,1'-binaphthyl)-2,2'-diamine as chiral auxilary gives the cyclic amide 16, which on subsequent deacetylation and lactonization yields lactone 17 in 97% diastereomeric excess. The absolute configuration of 17 has been determined and it has been converted to the pheromone (R)-6-undecyltetra-hydropyran-2-one (Y. Yamamoto et al., J. Org. Chem., 1991, 56, 1112).

(16)

(17)

The enantiomers of 3,6-dimethyltetrahydropyran-2-one (18) have been synthesized from (R)- and (S)-2-(2-hydroxy-propyl)-1,3-dithiane (19) (R. Bernardi and D. Ghiringhelli, Synthesis, 1989, 938). 6-Alkyl-3-methyl-and 3-alkyl-6-methyltetrahydropyran-2-ones 20 (R^1 = Me, R^2 = C_{2-10}alkyl;

$R^1 = C_{2-10}$alkyl, $R^2 = Me$) as odourous substances have been prepared by the Baeyer-Villiger oxidation of cyclopentanones 21 (T. Keil and K. Schulze, Eur. Pat. Appl. EP 401,491, 1990). The preparations of 3-ethylidene-6-vinyl-tetrahydropyran-2-one by telemirization of $CH_2 = CH\text{-}CH = CH_2$ and CO_2 have been reported (P. Braunstein, D. Matt and D. Nobel, J. Amer. Chem. Soc., 1988, <u>110</u>, 3207; Fr. Demande FR 2,617,163, 1988).

(18) $R^1 = R^2 = Me$ (19)   (21)
(20)

New approaches to the synthesis of 3-alkyl-6-methylenetetrahydropyran--2-ones 22 ($R = Pr^i$ CH_2CHMe_2, CHMeEt, Pr), inhibitors of the serine protease elastase (W. Dai and J.A. Katzenellenbogen, J. Org. Chem., 1993, <u>58</u>, 1900) and the extension of the intramolecular Stille cross coupling reaction of vinylstannyl chloroformates to the formation of 6-substituted 3--methylenetetrahydropyran-2-ones (R.M. Adlington et al., J. Chem. Soc., Perkin 1, 1994, 1677) have been discussed.

(22)   (23)   (24)

196

4-Substituted 3-(tributylstannylmethylene)tetrahydropyran-2-ones (o-tri-butylstannylmethylene-δ-valerolactones) 23 (R = Ph, 2,6-Me$_2$C$_6$H$_3$) are readily obtained *via* intramolecular radical cyclization of propiolates and acetylenic ketones mediated by reaction with Bu$_3$SnH/AIBN (E. Lee, C.U. Hur and C.M. Park, Tetrahedron Letters, 1990, <u>31</u>, 5039. Homoallylic diazoacetates, for example, CH$_2$ = CHCH$_2$CH$_2$O$_2$CCH = N$_2$ undergo enantioselective intramolecular cyclopropanation with the rhodium catalyst Rh$_2$(5S-MEPY)$_4$ (MEPY = methyl 2-pyrrolidinone-5S-carboxylate) to yield oxabicyclo[4.1.0]-heptanes, for example, 24 (S.F. Martin, C.J. Oalmann and S. Liras, *ibid.*, 1992, <u>33</u>, 6727). The effect of conformational mobility and hydrogen-bonding interactions on the selectivity of some 3- and 4-(4-guanidinophenyl)--6-methylene- and -6-(iodomethylene)tetrahydropyran-2-ones, mechanism--based inhibitors of trypsin-like serine proteases, has been reported (R. Rai and J.A. Katzenellenbogen, J. Med. Chem., 1992, <u>35</u>, 4297). The palladium(O)-mediated intramolecular lactonization of allylic alcohol derivatives and the unusual substituent effect of the trifluoromethyl group in the homoallylic position on the formation of tetrahydropyran-2-ones has been discussed (Y. Hanzawa *et al.*, Chem. Comm., 1990, 394).

A suspension of HO(CH$_2$)$_3$CHMeCH(OH)Et and platinum black in water containing 3 drops of concentrated hydrochloric acid, on heating under oxygen at 55°C with stirring yields (5S, 6R)-6-ethyl-5-methyltetrahydro-pyran-2-one (25) (S. Takano *et al.*, Jpn. Kokai Tokkyo Koho JP 02 250,880 [90 250,880], 1990). For the preparation of 6-ethyl-6-methyltetrahydro-pyran-2-one, see D.J. Ramon and M.Yus, Tetrahedron Letters, 1990, <u>31</u>, 3767. Tetrahydropyran-2-ones 26 with one to three alkyl substituents in the 3-, 4-, and 6-positions are conveniently prepared from saturated alcohols 27 (R^1-R^3 = H, alkyl) and carbon monoxide using a 1-electron oxidation system of lead tetraacetate in benzene and heating under pressure (S. Tsunoi, I. Ry, and N. Sonoda, J. Amer. Chem. Soc., 1994, <u>116</u>, 5473).

(25)   (26)   (27)

Boron enolate 28 undergoes a highly diastereoselective aldol reaction with chiral aldehyde 29, followed by lactonization to give 6-(but-3-one-2-yl)--3,5-dimethyltetrahydropyran-2-one (30) (R.P. Short and S. Masamune, Tetrahedron Letters, 1987, **28**, 2841). For the preparations of 4-allyl-6--methyl-5-methylenetetrahydropyran-2-one, see Y. Sasaki, Jpn. Kokai Tokkyo Koho JP 02 62,872 [90 62,872], 1990; and substituted 6-(phenyl-vinyl)tetrahydropyran-2-ones, S. Takano *et al.*, *ibid.*, 06 41,114 [94 41,114], 1994.

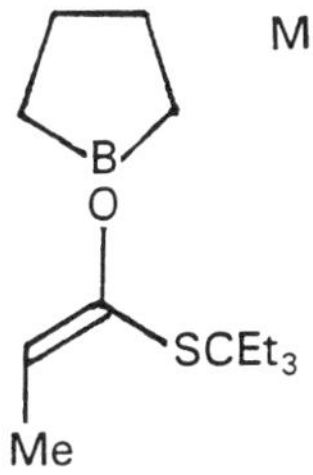

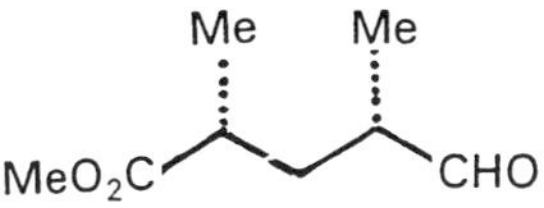

(29)

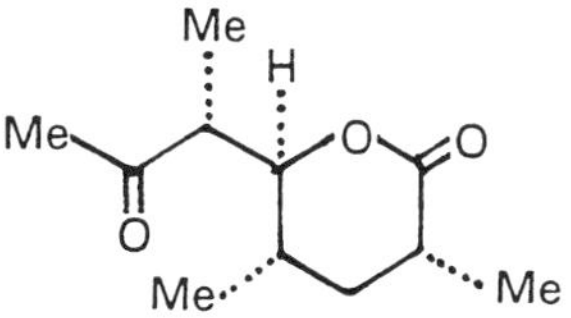

(30)

(28)

Photoadducts 1,3,3-trimethyl-2,4-dioxabicyclo[4.2.0]octan-5-ones 31 [R^1, R^2, R^3 = Me; R^1R^2 = $(CH_2)_4$, R^3 = H] derived from 2,2,6-trimethyl-4H-1,3--dioxin-4-one, on di*iso*butylaluminium hydride reduction afford 4,4,5,5-tetra--substituted 6-methyltetrahydropyran-2-ones 32 (G.L. Lange and M.G. Organ, Synlett. 1991, 665). The preparation of 3,3-dialkyl-5,6,6-trimethyl-tetrahydropyrans 33 [R^1, R^2 = Me,, Et; R^1R^2 = $(CH_2)_5$] by the acid-catalyzed lactonization of 3-cyclopropylalkanoic acids has been reported (K. Sugahara *et al.*, Synthesis, 1990, 783).

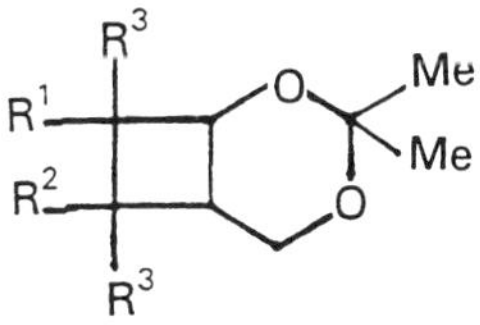

(31) 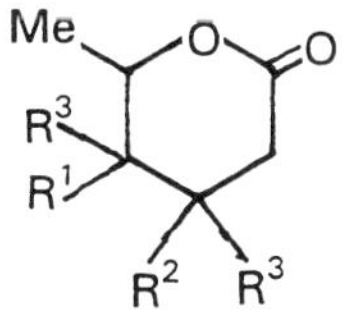   (32) 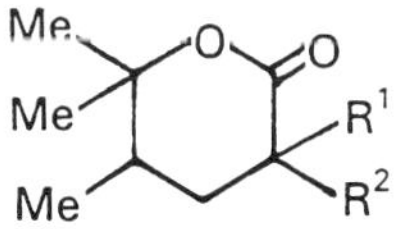   (33)

198

6-Phenyltetrahydropyran-2-one (34) has been obtained along with
HOCHPh(CH$_2$)$_3$CO$_2$Et by the chlorotri*iso*propoxytitanium-mediated aldol-type
condensation of γ-zinc ester ZnCH$_2$(CH$_2$)$_2$CO$_2$Et with PhCHO (H. Ochiai *et
al.*, J. Org. Chem., 1988, **53**, 1343). The resolution of racemic 6-phenyl-
tetrahydropyran-2-one (34) by chromatography on cellulose triacetate (CTAI)
resulted in one of the best separations of optical antipodes observed so far
on this chiral stationary phase (R.M. Wolf, E. Francotte, and J. Hainmueller,
Chirality, 1993, **5**, 538). Enantioselective hydrolysis of the appropriate
racemic tetrahydropyran-2-ones using esterase affords optically active (*R*)-4-
-methyl-, (*R*)-4-phenyl-, or (*R*)-6-phenyltetrahydropyran-2-ones and the
corresponding (*S*)-δ-hydroxypentanoic acid derivatives (T. Izumi, F. Tamura,
and M. Akutsu, J. Heterocyclic Chem., 1994, **31**, 441). The preparations of
a number of 6-aryl- and aralkyltetrahydropyran-2-one derivatives as antiper-
cholesterolemics (G.E. Stokker, Eur. Pat. Appl. EP 283,217, 1988) and of
trans-3,4-diphenyl- and *trans,trans*-2,3,4-triphenyltetrahydropyran-2-ones
(J.A. Stanley *et al.*, Org. Prep. Proced. Int., 1991, **23**, 193) have been
reported.

(34)

(35)

The synthesis of high specific activity tritium labelled (*E*)-6-(bromo-
methylene)-3-([4-^{3}H]naphth-1-yl)tetrahydropyran-2-one (35) and
investigations for its use as a selective probe for calcium-independent
phospholipase A$_2$ have been described (R.C. Durley, B.L. Parnas, and R.H.
Weiss, J. Labelled Compd. Radiopharm., 1992, **31**, 685). For the
preparations of some optically active 2-(substituted aryl)-5-substituted-tetra-
hydropyran-2-ones, see K. Sakashita, T. Ikemoto, and K. Mori, Jpn. Kokai
Tokkyo Koho JP 03 31,274 [91 31,274], 1991).

7-(4-Cyanophenyl)-3-hydroxy-1-phenylheptan-1-one has been converted to ($\pm$)-*cis*-6-[4-(4-amidinophenyl)butyl-4-phenyltetrahydropyran-2-one (36) and a number of related derivatives useful as platelet aggregation inhibitors have been prepared (N.A. Abood, R.E. Manning, and M. Miyano, PCT Int. Appl. WO 94 00,424, 1994).

(36)   (37)   (38)

In order to study the enantioselectivity of alternate substrate inhibition of chymotrypsin by chiral α- and ß-aryl-substituted enol lactones, four of these lactones have been prepared in optically pure form; 6-methylene-3-phenyl-(37), and 6-methylene-3-(1-naphthyl)-, 6-methylene-4-phenyl- (38), and 6--methylene-4-(1-naphthyl)tetrahydropyran-2-ones. The syntheses utilise the acetylenic acid precursors, α- and ß-arylhex-5-ynoic acids, resolved by silica gel chromatographic separation of the corresponding diastereoisomeric phenylglycinol amide derivatives (D.J. Baek *et al.*, J. Org. Chem., 1989, <u>54</u>, 3963). Both enantiomers of a wide variety of 6-alkylated tetrahydropyran-2-ones have been prepared by enzymic kinetic resolution (B. Haase and M.P. Schneider, Tetrahedron Asymmetry, 1993, <u>4</u>, 1017).

The reactions of tetrahydropyran-2-one with Gringard reagents, followed by phase-transfer oxidation afford the 6,6-disubstituted tetrahydropyran-2--ones 39 (R = Me, Et, Pr, pentyl, hexyl, Ph, 4-ClC$_6$H$_4$, 4-FC$_6$H$_4$) (J. Lehmann and N. Marquardt, Ann., 1988, 827).

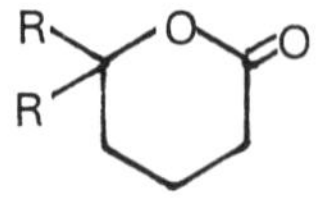

(39)
(40) 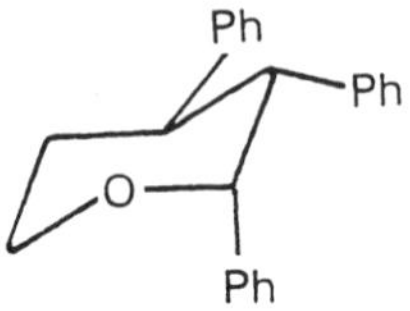

A new interpretation of the procedure of Meerwein for the addition of deoxybenzoin to cinnamic aldehyde, shows that the crude product is not the corresponding aldehyde but its hemiacetal, 4,5,6-triphenyltetrahydropyran--2-one (3,4,5-triphenyl-5-pentanolide) (40) (T. Gospodova and Y. Stefanovsky, Bulg. Chem. Comm., 1992, $\underline{25}$, 358).

3,4-Diaryl-6-methyltetrahydropyran-2-ones (41 and 42) (R^1 = Cl, R^2, R^3 = H, OMe, OH, R^2R^3 = OCH_2O; R^1 = R^2 = H, R^3 = OMe) have been prepared by hydrolysis-lactonization of the corresponding $HOCHMeCH_2CH(C_6H_4R^1$-4)-$CH(C_6H_3R^2R^3$-3,4)CN and tested for psychotropic activity. 42 (R = Cl, R^2 = OH, R^3 = H; R^1 = Cl, R^2 = H R^3 = OMe, OH) showed psychostimulant activity without antidepressant activity. The relative configurations and conformations of 41 and 42 have been determined by NMR and IR spectroscopy and the relationship between their stereochemistry and psychostimulant activity discussed (S. Axiotis et al., Eur. J. Med. Chem., 1987, $\underline{22}$, 293). The preparations of (4R)-substituted (6S)-phenoxymethyl-, (6S)-ß-phenylethyl-, and (6S)-ß-styryltetrahydropyran-2-ones as antihypercholesteremics and antiartheriosclerotics have been reported (H. Jendralla et al., Ger. Offen. DE 3,615,620, 1987).

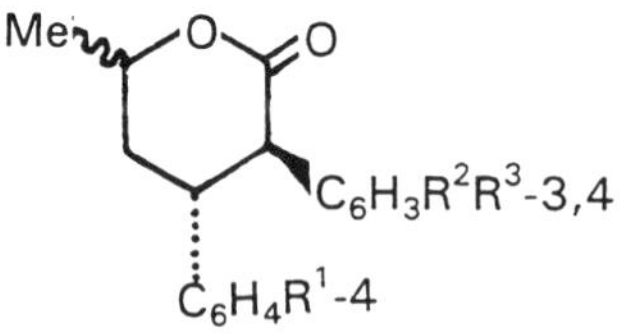

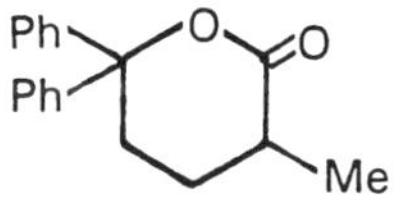

(41) 4-Me(β)
(42) 4-Me(α)

(43)

Catalytic hydrogenation of 3-(dimethylaminomethylene)-6,6-diphenyl-tetrahydropyran-2-one yields 3-methyl-6,6-diphenyltetrahydropyran-2-one (43), $Ph_2CHCH_2CH_2CH(CO_2H)CH_2NMe_2$, and $Ph_2CH(CH_2)_4NMe_2$ [J. Lehmann, M. Neugebauer, and N. Marquardt, Arch. Pharm. (Weinheim, Ger.), 1990, $\underline{323}$, 117).

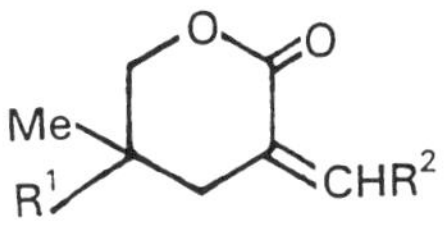

(44) 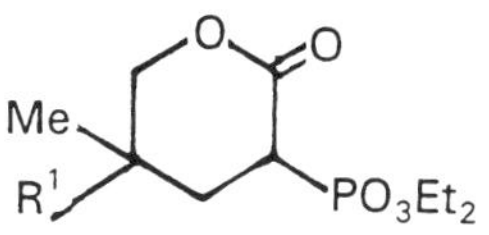   (45) 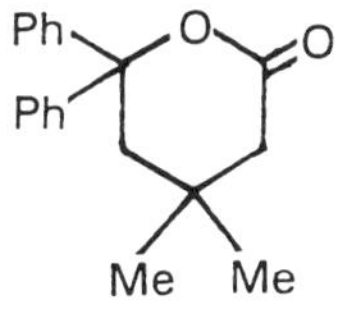   (46)

For the synthesis of 5-substituted 5-methyl-3-(monosubstituted methylene)tetrahydropyran-2-ones 44 [R^1 = Pr, R^2 = 2-naphthyl, 5-methyl-2-thienyl, CHMePr; R^1 = Me, R^2 = 4-$O_2NC_6H_4$, $(CH_2)_3CO_2Me$, $(CH_2)_6Me$] from the appropriate a-phosphonolactones 45 and the aldehydes R^2CHO, see G. Falsone, U. Wingen, and D. Wendisch, Tetrahedron Letters, 1989, $\underline{30}$, 675. 4,4-Dimethyl-6,6-diphenyltetrahydropyran-2-one (46) has been prepared from $ClCOCH_2CMe_2CH_2COCl$, C_6H_6, and $AlCl_3$ and its crystal structure data reported (J. Novotny et al., Collect. Czech. Chem. Comm., 1989, $\underline{54}$, 1661).

Hydroxytetrahydropyran-2-ones. Reaction of 5-butyl-2-hydroxytetra-hydrofuran with KCN in EtOH-AcOH at 0°C affords 2,5-dihydroxynonane-nitrile, which on treatment with 12N HCl at 100-115°C gives 6-butyl-3--hydroxytetrahydropyran-2-one (1). Treatment of derivative 1 with mesyl chloride and Et_3N in DMF yields 6-butyl-5,6-dihydro-2*H*-pyran-2-one, which along with other 6-alkyl-5,6-dihydro-2*H*-pyran-2-ones, are useful as butter- or milk-like flavouring materials (T. Kawanobe, Jpn. Kokai Tokkyo Koho JP 63 222,164 [88 222,164], 1988). The synthesis of 6-alkyl 3-hydroxytetra-hydropyran-2-ones 2 (R = C_{1-16}alkyl) (J. Nakauchi et al., *ibid.*, 63 258,872 [88 258,872], 1988) and 6-(substituted aryl)-3-hydroxytetrahydropyran-2--ones 3 (R = C_{1-16} linear or branched alkyl, C_{1-16} linear or branched alkoxy, OH, CH_2OH, CO_2H; A = arylene) (K. Sakashita, T. Ikemoto, and K. Mori, *ibid.*, 03, 31,276 [91 31,276], 1991) has been reported.

(1)

(2)

(3)

Several halogenoenol lactones, for example, 6-bromomethylene-3-
-phenoxytetrahydropyran-2-one (4) have been prepared from arylacetic acid
derivatives, for example, $PhCH_2CO_2H$. They are potent mechanism-based
inhibitors of a novel class of calcium-independent phospholipases A_2, which
have been implicated as the enzymic mediators of membrane dysfunction
during myocardial ischemia (L.A. Zupan et al., J. Med. Chem., 1993, 36,
95).

(4)

(5)

(6)

4-Hydroxy-4-[^{13}methyl]-[5-^{13}C]-tetrahydropyran-2-one (5) has been
prepared in 4 steps from [1,3-^{13}C]-acetone (L.O. Zamir, M. Lin, and Nguyen
Cong Danh, J. Labelled Compd. Radiopharm., 1988, 25, 1081). 4-Ethyl-4-

-hydroxytetrahydropyran-2-one (6) is obtained by the cyclization, on treatment with $Ag(OAc)_2$, of $PhSiMe_2CH_2CH_2C(OH)EtCH_2CO_2Bu^t$, prepared by reacting $PhSiMe_2CH_2CH_2COEt$ with $AcOBu^t$ in THF containing $(Pr^i)_2NLi$ (I. Fleming, S.K. Armstrong, and R.J. Pollitt, J. Chem. Res., S, 1989, 19). A convenient synthesis of 5-substituted *cis*- and *trans*-4-hydroxytetrahydro-pyran-2-ones 7 and 8 ($R^1 = CH_2Ph$, $4-CH_2C_6H_4Ph$) (J.D. Prugh and A.A. Deana, Tetrahedron Letters, 1988, _29_, 37) and preparations of optically active forms of 6-pentyl-4-hydroxytetrahydropyran-4-one (T. Sato, Canad. J. Chem., 1987, _65_, 2732) and asymmetric 6-benzyloxy-4-hydroxytetrahydro-pyran-2-one (9) (L. Shao *et al.*, Tetrahedron, 1993, _49_, 1997) has been reported. Also reported are the preparations of several 6-alkyl-, 6-alkenyl-, and 6-hydroxymethyl-4-hydroxytetrahydropyran-2-ones 10 ($R = C_{3-15}$alkyl, alkenyl, CH_2OH) as pure stereoisomers (P. Hammann *et al.*, Eur. Pat. Appl. EP 469,480, 1992).

(7) 4-OH(β)   (9)   (10)
(8) 4-OH(α)

For the preparation of a number of 6-aryloxymethyl-4-hydroxytetrahydro-pyran-4-ones and related compounds as HMG-CoA reductase inhibitors and biological antioxidants, see H. Jendralla *et al.*, J. Med. Chem., 1991, _34_, 2962; Eur. Pat. Appl. EP 418, 648, 1991) and for optically active *trans*- and *cis*-benzyloxymethyl-4-hydroxytetrahydropyran-2-ones, such as 11, *via*, 3--hydroxyalkenyl phenyl sulphides, S. Takano, Y. Sugihara, and K. Ogasawara, Tetrahedron Asymmetry, 1993, _4_, 1795.

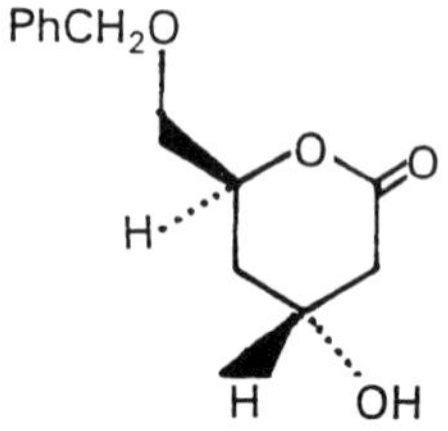

PhCH$_2$O

(11)

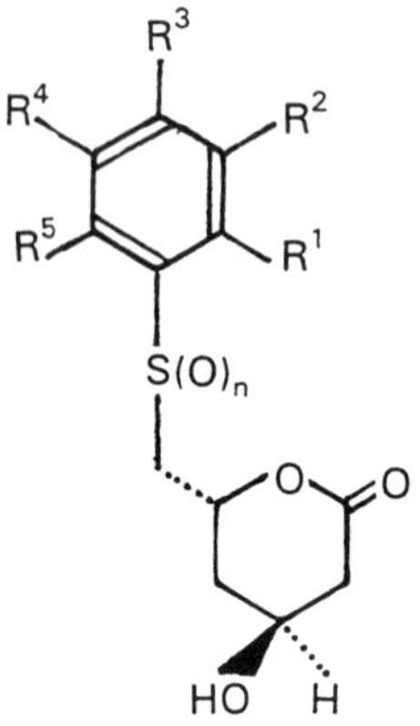

R^3, R^4, R^2, R^5, R^1

S(O)$_n$

(12)

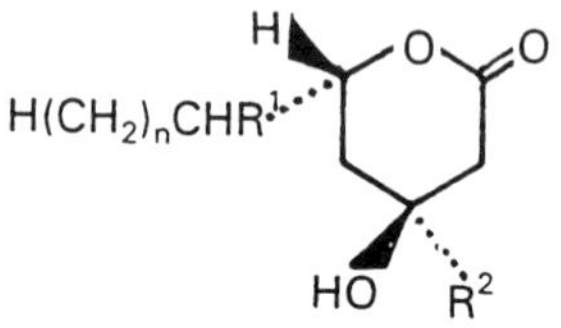

H(CH$_2$)$_n$CHR^1

HO R^2

(13)

HO H

6-(*S*)-Arylthiomethyl-4(*R*)-hydroxytetrahydropyran-2-ones and their corresponding sulphoxides and sulphones 12 [n = 0-2; R^1, R^5 = H, halogeno, (un)substituted Ph or alkyl; R^2, R^4 = H, halogeno, Me, Et, MeO, EtO, PhCH$_2$O; R^3 = H, halogeno, CF$_3$, Me, Et, MeO, EtO] and the corresponding dihydroxycarboxylic acids, their salts with bases and their esters, useful in treating hypercholesterolemia, atherosclerosis, and tumors have been prepared (W. Bartmann *et al.*, Ger. Offen. DE 3,632,893, 1988). The preparations of a number of 4-substituted *trans*-6-(1-halogenoalkyl)-4-
-hydroxytetrahydropyran-2-ones (*trans*-5-halogenoalkylmevalonolactones) 13 [R^1 = F, Cl, Br, iodo; R^2 = H, Me(CH$_2$)$_m$; m = 0-2; n = 0-12] have been reported (C.J. Morrow and J.M. Eridon, U.S. US 5,149,835, 1992).

The iodolactonization of (*Z*)- or (*E*)-3-siloxy-5-alkenoic acids 14 and 15 (R = Et, Bu) yields the *trans*-disubstituted 6-substituted 4-tri*iso*propylsiloxy-tetrahydropyran-2-ones 16 and 17, which differ in stereochemistry of the iodine substituent (S.B. Bedford *et al.*, Tetrahedron Letters, 1992, 33, 6505).

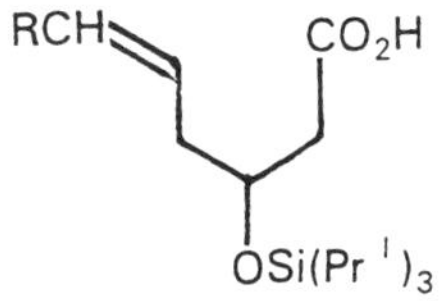

RCH=... CO$_2$H OSi(Pr^i)$_3$

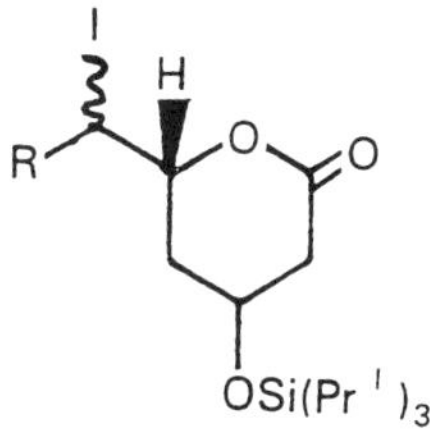

(14) *(E)*
(15) *(Z)*

(16) *(R)*
(17) *(S)*

The 6-substituted 4-hydroxy-2-oxotetrahydropyran-4-acetic acids (substituted glutaric acid lactones) 18 (R = alkyl, cycloalkyl, aryl, MeCHR^1; R^1 = cycloalkyl, Ph; n = 9-13), useful as HMG-CoA reductase inhibitors for treating hyperlipidemia, have been prepared (J.S. Baran, T.J. Lindberg, and H.S. Lowrie, U.S. US 4,895,973, 1990).

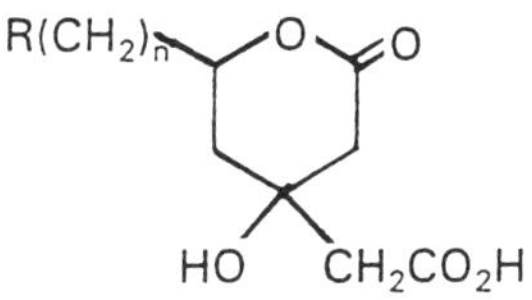

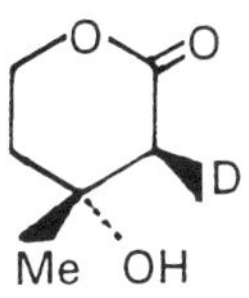

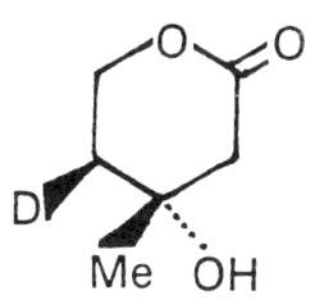

(18)

(19)

(20)

The synthesis of (3S, 4R)-3-deuterio-4-hydroxy-4-methyltetrahydropyran--2-one [(2S, 3R)-mevalonolactone-2-*d*] (19) and (4S, 5S)-5-deuterio-4--hydroxy-4-methyltetrahydropyran-2-one [(3S, 4S)-mevalonolactone-4-*d*] (20) starting from 4-methyl-5,6-dihydro-2H-pyran has been reported (S. Ohta *et al.*, J. Sci. Hiroshima Univ., Ser. A, Phy. Chem., 1989, 49).

Treatment of (*R*)-methyl-6-benzyloxy-5-hydroxyhex-2-ynoate with LiCuMe$_2$ affords (*R*)-6-benzyloxymethyl-4-methyl-5,6-dihydro-2*H*-pyran-2-
-one, which on oxidation with 30% H$_2$O$_2$ yields epoxide 21. Reduction of epoxide 21 followed by deprotection furnishes 4-hydroxy-6-hydroxymethyl-
-4-methyltetrahydropyran-2-one (22). A number of related derivatives 23 (R^1 =H, OH-protecting group; R^2 =alkyl, alkenyl, Ph; R^3 =H, OH; R^4 =H; R^{3}R^4 may form epoxy group) have been prepared (S. Takano and K. Ogasawara, Jpn. Kokai Tokkyo Koho JP 03 246,289 [91 246,289], 1991).

(21)   (22)   (23)

The cyclization of a series of ß-halogenoacetoxy ketones and aldehydes, for example, 24 (R^1 =H, Me, Et, Ph, Pr^i, Bu^t; R^2 =H, Me; R^3 =H, Me; R^4 =H, Me, Ph, Pr; R^5 =H, EtPh; R^6 =H, Me), promoted by samarium (II) iodide, to give 4-hydroxytetrahydropyran-2-ones 25, with up to six substituents has been reported (G.A. Molander *et al.*, J. Amer. Chem. Soc., 1991, <u>113</u>, 8036). For the preparation of 6-phenoxymethyl- and 6-thio-phenoxymethyl-4-hydroxytetrahydropyran-2-ones and related aryl-, biphenyl-, biaryloxy- and biarylthio derivatives, their acid derivatives, salts, and esters as antihypercholesterinemics, see H. Jendralla *et al.*, Eur. Pat. Appl. EP 341,681, 1989.

(24)   (25)

A large number of 6-(2-substituted ethyl and 2-substituted ethenyl)-4-
-hydroxytetrahydropyran-2-one derivatives have been prepared, some of
which have been reported as HMG-CoA reductase inhibitors useful as anti-
hypercholesterolemics, others as intermediates for their production and some
as antiatheroselerotics. The following references relate to derivatives
containing a substituted naphthyl or a substituted hexa-, octa-, or
decahydronaphthyl group attached to the 2-position of the 6-ethyl group:
W.F. Hoffmann *et al.*, Eur. Pat. Appl. EP 245,990, 1987; Merck and Co.,
Inc., Jpn. Kokai Tokkyo Koho JP 63 203,675 [88 203,675], 1988; G.D.
Hartman and R.L. Smith, Eur. Pat. Appl. EP 306,264, 1989; A.E. DeCamp,
T.R. Verhoeven, and I. Shinkai, *ibid.*, 306,210, 1989; T.J. Lee, G.E.
Stokker, and R.L. Smith, *ibid.*, 306,263, 1989; R.A. DiPietro, J.I. Tu, and
N.Z. Turabi, U.S. US 4,857,522, 1989; W.F. Hoffman, R.L. Smith, and T.J.
Lee, Eur. Pat. Appl. EP 323,867, 1989; A. Endo *et al.*, PCT Int. Appl. WO
89 08,094, 1989; W.F. Hoffman and T.J. Lee, Eur. Pat. Appl. EP 325,817,
1989; U.S. US 4,855,456, 1989; Eur. Pat. Appl. EP 331,240, 1989; A.S.
Thompson, T.R. Verhoeven, and I. Shinkai, U.S. US 4,866,186, 1989; M.
Duggan and G.D. Hartman, Eur. Pat. Appl. EP 323,866, 1989; T.J. Lee and
W.F. Hoffmann, *ibid.*, 331,239, 1989; L.R. Treiber, *ibid.*, 351,918, 1990;
A. Goldstein, J.A. McLane, and M.R. Uskokovic, *ibid.*, 369,288, 1990; E.S.
Inamine *et al.*, U.S. US 4,940,727, 1990; D.S. Karanewsky, Eur. Pat. Appl.
EP. 419,856, 1991; M.E. Duggan, W. Halczenko, and G.D. Hartman, *ibid.*,
415,488, 1991; M. Matsumoto *et al.*, Jpn. Kokai Tokkyo Koho JP 03
184,970 [91 1184,970], 1991; and A. Sato, A. Terahara, and Y. Tsujita,
ibid., 04 217,971 [92 217,971], 1992.

The following refer to 6-(2-substituted ethyl and 2-substituted ethenyl)-4-
-hydroxytetrahydropyran-2-ones and the 2-substituents attached to the ethyl
or ethenyl group are:- substituted phenyl group, T.J. Commons, R.E.
Mewshaw, and D.P. Strike, U.S. US 4,812,583, 1989; B. Dreckmann, R.

Heck, and J. Pill, Eur. Pat. Appl. EP 344,602, 1989; R. Heck, B. Dreckmann, and J. Pill, Ger. Offen. DE 3,821,538, 1990; G.B. Reddy *et al.*, J. Org. Chem., 1991, <u>56</u>, 5752; T. Hiyama *et al.*, Eur. Pat. Appl. 475,627, 1992; B. Henkel, A. Kunath, and H. Schick, Tetrahedron Asymmetery, 1993, <u>4</u>, 153: substituted biphenyl groups, R.L. Smith and G.E. Stokker, Eur. Pat. Appl. EP 232,997, 1987: substituted naphthyl and reduced naphthyl groups, J.D. Prugh *et al.*, J. Med. Chem., 1990, <u>33</u>, 758; C.N. Lewis *et al.*, PCT Int. Appl. WO 91 00,280, 1991; J.N. Barton U.S. US 5,114,964, 1992: substituted and cycloalkyl groups, B.D. Roth, *ibid.*, 4,906,657, 1990; J.R. Regan, J.G. Bruno, and K.W. Neuenschwander, *ibid.*, 4,939,143, 1990; J.R. Regan *et al.*, Eur. J. Med. Chem., 1992, <u>27</u>, 735: substituted 1*H*-pyrrol-1-yl groups, B.D. Roth and D.R. Sliskovic, U.S. US 4,735,958, 1988; D.E. Butler *et al.*, PCT Int. Appl. WO 89 07,598, 1989; B.D. Roth *et al.*, J. Med. Chem., 1990, <u>33</u>, 21; 1991, <u>34</u>, 357; D.E. Butler, T.V. Le, and T.N. Nannings, U.S. US 5,298,627, 1994: and substituted pyridinyl and pyrimidinyl groups, G. Wess *et al.*, Eur. Pat. Appl. EP 319,847, 1989; A.W. Chucholowski, B.D. Roth, and D.R. Sliskovic, U.S. US 4,868,185, 1989; and Y. Saito *et al.*, Eur. Pat. Appl. EP 535,548, 1993.

The substituent in the following cases is a cyclic alkyl system with an aryl substituent attached as the end group to 6-(2-substituted ethyl and ethenyl, and 3-substituted propyl)-4-hydroxytetrahydropyran-2-ones. In some instances there are also other substituents at various positions on the ethyl, ethenyl, and propyl groups: K.W. Neuenschwander *et al.*, PCT Int. Appl. WO 89 05,639, 1989; J.R. Regan and K.W. Neuenschwander U.S. US 4,900,754, 1990; and K.W. Neuenschwander and A.C. Scotese, *ibid.*, 4,892,884, 1990. Compounds containing groups similar to those quoted above, but possessing a 6-(2-substituted acetylenic) group have been reported (J.W. Ullrich and J.R. Ragan, PCT Int. Appl. WO 92 14,692, 1992).

5-Hydroxy-4-methyl-, butyl-, and benzyltetrahydropyran-2-ones 26 (R^1 = Me, Bu, PhCH$_2$) have been obtained by the ring expansion of the 4--substituted 5-(halogenomethyl)tetrahydrofuran-2-ones 27 (R^2 = iodo, Br) with (Bu$_3$Sn)$_2$O. Treatment of tetrahydrofuran-2-one 27 (R^1 = H, R_2 = Br) yielded 5-hydroxytetrahydropyran-2-one (26; R^1 = H) (I. Shibata *et al.*, J. Org. Chem., 1991, <u>56</u>, 475). For the preparations of 3-substituted (alkyl or CO$_2$H) 6-alkyl-5-hydroxytetrahydropyran-2-ones and 6-(cyclohexylmethyl-3--*iso*propyl-5-(*tert*-butyldimethylsiloxy)tetrahydropyran-2-ones (28 and 29), see M. Shiosaki, Jpn. Kokai Tokkyo Koho JP 02 149,574 [90 149,574], 1990).

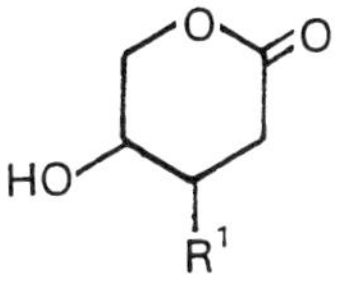

(26)

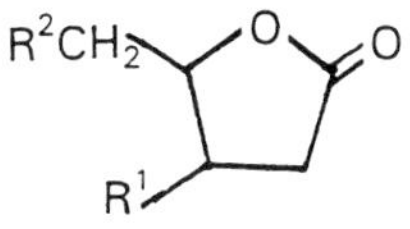

(27)

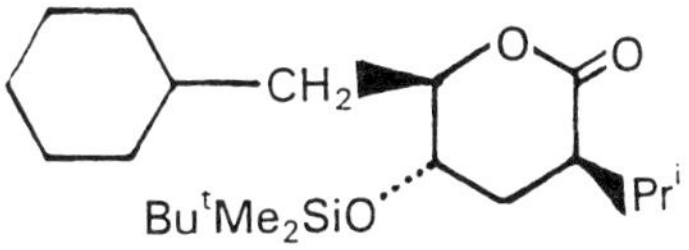

(28)

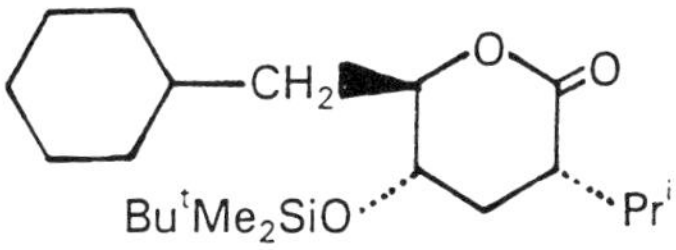

(29)

Oxidation by V_2O_5, of 4-(4-chlorobenzyl)-6,7-dimethoxyisoquinoline methiodide (30) in aqueous H_2SO_4 affords a mixture of 4-$ClC_6H_4COCH_2OH$, 4-$ClC_6H_4CH_2CR^1R^2CO_2H$ ($R^1 = R^2 = H$; $R^1R^2 = O$), 4-$ClC_6H_4C(CH_2NHMe) = CH$-CO_2H, and the 5-arylmethylene-2-oxo-6-hydroxytetrahydropyran-3-acetic acid and its methyl ester 31 (R = H, Me, respectively) (E. Postaire et al., Talanta, 1987, 34, 799). The preparations of a number of 4,4-disubstiuted 6-benzyloxytetrahydropyran-2-ones 32 (R^1 = Me, Et, Pr^i, Ph, vinyl; R^2 = H, Me, Et, Ph) have been reported (S. Takano et al., Chem. Letters, 1990, 1177).

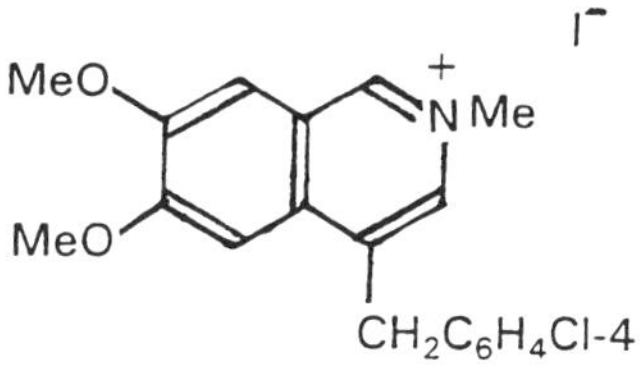

(30)

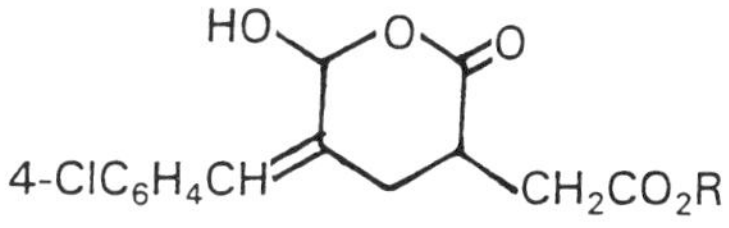

(31)

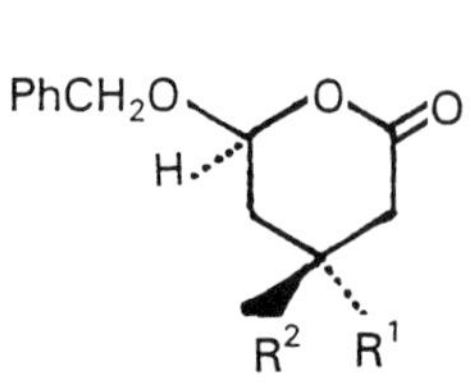

(32) 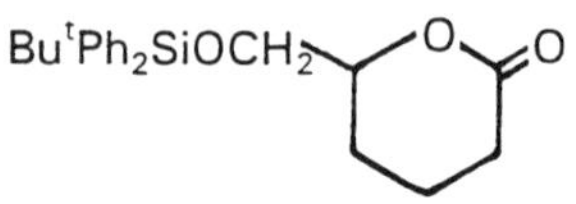 (33) 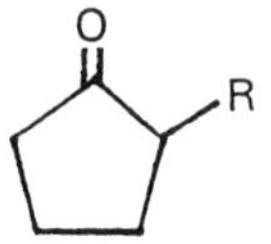 (34)

6-(*tert*-Butyldiphenylsiloxy)methyltetrahydropyran-2-one (33) has been
obtained in five steps from ethyl 2-oxocyclopentanecarboxylate (34;
R = CO$_2$Et). The key step is the Bayer-Villiger oxidation of derivative 34
(R = Bu^{t}Ph$_2$SiOCH$_2$) with 3-ClC$_6$H$_4$CO$_3$H (R.J.K. Taylor, K. Wiggins, and D.H.
Robinson, Synthesis, 1990, 589). For the preparation of 6-benzyloxy-4-
-methyltetrahydropyran-2-one (35), see S. Takano and K. Ogasawara, Jpn.
Kokai Tokkyo Koho JP 04 49,287 [92 49,287], 1992; and 6-arylthiomethyl-
tetrahydropyran-2-ones 36 (R = Pr, Ph, 4-MeC$_6$H$_4$), Z.K.M. Abd El Samii, M.I.
Al Ashmawy, and J.M. Mellor, J. Chem. Soc., Perkin 1, 1988, 2523.

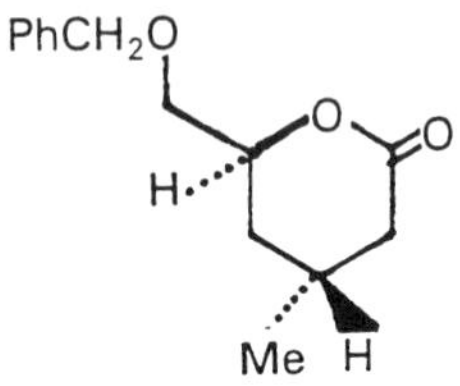

(35) 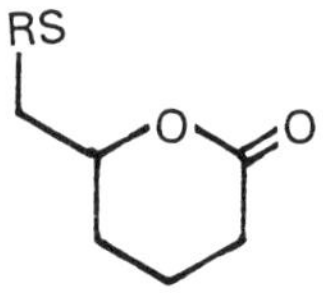 (36)

Tetrahydropyran-2-one carboxylic acids and related derivatives. The
[2,3] sigmatropic rearrangement of the cyclic propargylsulphonium ylide
generated from ethyl 5-phenylthiopent-3-ynyl diazomalonate furnishes the
novel allenic lactone, ethyl 2-oxo-3-phenylthiotetrahydropyran-4-vinylidene-
tetrahydropyran-3-carboxylate (2-ethoxycarbonyl-2-phenylthio-3-vinylidene-5-
-pentanolide) (1) (F. Kido, T. Abiko, and M. Kato, Bull. Chem. Soc. Jpn.,
1992, 65, 2471).

(1)   (2)

The rhodium (II) acetate-catalyzed [2,3] sigmatropic rearrangement of the
diazomalonates of (Z)-5-phenylthiopent-3-en-1-ol derivatives affords the 6-
substituted ethyl 2-oxo-3-phenylthio-4-vinyltetrahydropyran-3-carboxylates 2
(R = Me, Bu, PhOCH$_2$), in a highly stereoselective manner (F. Kido *et al.*,
Tetrahedron, 1990, 46, 4887). The preparations of a number of 3,6-
disubstituted 3-(arylamido and arylthioamido)tetrahydropyran-2-ones 3
[R^1 = H, C$_{1-20}$alkyl, C$_{3-6}$cycloalkyl, (un)substituted Ph, 2-, 3-, 4-pyridyl;
R^2 = H, C$_{1-6}$alkyl, (un)substituted Ph, (CH$_2$)$_n$Ph; X = O, S; Ar = (un)-
substituted Ph] and related compounds, useful as therapeutic agents to
lower blood cholesterol levels have been reported (C.E. Augelli-Szafran, U.S.
US 5,185,349, 1993).

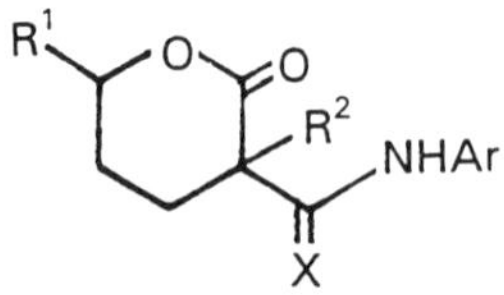

212

(3) 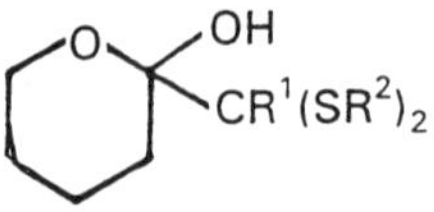 (4) 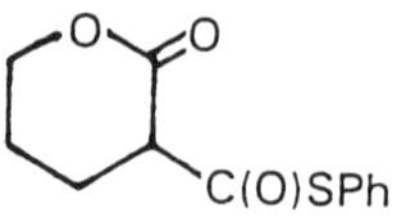 (5)

Tetrahydropyran-2-one reacts with tris(methylthio)methyllithium, $(MeS)_3$-
CLi to yield tetrahydropyran-2-ol derivative 4 (R^1 = H, R^2 = Me) and $HO(CH_2)_4$-
$COCH(SMe)_2$. When $(PhS)_3CLi$ is used 3-(phenylthiocarbonyl)tetrahydro-
pyran-2-one (5) is obtained. Lithiated 2-methylthio-1,3-dithione gives 4
[R^1 = SMe, R^2R^2 = $(CH_2)_3$], 4 [R^1 = H, R^2R^2 = $(CH_2)_3$], and $HO(CH_2)_4COCR^1$-
$(SR^2)_2$ [R^2R^2 = $(CH_2)_3$] (R^1 = SMe, H) (P. Yates, J.J. Krepinsky, and A. Seif-el-
Nasr, Chem. Comm., 1989, 177).

The sodium salt of 3-hydroxy-methylenetetrahydropyran-2-one (6)
(R = ONa), obtained by the Claisen condensation of tetrahydropyran-2-one
with ethyl formate, has been converted into its sulphonates and
carboxylates, for example, 6 (R = OSO_2Ph, OCOMe), obtained either as pure
E-isomers or as mixtures of *E*- and *Z*-isomers, separable by chromatography
(C. Mazal and J. Jonas, Coll. Czech. Chem. Comm., 1993, <u>58</u>, 1607).

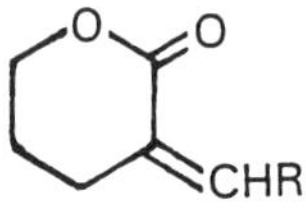

(6) 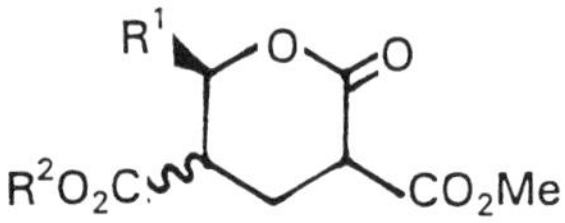 (7) 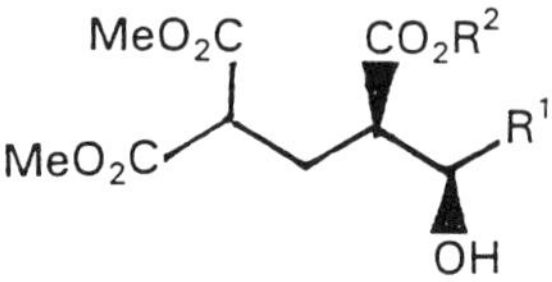 (8)

The Michael addition of $CH_2(CO_2Me)_2$ to $R^1CH(OH)C=(CH_2)CO_2R^2$ (R^1 = Ph, 3-pyridyl, 2-furyl, 1-naphthyl, 4-$NO_2C_6H_4$, R^2 = Me, CH_2Ph) in the presence of NaH in THF gives a mixture of 6-substituted tetrahydropyran--3,5-dicarboxylic acid esters 7, whereas the Michael reaction in the presence of KF/18-crown-6 in MeCN affords adducts 8 as major products (R.M. Lawrence and P. Perlmutter, Chem. Letters, 1992, 305). The manufacture of (2R, 3S, 4R, 5R)-4-acetylamino-2,5-dimethyl-6-oxotetrahydropyran-3-carboxylic acid (9), useful as an intermediate for 1ß-methylthienamycins has been reported (T. Honda *et al.*, Jpn. Kokai Tokkyo Koho JP 03 246,290 [91 256,290], 1991).

(9)

(10)

Polyfunctional tetrahydropyran-5-carboxylic acid esters have been prepared conveniently by successive additions of a-alkylacrylates and aldehydes to Grignard reagents, for example, the reaction of EtMgBr with $CH_2=CEtCO_2Me$, followed by Stobbe condensation with EtCHO affords methyl 3,5,6-triethyl-3-propyl-2-oxotetrahydropyran-5-carboxylate (10) in 2.3:1 cis/trans ratio (K. Yamaguchi, Y. Kurasawa, and K. Yokota, Chem. Letters, 1990, 719).

The ring-opening aldol-type reaction of the 2-methoxy-2-(trimethylsiloxy)-cyclobutanecarboxylates 11 (R^1 = H, Me) with aldehydes R^2CHO (R^2 = $PhCH_2CH_2$, nonyl, Me_2CHCH_2, Pr^i, cyclohexyl, Bu^t, Ph) proceeds in the presence of a Lewis acid, optimally Ti(OPr^i)Cl$_3$. Treatment of the crude products with catalytic 4-TsOH yields the 6-substituted methyl 3-methyl- and 3,3-dimethyl-2-oxotetrahydropyran-5-carboxylates 12. Similar reactions with unsymmetrical ketones have been reported (S. Shimada *et al.*, *ibid.*, 1993, 1117).

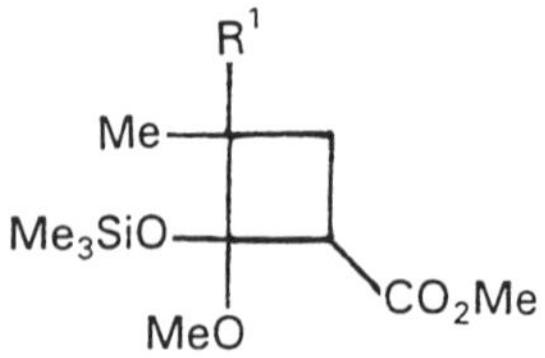

(11)

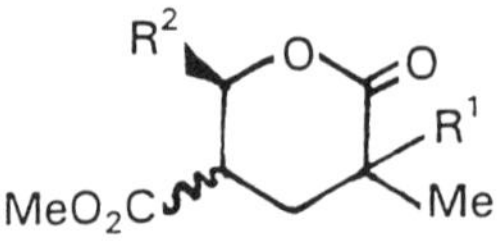

(12)

The (4*S*)-*iso*propyl-1,3-thiazolidine-2-thione diamides of 3-substituted glutaric acids, for example, $MeCH(CH_2CO_2H)_2$, have been submitted to enolization with $Sn(CF_3SO_3)_2$ and *N*-ethylpiperidine. The resultant tin(II) enolate on treatment with RCHO (R = Ph, Me, Pr^i, *sec*-Bu, *n*-hexyl) yields the aldol; which readily undergoes basic lactonization furnishing the corresponding tetrahydropyran-2-one, for example, 13, possessing three contiguous asymmetric centres (Y. Nagao *et al.*, *ibid.*, 1992, 335).

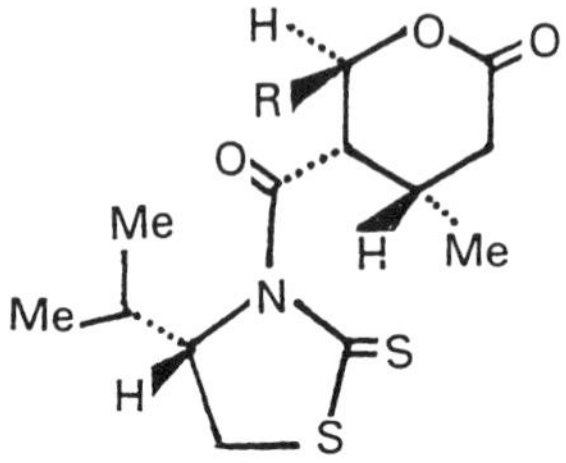

(13)

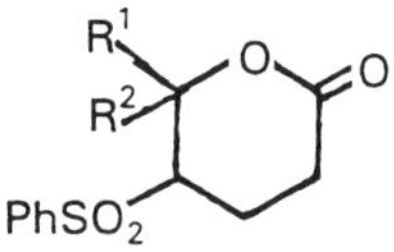

(14)

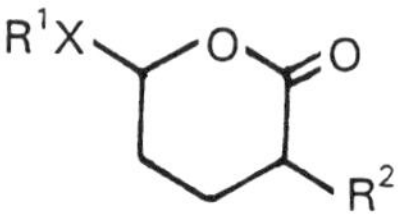

(15)

6,6-Disubstituted 5-sulphonyltetrahydropyran-2-ones 14 [R^1 = Bu, Ph, PhCH = CH, Me(CH$_2$)$_{10}$, R^2 = H; R^1 = R^2 = Me, Et; R^1R^2 = (CH$_2$)$_5$; R^1 = Ph, R^2 = Me] are obtained by the stepwise treatment of ketones and aldehydes R^1COR^2 with the dianion of 4-phenylsulphonylbutanoic acid, PhSO$_2$(CH$_2$)$_3$-CO$_2$H and (CF$_3$CO)$_2$O (C.M. Thompson, Tetrahedron Letters, 1987, _28_, 4243). For the preparations of some optically active derivatives of 2-oxo-tetrahydropyran-3- and -6-carboxylic acids 15 (R^1 = C$_{1-16}$linear or branched alkyl, alkoxy, OH, CH$_2$OH, CO$_2$H; R^2 = CO$_2$H, CH$_2$OH; X = direct bond, arylene) and 2-oxotetrahydropyran-3,6-dicarboxylic acid (15; R^1 = R^2 = CO$_2$H; X = direct bond), see K. Sakashita and T. Ikemoto, Jpn. Kokai Tokkyo Koho JP 03 27,374 [91 27,374], 1991.

Halogeno and aminotetrahydropyran-2-ones. Electrochemical reduction of glutaryl dichloride at a mercury electrode in MeCN containing 0.1M tetra-ethylammonium perchlorate results in the formation of 6-chlorotetrahydro-pyran-2-one (5-chlorovalerolactone) (1) and polymeric material (G.A. Urove and D.G. Peters, Tetrahedron Letters, 1993, _34_, 1271). The reaction of HO$_2$CCH$_2$(CH$_2$)$_2$COC(= PPh$_3$)CO$_2$Et with bromine/Et$_3$N results in a novel bromo lactonization to give a mixture of (_E_)- and (_Z_)-bromo enol lactones, 6--[bromo(ethoxycarbonyl)methylene]tetrahydropyran-2-ones (2) (A.D. Abell and J.O. Trent, Chem. Comm., 1989, 409). 3-Benzyl-6,6-difluorotetra-hydropyran-2-one (2-benzyl-5,5-difluoro-δ-valerolactone) (3) has been prepared by treating F$_2$C = CHCH$_2$CH(CH$_2$Ph)CO$_2$H with Hg(OAc)$_2$ followed by reaction with Bu$_3$SnH (D.A. Kendrick and M. Kolb, J. Fluorine Chem., 1989, _45_, 273).

| (1) | (2) | (3) |

tert-Butyl _N_-(2-oxo-5,6-dihydro-2_H_-pyran-5-yl)carbamate (4) prepared in 5 steps from γ-ethyl L-glutamate, reacts with R$_2$CuLi (R = Bu, Me) to give only _tert_-butyl _N_-(4-butyl- and 4-methyl-2-oxotetrahydropyran-5-yl)-carbamates, respectively (H. Yoda _et al._, Chem. Express, 1989, _4_, 585).

216

(4)

(5)

Tetrahydropyran-3-ones. Addition of pyridine dropwise to a solution of 6-hydroxy-3,6-dihydro-2H-pyran-3-one (2; $R^1 = R^2 = H$) in Ac_2O at 0°C gives derivative 2 ($R^1 = H$, $R^2 = Ac$), which on treatment with Pd/C under hydrogen in H_2O/THF yields tetrahydropyran-3-one (1; $R^1 = H$). The preparation of a number of 2-substituted tetrahydropyran-3-ones 1 ($R^1 = C_{1-10}$linear or branched alkyl, C_{3-6}cycloalkyl, Ph, biphenylyl, CH_2Ph, acyloxymethyl, CH_2OH), useful as intermediates for perfumes, drugs, and agrochemicals have been reported (T. Numata, M. Hatanaka, and J. Watanabe, Jpn. Kokai Tokkyo Koho JP 63 170,372 [88 170,372], 1988). For the preparation of several 2,2-disubstituted tetrahydropyran-3-ones see T. Satoh *et al.*, Bull. Chem. Soc. Jpn., 1988, <u>61</u>, 2109.

(1)

(2)

Lithiation of the methoxy enol ether 3 (R=H) provides the ß-lithiated enol ether 3 (R=Li) and although this derivative does serve as an equivalent of the tetrahydropyran-3-one enolate 4 (R=O⁻), and can be alkylated to give 2--alkyl derivatives 3 [R=(CH$_2$)$_3$OR^1, CHPhOH; R^1=CH$_2$Ph, tetrahydropyran-2--yl] is controlled by the endo enol ether function. The isomerically pure silyl enol ether 4 (R=OSiMe$_2$Bu^t) has also been prepared and this reagent represents a more general solution to the synthetic problems associated with access to the enolate 4 (R=O⁻) and reacts with a range of electrophiles in Lewis acid-mediated processes (P. Cox, S. Lister, and T. Gallagher, J. Chem. Soc., Perkin 1, 1990, 3151).

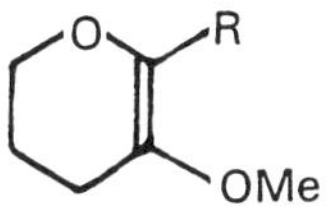

(3) 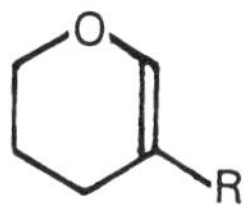   (4)

Treatment of   5,5-dimethyltetrahydropyran-3-one (5) with LiN(Pr^i)$_2$ in MeOCH=CHOMe at -78°C, followed by quenching with Ac$_2$O yields enol acetates 6 and 7 in a 5.7:1 ratio but treatment with HClO$_4$ in Ac$_2$O-CCl$_4$ gives only 6 (D.J. Goldsmith, C.M. Dickinson, and A.J. Lewis, Heterocycles, 1987, <u>25</u>, 291).

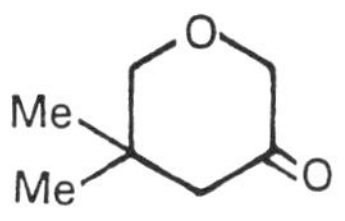

(5) 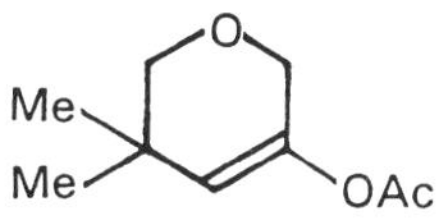   (6) 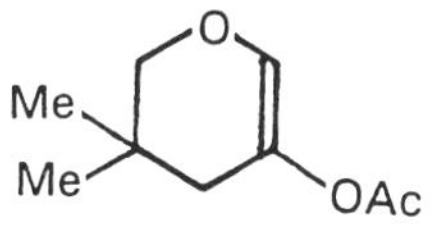   (7)

Addition of chiral imine 8, obtained from 2-methyltetrahydropyran-3-one and (R)-(+)-1-phenylethylamine, to methyl acrylate leads to the (S)-keto ester, methyl (S)-2-methyltetrahydropyran-2-propionate (9) (D. Desmaele, G. Pain, and J. D'Angelo, Tetrahedron Asymmetry, 1992, 3, 863).

(8)

(9)

(10) 2-Ph (α)
(11) 2-Ph (β)

The *a-O*-allyl-*a'*-diazopropanone derivative, $CH_2 = CHCH_2OCHPhCOCHN_2$ on treatment with cupric acetylacetonate in CH_2Cl_2 at room temperature affords a mixture of 4-phenyl-3-oxabicyclo[4.1.0]heptan-5-ones (10 and 11) in the ratio 93:7 (Y.S. Hon and R.C. Chang, Heterocycles, 1991, 32, 1089).

6-Methoxytetrahydropyran-3-one (12) has been treated with $PhNHNH_2$ to afford annelated pyran derivatives. Regioselective reactions of pyranone 12 are successful after conversion to the respective enamine, silyl enol ether, or lithio enamine. Cycloaddition, cyclocondensations, and electrophilic aldol reactions have also been investigated [F. Eiden and B. Wuensch, Arch. Pharm. (Weinheim, Ger.), 1990, 323, 393].

(12)

(13)

(14)   (15)

The Wadsworth-Emmons reactions of 6-methoxytetrahydropyran-3-one
(12) and 5-arylthio-6-methoxytetrahydropyran-3-ones 13 (R^1 = H, 4-Cl, 4-Me,
2-Pr^i, R^2 = H; R^1 = 2-Cl, R^2 = Cl) have been investigated and the ratio of (E)-
and (Z)-α,ß-unsaturated esters 14 (R^3 = Me, Et) and 15 formed shown to be
affected by the presence and type of arylthio substituent (P.L. Lopez
Tudanca, K. Jones, and P. Brownbridge, J. Chem. Soc., Perkin 1, 1992,
533).

(16)

Allenyl aldehydes and ketones are oxidatively cyclized by dimethyl-
dioxirane (DDO) to provide cyclic acetals and hemiacetals, for example,
oxidation of $Me_2C = C = CHCMe_2CHO$ with rigorously dried DDO in MeOH/
anhydrous K_2CO_3 affords $trans$-4-hydroxy-6-methoxy-2,2,5,5-tetramethyl-
tetrahydropyran-3-one (16) (J.K. Crandall and E. Rambo, Tetrahedron
Letters, 1994, <u>35</u>, 1489).

Tetrahydropyran-4-ones. Olefination of **tetrahydropyran-4-one** (1) with $(EtO)_2P(O)CH_2CO_2Et$ gives 4-(ethoxycarbonyl)methylenetetrahydropyran (2). Addition of $4\text{-}MeOC_6H_4CH_2SH$ to derivative 2 affords 4-(4-methoxy)benzyl-thiotetrahydropyran-4-acetic acid (3) (B. Lammek, I. Derdowska, and P. Rekowski, Pol. J. Chem., 1990, **64**, 351).

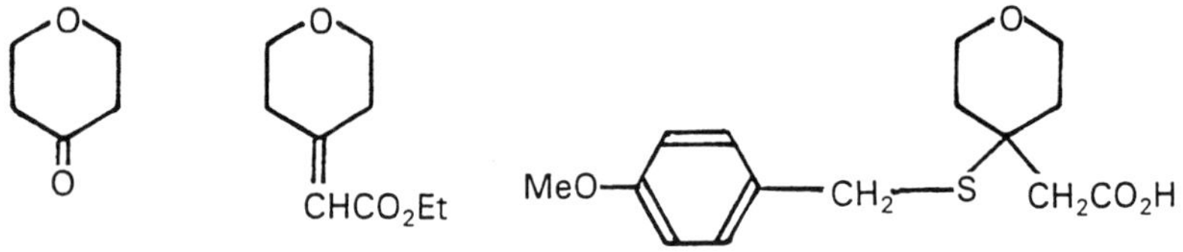

(1)   (2)   (3)

Condensation of 2,2-dimethyltetrahydropyran-4-one (1) with benzaldehyde gives 2,2-dimethyl-5-benzylidenetetrahydropyran-4-one (4; R = H) (N.S. Arutyunyan *et al.*, Arm. Khim. Zh., 1987, **40**, 570). For the preparations of 2,2-dimethyl-5-arylidenetetrahydropyran-4-one 4 [R = 4-OH-3-
-OMe, 4-OMe, 4-OPr^i, 3,4-(OMe)$_2$, 4-OH], see G.I. Manucharyan *et al.*, *ibid.*, 1988, **41**, 487).

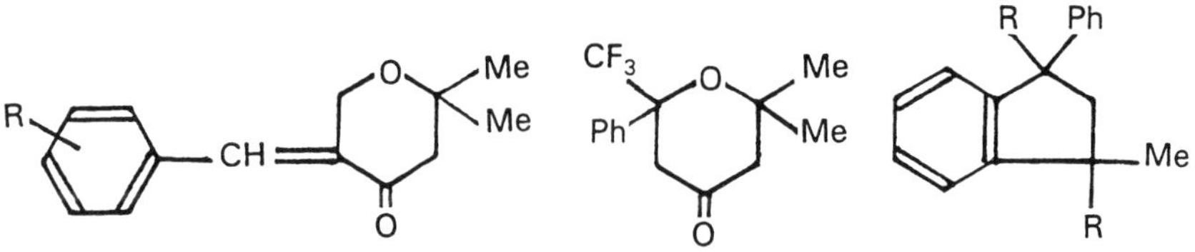

(4)   (5)   (6)

Me$_2$C=CHCOCH$_2$C(OH)PhR (R=CF$_3$) cyclizes in CF$_3$CO$_2$H to afford 2,2-
-dimethyl-6-phenyl-6-trifluoromethyltetrahydropyran (5). Other starting
materials (e.g. when R=Me, Ph) decompose in CF$_3$CO$_2$H to give
Me$_2$C=CHCO$_2$H and indanes 6 (V. Ya. Sosnovskikh and I.S. Ovsyannikov,
Izv. Vyssh. Uchebn. Zaved., Khim. Khim. Takhnol, 1989, _32_, 121).

RCHO (R=Ph, 2-MeOC$_6$H$_4$, 2-ClC$_6$H$_4$, 4-MeC$_6$H$_4$, 4-MeOC$_6$H$_4$, 2,5-
-Me$_2$C$_6$H$_3$) cyclocondenses with (HO$_2$CCH$_2$)$_2$CO in the presence of HCl to
give stereoselectively _cis_-2,6-diaryltetrahydropyran-4-ones 7. _trans_-2,6-
-Diaryltetrahydropyran-4-ones 8 (R=Ph, 2-MeOC$_6$H$_4$, 2-ClC$_6$H$_4$) have been
isolated as minor products (T.M. Fakler and K.D. Berlin, Org. Prep. Proced.
Int., 1989, _21_, 327).

(7)

(8)

A variety of ß-silyloxy silyl enol ethers, for example, Me$_3$SiOCH$_2$CH$_2$C-
(=CH$_2$)OSiMe$_3$, is easily produced by double deprotonation of readily
available ß-hydroxyketones. These substances undergo cyclization reactions
with dioxenium cations to yield 2-alkoxytetrahydropyran-4-ones, for example
2-ethoxytetrahydropyran-4-one (9) (V.A. Martin, F. Perron, and K.F. Albizati,
Tetrahedron Letters, 1990, _31_, 301).

(9)

(10)

(11)

Cycloadditions of 1-methoxy-3-trimethylsilyloxybuta-1,3-diene with halogenated carbonyl compounds R^1COR^2 ($R^1 = H$, $R^2 = CCl_3$, CH_2Cl; $R^1 = Ph$, $R^2 = CF_3$, $CHCl_2$) afford the corresponding 6,6-disubstituted 2-methoxytetrahydropyran-4-ones 10 and 2,2-disubstituted 3,4-dihydro-2*H*-pyran-4-ones 11. Lewis acids are effective as catalysts (A. Sera *et al.*, Heterocycles, 1991, 32, 273).

A 2-morpholinopenta-1,3-diene derivative on treatment with RCHO ($R = Ph$, tolyl, ClC_6H_4, furyl, thienyl) and $MgBr_2.Et_2O$, followed by treatment of the mixture with aqueous AcOH affords 6-substituted 2-methoxymethyl--3-methyltetrahydropyran-4-ones 12. The diene and PhCHO yield, after treatment with NaOAc in aqueous AcOH, 2-methoxymethyl-3-methyl-4-(1--morpholino)-6-phenyl-5,6-dihydro-2*H*-pyran (13) (J. Barluenga *et al.*, Chem. Comm., 1988, 1247).

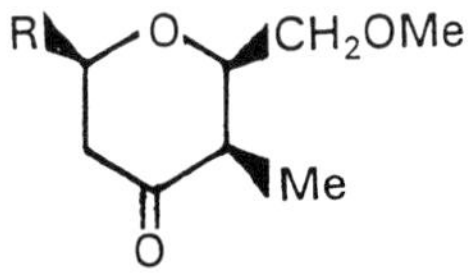

(12) 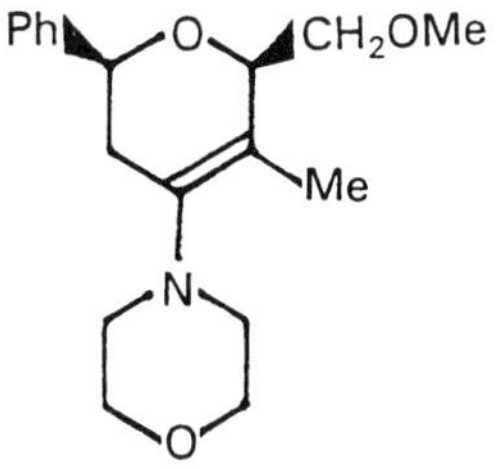 (13)

The hetero-Diels-Alder reaction of 2-trimethylsiloxybuta-1,3-diene, $CH_2 = C(OSiMe_3)CH = CH_2$, with *n*-butyl glyoxylate, BuO_2CCHO, yields a mixture of *n*-butyl 4-oxotetrahydropyran-2-carboxylate (14) and *n*-butyl 4--trimethylsiloxy-3,6-dihydro-2*H*-pyran (15). The dihydropyrancarboxylates are important byproducts in sugar synthesis (H. Melanson, D. Labrecque, and C. Jankowski, Rev. Latinoam. Quim., 1989, 20, 21).

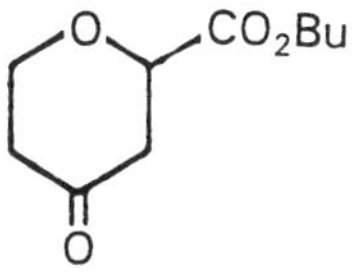

(14)

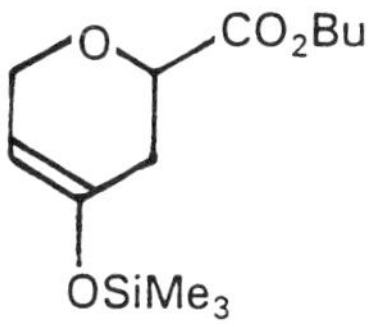

(15)

Tetrahydropyran-4-one on treatment with KCN and amines RH
(R = pyrrolidino, piperidino, 1,2,3,4-tetrahydroisoquinoline) in H_2O containing
HCl affords the 4-cyanotetrahydropyran derivatives 16, which react with
Grignard reagents R^1MgBr (R^1 = Ph, 3- and 4-anisyl, 2-thienyl) to give
oxaphencyclidines, possessing a better analgesic activity/excitability
relationship than phencyclidine. The same observations applied to similar
derivatives 17 obtained from tetrahydropyran-3-one. Related compounds
have been obtained from tetrahydrothiopyran-4-one [F. Eiden, M. Schmidt,
and H. Buchborn, Arch. Pharm. (Weinheim, Ger.), 1987, 320, 348].

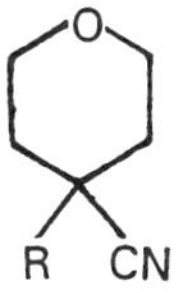

(16)

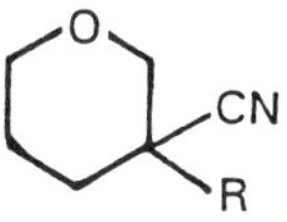

(17)

Treatment of 2,2-dimethyltetrahydropyran-4-one (18; X = O) with R^1NH_2
[R^1 = $Et_2N(CH_2)_3$] yields 18 (X = NR^1), which on acylation with R^2COCl
(R^2 = Et, Ph, 4-ClC_6H_4, 2-BrC_6H_4) affords mixtures containing equimolar
amounts of amides 19 and 20 (B.U. Minbaev, G.S. Bekenova, and S.M.
Shostakovskii, Vestn. Akad. Nauk Kaz. SSR, 1989, 46).

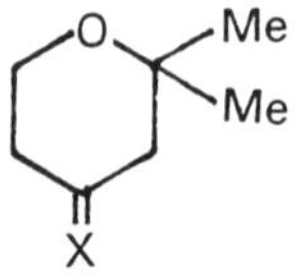

224

(18) 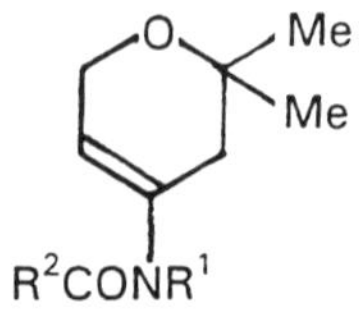   (19) 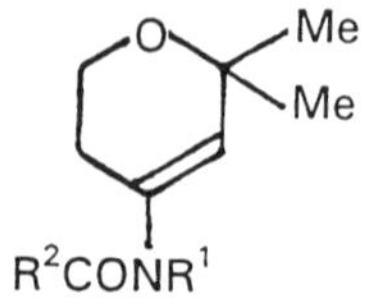   (20)

A mixture of ClCOCH:CRMe (R = H, Me) and 1-morpholinocyclopentane
in CHCl$_3$ on treatment with Et$_3$N and then HCl gives 2-methyl- and 2,2-
-dimethyl-3,4-dihydro-2*H*-cyclopenta[*b*]pyran-4-one (21), respectively, which
on hydrogenation over Raney Ni yield the isomeric cyclopentatetrahydro-
pyranols 22 and 23. Jones oxidation of pyranols 22 and 23 affords 2-
-methyl- and 2,2-dimethylcyclopenta[b]tetrahydropyran-4-one (24),
respectively (R. Dolmazon, Bull. Soc. Chim. Fr., 1987, 873).

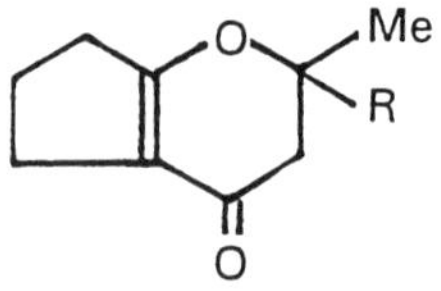

(21) 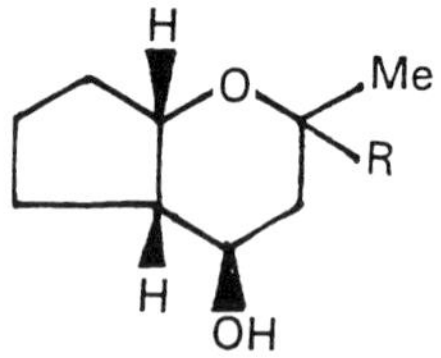   (22)

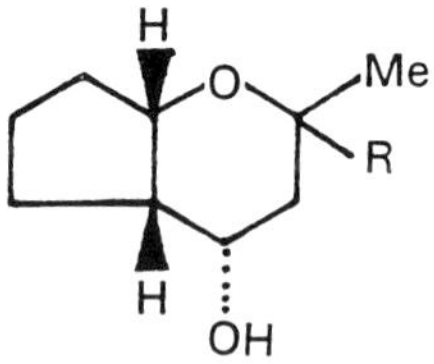

(23) 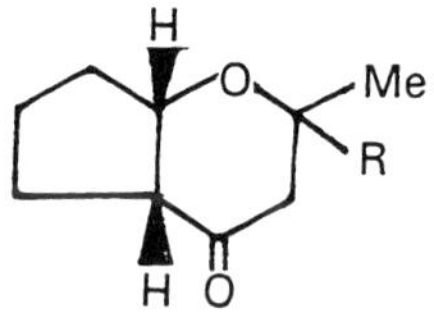   (24)

Tetrahydropyran-2,4-, -2,5-, -2,6-, and -3,5-diones. Tetrahydropyran-
-2,4-dione (ß-keto-δ-valerolactone) (1) has been prepared in 5 steps from
$HC \equiv CCH_2CH_2OH$. Thus, it was initially converted to a ketal derivative,
which upon sequential condensation with $ClCO_2CH_2Ph$ and hydrolysis gave
$PhCH_2O_2C \equiv CCH_2CH_2OH$. Benzyloxymercuration of the latter yielded the
dihydropyran-2-one 2, which underwent hydrogenolysis to give (1) (J.
D'Angelo and D. Gomez-Pardo, Tetrahedron Ledtters, 1991, _32_, 3063).

(1) (2) (3)

Bromozincates, $BrZnOCHR^2CBrR^1COCMe_2CO_2Et$, have been transformed
to $R^2CH = Cr^1COCMe_2CO_2Et$ ($R^1 = H$, $R^2 = Pr$, Ph, $4\text{-}ClC_6H_4$) and 5,6-
disubstituted 5-bromo-3,3-dimethyltetrahydropyran-2,4-diones 3 ($R^1 = H$, Me,
$4\text{-}ClC_6H_4$, $R^2 = CCl_3$, $4\text{-}FC_6H_4$, $4\text{-}ClC_6H_4$) (V.V. Shchepin *et al.*, Zh. Org.
Khim., 1992, _28_, 2033). 5-Bromo-5-ethyl-3,3-dimethyl-6-(4-chlorophenyl)-
tetrahydropyran-2,4-dione, 5-ethyl-3,3,5-trimethyl-6-(4-chlorophenyl)tetra-
hydropyran-2,4-dione, 5-*iso*propyl-3,3-dimethyl-6-(4-chlorophenyl)tetra-
hydropyran-2,4-dione (V.V. Shchepin V.V. and G.E. Gladkova, *ibid.*, 1993,
_29_, 474) and a number of 3-arylcarbonyltetrahydropyran-2,4-ones with upto
four substituents (C.G. Knudsen *et al.*, Eur. Pat. Appl. EP 249,812, 1987)
have been prepared.

Hypervalent iodine oxidation of diketo acid $PrCOCH_2COCH_2CH_2CO_2H$
using [hydroxy(tosyloxy)iodo]benzene in boiling CH_2Cl_2 gives 6-propionyl-
tetrahydropyran-2,5-dione (4) (R.M. Moriarty *et al.*, Tetrahedron Letters,
1990, _31_, 201).

(4)

(5)

Tetrahydropyran-2,6-dione (glutaric anhydride) (5) has been obtained by reacting glutaric acid with the catalyst montmorillonite KSF in the presence of *iso*propenyl acetate under microwave irradiation (D. Vilemin, B. Labiad, and A. Loupy, Synth. Comm., 1993, $\underline{23}$, 419). The 2,6-dione 5 and derivatives have been prepared by carbonylation of the appropriate (un)substituted alkenoic acid using various catalysts (E. Drent, Eur. Pat. Appl. EP 293,053, 1988). The reduction of tetrahydropyran-2,6-dione (5) with an alcohol (e.g. propan-2-ol) in the presence of H_2O-containing Zr oxide as a catalyst affords tetrahydropyran-2-one (K. Takahashi, M. Shibagaki, and H. Matsushita, Jpn. Kokai Tokkyo Koho JP 04 154,776 [92 154,776], 1992).

In organic solvents, lipase Amano P catalyzes the asymmetric ring opening of 4-substituted tetrahydropyran-2,6-ones 6 (R = CH_2Cl, CH_2F, OMe) with MeOH and BuOH to yield the (3R)-$HO_2CCH_2CHRCH_2CO_2R^1$ (R^1 = Me, Bu) in 70-86% enantiomeric excess. The esters have been converted into the 4-substituted tetrahydropyran-2-ones 7. The R-configuration has been assigned to the starting material and pyran-2-ones 7 by means of CD analytical and chemical correlation (Y. Yamamoto *et al.*, Agric. Biol. Chem., 1990, $\underline{54}$, 3269). Similar catalyzed ring opening of 3,3-disubstituted tetrahydropyran-2,6-pyrans 8 (R^1 = Me, Pr^i, Ph; R^2 = H, Me, or Et) in Pr^i_2O at 20° occurs preferentially at the less hindered carbonyl group to afford monoesters with high regioselectivity (J. Hiratake *et al.*, Tetrahedron Letters, 1989, $\underline{30}$, 1555).

(6) X = O
(7) X = H$_2$

(8)

The highly diastereoselective alcoholysis of 4-*iso*propyltetrahydropyran-
-2,6-dione (9) has been performed using 1-phenyl-3,3-bis(trifluoromethyl)-
propane-1,3-diol to give the chiral half acid ester 10 (Y. Suda *et al.*, Chem.
Letters, 1992, 389).

(9)

(10)

The Wittig reaction of Ph$_3$P = CMeCO$_2$R (R = Et, Bu^t) with tetrahydro-
pyran-2,6-diones 11 (R^1, R^3, R^4 = H, Me; R^2 = H, Me, Ph) has been
investigated (S. Tsuboi *et al.*, Bull. Chem. Soc. Jpn., 1987, <u>60</u>, 689). The
condensation of 2,6-bis(trimethylsiloxy)-4*H*-pyran with CH$_2$(OMe)$_2$ gives 3,5-
-di(methoxymethyl)tetrahydropyran-2,6-dione (12). A number of related
derivatives have been prepared (F. Mezger, G. Simchen, and P. Fischer,
Synthesis, 1991, 375).

(11) (12) (13)

A number of 4-aroyltetrahydropyran-3,5-diones 13 (R^1, R^4 = H, Me; R^2 = H, Me, MeS; R^3 = H, Me, Et; R^5 = Cl, O_2N; R^6 = H, Cl, MeO; R^7 = $MeSO_3$, $EtSO_3$, $PhSO_3$, Me_2CHSO_3) useful as acaricides and herbicides (R.J. Anderson *et al.*, Eur. Pat. Appl. EP 336,898, 1989) and (R^1-R^4 = H, C_{1-6}alkyl; R^5 = halogeno, C_{1-8}alkyl, alkoxy, NO_2; R^6,R^7 = any group defined by R^5, alkylsulphonyloxy) as rice paddy herbicides (T. Nishisaka and K. Komatsubara, Jpn. Kokai Tokkyo Koho JP 03 120,202 [91 120,202], 1991), and 4-(3-methoxy-4-methylsulphonyloxy-2-nitrobenzoyl)-2,2,6,6--tetramethyltetrahydropyran-3,5-dione (13; R^1-R^4 = Me, R^5 = O_2N, R^6 = OMe, R^7 = $MeSO_3$) and related derivatives (S.F. Les *et al.*, U.S. US 5,089,046, 1992) have been prepared.

3. Benzo[*b*]pyrans (1-benzopyrans, 5,6-benzopyrans, chromenes) and their derivatives

Benzo[*b*]pyran is used as the parent name for compounds in this section, but some trivial names still acceptable according to IUPAC rules are used when appropriate, for example, coumarin, chromone, and flavone.

(a) Benzo[b]pyrans and their oxidation products

(i) Alkyl- and arylbenzo[b]pyrans and flavenes
Substituted 2H-benzo[b]pyrans (3-chromenes, 2H- or α-chromenes).
Cyclization of phenol with HC≡CCR^{1}R^{2}OAc (R^1 = Me, R^2 = Ph, cyclopropyl, Et, 4-MeOC$_6$H$_4$, R^1R^2 = cyclohexylidine, adamantylidene) in 1,2-dichloro-benzene affords 2,2-disubstituted benzo[*b*]pyrans 1, for example, 2-cyclo-propyl-2-methyl-2*H*-benzo[*b*]pyran (1; R^1 = Me, R^2 = cyclopropyl). Similarly naphthopyrans can be obtained from 1- and 2-naphthols (H.A. Al Sehaibani, J. Indian Chem. Soc., 1993, 70, 168).

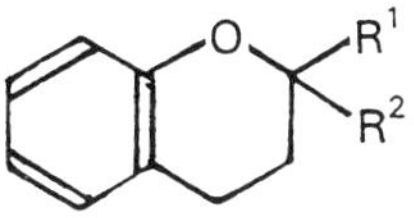

(1)

The cyclization of aryl allenic (buta-2,3-dien-1-yl) ethers, for example, 2 (R = H, Me, Cl, OMe, NO_2) to 6-substituted 4-methyl-2*H*-benzo[*b*]pyrans 3 and 6-substituted 4-methylene-3,4-dihydro-2*H*-benzo[*b*]pyrans 4 has been achieved by use of mercury (II) trifluoroacetate. Derivatives 4 can be isomerized to benzopyrans 3 by 4-toluenesulphonic acid in benzene (T. Balasubramanian and K.K. Balasubramanian, Tetrahedron Letters, 1991, <u>32</u>, 6641). The thermal Claisen rearrangement of aryl allenylmethyl ethers in diethylene glycol to yield 2-(2-hydroxyaryl)buta-1,3-dienes and cyclization products 4-methyl-2*H*-benzo[*b*]pyrans has been reported (*idem*, Chem. Comm., 1992, 1760).

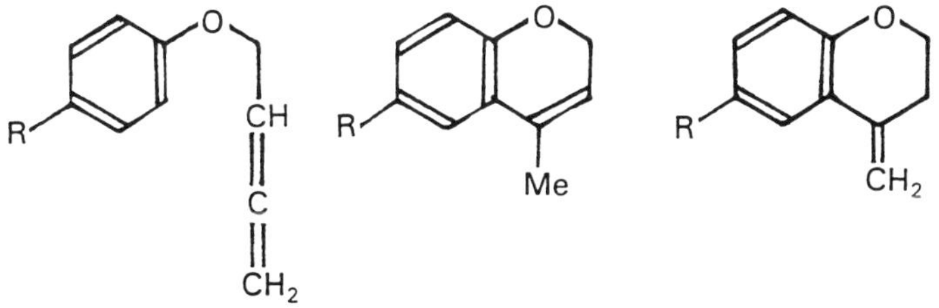

(2)   (3)   (4)

2,6,8-Trisubstituted 2-methyl-2*H*-benzo[*b*]pyrans 5 (R^1 = Me, CH$_2$CH$_2$-CH = CMe$_2$, CH$_2$CH$_2$CH = CMeCH$_2$CH$_2$CH = CMe$_2$; R^2 = CHO, OH; R^3 = H, OMe) are obtained by the Pd-catalyzed heteroannulation of iodophenols 6 with tertiary allylic alcohols CH$_2$ = CHC(OH)R^{1}Me (X. Garcias, P. Ballester, and J.M. Saa, Tetrahedron Letters, 1991, *32*, 7739). For the preparation of 1-(2,2-dimethyl-2*H*-benzo[*b*]pyran-6-yl)alkan-1-ols, see J. Kahovcova *et al.*, Czech. CS 250,585, 1988), and for the reduction of 2,2-dimethyl- and 2,2,4-trimethyl-2*H*-benzo[*b*]pyran with sodium or lithium in liquid ammonia, M. Aniol, P. Lusiak, and C. Wawrzenczyk, Heterocycles, 1994, *38*, 991).

(5)   (6)

2,2-Pentamethylenechroman-4-one on reduction with NaBH$_4$ in boiling Bu^{t}OH yields the corresponding alcohol, which on treatment with 4-MeC$_6$H$_4$-SO$_3$H affords the spiro compound 2,2-pentamethylene-2*H*-benzo[*b*]pyran (A. Ascaval, Rom. RO 100,570, 1991). Addition of halogens to the 2,2-spiro--2*H*-benzo[*b*]pyrans 7 (n = 2,3,4) gives the 3,4-dihalogeno derivatives 8 (R^1 = R^2 = Cl, Br). Hydrolysis of the dihalogeno derivatives furnishes the corresponding 3-halogeno-4-hydroxy derivatives 8 (R^1 = Cl, Br; R^2 = OH), converted on oxidation to the respective 3-halogeno-3,4-dihydro-2*H*-benzo-[*b*]pyran-4-ones (3-halogenochroman-4-ones) (H.A. Alsehaibani, Asian J. Chem., 1993, *5*, 416).

(7)   (8)

2-Ethoxy- and 2-benzyloxy-2*H*-benzo[*b*]pyrans (9; R = Et, PhCH$_2$) have been prepared by reduction of coumarin with diisobutylaluminium hydride and subsequent *O*-alkylation of the resulting lactols. Separation or enrichment of enantiomers has been achieved by liquid chromatography on triacetylcellulose and tribenzoylcellulose (L. Loncar *et al.*, Croat. Chem. Acta., 1933, <u>66</u>, 209).

(9) (10) (11)

The carbene 10 generated in methanol/sodium methoxide by photolysis of 2-methylchromone tosylhydrazone does not accept methanol at the carbenic site, but affords 2-methoxy-2-methyl-2*H*-benzo[*b*]pyran (11) (G. Hoemberger, W. Kirmse, and R. Lelgemann, Ber., 1991, <u>124</u>, 1867). Preparations of 3,6-disubstituted 2-methoxy-2*H*-benzo[*b*]pyrans 12 (R^1 = H, Me, OMe; R^2 = Me, Bu, Pr, Pr^i) (F.W. Ulrich and E. Breitmaier, Synthesis, 1987, 951), and 3,6,8-trisubstituted 2*H*-benzo[*b*]pyran-2-ols 13 (R^1 = Me, R^2 = R^3 = Br; R^1 = Me, R^2 = Br, R^3 = NO$_2$; R^1 = Et, R^2 = OMe, NO$_2$, R^3 = H; R^1 = Et, R^2 = NO$_2$, R^3 = Br; R^1 = Et, R^2 = Me, R^3 = NO$_2$) (Yu. M. Chunaev, N.M. Przhiyalgovskaya, and L.N. Kurkovskaya, Khim. Geterotsikl. Soedin., 1988, 460) have been reported.

(12) (13)

The cyclocondensation of trialkyl(4-halogeno-3-methylbut-2-enyl)-quinones 14 (R^1 = alkyl, R^2 = halogeno) has been used to obtain the 2,2,5,7,8-pentaalkyl-2*H*-benzo[*b*]pyran-6-ols 15 (R^1 = alkyl, R^3 = alkyl, Ph), for example, 2-[(4-nitrophenoxy)methyl]-2,5,7,8-tetramethyl-2*H*-benzo[*b*]-pyran-6-ol (15; R^1 = Me, R^3 = 4-$O_2NC_6H_4$) is an intermediate for the preparation of pharmaceuticals for the treatment of hyperlipidemia (D. Laffan, Eur. Pat. Appl. EP 543,346, 1993). For the preparation of 3-(*N*--hydroxyaminomethyl)-6-phenoxy-2*H*-benzo[*b*]pyran-HCl and related derivatives as 5-lipoxygenase inhibitors see, J.L. Stanton, Y. Satoh, and A.J. Hutchinson, *ibid.*, 392,510, 1989.

(14)   (15)

A two-step synthesis of 6-alkyl-4-hydroxymethyl-2*H*-benzo[*b*]pyrans 16 (R = Me, Et) *via* addition of phenyl vinyl sulphoxide carbanion to 5-alkyl-2--hydroxyacetophenone and the synthesis of 4-methyl-2*H*-benzo[*b*]pyran and 4,4-dimethylchroman have been described (G. Solladie and A. Girardin, Synthesis, 1991, 569).

The photoirradiation of 4-acetoxy-2*H*-benzo[*b*]pyran (17) yields 2-$AcOC_6$-$H_4COCH = CH_2$, chroman-4-one (18; R = H), chromone (19) 3-acetoxy- and 3-hydroxychroman-4-one (18; R = OAc, OH) and the cyclobutane dimer 20, with anti head to head structure (J. Climent *et al.*, Tetrahedron, 1987, <u>43</u>, 999).

(16)   (17)   (18)

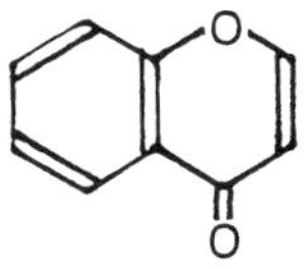

(19) 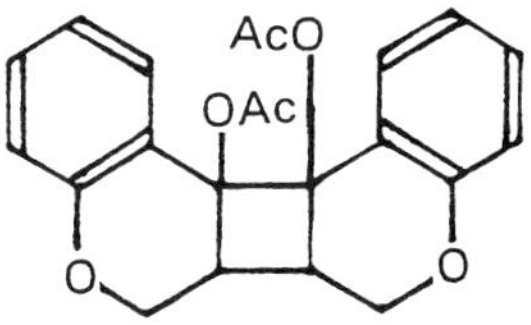   (20)

Treatment of 7,8-diacetoxy-2,2-dimethylchroman (21) with *N*-bromo-succinimide in the presence of benzoyl peroxide gives 7,8-diacetoxy-3,4--dibromo-2,2-dimethyl-2*H*-benzo[*b*]pyran (22) (D. Hu, D. Jin, and S. Chen, Youji Huaxue, 1992, 12, 384).

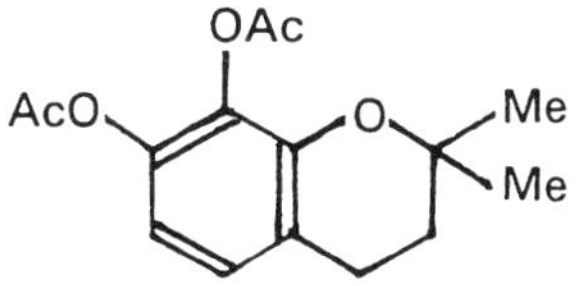

(21) 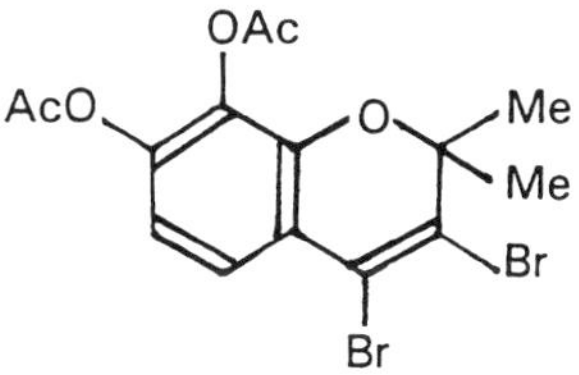   (22)

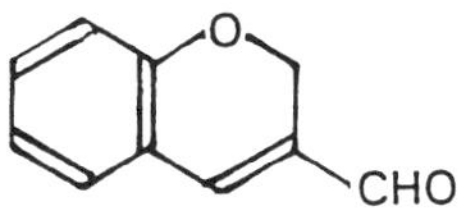

(23) 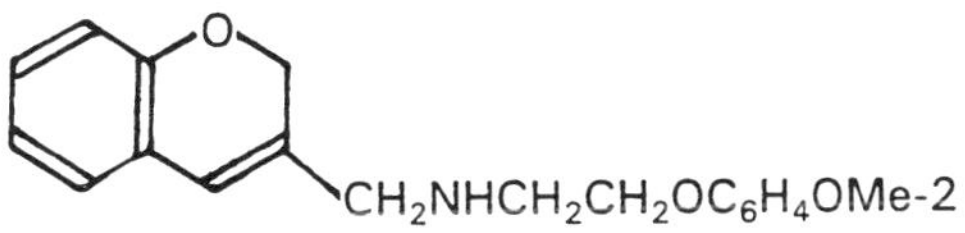   (24)

The addition of 2-MeOC$_6$H$_4$OCH$_2$CH$_2$NH$_2$ to a solution of 2*H*-benzo[*b*]-pyran-3-carboxaldehyde (23) in MeOH, followed by treatment with NaBH$_4$ gives 3-[2-(2-methoxyphenoxy)ethylamino]methyl-2*H*-benzo[*b*]pyran (24) (H. Matsuda *et al.*, Jpn. Kokai Tokkyo Koho JP 62 67,079 [87, 67,079], 1987).

6-Cyano-4-(1,2-dihydro-2-oxopyrid-1-yl)-2,2-dimethyl-2*H*-benzo[*b*]pyran-3-carboxaldehyde
(25) has been obtained by reacting 4-bromo-6-cyano-2,2-dimethyl-2*H*-
benzo[*b*]pyran-3-carboxaldehyde (26) with 1*H*-2-pyridone. A number of
related derivatives have been prepared as cardiovascular agents and smooth
muscle relaxants (R. Gericke *et al.*, Eur. Pat. Appl. EP 410,208, 1991).

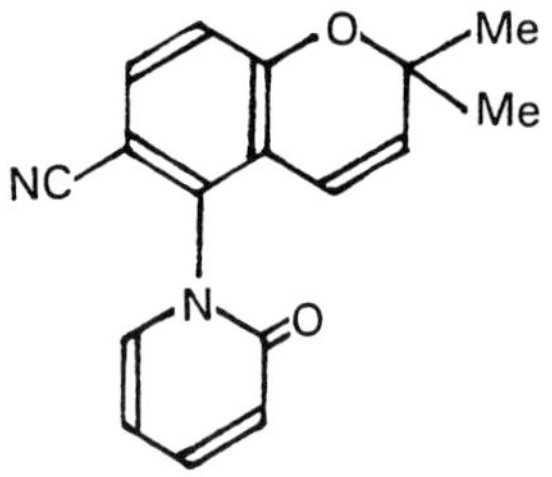

(25)

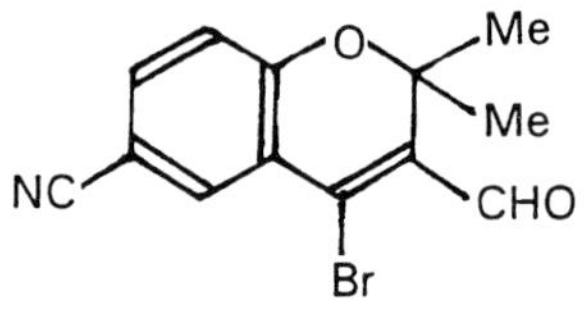

(26)

2-Substituted derivatives of 8-chloro-5-hydroxy-2,7-dimethyl-2*H*-benzo-
[*b*]pyran-6-carboxaldehyde, for example, 27, useful as anticancer agents
have been prepared (T. Hyama and H. Saimoto, Jpn. Kokai Tokkyo Koho JP
62 181,276 [87 181,276], 1987).

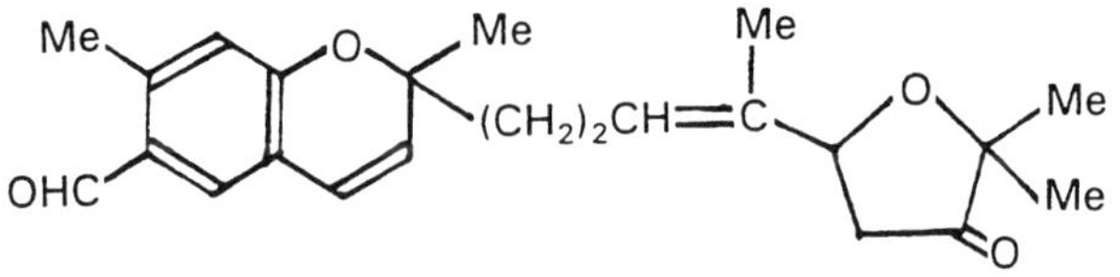

(27)

A number of substituted 2*H*-benzo[*b*]pyrans 28 (R^1-R^4 = H, halogeno, alkyl, alkenyl; R^5 = carboxy, alkoxycarbonyl, alkenoyloxycarbonyl) useful as anticancer agents have been prepared from the appropriate 3-acetyl-2*H*--benzo[*b*]pyrans (J.D. Brion *et al.*, Eur. Pat. Appl. EP 337,885, 1989).

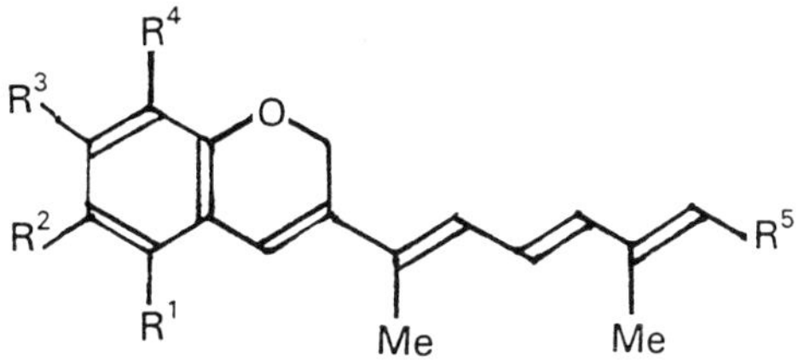

(28)

Claisen rearrangement of aryloxy(aryloxymethyl)propenoates 29 (R^1 = H, Cl; R^2 = H, Me, Et, Ph, Cl) in boiling PhNEt$_2$ affords hydroxyphenyl-methylenedihydrobenzopyranones 30 and methyl 6,8-disubstituted 2*H*--benzo[*b*]pyran-3-carboxylates 31 (D. Gopal and K. Rajagopalan, Tetrahedron Letters, 1987, <u>28</u>, 5327).

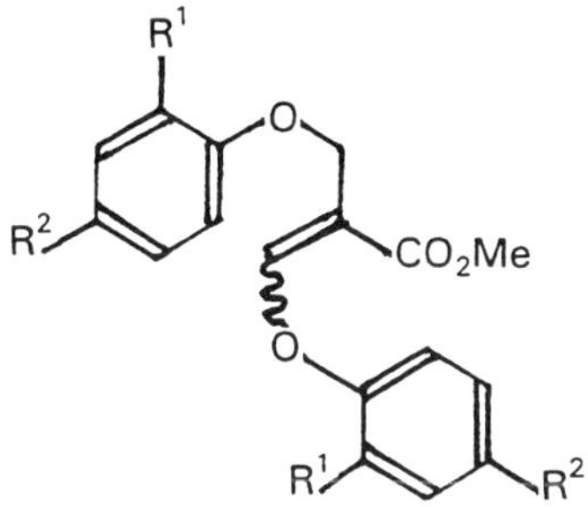

(29)

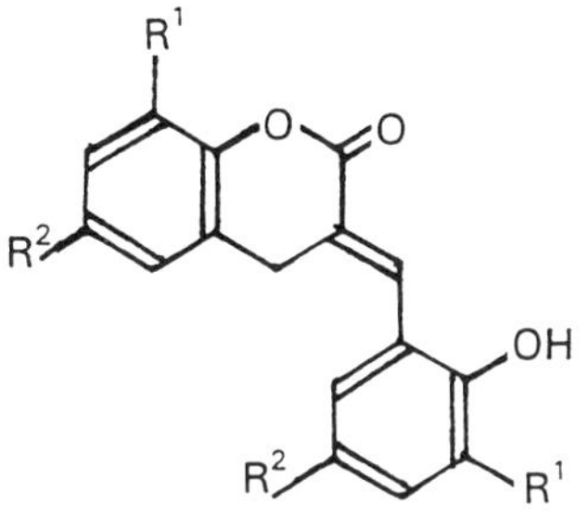

(30)

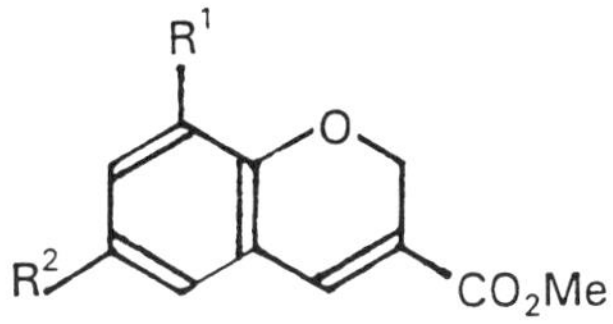

(31)

2*H*-Benzo[*b*]pyran-3-carboxylic acid esters 32 (R^1 = alkyl; R^2-R^5 = H, alkoxy, alkyl, halogeno) as intermediates for polymethine dyes and photographic sensitizers are prepared by cyclocondensation of salicylaldehyde derivatives 33 with MeO$_2$CCH = CClCH$_2$CO$_2$Me in boiling MeOH containing Et$_3$N (D. Greif *et al.*, Ger. (East) DD 272,8846, 1989).

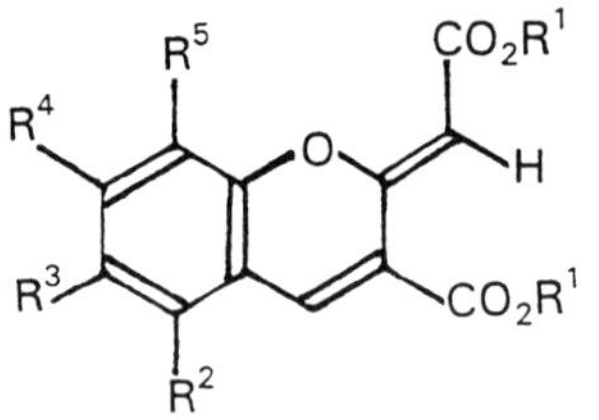

(32) 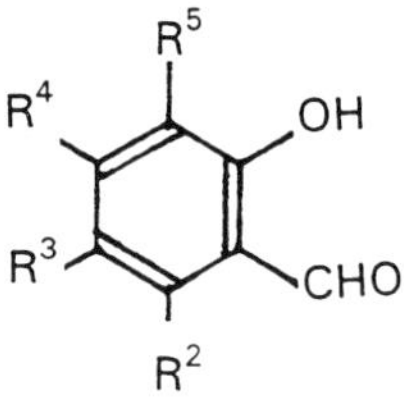 (33)

Cyclocondensation reactions of salicylaldehyde with conjugated olefins, Me$_2$C = CHR^1 (R^1 = CO$_2$Et, CN, COMe) affords 2,2-dimethyl-2*H*-benzo[*b*]pyran derivatives 34, while the condensations with Me$_2$C = CR^{1}CO$_2$Et or 35 yields 2-methyl-2*H*-benzo[*b*]pyran-2-acetic acid derivatives 36 [R^2 = Me, R^3 = H, R^1 = CO$_2$Et; R^{2}R^3 = (CH$_2$)$_4$, R^1 = CO$_2$Et, respectively] (S. Yamaguchi *et al.*, J. Heterocyclic Chem., 1992, <u>29</u>, 755). 6,7-Dimethoxy-2,2-dimethyl-2*H*--benzo[*b*]pyran (37) present in *Ageratum houstonianm* has been synthesized. Other benzo[*b*]pyrans have been obtained by the reaction of 2-hydroxybenzaldehydes with allyl bromide in the presence of powdered tin (B. Huang *et al.*, Youji Huaxue, 1989, <u>9</u>, 552).

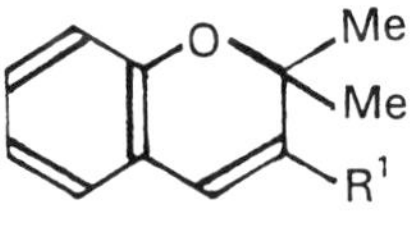

(34) 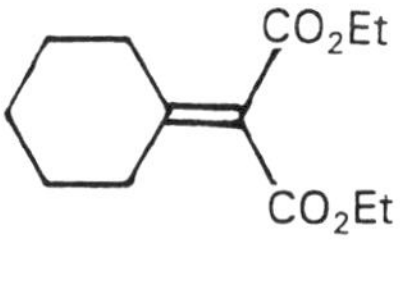 (35) 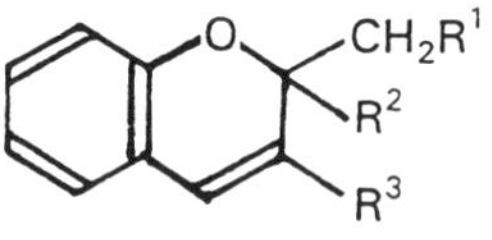 (36)

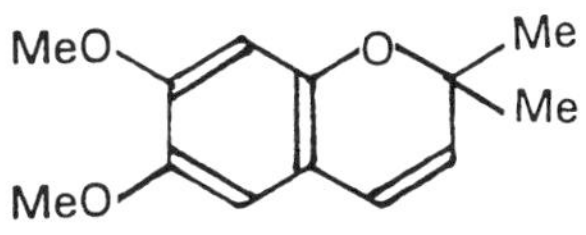

(37)

2-(2-Hydroxyethylthio)ethyl esters of 1-(2,2-dimethyl-2*H*-benzo[*b*]pyran-
-6-yl)-1-(alkoxycarbonyl)propanoic acids (J. Kahovcova, Czech. CS 269,383,
1991) and 2,2-dimethyl-6-[1-(2-methyl-2-acetoxy)propionyloxy]alkyl-2*H*-
-benzo[*b*]pyrans (*idem, ibid.*, 269, 387, 1991) have been prepared as
pesticides for application against termites *Prorhinotermes simplex*. The
synthesis, x-ray structure, and laser properties of 4-(trifluoromethyl)-7-
-hydroxy-2(cyanoethoxycarbonylmethylene)-2*H*-benzo[*b*]pyran (38) have
been reported (Ya. V. Voznyi *et al.*, Izv. Akad. Nauk SSSR, Ser. Khim.,
1989, 913).

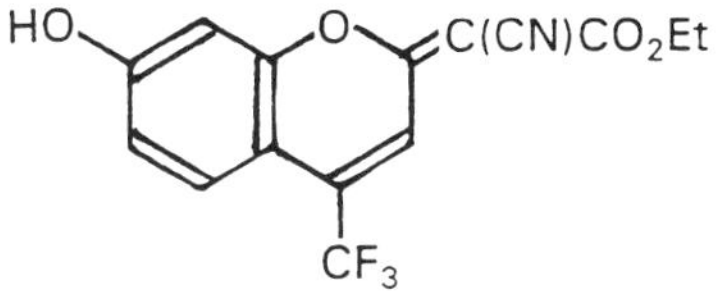

(38)

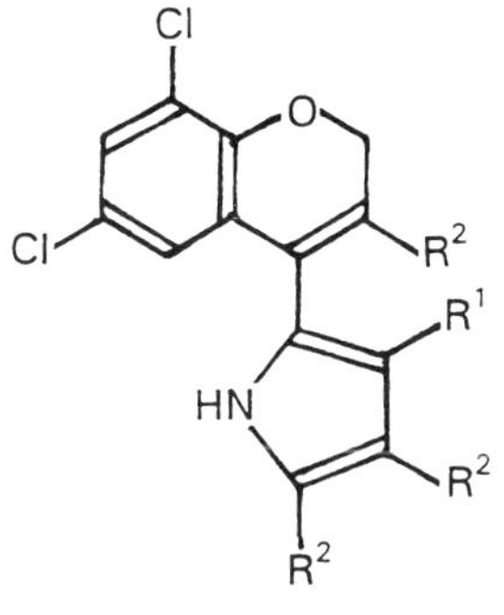

(39)

238

The preparations of 6,8-dichloro-4-(pyrrol-2-yl)-2*H*-benzo[*b*]-pyran derivatives 39 (R^1 = H, halogeno; R^2 = halogeno), as medical and agrochemical microbicides and antiseptics, have been reported (N. Shioiri *et al.*, Jpn. Kokai Tokkyo Koho JP 63 183,579 [88 183,579], 1988). The 7--alkoxy-4-halogeno-2,2-dimethyl-2*H*-benzo[*b*]pyrans 40 (R^1 = Cl, R^2 = Me, Et, Pr; R^1 = Br, R^2 = Me) have been synthesized from the corresponding chroman-4-ones and phosphorus trihalides (T. Eszenyi, T. Timar, and P. Sebok, Tetrahedron Letters, 1991, <u>32</u>, 827). Some 6-alkyl-7-alkoxy-4--chloro-2,2-dimethyl-2*H*-benzo[*b*]pyrans 41 (R^1 = H, Me, Et, Pr; R^2 = CHMeOH, CHEtOH) have also been synthesized (T. Eszenyi and T. Timar, Synth. Comm., 1990, <u>20</u>, 3219).

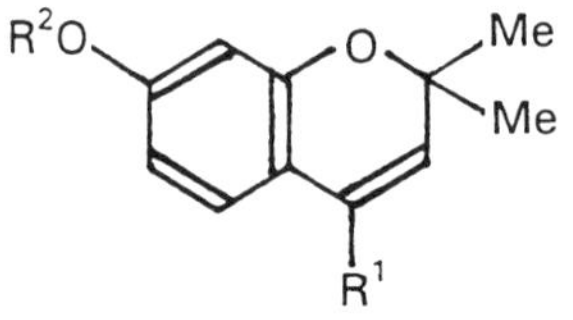

(40)

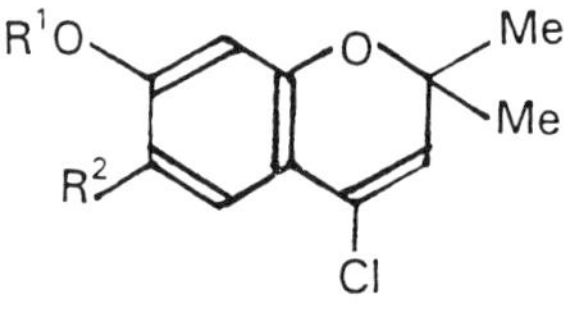

(41)

A number of 4-halogeno-2*H*-benzo[*b*]pyrans and 2*H*-benzo[*b*]thiopyrans 42 [R^1 = Me, H; R^2 = Me, Et, Pr^i, Bu, Ph; R^{1}R^2 = (CH$_2$)$_5$, (CH$_2$)$_6$; R^3 = H, 7--OMe; R^4 = Br, Cl; X = O, S, SO$_2$] have been prepared. Conversion of the bromo compounds to the lithio derivatives provides access to a wide range of novel 4-substituted 2*H*-benzo[*b*]pyrans and 2*H*-benzo[*b*]thiopyrans (C.D. Gabbutt *et al.*, Tetrahedron, 1994, <u>50</u>, 2507).

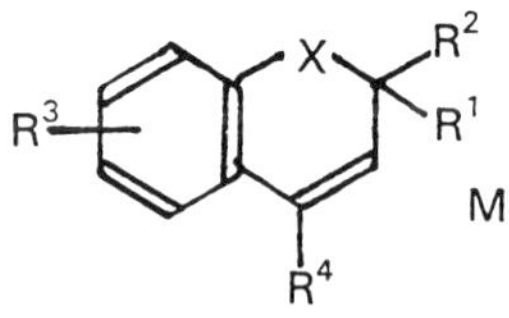

(42)

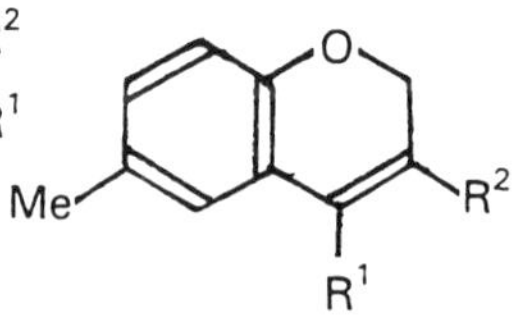

(43)

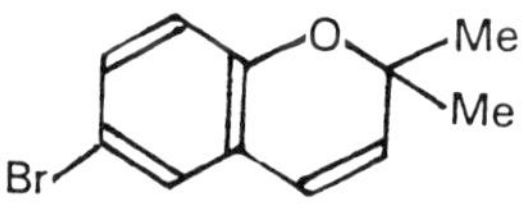

(44)

Aryl γ-bromopropargyl ethers, for example, $4\text{-MeC}_6\text{H}_4\text{OCH}_2\text{C}\equiv\text{CBr}$, undergo transformation in PhNEt_2 at 215°C to afford a mixture of 3- and 4-
-bromo-2*H*-benzo[*b*]pyran derivatives, for example, 43 ($R^1 = H$, $R^2 = Br$; $R^1 = Br$, $R^2 = H$) and *trans*-2,3-dibromoallyl aryl ethers. Under similar conditions aryl γ-chloropropargyl ethers, give 4-chloro-2*H*-benzo[*b*]pyrans, such as 43 ($R^1 = Cl$, $R^2 = H$) (G. Ariamala and K.K. Balasubramanian, *ibid.*, 1989, 45, 309). 3-, 4-, And 6-Halogeno-2*H*-benzo[*b*]pyrans are obtained by reacting the related halogenated coumarins with Li, Mg, or Al organo-metallics, and subsequent cyclization of the resulting 2-hydroxycinnamyl alcohols, for example, the preparation of 6-bromo-2,2-dimethyl-2*H*-benzo-[*b*]pyran (44). The 3- and 4-halogeno-2*H*-benzo[*b*]pyran derivatives yield nonmetalated products when subjected to metal-halogen interchange with butyllithium, whereas the 6-bromo derivatives give 6-lithio-2*H*-benzo[*b*]-pyrans (A. Alberola *et al.*, Heterocycles, 1994, 38, 819). The anionic cleavage of 3-bromo-2*H*-benzo[*b*]pyrans resulting in the formation of allenes has been investigated (C.D. Gabbutt *et al.*, J. Chem. Soc., Perkin 1, 1994, 1733).

The reaction of 2,2,4-trimethyl-2*H*-benzo[*b*]pyran with various amounts of bromine affords 3-bromo-4-bromomethyl-2,2-dimethyl-2*H*-benzo[*b*]pyran (45), 3,6-dibromo-4-bromomethyl-2,2-dimethyl-2*H*-benzo[*b*]pyran (46), and 3,6-dibromo-4-dibromomethyl-2,2-dimethyl-2*H*-benzo[*b*]pyran (47). The end product is 3,6,8-tribromo-4-dibromomethyl-2,2-dimethyl-2*H*-benzo[*b*]pyran (48) (E. Bradley *et al.*, J. Chem. Res., S, 1993, 314).

(45) R = H
(46) R = Br

(47) R = H
(48) R = Br

The following 2,2-dimethyl-2*H*-benzo[*b*]pyran derivatives containing a 6-
-cyano substituent plus another substituent in the 4-position and in some instances a substituent in the 3-position have been prepared and their properties reported; (i) a reinvestigation by NMR spectroscopy of the dynamic behaviour of cromakalim and related potassium channel activators

240

49 (R=H, n=1,2; R=Me, n=1) and 50 shows that previous evidence for a single rigid conformation in solution is incorrect (W.A. Thomas and I.W.A. Whitcombe, Chem. Comm., 1990, 528); (ii) structural modifications of cromakim have been described in which the amide moiety at C-4 had been replaced by carboxamide and thiocarboxamide functions to furnish compounds 51 (R^1=OH, R^2=Me, Bu^t, Ph, 4-FC$_6$H$_4$, 4-MeOC$_6$H$_4$, X=S; R^2=Ph, 4-MeOC$_6$H$_4$, X=O; R^1=H, R^2=Me, Bu^t, X=S; R^2=Ph, X=O) (J.R.S. Arch et al., J. Med. Chem., 1991, 34, 2588); (iii) a pharmacophore model, which explains rationally structure-activity relationships of chemical diverse potent K^+ channel openers, has been constructed and potent benzo-pyran derivatives 52 (X=S, O, NCN), with thioamide, amide, and (N-cyano)-amidine groups, at the 4-position have been designed using the model (H. Koga et al., Bioorg. Med. Chem. Letters, 1993, 3, 625); (iv) investigation of the structure-activity relationships of 6-substituted 2H-benzo[b]pyran-4-carbothioamide potassium channel openers 53 (R=CN, PhSO$_2$, CF$_3$, Cl) T. Ishizawa et al., ibid., p.1659); (v) a series of benzopyran derivatives of rolipram has been investigated, one of which 54, proved to be a potent inhibitor of the phosphodiesterase (PDE) IV isoenzyme (I.L. Pinto et al., ibid., p.1743); and (vi) the devicing of a practical and highly effective process for the asymmetric epoxidation of several 2,2-dimethyl-2H-benzo[b]pyrans 55 (R^1=CN, NO$_2$, H, R^2=R^3=R^4=H; R^1=Ac, R^2=MeO, R^3=R^4=H; R^1=R^2=R^4=H, R^3=Me; R^1=CN, R^2=R^3=H, R^4=Me) (N.H. Lee, A.R. Muci, and E.N. Jacobsen, Tetrahedron Letters, 1991, 32, 5055).

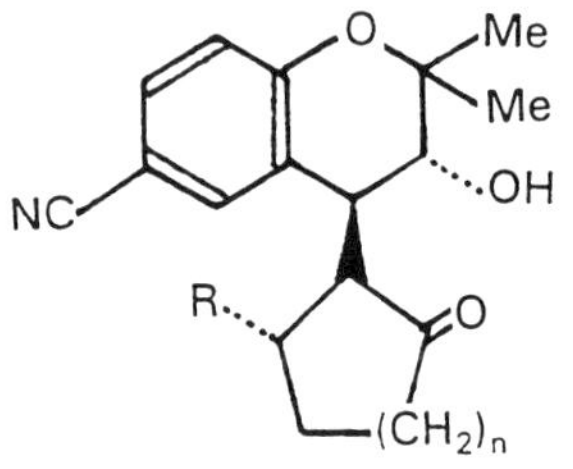

(49) 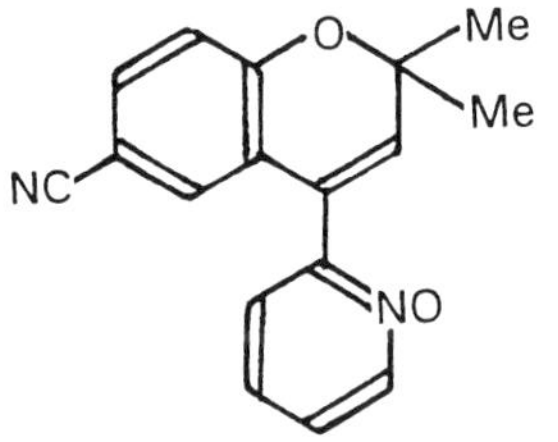   (50) 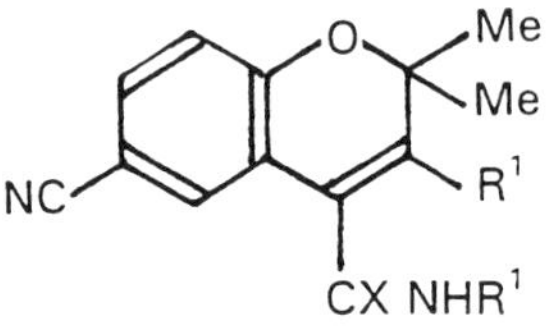   (51)

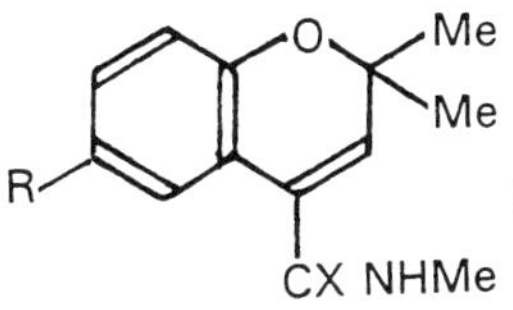

(52) R = CN
(53) R (see text), X = S

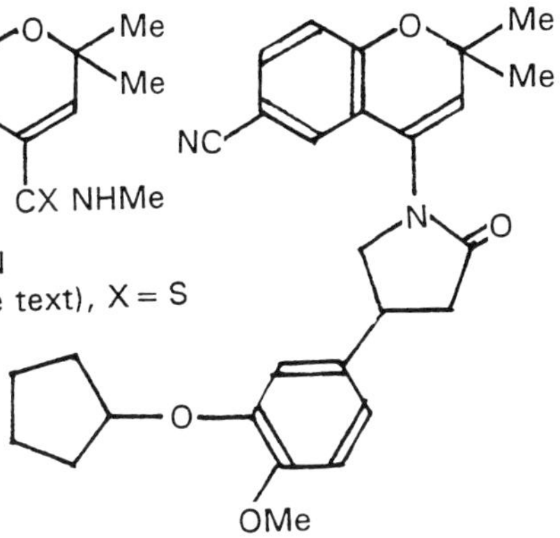

(54)

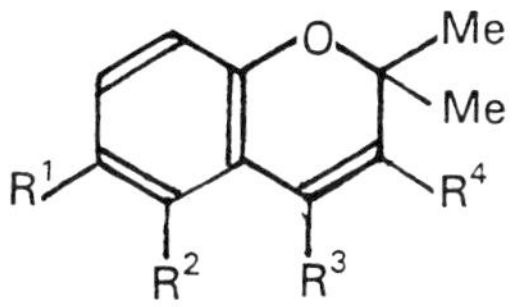

(55)

3-Nitro-2*H*-benzo[*b*]pyrans, for example, 56, are obtained by the
treatment of 2-hydroxybenzaldehydes by nitroethylene generated in situ, in
the presence of dibutylamine (M. Al Neirabeyeh, R. Koussini, and G.
Guillaumet, Synth. Comm., 1990, 20, 783). Some 3-nitro-2*H*-
-benzo[*b*]pyrans 57 have been prepared by the cyclization of salicylaldehydes
58 (R^2 = H, Cl; R^3 = H, OMe) with R^{1}CH = CHNO$_2$ [R^1 = Ph, 4-anisyl, 2-
-thienyl, 3,4-(MeO)$_2$C$_6$H$_3$] (N. Ono *et al.*, Chem. Comm., 1993, 1781).

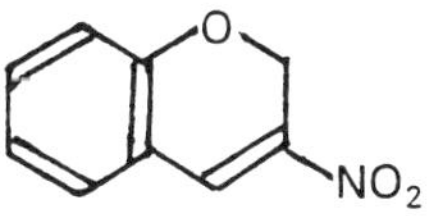

(56)

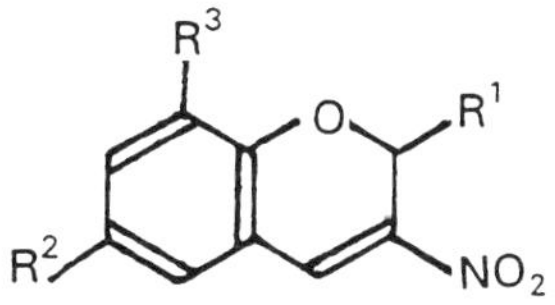

(57)

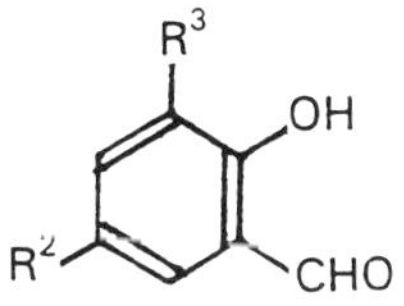

(58)

The *N*-substituted 2,2-dimethyl-6-nitro-2*H*-benzo[*b*]pyran-4-carboxamides 59 (R = H, alkyl) have been synthesized and some of them exhibited potent vasorelaxant activity (H. Koga *et al.*, Bioorg. Med. Chem. Letters, 1993, 3, 1115). It has been reported that 2,2-di(fluoromethyl)-6-nitro-2*H*-benzo[*b*]-pyran-4-(*N*-2-cyanoethyl)thiocarboxamide (KC-399) (60) showed a highly potent, slow and long-lasting antihypertensive effect with reduced reflex tachycardia (*idem, ibid.*, p.2005). For the syntheses of the stable epoxides 61 (R^1, R^2 = H, NO_2; $R^1 = R^2 = NO_2$; $R^1 = NO_2$, $R^2 = NHCO_2Et$) of 5- and 7-, and 5,7-dinitro-2,2-dimethyl-2*H*-benzo[*b*]pyrans and of 7-(ethoxycarbonyl-amino)-2,2-dimethyl-5-nitro-2*H*-benzo[*b*]pyran, see P.E. Brown and R.A. Lewis, J. Chem. Soc., Perkin 1, 1992, 573.

(59)

(60)

(61)

(62)

The [bis(salicylidene)ethylenediamine]manganese (III) complex has been found to catalyze the epoxidation of 2,2-dimethyl-2*H*-benzo[*b*]pyrans, for example, 6-acetylamino-2,2-dimethyl-7-nitro-2*H*-benzo[*b*]pyran (62), with high enantioselectivity (86-92% ee). Thus, treatment of benzopyran 62 with PhIO/complex in MeCN at -20°C affords the epoxide in 82% yield and 90% ee (A. Hatayama *et al.*, Synlett, 1992, 407).

The preparations of photochromic indoline-2-spiro-2'-(6-methoxy-8-nitro--2*H*-benzo[*b*]pyrans) 63 ($R^1 = C_{1-30}$alkyl, $R^2 = NO_2$), useful in optical storage media have been reported (J. Hibino and E. Ando, Eur. Pat. Appl. EP 385,407, 1990). For the preparation of related compounds and their properties see, M. Inouye, M. Ueno, and T. Kitao, J. Org. Chem., 1992, _57_, 1639.

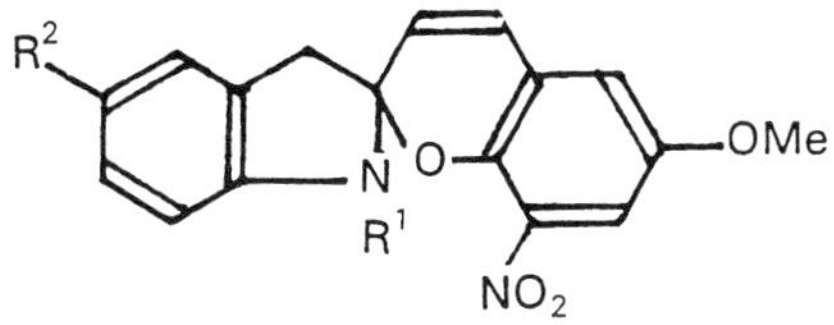

(63)

2,2,6-Trisubstituted 2*H*-benzo[*b*]pyrans and 2*H*-benzothipyrans 64 (R^1-=C$_{2-5}$alkyl, Ph; R^2=alicyclic *tert*-amino having 5-6 atoms consisting of 1N and C atoms, wherein at least one of the C atoms may be substituted with O; R^3=electron withdrawing group; X=O, S), possessing photochromicity with improved thermal stability in the colouration state, have been prepared by cyclocondensation of appropriate benzaldehyde derivatives with amines (R. Asakawa and I. Arai, Jpn. Kokai Tokkyo Koho JP 05 186,457 [93 186,457], 1993).

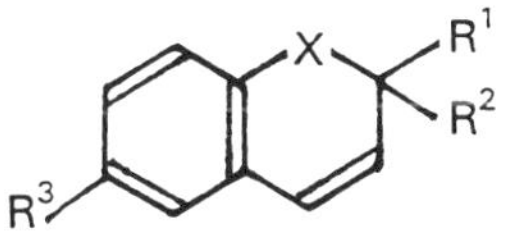

(64) 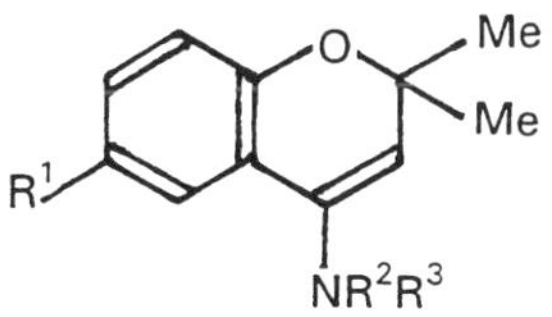   (65)

Several iodinated 3-amino-2*H*-benzo[*b*]pyrans and related compounds
have been investigated in the development of selective ligands for dopamine
D2 and D3 receptors (S. Chumpradit *et al.*, J. Med. Chem., 1994, $\underline{37}$,
4245). For the preparation of some antihypertensive 6-substituted 4-amino-
-2,2-dimethyl-2*H*-benzo[*b*]pyrans 65 [$R^1 = F_3CO$, F_3CCH_2O; R^2, $R^3 = H$,
C_{1-5}alkyl, C_{5-8}cycloalkyl, (substituted) $PhCH_2$, Bz, furoyl; $R^2R^3 = (CH_2)_n$,
n = 4-7, $(CH_2)_m CO$, m = 3,6] and their salts, see D.A. Quagliato, U.S. US
5,171,857, 1992; and of 3-(*N*-carbamoyl-*N*-hydroxyaminomethyl)benzo-
pyrans and analogues as 5-lipoxygenase inhibitors, J.L. Stanton, Y. Satoh,
and A.J. Hutchison, *ibid.*, 5,155,130, 1992; and for the chemistry of 2-*tert*-
-butyl-4-diazobenzopyran, S.E. Mulhall, Diss. Abstr. Int. B, 1990, $\underline{51}$, 216.

Flavenes (2-aryl-2H-benzo[b]pyran) and diaryl-2H-benzo[b]pyrans.
Heating aryl *o*-ethynylbenzyl ethers 1 ($R^1 = H$, OMe; $R^2 = H$, OMe; $R^3 = H$,
Me, Cl, OMe, NO_2; $R^4 = H$, Me) in $PhNEt_2$, polyethylene glycol, or 1,2-
-$Cl_2C_6H_4$ affords the 6,8-disubstituted flavenes (6,8-disubstituted 2-aryl-2*H*-
-benzo[*b*]pyrans) 2. 2,4-Diphenyl-2*H*-benzo[*b*]pyran has been obtained from
$PhOCPh_2C \equiv CPh$ (R.S. Subramanian and K.K. Balasubramanian, Tetrahedron
Letters, 1988, $\underline{29}$, 6797).

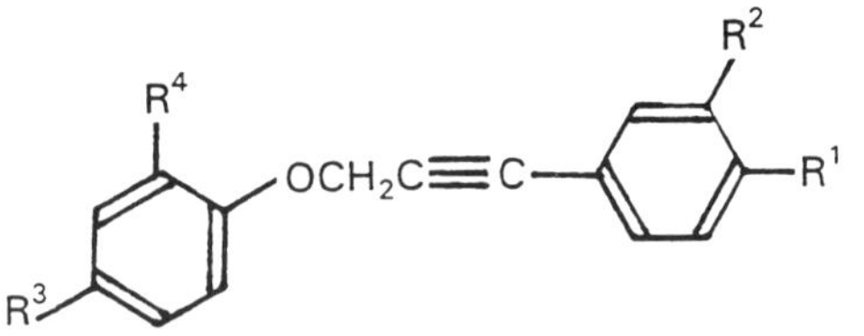

(1)

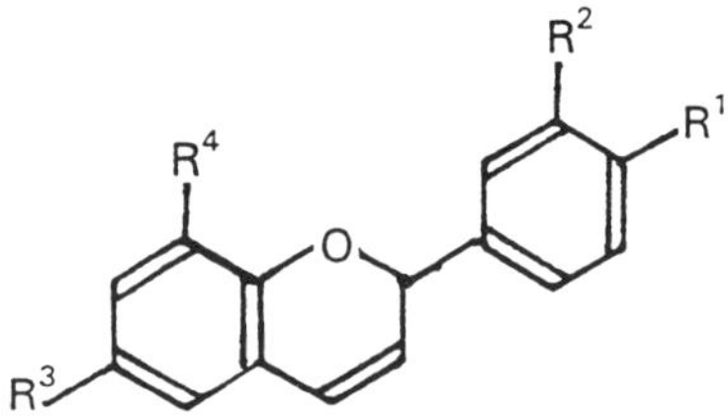

(2)

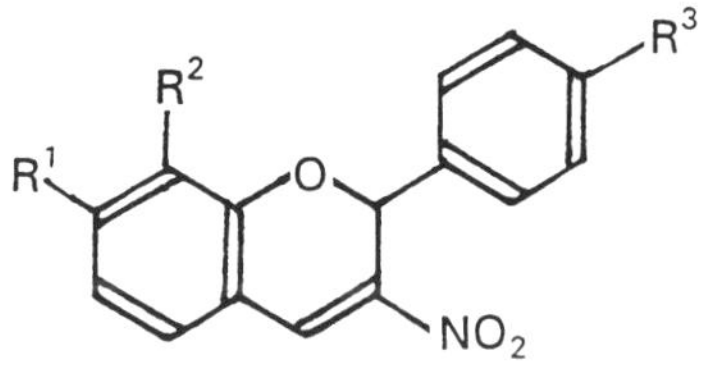

(3)

7,8-Disubstituted 3-nitroflavenes 3 (R^1, R^2=H, OMe; R^3=H, Cl, OMe, Me) when treated with KO_2 in Me_2SO are degraded mainly to the corresponding salicylic acids and benzoic acids. The formation of related flavonols as a minor product is also observed (T.S. Rao and G.K. Trivedi, Heterocycles, 1987, <u>26</u>, 2117). Reactions of the photosensitized ethylenic compounds, including diflavylene and dithioflavylene have been investigated (W.M. Abdou, Y.O. El Khoshnieh, and M.M. Sidky, J. Chem. Res.,, S, 1994, 184).

7-Substituted 2,3-diaryl-2*H*-benzo[*b*]pyrans 4 (R^1, R^2=H, HO, C_{1-17}-alkoxy, C_{2-18}alkoxycarbonyl, R^3=piperidinoethoxy, 2-pyrrolinylethoxy) useful as antiestrogens in conditions such as breast cancer have been prepared (R.S. Kapil *et al.*, Eur. Pat. Appl. EP 470,310, 1992) and studies on the structure-activity relationship of antiestrogens, 2-aryl-3-phenyl-2*H*-benzo[*b*]-pyrans 5 (R=H, MeO) (A. Saeed *et al.*, J. Med. Chem., 1990, <u>33</u>, 3210), 2,3-diaryl-2*H*-benzo[*b*]pyrans 6 (R^1=H, OH, OMe; R^2=H, OH; R^3=H, piperidinoethyl) (A.P.Sharma *et al.*, *ibid.*, 1990, <u>33</u> 3222), and 4,7-disubstituted 2-aryl-3-phenyl-2*H*-benzo[*b*]pyran 7 (R^1=tertiary aminoethoxy, in 2-, 3-, or 4-position; R^2=H, Me, Et; R^3=H, OMe) (*idem, ibid.*, p.3216) have been reported. For the preparation of some 3-alkyl-4-aryl-2*H*-benzo[*b*]pyrans with a number of substituents, for example, 8, see D. Festal *et al.*, Eur. Pat. Appl. EP 380,392, 1990.

(4)

(5)

(6)

(7)

246

(8)

The 7-substituted 4-aryl-3-(4-fluorophenyl)-2H-benzo[b]pyrans [4-aryl-3-
-(4-fluorophenyl)chrom-3-enes] 9 (R^1 = Me, R^2 = NEt_2, pyrrolidino, piperidino,
morpholino; R^1 = H, R^2 = pyrrolidino) have been prepared by the Grignard
reaction between isoflavanones 10 (e.g., R^1 = Me, tetrahydropyranyl) and 4-
-$BrC_6H_4OCH_2CH_2R^2$ as selective ligands for antiestrogen-binding sites (C.C.
Teo, O.L. Kon, and K.Y. Sim., J. Chem. Res., S, 1990, 4).

(9)   (10)

Substituted 4H-benzo[b]pyrans (2-chromenes, 4H- or ß-chromenes). 4H-
-Benzo[b]pyrans 1 (R^1 = H, Me; R^2 = CO_2Et, NO_2, CHO, COMe, COPh;
R^1R^2 = $CH_2CH_2CH_2CO$; R^3 = H, OMe; R^4 = H, OMe) have been prepared from
ß-functionalized enamines 2 and 2-hydroxybenzyl alcohols 3 in Ac_2O or
Ac_2O/AcOH (L. Rene, Synthesis, 1989, 69).

(1)

(2)

(3)

Heating a mixture of 2-(2,4-dihydroxyphenyl)-2,4,4-trimethylchroman-7-ol, toluene, H_2O, and H_2SO_4 in an autoclave at 140°C affords 2,4,4-trimethyl-4*H*-benzo[*b*]pyran-7-ol (4) (56%) and 2,4,4-trimethyl-3,4-dihydro-2*H*-benzo[*b*]pyran-2,7-diol (16%). The latter can be converted to the former derivative by distillation (I. Tomino *et al.*, Jpn. Kokai Tokkyo Koho JP 03 48,675 [91 48,675], 1991).

(4)

(5)

(6)

Substituted 2-amino-4*H*-benzo[*b*]pyrans 5 (R^1 =H, hexyl; R^2 =H, F, Cl, Br, NO_2, OH; R^3 =OH, NH_2; R^4 =CO_2Et, CN) have been obtained by the

248

cycloaddition of phenols 6 with $R^2C_6H_4CH=CR^4CN$ (G.V. Klokol *et al.*, Zh. Org. Khim., 1987, <u>23</u>, 412). 3-Substituted 2-amino-4-aryl-7-hydroxy-4*H*--benzo[*b*]pyrans 7 (R^2 = aryl, R^1 = CN, CO_2Et) have been prepared by the reaction of cinnamonitrile with resorcinol (A.G.A. Elagamey and F.M.A.A. El--Taweel, Indian J. Chem., 1990, <u>29B</u>, 885). 2-Amino-3-cyano-4-phenyl- and -4-(4-methoxyphenyl)-4*H*-benzo[*b*]pyrans (8; R = H and MeO, respectively have been prepared (A.G.A. Elagamey *et al.*, Coll. Czech. Chem. Comm., 1988, <u>53</u>, 1534).

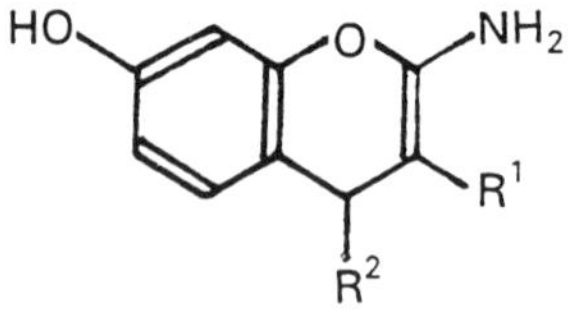

(7) 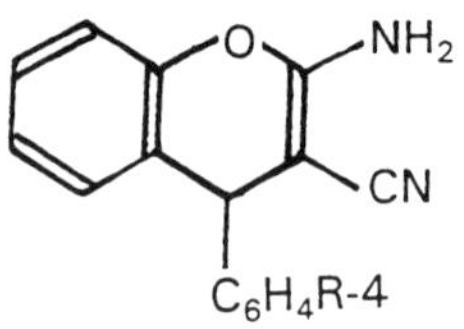   (8)

2,4-Diphenyl-4*H*-benzo[*b*]pyran (9) under Vilsmeier and Mannich reaction conditions affords 2,4-diphenyl-4*H*-benzo[*b*]pyran-3-carboxaldehyde (10), which when treated with $HClO_4$ in Ac_2O cyclizes to yield benzo[*b*]indeno-[2,1-*e*]pyrylium perchlorate (11) (A.V. Koblik and K.F. Suzdalev, Zh. Org. Khim., 1989, <u>25</u>, 171).

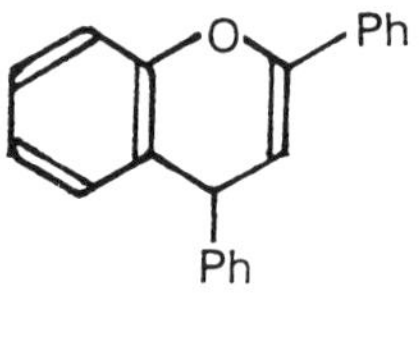

(9) 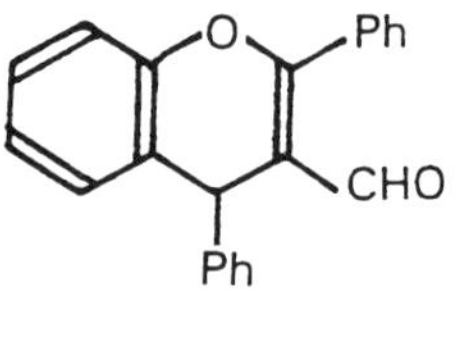   (10) 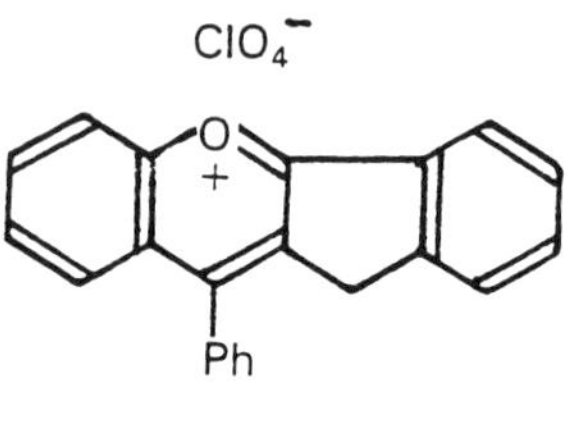   (11)

Acylation of 2,4-diphenylbenzopyran 9 by Ac_2O-$HClO_4$ gives 3-acetyl-
-2,4-diphenyl-4*H*-benzo[*b*]pyran (12) (36%) along with the indenobenzo-
pyrylium perchlorate 13 (29%), which on reduction with $LiAlH_4$ yields the
indenobenzopyran 14 (K.F. Suzdalev and A.V. Koblik, Khim. Geterotsikl.
Soedin., 1990, 1042).

(12) (13) (14)

2-*tert*-Butyl-4*H*-benzo[*b*]pyran (15) reacts with Cl_2C:, generated in situ
from $CHCl_3$, Bu^tOK, in hexane at 0°C to give 4-(dichloromethyl)-2-*tert*-butyl-
-4*H*-benzo[*b*]pyran (16) (80%) and the cyclopropanobenzopyran 17 (R = H)
(12%). Similarly the reaction of 15 and Cl_2C:, generated from $CHCl_3$ and
50% aqueous NaOH containing benzyltriethylammonium chloride at 40°C
affords derivative 17 (R = $CHCl_2$) (6%) and the spiro derivative 18 (7%) (R.R.
Kostikov, A.P. Molchanov, and K.F. Suzdalev, Zh. Org. Khim., 1990, 26,
355). Carbene adducts 19 (R = H, Ph) obtained from 3-methyl-2-phenyl- and
3-methyl-2,4-diphenyl-4H-benzo[b]pyrans, respectively, on boiling in DMF
furnish benzoxepins 20 (A.V. Koblik, K.F. Suzdalev, and A.A. Loktionov,
Khim. Geterotsikl. Soedin., 1987, 188).

(15) R = H (17) (18)
(16) R = $CHCl_2$

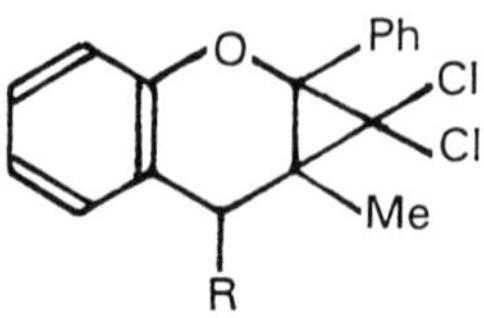

(19)

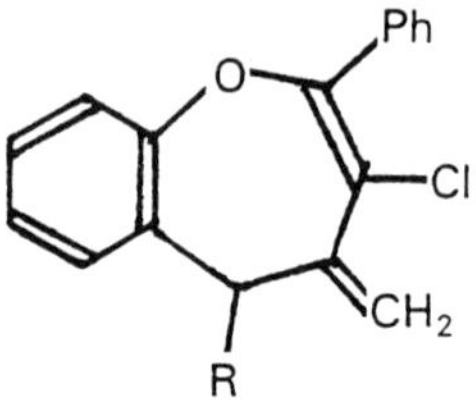

(20)

Miscellaneous compounds related to benzopyran. 2,5,7-Trisubstituted 4-
-ethoxycyclopentapyrans 21 (R^1 = 1-ethoxycyclopropyl, CMe_2OEt, Bu^t, Ph;
R^2 = Ph, Pr), related to benzopyran, have been obtained by treating
$(CO)_5Cr = C(OEt)CH = CR^1R^3$ ($R^3 = Me_2N$, EtO) with $R^2C \equiv CH$. The structure
of 21 (R^1 = ethoxycyclopropyl, R^2 = Ph) has been determined by x-ray
analysis (F. Stein *et al.*, Angew. Chem., 1991, <u>103</u>, 1669).

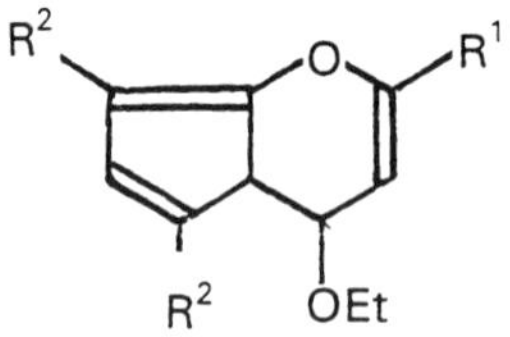

(21)

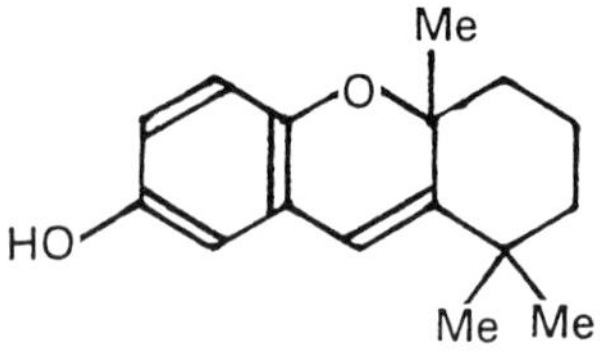

(22)

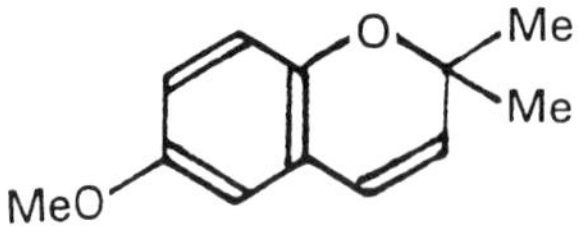

(23)

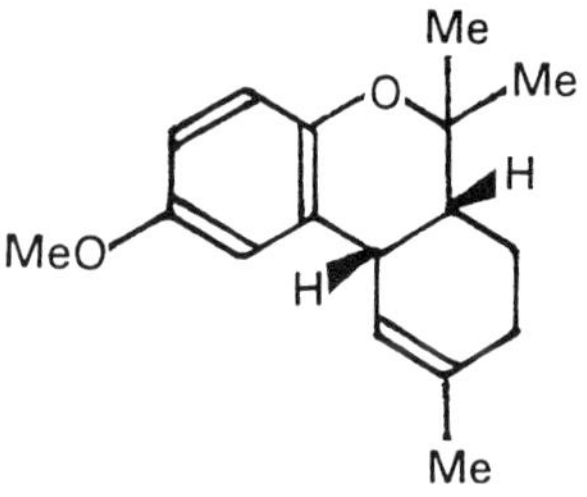

(24)

The synthesis of 2,2-dialkyl-2*H*-benzo[*b*]pyrans, for example, 22, 23, and 24, have been achieved by employing the 1,2-addition reaction of aryllithium salts to cyclocitral, 3-methyl-2-butenal, and farnesal, respectively (R. Cruz-Almanza, F. Perez-Flores, and C. Lemini, Heterocycles, 1994, <u>37</u>, 759).

The synthesis of 2,2-dimethylchromene-5,8-quinone (25), 2,2-dimethyl-chromene-6,7-quinone (26), and 6,7-dimethoxy-2,2-dimethylchromene-5,8--quinone have been reported. Since they are easily bioreducible to electron-rich chromenes, they and their oxirane derivatives are of interest as possible antitumor alkylating agents (P.E. Brown, R.A. Lewis, and M.A. Waring, J. Chem. Soc., Perkin 1, 1990, 2979).

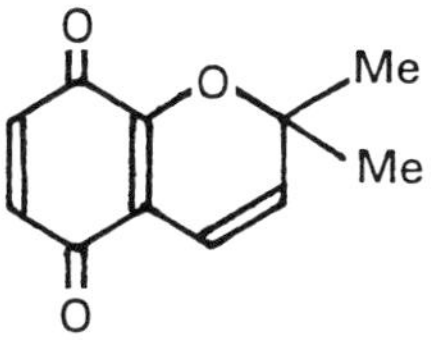

(25)

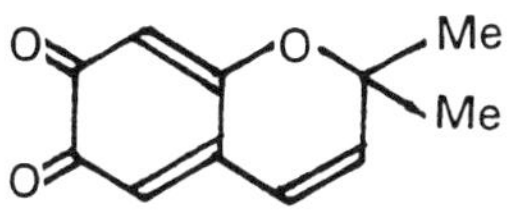

(26)

(ii) Naturally occurring derivatives of 2,2-dimethyl-2H-benzo[b]pyran
Cyclic analogues of precocene 1 (2-spirobenzopyrans) have been obtained by the Friedel-Crafts reaction of various 1,3-dihydroxybenzenes and cycloalkylideneacetic acid in the presence of a Lewis acid (S.Y. Dike, J.R. Merchant, and N.Y. Sapre, Tetrahedron, 1991, <u>47</u>, 4775).

Natural pro-allatocidin precocene II (1) has been prepared from 1,4-(HO)-3-MeOC$_6$H$_3$ *via* Fries condensation with Me$_2$C$=$CHCO$_2$H, followed by *O*--methylation, reduction by NaBH$_4$ in MeOH, and dehydration of the resulting hydroxychromanone 2 (P. Teixidor, F. Camps, and A. Messeguer, Heterocycles, 1988, <u>27</u>, 2459).

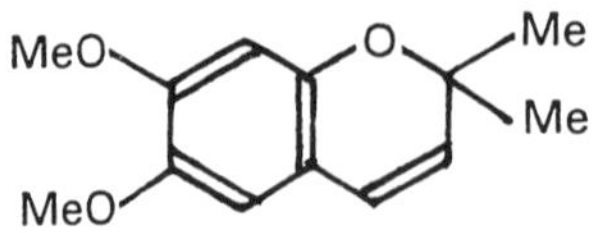

252

MeO— ... Me, Me (ring structure)

(1)

MeO— ... Me, Me; HO (ring structure)

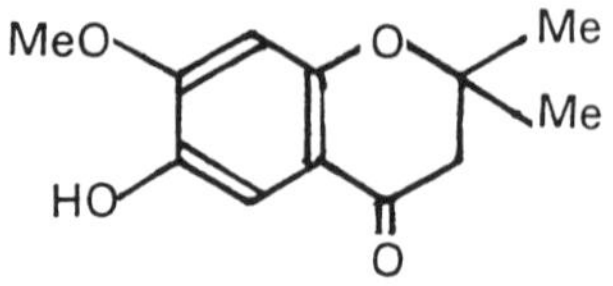

(2)

Analogues 3 (R = Pr, Pr^i, Bu, MeCHEt, Me_2CHCH_2, cyclopentyl, allyl, prenyl, HC≡CCH_2, PhCH_2) of precocene 3 containing various 7-*O*-substituents have been prepared from 6,7-dihydroxy-2,2-dimethylchroman-4-one (4) by sequential regioselective alkylation, followed by methylation, reduction, and dehydration (T. Timar, J.C. Jaszberenyi, and S. Hosztafi, Acta Chim. Hung., 1989, 126, 149). Related derivatives with the methoxy group in other positions of the benzene ring have been reported (*idem, ibid.*, p.487) and the precocene analogues 5 (R¹, R³ = H, Me; R² = H, OMe; R⁴ = 4-Me, 2-F) and 6 (R¹, R³ = H, Me; R² = H, OMe; R⁵ = Me, Et) have been prepared (A. Levai and T. Timar, Pharmazie, 1990, 45, 660).

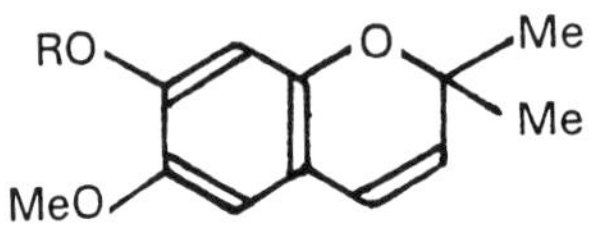

RO— ... Me, Me; MeO (ring structure)

(3)

HO— ... Me, Me; HO (ring structure)

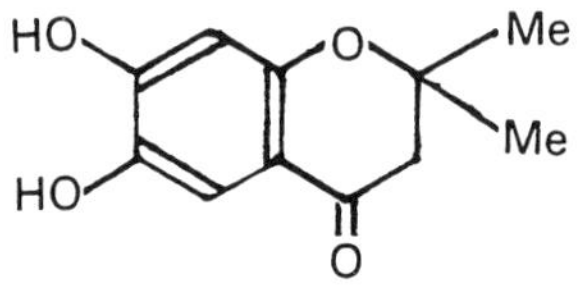

(4)

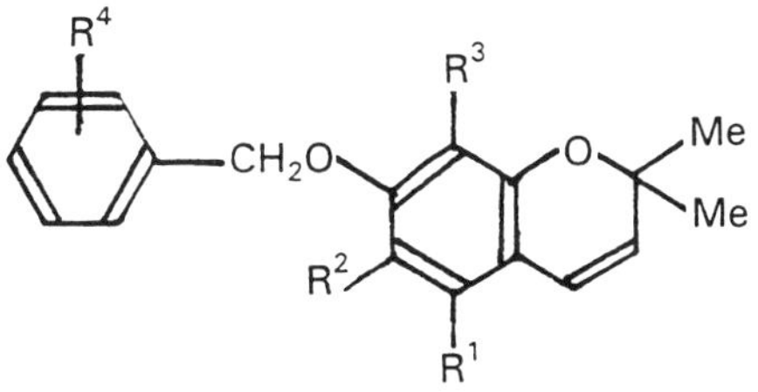

Phenylselenylation of precocenes, for example, 7, with PhSeCl in CH_2Cl_2 affords a mixture of phenylseleno derivatives (e.g., 8), chloro derivatives (e.g., 9), and dimers (e.g., 10) instead of the expected addition products. For 6-unsubstituted precocenes, aromatic substitution (of PhSe) at C-6 also occurs (P. Wagner *et al.*, J. Chem. Soc., Perkin 1, 1989, 2128).

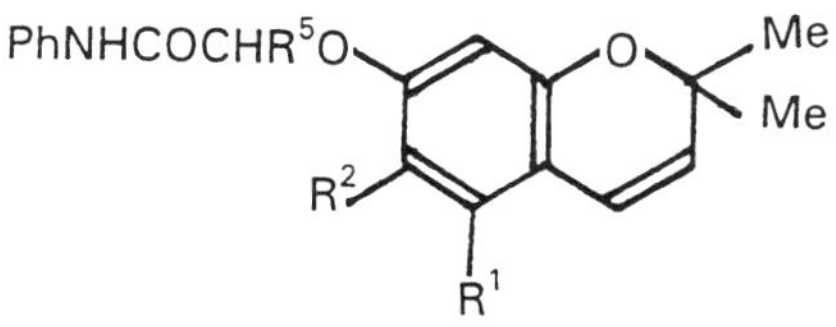

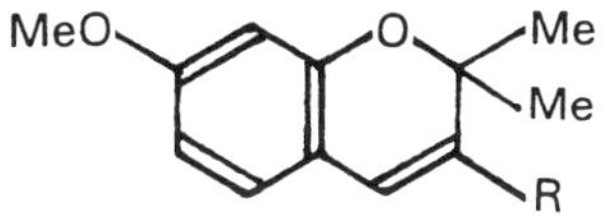

(5)

(6)

(7) R = H
(8) R = SePh
(9) R = Cl

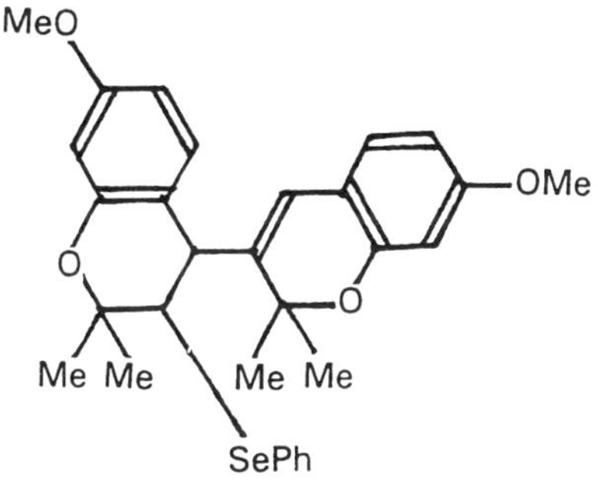

(10)

(iii) Benzopyrylium or chromylium and flavylium salts

The condensation of salicylaldehyde with with $MeCOBu^t$ in Me_2SO containing KOH, followed by cyclization with $HClO_4$ yields 2-*tert*-butylbenzo-[*b*]pyrylium perchlorate (1), which on reduction with a variety of reducing agents (e.g. $LiAlH_4$, $NaBH_4$, 1,3-dimethylbenzimidazoline) affords mixtures of 2-*tert*-butyl-2*H*- and -4*H*-benzo[*b*]pyrans (K.F. Suzdalev and A.V. Koblik, Khim. Geterotsikl. Soedin., 1989, 313).

$$ClO_4^-$$

(1)

The reaction of benzopyrylium salts with trialkyl phosphite gives benzopyranphosphoric acid derivatives, for example, 2, which undergoes a Wittig reaction with RCHO (R = Ph, 4-ClC$_6$H$_4$) to yield the benzylidene substituted derivative 3. Treatment of 3 (R = Ph) with CF$_3$SO$_3$H and Me$_2$C = C(OSiMe$_3$)OMe furnishes methyl 2-phenylmethyl-4*H*-benzo[*b*]pyran-4- -dimethylacetate (4) (H. Iwasaki and K. Akiba, Heterocycles, 1987, <u>28</u>, 2857).

(2)   (3)   (4)

4-*tert*-Butyldimethylsiloxybenzopyrylium salt (5) has been prepared in situ from chromone and F$_3$CSO$_3$SiMe$_2$Bt^t and it reacts with enol silyl ethers, ketene silyl acetals, and active methylene compounds to afford 2-substituted 4-*tert*-butyldimethylsiloxybenzopyrans 6 or chroman-4-ones 7 (R^1 = H, Me, Ph, CO$_2$Me, CN; R^2 = H, Me, COCHMe$_2$, CN, CO$_2$Me, Bz, CO$_2$Et; R^3 = H,

$COCHMe_2$, COEt, Ac, COC_6H_4Me-4, CO_2Me). Some derivatives of 6 have been converted to chroman-4-ones (H. Iwasaki *et al.*, Tetrahedron Letters, 1987, <u>28</u>, 6355).

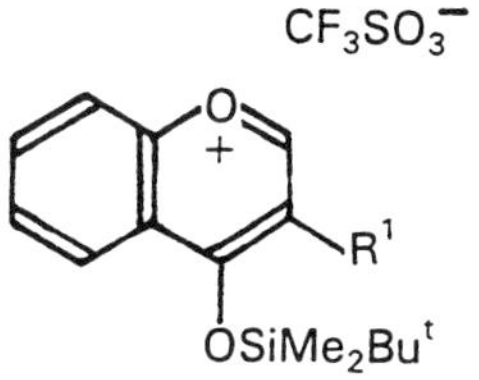

$CF_3SO_3^-$

(5) $R^1 = H$

(8) R^1 (see text)

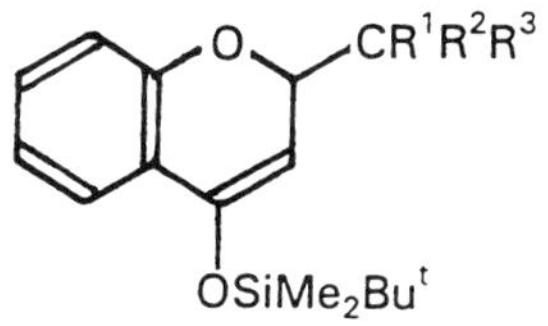

(6)

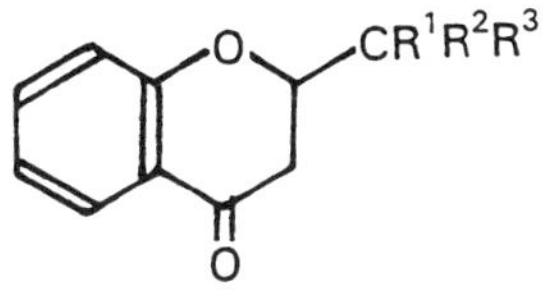

(7)

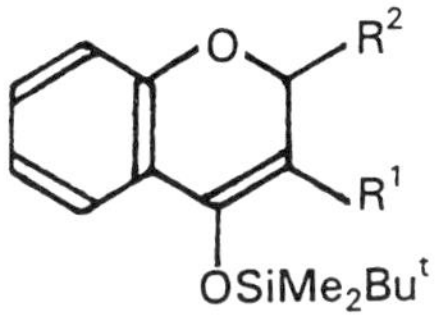

(9)

Reactions of 3-substituted 4-*tert*-butyldimethylsiloxybenzopyrylium triflates 8 ($R^1 = H$, Me, CO_2Me) with silyl enol ethers or allyl organometallic reagents give the corresponding 2,3-disubstituted 4-siloxy-2*H*-benzo[*b*]-pyrans 9 ($R^1 = H$, Me, CO_2Me; $R^2 = CH_2CH = CH_2$, CH_2COR^3, $R^3 = OEt$, Pr^i, Ph) along with 2,3-dihydrobenzopyrone derivatives. The reaction of benzo-pyrylium salts 8 with α,β-unsaturated ketones in the presence of *tert*-butyldi-methylsilyl triflate and 2,6-lutidine has been investigated (Y.G. Lee *et al.*, J. Org. Chem., 1991, <u>56</u>, 2058).

2-(Benzo[*b*]pyran-2-ylidenemethyl)benzopyrylium salts 10 [R^1-$R^3 = H$, alkyl, OH, alkoxy, NO_2, halogeno, (un)saturated aryl; R^1R^2, $R^2R^3 = $ atoms of a complete fused (heterocyclic) ring; $X^- = $ strong acid anion] as intermediates for thermo- and photochromic compounds have been prepared [R. Dusi and H. Hartmann, Ger. (East) DD 265,622 1989].

(10)

4-Chloromethylflavylium perchlorate (11; R = H) and 4-chloromethyl-6-
-methylflavylium perchlorate (11; R = Me) are prepared by the condensation
of 5,2-R(OH)$C_6H_3COCH_2Cl$ with PhCOMe in AcOH containing 70% $HClO_4$.
The related 4-chloromethyl-2-methyl- and -2,6-dimethylbenzothiopyrylium
perchlorates have also been prepared (S.V. Tolkunov and V.I. Dulenko, Khim.
Geterotsikl. Soedin., 1987, 766).
The reaction of 4,3-$R^1R^2C_6H_5$OAc (R^1 = H, Me, Cl; R^2 = H, Me) with
$SnCl_4$ gives mixtures of bis(aryloxy)stannates 12 and benzopyrylium pentachlorostannates (favylium pentachlorostannates) 13. Compound 12 and 13
have been separated and flavylium salts 14 and 15 have been prepared (H.
Jolibois et al., Helv., 1988, 71, 812). Flavylium salts 16 (R^1, R^2 = H, YO,
Y = H, C_{1-5}α-amino acyl or α-amino alkyl; R^3 R^4, R^7 = H, C_{1-5}alkyl; R^5 = H,
C_{1-5}αamino acyl, or α-amino alkyl; R^6 = H, HO, C_{1-5}alkyl; X = anion) useful as
antiviral, particularly against HIV, have been prepared (G. Sulyok et al., PCT
Int. Appl. 92 9,203,000, 1992).

ClO_4^-

(11)

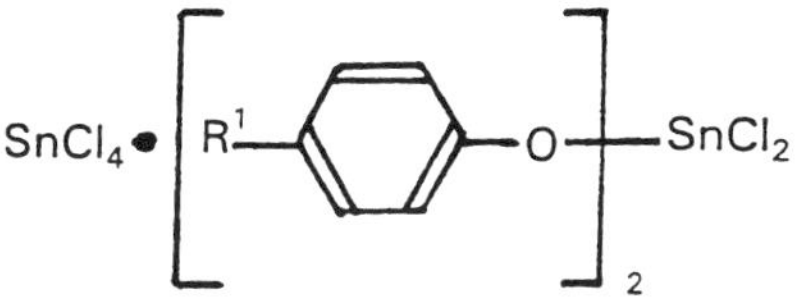

(12)

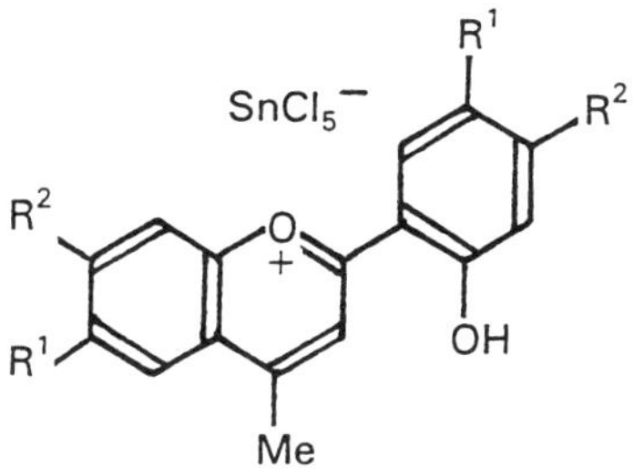

(13)

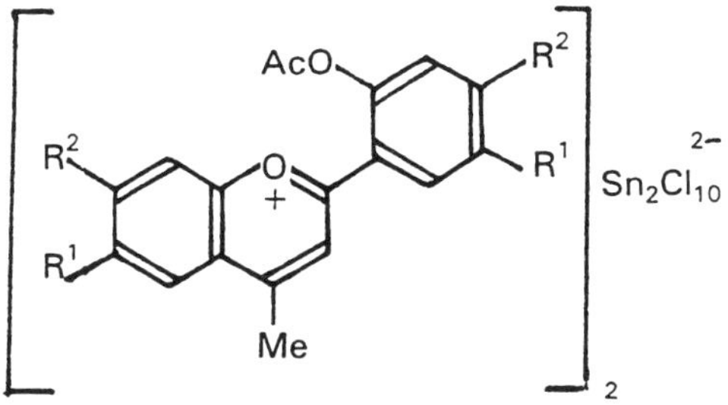

(14)

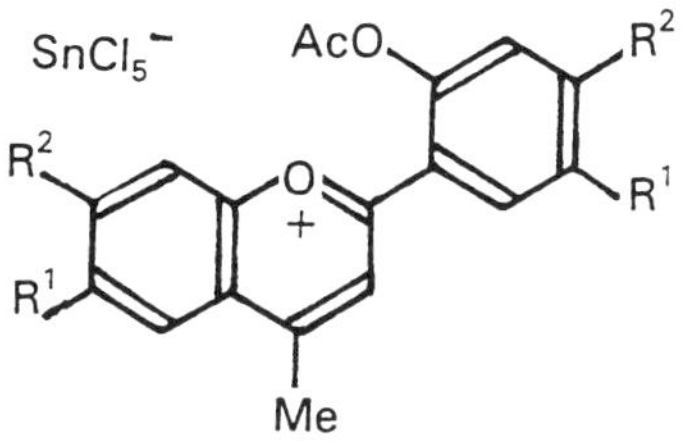

(15)

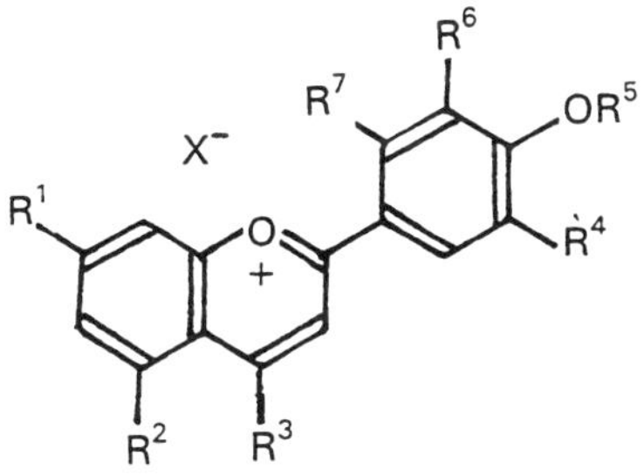

(16)

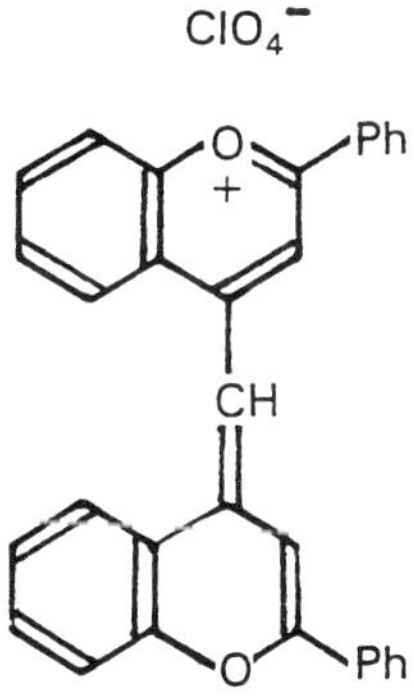

(17)

Cyanines, for example, **17**, have been obtained from the appropriate benzopyrylium perchlorates and an aliphatic acid anhydride (A.I. Pyshchev *et al.*, Khim. Geterotsikl. Soedin., 1989, 1563).

The reaction of phenylpropynal diethyl acetal, PhC≡CCH(OEt)$_2$ with dimedone gives the intermediate **18**, which on subsequent cyclization with HClO$_4$ affords 7,7-dimethyl-5-oxo-5,6,7,8-tetrahydroflavylium perchlorate (**19**) (E.P. Olekhnovich *et al.*, Zh. Org. Khim., 1990, <u>26</u>, 2609). 4-Hydroxy- -7,7-dimethyl-5-oxo-5,6,7,8-tetrahydrobenzopyrylium perchlorate (**20**) in the presence of NaOAc is converted to 7,7-dimethyl-5-oxochromone (**21**) (E.P. Olekhnovich, I.V. Korobka, and V.V. Mezheritskii, *ibid.*, p.173). For the synthesis of 2-(polyfluoroaryl)-4-phenyl-5,6,7,8-tetrahydrobenzopyrylium perchlorates from polyfluorochalcones, see N.A. Orlova *et al.*, *ibid.*, p1313.

(18)

(19) R = H
(20) R = OH, Ph = H

(21)

(22)

The reaction of 3-acetyltropolone with 2-hydroxybenzaldehyde, 2- -hydroxy-5-methylbenzaldehyde, and 5-chloro-2-hydroxybenzaldehyde in the presence of ethyl formate and perchloric acid affords the corresponding benzopyrylium tropolone perchlorates **22** (R = H, Me, Cl) (D.L. Wang and Z.T. Jin, Chin. Chem. Letters, 1993, <u>4</u>, 877).

The trihydroxybenzopyrylium chlorides 23 ($R^1 = 3\text{-HOC}_6\text{H}_4$, $R^2 = Me$; $R^1 = Ph$, $R^2 = 2\text{-HOC}_6\text{H}_4$) have been obtained *via* the cyclocondensation of benzene-1,2,4-triol triacetate (24) with $3\text{-HOC}_6\text{H}_4\text{COCH}_2\text{COMe}$ and $PhCOCH_2COC_6H_4OH\text{-}2$, respectively (G.F. Tantsyura, L.Ya. Grinskaya, and L.P. Kharchenko, Ukr. Khim. Zh., 1989, 55, 1082).

(23)   (24)

(iv) Coumarin (2H-benzo[b]pyran-2-one, 5,6-benzo-2-pyrone, 5,6-benzo--a-pyrone) and its derivatives

Alkyl- and arylcoumarins. The fused benzopyranone derivative 1, obtained by the Pd-catalyzed reaction of $2\text{-IC}_6\text{H}_4\text{OH}$, CO, and norbornadiene, undergoes a retro Diels-Alder reaction to yield coumarin (2) and cyclopentadiene (Z.W. An, M. Catellani, and G.P. Chiusoli, J. Organomet. Chem., 1989, 371, C51).

(1)   (2)

260

Coumarin (2) has also been obtained by the dehydrogenation of dihydrocoumarin with sulphur and desulphurizing agents, for example, Raney-Ni (W2) (K. Okumura, Jpn. Kokai Tokkyo Koho JP 02 96,574 [90 96,574], 1990; treating salicylaldehyde and acetic anhydride with anhdrous sodium fluoride as catalyst (C. Zhou and J. Hu, Xiangtan Daxue Ziran Kexue Xuebao, 1990, _12_, 79); heating salicylaldehyde with Ac_2O and polyethylene glycol to give $2\text{-}AcOC_6H_4CHO$ and then heating the product with Ac_2O and KF (C. Zhou, G. Xie, and F. Lin, Yingyoung Huaxue, 1992, _9_, 79); the palladium-catalyzed intramolecular vinylation of 2-bromophenyl acrylate (L. Cao, M. Bei, and L. Hua, Fenzi Cuihua, 1991, _5_, 296); the condensation of $2\text{-}HOC_6H_4CHO$ with Ac_2O in the presence of a quaternary salt or a trialkyl-amino catalyst (T.L. Simandan _et al._, Rom. RO 94,549, 1988); and the use of dibenzo-18-crown-6 as an additional catalyst to potassium fluoride in the above condensation reaction (G. Xie, Xiangtan Daxue Ziran Kexue Xuebao, 1993, _15_, 87).

Coumarin on heating in tetralin at 400°C affords PhOH (3.7%), $2\text{-}MeC_6H_4OH$ (7.8%), and $2\text{-}EtC_6H_4$ (8.9%). Dihydrocoumarin and $2\text{-}HOC_6H_4CH_2CH_2CO_2H$ are also isolated. A mechanism for the formation of the phenols has been proposed (C. Huang, H. Yang, and D. Han, Wul Huaxue Xuebao, 1988, _4_, 194). A reinvestigation of a sensitized photochemical cycloaddition of coumarin with cyclopentene indicated that the cycloaddition gave a sterically disfavoured _cis_-cisoid-_cis_-adduct 3 (18%) together with a _cis_--transoid-_cis_-adduct 4 (25%) and a head-to-head coumarin dimer 5 with a _cis_-transoid-_cis_ disposition (H. Suginome and K. Kobayashi, Bull. Chem. Soc. Jpn., 1988, _61_, 3782). The transient absorption spectra and quenching of coumarin excited states by nucleic acid bases have been investigated (H.K. Kang, E.J. Shin, and S. Shim., J. Photochem. Photobiol., B, 1992, _13_, 19).

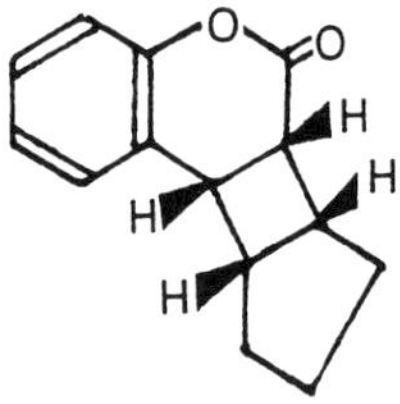

(3) 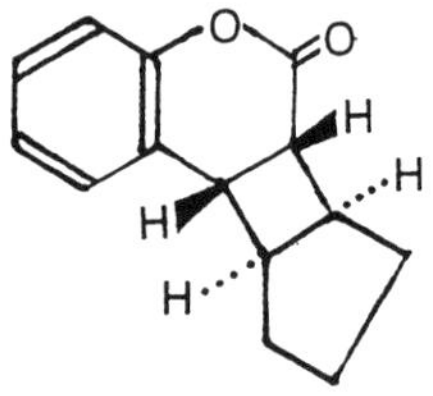   (4) 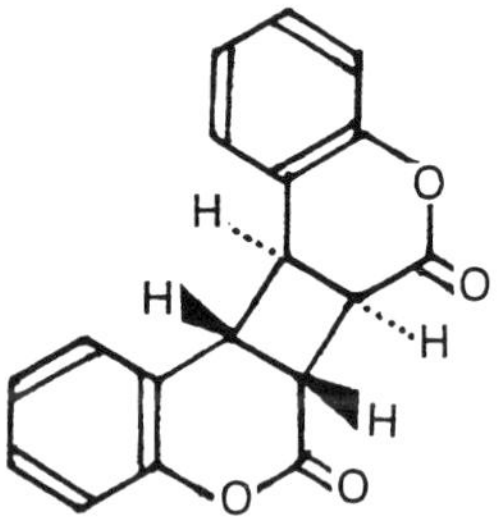   (5)

3-Methylcoumarin (6) has been obtained, following a number of steps from (Z)-oxazolinyl(phenylthio)styrene (J.C. Clinet, Tetrahedron Letters, 1988, **29**, 5901) and some 3-alkylcoumarins have been synthesized (G. Falsone *et al.*, Z. Naturforsch., B, Chem. Sci., 1993, **48**, 1391). Reaction of a preformed complex of amides (e.g., $PhCH_2CH_2CONEt_2$) and $POCl_3$ with substituted salicylaldehyde affords derivatives of 3-benzylcoumarin (7) (S.Y. Deshmukh, S.L. Kelkar, and M.S. Wadia, Synth. Comm., 1990, **20**, 855).

(6) R = Me
(7) R = CH_2Ph

(8)

It has been found that montmorillonite clay is an effective condensing agent in the Pechmann reaction for the synthesis of coumarin derivatives, for example, heating a mixture of PhOH, $MeCOCH_2CO_2Et$, and montmorillonite clay affords 4-methylcoumarin (8) (~70%) (G.K. Biswas *et al.*, Indian J. Chem. 1992, **31B**, 628). 4-Methylcoumarin (8) has also been obtained by the cyclization reaction of phenol and acetic anhydride over CeNaY zeolite, other zeolites, CuNaY, CrNaY, LaNaY, and HY have also been tested (Y.V.S. Rao *et al.*, Chem. Comm., 1993, 1456), and by the bis(dibenzylidene-acetone)palladium complex catalyzed Heck reaction of $2\text{-}IC_6H_4OCOCH_2\text{-}CH=CH_2$ in the presence of potassium butyrate/benzonitrile/CO/PPh_3 in anisole at 80°C (M. Catellani *et al.*, Tetrahedron Letters, 1994, **35**, 5919). The Grignard addition of alkylmagnesium halides to 3-ethoxycarbonyl-coumarin in the presence of CuBr, followed by hydrolysis, decarboxylation, and dehydrogenation of the resulting dihydrocoumarins yields the 4-alkyl-coumarins 9 (R = Me, Et, CH_2Ph) (A. Patra and S.K. Misra, Indian J. Chem., 1988, **21B**, 272). 4-Substituted coumarins 9 (R = Me, Et, Pr, Ph) have been obtained by the Baeyer-Villiger oxidation of indanones 10, followed by dehydrogenation of the resulting dihydrocoumarin (*idem, ibid.*, 1990, **29B**, 66).

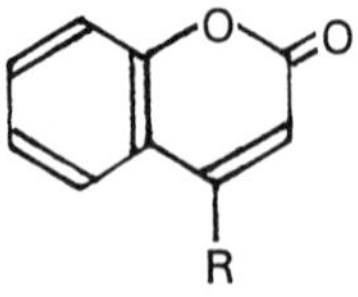

262

(9)

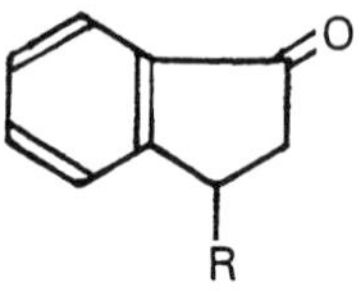

(10)

Several dialkylcoumarins 11 (e.g., R^1-R^4=H, C_{1-4}alkyl, with at least no substitutent on two of these positions along with related 3,4-dihydro-coumarins have been prepared by heating an appropriate 3-(2-cyclohexano-yl)propionic ester in presence of a Pd catalyst (Y. Nishida *et al.*, Jpn. Tokkyo Koho JP 05 78,345 [93 78,345], 1993; Eur. Pat. Appl. EP 526,231, 551,019, 1993) or a catalyst comprising a carrier supporting Pd and either chromium oxide or chromium hydroxide (T. Shirafuji, K. Sakai, and K. Okasako, *ibid.*, 434,410, 1991).

Dialkylcoumarins 11 have been obtained in good yield by dehydrogenation of the corresponding 3,4-dihydrocoumarins, either on heating them with 50% hydrated Pd/C under nitrogen at 230°C and 450 torr (Y. Nishida, K. Sakai, and T. Shirafuji, *ibid.*, 539,181, 1993) or on heating over 5% Pd/C in Ph_2O (K. Okumura, A. Imamura, and H. Mizumoto, Jpn. Kokai Tokkyo Koho JP 03 197,478 [91 197,478], 1991).

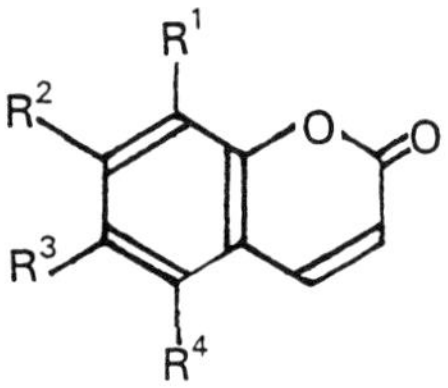

(11)

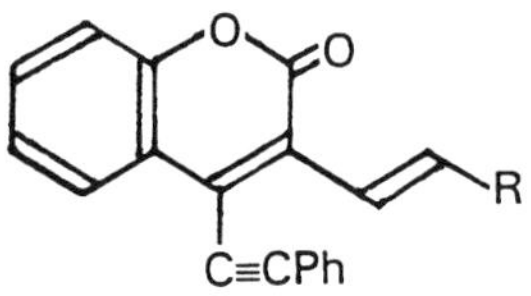

(12)

Pd(O) has been found to catalyze the reactions of an aryl iodide with CO, the triple bond of alkynes and the allylic C of an alkenoic chain, to form 3,4-
-disubstituted coumarins, for example, 3-(2-substituted ethenyl)-4-(2-phenyl-
ethynyl)coumarins 12 (R = Me, Ph) (M. Catellani *et al.*, Tetrahedron Letters, 1994, 35, 5923). The acylation of 4-cresol with Ac_2O at 200-380°C over a zeolite containing aluminosilicate, Al_2O_3, $AlCl_3$, Ca-Ni phosphate, or modenite catalyst affords a mixture of $4-MeC_6H_4OAc$, $5,2-Me(HO)C_6H_3COMe$, 4,6-
-dimethylcoumarin (22% at 380-400°C) and 2,6-dimethylchromone (A.A. Medzhidov and Sh.G. Gamzaeva Sb. Tr. - Akad. Nauk Az. SSR, Inst. Neftekhim. Protsessov im Akad. Yu. G. Mamedalieva, 1986, 15, 41).

Both 3- and 4-alkyl-7-substituted coumarins 13 (R^1 = H, alkyl, Ph; R^2 = H, OMe) on treatment with SeO_2 yield the corresponding 3- or 4-acylcoumarins 14. 4-Methyl-2*H*-benzo[*b*]-pyran-2-thione on reaction with SeO_2 affords 4-
-methylcoumarin (8) (K. Ito and K. Nakajima, J. Heterocyclic Chem., 1988, 25, 511).

(13) R = CH_2R^1
(14) R = COR^1

(15)

(16)

Coumarins 15 (R^1 = H, Me, Cl, R^2 = H, Me, R^3 = H, Me, OMe) reacted at room temperature during 2 days with hexadiynediols 16 (R^4 = H, Cl) in $CHCl_3$ or MeOH to afford 1:2 inclusion complexes (J.N. Moorthy and K. Venkatesan, J. Org. Chem., 1991 56, 6957).

264

The Wittig reaction of dihydroxybenzils 17 (R^1 = H, OMe, R^2 = H, Me)
with Ph_3P = $CHCO_2Et$ gives biscoumarins 18. Compound 18 (R^1 = R^2 = H) has
also been prepared in a number of stages from 2-$HOC_6H_4COCH_2C_6H_4OMe$-2
(R.S.K. Deshmukh and M.V. Paradkar, Synth. Comm., 1988, _18_, 589).

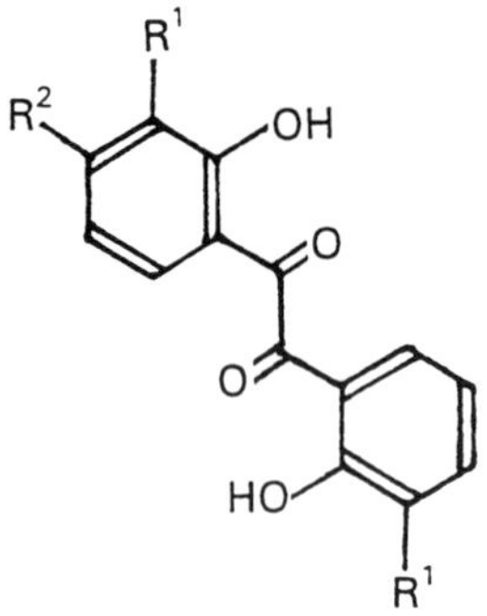

(17) 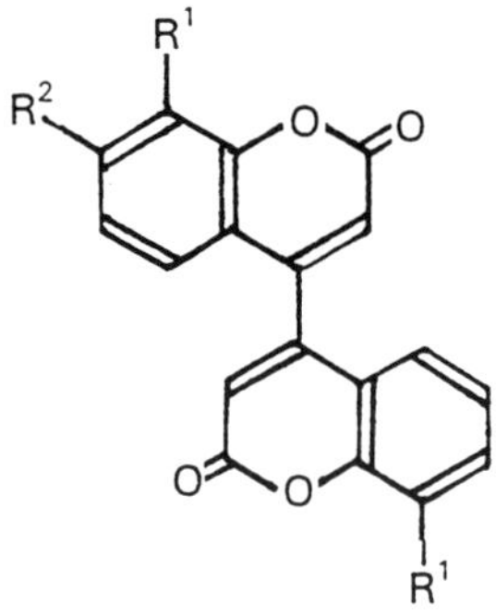   (18)

Several cyclopentanocoumarins (cyclopentabenzopyran-2-ones) 19
[R^1 = H, OH, OMe; R^2 = H, OH, Me, *n*-pentyl; R^3 = H, Cl, $Me(CH_2)_n$, Bu^t;
R^4 = H, OH; n = 0-5] have been obtained by the condensation of
appropriately substituted phenols with ethyl cyclopentanone-2-carboxylate in
the presence of phosphorus oxychloride or 73% sulphuric acid (B.S. Verma
et al., Chim. Acta Turc. 1989, _17_, 433).

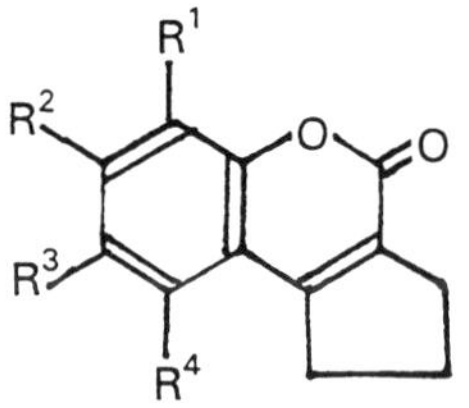

(19) 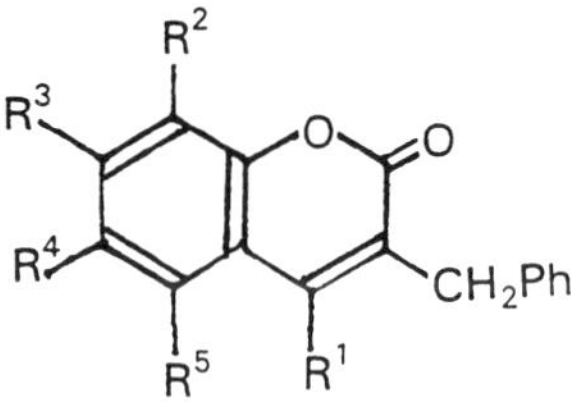   (20) 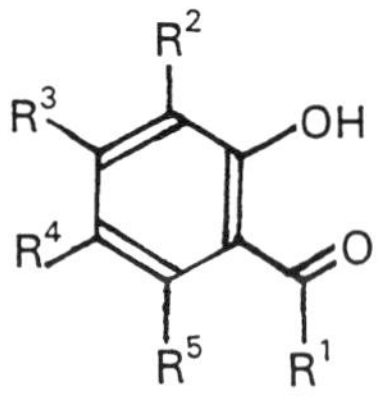   (21)

The substituted 3-benzylcoumarins 20 (R^1 =H, Me, Ph; R^2,R^3 =H, Me; R^4 =H, Me, MeO, Cl; R^5 =H; R^2R^3 = R^4R^5 =CH=CHCH=CH) are prepared from phenols 21 and Ph_3P=C(CO_2Et)CH_2Ph, for example, 3-benzyl-4-phenyl-coumarin (20; R^1 =Ph, R^2-R^5 =H) is obtained by heating 2-HOC$_6$H$_4$COPh with Ph_3P=C(CO_2Et)CH_2Ph at 210°C for several hours (N. Britto *et al.*, Synth Comm., 1989, *19*, 1899).

Cyclocondensation of phenols 22 (R^1 =H, Me, Et, Ph; R^2 =H, Me, Cl, Br; R^3 =H, $PhCH_2CO_2$, HO, 4-BrC$_6$H$_4$CH$_2$CO$_2$; R^4 =H, Me) with 4-R^5C$_6$H$_4$CH$_2$-COCl; R^5 =H, Br, MeO) under phase-transfer conditions in the presence of Bu_4N^+ HSO$_4^-$ affords the 3-arylcoumarins 23 (G. Sabitha, Mrs., and A.V. Subba Rao, *ibid.*, 1987, *17*, 341). Similarly, salicylaldehydes react with phenylacetic anhydride and its methoxyphenyl derivatives under solid-liquid phase transfer-catalyzed conditions using $Et_3N^+CH_2$PhCl$^-$ to give 3-phenyl and 3-arylcoumarins (S. Mohanty, J.K. Makrandi, and S.K. Grover, Indian J. Chem., 1989, *28B*, 766).

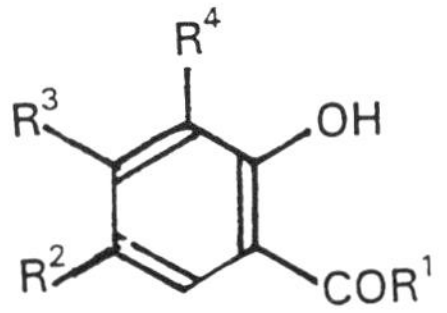

(22)

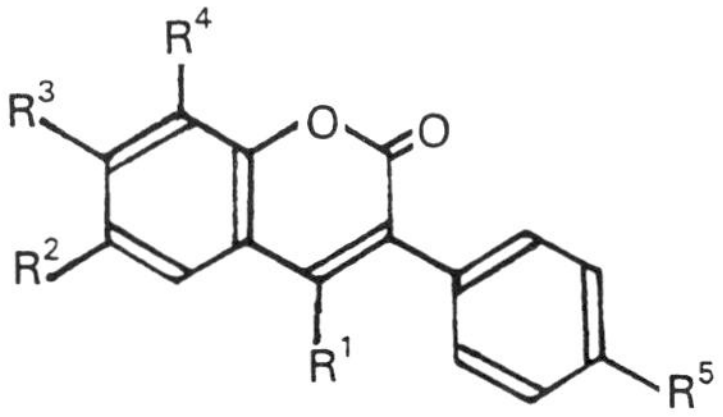

(23)

3-Arylcoumarins 24 (R^1 =H, OMe; R^2 =H, Me; R^3 =H; R^2R^3 =benzo; R^4 =H, $PhCH_2O$; R^5 =OMe, H; R^6 =H, OMe) have also been obtained by treating salicylaldehydes with 2-phenylthioacetamides and $POCl_3$. 3-Aryl-coumarin 24 (R^1 =R^2 =R^3 =R^5 =R^6 =H, R^4 =$PhCH_2O$) has been converted into coumestan (25) (M.S. Phansalkar *et al.*, Indian J. Chem., 1987, *26B*, 562). Fluorescent 3-arylcoumarins 26 (R=Ph, 1-naphthyl, 9-phenanthryl, 2-furyl, 1-methylpyrrol-2-yl, 1-methylindol-2-yl) are prepared by the photochemical coupling of 3-bromocoumarin (26; R=Br) with the appropriate aromatic or heteroaromatic compound (J. Meng *et al.*, Synthesis, 1990, 719).

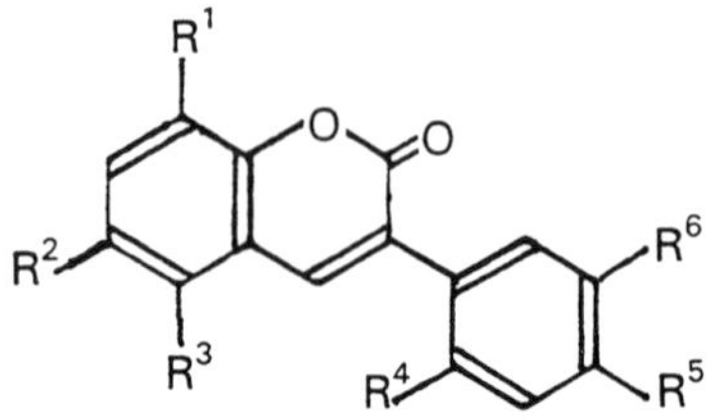

266

(24) 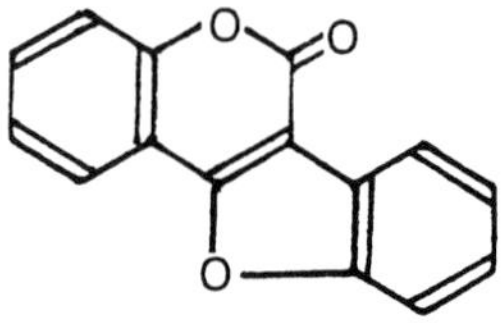   (25)

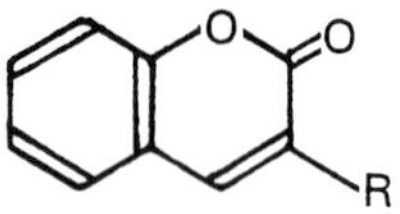

(26) 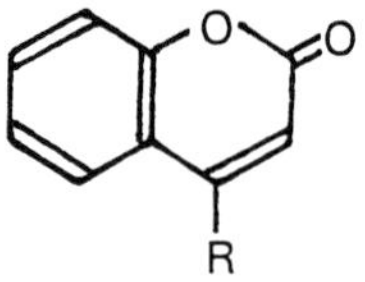   (27)

4-(4-Fluorophenyl)- and 4-(pyrid-3-yl)coumarins 27 (R = 4-FC$_6$H$_4$, pyrid-3-
-yl, respectively) have been prepared by the (PPh$_3$)$_4$Pd catalyzed coupling
reaction of coumarin 27 (R = triflate) with (R^1)$_3$R^{2}Sn (R^1 = Bu, R^2 = 4-FC$_6$H$_4$;
R^1 = Me, R^2 = pyrid-3-yl, respectively) (S. Wattanasin, Synth. Comm., 1988,
18, 1919). 3,4-Disubstituted coumarins 28 (R^1 = Me, Ph; R^2 = H, R^3 = H,
Me, OMe; R^{2}R^3 = benzo; R^4 = Me, Ph), for example, 3,4-dimethyl- and 3,4-
-diphenylcoumarins 28 (R^2 = R^3 = H, R^1 = R^4 = Me, Ph, respectively), are
prepared by the reaction of 2-hydroxyphenyl ketones 29 with R^{4}CH$_2$CONEt$_2$
and POCl$_3$ (V.O. Patil, S.L. Kelkar, and M.S. Wadia, Indian J. Chem., 1987,
26B, 674). For the preparation of 6,7-disubstituted 4-aryl-3-phenyl-
coumarins 30 (R^1 = Me, H; R^2 = H, OMe; R^3 = H, OAc, EtCO$_2$), see I. Sharma
and S. Ray, ibid., 1988, 27B, 374); and 3-(naphth-1-yl)coumarins 31
(R^1 = R^2 = H, R^3 = H, Me, Cl; R^1 = H, R^2 = MeO, R^3 = H; R^1 = MeO,
R^2 = R^3 = H), R.B. Tejwani et al., ibid., 1989, 28B, 414. Some coumarins on
treatment with Me$_2$S-BH$_3$ have been converted to the corresponding
chromans (P. Verma et al., Synthesis, 1988, 68).

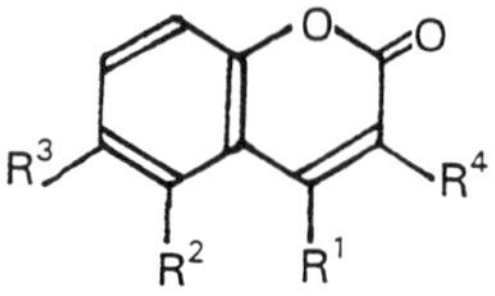

(28) 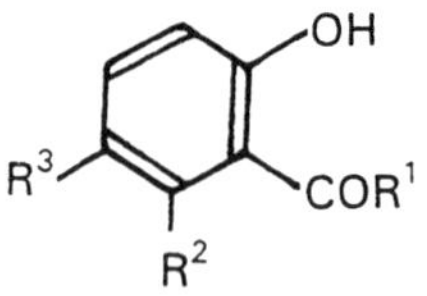   (29)

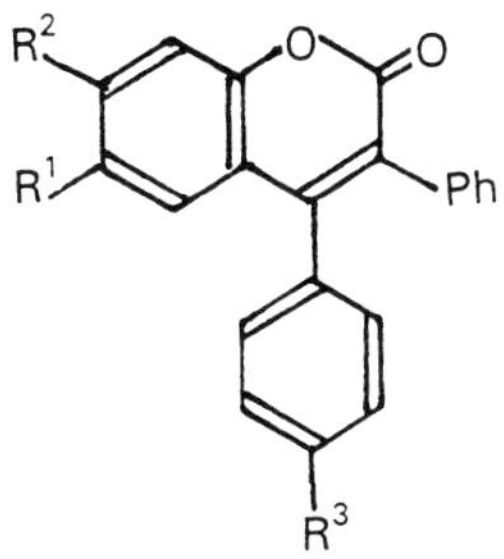

(30)

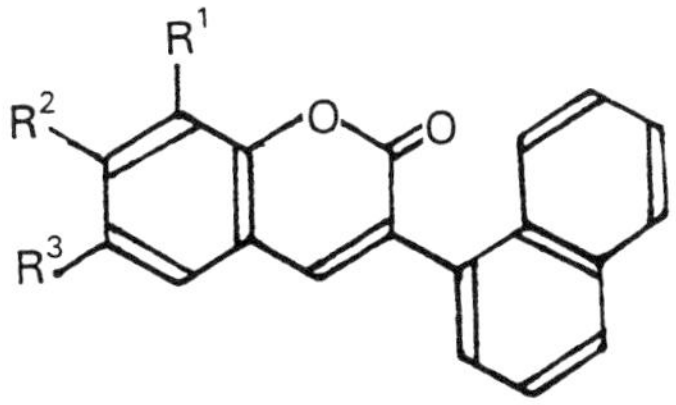

(31)

3-Styrylcoumarins 32 (e.g. R = H, OMe, Cl, NMe$_2$) have been prepared by condensing coumarin-3-acetic acid with 4-RC$_6$H$_4$CHO in the presence of piperidine (N.K. Chodankar *et al.*, Indian J. Chem., 1987, 26B, 427). For the preparations of 3-phenyl-4-styrylcoumarins, see P.D. Lokhande and B.J. Ghiya, J. Indian Chem. Soc., 1989, 66, 314.

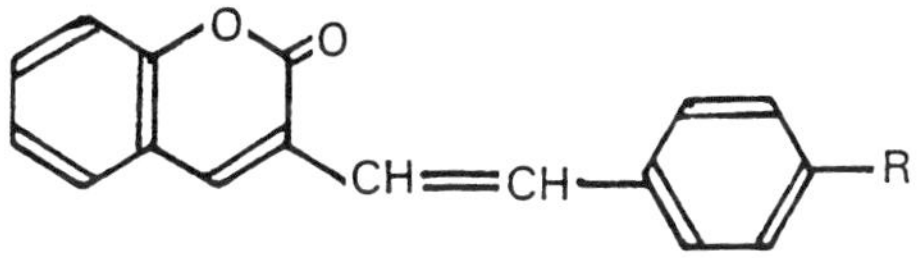

(32)

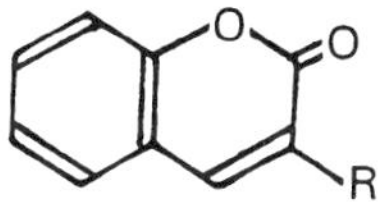

(33)

The preparations of the following coumarins possessing a 3-heteroaryl substituent have been reported; coumarins 33 (R = heteroaryl containing one or more hetero atoms) (S.S. El-Morsy, A.A. Fadda, and M.S. El-Hossini, *ibid.*, 1988, 65, 699), 3-(benzofuran-3-yl)coumarins (A.R. Deshpande and M.V. Paradkar, Synth. Comm., 1990, 20, 809), and 3-(2-aminothiazol-4-yl)- and 3-(indol-2-yl)coumarins (L. Rao and A.K. Mukerjee, Indian J. Chem., 1994, 33B, 166). A number of 6-azolylcoumarins have been prepared as herbicides (M. Ganzer *et al.*, Ger. Offen. DE 3,810,706, 1989).

268

Acylcoumarins and coumarincarboxylic acids. 3-Acetylcoumarin derivatives 34 (R^1 = H, R^2 = OMe; R^1 = Br, R^2 = H, Br; R^3 = H) on treatment with bromine afford bromoacetylcoumarin derivatives 34 (R^1R^2 same R^3 = Br), which react with thiourea or amines to yield coumarinylthiazoles 35 or derivatives 34 (R^3 = NHR^4; R^4 = substituted Ph), respectively. Derivative 34 (R^3 = Br) has been converted into 3-(2-arylamino-1-aryliminoethyl)-coumarin 36. The antibacterial activity of derivatives 34, 35, and 36 has been studied and they are all active against *Escherichia coli* (O.H. Hishmat *et al.*, Arch. Pharmacal Res., 1989, 12, 181).

(34)

(35)

(36)

The reaction of 3-acetylcoumarin with aqueous NH_3 yields oxaazatri-cyclotridecatrienone 37 (R^1 = coumarin-3-yl, R^2 = OH), whereas reaction with NH_4OAc affords compound 37 (R^1 = coumarin-3-yl, R^2 = NH_2). The former derivative with NH_4OAc furnishes the latter derivative, and mild hydrolysis of latter gives the former (C.N. O'Callaghan and T.B.H. McMurry, J. Chem. Res., S, 1989, 329).

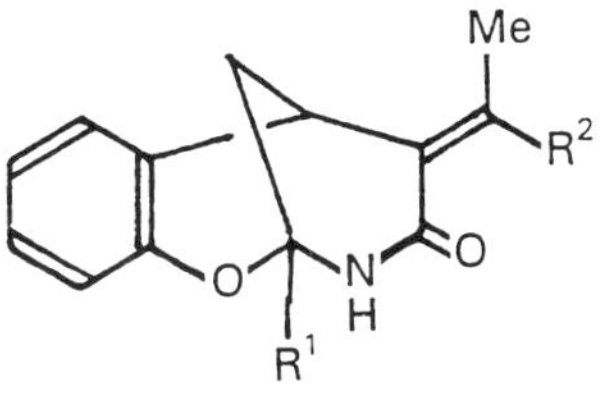

(37)

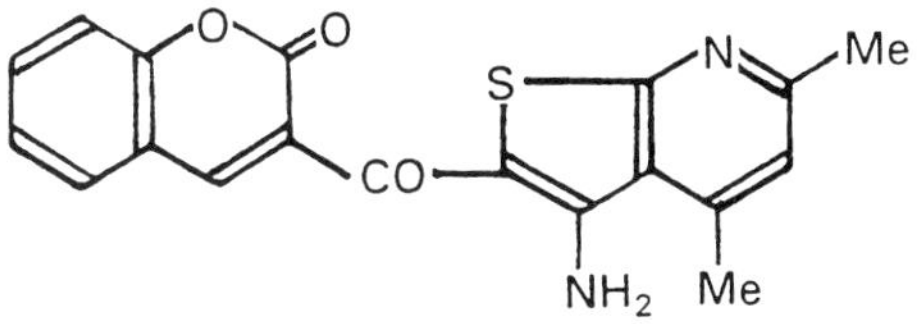

(38)

The reaction of 3-acetyl and 3-benzoylcoumarins with acid anhydrides in the presence of AcONa and Et_3N has been investigated (A. Bodzhilova and Kh. Ivanov, Indian J. Chem., 1987, 26B, 731). The reaction of 3-(bromo-acetyl)coumarin with 3-cyano-4,6-dimethylpyridine-2-thiol in the presence of K_2CO_3 gives the coumarin derivative 38 (A.M. El-Agrody et al., Pak. J. Sci. Ind. Res., 1993, 36, 175).

(Benzopyranylidene)(methoxycarbonyl)ethylidenedimethylammonium perchlorates under hydrolysis conditions give retro-Knoevenagel products, i.e. methyl coumarin-3-carboxylates (M. Weissenfels et al., Z. Chem., 1990, 30, 19). The interaction of esters 39 (R = CO_2Me, CO_2Et, CO_2Ph) of coumarin-3-carboxylic acid and 3-cyanocoumarin (39; R = CN) with CCl_3-CO_2H yields the chroman-2-ones 40, cyclopropanes 41, and dihydrobenzo-cyclopropanones 42 (A. Bozhilova, Synth. Comm., 1990, 20, 1967).

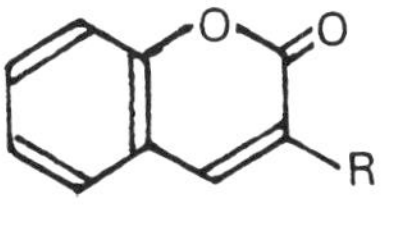

(39)

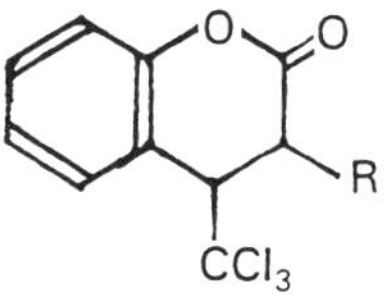

(40)

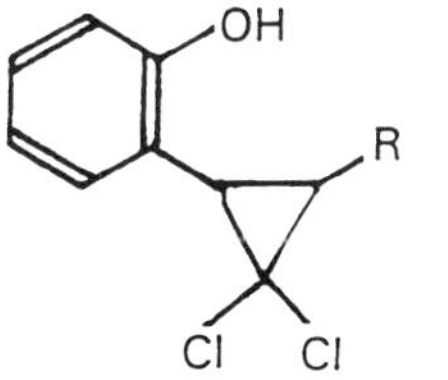

(41)

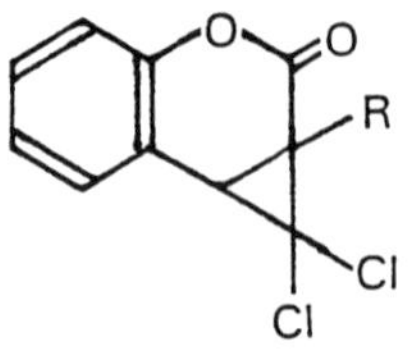

(42)

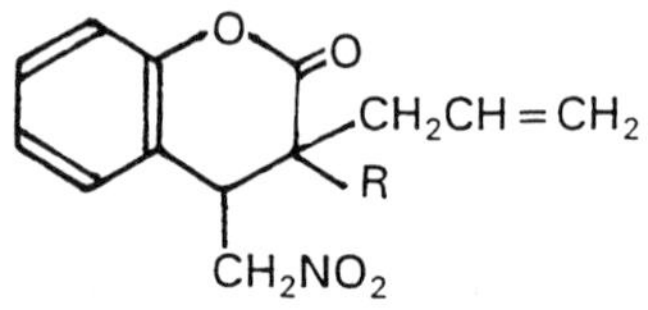

(43)

The addition of $MeNO_2$ to ester 39 ($R = CO_2R^1$; $R^1 = Me$, Et, Bu^t, $CH_2C_6H_4NO_2$-4, Ph) and 3-cyanocoumarin 39 ($R = CN$) by the action of 1,8--diazabicyclo[5.4.0]undec-7-ene proceeds only in the presence of allyl bromide to give chroman-2-ones 43 (A. Bozhilova, T. Kostadinova, and Kh. Ivanov, Ann., 1989, 1041). These essentially complete (>99%) asymmetric Diels-Alder reactions have been achieved between 3-(8-phenyl-menthyloxy-carbonyl)coumarin (44) and buta-1,3-dienes in high yields to afford the adducts 45 (R^1, $R^2 = alkyl$) (K. Ohkata et al., Tetrahedron Letters, 1993, 34, 6575).

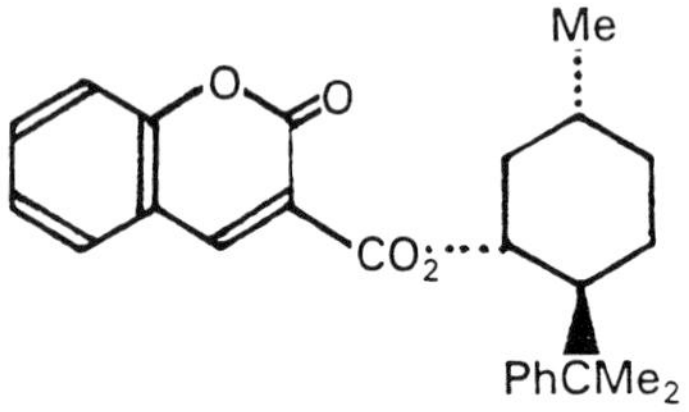

(44)

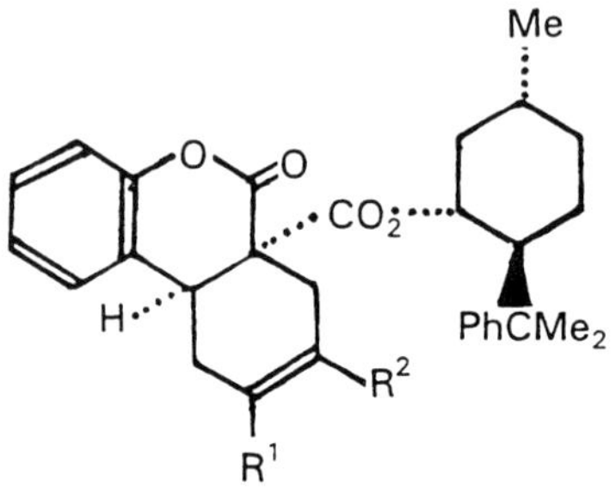

(45)

The cyclopropanation of ethyl coumarin-3-carboxylate (39; $R = CO_2Et$) with phenacyl bromide in the presence of benzyltriphenylphosphonium chloride phase transfer catalyst gives the *trans* and *cis*-substituted cyclopropane derivatives 46 in a ratio of 80:20 and an overall yield of 92% (A. Bojilova, A. Trendafilova, and C. Ivanov, Tetrahedron, 1993, <u>49</u>, 2275). A study has been made of the reactions of coumarin-3-carboxylic acid chloride with some nucleophilic agents (D.I. Gezalyan *et al.*, Khimiya i Khim. Tekhnol., Erevan, 1989, 127. From Ref. Zh., Khim. 1990, Abstr. No 12Zh236).

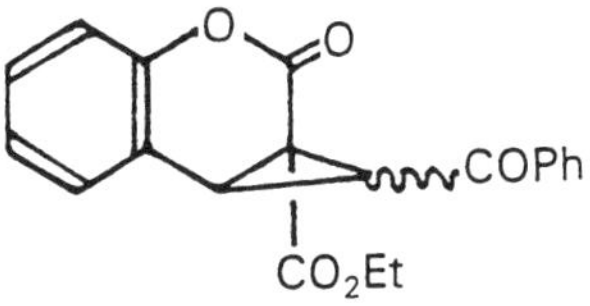

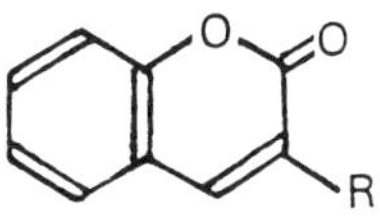

$R = (CH_2)_{n-1}CO_2H$

(46) (47) (48)

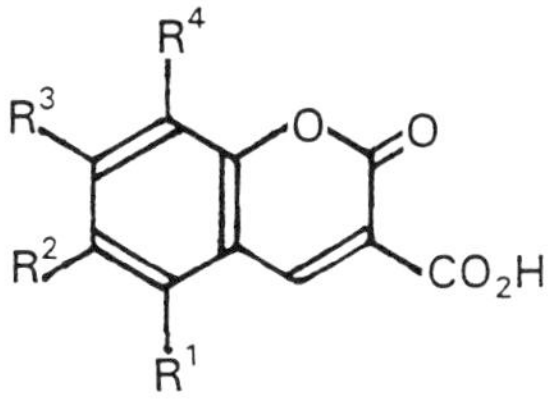

ω-(Coumarin-3-yl)alkanoic acids 47 (n = 3-6) have been prepared by the cyclization of the corresponding ethyl 2-formylphenyl alkanedioate, 2--OHCC$_6$H$_4$O$_2$C(CH$_2$)$_n$CO$_2$Et with DBU, followed by hydrolysis (S. Yamaguchi, M. Saitoh, and Y. Kawase, J. Heterocyclic Chem., 1991, <u>28</u>, 125). The reaction between 2-methoxybenzaldehydes and Meldrum's acid gives benzylidene derivatives, which on cyclization in H_2SO_4 afford coumarin-3--carboxylic acid (48; R^1-R^4 = H) or substituted coumarin-3-carboxylic acids, for example, 48 (R^1-R^4 = OMe) (J.A. Valderrama and R. Tapia, Synth. Comm., 1988, <u>18</u>, 717).

The 7-aminocoumarin-4-carboxylates 49 (R^1 = H, Me, CH$_2$PH; R^2 = H, Ac) and the related derivative 50 (R = H, Me, Ac) have been synthesized and their fluorescence characteristics investigated and compared with 7-amino-4--methylcoumarin (T. Besson, G. Coudert, and G. Guillaumet, J. Heterocyclic Chem., 1991, <u>28</u>, 1517).

272

(49)

(50)

The reaction of alkoxycarbonylmethylene(triphenyl)phosphoranes with 4-
-(triphenylmethyl)-1,2-benzoquinone in dry benzene at room temperature
gives compound 51 (R = alkyl) along with triphenylphosphine oxide. Boiling
compound 51 in toluene affords a mixture of alkyl 6-triphenylmethyl-
coumarin-4-carboxylate 52, 5-triphenylmethyl-3-alkoxycarbonylmethylene-
benzo[*b*]furan-2(3*H*)-ones and triphenylphosphine (F.H. Osman, N.M. Abd
El-hahman, and F.A. El-Samahy, Tetrahedron, 1993, _49_, 8691). The
synthesis and some reactions of 7-methylcoumarin-4-acetic acid have been
discussed (A.Y. Soliman *et al.*, Indian J. Chem., 1991, _30B_, 477).

(51)

(52)

Methyl coumarin-6-carboxylate (53; R = 6-CO$_2$Me) (96%) and related
derivatives (R = Br, OH, OMe, CO$_2$Me) are obtained in high yield by the Wittig
reaction of salicylaldehydes with Ph$_3$P = CHCO$_2$Et and cyclization of the
resulting *trans*-olefins 54 (T. Harayama *et al.*, Chem. Express, 1993, _8_,
245).

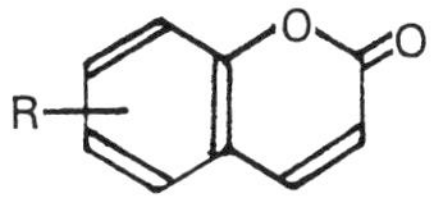

(53) 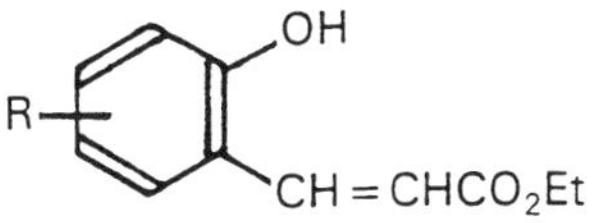 (54) 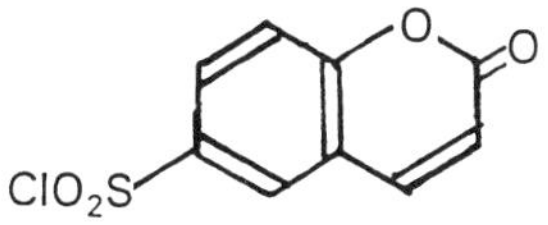 (55)

The reaction of coumarin with $ClSO_3H$ affords coumarin-6-sulphonyl chloride (55) and it has been shown that this derivative can be used as a luminescence labelling reagent for amino acids, amines, etc. which have reactivity with the SO_2Cl group (S. Yang and L. Li, Fenxi Huaxue, 1992, 20, 202). The fluorogenic properties of 4-methylumbelliferone-6-sulphonyl chloride, 4-methyl-7-acetoxylcoumarin-6-sulphonyl chloride, 4-methyl-7-methoxycoumarin-6-sulphonyl chloride, and 4-methylcoumarin-6-sulphonyl chloride have been investigated (Q.-E. Cau Q.-H. Xu, Gaodeng Xuexiao Huaxue Xuebao, 1994, 15, 675). For reactions of coumarin-6-sulphonyl chloride with 2-, 3-, and 4-phenylenediamines to give the sulphonamides, and some reactions of the sulphonamides see, H. Abd El-Bary *et al.*, Afinidad, 1994, 51, 311; and for the preparation and chemistry of several coumarin-6-sulphonamides, R.J. Cremlyn and S.M. Clowes, J. Chem. Soc. Pak., 1988, 10, 97.

Halogenocoumarins. 3-Chlorocoumarin (56) has been prepared by treating coumarin with hydrogen chloride and 3-chloroperbenzoic acid in DMF at room temperature (K.M. Kim *et al.*, Synthesis, 1993, 283). Coumarin and its derivatives have been halogenated selectively at the C-3 position with copper (II) halides adsorbed on neutral alumina and halogenobenzenes (P.C. Thapliya, P.K. Singh, and R.N. Khanna, Synth. Comm., 1993, 23, 2821).

(56)   (57)   (58)

3-Halogenocoumarins 57 (R^1 = H, Me; R^2 = Cl, Br) react with magnesium, lithium, aluminium, and copper derivatives to yield 3,4-dihydrocoumarins (chroman-2-ones) 58 (R^1 = Et, Bu, Pr^i; R^2 = Cl, Br) and 3-(2-hydroxyphenyl)-propenols, e.g., 2-HOC_6H_4CH = $CClCHMeCH_2OH$, as major products. The tandem alkylation-dehydrogenation of 3-halogenocoumarins provides a versatile route to 4-alkylcoumarins (A. Alberola *et al.*, J. Chem. Soc., Perkin 1, 1991, 203). Salicylaldehydes on heating with $ClCH_2CONEt_2$ and $POCl_3$ in $CHCl_3$ give the substituted 3-chlorocoumarins 59 (R^1 = H, OMe; R^2 = H, OMe; R^3 = H, Me; R^4 = H, or R^3R^4 = benzo). Similarly, coumarins 60 (R^1 = H, Me; R^2 = H, or R^1R^2 = benzo; R^3 = H, OH) are obtained from 2-hydroxybenzo-phenones and $ClCH_2CONEt_2$. Coumarin 60 (R^1 = R^2 = H, R^3 = OH) on treatment with KOH furnishes isocoumestan (61) (A.P. Borse, S.L. Kelkar, and M.S. Wadia, Indian J. Chem., 1987, <u>26B</u>, 1180).

(59)   (60)   (61)

Reactions of morpholine with 4- or 3-halogenocoumarins and 2-halogeno-
-1,4-naphthoquinones yield two different products, one where halogen is
replaced by nucleophile at the same carbon and the other where the
nucleophile is attached to carbon vicinal to that bearing the halogen (C. Oh,
I. Yi, and K.P. Park, J. Heterocyclic Chem., 1994, $\underline{31}$, 841).

4-Chlorocoumarin (62) is a highly versatile starting material and when
treated with organometallic reagents, for instance, it allows the selective
synthesis either directly, or through simple additional transformations, of 4-
-alkylcoumarins 63 (R = Me, Bu, Pr^i) (with R$_2$CuLi in Et$_2$O or Me$_2$CHMgBr in
THF), 2-chloro-2-(2-hydroxyphenyl)allyl alcohols, e.g., 2-HOC$_6$H$_4$CCl = CHCR$_2^1$-
OH (R^1 = Me, Et, Pr, Bu, Pr^i) or their 4-chloro-2H-benzo[b]pyran derivatives
64 (R^1 = Me, Et, Pr, Bu) [with R^{1}MgX (X = halogeno) in THF]. 2,2,4-Trialkyl-
-2H-benzo[b]pyrans 65 are obtained when excess of reagents, R^{1}MgX or R$_3^1$Al
are used (A. Alberola et $al.$, J. Chem. Soc., Perkin 1, 1992, 3075).

(62) R = Cl
(63) R (see text)

(64)

(65)

4-Chloro-7-dialkylaminocoumarins 66 (R^1 = Et, R$_2^1$N = piperidino,
morpholino, R^2 = H, Et, cyclohexyl, CH$_2$Ph) have been prepared by heating
aminophenols 67 with R^{2}CH(CO$_2$H)$_2$ in excess POCl$_3$. Reacting derivatives
66 (R^1 = Et, R^2 = H) with NaOEt, Bu$_2$NH, and N$_2$H$_4$.H$_2$O affords coumarins 68
(R^3 = OEt, Bu$_2$N, NHNH$_2$) in 45, 63, and 17% yields, respectively (M.A.
Kirpichenok, S.K. Gorozhankin, and I.I. Grandberg, Khim. Geterotsikl.
Soedin., 1990, 830). Methods of producing 4-chloro-7-dialkylamino-
coumarins have been discussed ($idem$, U.S.S.R. SU 1,594,177, 1990).

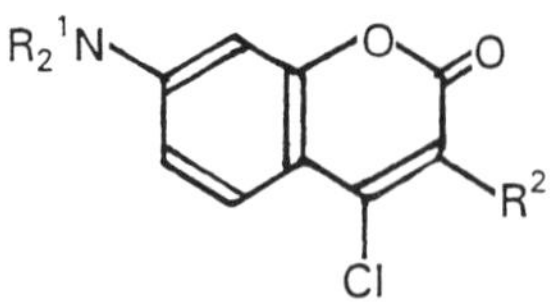

(66) 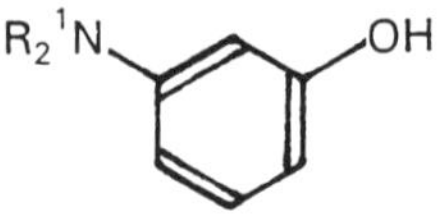   (67) 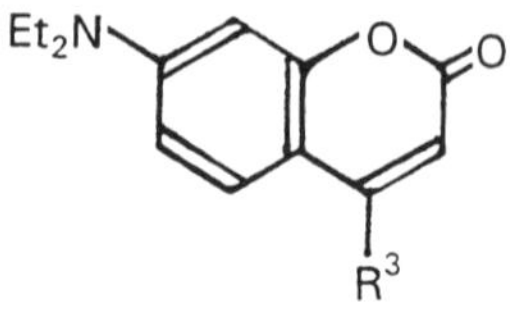   (68)

The structure of 4-chloro-7-diethylaminocoumarin (66; R^1 = Et, R^2 = H) has been confirmed by x-ray analysis (D.S. Yufit *et al.*, Izv. Akad. Nauk, Ser. Khim., 1992, 445) and the reactions of 4-chloro-7-dialkylamino- and 3-alkyl--4-chloro-7-dialkylaminocoumarins with primary and secondary alkylamines have been investigated (M.A. Kirpichenok, S.K. Gorozhankin, and I.I. Grandberg, Khim. Geterotsikl. Soedin., 1990, 836). For the preparations and reactions of 3-acylamino-6,8-dichlorocoumarin, see M.M. Ismail and M.M. Kandeel, Rev. Roum. Chim., 1994, *39*, 183.

The reaction of 4-chloromethylcoumarin 69 (R^1 = Bu^t, R^2 = R^3 = H) with 4--ClC$_6$H$_4$OH in Me$_2$CO containing K$_2$CO$_3$ gives 4-(4-chlorophenoxymethyl)-6--*tert*-butylcoumarin (70). Coumarins 69 (R^1 = Bu^t, R^2 = R^3 = H; R^1 = H, R^2 = R^3 = Me) have been treated with substituted phenols and the resulting phenoxy derivatives tested for toxicity towards mycelial growth of 7 plant--pathogenic fungi in culture. 7,8-Dimethyl-4-(4-nitrophenylmethyl)coumarin was the most active compound (R. Singh *et al.*, Indian J. Chem., 1989, *28B*, 996).

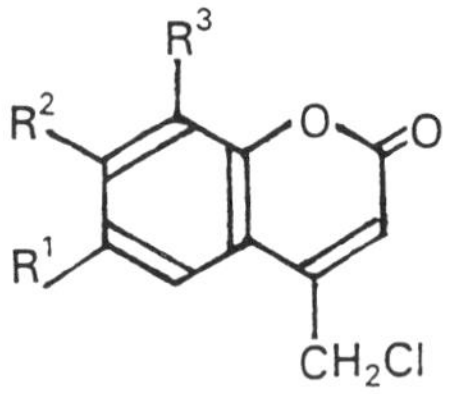

(69) 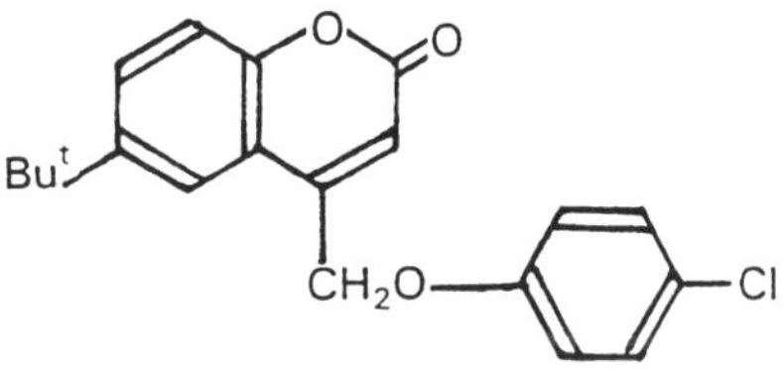   (70)

Racemic-threo-2-dichloroacetamidol-1-[3-(6-bromocoumarinyl)]propane-1,3-diol
(71) an analogue of chloramphenicol has been synthesized (R.V. Joshi and
V.V. Badiger, J. Heterocyclic Chem., 1988, <u>25</u>, 45). The reaction of 3-(α-
-bromobenzyl)coumarin (72; R = Br) with alcohols affords the alkyl ethers 72
(R = OR^1; R^1 = Me, Et, Pr, Pr^i, CH$_2$CHMe, CHMeEt, Bu^t, CH$_2$Ph, cyclohexyl)
of 3-(α-hydroxybenzyl)coumarin (A. Bozhilova and Kh. Ivanov, Doklady Bolg.
Akad. Nauk, 1989, <u>42</u>, 83). For the preparation of 3-(2-bromoethyl-
coumarins 73 (R^1 = H, Me, CMe = CH$_2$; R^2 = R^3 = H; R^2 = MeOCH$_2$O, R^3 = H,
pentyl), see T. Minami *et al.*, J. Org. Chem., 1992, <u>57</u>, 167.

(71)

(72)

(73)

Substituted 7-dialkylamino-3-iodocoumarins 74 [R^1 = Cl, Me, morpholino;
R^2 = Et; R^3 = H; R^{2}R^3 = (CH$_2$)$_3$] have been obtained by the reaction of the
corresponding 7-aminocoumarin with iodine in dioxane or THF in the
presence of pyridine as the catalyst (N.A. Gordeeva, M.A. Kirpichenok, and
I.I. Grandberg, U.S.S.R. SU 1,578,126, 1990). The reaction of 7-dialkyl-
amino-3-iodocoumarins 75 [R^1 = H, Me; R$_2^2$ = Et$_2$, (CH$_2$)$_3$] with secondary
amines (Et$_2$NH, piperidine, morpholine, imidazole,, benzimidazole) (M.A.
Kirpichenok, L. Yu. Fomina, and I.I. Grandberg, Khim. Geterotsikl. Soedin.,
1991, 609) and the photochemical reactions of 7-diethylamino-3-iodo-4-
-methylcoumarin with olefins (N.A. Gordeeva *et al.*, *ibid.*, 1990, 1033) have
been investigated.

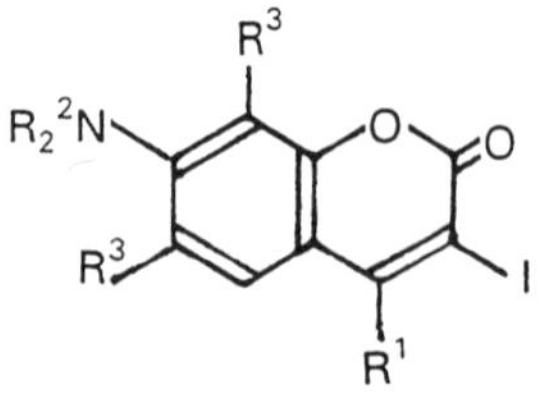

(74)

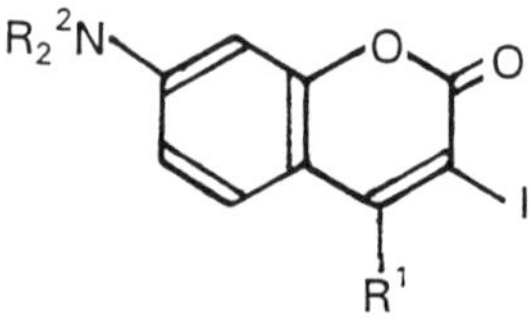

(75)

6-Aminocoumarin hydrochloride (76) in EtOH at 60°C on treatment with iodine monochloride affords 6-amino-5-iodocoumarin (77; R^1-R^4 = H), which inhibits cell growth and also inhibits HIV proliferation in human lymphoblasts. The preparations of a number of substituted 6-amino-5-iodocoumarins 77 [R^1, R^2, R^3, R^4 = H, HO, H_2N, alkyl, alkoxy, cycloalkyl, halogeno, (substituted) Ph] or a salt thereof, have been reported (E. Kun and J. Mendeleyev, PCT Int. Appl. WO 92 06,687, 1992).

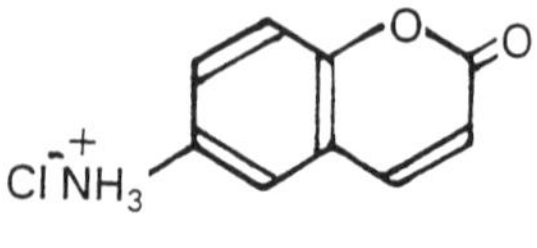

(76)

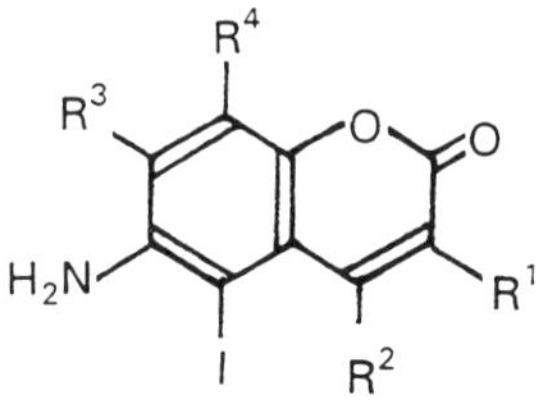

(77)

A number of 3-perfluoroalkylated coumarins (B. Huang, J. Liu, and W. Huang, Chem. Comm., 1990, 1781; M. Matsui *et al.*, Synlett, 1991, 113; H. Muramatsu *et al.*, Jpn. Kokai Tokkyo Koho JP 04 124,184 [92 124,184], 1992; H. Sawada and M. Nakayama, *ibid.*, 04 128,280 [92 128,280], 1992; B.N. Huang, J.T. Liu, and W.Y. Huang, J. Chem. Soc., Perkin 1,

1994, 101) and 3-aryl-7-hydroxy-4-trifluoromethyl-, 3-aryl-5-hydroxy-7-
-methyl-4-trifluoromethyl-, and 3-aryl-7-methoxy-4-trifluoromethylcoumarins
(T. Nishiwaki and H. Kikukawa, J. Heterocyclic Chem., 1994, _31_, 889) have
been prepared. It has been found that in sharp contrast to the packing and
photobehaviour of unsubstituted coumarin, molecules of both 6- and 7-
-fluorocoumarins attain an assembly, which enables them to undergo
[2 + 2]photodimerization in the solid state, leading stereospecifically to a
mirror symmetrical dimer (V.A. Kumar, N.S. Begum, and K. Venkatesan, J.
Chem. Soc., Perkin 2, 1993, 463).

*Amino-, cyano-, and nitrocoumarins and coumarincarboxamides and
related derivatives*. 2-Hydroxyaryl ketones, for example, $3,6\text{-Cl(HO)}C_6H_3Ac$,
condense with ethyl cyanoacetate in the presence of piperidine or
ammonium acetate as a catalyst to give 3-cyano-4-methylcoumarins, for
example, 6-chloro-3-cyano-4-methylcoumarin (78). The behaviour of 3-
cyano-4-methylcoumarin (79; R = Me) towards Grignard and Michael
reactions has been investigated (H.M.F. Madkour, Heterocycles, 1993, _36_,
947). The reaction of 3-cyanocoumarin with Grignard reagents, RMgX
(X = halide) gives 4-substituted 3-cyanocoumarins 79 (R = Et, 4-tolyl) (A.M.
El-Agrody *et al.*, J. Chem. Soc. Pak., 1993, _15_, 261). 3-Cyano-4-methyl-
coumarin (79; R = Me) condenses with aromatic aldehydes, ArCHO, to give
4-(2-arylethenyl)-3-cyanocoumarins 80 (A.A. Avetisyan *et al.*, Arm. Khim.
Zh., 1990, _43_, 59). Also reported are the preparations of ethenylcyano-
coumarins 80 [Ar = Ph, 4-MeC_6H_4, 4-MeOC_6H_4, $4\text{-(Me}_2\text{N)}C_6H_4$, 4-NCC_6H_4, 4-
-ClC_6H_4, 1-naphthyl, 9-anthryl, 3,4-(methylenedioxy)phenyl, 4-pyridyl] (I.
Angelova, and E. Dimitrova, Org. Prep. Proced. Int., 1989, _21_, 341).

(78)　　　　　　(79)　　　　　　(80)

A series of nitrocoumarin and nitrobenzo[*b*]pyran derivatives have been
prepared and shown to inhibit the phosphatidylinositol-specific phospholipase
C (PLC) isolated from human melanoma. The inhibition of PLC by 3-nitro-
coumarin (81) has been shown to be time-dependent and irreversible (F.W.

Perrella *et al.*, J. Med. Chem., 1994, _37_, 2232). The interaction of 4-chloro-
-3-nitrocoumarin (82; R = Cl) with thiolating reagents, for example, NaSH,
Na_2S, KSCN, $Na_2S_2O_3$, PhCOSNa, yields mostly 4-hydroxy-3-nitrocoumarin
along with some dimer 83. The expected product 4-mercapto-3-nitro-
coumarin is not formed (E.A. Parfenov and L.D. Smirnov, Khim. Geterotsikl.
Soedin., 1991, 1032). However the reaction of 4-chloro-3-nitrocoumarin
(82; R = Cl) with an equimolar amount of the appropriate benzylthiol yielded
4-benzylthio-3-nitrocoumarin (82; R = SCH_2Ph) and 4-(2-carboxyphenylthio)-
-3-nitrocoumarin (82; R = $SC_6H_4CO_2H$-2). With excess $PhCH_2SH$ 3,4-
-dibenzylthiocoumarin (84) was obtained and 4-methoxy-3-nitrocoumarin
(82; R = OMe) with an equimolar amount of $PhSCH_2SH$ afforded a mixture of
derivatives 82 (R = SCH_2Ph) and 84 (*idem, ibid.*, p.1476).

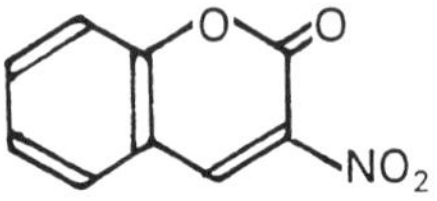

(81) 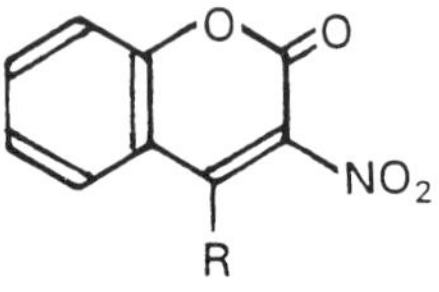   (82) 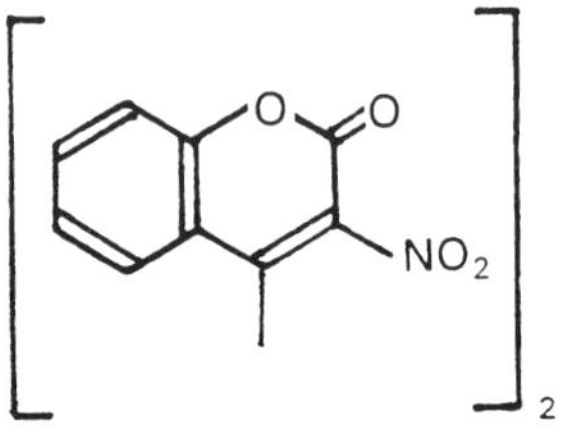   (83)

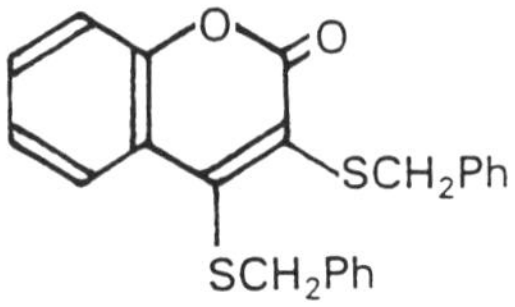

(84) 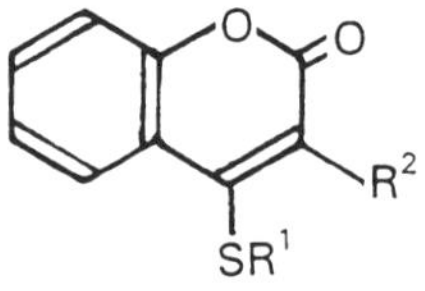   (85)

The 4-(substituted thio) derivatives 85 (R^1 = $PhCH_2$, 2-$HO_2CC_6H_4$;
R^2 = NO_2) nitrocoumarin on reduction with $Na_2S_2O_4$ afford 21-26% 85
(R^2 = NH_2), 25-42% 85 (R^2 = H), and 19% 85 (R^1 = 2-$HO_2CC_6H_4$, R^2 = SH).
An $S_{RN}1$ mechanism involving neighbouring-group participation has been
proposed (*idem, ibid.*, 1990, 1135). 4-Benzylthio-3-nitrocoumarin (82;
R = SCH_2Ph) on treatment with SO_2Cl_2 in $CHCl_3$ containing DMF and then

with 4-toluidine in Et_2N affords 3-nitro-4-(4-tolylamino)coumarin (86) (*idem,
ibid.*, 1992, 888). The stabilization of 3-nitrocoumarin-4-ylthiolate by
ammonium counterions has been discussed (*idem, ibid.*, 1993, 459).
Heating spirocoumarin salts 87 (R^1 =H, NO_2) in MeCN give the 4-mercapto-
-3-nitrocoumarins 88 (R^2 = S⁻ Et_3 NH⁺), which are alkylated with EtI to yield
derivatives 88 (R^2 = SEt), also obtained on treating 4-chloro-3-nitro- and 3,6-
-dinitrocoumarins with EtSH (T. Ya. Mozhaeva *et al.*, Zh. Org. Khim., 1990,
<u>26</u>, 1597).

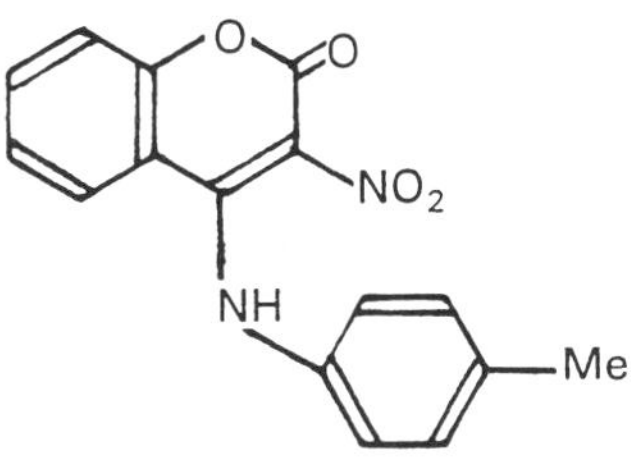

(86)

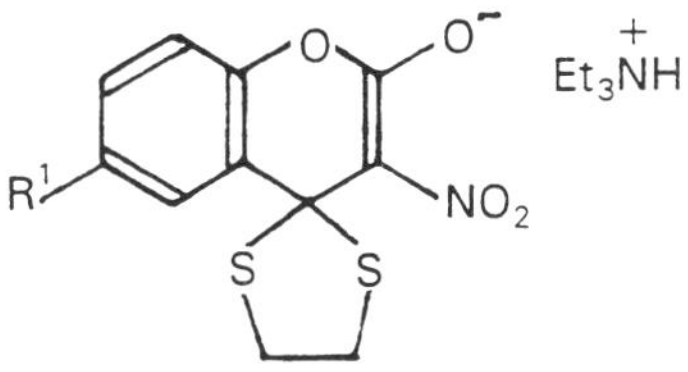

(87)

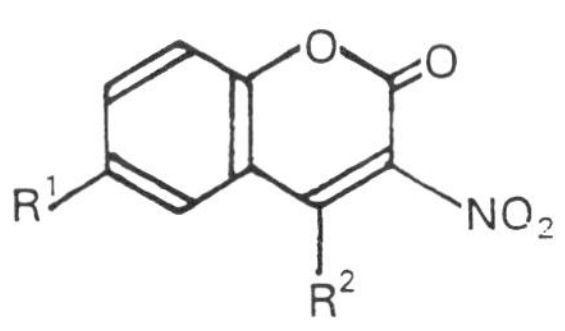

(88)

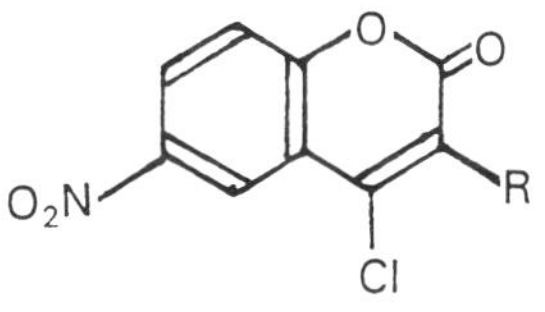

(89)

Treatment of 4-chloro-3,6-dinitrocoumarin (89; R = NO_2) with $POCl_3$ in
DMF gives substitution products 89 (R = CHO, Cl). Similar results are
obtained with 4-hydroxy-3,6-dinitrocoumarin, but 4-chloro-3-nitrocoumarin is
essentially unreactive (T. Ya. Mozhaeva, O.L. Samsonova, and V.L. Savel'ev,
Khim, Geterotsikl. Soedin., 1988, 1287).

The 3-acetamidocoumarins 90 (R = NHPh, substituted anilino, piperidino, NHNHPh) and 3-imino-4-(phenylhydrazono)dihydrocoumarins 91 (R = H, Me, NO_2) have been prepared from 3-aminocoumarin (H.M. Hassan *et al.*, Pak. J. Sci. Ind. Res., 1987, 30, 573).

(90) R = $NHCOCH_2R$
(92) R = NMe_2

(91)

The reaction between salicylaldehyde, betaine, and acetic anhydride gives 3-(*N,N*-dimethylamino)coumarin (92) and not 2-(*N,N*-dimethamino)-indan-1,3-dione as previously reported (P. Hrnciar, A. Gaplovsky, and J. Donovalova, Tetrahedron Letters, 1989, 30, 1709; Chem. Pap., 1989, 43, 669). Confirmation of this has been obtained from IR (A. Perjessy *et al.*, Coll. Czech. Chem. Comm., 1990, 55, 185) and NMR spectral data (T. Liptaj *et al.*, Chem. Pap., 1990, 44, 71). The above method has been used to prepare substituted 3-(*N,N*-dimethylamino)coumarins 93 (R^1 = H, Cl, NO_2; R^2 = H, Br, MeO; R^3 = H, Cl, MeO), colourless light stabilizers for plastics, from the corresponding salicylaldehydes (P. Hrnciar, A. Gaplovsky, and J. Donovalova, Czech. CS 275,270, 1992). 3-(*N*-Methylaminocoumarins 94 (R^1 = H, Cl, NO_2; R^2 = H, Br, MeO; R^3 = H, Cl, MeO) have been prepared by photochemical monodemethylation of the corresponding 3-(*N,N*-dimethyl-amino)coumarins (J. Donovalova, A. Gaplovsky, and P. Hrnciar, *ibid.*, 276,460, 1992). The preparations of 3-acetyl, 3-benzoyl, and 3-(4-methyl-benzoyl)aminocoumarins and their 8-hydroxy derivatives have been reported (A. Jain and A.K. Mukerjee, J. Prakt. Chem., 1989, 331, 493).

(93) R = NMe$_2$
(94) R = NHMe

(95)

(96)

3,4-Diaminocoumarins 95 (R^1, R^2 = H, alkyl, alkanediyl, aryl; R^3 = NH$_2$) have been prepared by reduction of the 4-amino-3-nitrocoumarins 95 (R^3 = NO$_2$) with Na$_2$S$_2$O$_4$. Similarly 4-hydroxy-3-nitrocoumarin afforded 3-amino-4-hydroxycoumarin, but Na$_2$S$_2$O$_4$ reduction of 4-methoxy-3-nitro- coumarin gave, along with the anticipated 3-amino-4-methoxycoumarin, 4-methoxycoumarin. The former is the preferred product in aqueous MeOH/NaOAc, whereas the latter is preferred in the presence of bicarbonate (E.A. Parfenov and L.D. Smirnov, Khim. Geterotsikl. Soedin., 1992, 329).

4-Aminocoumarins 96 (R = NH$_2$, alkylamino) are prepared by treating 4-hydroxycoumarin (96; R = HO) with NH$_4$OAc or the corresponding primary amine in AcOH [I. Ivanov, S. Karagiozov, and I. Manolov, Arch. Pharm. (Weinheim, Ger.), 1991, 324, 61]. A series of 4-aminocoumarins 97 [R = H, C(NH$_2$) = NH, COMe, azolyl] and 98 (R^1 = azolyl, arylamino, R^2 = H) and related derivatives 98 (R^1 = OH, R^2 = azolylmethyl, arylaminomethyl) have been prepared and the preliminary evaluation of their in vitro antimicrobial activity described (M.M. Badran, A.K. El-Ansari, and S. El-Meligie, Rev. Roum. Chim., 1990, 35, 777).

(98)

(97)

284

6-Aminocoumarin (99; $R = NH_2$) has been prepared by the slow addition of KBH_4 to a suspension of 10% Pd/C in H_2O, followed by the introduction of 6-nitrocoumarin (99; $R = NO_2$). A number of related derivatives 100 [R^1-R^5 = H, OH, NH_2, alkyl, alkoxy, cycloalkyl, (substituted)Ph], useful as antiviral agents, especially effective against AIDS, have been prepared (E. Kun and L. Aurelian, PCT Int. Appl. WO 91 04,663, 1991).

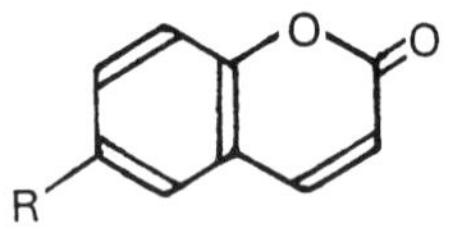

(99)

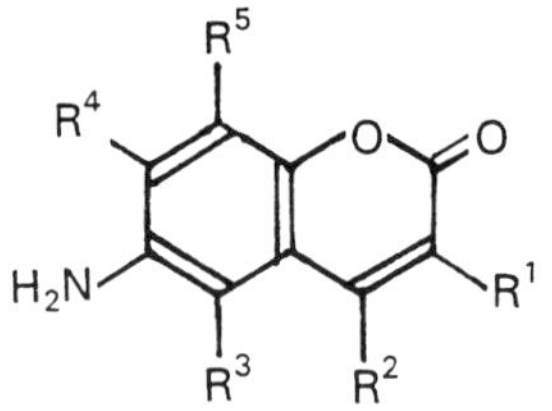

(100)

Several 7-acylamino-4-methylcoumarins 101 (R^1 = H, acyl; R^2 = H, alkyl, aryl), as well as their protonated, or alkali or NH_4 salts, have been prepared for use as substrates in determining α-adipinylamidase activity. Compounds 101 upon hydrolysis afford the strongly fluorescing 7-amino-4-methyl-coumarin (102) (W. Aretz and U. Hedtmann, Ger. Offen. DE 3,834,597, 1990).

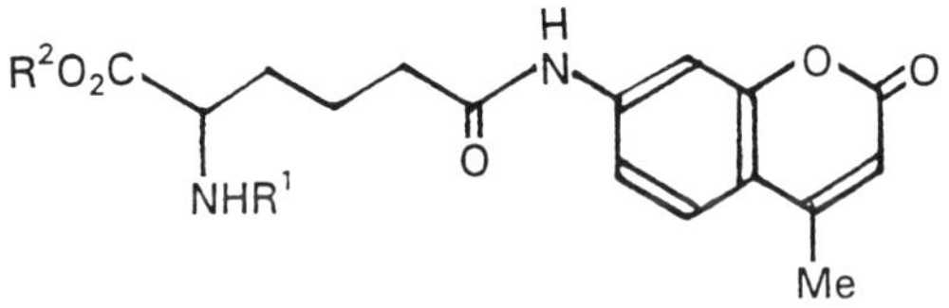

(101)

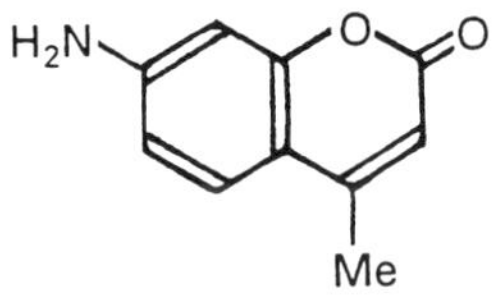

(102)

7-Amino-4-methylcoumarin (102) is also prepared by reacting 3-
-$H_2NC_6H_4OH$ with an alkyl chloroformate in an organic solvent in the
presence of an alkaline agent and subsequent cyclocondensation of the
resulting N-(alkoxycarbonyl)-3-aminophenol with acetoacetic ester in a
H_2SO_4 medium, followed by heating the resulting 7-[N-(alkoxycarbonyl)-
amino]-4-methylcoumarin. The yield of 102 is increased and the procedure
is simplified by using $ClCO_2Me$ methyl chloroformate and finally carrying out
the hydrolysis of the carbamate group in the presence of 3-4 equivalents of
concentrated alkali (e.g., KOH) at 80-100°C (V.F. Pozdnev, U.S.S.R. SU
1,325,050, 1987; Khim. Geterotsikl. Soedin., 1990, 312). Reactions of 7-
-(diethylamino)-4-methylcoumarin with different electrophiles, for example,
Ac_2O, complex unsaturated compounds ($PhCH = CH_2$, $PhC \equiv CH$, dihydropyran), and Lewis acids (copper halides), yield 3-, 6-, and 8-substituted
coumarins (M.A. Kirpichenok, S.L. Levchenko, and I.I. Grandberg *ibid.*,
1987, 1324).

Treatment of 7-diethylaminocoumarins 103 (R^1 = H, Me) with $Et_3O^+BF_4^-$ in
CH_2Cl_2 at 20°C, followed by $NaCHR^2R^3$ ($R^2 = R^3 = Ac$, CO_2Et, CN; $R^2 = Ac$,
$R^3 = CO_2Et$; $R^2 = CO_2Et$, $R^3 = CN$; $R^2R^3 = $ phthaloyl) yields the 7-diethylamino-
-2-(substituted methylene)-2H-benzo[b]pyrans 104, *via* an intermediate
benzopyrylium tetrafluoroborate (M.A. Kirpichenok, S.K. Gorozhankin, and
I.I. Grandberg, *ibid.*, 1988, 751). A number of 4-(functionally substituted
alkyl)-7-(dialkylamino)coumarins have been synthesized (S.K. Gorozhankin,
M.A. Kirpichenok, and I.I. Grandberg, *ibid.*, 1990, 1326).

(103)

(104)

The photochemical addition of *trans*-stilbene to 7-diethylaminocoumarins
103 (R^1 = H, Me, morpholino) yields adducts 105 *via* a stereoselective
[2 + 2]-cycloaddition at the C-3-C-4 bond. Evidence for structure of adducts
105 is based on NMR and crystal structure data (M.A. Kirpichenok *et al.*,
ibid., 1990, 1022). Adducts 106 (R^1 = H, R^2 = Me, R^3 = H, Et; R^2 = H,

R_2^3N = morpholino; R^1 = EtO$_2$CCH$_2$, R^2 = Me, R^3 = Et) have been obtained from the appropriate 7-aminocoumarins and styrene (*idem, ibid.*, 1988, 1176). Related [2 + 2]-cycloadducts have been reported from the photochemical reaction between 7-aminocoumarins and *trans,trans*-1,4-diphenylbuta-1,3--diene (*idem, ibid.*, 1990, 1319).

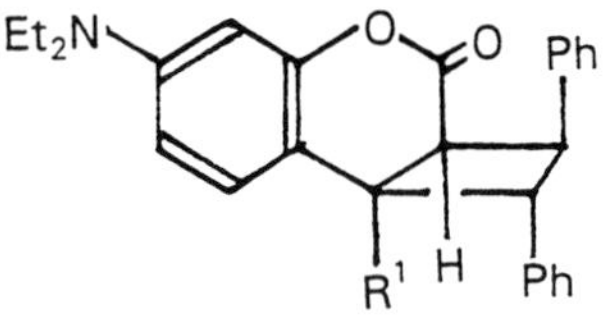

(105)

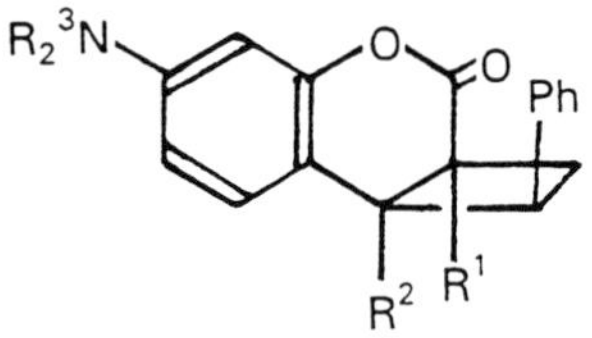

(106)

3-Alkenyl- and 3-alkynyl-7-diethylamine-4-methylcoumarins 107 (R = CH = CHPh, CH = CHCN, C≡CPh, inden-2-yl) have been obtained by the alkenylation and alkynylation of 7-diethylamino-3-iodo-4-methylcoumarin with RH in UV-irradiated solutions (*idem*, U.S.S.R. SU 1,505,941, 1989).

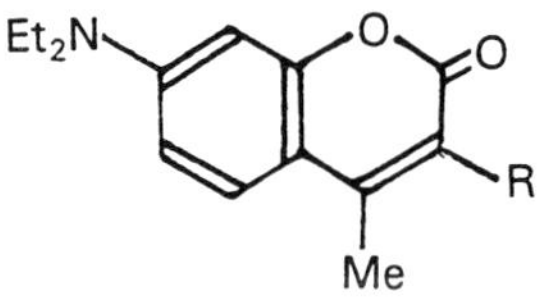

(107)

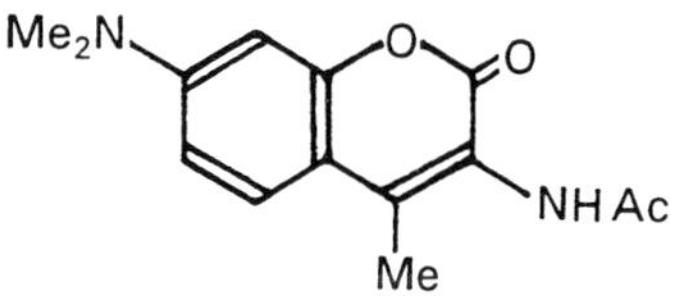

(108)

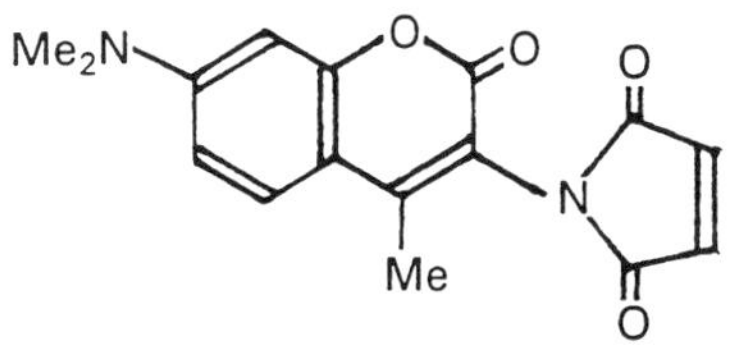

(109)

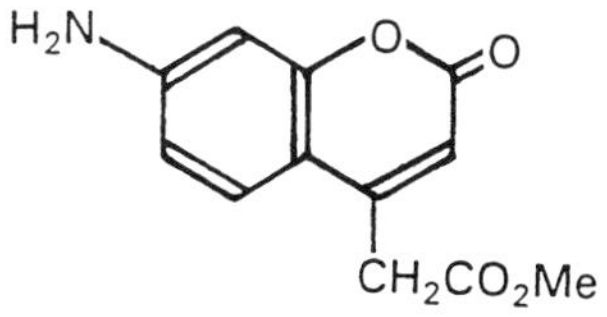

(110)

The Pechmann condensation of 3-(Me$_2$N)C$_6$H$_4$OH and MeCOCH(NHAc)-CO$_2$Et leads directly to 3-acetamido-7-dimethylamino-4-methylcoumarin (108), which is readily converted to the fluorescent, thioreactive *N*-[7-(dimethylamino)-4-methylcoumarin-3-yl]maleimide (109) (J.E.T. Corrie, J. Chem. Soc., Perkin 1, 1990, 2151). The fluorescent properties of new heterobifunctional fluorescent probes, including methyl 7-amino-coumarin-4--acetate (110) have been investigated (T. Besson *et al.*, Heterocycles, 1992, 34, 273) and the synthesis and reactivity of 7-dimethylaminocoumarin-3--isocyanate (111) as a fluorescent derivatization reagent for alcohols have been reported (H. Fujino, M. Eguchi, and S. Goya, Yakugaku Zasshi, 1990, 110, 155). The dipole moments of some derivatives of 7-aminocoumarin have been determined (L.I. Kuznetsova *et al.*, Izv. Timiryazevsk. S-kh. Akad., 1991, 169).

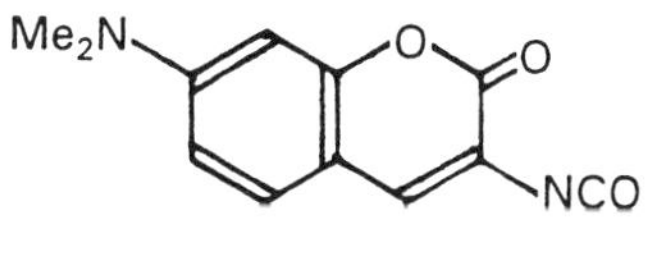

(111)

288

The preparations of water-soluble coumarin derivatives 112
[R^1 = $(OCH_2CH_2)_nOCH_2R^7$ or $(OCHMeCH_2)_nOCH_2R^7$, n = 1-30, R^7 = H or a
variety of groups, with a substituted amino group at C-7 and a substituent at
C-4 and their use as enzyme substrates or for the preparation of such
compounds have been discussed. Compound 113 has been used as a
water-soluble substrate for leucine aminopeptidase (J. Chalom *et al.*, Eur.
Pat. Appl. EP 516,532, 1992).

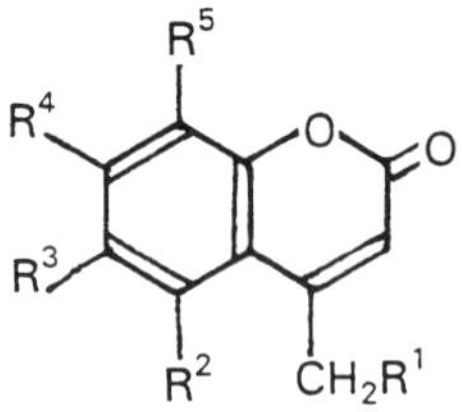

(112)

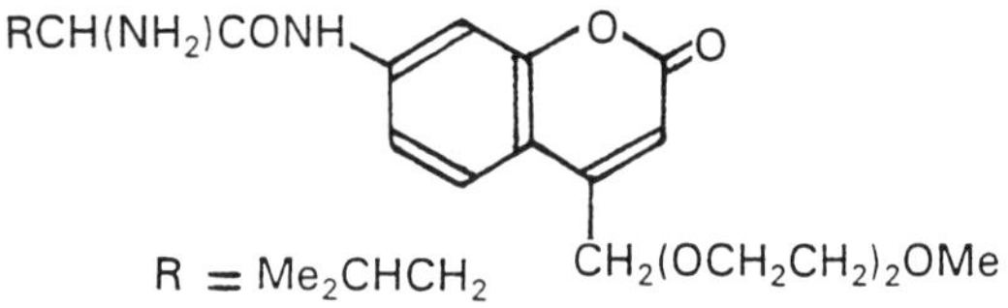

R = Me_2CHCH_2 $CH_2(OCH_2CH_2)_2OMe$

(113)

The reaction of 3-(1-propanoyloxybenzyl)coumarins with the nitriles of 4-
-(dimethylamino)benzoic, 4-(diethylamino)benzoic, nicotinic, isonicotinic, and
3-(dimethylamino)propionic acids in concentrated H_2SO_4 gives the 3-(1-acyl-
aminobenzyl)coumarins 114 (R = $4\text{-}Me_2NC_6H_4$, $4\text{-}Et_2NC_6H_4$, 3-pyridyl, 4-
-pyridyl, $Me_2NCH_2CH_2$) [A. Bozhilova and D. Ivanova, Arch. Pharm.
(Weinheim, Ger.) 1991, <u>324</u>, 483].

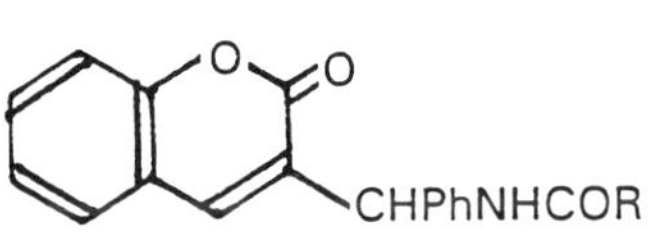

(114)

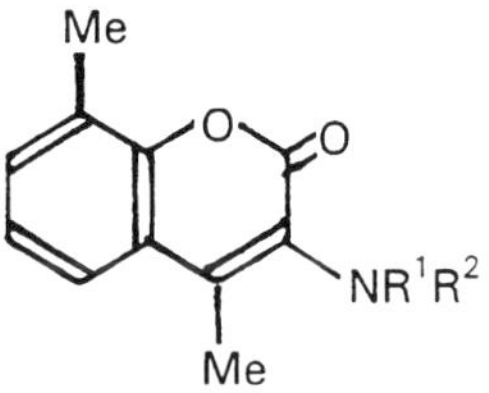

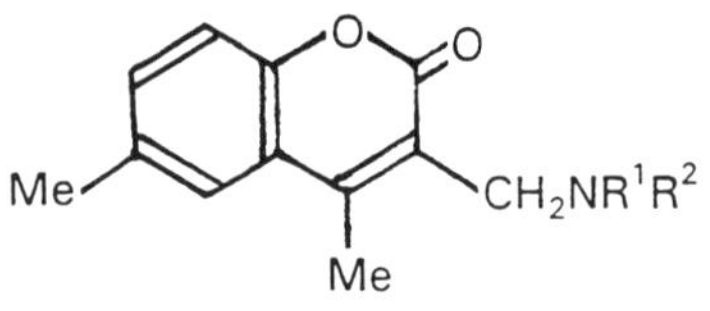

(115)

(116)

3-Chloromethyl-4,6-dimethylcoumarin reacts with amines to afford the 3-
-aminomethyl-4,6-dimethylcoumarins 115 (R^1 = H, alkyl, $PhCH_2$; R^2 = alkyl,
Ph, $PhCH_2$, $PhCH_2CH_2$, tolyl, ClC_6H_4, pyridylmethyl, methylpyridyl;
NR^1R^2 = morpholino, piperidino), many of which show bactericidal activity
(M. Nagesam, M.S. Raju, and K.M. Raju, J. Indian Chem. Soc., 1987, 64,
418). The preparations of antibacterial aminomethylcoumarins 115 (R^1 = 3-
-methylpyrid-2-yl, 4-ClC_6H_4, 1-naphthyl, 2-tolyl, 2,5-$Cl_2C_6H_3$, R^2 = H;
R^1 = R^2 = Ph), tested against *Xanthomonas citri*, *Bacillus subtilis*, *Escherichia
coli*, and *Pseudomonas viticola* (*idem*, Chim. Acta Turc., 1989, 17, 255) and
3-aminomethyl-4,5-dimethylcoumarins 116 (R^1 = H, alkyl, $PhCH_2$; R^2 = tolyl,
Ph, alkyl, methylpyridyl, naphthyl, $PhCH_2$; NR^1R^2 = morpholino, piperidino, 2-
-phenylindol-1-yl, 3,5-dimethylpyrazol-1-yl) (*idem*, J. Indian Chem. Soc.,
1988, 65, 380), have been reported. For the preparation of 3-(2-diethyl-
amino)ethyl-7-(ethoxycarbonyl)methoxy-4-methylcoumarin which acts as a
vasodilator, see R. Mazurkiewicz *et al.*, Pol. PL 149,196, 1990; and for its
hydrochloride, *idem*, *ibid.*, 149,194, 1990.

Several 6-(substituted phenylazo)-3-(substituted amido)coumarins 117
(R^1 = 2-, 4-EtO, 4-Br, R^2 = H, 2-, 3-, 4-Me, 2-, 3-, 4-Cl, 2-, 3-, 4-MeO) have
been prepared by cyclocondensation of (phenylazo)salicylaldehydes 118 with
$R^2C_6H_4NHCOCH_2CO_2H$. Some of the compounds have been tested for
fungicidal and bactericidal activities (V.S. Jolly, L. Singh, and S. Pendse, J.
Indian Chem. Soc., 1992, 69, 105). The synthesis and chemotherapeutic
evaluation of some diazocoumarins, as antimicrobial and potential
carcinostats, have been reported (A. Pradhan, V. Pradhan, and V.S. Jolly,
Orient. J. Chem., 1994, 10, 73).

(117)

(118)

6-Nitrocoumarin on reduction by $SnCl_2$ gives the related hydroxylamine, which on oxidation in situ by $FeCl_3$ affords the corresponding 6-nitroso derivative. Condensation of the nitroso derivative with cyanamide in the presence of iodobenzene diacetate yields 6-cyanoazoxycoumarin (119), which has a particularly high activity against *Plasmopara viticola* in a translaminar protectant test. A number of related biocidal coumarinylazo-carboxylic acid derivatives have been prepared (M. Pearson *et al.*, Eur. Pat. Appl. EP 371,560, 1990).

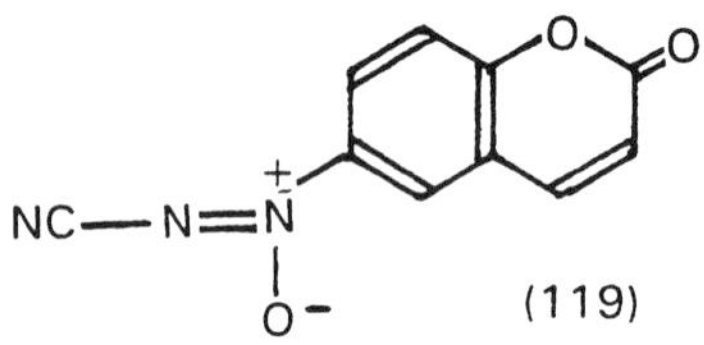

(119)

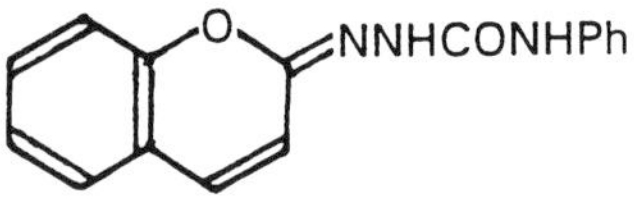

(120)

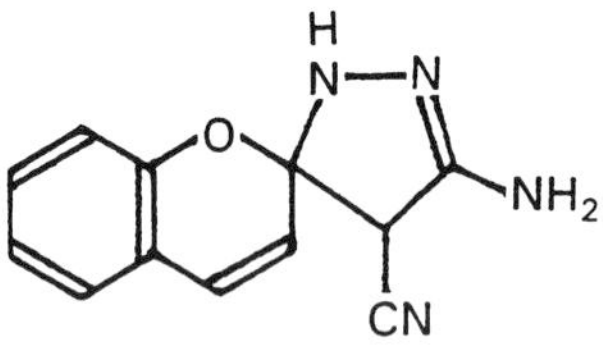

(121)

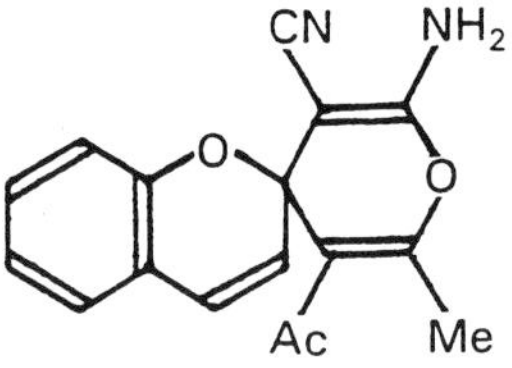

(122)

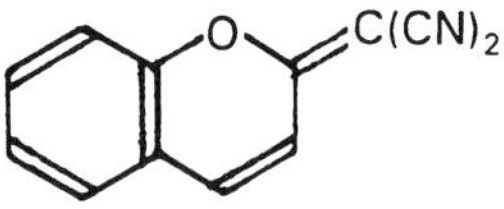

(123)

The coumarin derivatives 120 (coumarin phenylsemicarbazone), 121 and 122 have been prepared in order to evaluate their activity against *Biomphlaria alexandrina* snails. The benzopyranylidenemalononitrile derivative 123 underwent cyclocondensation with acetylacetone to furnish compound 122 [I. Zeid *et al.*, Arch. Pharm. (Weinheim, Ger.), 1991, <u>324</u>, 589).

3-Acetylcoumarin on treatment with thiosemicarbazides, $H_2NNHC(S)NHR$ (R = Et, Bu, cyclohexyl, Ph) in EtOH yields the semicarbazones 124. Cyclocondensation of 3-(bromoacetyl)coumarin with $R^1NHC(SH) = NN = CHR^2$ (R^1 = Et, cyclohexyl, Ph, R^2 = substituted Ph) in $CHCl_3$-EtOH affords thiazolone derivative 125. None of the compounds exhibit any significant bactericidal, antifungal, insecticidal, herbicidal, or plant growth regulator activities (A. Gursoy, N. Karali, and G. Otuk, Acta Pharm. Turc., 1992, <u>34</u>, 9). The synthesis of semicarbazones 124 (R = H, Ph) and related acylhydrazones 124 (S = O, NHR = Me, Ph, $2\text{-HOC}_6\text{H}_4$, $2\text{-AcOC}_6\text{H}_4$) and their transformation to 5-substituted 3-acetyl-2-(coumarin-3-yl-2-methyl-1,3,4--thia(oxa)diazolines under acetylating conditions have been reported (L. Somogyi, Ann., 1994, 623).

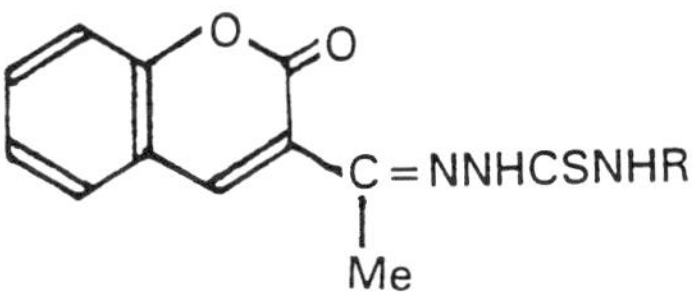

(124)

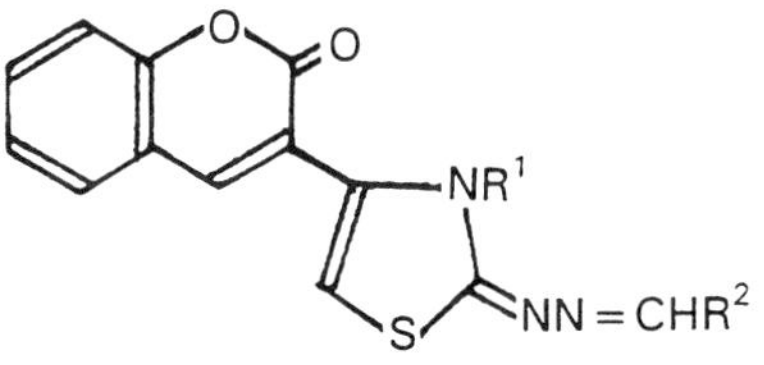

(125)

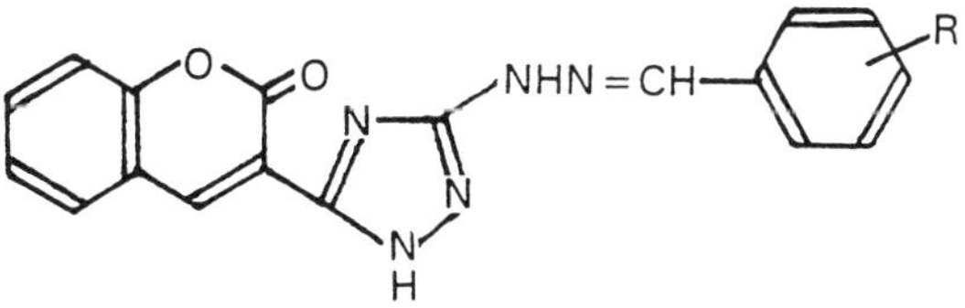

(126)

292

The electronic absorption spectra of 3-(3-*N*-arylidenehydrazino-1,2,4-
-triazol-5-yl)coumarins 126 (R = 4-O_2N, 3-HO, 4-MeO) have been investigated
in organic solvents of varying polarities and buffer solutions at different pH.
The important bonds in the IR spectra and the main ^{1}H NMR signals have
been assigned and discussed in relation to molecular structure (F.A. Yassen,
A.A.F. Wasfy, and M.E. Moustafa, Spectrosc. Letters, 1993, 26, 1855).
The photoreactions of 3- and 4-azidocoumarins in the presence of nucleophiles such as alcohols, thiols, and amines have been discussed and plausible
mechanisms suggested for these reactions (K. Ito *et al.*, Heterocycles, 1993,
35, 937).

Coumarin-3-carboxamides 127 (R^1 = H, NH_2, R^2 = H, Me, CO_2Et,
CH_2CO_2H, $CONHCH_2CO_2H$, $CONHC_6H_4Me$-4, $CONHCH_2CO_2H$) and 128 have
been prepared by amidation of 3-ethoxycarbonylcoumarin with aniline
derivatives. Compound 128 showed both bactericidal and fungicidal
activities (A.H. Bedair, J. Prakt. Chem., 1987, 329, 359). The antimicrobial
and insecticidal activities of coumarin-3-*N*-substituted carboxamides 129
(R = CH_2CH_2OH, $CH_2CHOHCH_2OH$, CH_2CH_2Cl, $CH_2CH_2NMe_2$) have been
determined (A.M. El-Agrody *et al.*, Afinidad, 1988, 45, 447) and test data
for the diuretic, analgesic, and myorelaxant activity of some derivatives, for
example, 7-methylcoumarin-3-*N*-phenylcarboxamide (130) have been
reported and discussed (L. Bonsignore *et al.*, Eur. J. Med. Chem., 1993, 28,
517). The preparation of carboxamides 129 (R = CH_2Ph, 4-anisyl) and their
reactions with active methylene compounds, ketones, Grignard reagents, and
aromatic amines have been described [A.F. El-Farargy *et al.*, Egypt. J.
Chem., 1987 (Pub. 1989), 30, 497].

(127)

(129)
(130) R = Ph

(128)

The reaction of 6-bromo- and 6,8-dibromo-3-ethoxycarbonylcoumarins with aniline derivatives results in condensation products, for example, derivatives 131 ($R^1 = R^2 = Br$, $R^3 = R^4 = Cl$, $R^5 = H$) and cyclocondensation products, for example, derivatives 132 ($R^1 = H$, $R^2 = Br$, $R^3 = R^4 = H$) (M.R. Selim, Sci. Phys. Sci., 1992, _4_, 34). Friedel-Crafts reaction of coumarin 133 with benzene affords derivative 134 (A.M. El-Agrody *et al.*, J. Chem. Soc., Pak., 1993, _15_, 261). The preparations of oxadiazoles 135 [R = H, 4-OH, 3--MeO, 3,4-(MeO)$_2$] and coumarin derivative 136 [R^1 = H, 4-OH, 3-MeO, 3,4--(MeO)$_2$, R^2 = 3-, 4-Cl, 3-, 4-Me] have been reported and tested for their antidepressant activity. Compound 136 [R^1 = 4-OH, 3-MeO, R^2 = 3-, 4-Cl; R^1 = 3,4-(MeO)$_2$, R^2 = Cl] had greater activity than imipramine with less toxicity (V. Singh *et al.*, Arzneim.-Forsch., 1992, _42_, 993). A number of coumarin-3-(*N*-aryl and *N*,*N*-diaryl)carboxamides and related compounds with many substituents have been prepared as UV-absorbents and for use in thermoplastin resin compositions and moldings (A. Ogiso *eg al.*, Jpn. Kokai Tokkyo Koho JP 06 145,164 [94 145,164], 1994).

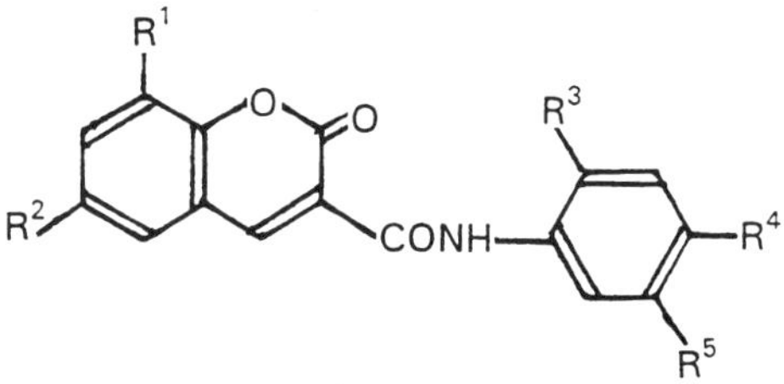

(131)

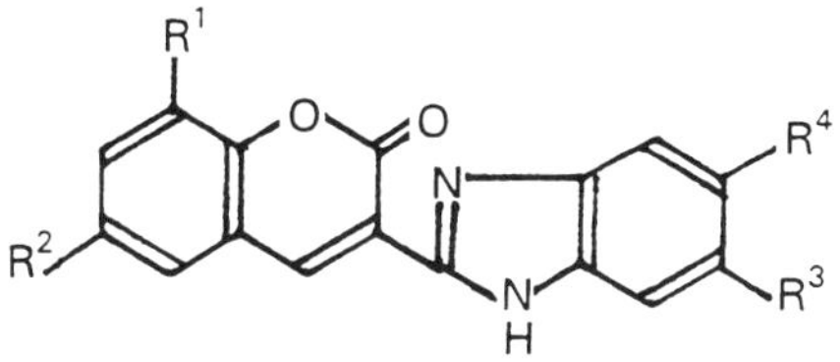

(132)

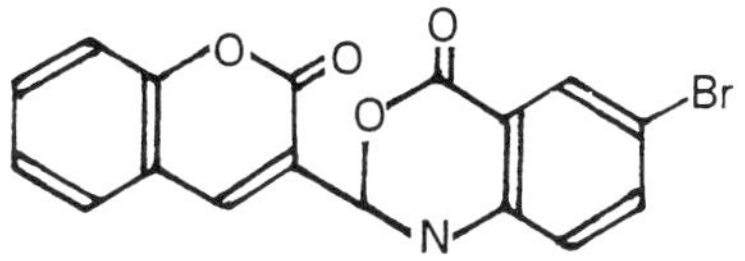

(133)

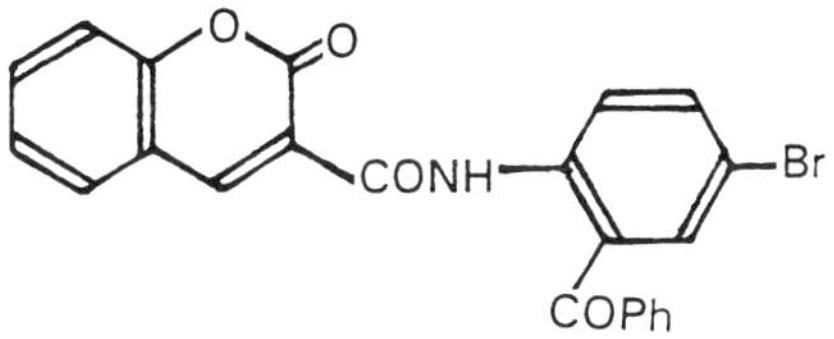

(134)

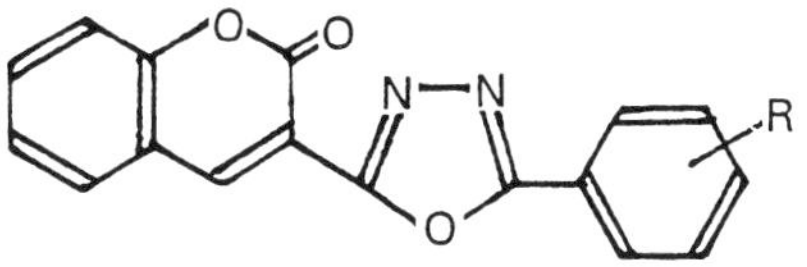

(135)

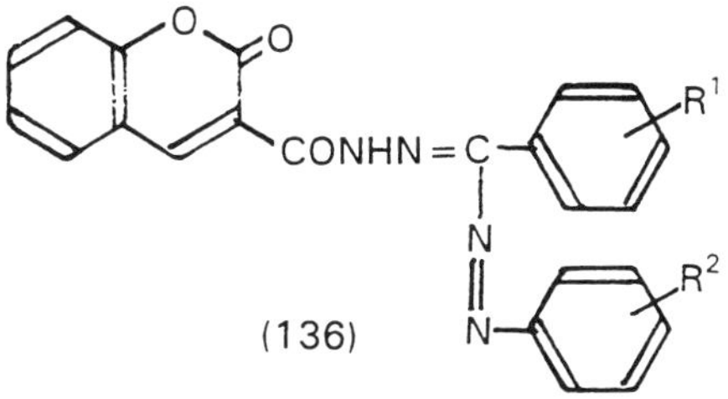

(136)

Coumarin-3-(4-aminosulphonyl)carbanilide derivatives 137 (R^1 = H, Br, NO_2, R^2 = H; R^1R^2 = CH = CHCH = CH; R^3 = H, Ac, 2-pyrimidyl, 2-thiazolyl, 5--methyl-3-isoxazolyl) have been prepared by cyclocondensation of $EtO_2CCH_2CONHC_6H_4SO_2NHR^3$-4 with 5,6-$R^1R^2C_6H_3CHO$ and their IR and ^1H NMR spectroscopic data reported (M.A.A. Moustafa, Sci. Pharm., 1991, <u>59</u>, 213).

(137)

The acrylate ester of salicylaldehyde, in the presence of DABCO, yields a crystalline coumarin salt 138. Salicylaldehyde, suitably protected, reacts with methyl acrylate to give 3-(*N,N*-dimethylaminomethyl)coumarin (139) (S.E. Drewes *et al.*, Synth. Comm., 1993, <u>23</u>, 2807).

(138)   (139)

Substituted 2*H*-benzo[*b*]pyran-2-thiones 140 (R^1 = H, R^2-R^5 = H, Me; R^1R^2 = CH = CHCH = CH; Z = S) and benzothiopyranthiones 141 (R^1,R^2 = H, Me, Z = S) undergo oxidation to 140 and 141 (Z = SO), which on further oxidation afford compounds 140 (Z = O, substituted coumarins) and 141 (Z = O) or polymeric products (K. Buggle and B. Fallon, Monatsh. Chem., 1987, <u>118</u>, 1197).

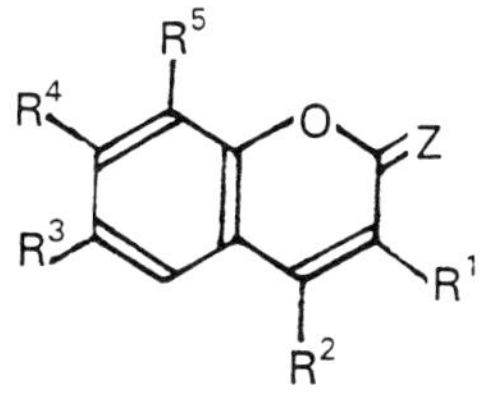

(140)

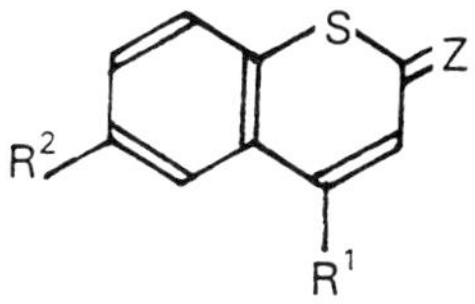

(141)

^{17}O NMR at natural abundance has been used effectively for the differentiation between coumarins 142 (R^1 =H, CO_2H, NO_2, $CONHCO_2Et$; R^2 =H, Me, Cl; R^3 =H, Me; R^4 =H, Br; R^3R^4 = CH = CHCH = CH; R^5 = H, Me; R^6 =H, MeO) and chromones 143 (R^1 =H, Me, CO_2H; R^2 =H, CN, CHO; R^3 =H, Me; R^4 =H, Br; R^3R^4 = CH = CHCH = CH; R^5 =H, Me; R^6 =H) (K. Nagasawa *et al.*, Chem. Pharm. Bull., 1993, **41**, 211).

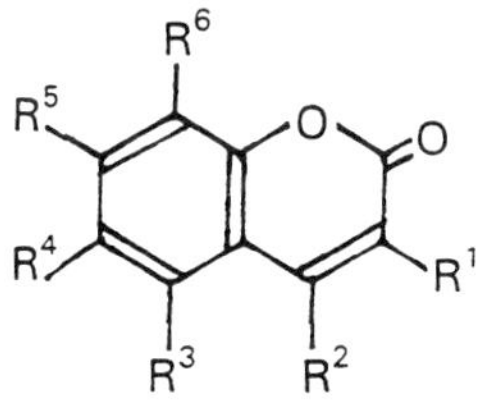

(142)

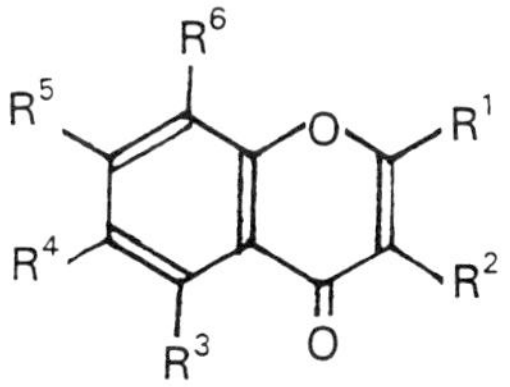

(143)

(v) Hydroxycoumarins

3-Hydroxycoumarins. The phase-transfer catalyzed alkylation of 3--hydroxycoumarin (1) with RBr (R = $PhCH_2$, allyl, crotyl, prenyl, propargyl)

296

yields the corresponding *O*-alkylation products 2, *C*-alkylation products 3 and
in one instance the C,O-dialkylated product 4 (K.C. Majundar, A.T. Khan,
and S.K. Chattopadhyay, Heterocycles 1989, 29, 1573).

OH

(1)

OR

(2)

OH

R

(3)

$OCH_2C{\equiv}CH$

$CH=C=CH_2$

(4)

3-Allyloxycoumarin (5; $R^1 = OCH_2CH = CH_2$, $R^2 = H$) and 4-allyl-3-
-methoxycoumarin (5; $R^1 = OMe$, $R^2 = $ allyl) have been converted into their
carbonyl derivatives 5 ($R^1 = OCH_2COMe$, $R^2 = HO$) and 5 ($R^1 = OMe$,
$R^2 = CH_2COMe$), respectively, in the presence of a catalytic amount of
palladium (II) chloride in an atmosphere of oxygen. However, when a free
adjacent hydroxy group is present cyclization occurs, as in the case of
derivative 5 ($R^1 = OH$, $R^2 = CMe_2CH = CH_2$) when furocoumarin 6 is formed
(J. Mitra and A.K. Mitra, Indian J. Chem., 1992, 31B, 693).

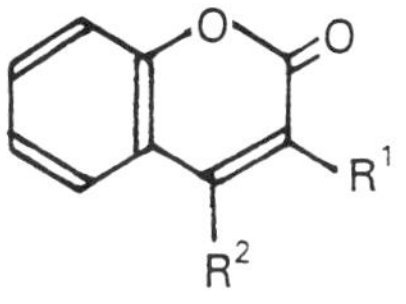

(5)

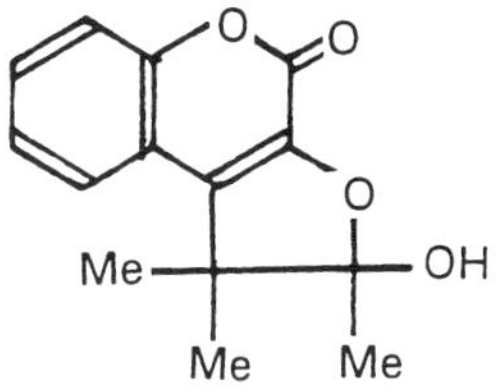

(6)

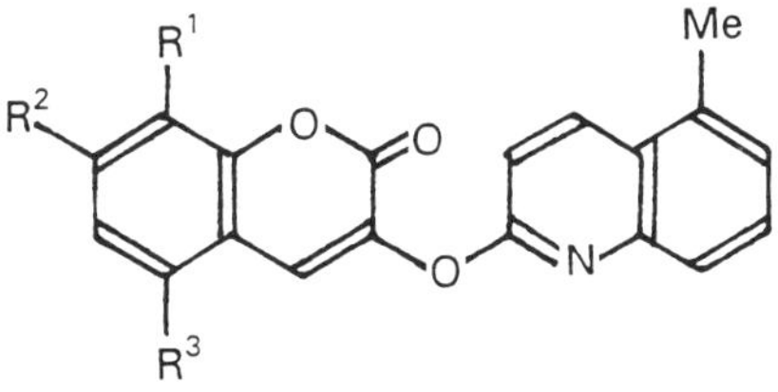

(7)

3-[(5-Methylquinolin-2-yl)oxy]coumarins (3-[(5-methylquinolin-2-yl)oxy]-
-2*H*-benzo[b]pyran-2-ones) 7 (R^1-R^3 = H, Me, OMe) have been prepared from
[(5-methylquinolin-2-yl)oxy]acetic acid and appropriate benzaldehyde
derivatives and tested against encephalomyocarditis virus (M. Kidwai, N.
Gupta, and K.C. Srivastava, Indian Drugs, 1993, 30, 377). The stigmatropic
rearrangement of 3-[(aryloxy)methyl]coumarins has been investigated and a
simple synthesis of hydroxylate 3-benzylcoumarins reported (K.C. Majumdar
et al., J. Chem. Soc., Perkin 1, 1993, 715).

The synthesis of 3-hydroxycoumarin phthalimido derivative 8 has been
reported along with those of the tosylamino coumarin derivative 9 and *N*-(7-
-hydroxy-4-methylcoumarin-6-sulphonyl)amino acids. Several of these
compounds possessed specific antimicrobial activities (T.M. Ibrahim *et al.*,
Proc. Indian Natl. Sci. Acad., Part A, 1993, 59, 189).

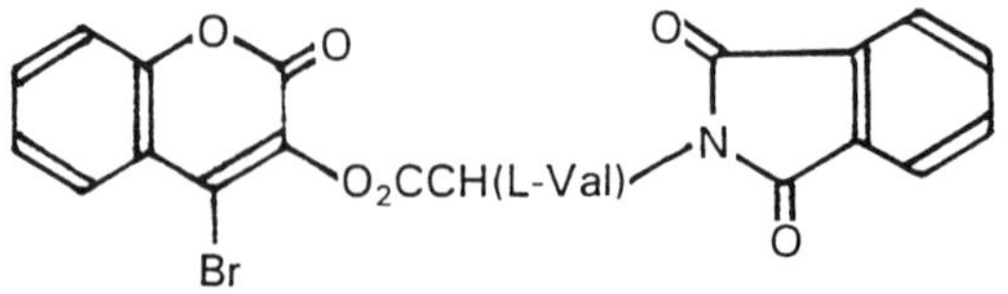

(8)

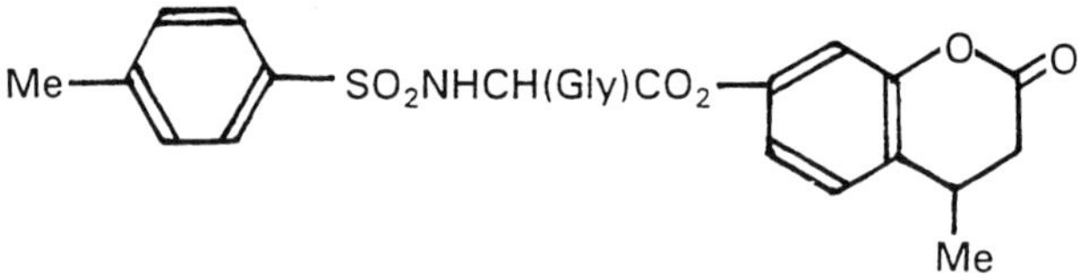

(9)

Treatment of coumarin with thallium(III) nitrate, Tl(NO$_3$)$_3$, in MeOH at
room temperature for 7 days affords 3-methoxy-6-nitrocoumarin (10) (18%).
However, treatment of 3,4-dihydroxycoumarin (11; R^1=R^2=H) with
Tl(NO$_3$)$_3$ in AcOH-EtOH for 5 days yields derivative 11 [R^1=H,
R^2=Tl(OAc)$_2$], which reacts with HClO$_4$-NaClO$_4$ in MeOH to give 3-hydroxy-
-4-methoxycoumarin (11; R^1=Me, R^2=H) (A. Banerji and G. Nandi,
Heterocycles, 1987, 26, 1221).

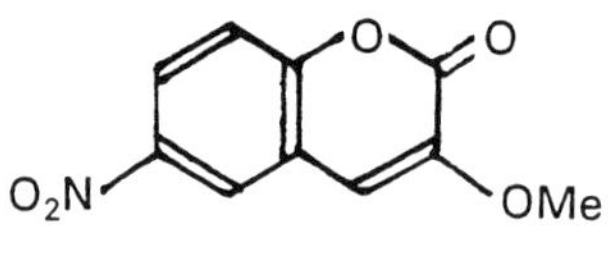

(10)

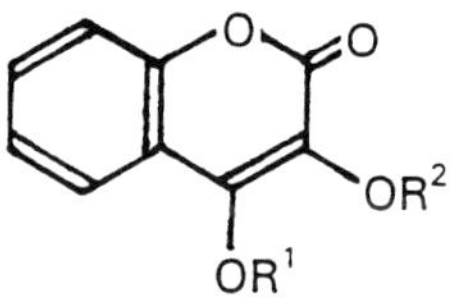

(11)

4-Hydroxycoumarins. 4-Hydroxycoumarin (1) useful as an intermediate for drugs, agrochemicals, and especially rodenticides has been prepared by the alkali-catalyzed reaction of acetylsalicyloyl halides with acetoacetate esters in ketones (e.g. methyl *iso*butyl ketone), cyclization of the resulting alkali metal a-(2-hydroxybenzoyl)acetoacetates by mineral acid (e.g. HCl) in the above mixture, and deacetylation of the ensuing 3-acetyl-4-hydroxy-coumarin in the presence of H_2SO_4 (N. Numamoto *et al.*, Jpn. Kokai Tokkyo Koho JP 02 36,178 [90 36,178], 1990).

(1)   (2)   (3)

4-Hydroxycoumarin (1) has also been obtained by initially adding 40% aqueous NaOH to a solution of $MeCOCH_2CO_2Me$ in $MeCOCH_2CHMe_2$ at 15-
-20°C, followed by acid chloride 2 and $MeCOCH_2CHMe_2$ at 30-35°C and the mixture kept at 40°C to yield a slurry of salt 3, which was then heated with 97% H_2SO_4 at 90-100°C (T. Yamamoto *et al.*, *ibid.*, 02 06,481 [90 06,481], 1990). A method for synthesizing 4-hydroxycoumarin (1), involving the esterification of $CH_2(COCl)_2$ with PhOH, and cyclization of the $CH_2(CO_2Ph)_2$ has been developed (E. Ozcan, C. Bayat, and A.M. Acar, Doga: Turk Kim. Derg., 1991, _15_, 123; 258). Malonic diesters have been treated with $MgCl_2$ and acetylsalicylic chloride (2) and the resulting dialkyl 2-(2-
-acetoxybenzoyl)malonates cyclized by alkali to afford 4-hydroxycoumarin (1) (T. Kakimoto and T. Hirai, Jpn. Kokai Tokkyo Koho JP 05 255,299 [93 255, 299]; 05 262,756 [93 262,756], 1993).

N-(Methylene-4-oxocoumarinyl)carbamates 4 (R = Me, Et, Bu, Bu^t, CH_2Ph, Ph, CH_2CH_2OH) are prepared by the condensation of H_2NCO_2R with 4-hydroxycoumarin (1) in the presence of ethyl orthoformate (V. Speziale, R. Sakellariou, and M. Hamdi, Synthesis, 1992, 921). Similarly a series of 3-
-ureidomethylenecoumarins 5 (X = O, S, R = H, allyl; X = O, R = Me, Bu, cyclohexyl, Ph, Ac) are obtained by condensation of substituted ureas with 4-hydroxycoumarin in the presence of ethyl orthoformate (R. Sakellariou, V. Speziale, and M. Hamdi, Synth. Comm., 1990, _20_, 3443).

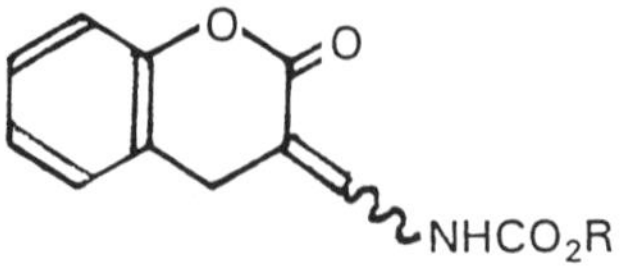

300

(4) 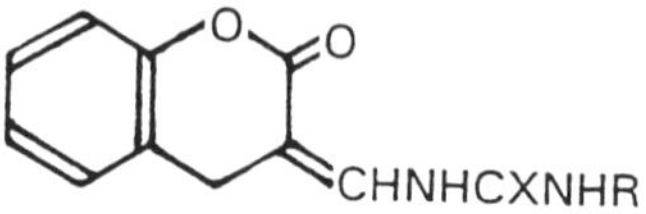   (5)

The reaction of 4-hydroxycoumarin (1) with aldehydes, RCHO (R = Ph, 2-, 3-$O_2NC_6H_4$, 2-ClC_6H_4, 4-$MeOC_6H_4$, Et, Pr, Pr^i) under Knoevenagel conditions initially yields intermediates 6, which subsequently react with addition 4-hydroxycoumarin or PhSH to give biscoumarins 7 or 3-aryl(phenylthio)-methyl-4-hydroxycoumarins 8 (R = Ph, 2-, 3-$O_2NC_6H_4$, 2-ClC_6H_4), respectively (K. Nawaz, M.A. Munawar, and M. Siddiq, J. Chem. Soc. Pak., 1991, 13, 272). Some related biscoumarins with substituents on the benzene rings (e.g., OH, Me) have been obtained by reacting the appropriate 4-hydroxycoumarin with the necessary aldehyde (A.K. Shah *et al.*, Sci. Cult., 1986, 52, 228). Phase transfer-catalyzed alkylation of 4-hydroxycoumarin (1) with active halides affords *O*-alkylated products, *C,O*-dialkylated products, and α,α-disubstituted 2-HOC_6H_4COMe derivatives. Only with Me_2C = CHCH$_2$Br is a *C,C*-dialkylated product obtained (K.C. Majumdar, A.T. Khan, and S.K. Chattopadhyay, Indian J. Chem., 1990, 293, 483).

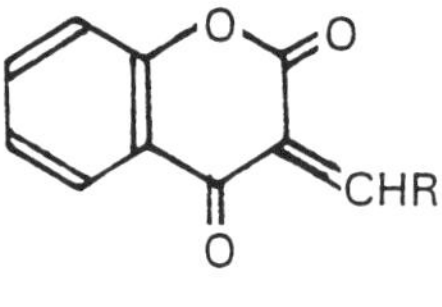

(6) 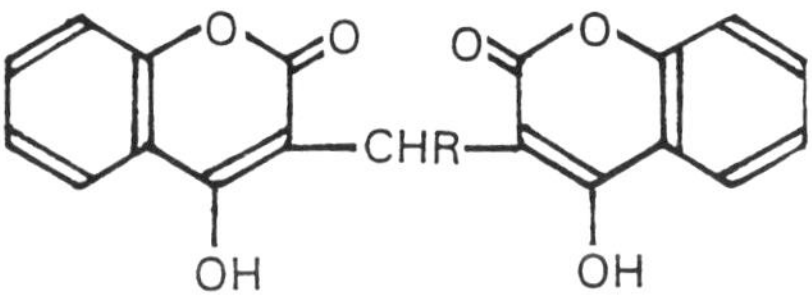   (7)

(8)

The reductive cleavage of biscoumarins 7 [R = (*E*)-CH = CHPh, $(CH_2)_5Me$, 4-$(Me_2N)C_6H_4$, 2-$MeOC_6H_4$, 1,3-benzodioxol-4-yl, 2-furyl, cyclohexyl] by $NaBH_3CN$ in boiling MeOH furnishes 3-alkyl-4-hydroxycoumarins 9 and 4--hydroxycoumarin (1). It has been suggested that the reaction might take place *via* hydride trapping of alkylidenechromandiones 6 formed from the biscoumarins in a *retro*-Michael reaction and that this reaction might have biological relevance (G. Appendino *et al.*, Helv., 1991, <u>74</u>, 1451).

(9)

(10)

Warfarin (10; R = H) and acenocoumarol (10; R = NO_2) have been prepared by Michael addition of 4-hydroxycoumarin (1) to 4-MeCOCH = CH-C_6H_4R-4 in the presence of NaF, KF, or a tetraalkylammonium halide (e.g., $PhCH_2N^+Et_3Cl^-$) as catalyst [I. Ivanov, L. Manolov, and L. Alexandrova, Arch. Pharm. (Weinheim, Ger.), 1990, <u>323</u>, 521]. Factors affecting the yield of warfarin have been examined (E. Ozcan, C. Bayat, A.M. Acar, Marmara

Univ. Fen Bilimleri Derg., 1989, $\underline{6}$, 155) and a process for obtaining it and its alkali metal salts from 4-hydroxycoumarin and PhCH=CHCOMe reported (W. de C. Silveira and B. Alves de Souza Netto, Braz. Pedido PI BR 87 00,724, 1988).

4-Alkoxycoumarins 11 (R^1 = Me, Et, allyl, Pr^i, Bu, MeCHEt, amyl, hexyl, octyl, octadeyl, R^2 = H, Me, Cl, R^3 = H, Me) with methyl groups at either the 6- or 7-position or at both and in some cases a Cl at the 6-position have been prepared by alkylation of the corresponding 4-hydroxycoumarin with R^1Br. The lower alkyl ethers are more active towards tested bacteria, *Xanthomonas citri*, *Bacillus subtilis*, and *Escherichia coli*, than the higher alkyl ethers of the corresponding 4-hydroxycoumarins (C.R. Prasad *et al.*, Indian J. Pharm. Sci., 1989, $\underline{51}$, 1). The sulphur-assisted carbonylation of 2-hydroxyacetophenones 12 (R^1, R^2 = H, Me; R^3 = H, OH, MeO) yields 4-hydroxycoumarins 13 with up to three substituents at positions -3, -6, and -7 (T. Mizuno *et al.*, Synthesis, 1988, 257).

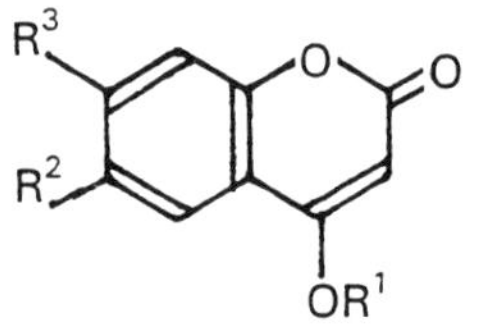

(11) 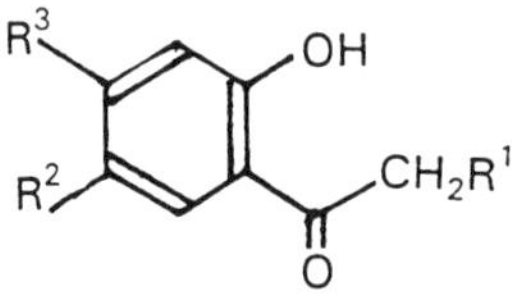   (12) 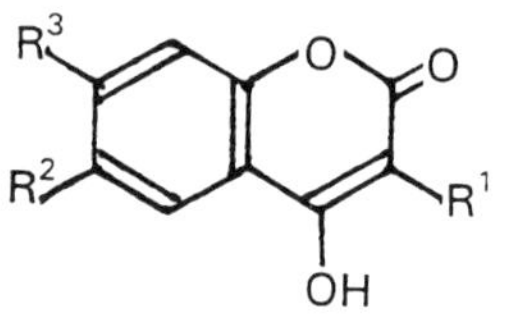   (13)

4-Hydroxy-3-phenylcoumarin (14) and related derivatives with substituents on the benzene ring have been synthesized in two steps by the titanium(III)-mediated reaction of methyl benzoylformate with unsubstituted and substituted salicylaldehyde, followed by lactonization, under acid conditions, of the resulting methyl 2,3-dihydroxy-3-(2-hydroxyaryl)-2-phenyl-propanoates (A. Clerici and O. Porta, *ibid.*, 1993, 99). 4-Hydroxy-3-phenyl-coumarin (14) has also been obtained on boiling a mixture of $2\text{-HOC}_6\text{H}_4\text{CO}_2\text{H}$ and phenylacetyl chloride, $PhCH_2COCl$, in acetone containing K_2CO_3. The method can also be used to prepare 3-aryl-4-hydroxycoumarins (R. Gandhidasan *et al.*, Indian J. Chem., 1988, $\underline{27B}$, 849). Related derivatives 15 (R^1-R^3 = H, R^4 = OMe; R^1 = R^2 = H, R^3 = R^4 = OMe; R^1 = OMe, R^2-R^4 = H; R^1 = H, R^2 = OMe, R^3R^4 = OCH_2O) have been obtained from suitably substituted methyl salicylates and phenylacetyl chlorides (S.K. Sripathi, R. Gandhidasan, and P.V. Raman, Indian J. Heterocyclic Chem., 1992, $\underline{1}$, 155).

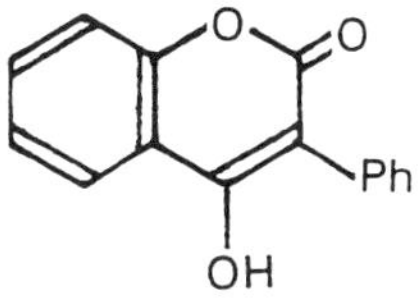

(14) 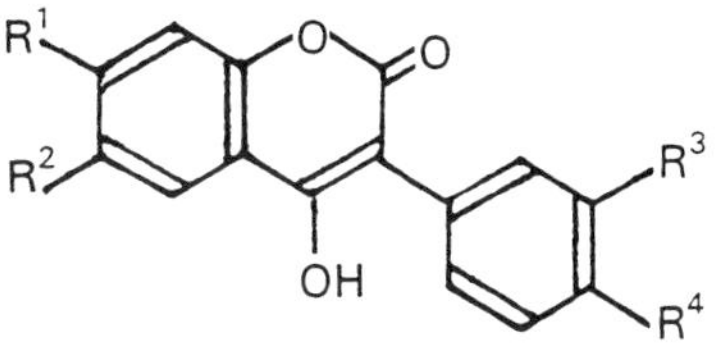 (15)

4-Hydroxycoumarin has been boiled in EtOH with acrolein, $CH_2 = CHCHO$, or MeCHBrCHO to yield derivative 16 (R = H), which has been converted to the 4-acetoxy and 4-methoxyderivatives 16 (R = Ac, Me, respectively). A similar reaction of 4-hydroxycoumarin with 3-ethoxy-2--methylacrolein, EtOCH = CMeCHO, or 3-amino-2-methylacrolein, $H_2NCH = CMeCHO$, afforded derivative 17, which has also been acetylated. Other related reactions and resulting products have been reported (M. Eckstein and H. Pazdro, Acta Pol. Pharm., 1988, 45, 8).

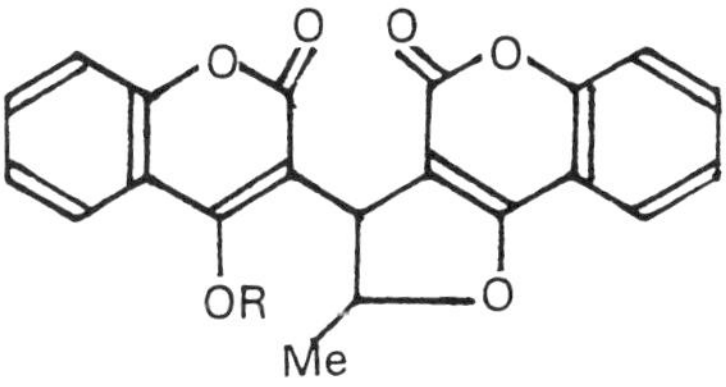

(16) 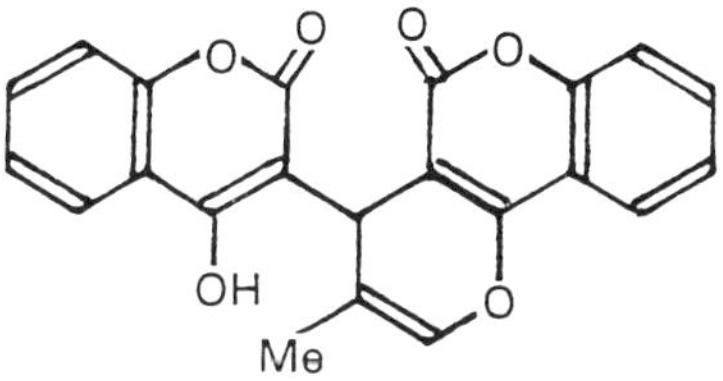 (17)

304

4-(Substituted phenoxymethyl)-6,7-dimethylcoumarins 18 (R^{1-4} = H, Me, Cl, NO_2) have been synthesized by the condensation of 4-chloromethyl-6,7--dimethylcoumarin with various phenols in acetone in the presence of K_2CO_3 and tested for their antimicrobial activity (R.V. Singh, O.P. Malik, and J.M. Makrandi, Chim. Acta Turc., 1991, _19_, 233).

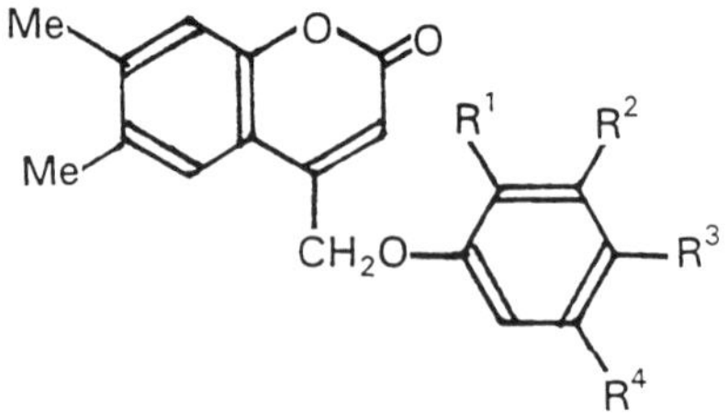

(18) 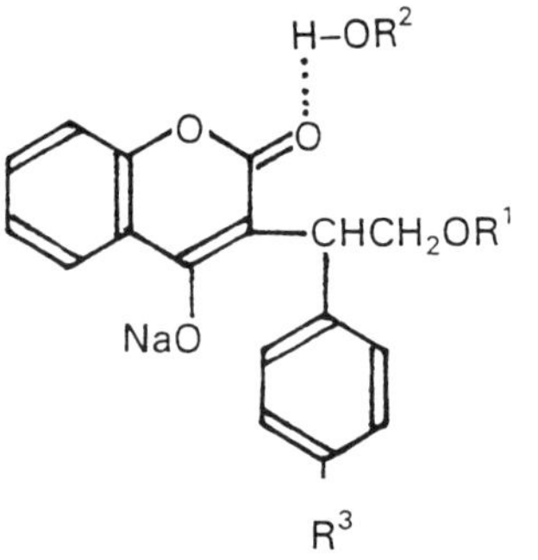   (19)

The preparations of the sodium salts of coumarin-alcohol 'dimers' 19 (R^1 = Me, Et, R^2 = Pr, Pr^i, Bu, Me_2CHCH_2, _sec_-octyl, Bu^t, _tert_-heptyl, EtCHMe, R^3 = H, OMe, OEt, NO_2) including phepromaron and nitrofarin, useful as injectable anticoagulants, have been reported (T.V. Smirnova _et al._, Tr. Inst.-Mosk. Khim.-Tekhnol. Inst. im. D.I. Mendeleeva, 1987, _149_, 92).

The antithrombotic 4-hydroxy-3-[1-(4-nitrophenyl)-3-oxobutylcoumarin [3-(_a_-acetonyl-4-nitrobenzyl)-4-hydroxycoumarin] has been obtained by the condensation reaction between 4-hydroxycoumarin and 4-nitrobenzylidene-acetone (P. Stepniewski _et al._, Pol. PL 156,096, 1992; S. Kotlicki, B. Sokolowska, and J. Szlukier, _ibid._, 156,117, 1992). The preparation of the 4-hydroxycoumarin-derived compounds 20 [R = H, (un)substituted Me, Et, Cl, Br, F, NO ; X = (CH) , (CH) O, R^1CHO, R^1CO; R^1 = Me, Et; n = 1, 2], useful as antithrombotic agents, have been reported (L. Lopez Belmonte, Eur. Pat. Appl. EP 553,590, 1993). For the preparations of a number of 3-alkoxycarbonyl-4-hydroxycoumarins, useful as intermediates for rodenticides, see T. Kakimoto and T. Hirai, Jpn. Kokai Tokkyo Koho JP 05 86,048 [93 86,048], 1993).

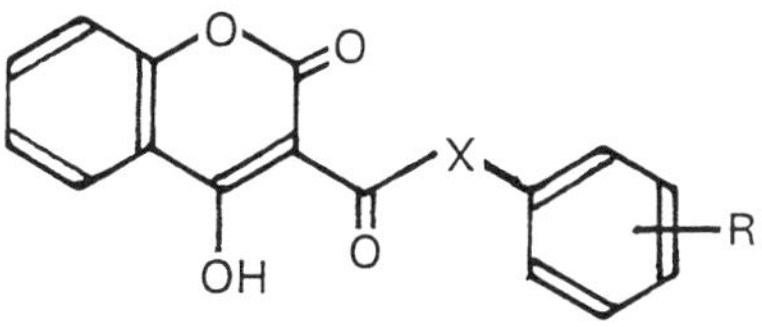

(20)

3-Bromo-4-hydroxycoumarin (21) reacts with formaldehyde, HCHO, in alcoholic solution to yield 2,3-dihydro-2-(2-hydroxybenzoyl)-4*H*-furo[3,2-*c*]-benzo[*b*]pyran-4-one and 2-(2-hydroxybenzoyl)-4*H*-furo[3,2-*c*]benzo[*b*]pyran--4-one. With salicylaldehyde the product is 7-(4-hydroxycoumaran-3-yl)--6*H*,7*H*-benzo[*b*]pyrano[4,3-*b*]benzo[*b*]pyran-6-one (Mujeeb-ur-Rahman *et al.*, Indian J. Chem., 1990, 29B, 941). The reactions of 3-bromo-4-hydroxy-coumarin with primary and secondary amines, ethylenediamine, and thiols and of 4-hydroxycoumarin-3-sulphonamide with aldehydes to give Schiff bases, have been discussed [Z.M. Nofal, A.H. Mandour, and M.I. Nassar, Egypt. J. Chem., 1990 (Pub. 1991), 33, 509].

6-Chloro-4-hydroxycoumarin on treatment with SF_4 dimerizes to 6,8'-di -chloro-2,2,4,4-tetrafluoro-4'-oxospiro[2*H*-benzo[*b*]pyran-3(4*H*),2'(4*H*)--benzo[*b*]pyrano[3,4-*d*][1,3]oxathiole]. 4-Fluorocoumarins 22 (R^1 = H, Cl, F, Me, R^2 = H; R^1 = H, R^2 = Cl) have been obtained from 4-chlorocoumarins by halogen exchange (H.J. Bertram, S. Boehm, and L. Born, Synthesis, 1991, 937).

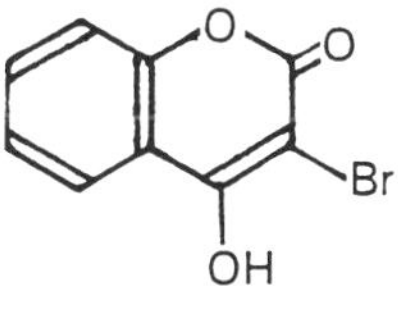

(21) 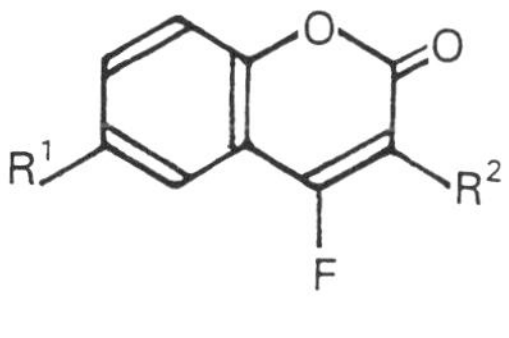   (22)

306

The $Na_2S_2O_4$ reduction of 4-hydroxy-3-nitrocoumarin (23; $R^1 = NO_2$,
$R^2 = OH$) in the presence of base affords 3-amino-4-hydroxycoumarin (23;
$R^1 = NH_2$, $R^2 = OH$); of 4-methoxy-3-nitrocoumarin (23; $R^1 = NO_2$, $R^2 = OMe$)
in aqueous MeOH containing AcONa or $NaHCO_3$, 3-amino-4-methoxy-
coumarin (23; $R^1 = NH_2$, $R^2 = OMe$) or 4-methoxycoumarin (23; $R^1 = H$,
$R^2 = OMe$) as respective main products; of 4-chloro-3-nitrocoumarin (23;
$R^1 = NO_2$, $R^2 = Cl$) in aqueous medium containing pyridine, 4-hydroxycoumarin
(23; $R^1 = H$, $R^2 = OH$) (E.A. Parfenov, V.L. Savel'ev, and L.D. Smirnov, Khim.
Geterotsikl. Soedin., 1989, 423).

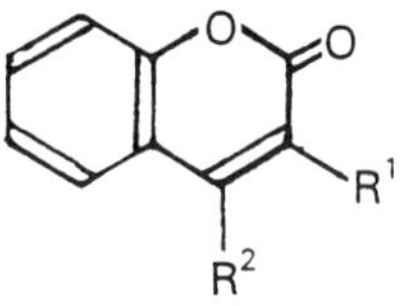

(23) 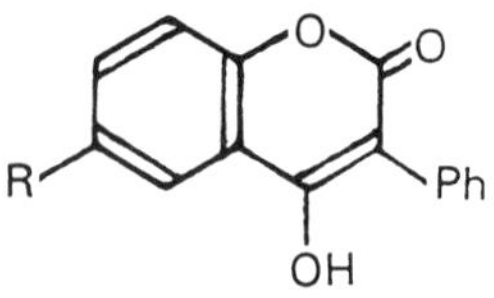　　　　　(24) 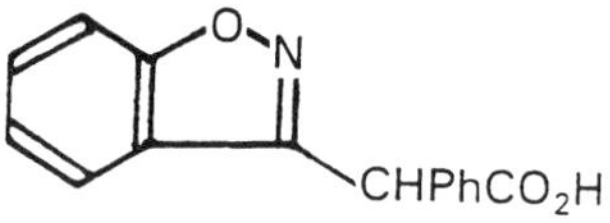　　　　　(25)

4-Hydroxy-3-phenyl- and 4-hydroxy-6-nitro-3-phenylcoumarins 24 (R = H,
NO_2, respectively) react with hydroxylamine hydrochloride, $HONH_2.HCl$, to
give benzisoxazole-3-phenylacetic acids 25 (P.D. Lokhande and B.J. Ghiya,
J. Indian Chem. Soc., 1989, 66, 314). The interaction of 3-nitro- and 3,6-
-dinitro-4-(2-hydroxyphenoxy)coumarins (26; $R^1 = H$, $R^2 = H$, Cl; $R^1 = NO_2$,
$R^2 = H$, Cl) with NEt_3 leads to the formation of triethylammonium salts of the
corresponding anionic spirocyclic compounds 27. The salts result from
intramolecular nucleophilic attack of the phenate ion on position-4 of the
coumarin (V.N. Drozd et al., Tetrahedron, 1992, 48, 469).

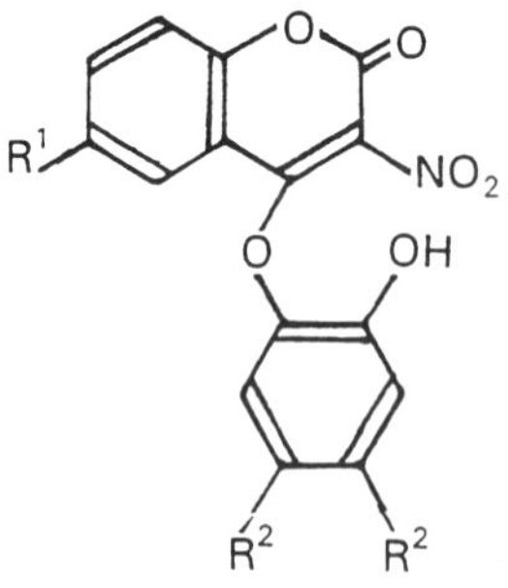

(26) 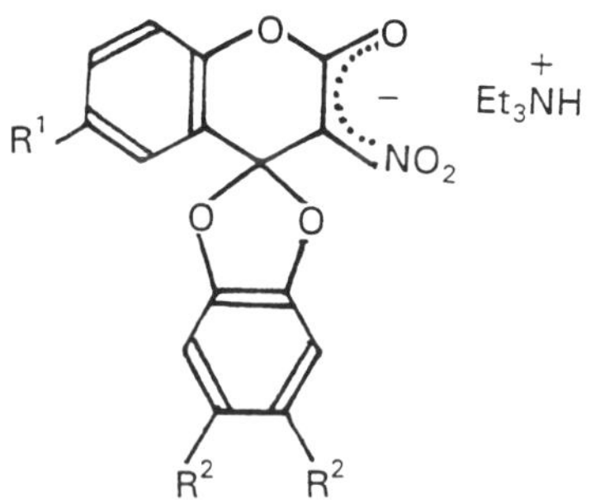　　　　　(27)

The hydroxyethylation of 4-hydroxycoumarin with ethylene carbonate and $Et_4N^+Br^-$ at 160°C affords 4-(2-hydroxyethoxy)coumarin, which reacts with HNO_3 and Ac_2O in AcOH-AcOEt to yield the nitrate ester 28. Related nitroxy coumarin derivatives with pharmaceutical properties have been prepared (M.A. Broekhoven *et al.*, PCT Int. Appl. WO 94 12,488, 1994).

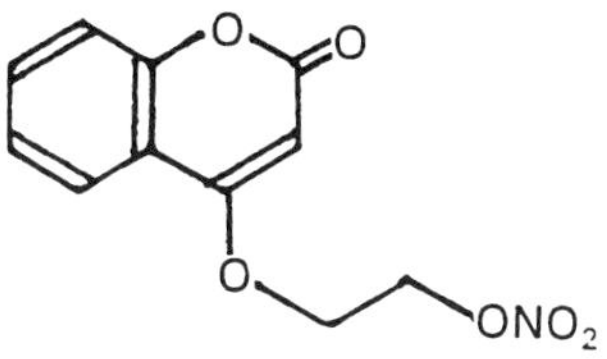

(28)

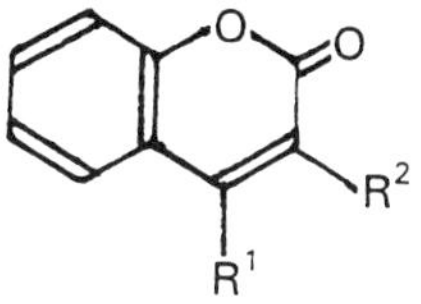

(29)

The synthesis of a series of aminocoumarins 29 (R^1, R^2 = H, OH, *N*--arylamino), some of which are 3-amino-4-hydroxycoumarins, has been reported. Some of the derivatives possessed moderate activity against *Bacillus subtilis* and *Bacillus pumilus* (M.M. Badran, A.K. El-Ansari, and S.El--Meligie, Egypt. J. Pharm. Sci., 1989, 30, 379). The Fe(II), Co(II), and Ni(II) complexes of 4-hydroxy-3-phenylazocoumarin (30) and coumarin derivative 31 have been prepared in a metal to ligand ratio of 1:2. The chelates showed increased fungicidal activity over ligands alone (B. Bharat Kumar, B. Jaya Tyaga Raju, and V. Ranbaore, J. Archaeol. Chem., 1986, 4, 35).

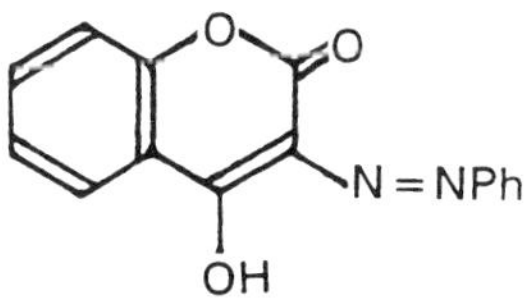

(30)

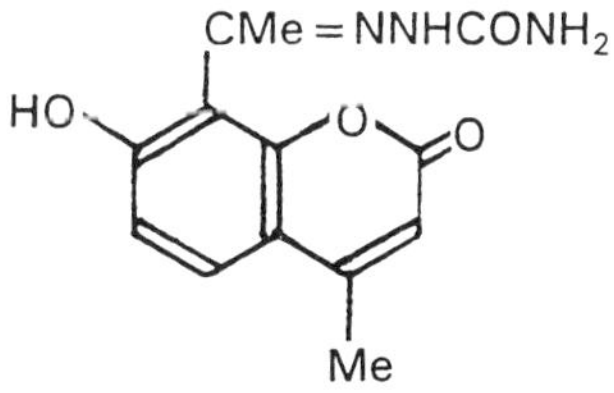

(31)

308

The preparations of 3-carbamido-4-hydroxycoumarins 32 [R^1 = H,
halogeno, alkyl, NO_2, CN, alkoxy; R^2 = (halogeno)alkyl, (un)substituted Ph,
naphthyl, pyridyl; XR^2R^3 = (halogeno)alkylenedioxy, (halogeno)oxyalkylene-
oxyalkylene; R^3 = R^4 = H, halogeno, CN, OH, NO_2, NH_2, mono- and
dialkylamioalkyl, (un)substituted aralkyl, aryl; X = O, S, SO, SO_2] have been
reported as anthelmintics, for instance derivative 32 (R^1 = R^2 = R^4 = H;
XR^2 = 4-F_3CO) has been obtained by adding 4-(F_3CO)C_6H_4NCO to 4-hydroxy-
coumarin in DMSO containing Et_3N (N. Mueller, E. Kranz, and P. Andrews,
Ger. Offen. DE 3,613,065, 1987). Also reported are the preparations of 3-
-carbamido-4-hydroxy coumarin derivatives 33 (R^1 = H, C_{1-6}alkyl, halogeno;
R^2 = H, halogeno, NO_2, NH_2, C_{1-6}alkoxy, C_{1-6}alkylthio; R^3 = H, halogeno, HO,
NO_2, NH_2, C_{1-6}alkoxy; R^4 = H, halogeno, C_{1-6}alkyl, C_{1-6}alkoxy, NH_2, NO_2;
R^5-R^8 = H, halogeno, C_{1-6}alkenyl, C_{1-6}alkyl, heterocyclyl), possessing
antibacterial activity (C.J. Dutton, J. Sutcliffe, and B. Yang, PCT Int. Appl.
WO 94 05,649, 1994). The synthesis of carbon-14-labelled 3-*N*-[(4-bromo-
phenyl)carbamido]-7-chloro-4-hydroxycoumarin and 3-*N*-[5-(trifluoromethyl)-
-1,3,4-thiadiazol-2-yl]carbamido-7-chloro-4-hydroxycoumarin have been
discussed (B.H. Lee and M.F. Clothier, J. Labelled Compd. Radiopharm.,
1993, <u>33</u>, 823).

(32)   (33)

The reaction of substituted cyclohex-2-en-1-ones 34 (X = O, S) with 3
equivalents of diethyl carbonate in the presence of sodium furnishes 5-fur-2-
-yl-4-hydroxy-7-thien-2-yl-5,6-dihydrocoumarin (35; X = O) and 4-hydroxy-
-5,7-dithien-2-yl-5,6-dihydrocoumarin (35; X = S) (G.A.M. Nawwar, B.M.
Haggag, and El S.M.A. Yakout, Z. Naturforsch., B, Chem. Sci., 1992, <u>47</u>,
1639).

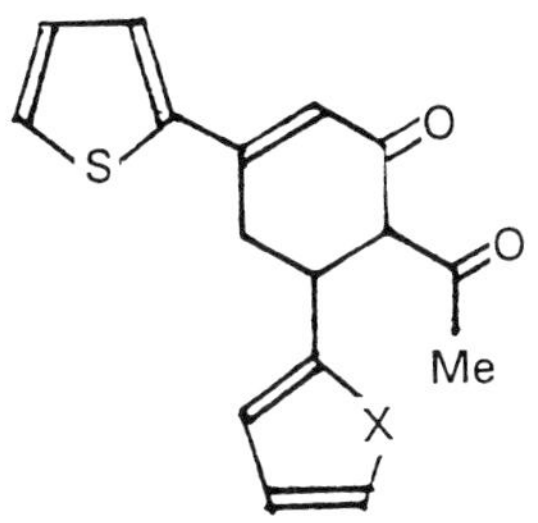

(34)

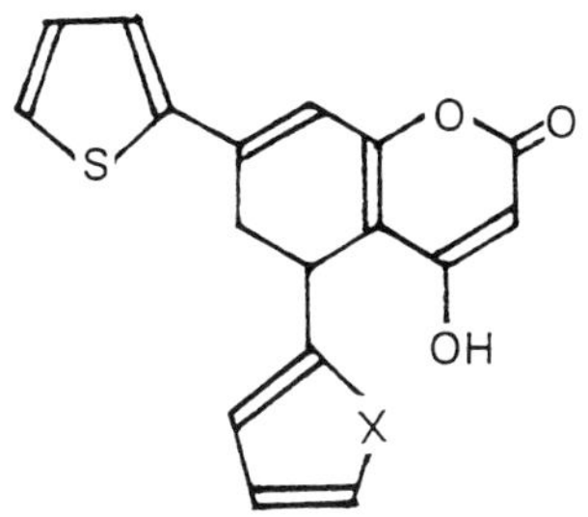

(35)

5- And *6-Hydroxycoumarins*. Benzaldehydes 1 (R = MeO, H) on boiling with 2,2-pentamethylene-1,3-dioxane-4,6-dione in AcOH in the presence of pyridine give 5-methoxycoumarin-3-carboxylic acid (2; R = MeO) and coumarin-3-carboxylic acid (2; R = H), respectively (F. Xu, W. Ding, and B. Da, Youji Huaxue, 1991, **11**, 624). It is reported that the abnormal cyclization of $PhCOC_6H_3(OH)_2$-2,4 with $EtOCH = C(Ac)CO_2Et$ affords 3-acetyl-6-benzoyl-5-hydroxycoumarin (3) (S. Xu, L. Li, and X. Huang, Chin. Chem. Letters, 1993, **4**, 951).

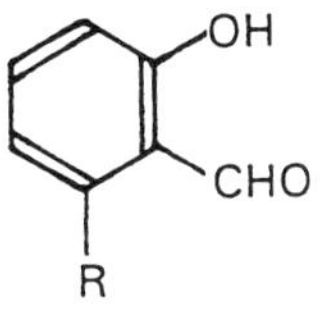

(1)

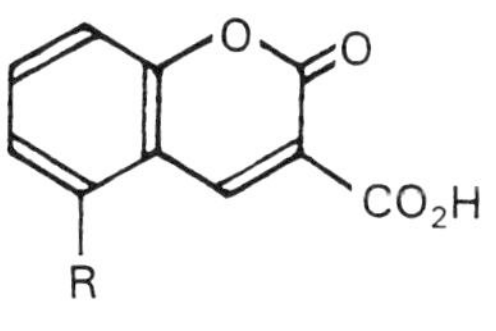

(2)

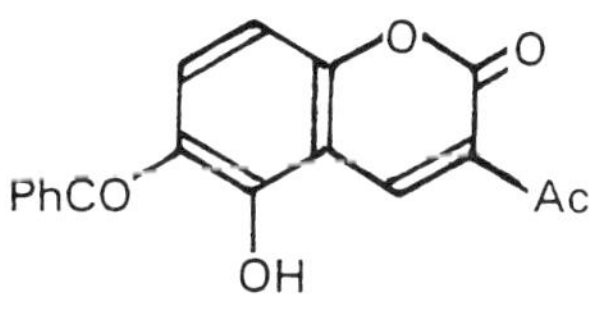

(3)

6-Hydroxy-4-methyl-8-*tert*-butylcoumarin (4) is prepared by the cyclocondensation of 2,5-di-*tert*-butyl-4-benzoquinone with $MeCOCH_2CO_2Et$ in H_2SO_4 and on nitration $(HNO_3$-AcOH) followed by reduction $(Na_2S_2O_4$--NH_3)$ it affords 5-amino-6-hydroxy-4-methyl-8-*tert*-butylcoumarin, which condenses with acetic acid in the presence of P_2O_5 to yield benzoxazole derivative 5. It also reacts with CS_2 in the presence of KOH/MeOH to give compound 6 (A.F. El-Farargy, J. Chem. Soc. Pak., 1991, 13, 59; Egypt. J. Pharm. Sci., 1991, 32, 625).

(4)   (5)   (6)

Some 6-(heterocyclylcarbonyl)alkoxycoumarins 7 $(R^1 = R^5COCHR^4$; R^2, R^3 = alkyl, aryl; R^4 = alkyl; R^5 = heterocyclic amine) have been prepared by the condensation of appropriate 6-hydroxycoumarins 7 $(R^1 = H)$ with haloketones R^5COCHR^4X $(X$ = halogeno), for example, 6-hydroxycoumarin 7 $(R^1 = H, R^2 = R^3 = Me)$ on boiling with N-(2-chloroacetyl)indoline in DMF containing K_2CO_3 furnished 6-(indolin-2-ylcarbonyl)methoxy)-4,7-dimethyl-coumarin (8) [K. Vogt, G. Kempter, and M. Klepel, Ger. (East) DD 272,464, 1989]. Some related derivatives have been prepared (G. Kempter, K. Vogt, and G. Sarodnick, *ibid.*, 255,343, 1988). Coumarin-6-sulphonates (2*H*--chromen-2-on-6-yl-sulphonates) 7 $[R^1 = R^4SO_2$; R^4 = (un)substituted alkyl, aryl; R^2, R^3 = alkyl, aryl] have been obtained by condensation of the appropriate 6-hydroxycoumarin with R^4SO_2Cl (K. Vogt, G. Kempter, and M. Klepel, *ibid.*, 286,172, 1991).

(7)

(8)

The synthesis of 6-bromoalkoxy-4-methylcoumarins 9 (R = Br, n = 2-6), 4-
-methyl-6-phenylthioalkoxycoumarins 9 (R = SPh, n = 2-6) and related
derivatives along with their antihistaminic activity have been reported
(D.B.Shinde and M.S. Shingare, Asian J. Chem., 1994, 6, 265).

(9)

The reaction involving PhNH$_2$, 2,5-dihydroxybenzaldehyde, and malonic
acid in pyridine at 60-125°C affords 6-hydroxy-3-phenylamidocoumarin (10),
chroman derivative 11 and ring-opened amide 12 in varying ratios (K.
Hayashi and M. Iinuma, Eur. Pat. Appl. EP 515,917, 1992).

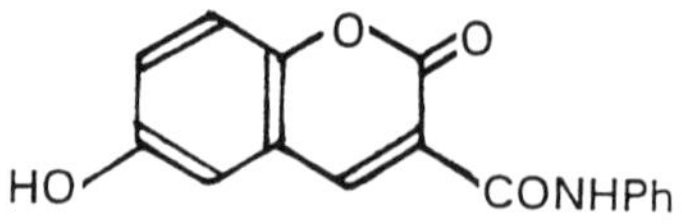

312

(10)

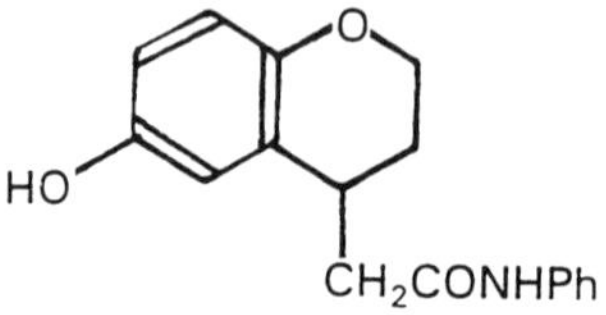

(11)

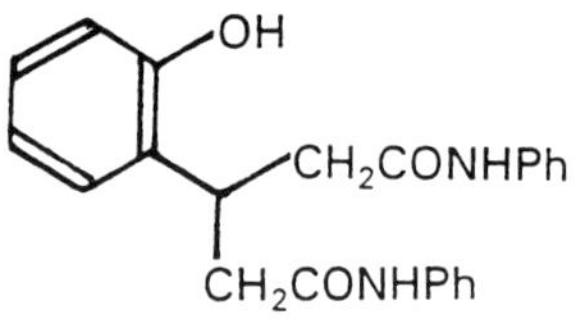

(12)

A series of 6- and 7-aryloxycoumarins and 1-(4-methylcoumarin-6-yloxy)-
-3-(substituted amino)propan-2-ols 13 (e.g., $R^1 = R^2 = Me$) have been
prepared. The latter compounds possessed hypotensive and antihistaminic
activity (S.L. El-Ansary *et al.*, Egypt. J. Pharm. Sci., 1992, <u>33</u>, 639).

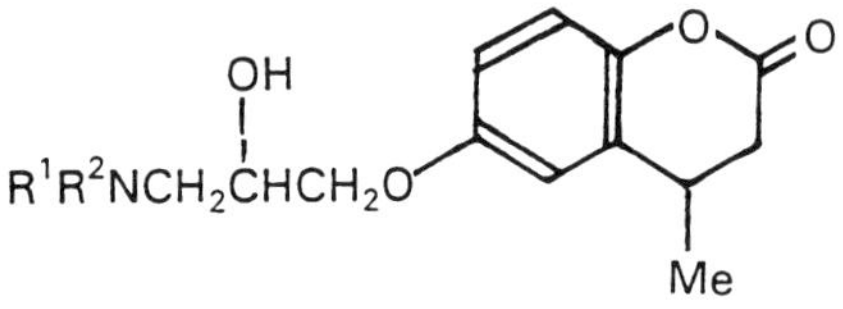

(13)

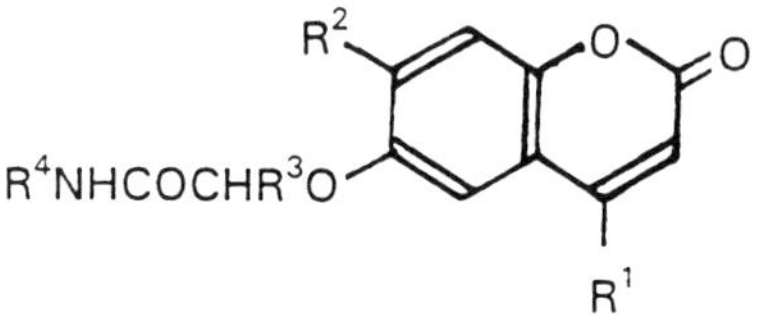

(14)

A number of *N*-substituted 4,7-dialkylcoumarin-6-yloxy carboxamides (4,7-dialkyl-2-oxo-2*H*-benzo[*b*]pyran-6-yloxy-carboxamides) 14 (R^1, R^2 = Me, Et; R^3 = Me, alkyl; R^4 = 4-ClC$_6$H$_4$, 4-BrC$_6$H$_4$, 4-MeC$_6$H$_4$, 1- or 2-naphthyl) have been synthesized [G. Kempter and K. Vogt, Ger. (East) DD 255,530, 1988].

The reaction of 4-EtO$_2$CC$_6$H$_4$N = CPhCl with 6-hydroxy-4-methyl-coumarin, followed by Chapman-Mumm rearrangement of the benzimidate product 15 yields 6-benzoyl(ethoxycarbonylphenyl)amino-4-methylcoumarin (16). Acidic hydrolysis removes the PhCO group to give 6-(4-ethoxy-carbonylphenyl)-amino-4-methylcoumarin (17). A similar sequence of reaction starting from 4-EtO$_2$CC$_6$H$_4$N = CPhCl and 8-benzoyl-7-hydroxy-4--methylcoumarin affords acridine derivative 18 (M.D. Bhavsar and U.G. Chavan, Man-Made Text. India, 1987, <u>30</u>, 159; 224). 4-Methylcoumarin-6--yl diethylglycinate (19), 6-(4,5-dihydro-5-thioxo-1,3,4-oxadiazol-2--yl)methoxy-4-methylcoumarin (20), and some related compounds derived from 6-hydroxy-4-methylcoumarin have been synthesized and evaluated for their pharmacological activity [S.L. El-Ansary *et al.*, Bull. Fac. Pharm. (Cairo Univ.), 1993, <u>31</u>, 203].

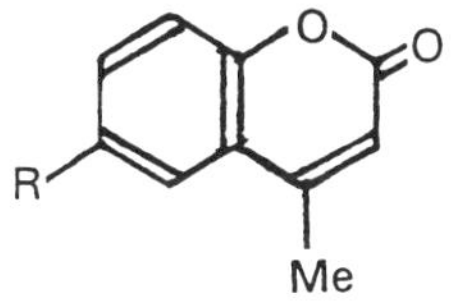

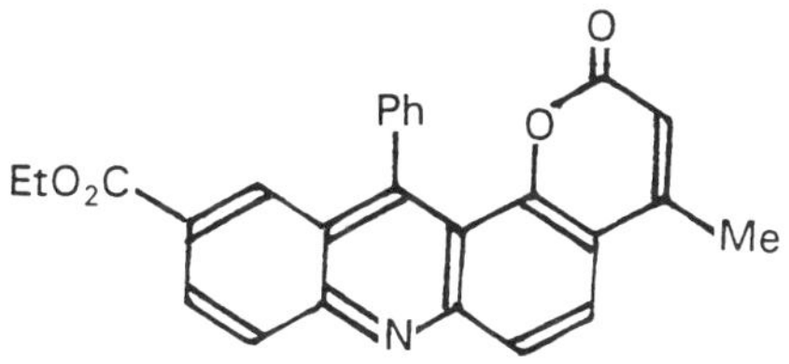

(15) R = 4-EtO$_2$CC$_6$H$_4$N = CPhO
(16) R = 4-EtO$_2$CC$_6$H$_4$NBz
(17) R = 4-EtO$_2$CC$_6$H$_4$NH

(18)

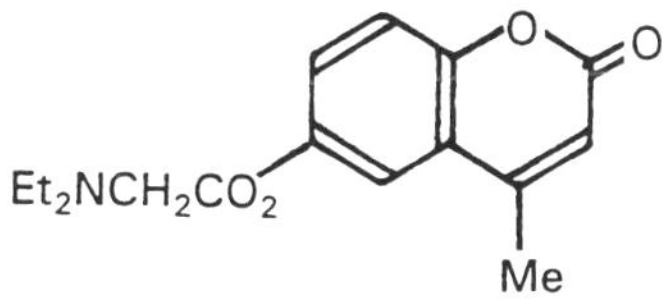

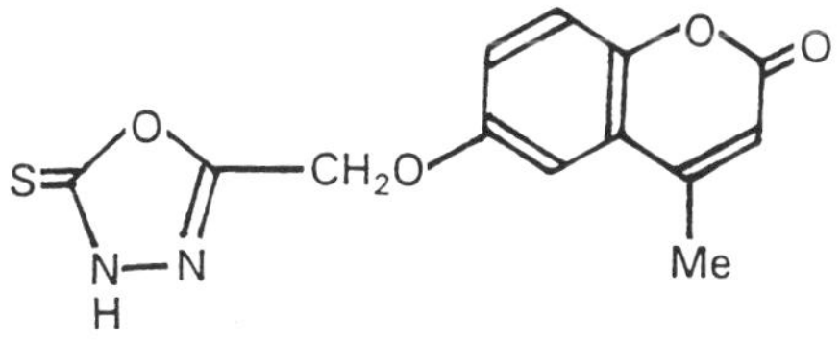

(19)

(20)

314

7-Alkyl-4-methylcoumarin-6-yloxy carboxylic acid hydrazides 21
(R^1 = alkyl, especially Me, Et; R^2 = H, Me) are prepared by reaction of the
corresponding ethyl esters with $N_2H_4.H_2O$ neat or in a solvent such as EtOH
[K. Vogt and G. Kempter, Ger. (East) DD 295,370, 1991]. 1-(4,7-Didi-
methylcoumarin-6-yloxyacetyl)-4-phenyl-thiosemicarbazide and related
-4-alkyl and -aryl-thiosemicarbazides are obtained by treating the acid
hydrazide with RNCS (R = Ph, alkyl, aryl) [K. Vogt, G. Kempter, and E.
Kleinpeter, *ibid.*, 294,021, 1991].

(21)

(22)

Cyclization of the aryl(4-methylcoumarin-6-yloxy)acetic acid hydrazides
23 (R = Ph, substituted Ph, PhCH = CH) gives the expected 4-acetyl-5-aryl-2-
(4-methylcoumarin-6-yloxymethyl)-1,3,4-oxadiazolines 24. N^4-Substituted-
-N^1-(4-methylcoumarin-6-yloxymethylcarbonyl)thiosemicarbazides 25 (R = Et,
allyl, Ph), 2-(*N*-substitued imino)-3-(4-methylcoumarin-6-yloxyacetamido)-4-
-oxothiazolidines 26 (R = Et, allyl, Ph) and related compounds (S.L. El-Ansary,
E.I. Aly, and M.A. Halem, Egypt. J. Pharm. Sci., 1992, <u>33</u>, 379), and the
4,7-dimethylcoumarins 27 (R^1 = H, Me, Et; R^2 = Me, Et) with a C-6 alkoxy
side chain containing a pyrazolyl end group (K. Spinkler, Chem.-Ztg., 1991,
<u>115</u>, 221) have been synthesized.

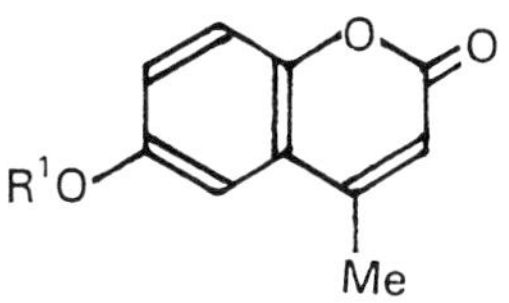

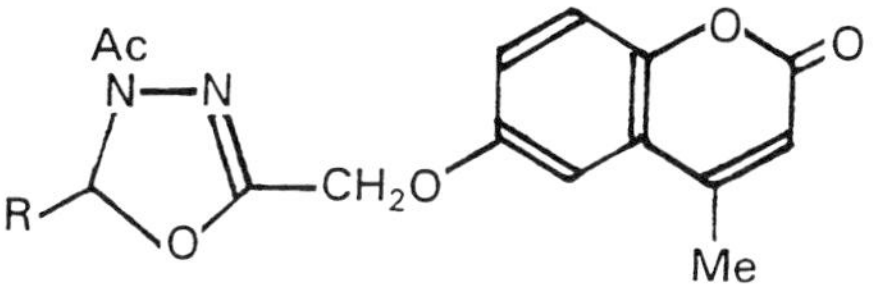

(23) $R^1 = RCH = NNHCOCH_2$
(25) $R^1 = RNHCSNHNHCOCH_2$

(24)

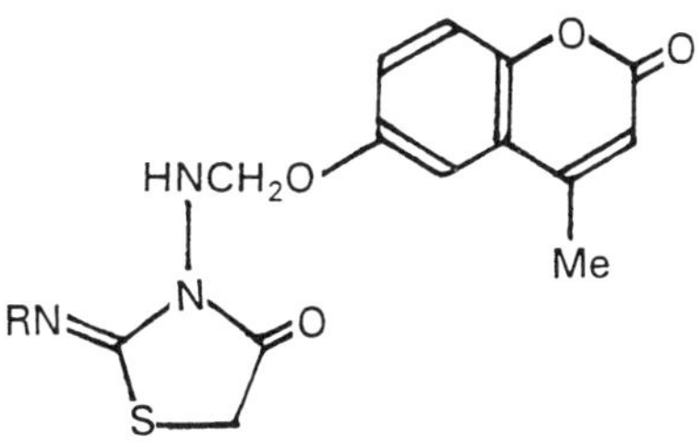

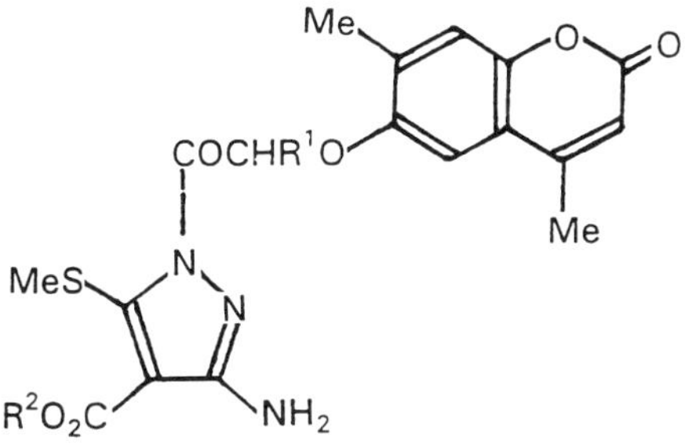

(26)

(27)

7-Hydroxycoumarins. The reaction of cinnamic acid derivatives with phenols in the presence of polyphosphoric acid yields coumarins in a single step, for example, the cyclization of $4,3-HO(MeO)C_6H_3CH = CHCO_2H$ with resorcinol monomethyl ether affords 7-methoxycoumarin (1) (J. Bhattacharjee and S.K. Paknikar, Indian J. Chem., 1989, 28B, 205). The condensation of umbelliferone (7-hydroxycoumarin) with hexamethylene dibromide gives 7,7'-[1,6-hexanediylbis(oxy)]biscoumarin (2) (N. Beuscher, H. Ritter, and C. Bodinet, Eur. Pat. Appl. EP 543,201, 1993).

(1)   (2)

7-Hydroxy-4-methylcoumarin (3) has been obtained by adding a mixture
of ethyl acetoacetate and resorcinol to boiling benzene containing H_2SO_4
with azeotropic removal of H_2O. The preparations of related derivatives have
also been reported (S. Balint *et al.*, Hung. Teljes HU 46,315, 1988). The
condensation of ethyl acetoacetate and resorcinol in the presence of
polyphosphoric acid yields 7-hydroxy-4-methylcoumarin (3) (63.5-67.8%)
and the coumarin dimer 4 (13-28%) (M. Parveen, N.U.-D. Khan, and M.K.
Logani, J. Indian Chem. Soc., 1991, 68, 617). A method for the
manufacture of 7-hydroxy-4-methylcoumarin from resorcinol and ethyl
acetoacetate (J. Leiner, D. Hesoun, and M. Rajsner, Czech. CS 259,633,
1989) and for its purification (H. Jozwiak *et al.*, Pol. PL 129,772, 1985)
have been described.

(3)   (4)

A number of coumarin-7-yl ethers **5** (R^1 = H, Me; R^2 = CH_2CH_2Br, $CH_2CH_2CH_2Cl$, adenin-9-ylethyl, benzimidazol-1-ylethyl) have been synthesized and studied for their biological activity (A.Z. Abyshev *et al.*, Khim.-Farm. Zh., 1993, <u>27</u>, 34). The synthesis of the first sample of a calixcoumarin analogue, 25,26,27-tris[(ethoxycarbonyl)methoxy]-2,8-(4-methyl--7-coumarinyloxycarbonylmethoxy)calix-4-arene (6) has been reported (H.M. Chawla and K. Srinivas, Indian J. Chem., 1993, <u>32B</u>, 1162).

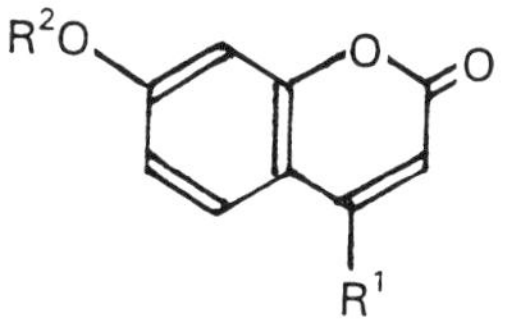

(5)

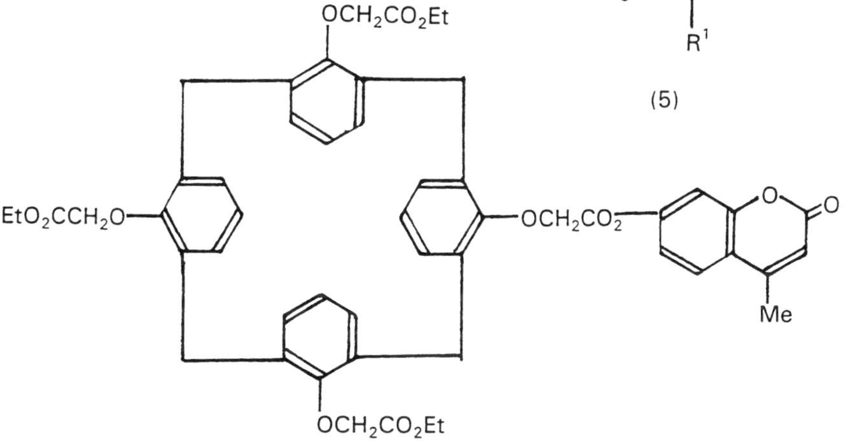

(6)

4-Substituted 8-allyl-7-methoxycoumarins **7** (R = H, Me, Et, allyl) are obtained, *via* a Claisen rearrangement, on heating *O*-allylsalicylaldehyde **8** with Ph_3P = $CRCO_2Et$ (R.S. Mali, A.R. Manekar, and S.G. Tilve *ibid.*, 1987, <u>26B</u>, 1007). Similarly appropriate substituted salicylaldehydes with ethoxycarbonylmethylenephosphorane in boiling diethylaniline afford coumarins **9** (R^1, R^2 = H, OMe; R^3, R^4 = H, Me, OMe, OPr^i) (H. Ishii *et al.*, Chem. Pharm. Bull., 1991, <u>39</u>, 3100). Substituted coumarin-7-yloxy-acetic acid (7--carboxymethoxycoumarin) and related derivatives **9** (R^3 = OCH_2CO_2H, CH_2CO_2H; R^1, R^2, R^4 = H, halogeno, lower alkyl), useful for treatment of complications of diabetes have been prepared (H. Nakayama *et al.*, Jpn. Kokai Tokkyo Koho JP 03 48,674 [91 484,674], 1991). Also the

318

preparations of a number of 7-carboxymethoxy-4-phenylcoumarins and related aldose reductase inhibitors have been reported (*idem, ibid.*, 02 191,269 [90 191,269], 1990).

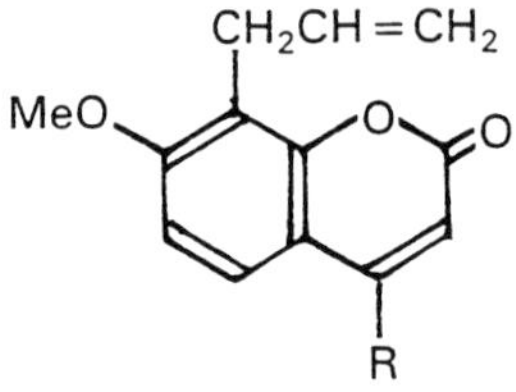

(7)

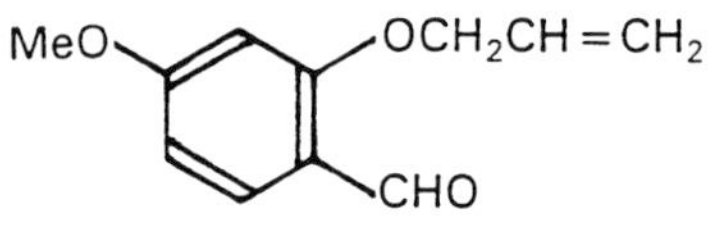

(8)

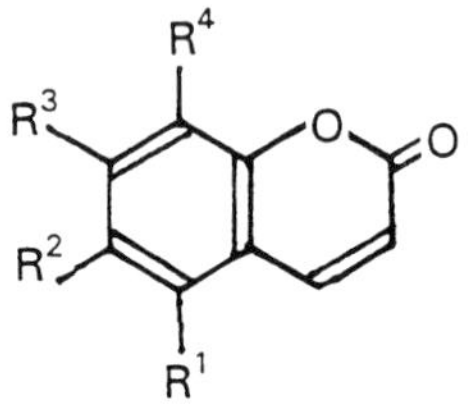

(9)

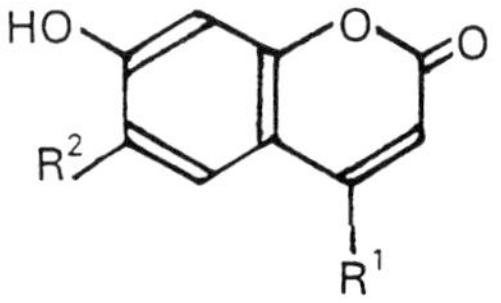

(10)

4,6-Disubstituted 7-hydroxycoumarins 10 (R^1 = Me, Ph; R^2 = Et, pentyl, hexyl, cyclohexyl, PhCHMe, PhCMe$_2$) are prepared by reaction of R^1COCH_2-CO_2Et with the appropriate substituted 4-alkylresorcinol in the presence of $BF_3.OEt_2$ at 105-108°C (S.P. Starkov, G.A. Goncharenko, and A.I. Panasenko, Zh. Obshch. Khim., 1993, 63, 1111). 7-Hydroxy-3,4-dimethylcoumarin (11) on treatment with NaH in DMF and then with 4-Pr^{i}C$_6$H$_4$CH$_2$Cl gives 7-(4-*iso*propylphenyl)methoxy-3,4-dimethylcoumarin (12). A number of related 7-(arylalkoxy)coumarins as central nervous system agents have been prepared (B. Rendenbach, H. Weifenbach, and H.J. Teschendorf, Ger. Offen. DE 3,834,861, 1990). The Fries rearrangement of esters of 5--hydroxy-4,7-dimethylcoumarin, 6-hydroxycoumarin, 6-hydroxy-4-methylcoumarin, 7-hydroxycoumarin, and 7-hydroxy-4-methylcoumarin with 2-, 3-, and 4-chlorobenzoic acid and 2-, 3-, and 4-methoxybenzoic acid have been studied to obtain corresponding hydroxyaroylcoumarins (M.D. Bhavsar and V. Chandini, Man-Made Text. India, 1989, 32, 183).

(11) R = H
(12) R = 4-Pr^{i}C$_6$H$_4$CH$_2$

4-Alkyl-3-phenylcoumarins and related compounds have been
synthesized and tested for their in vitro growth inhibitory activity against
several phytopathogenic fungi. 7-Acetoxy-4-hept-3-yl-3-phenylcoumarin
(13) has been found to be most effective against *Macrophomina phaseolina*
and *Rhizoctonia solani* (N.K. Sangwan *et al.*, Indian J. Chem., 1990, <u>29B</u>,
294).

(13)

(14)

The 7-alkoxy-3-arylcoumarins 14 (R^1 = H, NH$_2$, NO$_2$; R^2 = C$_{1-18}$alkyl) are
obtained by etherification of the corresponding 3-aryl-7-hydroxycoumarins
(T.C. Nallappa, K. Mohana Raju, and M. Subramanyam Raju, Acta Ciene.
Indica Chem., 1985, <u>11</u>, 47). The syntheses of 6-hydroxy-3,4-diphenyl-

coumarin, 7-hydroxy-3,4-diphenylcoumarin (15; R=H), and 7-methoxy-3,4-
-diphenylcoumarin (15; R=Me) have been effected by the condensation of
2-hydroxybenzophenones with phenylacetyl chloride in the presence of
K_2CO_3 in acetone. The preparation of 1,2-diphenyl-3*H*-naphtho[2,1-*b*]pyran
has also been reported (V. Narayanan and M. Natarajan, Indian J.
Heterocyclic Chem., 1992, <u>2</u>, 69). For the synthesis of 4-benzyl-7-methoxy-
-3-phenylcoumarin and related derivatives, see B.S. Verma, K.S. Dhindas,
and N.K. Sangwan, Indian J. Chem., 1993, <u>32B</u>, 239.

(15)

(16)

The cyclocondensation of 2,4-$(HO)_2C_6H_3CHO$ with $MeCOCH_2CO_2Et$ in
the presence of piperidine as catalyst in PrOH gives 3-acetyl-7-hydroxy-
coumarin (16). The preparations of a number of 3-cinnamoylcoumarins have
been reported (M.M. Hamad *et al.*, Pak. J. Sci. Ind. Res., 1990, <u>33</u>, 515).
Several 8-acyl-7-hydroxy-4-methylcoumarins 17 (R = CH = CHR[1], CH_2CH_2-
NHR[2]; R[1] = Ph, substituted Ph, CH = CHPh, indol-3-yl; R[2] = Ph, substituted
Ph), 18 (R = Ph, substituted Ph, indol-3-yl) and 19 have been synthesized
from 8-acetyl-7-hydroxy-4-methylcoumarin (17; R = Me) and along with
some related compounds screened for their anti-inflammatory activity (M.
Bhalla *et al.*, Indian J. Chem, 1992, <u>31B</u>, 183).

(17)

(18)

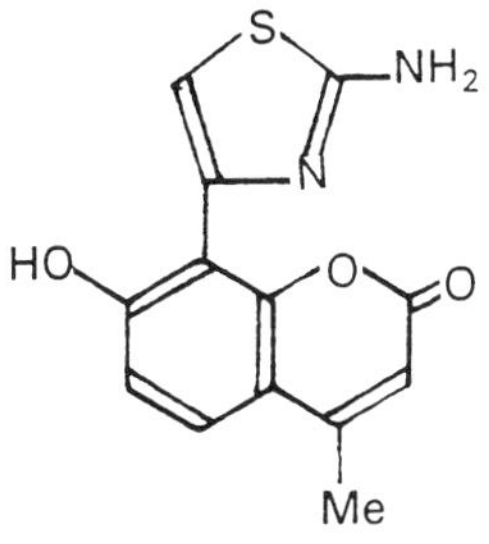

(19)

The reaction of alkylmagnesium halides with 8-acetyl-7-methoxy-coumarin (20; R = H) yields 3-acetyl-2-hydroxy-4-methoxy-Z-cinnamate esters 21 (R = Et, Pr^i, CH$_2$CH$_2$Ph). 8-Acetyl-7-methoxy-4-methylcoumarin (20; R = Me) affords the expected alkylated coumarin derivatives on reaction with alkyl and aryl Grignard reagents (L.N. Dutta, N.C. Sinha, and A.K. Sarkar, *ibid.*, 1991, 30B, 1112).

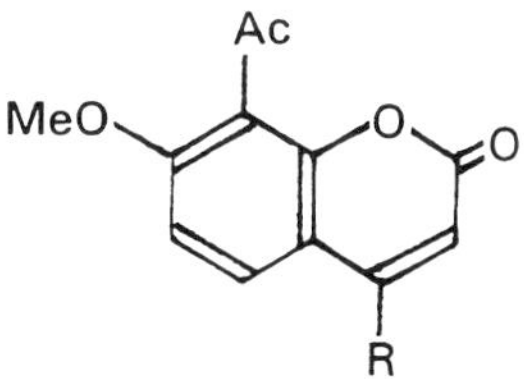

(20) 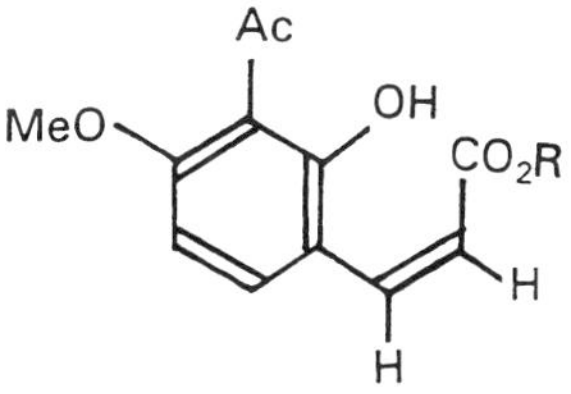　　　　　　　(21)

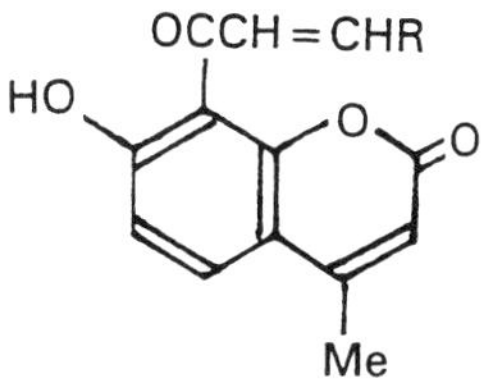

(22) 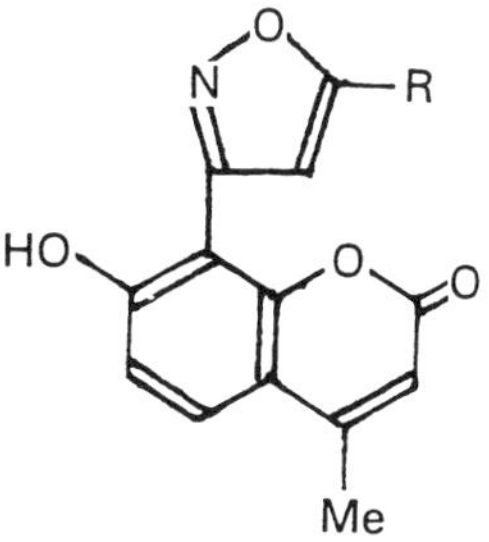　　　　　　　(23)

Condensation of 8-acetyl-7-hydroxy-4-methylcoumarin (20; R = Me) with aldehydes, R^1CHO [R^1 = 4-tolyl, 3-O_2NC$_6$H$_4$, 4,3-EtO(MeO)C$_6$H$_3$, 4,3-BuO-(MeO)C$_6$H$_3$, 4-Me$_2$NC$_6$H$_4$, 3,5,2-Br$_2$(HO)C$_6$H$_2$, 5,2-Me(HO)C$_6$H$_3$, 4-ClC$_6$H$_4$] gives 8-substituted 7-hydroxy-4-methylcoumarin-8-yl chalcones, which cyclize with NH$_2$OH to yield 7-hydroxy-8-isoxazol-3-yl-4-methylcoumarins 23. Derivatives 22 [R = 3-O_2NC$_6$H$_4$, 4,3-EtO(MeO)C$_6$H$_3$] and 23 (R = 4-ClC$_6$-H$_4$) possess fungicidal activity against *Aspergillus niger* comparable to carbendazim (P. Sharan, S. Giri, and Nizamuddin, J. Indian Chem. Soc., 1989, <u>66</u>, 393). Schiff bases of 8-acetyl-7-hydroxy-4-methylcoumarin and aniline derivatives have been prepared and their acid dissociation constants measured and their relationships discussed [G.P. Pokhariyal, J. Inst. Chem. (India), 1988, <u>60</u>, 181].

7-Hydroxycoumarin-6- and -8-carboxylic acids and their 4-methyl derivatives have been separated from mixtures by adsorption on cation exchange resin, followed by elution with acidic solvents [H.S. Rama, J. Inst. Chem. (India), 1990, <u>62</u>, 189]. 7-Hydroxycoumarin-4-acetic acid (24; R^1 = H) and 6-chloro-7-hydroxycoumarin-4-acetic acid (24; R^1 = Cl) are prepared from the appropriate 4-R^1C$_6$H$_4$OH and citric acid with H$_2$SO$_4$ as the condensing agent. The related 4-styrylcoumarins 25 (R^2 = H, MeO, NMe$_2$) are obtained by condensing derivatives 24 with 4-R^2C$_6$H$_4$CHO in piperidine (A.D. Gharde and B.J. Ghiya, J. Indian Chem. Soc., 1992, <u>69</u>, 397).

(24)

(25)

The 3-chloro-7-hydroxy (or alkoxy)-4-(substituted methyl)coumarins 26 (R^1 = H, alkyl; R^2 = R^1 = Cl, Br, aryl), useful as agrochemical intermediates, are prepared by the condensation of 3-(R^1O)C$_6$H$_4$OH with R^2CH$_2$COCH$_2$CO$_2$-R^3 (R^3 = alkyl) at elevated temperature in an inert solvent (e.g., octane) containing a catalyst (e.g., H$_2$SO$_4$) (R.H.S. Wang, U.S. US 4,788,298, 1988).

(26)   (27)

Treatment of 7-hydroxy-4-methylcoumarin (4-methylumbelliferone) with bromine at room temperature affords 3-bromo-7-hydroxy-4-methylcoumarin (27; $R^1 = R^2 = H$). Related derivatives 27 ($R^1 = H$, EtCO; $R^2 = H$, EtCO, MeCHBrCO, Me_2NCS_2CHMeCO, Et_2NCS_2CHMeCO, 2-pyrrolidino-, 2- -piperidino-, 2-morpholinocarbodithiopropionyl) having potential biological activity have been prepared (A. Cascaval *et al.*, Rom. RO 94,546, 1988). The Mannich reaction of 3-bromo-7-hydroxy-4-methylcoumarin (27; $R^1 = R^2 = H$) with various primary and secondary amines has been utilized to provide various 3-bromo-7-hydroxy-4-methyl-8-(substituted aminomethyl)-coumarins and related compounds (J.N. Gadre and V.B. Dubhashi, Indian J. Heterocyclic Chem., 1994, **3**, 181).

The radioiodination of 7-methoxy- and 6,7-dimethoxy-4-bromomethyl-coumarins 28 ($R^1 = R^2 = H$; $R^1 = MeO$, $R^2 = H$) with $F_3CCO_2{}^{125}I$ gives derivatives 28 ($R^1 = {}^{125}I$, $R^2 = H$, $R^2 = {}^{125}I$; $R^1 = MeO$, $R^2 = {}^{125}I$, respectively) (J. Baranowska-Kortylewicz and Z.P. Kortylewicz, J. Labelled Compd. Radiopharm., 1991, **29**, 1301).

(28)   (29)   (30) $R^1 = H$
(31) $R = Ac$, $R^1 = CF_3$

324

The reaction of 7-hydroxycoumarin with the hypervalent iodine reagent, (diacetoxyiodo)benzene gives 8-iodo-7-phenoxycoumarin (29) (U.K. Mallik and A.K. Mallik, Indian J. Chem., 1992, _31B_, 696).

3-Cyano-7-hydroxycoumarin (30; R=H) in toluene has been treated with Ag_2O, quinoline, and ethyl iodide to give 3-cyano-7-ethoxycoumarin (30; R=Et). Derivative 30 $[R=Me(CH_2)_4]$ has also been prepared and both compounds are useful in fluorometric assays of enzymes (I.N.H. White, Brit. UK Pat. Appl. GB 2,211,500, 1989). A method for preparing 7-hydroxy-4--trifluoromethylcoumarins, which contain electron-withdrawing groups (CN, CF_3CO) in the 3-position has been developed. The structure of 7-acetoxy-3--cyano-4-trifluoromethylcoumarin (31) has been supported by x-ray studies (Ya.V. Voznyi _et al._, Izv. Akad. Nauk, Ser. Khim., 1992, 1371). A number of 3-cyano- and 3-substituted 7-hydroxycoumarins and other hydroxy substituted coumarin derivatives and related heterocycles with a S atom or a NH group in place of the hetero O atom have been synthesized. Some of the compounds are useful as antitumor agents, which inhibit the activity of tyrosine kinase enzyme without influencing the activity of insulin-receptor tyrosine kinase (G. Keri _et al._, PCT Int. Appl. WO 93 16,064, 1993).

Some syntheses starting from 4-amino-7-methoxycoumarin (32) to yield products of types 33 (R=MeO, 4-MeC_6H_4NH) and 34 have been described (H. Braeuniger _et al._, Wiss. Z. Wilhelm-Pieck-Univ., Rostock, Naturwiss. Reihe, 1986, _35_, 34).

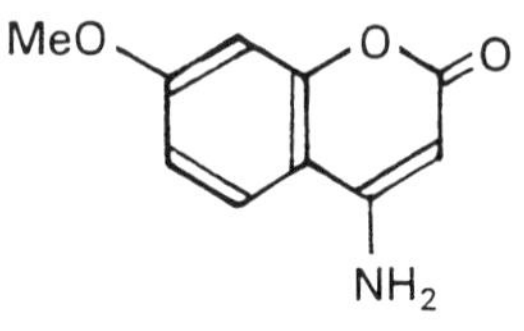

(32) 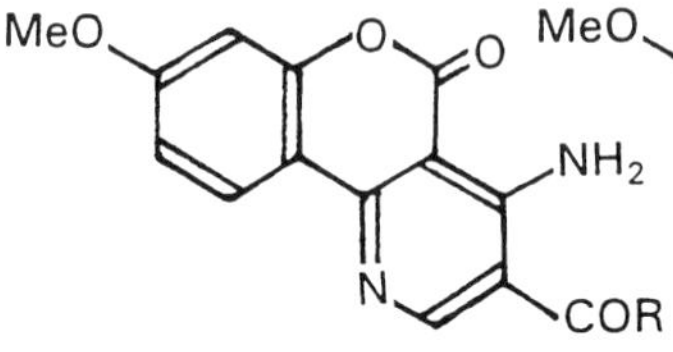   (33) 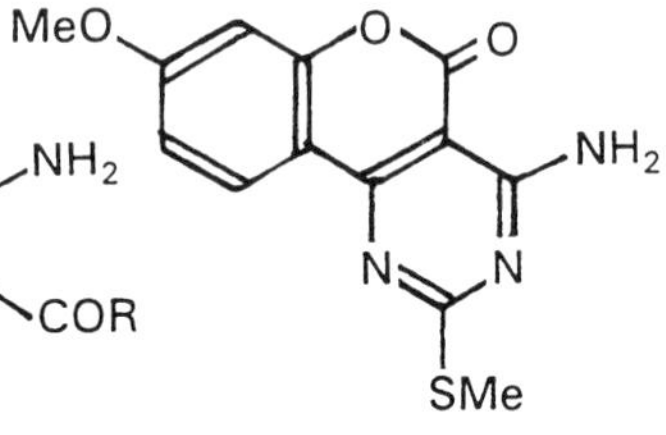   (34)

Some 6- and 8-amino-4-methyl-7-hydroxycoumarins 35 ($R^1\text{-NH}_2$, R^2=H; $R^2=NH_2$, R^1=H, respectively) have been _N_-acylated by $ClCH_2CH_2COCl$ and the resulting derivatives 35 (R^1, R^2=H, $NHCOCH_2CH_2Cl$) dehydrochlorinated. Acylation of coumarin 35 ($R^1=NH_2$, R^2=H) with $ClCH_2COCl$, followed by treatment with Et_3N afforded oxazinocoumarin derivative 36 (K.M. Amin, Egypt. J. Pharm. Sci., 1986, _27_, 333).

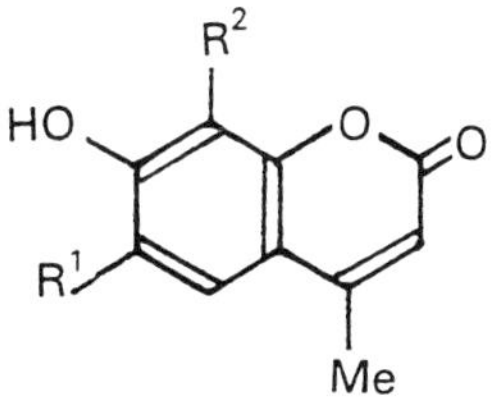

(35)

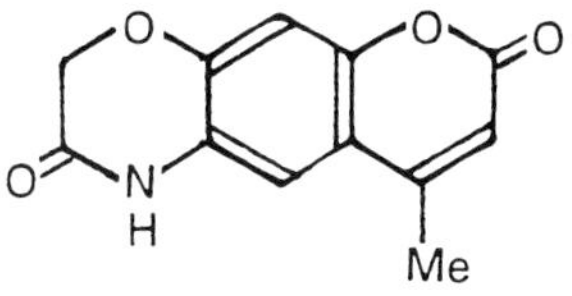

(36)

Treatment of 7-ethoxy-4-methyl-8-(thioureido)coumarins with α-halogeno ketones and α-halogenoacetic acids furnishes thiazolinylcoumarins 37 (R^1 = alkyl, allyl, cyclohexyl, PhCO; R^2 = Me, halogeno Ph, anisyl) and thiazolidinylcoumarins 38 (R^1 = alkyl, allyl, cyclohexyl; R^2 = H, CO$_2$Et) (M.Y. Ebeid, K.M. Amin, and M.M. Hussein, *ibid.*, 1987, <u>28</u>, 183).

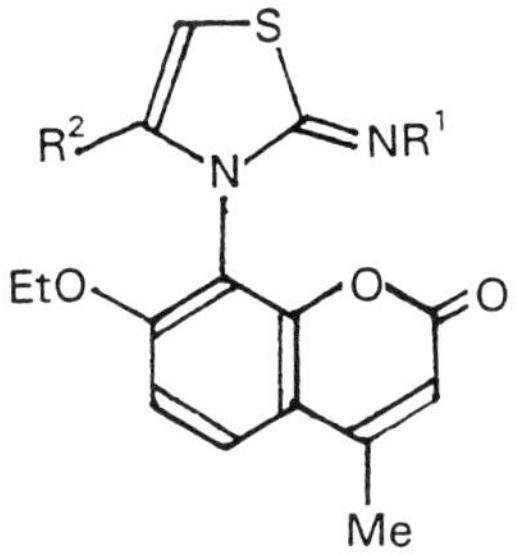

(37)

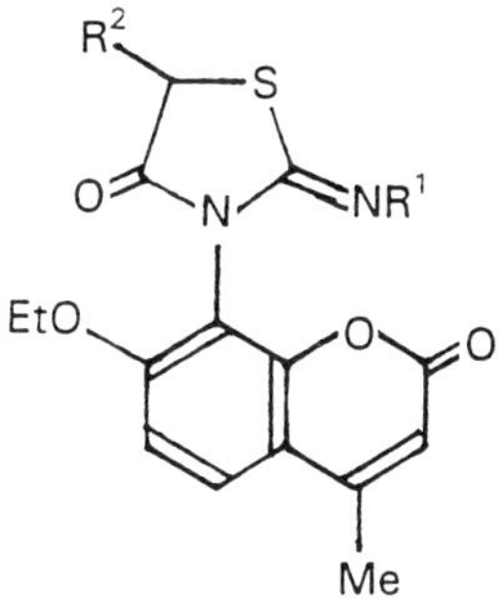

(38)

326

A method for the purification of crude 3-(2-diethylamino)-7-hydroxy-4-
-methylcoumarin has been described (B. Szczycinski and E. Boguszewska,
Pol. PL 136,525, 1986). 4-Formyl-7-ethoxycoumarin has been reduced to
the corresponding alcohol, which on treatment with PCl_5 gave 4-chloro-
methyl-7-ethoxycoumarin. The latter reacted with amines to afford 4-
-(substituted)aminomethyl-7-ethoxycoumarins 39 (R^1 = H, Et, Ph; R^2 = Ph,
methylpyridyl, pyridylmethyl, tolyl, $PhCH_2CH_2$; NR^1R^2 = piperidino,
morpholino), which exhibited bactericidal activity (M. Nagesam, M.K. Raju,
and M.S. Raju, Indian J. Pharm. Sci., 1988, _50_, 49). The preparations of 8-
-(substituted)aminomethyl-7-hydroxy-4-methylcoumarins 40 (R^1, R^2 = Et;
NR^1R^2 = morpholino, pyrrolidino, piperidino), useful as optical bleacher of
polymers, have been reported (A. Cascaval et al., Rom. RO 92,697, 1987).

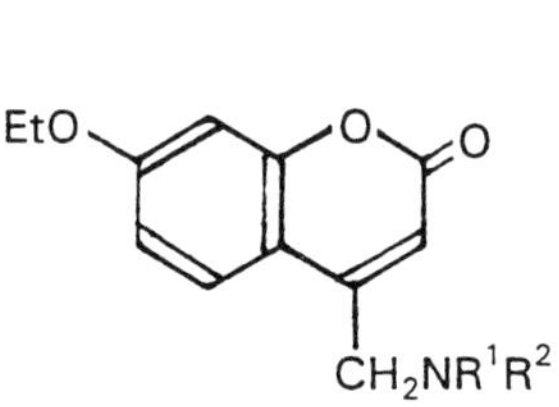

(39)

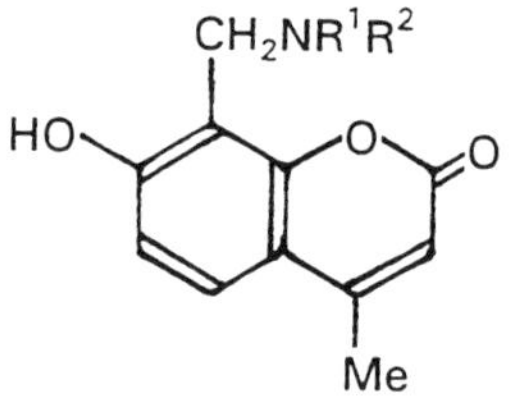

(40)

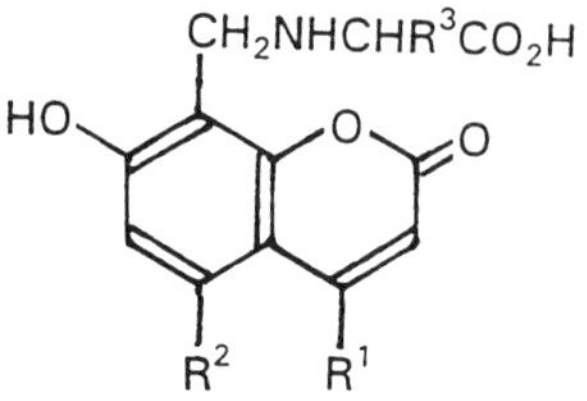

(41)

Several substituted derivatives 41 (R^1 = H, Ph; R^2 = H, Me; R^3 = H, Me,
Pr^i, CH_2OH, CH_2CH_2SMe, CH_2Ph) of 7-hydroxycoumarin have been prepared
by the Mannich reaction of 7-hydroxy-, 7-hydroxy-4-phenyl-, and 7-hydroxy-
-5-methylcoumarins with $H_2NCHR^3CO_2H$. It has been found that derivative
41 (R^1 = Ph, R^2 = H, R^3 = Pr^i) was active against S. typhosa and 41
(R^1 = R^2 = H, R^3 = CH_2Ph) was active against S. typhosa and S. albus (S.
Shah, R. Vyas, and R.H. Mehta, J. Indian Chem. Soc., 1991, _68_, 411). For

the preparation of *N*-hydroxysuccinimidyl 3-(7-hydroxycoumarin-4-yl)-propanoate as a fluorophore, see S.R. Bouma and J.E. Celebuski, Eur. Pat. Appl. EP 413,152, 1991.

An improved method for the preparation of *N*-(2-dialkylaminoalkyl)-substituted 2-oxo-2*H*-benzo(thio)pyran-3-carboxamidines 42 [X = O, S; R^1 = H, OMe, OCH$_2$Ph, OCH$_2$CO$_2$Et; NR$_2^2$ = NMe$_2$, NEt$_2$, morpholino] from the corresponding methyl(thio)coumarin-3-carboximidothiolates has been described (Y. El-Ahmad, J.D. Brion, and P. Reynaud, Heterocycles, 1993, <u>36</u>, 1979).

$$R^1 \text{–}\quad\text{(coumarin ring with X, }=O\text{)}\quad\text{C(= NH)CH}_2\text{CH}_2\text{NR}_2{}^2$$

(42)

7-Hydroxy-4-methylcoumarin has been converted into 7-(3-substituted amino-2-hydroxypropoxy)-4-methylcoumarins 43 (R = substituted amino) *via* etherification with epichlorohydrin. Derivative 43 (R = morpholino) exhibited antiinflammatory activity and derivative 43 (R = N$^+$HPrMe I$^-$) showed anti-hypertensive activity (M.S. Malik, N.K. Sangwan, and S.N. Rastogi, Chim. Acta Turc., 1986, <u>14</u>, 307). Related derivatives 44 [R^1 = CH$_2$CH(OH)CH$_2$R^5, R^5 = Pr^{i}NH, morpholino, piperidino, Me$_2$N, (HOCH$_2$CH$_2$)$_2$N; R^2 = H, Ph; R^3 = Me, Ph; R^4 = H, Me], as possible ß-blockers have been synthesized (S.M. Desai and K.N. Trivedi, J. Indian Chem. Soc., 1989, <u>66</u>, 415). The following derivatives have also been synthesized and tested for the same properties, 45 [A = morpholino, (substituted)alkylamino; R^1,R^2,R^3 = H, C$_{1-4}$alkyl, halogeno; n = 1,2] (H. Yanagisawa *et al.*, Eur. Pat. Appl. EP 411,912, 1991), 46 and 47 (R^{1}R^{2}N = BuNH, Me$_2$N, H$_2$C = CHCH$_2$NCH$_2$Ph, PhCH$_2$NH; R^3 = H, allyl) (S.L. El-Ansary and A.B. Hassan, Egypt. J. Pharm. Sci., 1992, <u>33</u>, 407).

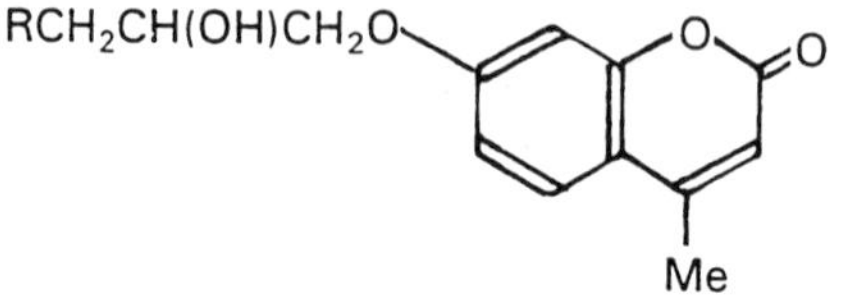
RCH2CH(OH)CH2O
O
O
Me

(43)

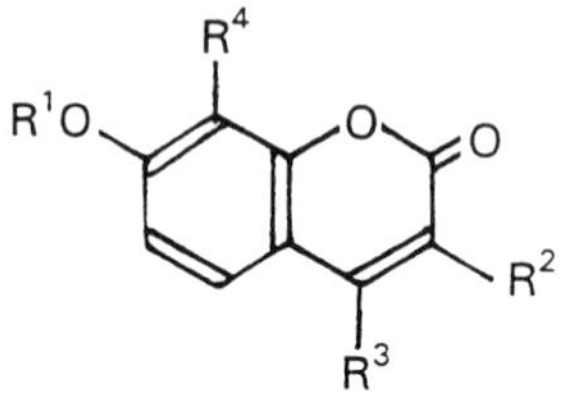
R4
R1O
O
O
R2
R3

(44)

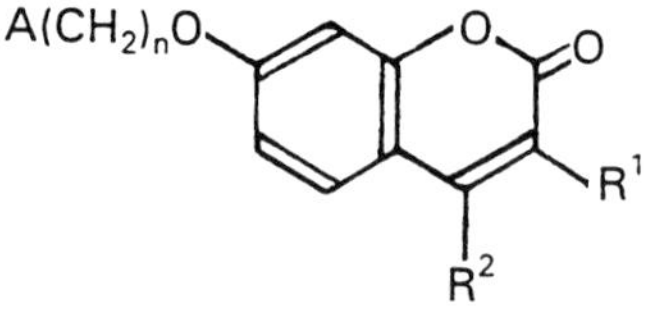
A(CH2)nO
O
O
R1
R2

(45)

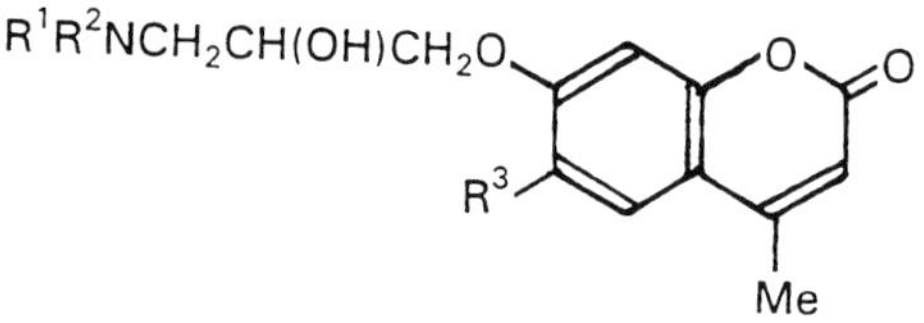
R1R2NCH2CH(OH)CH2O
O
O
R3
Me

(46)

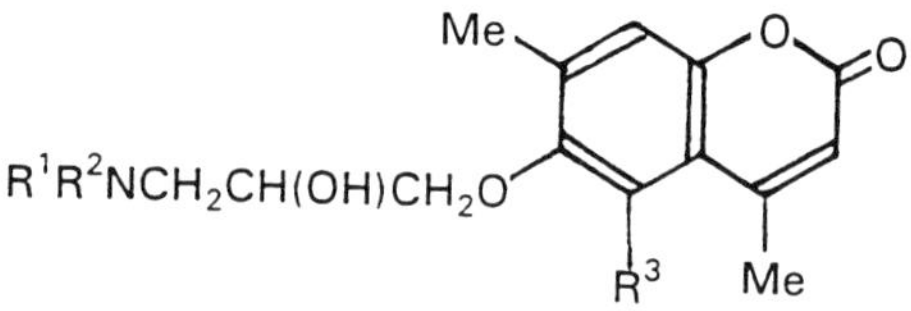
Me
O
O
R1R2NCH2CH(OH)CH2O
R3
Me

(47)

Condensation of 8-acetyl-7-hydroxy-4-methylcoumarin (48; R = Me) with $R^1C_6H_4CHO$ (R^1 = H, 3-MeO, 4-MeO) yields 48 (R = CH = $CHC_6H_4R^1$), which on cyclization with N_2H_4 in HCO_2H or AcOH affords benzopyranopyrazoles 49 (R^1 = H, 3-MeO, 4-MeO; R^2 = H, Me) (N.K. Sangwan, B.S. Verma, and K.S. Dhindsa, J. Prakt. Chem., 1988, 330, 137).

The synthesis of 7-hydroxycoumarin-4-acethydrazide (50) and its use as a convenient and sensitive fluorophoric derivatizing reagent for aldehydes and ketones has been described (N. Baggett *et al.*, Anal. Chim. Acta, 1992, <u>265</u>, 111).

8-Hydroxycoumarins. 4-Chloro-3,5-dimethyl-, 2-methoxy- and 4-(propen--2-yl)-2-methoxyphenols condense with ethyl 4-chloroacetoacetate in the presence of 73% H_2SO_4 to give substituted 4-chloromethylcoumarin 1 ($R^1 = Cl$, $R^2 = R^4 = Me$, $R^3 = Cl$, $R^5 = H$) and substituted 4-chloromethyl-8--methoxycoumarins 1 ($R^1 = Cl$, R^2-$R^4 = H$, $R^5 - OMe$; $R^1 = Cl$, $R^2 = R^4 = H$, $R^3 = CH_2CH = CH_2$, $R^5 = OMe$), respectively, which on treatment with substituted phenol afford the corresponding 4-phenoxycoumarin derivatives, for example, 1 ($R^1 = 4\text{-}ClC_6H_4$, R^2-R^5 as above) (R. Singh, S. Sharma, and O.P. Malik, J. Indian Chem. Soc., 1992, <u>69</u>, 338).

(1)

(2)

(3)

3-Bromoacetyl-8-methoxycoumarin (2) has been converted, by sequential substitution reaction with hexamethylenetetramine, acylation with Cl_2CHCO-Cl, and hydroxymethylation with HCHO, into the chloroamphenicol analogue, 2-dichloroacetamido-3-hydroxy-1-(8-methoxycoumarin-3-yl)propan-1-one 3, which showed bactericidal and fungicidal activity (R.V. Joshi and V.V. Badiger, Curr. Sci., 1987, 56, 811).

A number of 5-(substituted amino)methyl-8-methoxycoumarins 4 (R = Ph, tolyl, anisyl, $PhCH_2$, $4\text{-}ClC_6H_4$, $4\text{-}O_2NC_6H_4$, $4\text{-}H_2NC_6H_4$, naphthyl) have been prepared by reacting 5-chloromethyl-8-methoxycoumarin with RNH_2 and tested for antibacterial activity (S. Shah and R.H. Mehta, J. Indian Chem. Soc., 1989, 66, 802).

(4)

(5)

(6)

The condensation of 7-formyl-8-hydroxycoumarin (5) with RNH_2 ($R = Ph$, 4-BrC_6H_4, 4-$O_2NC_6H_4$, 2-ClC_6H_4CONH) affords the coumarin Schiff base derivatives, 7-(substituted imino)methyl-8-hydroxycoumarins 6, which have been examined for antibacterial activity (S. Shah, R. Vyas, and R.H. Mehta, *ibid.*, 1992, 69, 590). 3-[*N*-(2-aminophenyl)amido]-8-methoxycoumarin (7) condensed with RCHO (R = aryl, furfuryl) to yield Schiff bases 8, which on screening for bactericidal activity were found to be inactive (R.R. Vyas and R.H. Mehta, *ibid.*, 1991, 68, 294).

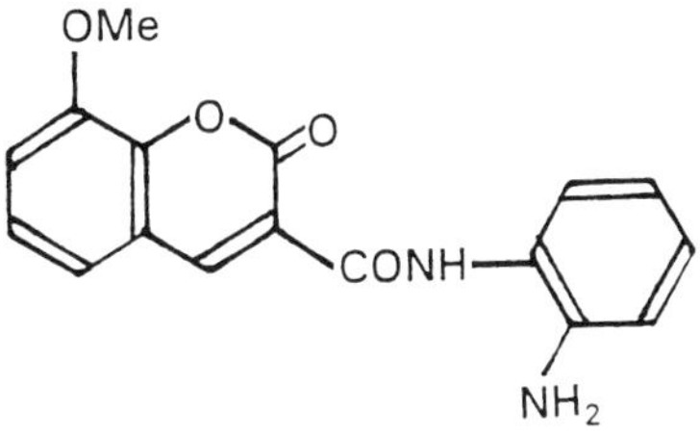

(7)

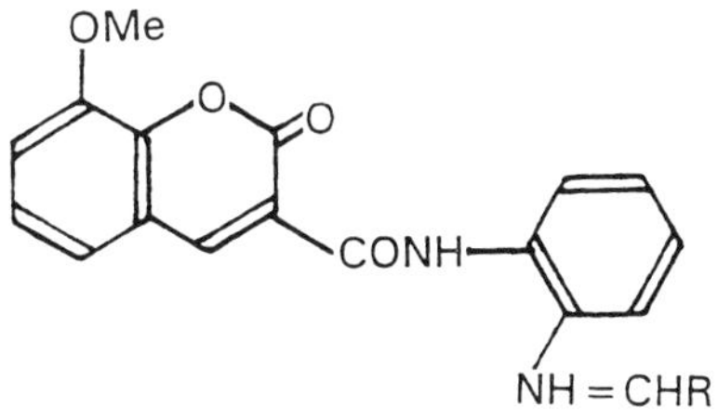

(8)

A number of 8-methoxycoumarin-3-carboxamides 9 ($R^1 = R^2 = H$, $R^3 = Ph$, substituted Ph, 1-naphthyl, 2-naphthyl, cyclohexyl, 4-phenylthiazol-2-yl; $R^1 = Br$, $R^2 = H$, $R^3 = 4$-BrC_6H_4; $R^1 = H$, $R^2 = R^3 = Ph$, Et; $R^1 = H$, $R^2 = Ph$, $R^3 = Et$; $NR^2R^3 = $ morpholino, 4-phenylpiperazino) have been prepared from the acid chlorides of the 6-substituted 8-methoxycoumarin-3-carboxylic acids 10. Compounds 9 ($R^1 = H$, Br, $R^2 = H$, $R^3 = 4$-BrC_6H_4) were active against some fungi (S. Shah and R.H. Mehta, *ibid.*, 1987, 64, 708).

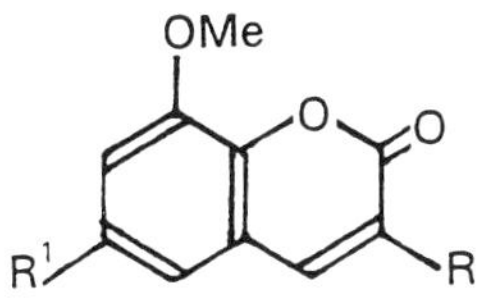

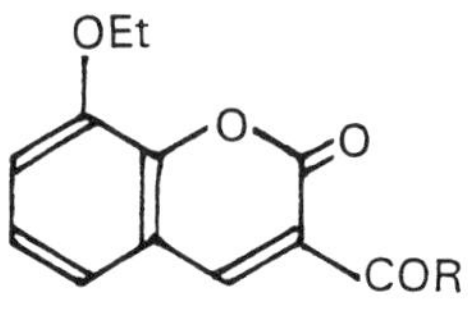

(9) R = $CONR^2R^3$
(10 R = CO_2H

(11)

2,3-(HO)(EtO)C$_6$H$_3$CHO on heating at 40°C with an equimolar amount of CH$_2$(CO$_2$E)$_2$ in EtOH containing piperidine affords ethyl 8-ethoxycoumarin-3--carboxylate (11; R = OEt), which on boiling with H$_2$NCH$_2$CH$_2$CH$_2$NMe$_2$ gives 8-ethoxycoumarin-3-(3-dimethylaminopropyl)carboxamide (11; R = NHCH$_2$CH$_2$CH$_2$NMe$_2$). Compounds of the latter type are potentially effective components of creams protective against UV radiation (J. Smidrkal and I. Hedrlin, Czech. CS 246,334, 1987).

(vi) Dihydroxycoumarins

It has been reported that the superoxide-mediated base catalyzed autoxidation of 3-hydroxy- and 3,4-dihydroxycoumarins 1 (R = H, OH, respectively) furnishes several products *via* a deprotonation-oxidation sequence. Unfortunately this method is complicated by a competing lactone saponification [A.A. Frimer, V. Marks, and P. Gilinsky-Sharon, Free Radical Res. Comm., 1991, 12-13 (Pt. 1), 93].

(1)   (2)   (3)

Synthetic procedures for *aci*-reductones belonging to the 6- and 7-mono- and disubstituted 3,4-dihydroxycoumarins (3,4-dihydroxy-2*H*-benzo[*b*]pyran--2-ones) 2 (R^1 = H, R^2 = Cl, H, OH, Ph; R^1 = Bu^t, R^2 = OH; R^{1}R^2 = OCH$_2$O) and their *cis*- and *trans*-4a,5,6,7,8,8a-hexahydro diastereomers 3 have been described. Redox potentials of all the *aci*-reductones have been determined by cyclic voltammetry and their antiaggregatory, antilipidemic, and redox properties compared with those of 4-substituted 2-hydroxytetronic acids (D.T.Witiak *et al.*, J. Med. Chem., 1988, 31, 1437).

Di and trihydroxycoumarins are arylated by 4-benzoquinone to yield 3--(4-benzoquinon-2-yl)-di and trihydroxycoumarins 4 (n = 1, 2). Their *O*-acetyl derivatives 5 have also been obtained and derivative 4 showed bactericidal activity (B.S.U. Rani and M. Darbarwar, J. Indian Chem. Soc., 1987, 64, 38).

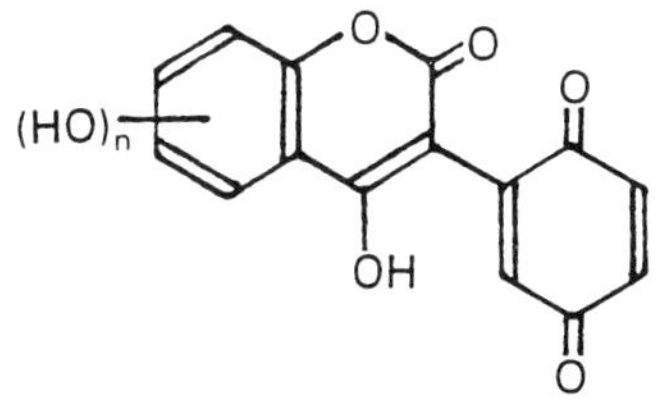

(4)

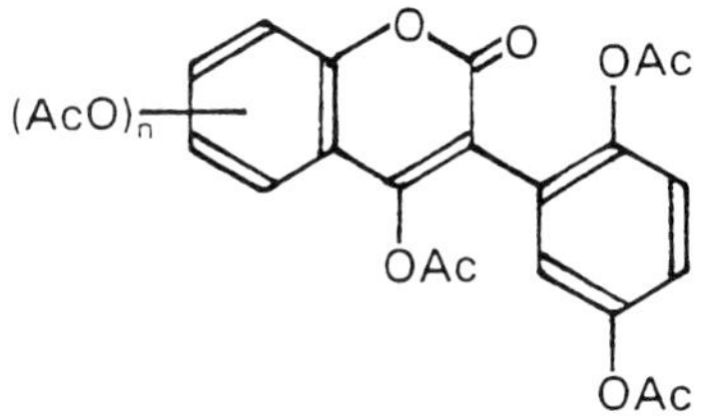

(5)

The cyclocondensation of ethyl acetoacetate with phenols 6 (R^1 = H, OH, R^2 = OH, R^3 = H; R^1 = H, R^2 = R^3 = OH; R^1 = R^3 = H, R^2 = OMe), catalyzed by $BF_3 \cdot Et_2O$, cationite KU2, or anhydrous phosphoric acid yields 6-hydroxy-, 6,8-dihydroxy-, and 5,6-dihydroxy-4-methylcoumarins, and 6-methoxy-4-methylcoumarin, 7, respectively (A.I. Panasenko, S.P. Starkov, and S.E. Sinyutina, Izv. Vyssh. Uchebn. Zaved., Khim., Khim. Tekhnol., 1994, 37, 16). The UV-absorption and fluorescence emission spectra of 5,7-diethoxy-4-methylcoumarin, 5-ethoxy-7-methoxy-4-methylcoumarin and 7,8-diethoxy-4-methylcoumarin have been reported in different polar and nonpolar organic solvents (R. Giri and M.M. Bajaj, Curr. Sci., 1992, 62, 522). A convenient two-step synthesis of 8-allyl-5,7-dimethoxycoumarins from 2-hydroxy-4,6-dimethoxybenzaldehyde has been reported (R.S. Mali and K.P. Sandhu, Synth. Comm., 1994, 24, 2883).

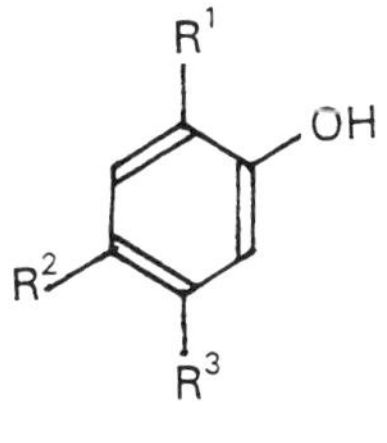

(6)

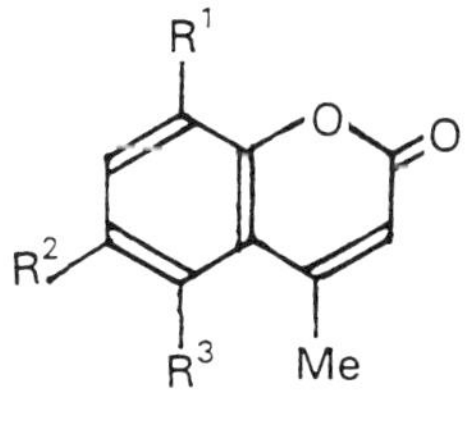

(7)

With reference to the structure elucidation of the so-called Reisch's coumarin, isolated from *Toddalia asiatica*, 6,7-dimethoxy- and 7,8-di-methoxy-5-[(*E*)-3-oxobut-1-enyl]coumarins 8 (R^1 = OMe, R^2 = H; R^1 = H, R^2 = OMe, respectively) have been synthesized from vanillin and 3,4,5-tri-methoxytoluene, respectively, *via* several steps. Comparison of the m.ps. and NMR spectral data of the synthesized coumarins 8 and of related coumarains, 5-methoxysuberenon, toddalenone, and Reisch's coumarin, suggested that the latter might be 6,7-dimethyl-5-[(*E*)-3-oxo-but-1-enyl]-coumarin (8; R^1 = OMe, R^2 = H) (H. Ishii *et al.*, Chem. Pharm. Bull., 1992, <u>40</u>, 2614).

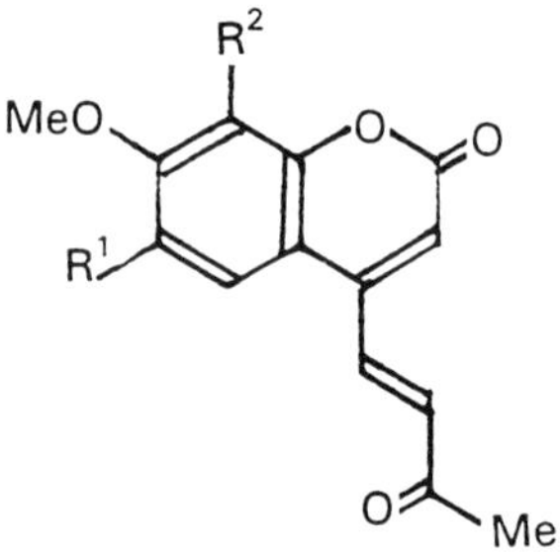

(8)
(9) 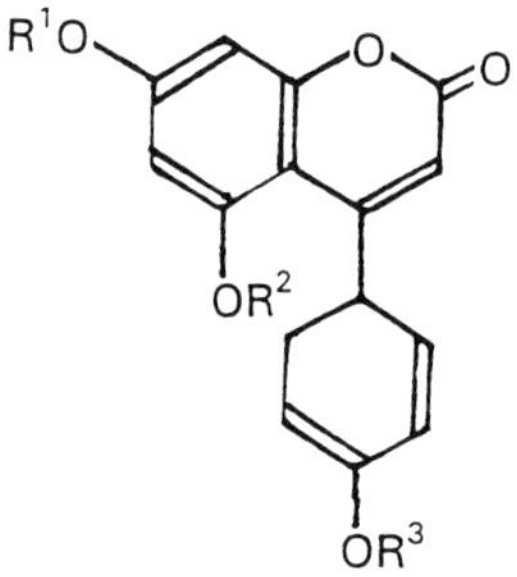

Several 5- or 7-hydroxy-5- or -7-methoxy-4-(4-substituted phenyl)-coumarins 9 (R^1 = H, R^2 = R^3 = Me; R^1 = R^3 = Me, R^2 = H; R^1 = R^3 = H, R^2 = Me; R^1 = Me, R^2 = R^3 = H; R^1 = H, R^2 = Me, R^3 = $PhCH_2$; R^1 = Me, R^2 = H, R^3 = $PhCH_2$) and two naturally occurring coumarins 9 (R^1 = R^2 = Me, R^3 = H; R^1 = R^2 = R^3 = Me) have been synthesized by the Pechmann condensation of 1,2-$(HO)_2C_6H_3OMe$-5 with 4-$R^3OC_6H_4COCH_2CO_2Et$ (R^3 = Me, $PhCH_2$) (P. Bose and J. Banerji, Indian J. Chem., 1990, <u>29B</u>, 422). The ^{13}C NMR resonances of some 5,7-dihydroxy- and 5,7-dimethoxy-4-arylcoumarins have been assigned by two-dimensional experiments (G. Delle Monache *et al.*, Magn. Reson. Chem., 1989, <u>27</u>, 1181).

Dialkyl 7-cyanomethoxy- and 7-hydroxy-5-methoxycoumarin-3,8-dicarboxylates 10 (R^1 = H, CH_2CN, alkali metal; R^2, R^3 = lower alkyl), useful as pharmaceutical intermediates have been prepared. Treatment of derivative 10 (R^1 = H, R^2 = Me, R^3 = Et) with NaH in DMF yielded 10 (R^1 = Na) (L.H. Schlager, Austrian AT 394,556, 1992).

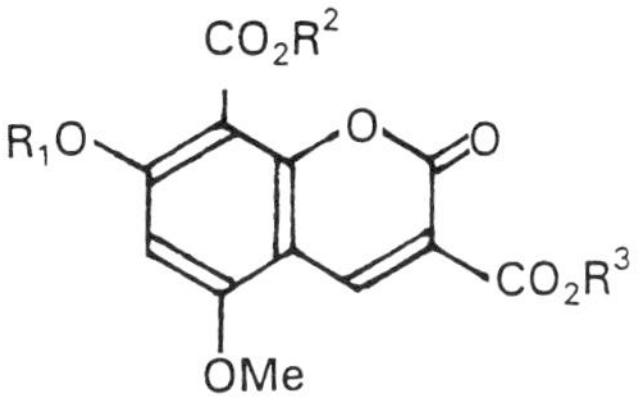

(10)

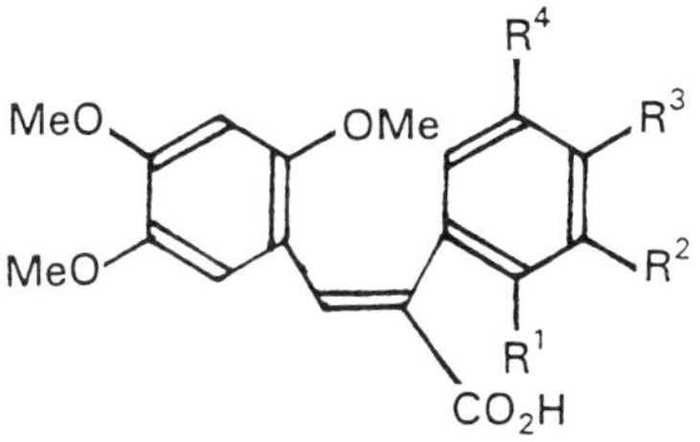

(11)

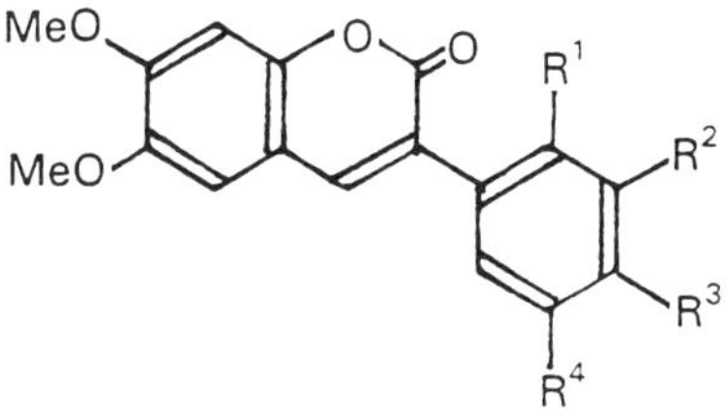

(12)

(*E*)-α-Phenylcinnamic acids 11 (R^1-R^4 = H, OMe) on heating with pyridine/POCl$_3$ give 3-aryl-6,7-dimethoxycoumarins 12 (A.S.R. Anjaneyule *et al.*, Indian J. Chem., 1991, 30B, 707). 2-Methoxybenzylidenemalonates 13 (R^1 = H, R^2 = H, 3-MeO, 4-MeO, 6-MeO, R^3 = Et; R^1 = MeO, R^2 = 4-MeO, R^3 = Et) have been hydrolyzed to afford 13 (R^3 = K), which on treatment with trifluoroacetic acid-trifluoroacetic anhydride yielded 8-, 7-, and 5-methoxy and 6,7-dimethoxycoumarin-3-carboxylic acids 14, respectively (O.E.O. Hormi, C. Peltonen, and R. Bergstrom, J. Chem. Soc., Perkin 1, 1991, 219). Cyclocondensation of 1,2,4-(HO)$_3$C$_6$H$_3$ with EtO$_2$CCH$_2$CH(COMe)CO$_2$Et affords ethyl 6,7-dihydroxy-4-methylcoumarin-3-acetate (15), which has been methylated and acetylated (V.S. Parmar, S. Singh, and J.S. Rathore, J. Indian Chem. Soc., 1987, 64, 254).

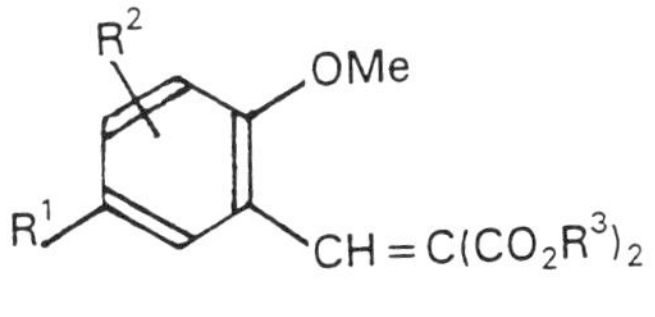

(13)

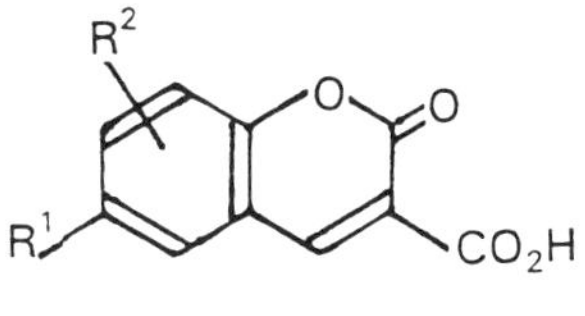

(14)

(15)

7-Methoxy-6-(oxiranylmethoxy)coumarin [7-methoxy-6-(oxiranyl-methoxy)-2*H*-benzo[*b*]pyran-2-one] reacts with 1-phenylpiperazine in EtOH to give 6-(2-hydroxy-3-piperazinylpropoxy)-7-methoxycoumarin (16). A number of related piperidinylalkoxy- and piperazinylalkoxycoumarins as neuro-protectants and anticonvulsants have been synthesized (S.S. Chatterjee, M. Noeldner, and H. Hauer, Ger. Offen. DE 4,111,861, 1992).

(16)

In the synthesis of bucumolol analogues, the reaction of 7-methoxy-coumarin with epichlorohydrin affords the epoxide 17, which reacts with amines to give 8-(3-amino-2-hydroxypropoxy)-7-methoxycoumarin hydro-chlorides 18 [R = Me_2CHNH, Me_2EtCNH, Bu^{t}NH, (Me_2CHCH$_2$)$_2$N, (Me_2CH)$_2$N, 4-benzyl-1-piperazinyl, 1,2,3,4-tetrahydro-2-isoquinolyl, 4-phenyl-1-piperazin-yl (base), AcNH(CH$_2$)$_2$NH (oxalate)], with ß-adrenolytic activity (T. Zawadowski and J. Kossakowski, Acta Pol. Pharm., 1990, 47, 49).

(17)

(18)

7,8-Dihydroxy-, 7-hydroxy-, 6-hydroxy(methoxy)-, 7-hydroxy-5-methyl-, and 7-hydroxy-6-pentyl-4-phenylcoumarins 19 ($R^1 = R^2 = H$, $R^3 = OH$, $R^4 = OH$, H; $R^1 = R^3 = R^4 = H$, $R^2 = OH$, OMe; $R^1 = Me$, $R^2 = R^4 = H$, $R^3 = OH$; $R^1 = R^4 = H$, $R^2 = pentyl$, $R^3 = OH$, respectively) are obtained by the condensation of the appropriate phenol with $PhC \equiv CCO_2Et$ in the presence of catalyst $BF_3.Et_2O$ (N.L. Polyanskaya, S.P. Starkov, and M.N. Volkotrub, Izv. Vyssh. Uchebn. Zaved., Khim. Khim. Tekhnol., 1992, <u>35</u>, 26). 7,8- -Diacetoxy-4-acetoxymethyl-, 7,8-diacetoxy-4-chloromethyl-, and 4-acetoxy- -7,8-dihydroxycoumarin 20 ($R^1 = Ac$, $R^2 = OAc$, Cl; $R^1 = H$, $R^2 = OAc$, respectively) have been synthesized *via* a phase transfer-catalyzed reaction from pyrogallol and ethyl 4-chloroacetoacetate (R.P. Singh, R.V. Singh, and O.P. Malik, Indian J. Chem., 1988, <u>27B</u>, 1031).

(19)

(20)

The condensation of 3-chloromethyl-7,8-dimethoxy-4-methylcoumarin (21; R = Cl) with various amines affords the 3-(substituted)aminomethyl-7,8--dimethoxy-4-methylcoumarins 21 (R = NMe$_2$, NMePh, NEtPh, piperidino, morpholino, arylamino). Some of the compounds have been found to be active against *E. coli* and *S. aureus* (R. Vyas, S. Bapat, and R.H. Mehta, J. Indian Chem. Soc., 1990, **67**, 482).

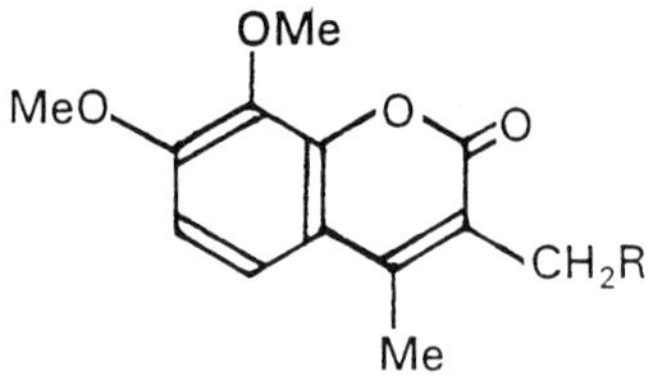

(21)

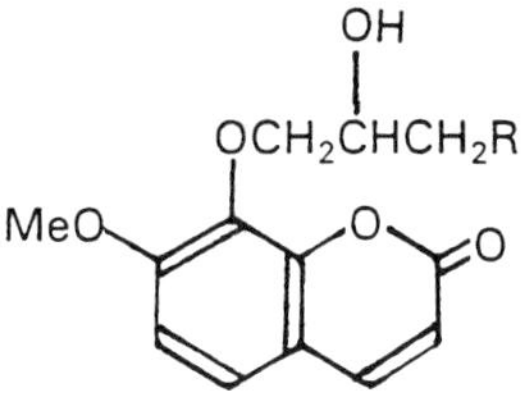

(22)

8-(3-Amino-2-hydroxypropoxy)-7-methoxycoumarins 22 (R = alkylamino, arylamino) have been prepared by etherification of 8-hydroxy-7-methoxycoumarin with epichlorohydrin and an oxirane ring cleavage of the resulting epoxypropyl ether by an amine (T. Zawadowski *et al.*, Pol. PL 145,013, 1988). For the preparation of derivatives 22 [R = 4-phenylpiper-azinyl, MeEtCHNH, Bu^{t}NH, Pr$^i_2$N, (Me$_2$CHCH$_2$)N, 4-benzylpiperazinyl, 1,2,3,4--tetrahydroisoquinolinyl], analogues of bucumolol, see T. Zawadowski and J. Kossakowski, Acta Pol. Pharm., 1933, **50**, 269.

(vii) Trihydroxycoumarins

3,4,7-Trihydroxycoumarin (1) and related trihydroxycoumarin derivatives 2 (R^1, R^2 = H, acyl, C$_{1-12}$alkyl, C$_{2-10}$alkenyl; R^3 = OH, acyloxy, C$_{1-10}$alkoxy, C$_{2-10}$alkenyloxy) have been prepared as antiallergic agents (H. Takagaki *et al.*, PCT Int. Appl. WO 92 13,852, 1992).

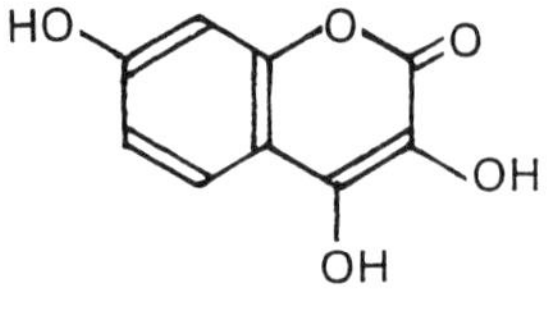

(1)

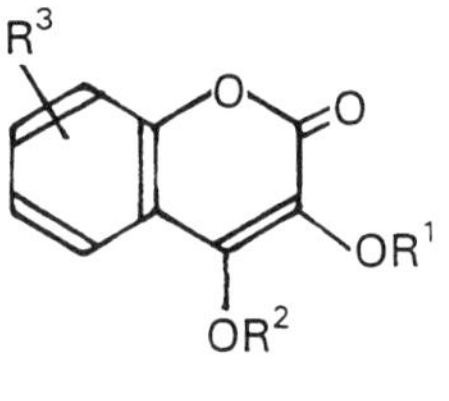

(2)

Methylation of 5,7-dimethoxy-6-hydroxy-4-methylcoumarin (3; R^1 = H, R^2 = Me) with Me_2SO_4 affords 5,6,7-trimethoxy-4-methylcoumarin (3; $R^1 = R^2$ = Me), which has been selectively demethylated with H_2SO_4 to give 5,6-dimethoxy-7-hydroxy-4-methylcoumarin (3; R^1 = Me, R^2 = H). Both dimethoxycoumarins have been converted into their respective, 6- and 7-(3-amino-2-hydroxypropoxycoumarins 3 [R^1 = R^3CHCH(OH)CH$_2$, R^2 = Me, R^3 = Et$_2$N, Pr^{i}NH, Bu^{t}NH, pyrrolidinyl, piperidinyl, 4-methylpiperazinyl, 4-morpholinyl; R^1 = Me, R^2 = R^3CH$_2$CH(OH)CH$_2$, R^3 as above less Bu^{t}NH]

These compounds have been prepared as potential circulatory system drugs (T. Zawadowski and J. Kossakowski, Acta Pol. Pharm., 1986, 43, 529). The related 8-(3-amino-2-hydroxypropoxy)-6,7-dimethoxycoumarins 4 [R = Pr^{i}NH, Me$_2$CHCH$_2$NH, Bu^{t}NH, MeEtCHNH, PhMeCHNH, 4-benzylpiperazinyl, 4-phenylpiperazinyl, 4-(2-hydroxyethyl)piperazinyl] have also been prepared (*idem, ibid.*, 1993, 50, 453).

(3)

(4)

Several 3-(coumarin-3-yl)thiophenes 5 (R^1 = H, OMe, R^2 = H, OH, OMe, NEt$_2$, R^3 = H, OH, OMe, Cl, NO$_2$, R^4 = H, OMe), containing a number of OH and OMe groups attacked to the benzene ring have been synthesized and applied on polyester fibres as fluorescent dispersed dyes. Their fluorescence and properties as dyes have been studied (R.W. Sabnis and D.W. Rangnekar, Sulphur Letters, 1993, 15, 263).

(5)

340

(viii) Miscellaneous coumarin derivatives

Salicylaldehydes and 2-hydroxybenzophenones on treatment with $Et_2NCO(CH_2)_3CONEt_2$ and $POCl_3$ yield di(coumarin-3-yl)methanes 1 (R^1 =H, Ph; R^2 =H, Me, Cl; R^3 =H, OMe) (J.R. Ahuja *et al.*, Synth. Comm., 1987, 17, 1951).

(1)

3-Cyano-4-methylcoumarin condenses with R^1CHO (R^1 = Ph, 4-HOC$_6$H$_4$, 2,4-(HO)$_2$C$_6$H$_3$, 4-MeOC$_6$H$_4$, 4-Me$_2$NC$_6$H$_4$, 2-ClC$_6$H$_4$) in the presence of AcONa to give 3-cyano-4-styrylcoumarins 2 and with phthalic anhydrides 3 (R^2 = R^3 =H; R^2 =H, R^3 = Cl) to yield 3-cyano-4-(phthalidylidenemethyl)-coumarins 4. However, phthalic anhydrides 3 (R^2 = R^3 = Cl, Br) condense with 3-cyano-4-methylcoumarin to afford 3-cyano-4-(1,3-dioxoindan-2-yl)-coumarins 5. Similar reactions of 3-cyanocoumarin-4-carboxamide and -4--carboxylic acid with aromatic aldehydes and phthalic anhydrides have been reported (F.M. Aly, A.H. Bedair, and M.R. Selim, Afinidad, 1987, 44, 489).

(2)   (3)

(4)

(5)

3-(2-Diethylaminoethyl)-7-hydroxy-4-methylcoumarin has been treated with KOH in EtOH and then with K_2CO_3 and $Br(CH_2)_3Br$ in butan-2-one to give the biscoumarin 6. The preparations of related biscoumarins 7 [R^2-R^5, R^7-R^{10} =H, halogeno, (esterified) carboxy, (esterified or etherified) OH, substituted (unsaturated) alkyl, monocycloalkyl, (substituted) monocycloaryl; R^1, R^4 = (unsaturated) (substituted) azaalkyl, azamonocycloalkyl; R^6, R^9 may also represent such units; X = (unsaturated) (substituted) (NH-, O-, S--interrupted) alkylene, monocycloarylalkylene, monocycloalkylalkylene, monocycloalkylene] as cardiovascular agents have been reported (A. Romeo and M. Prosdocimi, PCT Int. Appl. WO 92 22,545, 1992).

(6)

342

(7)

4-(2-Methylphenyl)-2-oxocyclopenta[*g*]benzo[*b*]pyran-3-carboxylic acid [4-(2-methylphenyl)cyclopenta[*g*]coumarin] has been converted into the 3--substituted tricyclic derivative 8 (K. Meguro, H. Tawada, and H. Ikeda, Eur. Pat. Appl. EP 481,243, 1992).

(8)

(ix) Furocoumarins
Cinnamates, 2-MeOC$_6$H$_4$CH=C[C$_6$H$_3$(OMe)R-2,4]CO$_2$Et (R=H, MeO) undergo cyclocondensation to give 3-(2-hydroxyphenyl)coumarin derivatives, which when treated with DDQ afford coumestans 1 (R.S. Mali and S.G. Tilve, Synth. Comm., 1990, 20, 1781).
Reactions of 3,3'-methylene-bis(5-hydroxyfuro[3,2-*g*]coumarins) [3,3'--methylene-bis(5-hydroxy-2*H*-furo[3,2-*g*]benzo[*b*]pyran-2-ones), bisallo-psoralens] 2 (R^1=H, OMe; R^2=OH) with PhCH$_2$NH$_2$, PhNH$_2$, 4-MeC$_6$H$_4$NH$_2$,

and H_2NNH_2 give the 5-amino or 5-hydrazino derivatives 2 (R^2 = $NHCH_2Ph$, NHPh, NHC_6H_4Me-4, $NHNH_2$), whereas reaction with $PrNH_2$ and $Me(CH_2)_5$-NH_2 results in ring cleavage giving hydroxybenzofurans 3 (R^3 = Pr, hexyl). The preparations of furocoumarins 4 (R^1 = H, R^4 = Pr; R^1 = OMe, R^4 = Pr, CH_2Ph) and benzofurans 5 from methylene-bis(5-hydroxyfurocoumarins) 2 have been described (O.H. Hishmat, H.I. El-Diwany, and N.M. Fawzy, Egypt. J. Pharm. Sci., 1989, 30, 317).

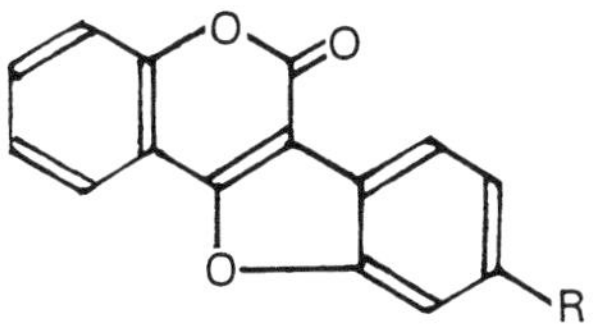

(1)

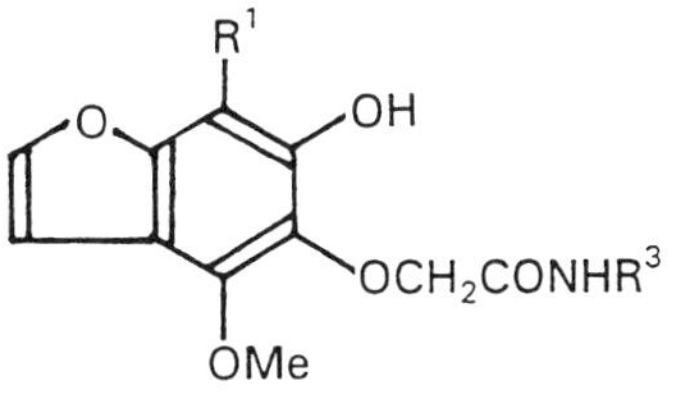

(3)

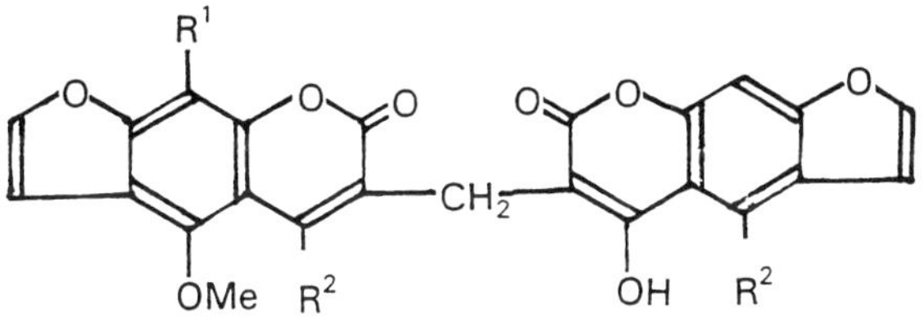

(2)

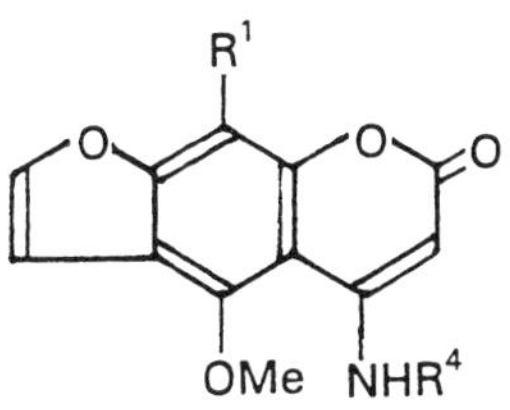

(4)

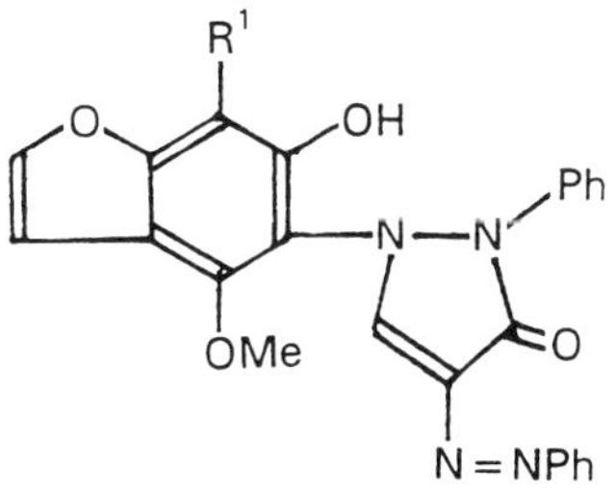

(5)

Treatment of 6-substituted-4,7,9-trimethylpsoralene 6 (R = H) in DMF with *N*-succinimidyloxybiotin at room temperature yields the biotinylated psorelen derivative 6 (R = biotinoyl), which at 1μg/mL followed by irradiation with near UV light inhibits lymphocyte proliferation by 99% in vitro (W.A. Saffran *et al.*, PCT Int. Appl. WO 87 05,805, 1987). For the aminomethylation of psoralens by hydroxymethylamides, see N.D. Heindel and M.D. Choudhuri, U.S. US 4,950,770; the preparations of dihydropsoralens, benzodipyranones and 7-alkoxycoumarins as photo-activated therapeutic agents and inhibitors of the epidermal growth factor; N.D. Heindel *et al.*, PCT Int. Appl. WO 90 08,529, 1990; and the preparation of di and trialkyl-(phthalimidomethyl)furocoumarins, G. Mikhail and H.J. Buysch, Ger. Offen. DE 3,940,597, 1991.

(6)

7-Hydroxycoumarin has been converted to 9-benzoyl-5,7-dimethylfuro-[3,2-*g*]coumarin (7) and 8-methyl-3-phenylfuro[2,3-*h*]coumarin (8), both of which showed photochemical antibacterial activity (K.M. Amin and A.N. Michael, Egypt. J. Pharm. Sci., 1987, *28*, 425).

(7)   (8)

Condensation of benzaldehyde 9 with $Ph_3P=CRCO_2Et$ (R =H, Me, Et, CH_2Ph, $CH_2CH=CH_2$) affords coumarins 10, also obtained by condensation of $2,3,4-HO(H_2C=CHCH_2)(MeO)C_6H_2CHO$ with $Ph_3P=CRCO_2Et$ followed by intramolecular cyclocondensation. Demethylative cyclization of coumarins 10 yields the 6-substituted 2,3-dihydro-2-methylfuro[2,3-*h*]coumarins 11 (R.S. Mali, A.R. Manekar, and S.G. Tilve, Indian J. Chem., 1990, <u>29B</u>, 113).

(9)

(10)

(11)

The reaction between 8-acetyl-7-hydroxy-6-methoxy-4-methylcoumarin (12) and $ClCH_2COMe$ in Me_2CO containing K_2CO_3 gives furocoumarin 13 (R = Me), which on demethylation with $AlCl_3$ in boiling chlorobenzene yields 2-acetyl-9-hydroxy-3,7-dimethylfuro[2,3-*h*]coumarin (13; R = H). Reaction of furocoumarin 13 (R = H) with epibromohydrin in EtCOMe affords derivative 14, which on treatment with amines in EtOH gives furocoumarins 13 [R = $CH_2CH(OH)CH_2R^1$; R^1 = 4-methyl-, 4-benzyl-, and 4-(2-hydroxyethyl)-1- -piperazinyl, 1-pyrrolidinyl, 4-morpholinyl, Bu^tNH, Pr^i_2N, EtMeCHNH]. The aminoalkanol derivatives have been investigated as potential ß-adrenolytics (J. Mazur *et al.*, Acta Pol. Pharm., 1989, <u>46</u>, 1). The syntheses of some 4- -(2-benzofuranyl)coumarins have been reported (A.R. Deshpande and M.V. Paradkar, Indian J. Chem., 1988, <u>27B</u>, 524).

346

(12)

(13) R (see text)
(14) R = 2-Ethene oxide

(x) Pyrancoumarins

The reaction of 4-hydroxycoumarin with 4-methoxycinnamaldehyde gives
the dihydropyrano[3,2-*c*]coumarin 1 (R^1 = 4-hydroxycoumarin-3-yl, R^2 = H,
R^3 = 4-MeOC$_6$H$_4$) and 3-[3-(4-methoxyphenyl)propenylidene]-3,4-dihydro-
-2*H*-benzo[*b*]pyran-2,4-dione (2). Reacting 4-hydroxycoumarin with alkyl
aldehydes, for example, propenal, affords dihydropyranocoumarin 1 (R^1 = OH,
R^2 = R^3 = H) and 2*H*-pyrano[3,2-*c*]coumarin 1 (R^1 = H, R^{2}R^3 = bond) (G.
Appendino *et al.*, Helv., 1990, 73, 1865).

(1)

(2)

4-[4-(2-Cresoxy)but-2-ynyloxy]coumarin 3 (R = 2-MeC$_6$H$_4$) and 4-[4-(4-
-cresoxy)but-2-ynyloxy]coumarin 3 (R = 4-MeC$_6$H$_4$) undergo a sigmatropic
rearrangement to give in addition to the normal products 4-aryloxymethyl-
-2*H*-pyrano[3,2-*c*]coumarins 4, products 5 and 6, respectively (K.C.
Majumdar *et al.*, Indian J. Chem., 1994, 33B, 120).

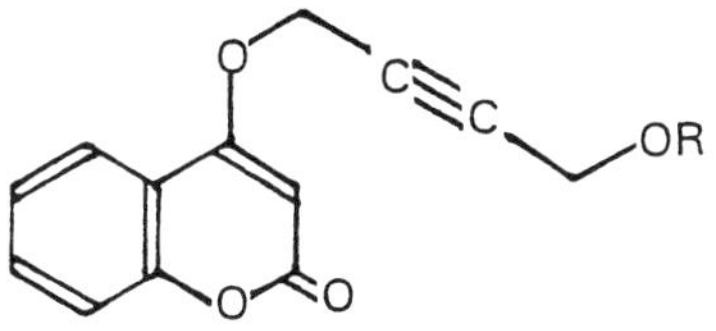

(3)

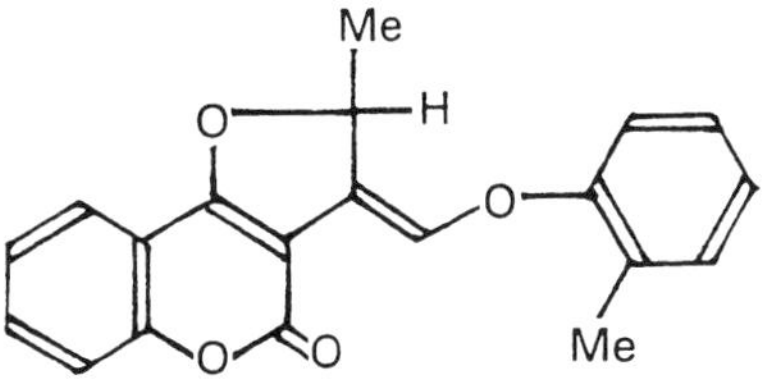

(5)

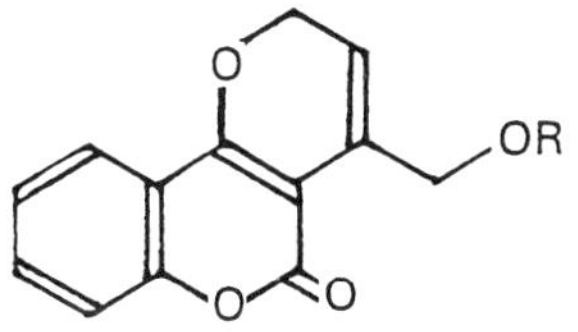

(4)

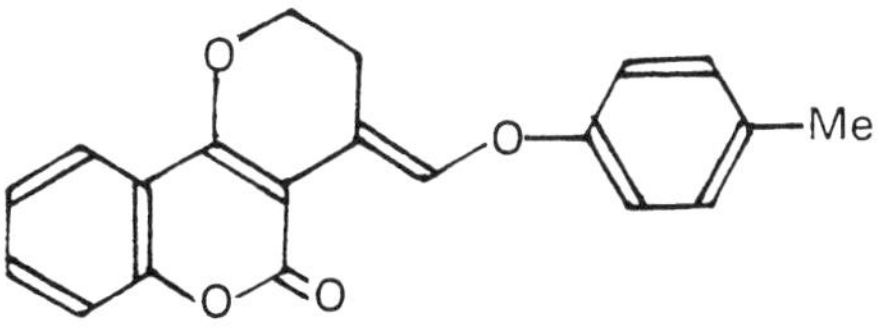

(6)

Cyclocondensation of 4-nitrobenzalacetone with 4-hydroxycoumarin
yields nitrowarfarin methyl ketal [2-methoxy-4-(4-nitrophenyl)-3,4-dihydro-
-2*H*-pyrano[3,2-*c*]coumarin] 7 (R^1 = NO$_2$, R^2 = H) (84:16 mixture of *trans-cis*
isomers), which on reduction of the nitro group affords amines 7 (R^1 = NH$_2$,
R^2 = H). 4-Nitrobenzalacetone on condensation with 6-bromo-4-hydroxy-
coumarin gives only *cis*-7 (R^1 = NO$_2$, R^2 = Br), which on reduction with tritium
over Pd/C affords ^{3}H-7 (R^1 = NO$_2$, R^2 = ^{3}H). Attempts to deuterate *cis*-7
(R^1 = NH$_2$, R^2 = Cl) failed (R.S. Obach and L.S. Kaminsky, J. Labelled Compd.
Radiopharm., 1992, 31, 763). For the preparation of dihydropyrano[5,4-*c*]-
coumarins 8 (R^1 = H, R^2 = Me, Et; R^1 = R^2 = Me; R^3 = Et, Pr^i), see A. Boilova
et al., Ann., 1991, 1279.

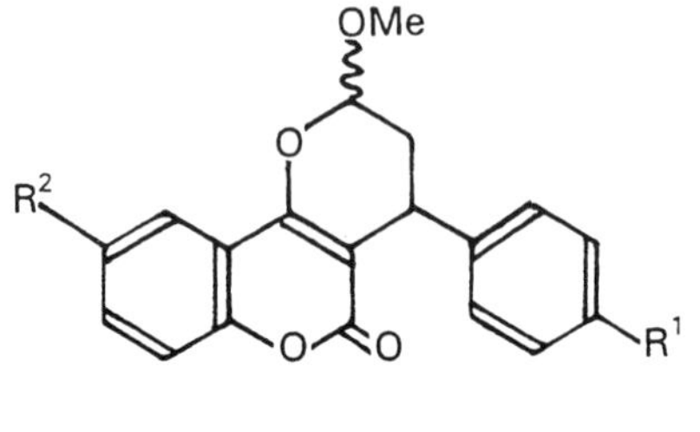

(7)

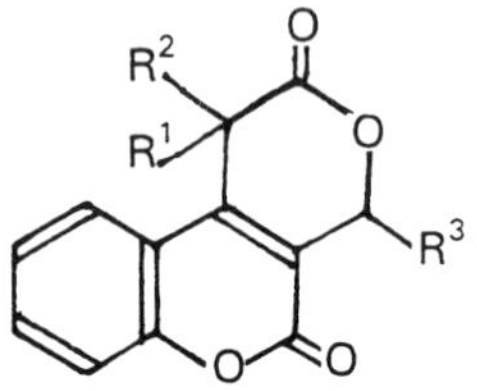

(8)

4-Methoxy- and 4,5-dimethoxy-8,8-dimethylpyrano[3,2-*g*]coumarins (9; R =H, MeO) have been prepared by cyclization of the 7-dimethylpropynyloxy- -8-iodo-4-methoxycoumarins 10 (R =H, MeO) in boiling Me_2NPh (M. Aslam Ansari and V.K. Ahuwalia, Curr. Sci., 1987, <u>56</u>, 145). 3-(1,1-Dimethylallyl)- -8,8-dimethylpyrano[2,3-*h*]coumarin [3-(1,1-dimethylallyl)seselin] has been obtained from 7-hydroxycoumarin *via*, prenylation, sigmatropic rearrangement, and further prenylation (R. Hernandez Galan *et al.*, Heterocycles, 1989, <u>29</u>, 297). 3,7-Disubstituted 4,6-dimethylpyrano[3,2-*g*]- benzopyran-2,8-diones (4,6-dimethylpyrano[3,2-*g*]coumarin-8-one) 11 (R^1 =H, Me; R^2 =H, Et) have been prepared, along with other products, *via*, the reaction of 4,6-diacetylresorcinol with appropriate alkoxycarbonylalkyl- idene(triphenyl)phosphoranes (D.N. Nicolaides *et al.*, J. Heterocyclic Chem., 1994, <u>31</u>, 173).

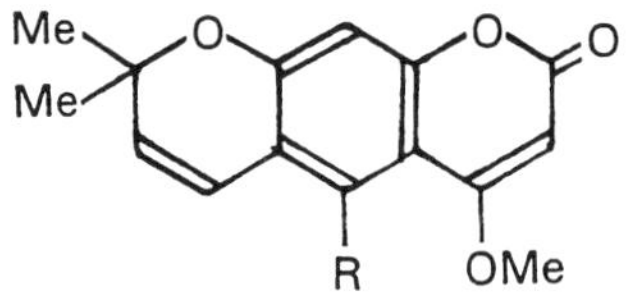

(9)

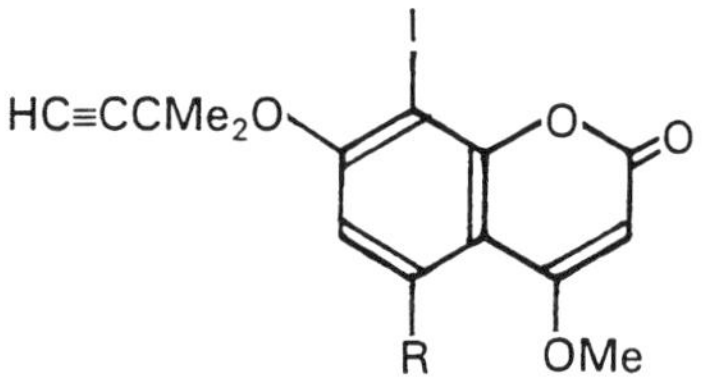

(10)

(11)

(12)

Four *o*-pyranoflavanones, for example, 12 (8-methyl-2-phenyl-3,4-
-dihydro-2*H*-pyrano[2,3-*h*]benzo[b]pyran-4,6-dione, 8-methyl-2-phenyl-3,4-
-dihydro[2,3-*h*]coumarin-4-one) have been prepared from 8-acetyl-7-hydroxy-
-4-methylcoumarin (M.S.Y. Khan and P. Sharma, Indian J. Chem., 1993,
<u>32B</u>, 374).

*(xi) Chromones (4H)-benzo[b]pyran-4-ones, 5,6-benzo-4-pyrones, 5,6-
-benzo-γ-pyrones)*
Alkyl- and arylchromones. The chemo-, regio-, and stereoselectivities of
hydrogenation of 2-methyl- and 3-methylchromones (2-methyl- and 3-methyl-
-4*H*-benzo[*b*]pyran-4-ones) (1 and 2) with group 8 metal catalysts, at
ambient temperature under ordinary pressure of hydrogen have been
examined. The hydrogenation of chromones 1 and 2 afforded predominantly
the respective cis saturated alcohols (K. Hanaya *et al.*, Nippon Kagaku
Kaishi, 1993, 774).

(1)

(2)

In a new efficient synthesis of 2-butylchromone (3) from salicylaldehyde, the key step involves the cyclization of $2\text{-HOC}_6\text{H}_4\text{COC}\equiv\text{CBu}$ with MeONa in MeOH (D. Pflieger and B. Muckensturm, Tetrahedron Letters, 1990, *31*, 2299). 2-Substituted 6-methylchromones 4 (R = Me, CO_2Et, Ph, 4--$MeOC_6H_4$, cyclopropyl) are prepared by the cyclocondensation of 2,5--$(HO)MeC_6H_3COCH_2Cl$ with RCOCl in the presence of Ph_3P in boiling toluene (K.J. Lee, S. Kim and H. Park, Bull. Korean Chem. Soc., 1991, *12*, 120).

(3) (4) (5)

The reactions of chromone (5; $R^1 = R^2 = H$) 2-methylchromone (5; $R^1 = Me$, $R^2 = H$) and 6-methylchromone (5; $R^1 = H$, $R^2 = Me$) with H_2NCH_2-CH_2NH_2 under various conditions yield several ring opened products, for example, 1,8-bis(2-hydroxybenzoyl)-2,7-octadiene-3,6-diamines 6 and dihydro(hydroxyphenyl)diazepines 7 (M. Owczarek and K. Kostka, Pol. J. Chem., 1991, *65*, 345).

(6)

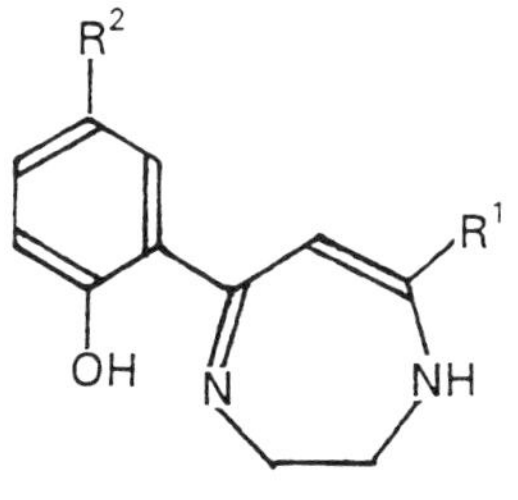

(7)

The syntheses of some 2-benzylchromones from 2-hydroxyacetophen-ones and phenylacetic acid *via* the Baker-Venkataraman rearrangement have been reported (A.K.D. Mazumdar *et al.*, J. Indian Chem. Soc., 1990, _67_, 148). 2,6-Dimethylchromone has been prepared and converted to various 2--styryl derivatives by reaction with aromatic aldehydes (I.S. Al Naimi and B.A. Hussain, Qatar Univ. Sci. J., 1992, _12_, 73). A number of [1-(2--hydroxybenzoyl)alkylidene]triphenylphosphoranes 8 ($R^1 = R^2 = R^3 = H$; $R^1 = R^3 = H$, $R^2 = Me$, $CH_2CH = CH_2$, Pr; $R^1 = H$, $R^2 = Me$, $CH_2CH = CH_2$, $R^3 = Me$; $R^1 = OMe$, $R^2 = CH_2CH = CH_2$, $R^3 = H$) have been obtained by acylation of $R^3CH = PPh_3$ with the corresponding methyl 2-hydroxybenzoates. Compounds 8 on cyclization with $ClCOCH = CHPh$ gave the 2-styrylchrom-ones 9 (F. Zammattio *et al.*, Synthesis, 1992, 375).

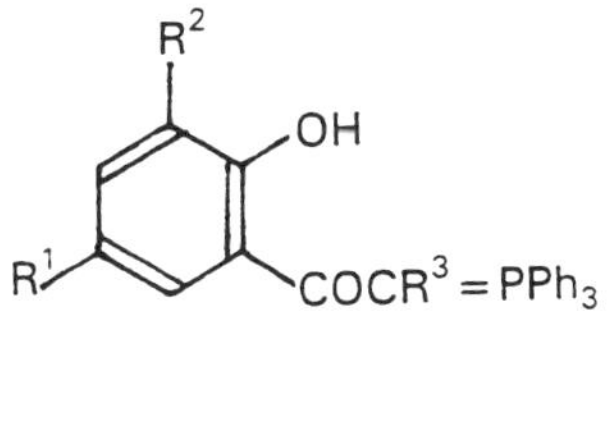

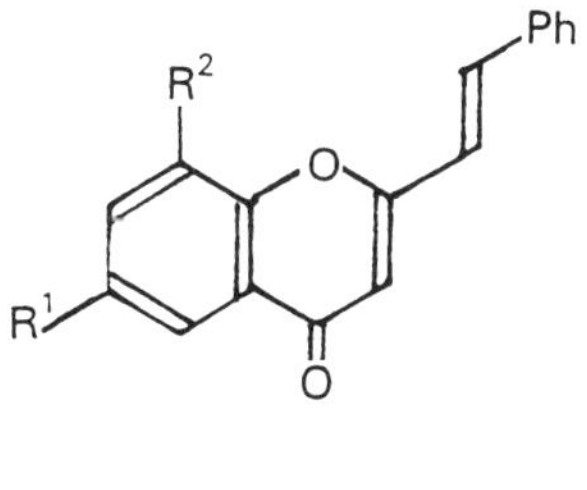

(8)

(9)

352

The reaction of 3-(2,2-diacetylvinyl)-6-methylchromone [1,1-diacetyl-
-2-(6-methyl-4-oxo-4*H*-benzo[*b*]pyran-3-yl)ethene] with phenyldiazomethane
affords a mixture of the *trans*-cyclopropane 10 (R = 6-methyl-4-oxo-4*H*-
-benzo[*b*]pyran-3-yl), 1,1-diacetyl-2-(4-hydroxy-6-methyl-2-phenylmethylene-
-2*H*-benzo[b]pyran-3-yl)ethene 11, pyrazole 12 and *cis*-dihydrofuran 13
(R^1 = H, R^2 = CHRPh) (C.K. Ghosh and S. Sahana, Tetrahedron, 1993, <u>49</u>,
4135).

(10)　　　　　　　　　　(11)

(12)　　　　　　　　　　(13)

2-Alkyl-, 2,6-dialkyl-, 2-thienyl-, and 6-alkyl-2-thienylchromones 14
(R^1 = H, lower alkyl;　R^2 = lower alkyl, thienyl) have been obtained by the
cyclization of 2,4-[(R^1)C_6H_3OH with HC≡CR^2 in the presence of Pd catalyst
at 90-120°C under 10-20 atmospheres CO in a secondary or tertiary amine
or in an organic solvent containing an amine. The yields of derivatives 14
were increased and the process simplified in the presence of a Cu halide
(V.N. Kalinin, M.V. Shostakovskii, and A.B. Ponomarev, U.S.S.R. SU
1,721,052, 1992). The reaction of 2-hydroxybiphenyl with ethyl acetoacetate in the presence of sulphuric acid affords 2-methyl-8-phenylchromone
(15) (A.Y. Soliman, M.R. Mahmoud, and H.M.F. Madkour, Rev. Roum.
Chim., 1993, <u>38</u>, 1117).

(14)   (15)

The preparations of some 6-substituted 2-heterochromones (J. Sun, Q.
Geng, and M. Cai, Huaxue Tongbao, 1990, 38) and 2-cyclopropylchromones
under phase transfer catalysis have been noted (D.M.S. Rao, E.V.S.B. Rao,
and A.V.S. Rao, Synth. Comm., 1990, 20, 817). Similarly some chromones
with a number of substituents for the treatment of osteoporosis (M.
Morimoto *et al.*, Jpn. Kokai Tokkyo Koho JP 05 112,552 [93 112,552],
1993), and some chromones and thiochromones, which induce gene
modulation in *Rhizobium* species (J.J. Berthelon and G. Quendo, Fr.
Demande FR 2,693,724, 1994) have been reported. The oxidation of
chroman-4-ones and 2-spirochroman-4-ones with [hydroxy(tosyloxy)iodo]-
benzene in boiling acetonitrile, under ultrasonification gives a convenient
route for the preparation of chromones, tetrahydroxanthones and their higher
homologues (D. Kumar *et al.*, Synth. Comm., 1994, 24, 2637). The
reaction of 2-propanoylcyclohexane-1,3-diones with dialkylformamide dialkyl
acetals provides an effective route, for example, to 3-methyl- and 3,7,7-
-trimethyl-5,6-7,8-tetrahydrochromon-5-ones (3-methyl- and 3,7,7-trimethyl-
-5,6,7,8-tetrahydro-4*H*-benzo[*b*]pyran-4,5-diones) 16 (R=H, Me) (A.L.
Mikhal'chuk, Zh. Obshch. Khim., 1991, 61, 261). For the preparation of 7-
-phenyl and 7-(2-phenylvinyl)chromone 17 (R=Ph, CH=CHPh), see D.
Obrecht, Helv., 1989, 72, 447.

(16)   (17)

354

6,8-Disubstituted 2-(2-fur-2-ylvinyl)chromone 18 (R^1 =H, R^2 =H, Me, Cl; R^1 =Me, R^2 =H, Br; R^1 =Br, R^2 =Cl, Me) are prepared by successive esterification of 2-hydroxyacetophenones with furylacrylic acid, Baker-
-Venkataraman rearrangement, and cyclization (A.K.D. Mazumdar *et al.*, J. Indian. Chem. Soc., 1990, _67_, 255).

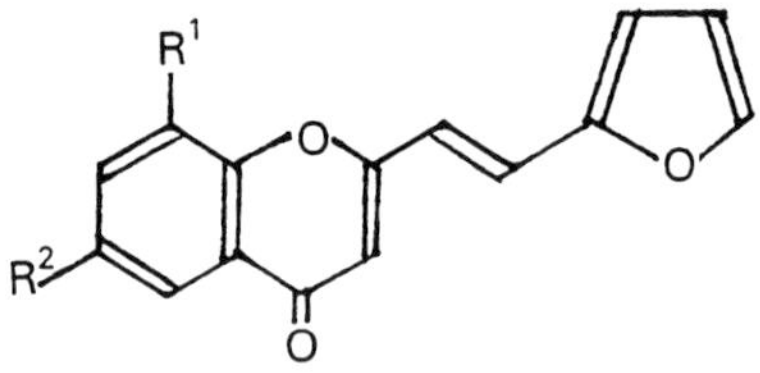

(18)

Hydroxychromones. Chromones 19 (R =H, Cl, Me) have been oxidized by 3-chloroperbenzoic acid to afford 3-hydroxychromones 20 and small amounts of 4-hydroxycoumarins 21 (C.P. Rao and G. Srimannarayana, Synth. Comm., 1987, _17_, 1507). S-Substituted derivatives have been obtained from 3-mercaptochromone by alkylation, acylation, and oxidation [W. Loewe and A. Kennemann, Arch. Pharm. (Weinheim, Ger.), 1988, _321_, 541].

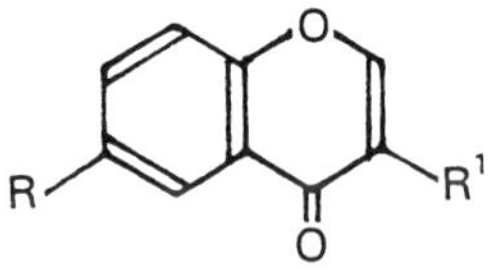

(19) R^1 =H
(20) R^1 =OH

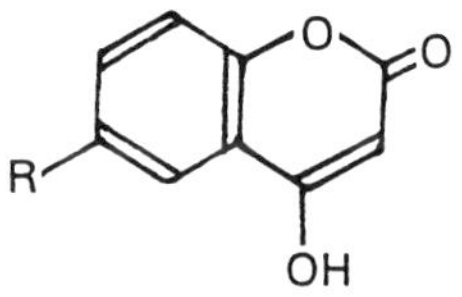

(21)

2-(3-Alkoxy-2-arylchromon-6-yl)propanoic acids 22 (R^1 = Me, CH_2Ph; R^2 = Ph, 4-MeOC$_6$H$_4$, 4-MeC$_6$H$_4$, 4-ClC$_6$H$_4$; R^3 = CHMeCO$_2$H) have been obtained and some selected derivatives inhibited the in vitro immuno-hemolysis of antibody coated sheep erythrocytes by guinea pig serum (O.V. Singh *et al.*, Indian J. Chem., 1992, <u>31B</u>, 248).

(22)

(23)

6,7-Disubstituted 2-hydroxymethylchromones 23 (R^1 = H, OMe; R^2 = H, F, Cl, Me) have been prepared by esterification of the corresponding carboxylic acids, followed by reduction with NaBH$_4$. 3-Hydroxymethyl-2*H*--benzo[*b*]pyran has also been prepared (G. Mouysset *et al.*, Eur. J. Med. Chem., 1988, <u>23</u>, 199).

3,5-Disubstituted 2-(fur-2-ylvinyl)-7-hydroxychromones 24 (R^1 = Me, Ph, OMe; R^2 = H, OH; R^3 = OH) have been obtained by the cyclocondensation of hydroxyacetophenones 25 with ClCOCH = CHfur-2-yl. Methylation of derivative 24 (R^3 = OH) with Me$_2$SO$_4$ yields methoxy derivative 24 (R^3 = OMe) (M.S. Reddy *et al.*, Indian J. Chem., 1989, <u>28B</u>, 1057).

(24)

(25)

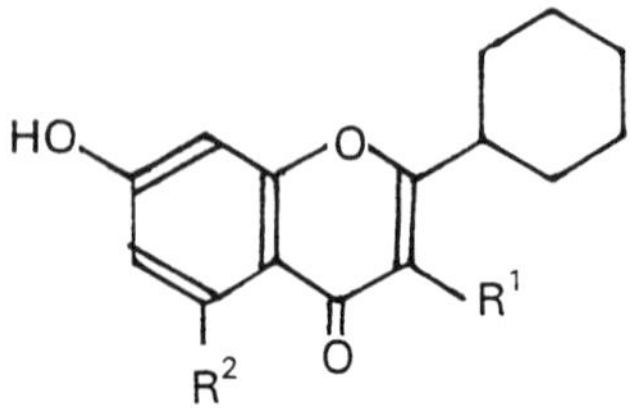

(26)

3-Substituted 2-cyclohexyl-7-hydroxy- and 3-substituted 2-cyclohexyl-
-5,7-dihydroxychromones 26 (R^1 =H, Me, Et, Ph; R^2 =H, OH) have been
obtained by the condensation of cyclohexanecarbonyl chloride with 2-alkan-
oylresorcinols and 2-alkanoylphloroglucinols 25 (R^2 = OH) respectively, in the
presence of tetrabutylammonium hydrogen sulphate and K_2CO_3 in H_2O/C_6H_6
(D.M. Rao and A.V.S. Rao, *ibid.*, 1992, <u>31B</u>, 335). 2-Alkanoylphenols 25
(R^1 =H, Me, Ph; R^2 =H, OH) have been treated with PhCH = CHCOCl and
K_2CO_3 in acetone to yield the corresponding 2-styrylchromones 27 (R^3 =H,
PhCH = CHCO$_2$), which have been saponified and the resulting hydroxychromones methylated by Me$_2$SO$_4$ (C.R. Reddy, G.L.D. Krupadanam, and G.
Srimannarayana, *ibid.*, 1987, <u>26B</u>, 974). A convenient synthesis of 2-
-styrylchromones 28 (R^1 =H, MeO; R^2 =H, Me, R^3 =H, MeO) by modified
Baker-Venkataraman transformation using phase-transfer catalysis has been
reported (J.K. Makrandi and V. Kumari, Synth. Comm., 1989, <u>19</u>, 1919).

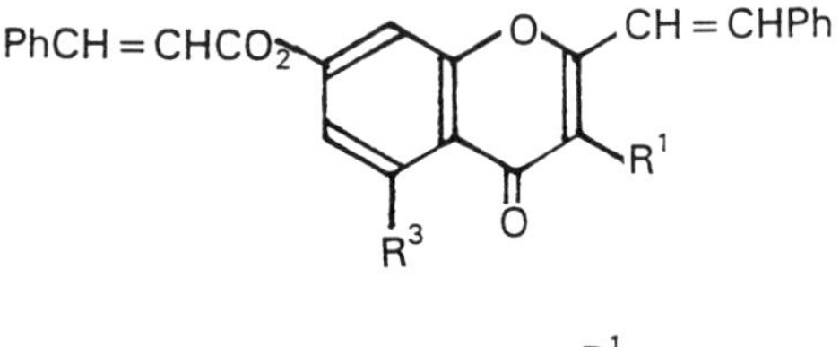

(27)

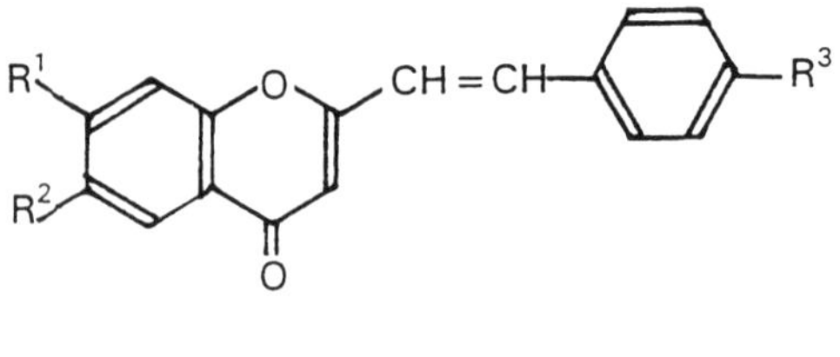

(28)

The 7-cinnamyloxychromones 29 (R^1 = H, Me, R^2 = H, R^3 = OH; R^1 = Me, R^2 = Ph, R^3 = H) undergo a Claisen rearrangement in boiling *N,N*-dimethyl-aniline to afford 7-hydroxy-8-(1-phenylprop-2-enyl)chromone derivatives 30, which have features similar to those of natural neoflavonoids (A.C. Jain, R.K. Gupta, and C. Chawla, Indian J. Chem., 1988, <u>27B</u>, 1032).

(29)

(30)

7-Hydroxy-, 5,7- and 7,8-dihydroxy and 5,7,8-trihydroxy-3-methylchrom-one derivatives 31 (R^1 = H, Me; R^2, R^3 = H, OH, OMe) have been prepared by treating 2-hydroxypropiophenones 32 with Me_2NCHO, $BF_3.EtO_2$, and $MeSO_2Cl$. Methylation of 31 (R^1 = H) give the corresponding methyl ethers 31 (R^1 = Me) (A.C. Jain, O.D. Tyagi, and R. Saksena, *ibid.*, 1989, <u>28B</u>, 678).

(31)

(32)

358

Six-coordinate cobalt-Schiff base complexes 33 [X = (CH$_2$)$_3$NH(CH$_2$)$_3$],
promote the highly selective conversion of 1-(2-hydroxyaryl)-1,3-diketones
34 (R^1 = Ph, 4-O$_2$NC$_6$H$_4$, 4-ClC$_6$H$_4$, 4-MeC$_6$H$_4$, 4-MeOC$_6$H$_4$, Me, Et, Pr^i;
R^2 = H, OMe, R^3 = H, OCH$_2$OMe) to 2-alkyl(aryl)-5- and 7-hydroxy- and -5,7-
-dihydroxychromone derivatives 35 under neutral conditions (A. Nishinaga *et
al.*, Synthesis, 1992, 839).

(33)

(34)

(35)

The preparations of the following 5,7-dihydroxychromone derivatives
have been reported; 5,7-dialkoxy-2-(4-hydroxyphenylthio)chromones 36 (R^1,
R^2 = H, lower alkyl) as aldose reductase inhibitors (Y. Igarashi, T. Yamaguchi,
and K. Hosaka, PCT Int. Appl. WO 93 24,477, 1993); 6,8-disubstituted 7-
-alkoxy-5-methoxy- and 5,7-dihydroxy-2-methylchromones 37
[X = PhNHCSNHN, 4-pyridylcarbonylhydrazono, 3,4-HO(HO$_2$C)C$_6$H$_3$N, PhN,
R^1 = H, R^2 = H, alkyl, arylmethyl; X = O, PhNHN, R^1 = PhN = N, 3-O$_2$NC$_6$H$_4$-
N = N, 3-MeC$_6$H$_4$N = N, 4-MeOC$_6$H$_4$N = N, R^2 = H], some of which were tested
for antibacterial, antitubercular, and spasmolytic activity (H. Abu-Shady, *et
al.*, Egypt. J. Pharm. Sci., 1988, 29, 471); 6,8-dicinnamyl- and 8-cinnamyl-
-5,7-dihydroxychromone 38 (R^1 = H, R^2 = R^3 = CH$_2$CH = CHPh; R^1 = H,
R^2 = CH$_2$CH = CHPh, R^3 = H) and their 2-methyl derivatives 38 (R^1 = Me)
which were evaluated as bactericides against *S. aureus* and *E. coli* (C.S.
Andotra and C. Chawla, Indian J. Pharm. Sci., 1990, 52, 281); 3-acetyl-
-5,7-dihydroxy-2-methylchromone derivatives 39 [R^1 = H, acetyl; R^2 = H,
alkyl, (CH$_2$)$_n$R^4; R^4 = OH, alkoxy, CO$_2$H; n = 1-6; R^3 = H, methoxymethyl,
alkyl] were tested as allergy inhibitors (S. Fujisawa *et al.*, Jpn. Kokai Tokkyo

Koho JP 02 255,672 [90 255,672], 1990); and 6-pyrazolylidene, imidazol-ylidine, and pyrimidinylidene derivatives of 7-hydroxy-5-methoxy-2-methyl-chromone by reaction of the respective 6-carboxaldehyde with the appropriate active methylene compound, compound 40 ($X = X^1$) was active against *Staphylococcus aureus*, and 40 ($X = X^2$) against *Escherichia coli* [A.K.M.N. Gohar, Egypt. J. Chem., 1988, (Pub. 1990), <u>31</u>, 367].

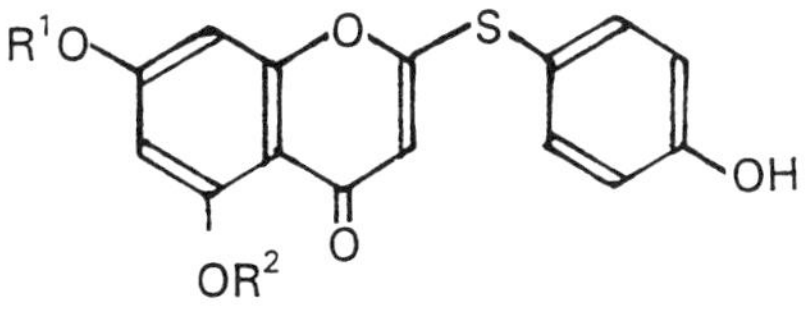

(36)

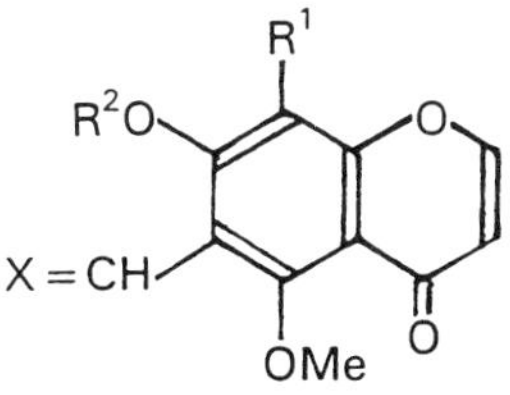

(37)

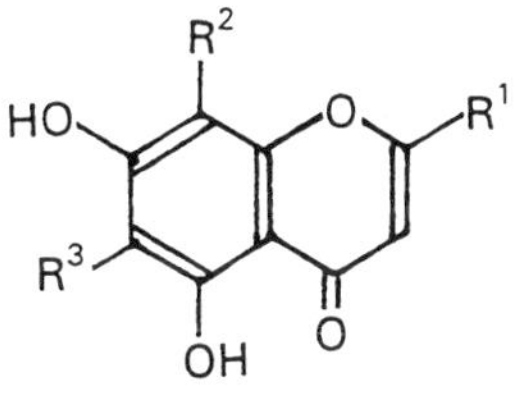

(38)

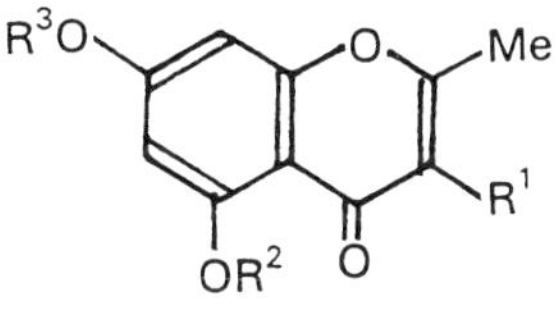

(39)

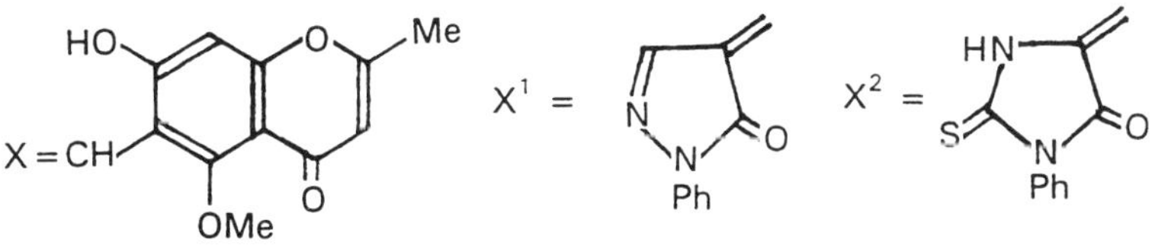

(40)

360

Chromonecarboxaldehyde (formylchromones) and acyclchromones.
Chromone-2-carboxaldehydes 41 (R¹=H, Me) have been obtained by
oxidation of the respective 2-methylchromone with SeO_2 (K. Ito and K.
Nakajima, J. Heterocyclic Chem., 1988, 25, 511). Others 42 (R¹ =H,
R²=H, Cl, Br, Me, R³=H, Cl, Me, R⁴=H, Cl, R¹R²=CH=CH-CH=CH) have
been obtained by hydrolysis of nitrones 43 (H.M. El-Shaaer *et al.*, Coll.
Czech. Chem. Comm., 1994, 59, 1673).

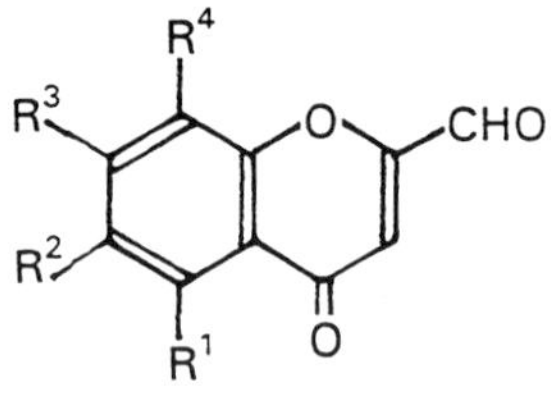

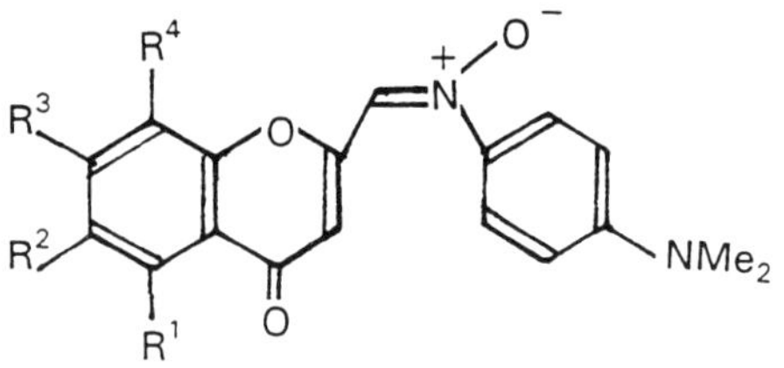

(41) R¹,R³=H,Me; R²=R⁴=H   (43)
(42) (See text)

Chromone-3-carboxaldehyde and methyl chromone-3-carboxylate (44;
R=CHO, CO_2Me, respectively) react with H_2C=CH-CH=CH_2 in the presence
of $TiCl_4$ to afford the corresponding [4 + 2] cycloadducts 45. Derivatives 44
undergo uncatalyzed cycloaddition to MeOCH=CHC(OSiMe₃)=CH_2 to give
the respective adducts 46 (P.T. Cremins, S.T. Saengchantara, and T.W.
Wallace, Tetrahedron, 1987, 43, 3075). The reactions of chromone-3-
-carboxaldehyde with pent-2-en-5-olide, but-2-en-4-olide, and coumarin
under Perkin reactions (G.S. Melikyan *et al.*, Chem. Pap., 1993, 47, 388)
and the condensation reactions of substituted chromone-3-carboxaldehydes
47 (R¹ =electron-donating and electron-accepting substituents) with MeCHO,
malonic acid, PhCOMe, and indandione to yield products with potential
biological activities (A.P. Polyakov *et al.*, Vestn. Khar'k. Unv., 1991, 359,
60) have been investigated.

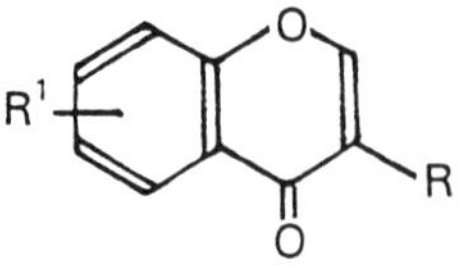

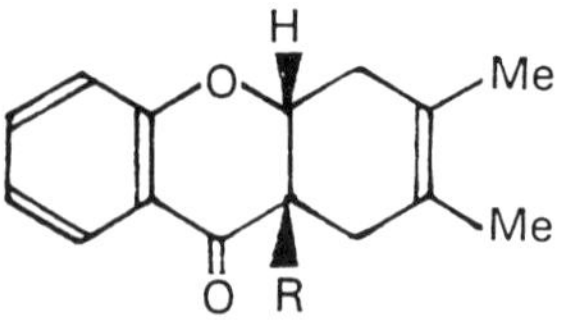

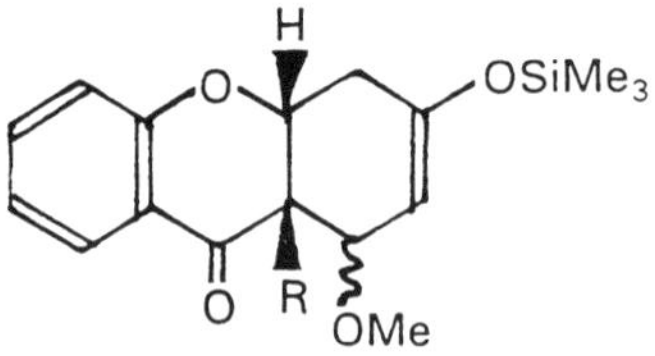

(44) R (See text), R¹=H   (45)   (46)
(47) R=CHO, R¹ (See text)

The 3-(cyclic ethylene acetal) of chromone-3-carboxaldehyde (48; R=H)
on treatment with CH_2N_2 in CH_2Cl_2-Et_2O affords fused pyrazolidine 49,
which on thermolysis in PhMe, followed by hydrolysis yields 2-methylchrom-
one-3-carboxaldehyde (48; R = Me). Compound 49 on heating at 160-170°
or percolating in AcOEt through an alumina column gives a small amount of
salicyloylpyrazole 50 (C.K. Ghosh *et al.*, J. Chem. Res., S, 1990, 117).

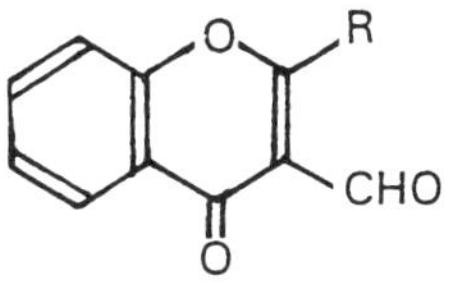

(48)

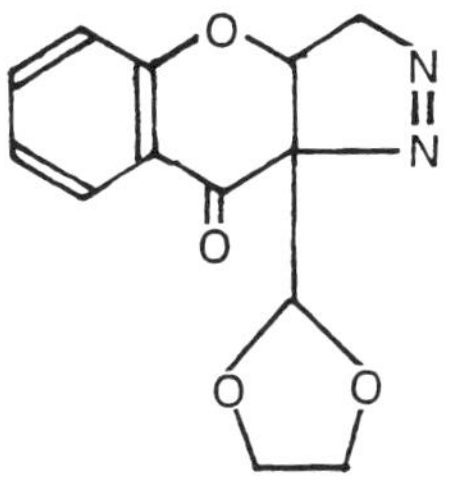

(49)

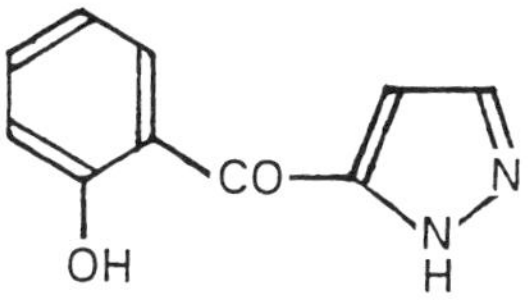

(50)

Chromone-3-carboxaldehyde condenses with RCOMe (R = Ph, ClC_6H_4,
tolyl, anisyl) to give chalcone analogues 51. 8-Cinnamoylchromones 52
[R = Ph, ClC_6H_4, tolyl, anisyl, $(MeO)_2C_6H_3$, $Me_2NC_6H_4$, HOC_6H_4, hydroxy-
naphthyl] are obtained from 8-acetyl-7-hydroxy-2,3-dimethylchromone and
RCHO (M.S.S. Shankar *et al.*, J. Indian Chem. Soc., 1989, 66, 30).

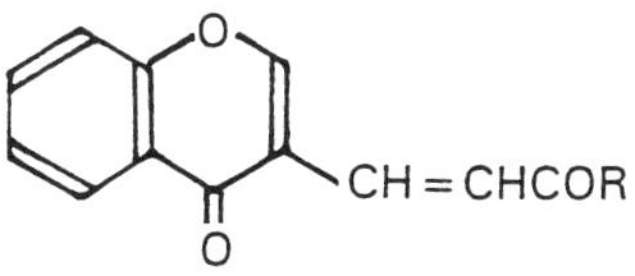

(51)

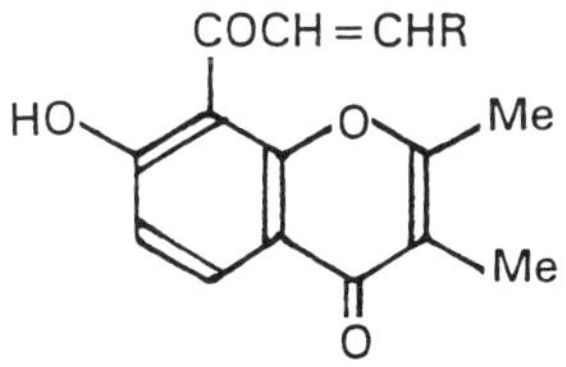

(52)

362

Substituted 3-(4-methoxycinnamoyl)-2-methylchromones 53 (R^1=Me, R^2=H, Br; R^1=H, R^2=Me, Cl; R^1=Br, R^2=Me, Cl) are prepared by ester-ification of 2-hydroxyacetophenones with 4-MeOC$_6$H$_4$CH=CHCO$_2$H, Claisen rearrangement and cyclization of the resulting diones with Ac$_2$O-AcONa (A.K.D. Mazumdar *et al.*, *ibid.*, 1990, <u>67</u>, 257). For the preparation of anisyl 3-acetyl-7-methoxychromones, see A.C. Jain and P.K. Bambah, Proc. Indian Natl. Sci. Acad., Part A, 1987, <u>53</u>, 564; and of 3-aroyl-2-(fur-2-yl)-chromones, R.T. Cummings, J.P. DiZio, and G.A. Krafft, Tetrahedron Letters, 1988, <u>29</u>, 69.

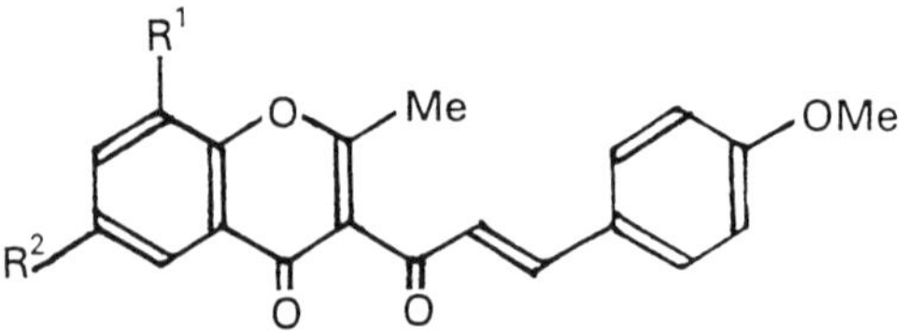

(53)

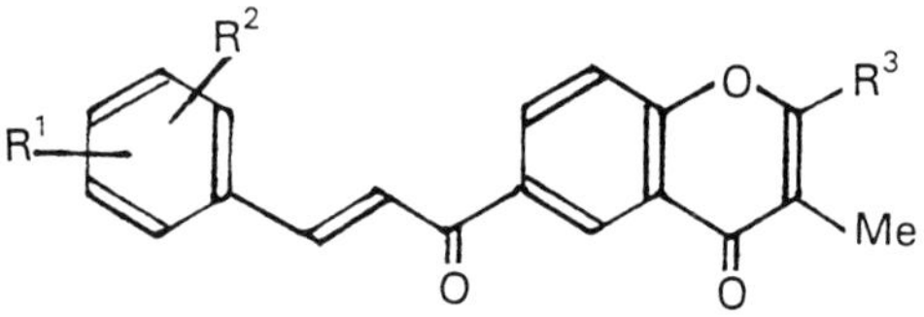

(54)

The preparations of 6-cinnamoyl-2,3-dimethylchromones, 6-cinnamoyl-3--methyl-2-phenylchromones, and 6-cinnamoyl-3-methylchromones 54 [R^1, R^2=H, 4-Cl, 4-Me, 4-MeO, 3,4-(MeO)$_2$; R^3=Me, Ph, H, respectively] from 5-cinnamoyl-2-hydroxypropiophenone have been reported (O.V. Singh *et al.*, *ibid.*, 1992, <u>69</u>, 147).

Chromonecarboxylic acids. The product obtained from the Claisen rearrangement of 2-(MeCO)C$_6$H$_4$OCH$_2$CH=CH$_2$ on cyclocondensation with (CO$_2$Et)$_2$ gives ethyl 8-allylchromone-2-carboxylate (55; R^1=Et, R^2=CH$_2$-CH=CH$_2$. The preparations of a number of 8-substituted chromone-2-carboxylic acids, esters and sodium salts have been described and it has been reported that sodium 8-hydroxychromone-2-carboxylate (55; R^1=Na, R^2=OH) inhibited the antigen-induced relase of SRS-A leukotrienes from human lung samples (L.M. Von Itzstein and H.F. White, PCT Int. Appl. WO 90 02,741, 1990). Some 6-methychromone-2-carboxylic acid derivatives have been synthesized and their local anesthetic and analgesic activities roughly estimated. The introduction of a methyl group into the 6-position appeared to enhance their pharmacological activities (W. Saito *et al.*, Annu. Rep. Tohoku Coll. Pharm., 1990, <u>37</u>, 23).

R²

CO_2R^1

(55)

R^4O_2C

OCH_2CHCH_2O
OR

(56) $R = CO(CR^1R^2)_nCOR^3$
(57) $R = H$

CO_2R^5

The dichromone derivatives (cromoglycates) 56 (R^1, $R^2 = H$, alkyl; $R^3 = $ alkyl; R^4, $R^5 = $ alkoxyalkyl, 1-alkanoyloxyalkyl; $n = 0$-4) useful as antiallergic agents have been prepared by esterification of hydroxy derivatives 57 with $R^3CO(CR^1R^2)_nCO_2H$ or diketene (S. Tamaki *et al.*, Jpn. Kokai Tokkyo Koho JP 62 81,380 [87 81,380], 1987). For the preparation of disodium cromoglycate 57 ($R^4 = R^5 = Na$), see M.U.J. Dahlstrom, Finn. FI 83,311, 1991; and for highly pure disodium cromoglycate tetrahydrate, T. Balogh *et al.*, Hung. Teljes HU 54,135, 1991.

Ethyl 5-(2-hydroxy-3-methylthiopropoxy)chromone-2-carboxylate (58; $R = $ Et, $n = 0$) on treatment in AcOH with 35% aqueous H_2O_2 affords ethyl 5--(2-hydroxy-3-methylsulphonylpropoxy)chromone-2-carboxylate (58; $R = $ Et, $n = 2$). Salts and esters of compound 58 ($R = H$, $n = 1$ or 2), useful as allergy inhibitors have been prepared (S. Boveri and W. Cautreels, Fr.Demande FR 2,592,653, 1987). The preparation of 7-(1-hydroxy-ω-phenoxyalken-2-yl-thio)chromone-2-carboxylates and analogues as leukotriene and phospho-lipase inhibitors have been reported (A. Von Sprecher and A. Beck, Eur. Pat. Appl. EP 335,315, 1989). Also reported are the preparations of 7-*a*--hydroxy-*a*-furyl-ω-phenoxyalkadienylthio)chromone-2-carboxylic acids (A. Von Sprecher and A. Beck, Eur. Pat. Appl. EP 419,410, 1991) and 7-[(4--acetyl-3-hydroxy-2-propylphenoxy)halogenoalkadienylthio]chromone-2-carboxylic acids (A. Von Sprecher, B. Schaub, and R.W. Lang *ibid.*, 419,411, 1991) and their salts and analogues as antiinflammatories and antiallergics.

CO_2R

$OCH_2CH(OH)CH_2S(O)_nMe$

(58)

R^3 R^5 R^2 R^4 R^1

$R^5 = CO_2Me$

(59)

R^3 OH R^2 R^5 R^1

(60)

Methyl chromone-2-carboxylate, methyl 3-methyl- and 5-hydroxychrom-one-2-carboxylates and dimethyl 7-hydroxychromone-2,6-dicarboxylate 59 ($R^1 = R^2 = R^3 = H$, $R^4 = H$, Me; $R^1 = OH$, R^2-$R^4 = H$; $R^1 = R^4 = H$, $R^2 = CO_2Me$, $R^3 = OH$, respectively) have been obtained on treating the appropriate pyruvic acid dimethylhydrazone, $R^4CH_2C = (NNMe_2)CO_2H$ with hydroxybenzoates 60, followed by hydrolysis and esterification (I. Tapia, V. Alcazar, and J.R. Moran, Canad. J. Chem., 1990, <u>68</u>, 2190).

Bromochromones 61 ($R^1 = Br$, $R^2 = H$; $R^1 = H$, $R^2 = Br$) on coupling with $CH_2 = CHR^3$ ($R^3 = CO_2Me$, Ph, CN) in the presence of $Pd(PPh_3)_2Cl_2$ and Et_3N regiospecifically yield the corresponding vinylchromones 61 ($R^1 = CH = CHR^3$, $R^2 = H$; $R^1 = H$, $R^2 = CH = CHR^3$) as predominantly the E-isomers. Similarly with 3,6-dibromochromone and $CH_2 = CHCO_2Me$, divinylchromone 62 (41%) and ring-opened cinnamate 63 (8%) are obtained (S.G. Davies, B.E. Mobbs, and C.J. Goodwin, J. Chem. Soc., Perkin 1, 1987, 2597).

(61)   (62)

(63)   (64)

Ethyl 5-hydroxychromone-2-carboxylates 64 (R = H) on treatment with alkyl halides and K_2CO_3 in N-methylpyrrolidinone with ultrasonication afford ethyl 5-alkoxychromone-2-carboxylates 64 (R = Me, Pr, Bu, CH_2Ph, allyl) in yields of 54-84%. Without ultrasonication yields are 28-59% (T.J. Mason *et*

al., *J. Chem. Res., S, 1988, 80)*. *This method has been compared with the*
O-*alkylation of 5-hydroxychromone 64 (R = H) using phase-transfer catalyst*
Aliquat 336 neat (idem, *Synth. Comm., 1990, 20, 3411)*.

6-Fluorochromone-2-carboxylic acid (65) obtained in 2 steps from 5-
-fluoro-2-acetophenone and $(CO_2Et)_2$ on hydrogenation over Pt/C in EtOH at
7 atmospheres and room temperature affords 6-fluoro-4-chromanone-2-
-carboxylic acid (66), useful as an intermediate for the synthesis of spiro
(4*H*-benzo[*b*]pyran-4,4'-imidazolidine)-2-carboxylic acid derivatives, which
are aldose reductase inhibitors (T. Koizumi and Y. Saito, Jpn. Kokai Tokkyo
Koho JP 02 218, 674 [90 218, 674]; 02 218,675 [90 218,675], 1990).
For the preparation of 6-fluorochromone-2-carboxylic acid esters and amides,
see T. Koizumi *ibid*., 03 17,075 [91 17,075], 1991). Several 6-chloro-
chromone-2-carboxylic acid derivatives have been synthesized and their local
anesthetic and analgesic activites estimated (Y. Okuyama *et al*., Annu. Rep.
Tohoku Coll. Pharm., 1990, 37, 31).

(65)   (66)

The cyclocondensation of thienoylacetophenone 67 with $(CO_2Et)_2$
affords ethyl 7,8-dichloro-6-thienoylchromone-2-carboxylate (68; R = Et),
which on hydrolysis yields the corresponding acid 68 (R = H), tested for
diuretic activity and found to be 80% as active as thienylic acid 69 (P.
Montanari, P. DaRe, and P. Valenti, J. Heterocyclic Chem., 1988, 25, 1277;
1991, 28, 2075).

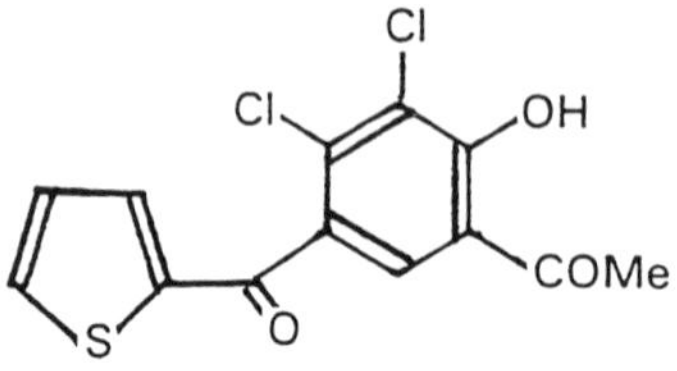

(67)

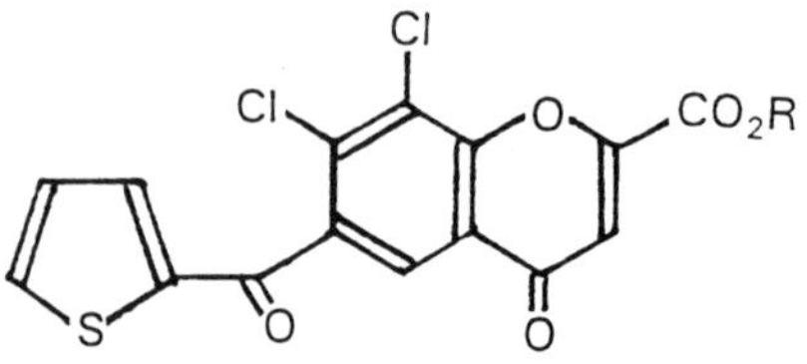

(68)

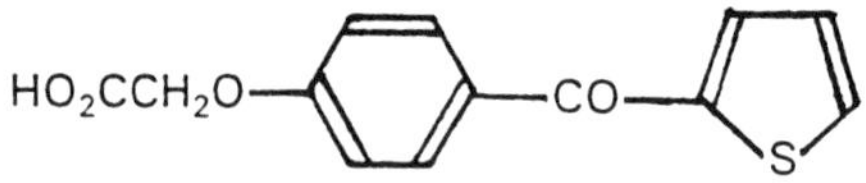

(69)

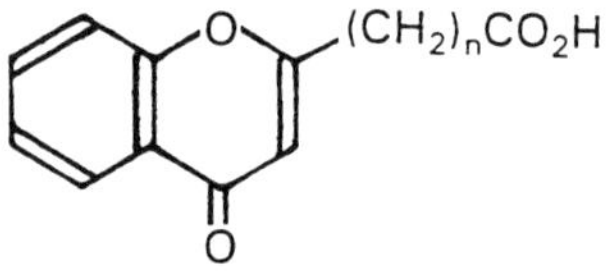

(70)

ω-(Chromone-2-yl)alkanoic acids 70 (n = 3-6) have been obtained by the cyclization of the corresponding 2-acetylphenyl ethyl alkandioate $2\text{-MeCOC}_6\text{-}H_4O_2C(CH_2)_nCO_2Et$ (S. Yamaguchi, M. Saitoh, and Y. Kawase, *ibid.*, p.125). Related ω-(chromone-3-yl)alkanoic acids have also been prepared (S. Yamaguchi *et al.*, *ibid.*, p.119).

Methyl chromone-3-carboxylate (71) on treatment with Me_2CuLi in THF--Et_2O gives 78% of the ketone 72 (R = CO_2Me) and enol 73, which on demethoxycarbonylation afford 2-methylchroman-4-one (72; R = H) (S.T. Saengchantara and T.W. Wallace, Tetrahedron, 1990, **46**, 3029).

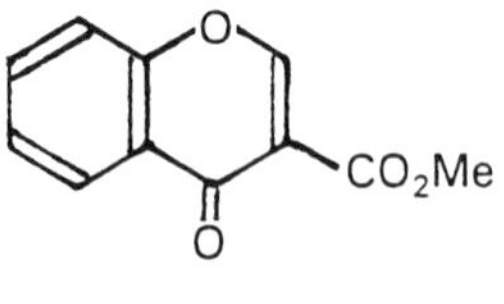

(71)

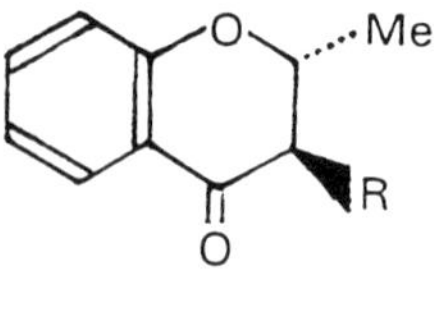

(72)

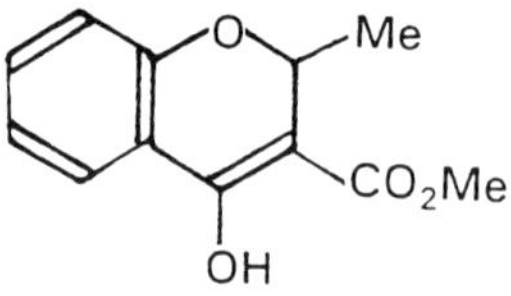

(73)

3-(Chromone-3-yl)-4-cyano-5-ethoxy-3-methyl-2,3-dihydrofurans [4-
-cyano-5-ethoxy-3-methyl-3-(4-oxo-4H-benzo[b]pyran-3-yl)-2,3-dihydrofurans]
74 (R = H, Me, Cl) obtained from the 6-substituted 3-(2-cyano-2-ethoxy-
carbonyl)vinylchromones 75 and CH_2N_2, thermally rearrange to yield
predominantly the Z-isomer of the ethyl 1-(6-substituted chromone-3-yl)-2-
-cyano-1-methylcyclopropane-2-carboxylates 76 (C.K. Ghosh and S. Biswas,
Chem. Comm., 1989, 1784). The addition of CH_2N_2 to compounds related
to compound 75 to afford dihydrofurans have been described, along with
their chemical properties (C.K. Chosh, S. Biswas, and S. Sahana, Indian J.
Chem., 1993, <u>32B</u>, 630).

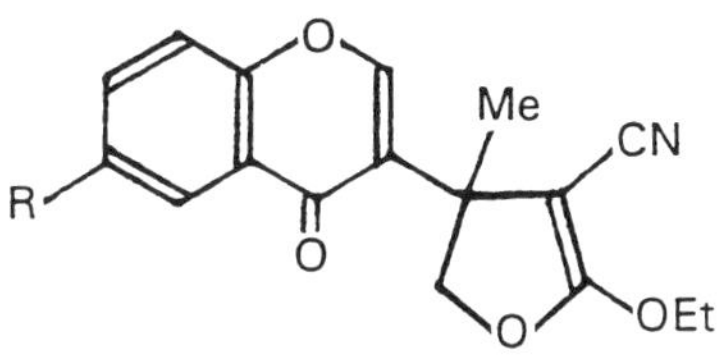

(74)

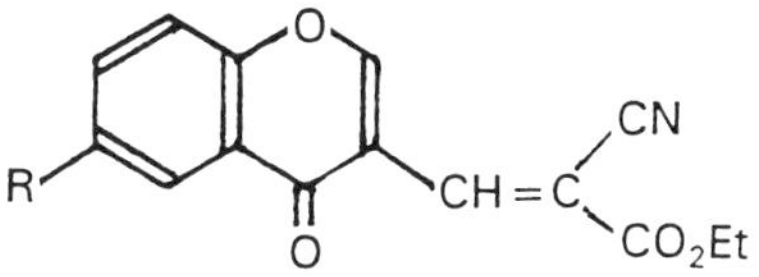

(75)

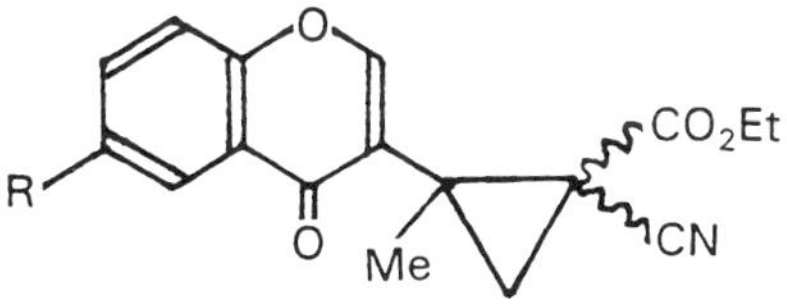

(76)

Methyl 2-methylchromone-3-acetates 77 (R^1 = H, R^2 = R^3 = Me;
R^1 = R^2 = R^3 = Me) and vinyl ketones 78 are obtained on irradiating enol
lactones 79 in MeOH. Irradiation in benzene affords ketones 78, chromone-
-3-acetic acid 77 (R^3 = H) and 2,3-dimethylchromone 80. The mechanistic
implications of these reactions have been discussed (L. Fillol *et al.*,
Heterocycles, 1989, <u>29</u>, 511).

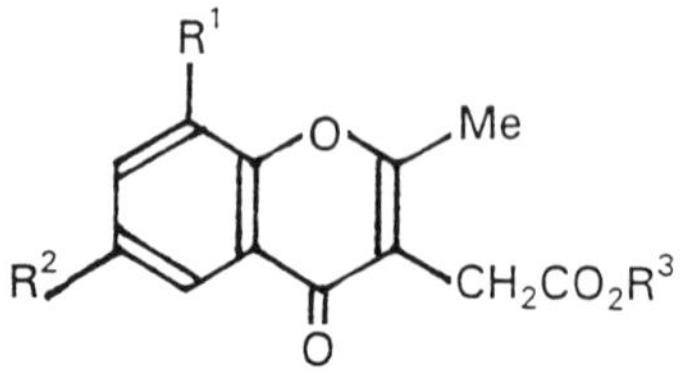

(77)

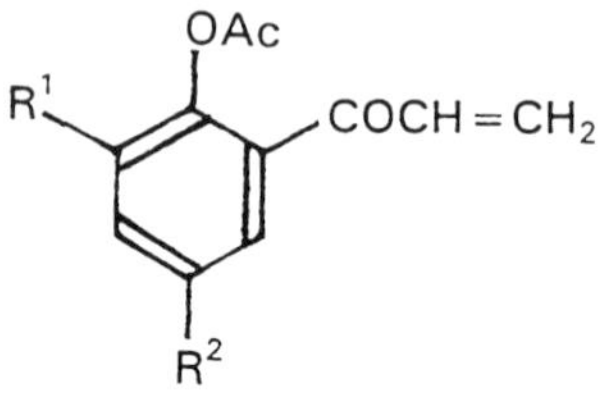

(78)

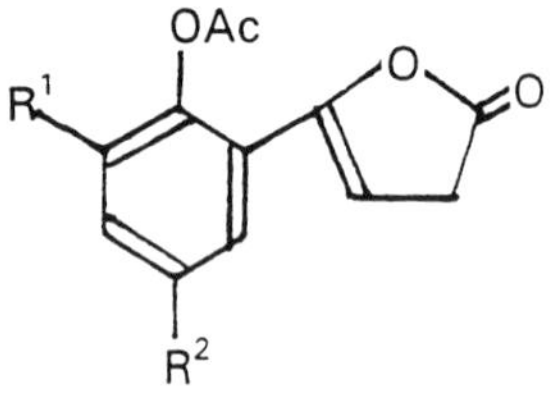

(79)

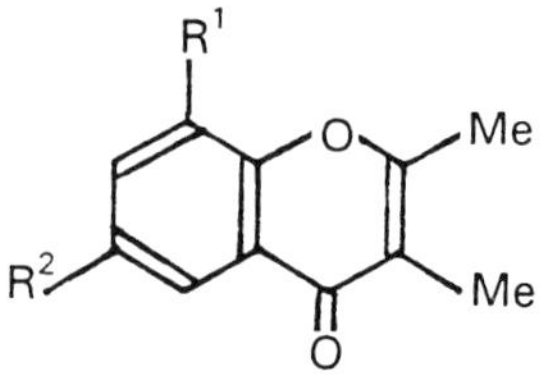

(80)

The reaction of benzoxathiin 81 (R = CHO) with NH_2OH in the presence of pyridine-acetic anhydride gives 2-aminochromone-3-sulphonic acid (82) along with the pyridine salt of 81 (R = CN). Derivative 81 (R = CH = NOH) is obtained from benzoxathiin 81 (R = CHO) and H_2NO_3SOH and is converted to sulphonic acid 82 on treatment with pyridine-acetic anhydride [W. Loewe, S. Goebel, and C. Mueller-Menke, Arch. Pharm. (Weinheim, Ger.), 1988, 321, 729].

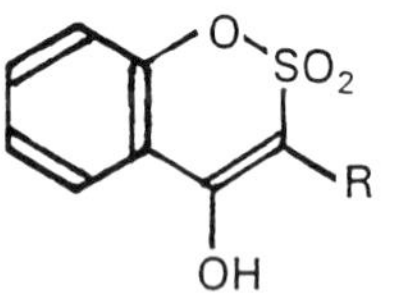

(81)

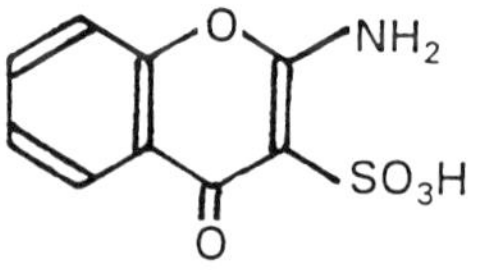

(82)

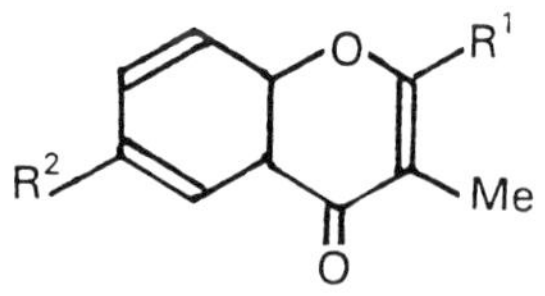

(83)

Methyl 5,7-dimethoxy-2-methylchromone-6-carboxylate has been condensed with PhCHO to give methyl 5,7-dimethoxy-2-styrylchromone-6--carboxylate (J.D. Brion *et al.*, Eur. Pat. Appl. EP 454,587, 1991). The preparation of 3-methylchromone-6-ylalkanoic acids 83 (R^1 = Me, Ph; R^2 = HO_2CCH_2, HO_2CCHMe) has been reported (O.V. Singh *et al.*, J. Indian Chem. Soc., 1989, <u>28B</u>, 814).

Bis(chromonylarene)dicarboxylic acids 84 (R = Me or H) have been obtained by the Baker-Venkataraman rearrangement of the appropriate esters 85 (G. Wurm and D. Schnetzer, Arch. Pharm. (Weinheim, Ger.), 1993, <u>326</u>, 167].

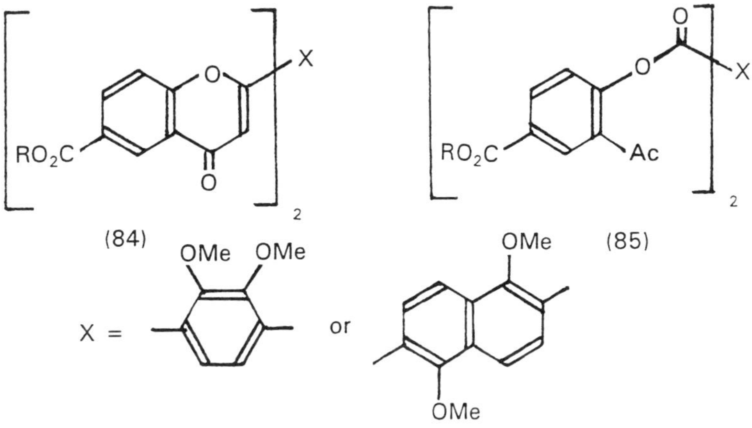

6-Hydroxy-2-methyl-5-nitrochromone-7-carboxylic acid (86) has been prepared from furochromone 87 (J. Kossakowski and T. Zawadwski, Pol. PL 161,393, 1993).

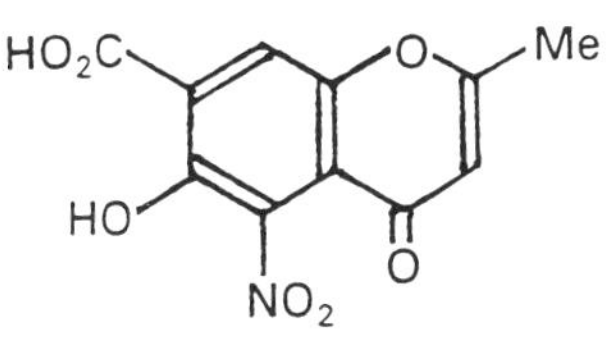

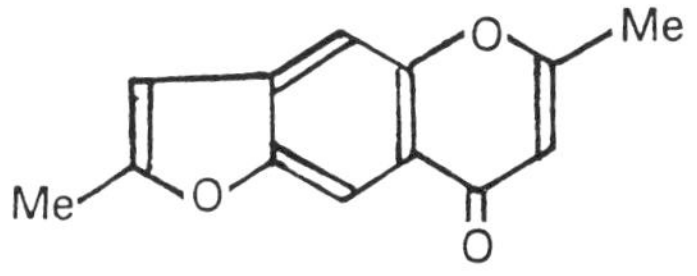

370

(Chromon-7-yl)oxyacetic acids 88 [R^1 =H, Me, Ph; R^2 = (un)substituted
Ph, Pr^i, $PhCH_2$, cyclopentyl; R^3 =Cl, Me, CF_3, F, I; R^4 =H, Cl, Me] and 5-
-chloro-3-phenyl-8,9-dihydrofuro[2,3-*h*]chromone-8-carboxylic acid (89) have
been synthesized, tested for natriuretic and uricosuric activities, and their
structure-activity relationships have been discussed (M. Kitagawa *et al.*,
Chem. Pharm. Bull., 1991, *39*, 2681).

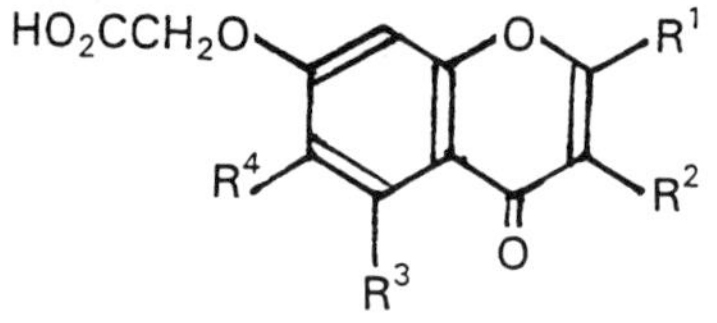

(88)

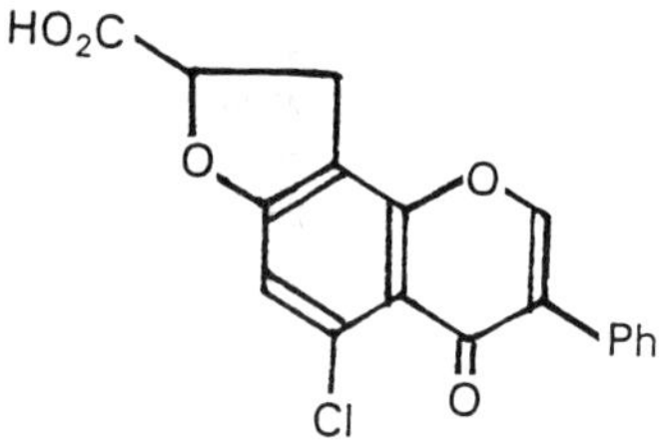

(89)

The preparations of (2-styrylchromon-8-yl)acetic acids (2-styrylchrom-
one-8-acetic acids) 90 [R^1 =H, Me; R^2 =H, MeO; R^3 =Ph, 3,4-$(MeO)_2C_6H_3$,
3,4,5-$(MeO)_3C_6H_2$, 4-$(HO_2C)C_6H_4$] (F. Zammattio *et al.*, J. Heterocyclic
Chem., 1991, *28*, 2013) and related derivatives 91 (R^1 =unsaturated
C_{3-7}alkyl, cycloalkyl; R^2 =H, halogeno; R^3 =H, alkyl; R^4 =CO_2H, its ester,
salt, amide) useful as antitumor agents (T. Aono and K. Mizuno, Eur. Pat.
Appl. EP 283,761, 1988) have been reported.

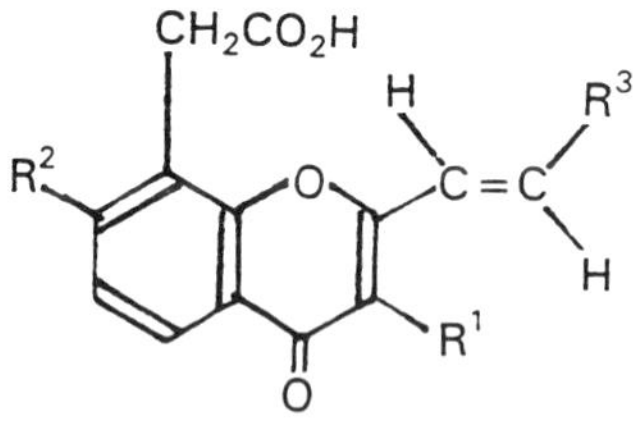

(90)

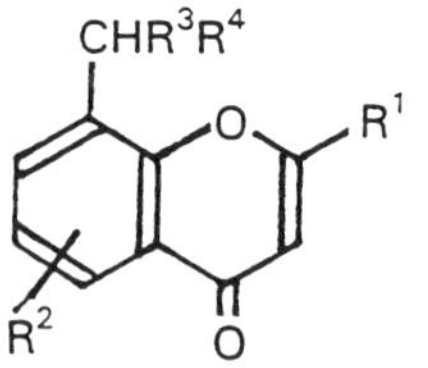

(91)

Halogenochromones. The cyclocondensation of $4\text{-}FC_6H_4O(CH_2)_2CO_2H$ in the presence of HF yields 6-fluorochromone (R. Hasegawa, Jpn. Kokai Tokkyo Koho JP 62 126,184 [87 126,184], 1987). 7-Fluorochromones **92** (R = H, alkyl, OH) containing a sulphur group in the 3-position have been prepared from the enaminone **93** and thionyl chloride [W. Loewe *et al.*, Arch. Pharm. (Weinheim, Ger.), 1994, <u>327</u>, 267]. 2-Fluoroalkylated chromones **94** [$R^1 = ClCF_2$, $Cl(CF_2)_3$, $Cl(CF_2)_5$; $R^2 = H$, Cl, Me; $R^3 = H$, Me; $R^4 = H$, Cl, Br, NO_2, Me, Ph] have been obtained by the Michael addition reactions of 2,2--dihydropolyfluoroalkanoates with phenols in the presence of Et_3N, followed by acid-catalyzed intramolecular ring-closure (W.Y. Huang, Y. Liu, and L. Lu, J. Fluorine Chem., 1994, <u>66</u>, 263).

(92)   (93)   (94)

The self-condensation of ethyl pentafluorobenzoylacetate gives 3-penta-fluorophenyl-6,7,8,9-tetrafluoro-1*H*-isopyrano[4,3-*b*]chromon-1-one (95), which on acidic hydrolysis yields 2-(2-pentafluorophenyl-2-oxoethyl)-5,6,7,8--tetrafluorochromone (96) (V.I. Saloutin *et al.*, Izv. Akad. Nauk, Ser. Khim. 1992, 2186).

(95)   (96)

Condensation of 6-chloro-2-methylchromone (97) with R^1CHO
[R^1 = (un)substituted Ph, fur-2-yl] gives 6-chloro-2-styrylchromones 98,
which undergo Diels-Alder reactions with maleic anhydride or *N*-arylmale-
imides to yield xanthone derivatives 99 [X = O, NR^1, R^1 = (un)substituted Ph]
(M.A.I. Salem *et al.*, J. Chem. Soc. Pak., 1990, 12, 189; Rev. Roum.
Chim., 1992, 37, 495). Some reactions of 6-chloro-2-methylchromone (97),
including those with hydroxylamine hydrochloride, ethyl oxalate, and
aromatic nitroso compounds have been investigated (*idem*, J. Chem. Soc.
Pak., 1992, 14, 24).

(97) R = Me

(98) R = CH=CHR^1

(99)

The photoirradiation of 3-benzyloxy-6-chloro-2-phenylchromones 100
(R = F, Cl) in benzene gives a mixture of cyclization products 101 and
derivatives with the suggested structure 102, but derivative 100 (R = H)
affords only the aromatic product 101 under the same conditions (N.S.
Yadav, S.N. Dhawan, and S.C. Gupta, J. Indian Chem. Soc., 1990, 67,
770). Some 2-(naphth-1-yl)halogenochromones have been synthesized and
a study made of their toxicity to fish (S.V. Kuberkar and Y.B. Vibhute, Asian
J. Chem., 1992, 4, 22).

(100)

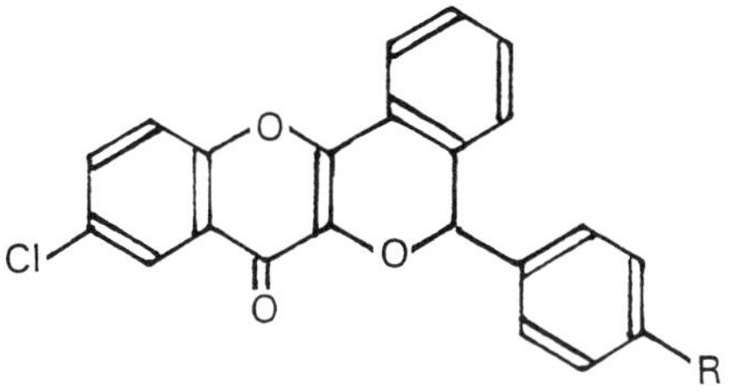

(101)

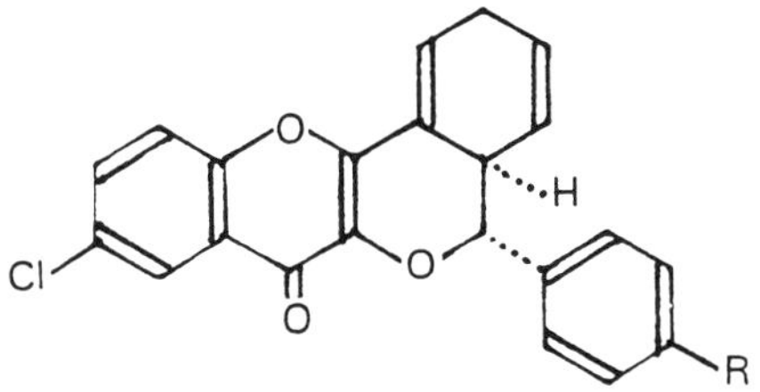

(102)

2-Bromo-1-(2-hydroxyphenyl)-5-phenylpent-4-ene-1,3-diones 103
($R^1 = R^2 = H$, Me, $R^3 = Br$) undergo cyclodehydration in $AcOH-H_2SO_4$ to give 3-
-bromo-2-styrylchromones 104, but with ethanolic KOH they undergo
rearrangement to afford 2-cinnamoyl-2,3-dihydrofuran-3-ones 105.
Bromination of compound 103 ($R^3 = H$) with NBS in CCl_4 yields a crude
product, which on cyclodehydration with $AcOH-H_2SO_4$ gives derivative 104
(B.D. Saraf and K.N. Wadodkar, Indian J. Chem., 1988, 27B, 771).

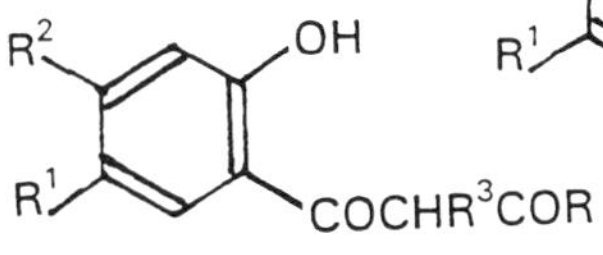

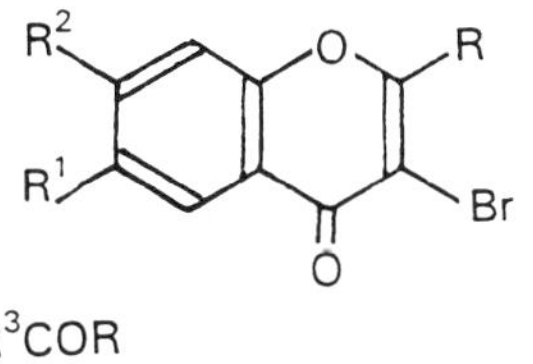

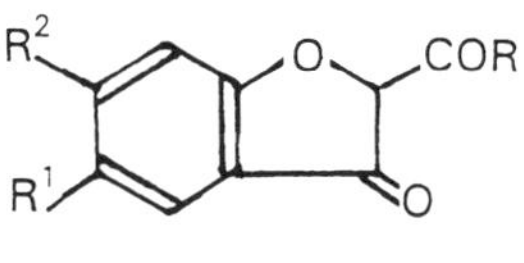

$R = CH{=}CHPh$

(103)

(104)

(105)

Amino- and nitrochromones and chromonecarboxamides and related derivatives. Treatment of $MeNO_2$ with *tert*-BuOK in DMF, followed by reaction with sodium 2,3-dimethylchromone-6-sulphinate in the presence of iodine, affords 2,3-dimethyl-6-nitromethylsulphonylchromone (106). Other analogues have been prepared and they have been investigated for use as aldose reductase inhibitors (D.R. Brittain and M.T. Cox, PCT Int. Appl. WO 90 08,763, 1990).

(106)   (107)   (108) R^1 = CONHR
                              (109) R^1 = CH=NR

Chromone-3-aldonitrones 107 (R = Me, Ph, $4\text{-}ClC_6H_4$) undergo rearrangement in the presence of AlI_3 to yield chromone-3-carboxamides 108 (80-84%) along with aldimines 109 (9-12%) (A.R. Mahajan, R.C. Boruah, and J.S. Sandhu, Chem. Ind., 1990, 261).

The boronate complex (110) of *N,N*-dimethylsalicylacetamide on hydrolysis with 1% aqueous $NaHCO_3$ yields 2-dimethylaminochromone (111) (84%) and complex 112 (J. Morris *et al.*, Tetrahedron Letters, 1993, <u>34</u>, 3817; J. Morris, D.G. Wishka, and Y. Fang, J. Org. Chem., 1992, <u>57</u>, 6502).

(110)   (111)

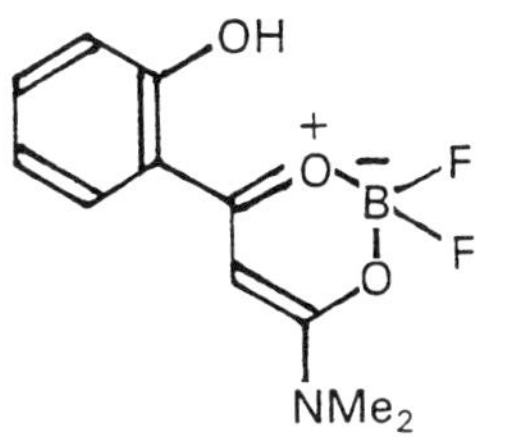

The structures at the top are labeled (112) and (113).

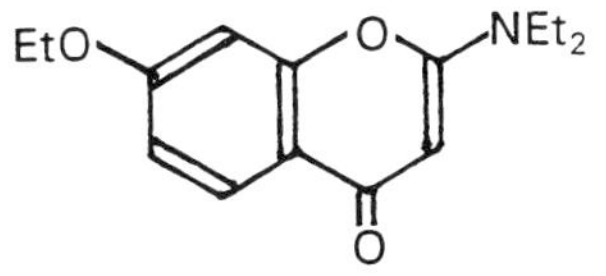

The cyclocondensation of $3\text{-EtOC}_6\text{H}_4\text{OH}$ with $\text{Et}_2\text{NCOCH}_2\text{CO}_2\text{Et}$ in the presence of $POCl_3$ gives 2-diethylamino-7-ethoxychromone (113). The preparations of a variety of related compounds have been reported and tested for their in vitro inhibition of human blood platelet aggregation. Compound 113 was found to be the most active (M. Mazzei *et al.*, Eur. J. Med. Chem., 1990, <u>25</u>, 617). The cyclocondensation of the appropriate salicylacetamides 114 with triflic anhydride yields the corresponding 2- -aminochromones 115 (R^1 =H, 6-, 7-, 8-MeO, 6-Br, 7-Cl, 7-, 8-Me; NR_2^2 =NMe_2, 4-morpholino) (J. Morris, D.G. Wishka, and Y. Fang, Synth. Comm., 1994, <u>24</u>, 849). For the synthesis and anti-platelet activity of some 2-dialkylaminochromones, see M. Mazzei *et al.*, Eur. J. Med. Chem., 1988, <u>23</u>, 237; for the preparation of 2-methylamino-6-*tert*-butylchromone-3- -carboxamide and derivatives as fungicides and agrochemicals, K. Konishi, A. Nohara, and K. Matsura, Jpn. Kokai Tokkyo Koho JP 62 228,001 [87 228,001], 1987; 2-amino-3-cyanochromone and related compounds as herbicide softeners, H. Hagen *et al.*, Ger. Offen. DE 4,039,281, 1992; 2- -(piperidin-1-yl)aminochromone, which showed antagonistic activity to α_1- and α_2-adrenergic receptors, G. Mouysset *et al.*, Eur. J. Med. Chem., 1987, <u>22</u>, 539; and antiatherosclerotic and antithrombotic 3-aminochromones and related compounds, R.B. Gammill, T.M. Judge, and J. Morris, PCT Int. Appl. WO 91 19,707, 1991.

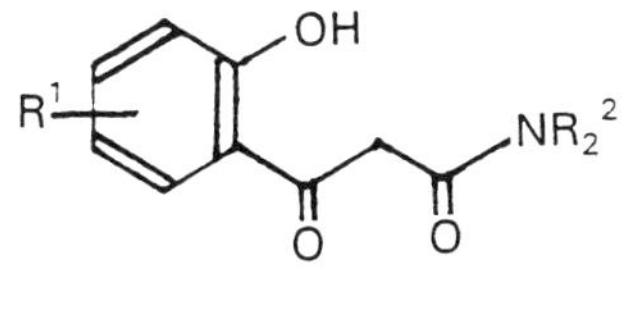

The structures at the bottom are labeled (114) and (115).

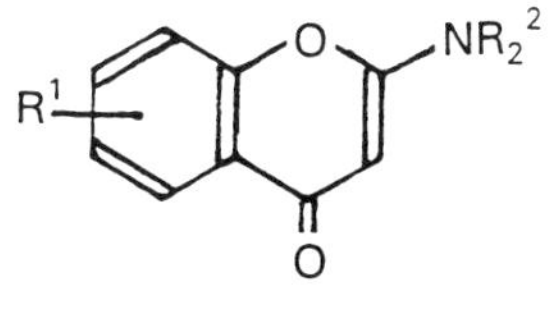

1-Benzylpiperazine reacts with 2-chloromethylchromone in THF to give 2-
-(1-benzylpiperazin-4-yl)methylchromone (116). Other related derivatives
have been prepared and tested for use as central nervous system agents (M.
Payard *et al.*, Eur. Pat. Appl. EP 580,503, 1994). A number of 2-dialkyl-
aminoethyl chromone-2-carboxylates including, 2-dialkylamino-1-methylethyl
chromone-2-carboxylate, 3-dialkylaminopropyl chromone-2-carboxylate, *N*-(2-
-dialkylaminoethyl)chromone-2-carboxamide, and *N*-(3-dialkylaminopropyl)-
chromone-2-carboxamides derivatives and their hydrochlorides have been
prepared. None of these compounds showed any local anesthetic activity,
but some of them showed analgesic activity approximately equivalent to that
of aspirin (K. Hisamichi *et al.*, Annu. Rep. Tohuku Coll. Pharm., 1989, 69).

Chromones 117 (R^1 = H, Me, Ph; R^2 = H, Me; R^3 = H, Me) react with
Ph_2S = NH to yield the azirinochromones 118 (K. Buggle and B. Fallon, J.
Chem. Res., S, 1988, 349).

(116)

(117)

(118)

(119)

Several 3-(acylamino)-7-alkylsulphonylamino-6-phenoxychromones, for
example, 3-formylamino-7-methylsulphonylamino-6-phenoxychromone (119),
and their salts, useful as anti-inflammatory agents have been prepared (T.
Takeno *et al.*, Jpn. Kokai Tokkyo Koho JP 05 97,840 [93 97,840], 1993;
T. Inaba *et al.*, *ibid.*, 05 125,072 [93 125,072], 1993; S. Takano *et al.*,
Ger. Offen. DE 3,834,204, 1989).

The Mannich between chromone, RNH_2 (R = 4-ClC$_6$H$_4$, 4-BrC$_6$H$_4$. 4-MeC$_6$H$_4$, 4-MeOC$_6$H$_4$, 3,4-Cl$_2$C$_6$H$_3$) and 37% aqueous HCHO in EtOH in the presence of HCl affords 3-arylaminomethylchromone 120 (H. Kiao and X. Xu, Beijing Shifan Daxue Xuebao, Ziran Kexueban, 1994, _30_, 1111).

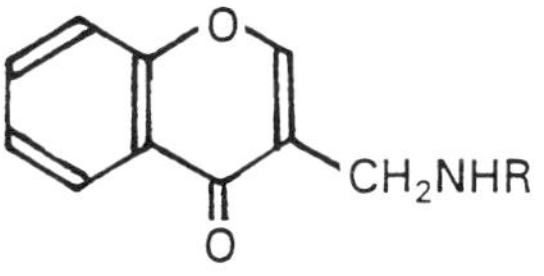

(120)

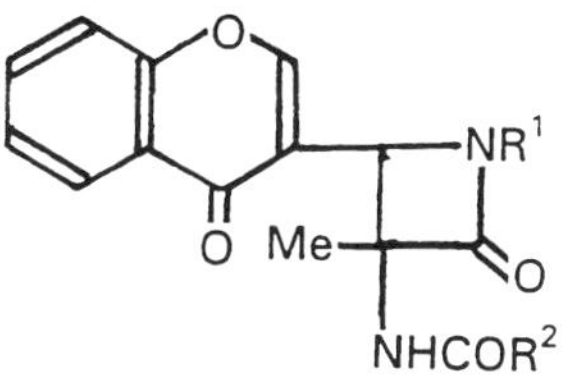

(121)

The cycloaddition reaction of 3-(aryliminomethyl)chromones with oxazolones generated directly from the corresponding _N_-benzoylamino acids, furnishes the 4-(chromon-3-yl)azetidin-2-ones 121 [R^1 = R^2 = (un)substituted Ph] (D. Prajapati, A.R. Mahajan, and J.S. Sandhu, J. Chem. Soc., Perkin 1, 1992, 1821).

The ester obtained from (±)-_cis_-3-acetoxy-1-methyl-4-(3-acetyl-4,6-dimethoxy-2-hydroxyphenyl)piperidine and thiophene-2-carboxylic acid in the presence of POCl$_3$ in pyridine, is rearranged by NaH in dioxane and cyclized by HCl(g) to yield (±)-_cis_-8-(3-acetoxy-1-methylpiperidin-4-yl)-5,7-dimethoxy-2-(thien-2-yl)chromone 122. A number of related derivatives have been prepared as antitumor agents (R.G. Naik _et al._, Ger. Offen. DE 3,836,676, 1990).

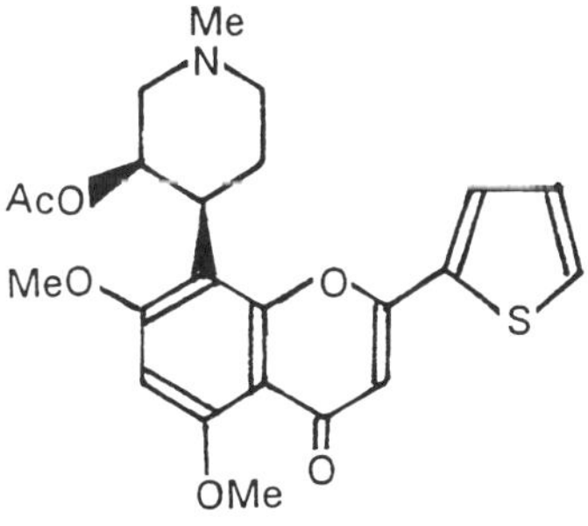

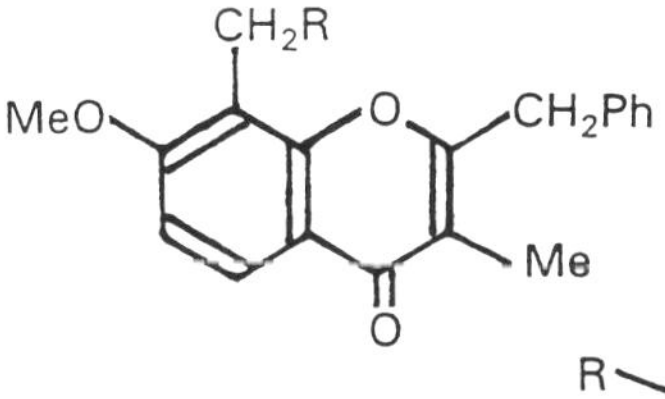

(122)

(123)

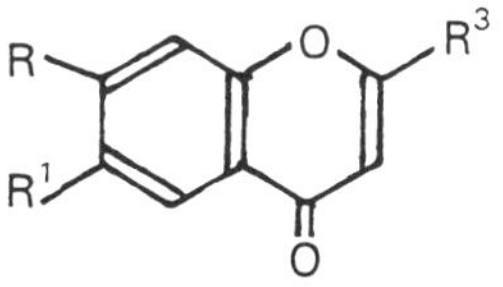

(124) R = H, R^3 = CONHR^2
(125) R^1 = H, R^3 = CONMe$_2$

8-Aminomethyl-2-benzyl-7-methoxy-3-methylchromones (8-aminomethyl-
-7-methoxy-3-methylhomoflavones) 123 (R = NMe$_2$, NEt$_2$, pyrrolidin-1-yl,
piperidin-1-yl, morpholino) have been obtained by the amination of 2-benzyl-
-8-chloromethyl-7-methoxy-3-methylchromone. Their analeptic properties
have been compared with those of dimefline and found to be less toxic (F.
Capolongo *et al.*, Coll. Czech. Chem. Comm., 1991, <u>56</u>, 2402).

The 6-substituted chromone-2-carboxamides 124 [R^1 = H, Cl, NO$_2$;
R^2 = 4,2-, 2,4-, 5,2-(O$_2$N)C$_6$H$_3$, 5-nitrothiazol-2-yl] have been prepard and
tested as protozoacides, amebicides, fungicides, and anthelmintics, but only
derivative 124 (R^2 = 5-nitrothiazol-2-yl) showed antiparasitic activity (A.
Taudou *et al.*, Eur. J. Med. Chem., 1987, <u>22</u>, 583).

7-Substituted *N*-dimethylchromone-2-carboxamides 125 (R = H, OMe,
NO$_2$, F, Cl, Br) are efficiently obtained without concomitant formation of
unwanted products, by treating the appropriate acid chloride with Me$_2$NH in
pyridine (D.N. Davidson and P.T. Kaye, Synth. Comm., 1990, <u>20</u>, 727).
6,7-Dimethylchromone-2-carboxamide derivatives 126 (NR$_2$ = NMe$_2$, NEt$_2$,
pyrrolidino, piperidino, morpholino, n = 2; NMe$_2$, NEt$_2$, n = 3) and their
hydrochlorides have been synthesized and examined for local anesthetic and
analgesic activities (M. Ohta *et al.*, Annu. Rep. Tohoku Coll. Pharm., 1991,
<u>38</u>, 35). Several *N*-dialkylaminoalkyl-6,8-dimethylchromone-2-carboxamides
have been prepared and similarly examined for pharmacological activity (H.
Ito *et al.*, *ibid.*, 1993, <u>40</u>, 91). 6,7-Methylene-dioxychromone-2-
carboxamide on treatment with POCl$_3$ gives 2-cyano-6,7-methylenedioxy-
chromone, which on heating with NaN$_3$ and NH$_4$Cl affords 2-(tetrazol-5-yl)-
-6,7-dimethylenedioxychromone (J. Go *et al.*, Jpn. Kokai Tokkyo Koho JP 06
41,140 [94 41,140], 1994).

N-(2-Dimethylaminoethyl)-5,7-dimethylchromone-2-carboxamide, *N*-(2-
-diethylaminoethyl)-5,7-dimethylchromone-2-carboxamide, *N*-(2-pyrrolidino-
ethyl)-5,7-dimethylchromone-2-carboxamide, *N*-(2-piperidinoethyl)-5,7-di-
methylchromone-2-carboxamide, *N*-(2-morpholinoethyl)-5,7-dimethylchrom-
one-2-carboxamide, *N*-(3-dimethylaminopropyl)-5,7-dimethylchromone-2-car-
boxamide, and *N*-(3-diethylaminopropyl)-5,7-dimethylchromone-2-carbox-
amide have been prepared and their local anesthetic and analgesic properties
determined (H. Ito *et al.*, Annu. Rep. Tohoku Coll. Pharm., 1992, <u>39</u>, 47).

N-Arylchromone-3-carboxamides (3-arylcarbamoylchromones) 127 (R^1-
-R^4 = H, halogeno, alkyl, alkenyl, alkynyl, R^5 = C$_{1-12}$alkyl, alkenyl, alkynyl,
halogenoalkyl, aryl; R^6-R^{10} = H, halogeno, alkyl, alkynyl, CN) have been
prepared and tested as plant growth inhibitors, for example, *N*-(2-butyl-6-
-ethylphenyl)-2-butylchromone-3-carboxamide (127; R^1-R^4 = R^7-R^9 = H,
R^5 = R^6 = Bu, R^{10} = Et) killed 100% rice plant and barnyard grass at 20ppm
(Y. Goto *et al.*, Jpn. Kokai Tokkyo Koho JP 03 130,278 [91 130,278],
1991).

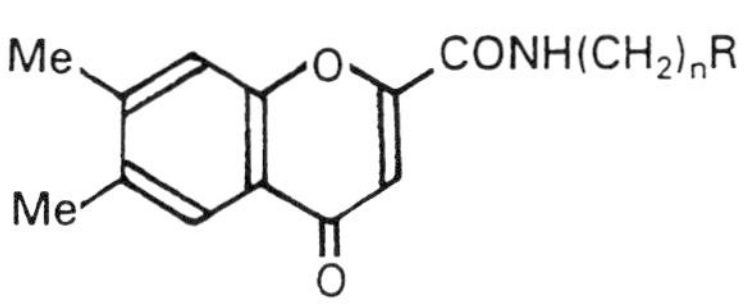

(126)

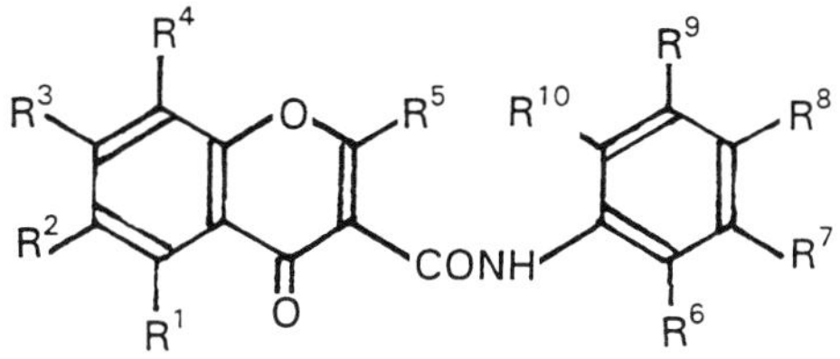

(127)

Unsubstituted and 6,8-disubstituted N-(ethoxycarbonyl)chromone-3-
-carboxamides 128 (R^1 =H, Cl, Br, R^2 =H, OMe, OEt, Br) have been prepared
by cyclocondensation of salicylaldehydes 129 with enamine 130 in hot EtOH
(G. Shaw, Chem. Comm., 1987, 1735).

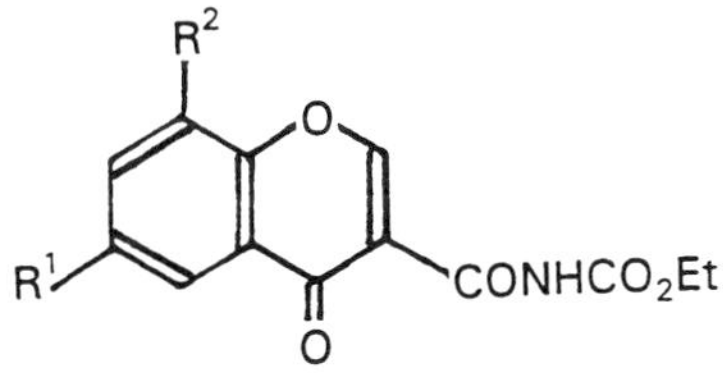

(128)

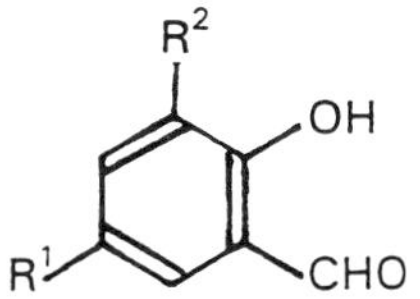

(129)

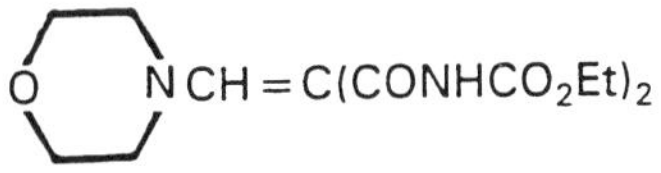

(130)

(xii) Miscellaneous compounds related to chromone
The preparations fo 3-methyl- and 3,7,7-trimethyl-5,6,7,8-tetrahydro-4H-
-benzo[b]pyran-4,5-diones (3-methyl- and 3,7,7-trimethyl-5,6,7,8-tetrahydro-
chromon-5-ones) 1 (R =H, Me) have been reported (A.L. Mikhal'chuk, Zh.
Obshch. Khim., 1991, 61, 261).

Ar = 4-MeC$_6$H$_4$

(1) (2) (3)

3-(4-Tolylsulphinyl)chromone (2) undergoes a diastereoselective conjugate addition of lithium dimethylcuprate to furnish a mixture of 2--methyl-3-(4-tolylsulphinyl)chroman-4-ones (3), which on heating to 140°C yields 2-methylchromone. Desulphurization of the mixed products from (S)-2 affords (S)-2-methylchroman-4-one in 90% enantiomeric excess (S.T. Saengchantara and T.W. Wallace, Tetrahedron, 1990, 46, 6553).

The enamines 4 (R^1 = E-CH = CHNMe$_2$; R^2 = H, Me, Ph) on treatment with NH$_3$ yields the fused pyridine 5 and with NH$_2$OH the corresponding pyridine 2-oxide. N$_2$H$_4$, PhNHNH$_2$, and guanidine undergo initial 1,6-addition to the enamines ultimately affording pyrazoles 6 and pyrimidines 7, respectively. Chromones 4 (R^1 = E-CH = CHNMe$_2$; R^2 = Me, Ph) with H$_2$NCH$_2$CO$_2$Et form esters, which on treatment with EtONa-EtOH yeld the azepines 8 (C.K. Ghosh et al. J. Chem. Soc., Perkin 1, 1988, 1489).

(4) (5) (6)

(7)

(8)

(xiii) Furochromones

Ethyl 4-chloro-6-hydroxy-5-phenylacetyl-2,3-dihydrobenzofuran-2-
-carboxylate on boiling in pyridine with the addition of HCO_2Et and piperidine
yields ethyl 4-chloro-5-oxo-6-phenyl-2,3-dihydro-5*H*-furo[3,2-*g*]benzo[*b*]-
pyran-2-carboxylate (1). A number of related derivatives useful as diuretics
and antihypertensives have been prepared (M. Kitagawa, Jpn. Kokai Tokkyo
Koho JP 63 83,089 [88 83,089], 1988). Thiation of khellin 2 (X = O) by
Lawesson's reagent to give 2 (X = S) is accompanied by 5-*O*-demethylation
of 2 (X = S) to afford 5-desmethylthiokhellin (T.S. Hafaz *et al.*, Phosphorus,
Sulphur Silicon Relat. elem., 1991, 56, 165).

(1)

(2)

(xiv) Pyranochromones

The condensation reaction between 4,6-diacetylresorcinol and R^1CHO
($R^1 = ClC_6H_4$, furyl, thienyl) affords derivative 1, which has been converted to
the 2,8-disubstituted 3,7-dihydroxy- and 3,7-dimethoxy pyrano[3,2-
g]chromon-6-ones (substituted 4,6-dioxo-4*H*,6*H*-benzo[1,2-b:5,4-b']pyrans)
2 ($R^2 = H$, Me) as potential insecticides (D. Ashok and P.N. Sarma, Indian J.
Chem., 1987, 26B, 900).

(1)

(2)

The self-condensation of ethyl pentafluorobenzoylacetate affords 6,7,8,9-tetrafluoro-3-pentafluorophenyl-1*H*-pyrano[4,3-*b*]chromon-1-one (3), which on hydrolysis yields 5,6,7,8-tetrafluoro-2-(pentafluorobenzoylmethyl)-chromone (4) (V.I. Saloutin *et al.*, J. Fluorine Chem., 1993, <u>65</u>, 37).

(3)

(4)

*(xv)Flavones (2-phenylchromones, 2-phenyl-4*H-*benzo[b]pyran-4-ones)*
and related derivatives

Heating 2-iodophenol, $PhC \equiv CH$, and Et_2NH at 120°C under 20 atmospheres CO in the presence of $PdCl_2$ [bis(diphenylphosphino)ferrocene] $(dppf)_2$ gives flavone 1. Similarly 2-iodophenols with arylacetylenes afford 2-aryl-chromones (V.N. Kalinin, M.V. Shostakovskii, and A.B. Ponomarev,

Tetrahedron Letters, 1990, 31, 4073) and 4-alkyl-2-iodophenols with arylacetylenes give 6-alkyl-2-arylchromones (*idem*, U.S.S.R. SU 1,719,402, 1992). The irradiation by UV light of the alkaline solutions of flavones and 2 (R=H, Cl) in benzene results in a rearrangement to yield the cyclobutane derivatives 3 [S. Jain, Natl. Acad. Sci. Letters (India), 1992, 15, 259].

(1) Ar=Ph   (3)
(2) Ar=4-ClC$_6$H$_4$

Lawesson's reagent converts flavone (1) into 2-phenyl-4*H*-benzo[*b*]-pyran-4-thione (T.S. Hafez *et al.*, Phosphorus, Sulphur Silicon Related Elem., 1991, 56, 165). Cyclization of dihydrochalcones, 5,2-R^1(AcO)C$_6$H$_3$COCHBr-CH(OMe)Ph (R=H, Me, Cl, NO$_2$), by aqueous ethanolic NaOH produces minor quantities of the 6-substituted flavones 4, besides the major products, aurones 5. Increasing the concentration of base increases the yield of flavones (J.A. Donnelly and C.L. Higginbotham, Tetrahedron, 1990, 46, 7219). Similarly, 4,2-R^2(AcO)C$_6$H$_3$COCHBrCH(OMe)Ph (R^2=H, OMe, Me) yields aurones 6 as the major products, together with small amounts of the 7-substituted flavones 7, but it is reported that when R=NO$_2$ or Cl the major product is the respective flavone 7 (*idem*, Monatsh., 1991, 122, 83). The palladium-catalyzed carbonylative coupling of 2-hydroxyaryl iodides with ethynylarenes has been carried out in DMF at 60°C under 1 atmosphere of CO pressure, using DBU as base and Pd(OAc)$_2$(dppf)$_2$ as catalyst to yield generally mixtures of flavones 8 (R^1=H, Me, Cl, OMe; R^2=H, Me; R^3=H, OMe; R^4=H, OMe, CO$_2$Me) and aurones 9 in varying yields, depending on substituent on both reactants (P.G. Ciattini *et al.*, Tetrahedron, 1991, 47, 6449).

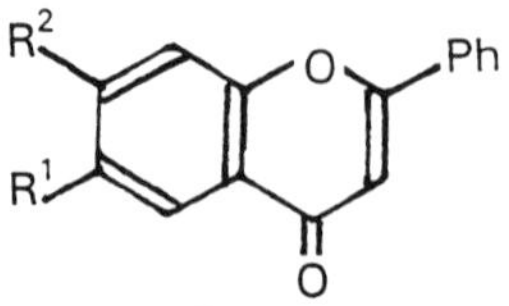

(4) $R^2 = H$
(7) $R^1 = H$

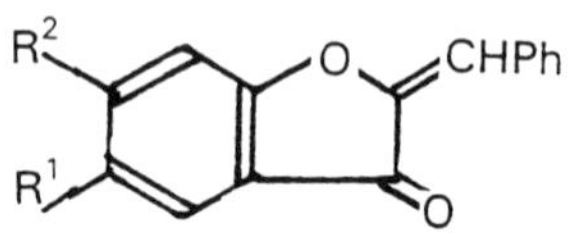

(5) $R^2 = H$
(6) $R^1 = H$

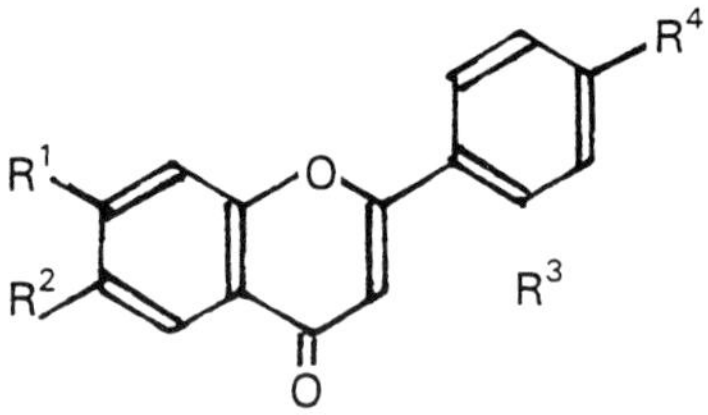

(8)

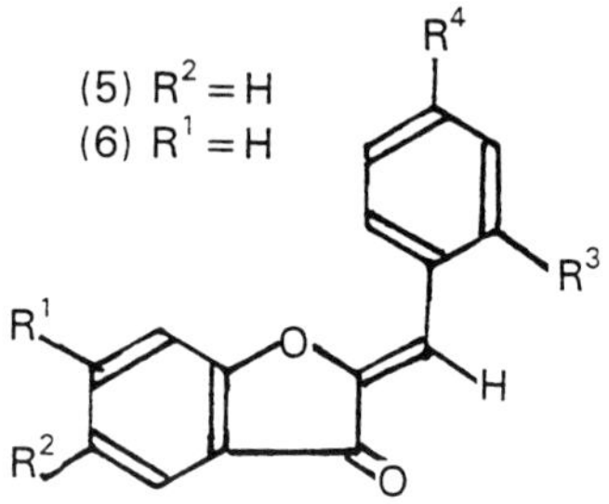

(9)

A number of 4-oxo- and 4-thioxo-4H-benzo[b]pyran-2-phenylalkanoic acids and esters 10 (R^1 = 3- or 4-CHR^{4}CO$_2$R^3; R^2 = H, 6-Me, 5,7-Me$_2$; R^4 = H, Me, R^3 = H, Me, Et, Pr, Pr^i, Bu, CH$_2$CHMe$_2$, octyl, cetyl) have been prepared and evaluated for their hypolipidemic activity (A. Orjales, A. Berisa, and L. Alonso-Cires, Indian J. Chem., 1994, <u>33B</u>, 27).

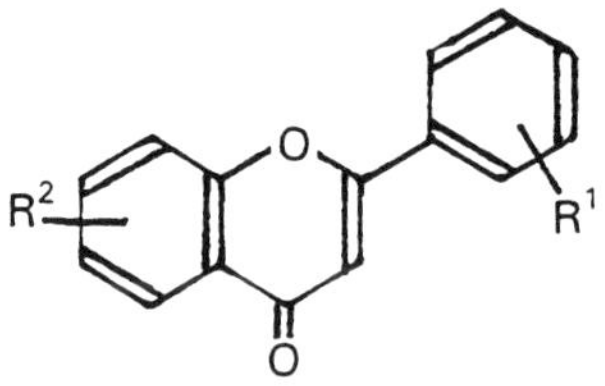

(10)

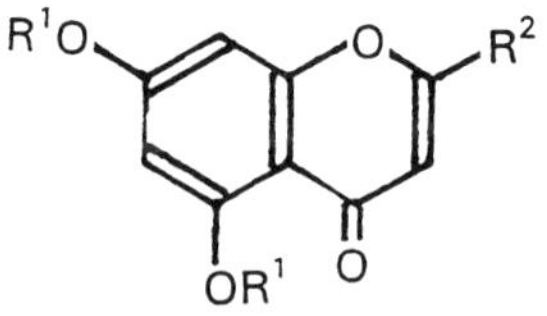

(11)

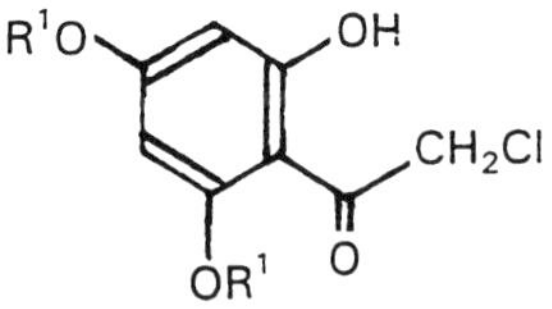

(12)

5-7-Dihydroxyflavones and their methylated derivatives 11 (R^1 = H, Me; R^2 = Ph, substituted Ph) have been obtained by the aldol condensation of α-chloro-2-hydroxyacetophenones 12 (R^1 = H, Me) with R^2CHO in aqueous alcohol solution containing 5-10% NaOH at room temperature (J. Chen and Y. Li, Xiamen Daxue Xuebao, Ziran Kexueban 1992, <u>31</u>, 651).

Flavones and 6-methylflavones 13 (R^1 = H, Me; R^2 = H, OMe; R^3 = H, OMe; R^4 = H, OMe, R^5 = H, OMe) containing up to two methoxy groups attached to the 7- and 8-positions and up to two methoxy groups at the 2- and 4-positions of the 2-phenyl group have been obtained by heating the appropriate 2-hydroxybenzoylmethane with Me_2SO_4-iodine (J.K. Makrandi and V. Kumari, Chem. Ind., 1988, 630). The mass spectra of flavones 14 (R^1 = H, Cl, NO_2; R^2 = H, Me) have been reported (J. Chen and Y. Li, Xiamen Daxue Xuebao, Ziran Kexueban, 1994, <u>33</u>, 123).

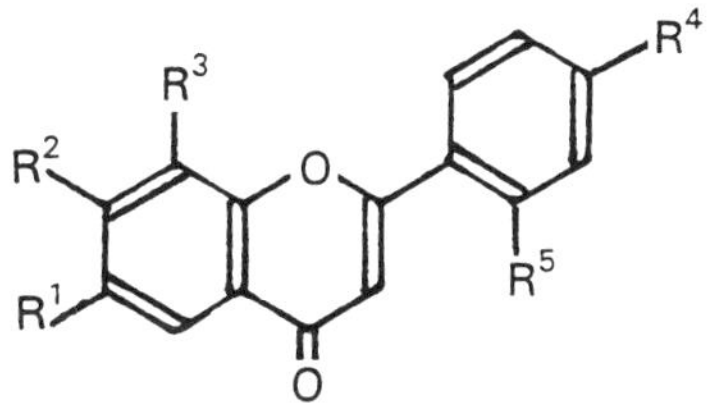

(13) 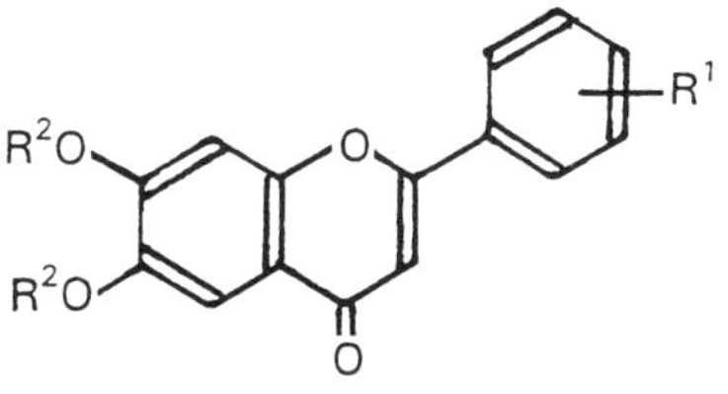   (14)

5-Benzoyl-2-hydroxy-3,4-dimethoxydibenzoylmethane on boiling with AcONa in AcOH gives 6-benzoyloxy-7,8-dimethoxyflavone, which on boiling with aqueous KOH in MeOH affords 6-hydroxy-7,8-dimethoxyflavone (15). Related derivatives 16 (R^1-R^3 = H, C_{1-3}alkyl) and 17 (R^1-R^3 = H, C_{1-3}alkyl) or their salts have been investigated for their use in the treatment of osteoporosis (Y. Kinoshita et al., Jpn. Kokai Tokkyo Koho JP 63 201,124 [88 201,124], 1988; 201,123 [88 201,123], 1988).

(15) $R^1 = R^2 = Me$, $R^3 = H$ (17)
(16) (see text)

The oxophilicity of cerium (IV) ammonium nitrate has been controlled by the addition of hydrogen peroxide and utilized to prepare the 3-aroylflavones 18 ($R^1 = R^2 = R^3 = R^4 = H$; $R^1 = R^3 = R^4 = H$, $R^2 = OMe$; $R^1 = R^2 = H$, $R^3 = R^4 = OMe$; $R^1 = OH$, $R^2 = R^3 = R^4 = H$) from the corresponding 3-arylideneflavones 19 (H.M. Chawla and S.K. Sharma, Bull. Soc. Chim. Fr., 1990, 656). A number of substituted flavones, for example, 3-(4-chloro-benzoyl)-4'-methoxy-6-methylflavone 18 ($R^1 = Me$, $R^3 = OMe$, $R^4 = Cl$), have been synthesized and tested for their antibacterial and antifungal activities and their IR and NMR spectral data have been reported (M.B.Hogale, B.N. Pawar, and B.P. Nikam, J. Indian Chem. Soc., 1987, 64, 486).

(18)   (19)

The reaction of 3-methylflavone-8-carbonyl chloride (20) with 2,2-di-methyl-3-[4-(2,3,4-trimethoxybenzyl)piperazin-1-yl]propanol in benzene affords the 3-methylflavone-8-carboxylic ester 21 (R^1-R^4 = H, X = NR, R = 2,3,4-trimethoxybenzyl). Other related derivatives 21 (R^1-R^4 = H, Me; X = O, CH_2, CH_2CH_2, NR; R = H, alkyl, acyl) have been prepared and derivative 21 (R^1-R^4 = H, X = NR, R = allyl) significantly inhibited bladder contraction in rats (H. Matsui *et al.*, Jpn. Kokai Tokkyo Koho JP 01 299,284 [89 299,284], 1989). Some related basic esters as flavoxate analgoues for example, ester 22, have been prepared and their activity as spasmolytics tested and compared to flavoxate (D. Nardi *et al.*, Arzneim-Forsch., 1993, <u>43</u>, 28).

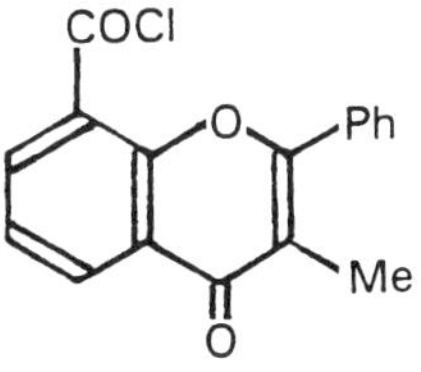

(20) 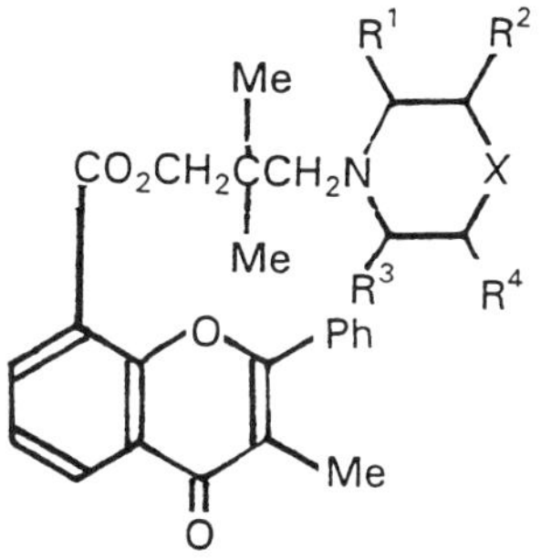   (21) 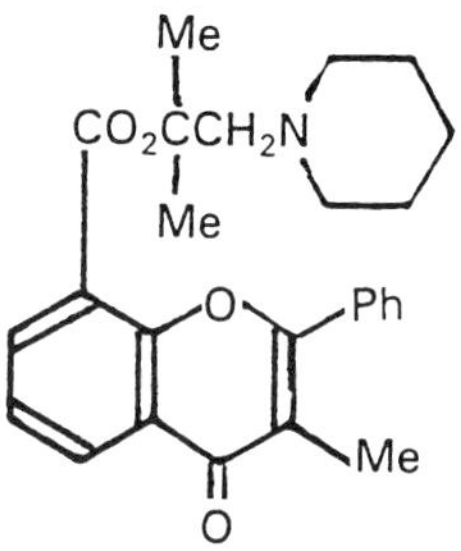   (22)

2-Acyloxybenzoic acids are readily cyclized by excess PPh_3-CCl_4 in CH_2Cl_2 at 24°C to 3-chloroflavones. Using Cl_3CCN instead of CCl_4 and heating to 180°C aspirin is cyclized to 3-cyano-2-methylflavone (H. Vorbruggen, B.D. Bohn, and K. Krolikiewicz, Tetrahedron, 1990, <u>46</u>, 3489). Flavones 23 (R^1, R^2 = Cl, $HOCH_2$; R^3 = Ph, 2-, 3-, 4-ClC_6H_4, 2-tolyl, 1-naphthyl), with chloro or hydroxymethyl substituents in the 6- and/or 8-positions have been prepared by the oxidative cyclization of the appropriate 1-(3,5-disubstituted 2-hydroxyphenyl)-3-arylprop-2-en-1-one 24 by SeO_2 in isoamyl alcohol (V.M. Gurav and D.B. Ingle, J. Indian Chem. Soc., 1988, <u>65</u>, 796).

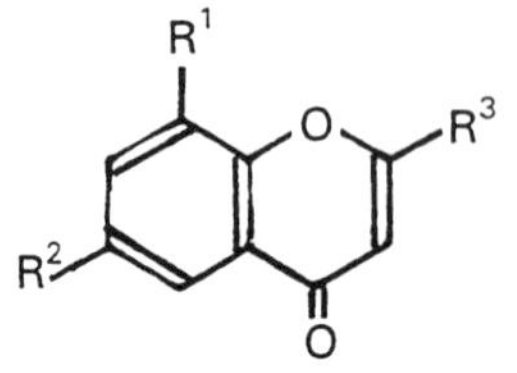

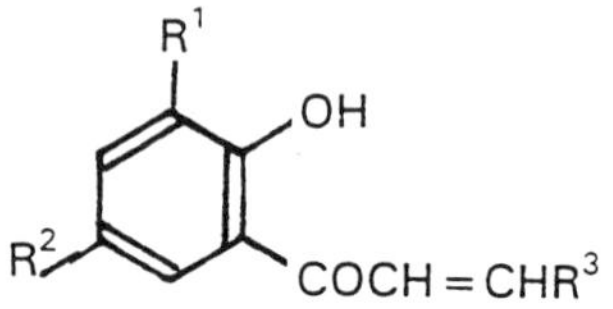

(23)

(24)

6,8-Disubstituted flavones 25 (R^1 =H, Me, Br; R^2 = Me, Cl, Br; R^3 =H) on bromination with $CuBr_2$ afford the 3-bromoflavones 25 (R^3 = Br) (A.K.D. Mazumdar *et al.*, *ibid.*, 1990, <u>67</u>, 845).

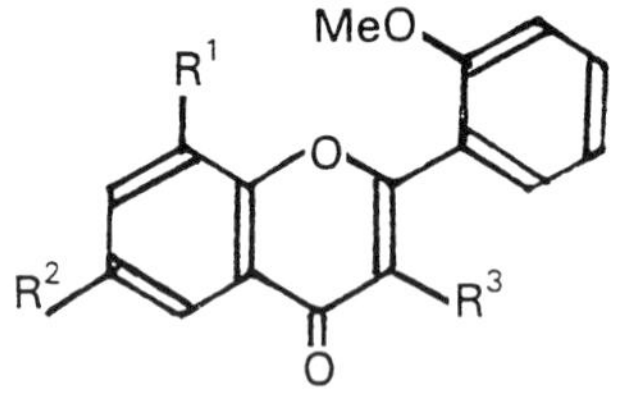

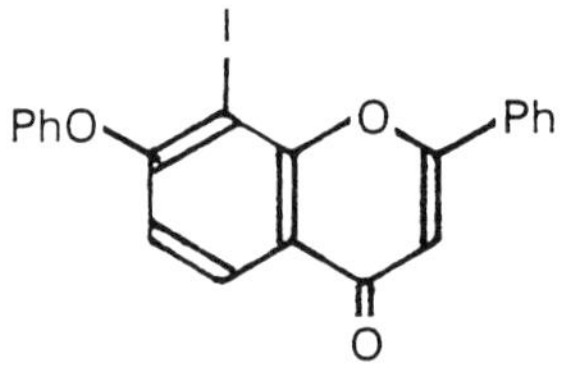

(25)

(26)

The reaction of the hypervalent iodine reagent, (diacetoxyiodo)benzene, with 7-hydroxyflavone yields 8-iodo-7-phenoxyflavone (26) (U.K. Mallik and A.K. Mallik, Indian J. Chem., 1992, <u>31B</u>, 696). Cyclization of iodochalcones 27 (R =H, MeO) with I_2-DMSO-H_2SO_4 gives the 8-iodo-5,7-dimethoxy-flavones 28 (Y. Xing, Z. Sun, and N. Chen. Zhongguo Yigao Gongye Zazhi, 1992, <u>23</u>, 325). Treatment of flavones 29 (R^1 =H, R^2, R^3 =H, alkoxy, halogeno) with the iodine-cerium(IV) ammonium nitrate system under mild conditions furnishes the corresponding 3-iodoflavones 29 (R^1 =I) (F. Zhang and Y. Li, Synthesis, 1993, 565).

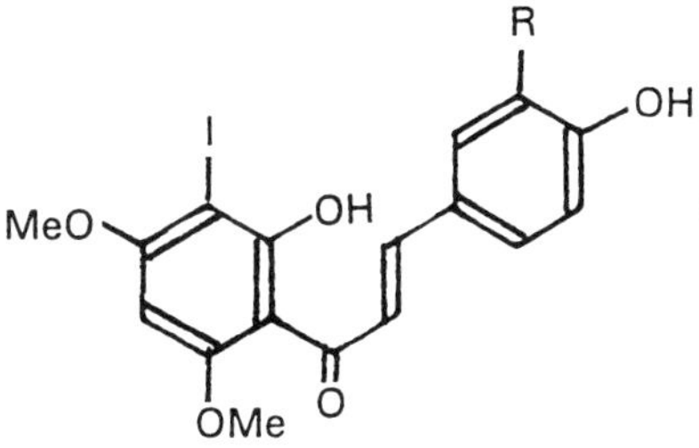

(27)

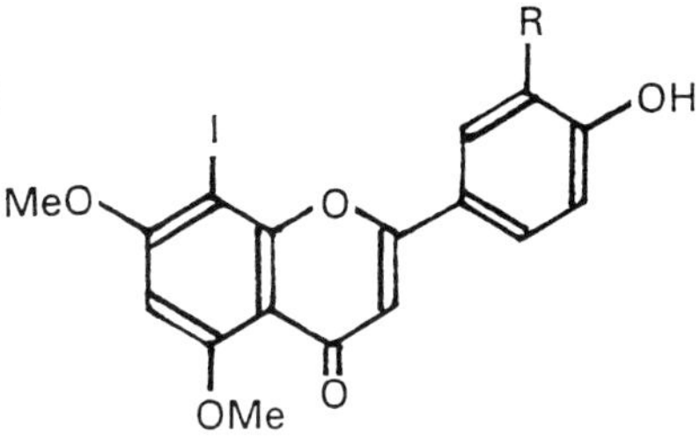

(28)

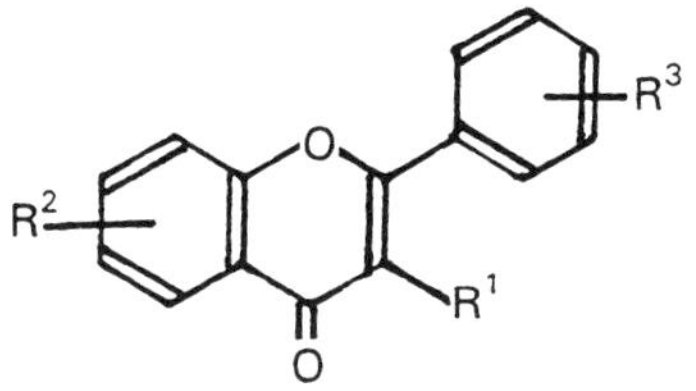

(29)

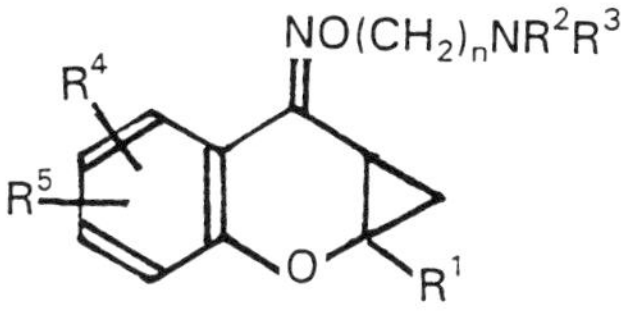

(30)

7-(Aminoalkoximino)-1a,7a-dihydrocyclopropa[*b*]chromens 30
[R^1 = (substituted) Ph; R^2, R^3 = H, (ar)alkyl, Ph; NR^{2}R^3 = (thio)morpholino,
pyrrolidino, piperidino; R^4, R^5 = H, OH, halogeno, alkoxy; n = 2-5] and other
related derivatives have been prepared from flavones for the treatment of
disorders from cerebral neurodegeneration for example, flavone was treated
with Me$_3$S(O)I and NaH and the product converted in 3 steps to derivative
30 (R^1 = Ph, R^2 = Me, R^3-R^5 = H, n = 2) (T. Tatsuoka, Eur. Pat. Appl. EP
460,358, 1991; T. Tatsuoka and K. Nomura, *ibid.*, 461,343, 1991).

2-Aryl-3-chloro-4-hydroxy-3-nitrochromans (2-aryl-3-chloro-4-hydroxy-3-
-nitro-3,4-dihydro-2*H*-benzo[*b*]pyrans) 31 (R^1, R^2, R^3 = H, Cl, NO$_2$, MeO; R^4,
R^5, R^7 = H, MeO; R^6 = H, Cl, Br, NO$_2$, MeO; R^8 = H, MeO, Br) have been
conveniently converted to the corresponding 3-nitroflavones 32.
Intermediates in the procedure were the 3-chloro-3-nitroflavanones (2-aryl-3-
-chloro-3-nitro-3,4-dihydro-2*H*-benzo[*b*]pyran-4-ones) 33 (D. Dauzonne and
C. Grandjean, Synthesis, 1992, 677).

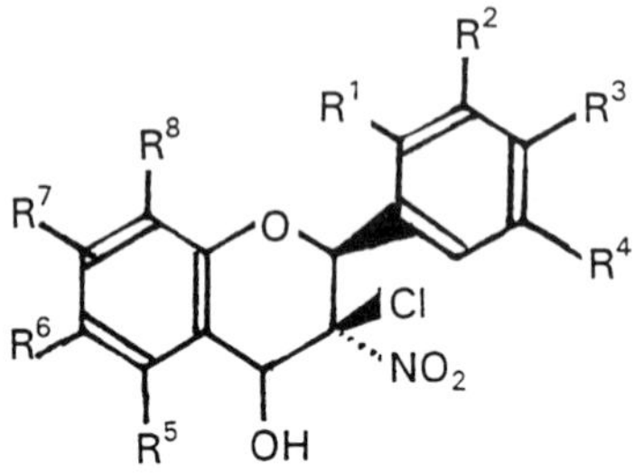

(31)

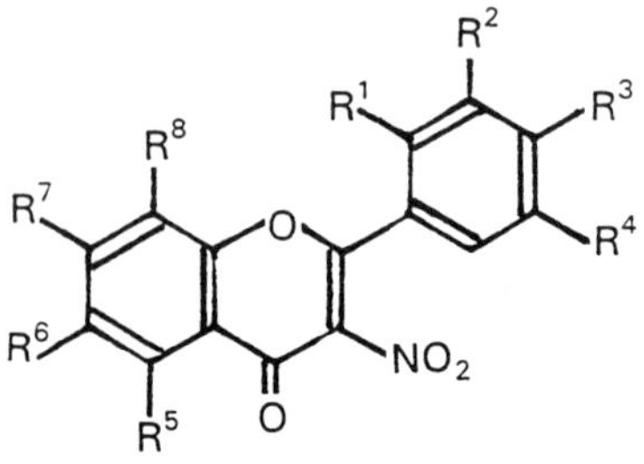

(32)

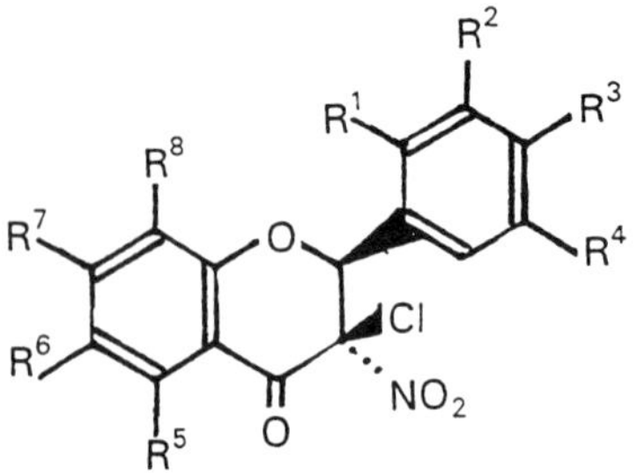

(33)

Treatment of chalcones 34 (R^1 = aryl, R^2 = NO_2, H) with SeO_2 gives 7-
-butoxy-6-nitroflavone (35; R^1 = aryl, R^2 = NO_2) and 7-butoxyflavone (35;
R^1 = aryl, R^2 = H), respectively, which along with related flavonols and
flavanones were tested for antimicrobial activity against *Staphylococcus
aureus*, *Escherichia coli*, and *Aspergillus niger* (M.D. Ankhiwala and H.B.
Naik, J. Indian Chem. Soc., 1989, <u>66</u>, 482). Some 6- and 8-nitroflavones
and their corresponding aminoflavones, for example, 36 (R^1 = NO_2, NH_2,
R^2 = H, R^3 = OH, MeO; R^2 = MeO, NO_2, OH, R^3 = H) have been synthesized
and the aminoflavones evaluated as potential inhibitors of protein-tyrosine
(M. Cushman *et al.*, J. Med. Chem., 1994, <u>37</u>, 3353).

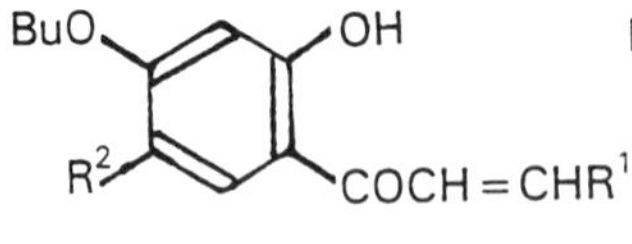

(34)

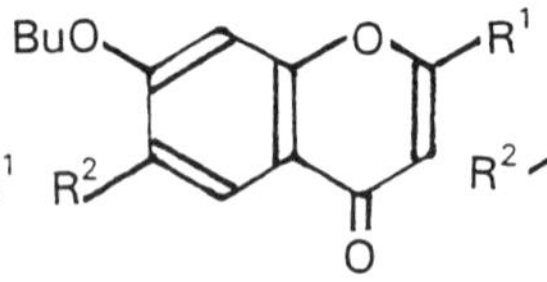

(35)

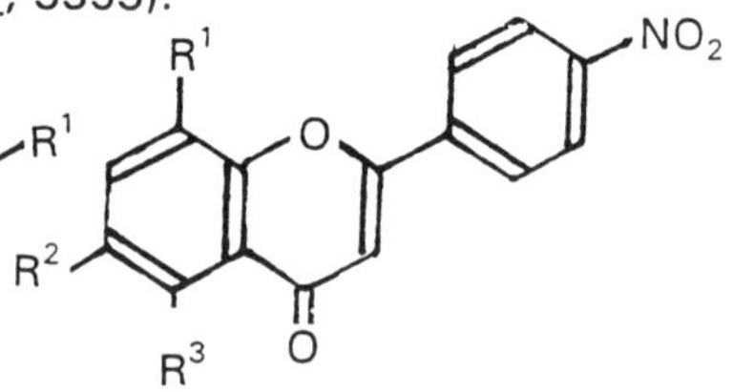

(36)

2-(4-Aminophenyl)chromone (37; R = H) has been converted into aryl-thiazolidinonylphenylbenzopyranones 38 (R = H, Cl, NO_2) and arylazetidinon-ylphenylbenzopyrans 39 (R = H, Me) by the reaction of the *N*-arylidene derivatives of 37 (R = H) with thioglycolic acid and chloroacetyl chloride, respectively. The *N*-chloroacetyl derivative 37 (R = $COCH_2Cl$) reacted with secondary amines and other nucleophiles to give derivatives, such as 37 (R = $COCH_2R^1$; R^1 = piperidino, morpholino, $NHNH_2$) (D.L. Coutinho and P.S. Fernandes, *ibid.*, 1992, <u>69</u>, 265). 3-Aminoflavone reacts with NaN_3 and $HC(OEt)_3$ in AcOH to give 3-(tetrazol-1-yl)flavone (G. Litkei *et al.*, Pharmazie, 1989, <u>44</u>, 791).

(37)

(38)

(39)

A number of 5-aminoflavone derivatives possessing antibacterial and antitumor activity (T. Akama *et al.*, Eur. Pat. Apl. EP 556,720, 1993) and *N*-substituted 3-methylflavone-8-carboxamides (flavoxate analogues), for example, derivative 40, as antagonists of α_1-adrenergic and 5-HT_{1A} receptors (A. Leonardi *et al.*, *ibid.*, 558,245, 1993) have been prepared.

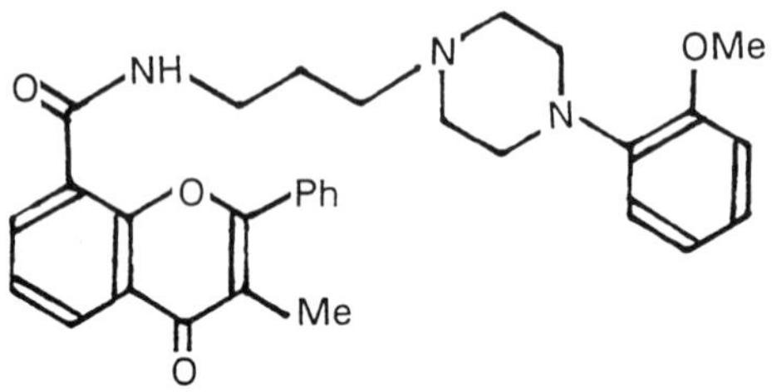

(40)

The 8-substituted 3-methylflavones 41 [R = (un)substituted (phenylalkyl)-
amino, (un)substituted dihydroisoquinolino; X = CO, CO_2, CONH, CONMe,
CH_2NH; n = 1-4], which exhibit powerful antispasmodic action and are
considerably more stable at physiological pH than flavoxate have been
synthesized (*idem, ibid.*, 566,288, 1993). For the synthesis of flavone and
flavanone-7-*O*-isopropanolamine derivatives and their antiarrhythmic activity,
see X. He *et al.*, Zhongguo Yaowu Huazue Zazhi, 1990, 1, 11.

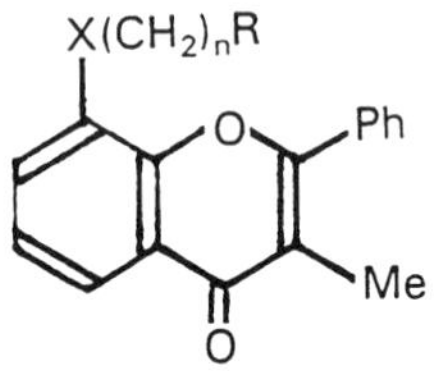

(41)

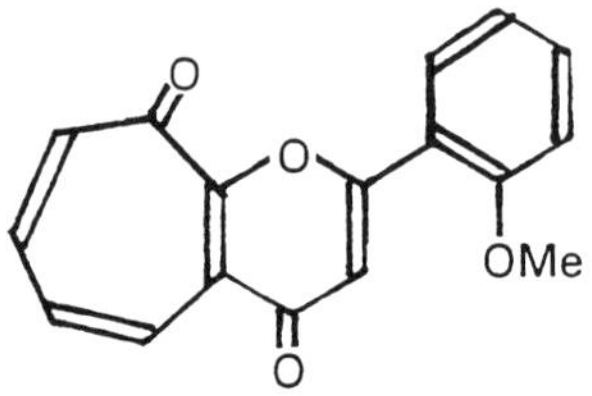

(42)

The reaction of 3-acetyltropolone with 2-methoxybenzaldehyde in the presence of ethyl orthoformate and perchloric acid gives 2-(2-methoxy-phenyl)-4,9-dihydrocyclohepta[*b*]pyran-4,9-dione (42) (D. Wang, Z. Jin, and K. Imafuku, J. Heterocyclic Chem., 1990, _27_, 891).

Artonin V has been isolated as pale yellow needles, m.p. 85-87°C from the root bark of *Artocarpus altilis* and shown to have the structure 43 [3,8--di(isoprenyl)-5,7,2',4',5'-pentahydroxyflavone] on the basis of spectroscopic evidence (Y. Hano, R. Inami, and T. Nomura, J. Chem. Res., S, 1994, 348).

(43)

(xvi) Flavonols (3-hydroxyflavones, 3-hydroxy-2-phenyl-4H-benzo[b]-
*   pyran-4-ones)*

3-Nitro-2-phenyl-2*H*-benzo[*b*]pyrans 1 (R^1, R^2 =H, OMe; R^3 =H, Cl, OMe, Me) are degraded on treatment with KO_2 in Me_2SO mainly to the corresponding salicylic acids and benzoic acids, along with flavonol 2 as a minor product. However the dihydro derivatives of 1 on treatment with KO_2 in C_6H_6 containing 18-crown-6 are converted to the corresponding flavonols 2 (T.S. Rao and G.K. Trivedi, Heterocycles, 1987, _26_, 2117).

(1)

(2)
(6) $R^1 = R^2 = H$, $R^3 = R$

The treatment of chalcones 3 (R^1 = aryl, R^2 = NO_2) with alkaline H_2O_2 and H_2SO_4 affords flavonols 4 and flavanones 5, respectively (M.D. Ankhiwala and H.B. Naik, J. Indian Chem. Soc., 1989, <u>66</u>, 482).

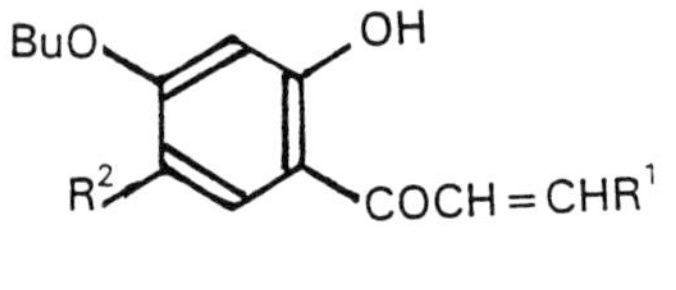

(3)

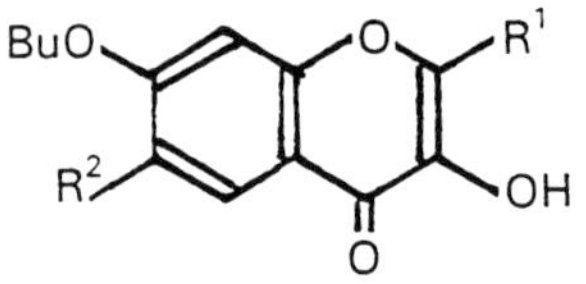

(4)

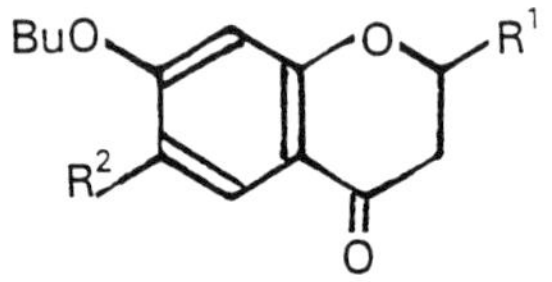

(5)

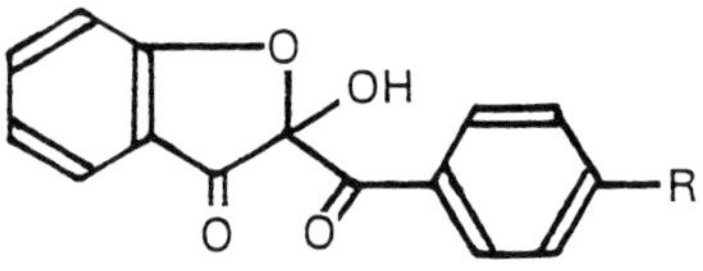

(7)

The reaction of flavonols 6 (R = H, OMe) with dioxygen in the presence of CuCl and $CuCl_2$ results in the formation of 2-benzoyl-2-hydroxy-2,3--dihydrobenzo[*b*]furan-3-one 7, along with other products including 2-hydroxybenzil, salicylic acid, and benzoic acid (E. Balogh-Hergovich and G. Speier, J. Mol. Catal., 1992, <u>71</u>, 1).

The selective oxidation of 6-propionylflavonols 8 (R = Ph, 4-MeOC$_6$H$_4$, 4--ClC$_6$H$_4$, 4-MeC$_6$H$_4$) and related compounds 8 (R = fur-2-yl, thien-2-yl) using hypervalent iodine reagnets, [bis(trifluoroacetoxy)iodo]benzene, [hydroxy-(tosyloxy)iodo]benzene, iodobenzene diacetate, and iodosylbenzene in methanol, results in the formation of 3-hydroxy-2,3-dimethoxy-6-propionyl-flavanones 9 and heteroylchromanones, respectively (M.S. Khanna *et al.*, Synth. Comm., 1992, <u>22</u>, 893).

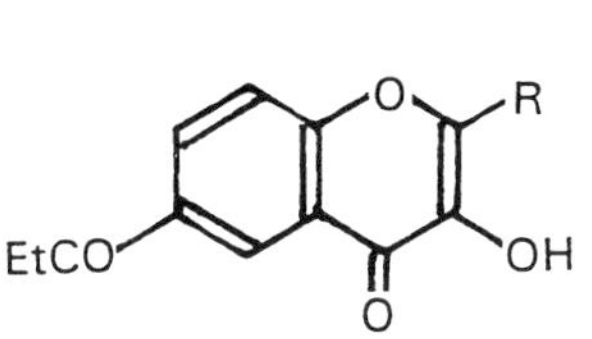

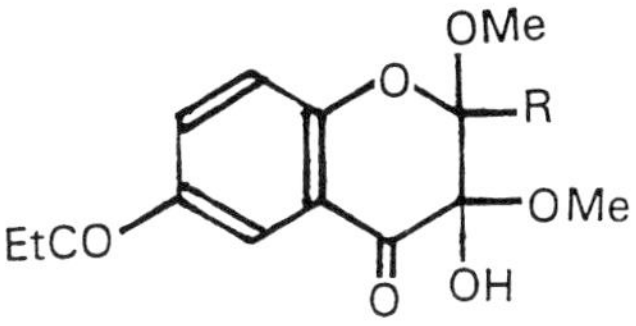

(8)

(9)

Hispidulin, useful as an antiasthmatic has been extracted from *Arnica montana* flowers (P. Potier, J. Benveniste, and B. Bourdillat, PCT Int. Appl. WO 87 02,981, 1987) and the preparation of magnesium quercetin-5-sulphonate has been reported (M. Kopacz and D. Nowak, Pol. PL 145,085, 1988).

The reaction of quercetin (10; $R^1 = R^2 = H$) with pivaloyl chloride in pyridine furnishes derivative 10 ($R^1 = H$, $R^2 = COBu^t$), which on methylation with diazomethane in Et_2O gives derivative 10 ($R^1 = Me$, $R^2 = COBu^t$). Hydrolysis of the latter derivative with HCl in MeOH yields azaleatin [3,7-dihydroxy-2-(3,4-dihydroxyphenyl-5-methoxy chromone, 7,3',4'-trihydroxy-5-methoxyflavonol] 10 ($R^1 = Me$, $R^2 = H$) (P. Pachaly and H.L. Jan, Arch. Pharm. (Weinheim, Ger.), 1994, 8 (327), 535).

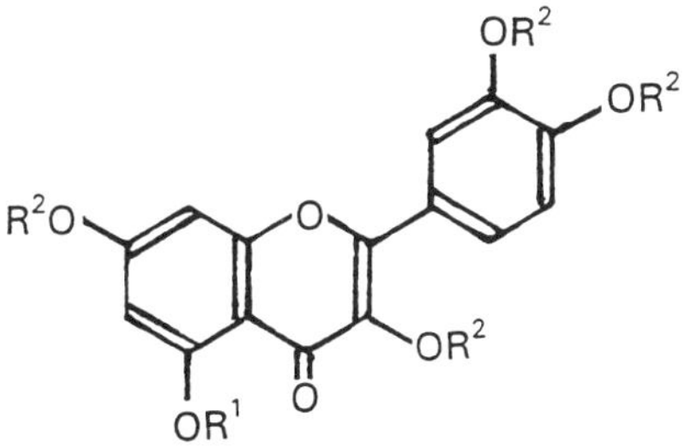

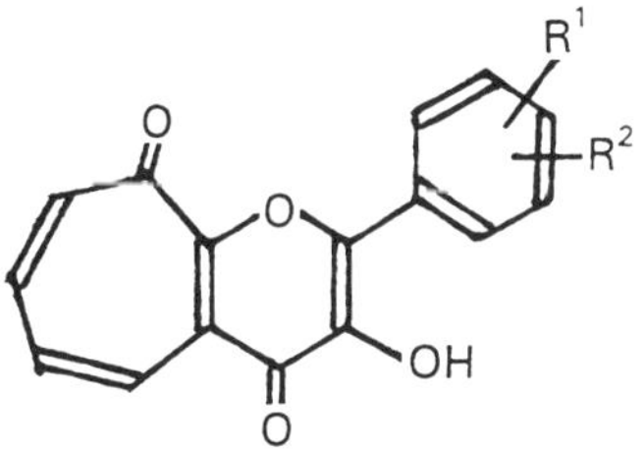

(10)

(11)

The reaction between 3-(2-bromoacetyl)tropolone and $R^1R^2C_6H_3CHO$ (R^1 = 4-MeO, 4-NMe$_2$, R^2 = H; R^1 = 2-OH, R^2 = H, 4-Me; R^1 = MeO, R^2 = 4-MeO, 4-OH; R^1R^2 = 3,4-OCH$_2$O) in the presence of alkali at low temperature affords the arylhydroxycyclohepta[*b*]pyrandiones 11 (Y. Jin, Z. Jin, and K. Imafuku, J. Heterocyclic Chem., 1991, <u>28</u>, 817).

(xvii) Isoflavones (3-phenylchromones, 3-phenyl-4H-benzo[b]pyran-4-ones)

Substituted isoflavones and 3-heteroarylchromones 1 [R^1 = H, lower alkyl, lower alkoxy, CO$_2$H; R^2 = (un)saturated Ph, 3,4-ethylenedioxyphenyl, 3,4-methylenedioxyphenyl, 5-ethoxycarbonylfur-2-yl, benzofuryl, 1-phenyl-pyrazol-4-yl] have been prepared by the reaction of *o*-substituted 2-hydroxy-acetophenone with HCO$_2$Ac in the presence of a tertiary alkylamine (V.P. Khilya, V.G. Pivovarenko, and N.V. Gorbulenko, U.S.S.R. SU 1,333,674, 1987).

(1)

4-Hydroxy-3-(2,5-dihydroxyphenyl)coumarins are cleaved by KOH to yield deoxybenzoins 2 (R^1 = OH, H; R^2 = H, Me; R^3 = Me, H, Cl), which cyclocondense with HCO$_2$Et to afford isoflavones 3. Some of the derivatives of 2 and 3 showed bactericidal and fungicidal activity (B.S.U. Rani and M. Darbarwar, I. Indian Chem. Soc., 1987, <u>64</u>, 555).

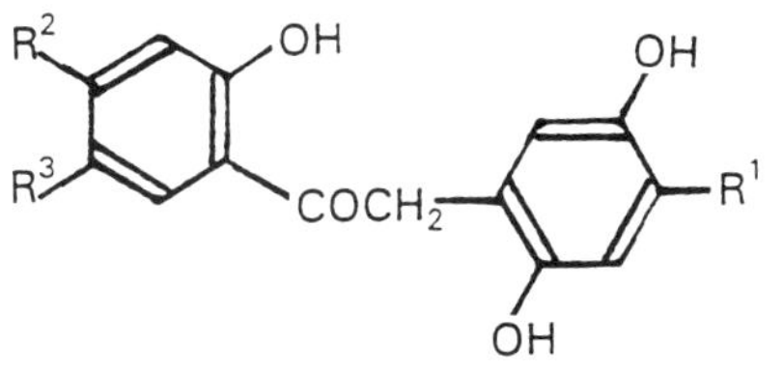

(2) 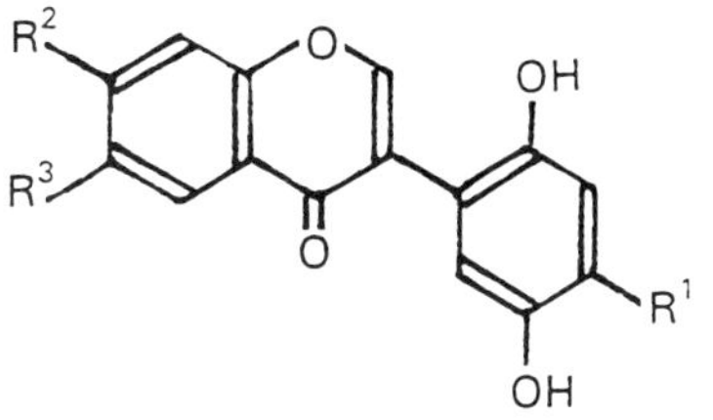   (3)

The conversion of 2-(carboxyalkoxy)benzils 4 ($R^1 = R^2 = R^3 = H$, OMe; $R^4 = H$, Me; $R^1 = R^3 = H$, $R^2 = $OMe, $R^3 = H$, Me) to the corresponding phenoxyketenes leads to an intramolecular [2 + 2] ketene cycloaddition, to ultimately give isoflavones 5 with up to four substituents in the 2-, 7-, 2'- and 4'-positions (OMe, Me) and/or 3-aroylbenzofurans 6 (W.T. Brady and Y.Q. Gu, J. Org. Chem., 1988, 53, 1353).

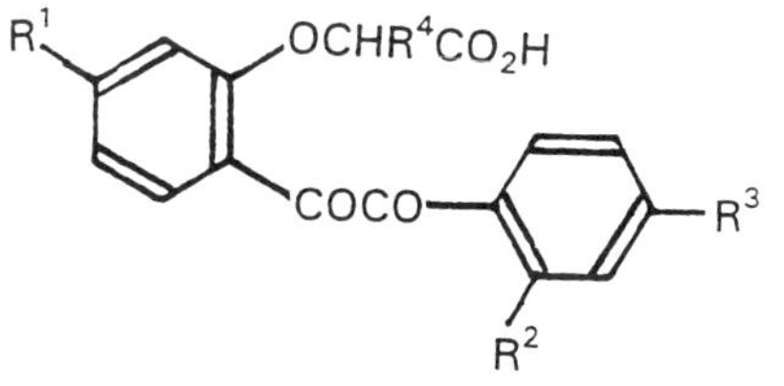

(4) 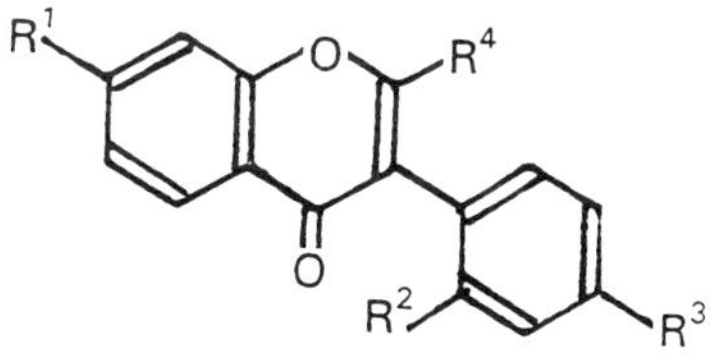   (5)

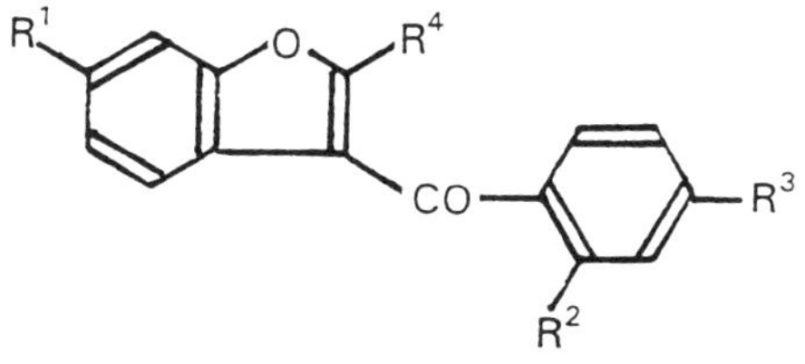

(6)

398

The reaction of isoflavones 7 [R^1 = H, 7-MeO, 7-AcO; R^2 = H, Me, Ph; R^3 = Ph, 2,4-(MeO)C_6H_3, 3,4-(MeO)$_2C_6H_3$, 4-$O_2NC_6H_4$, 4-MeOC_6H_4; X = O] and 3-benzylchromones 7 (R^3 = NC$C_6H_4CH_2$, 4-$O_2NC_6H_4CH_2$) with Lawesson's reagent in hot anhydrous toluene affords the corresponding 4-
-thioisoflavones 7 (X = S) and 3-benzyl-4H-benzo[b]pyran-4-thiones (A. Levai, J. Chem. Res., S, 1992, 163).

(7)

(8)

4'-Hydroxy-7-methoxyisoflavone (8) and 7-hydroxy-4'-methoxyisoflavone are prepared by condensation of the corresponding deoxybenzoin with DMF dimethylacetal. 7-Methoxy-2-methyl-4'-hydroxyisoflavone has also been prepared (Q.E. Ji and Y.L. Wei, Yaoxue Xuebao, 1989, 24, 906). For the preparation of 7-hydroxyisoflavone, see V.G. Pivovarenko and V.P. Khilya, Khim. Geterotsikl. Soedin., 1992, 595; and for 6-bromoisoflavone, B. Yue, Z. Zhou, and M. Cai, Gaodeng Xuexiao Huaxue Xuebao, 1990, 11, 99.

The irradiation in the presence of air of 2-(2-furylvinyl)-7-methoxyiso-
flavone [9; R = (E)-2-furylethenyl] in benzene containing I_2 results in photochemical oxidation to yield furylmethoxybenzoxanthone (10), 2-formyl-
-7-methoxyisoflavone (9; R = CHO) and furfural (M.S. Reddy, G.L.D. Krupadanam, and G. Srimannarayana, Indian J. Chem., 1991, 30B, 613).

(9)

(10)

2,8-Dimethyl-7-(1,1-dimethylprop-2-ynyloxy)isoflavone (11) on heating in dimethylaniline gives cyclopentabenzopyrandione 12. Other related 7-(1,1--dimethylprop-2-ynyloxy)isoflavones have been treated in a similar fashion (S.C. Joshi and K.N. Trivedi, Tetrahedron, 1992, $\underline{48}$, 563).

(11)

(12)

2-Hydroxy-4-*iso*propoxyacetophenone (13) on treatment with $HC(OEt)_3$ and morpholino in DMF affords a new fish-feed supplement, 7-*iso*propoxyiso-flavone (14) (C. Shi and S. Wang, Zhongguo Haiyang Yaowu, 1990, $\underline{9}$, 10).

(13)

(14)

2-Aminoisoflavones 15 (R^1 =H, Cl, SO_2NMe_2; NR_2^2 =NMe_2, NEt_2, morpholino) have been synthesized and 2-amino-6-chloro-3'-diethylamino-methyl-4'-hydroxyisoflavone (15; R^1 = Cl, R_2^2 =NEt_2) showed antihypoxia activity in mice (Y. Li and Q. Ji, Yaoxue Xuebao, 1987, $\underline{22}$, 655). The 3'-

400

-(substituted aminomethyl)-4'-hydroxy-7-methoxy- and 3'-(substituted aminomethyl)-4'-hydroxyl-7-methoxy-2-methyliso flavones 16 [R^1 = H, Me; R^2 = NMe$_2$, N(CH$_2$CH$_2$OH)$_2$, pyrrolidino, piperidino, piperazino] have been prepared from isoflavone 16 (CH$_2$R^2 = H) by the Mannich reaction. Derivative 16 (R^1 = H, R^2 = piperazino) showed antiarrhythmic and antihypoxic activities (H. Zhu, Q. Ji, and W. Xiao, Zhongguo Yiyao Gongye Zazhi, 1990, 21, 402).

(15)

(16)

(17)

Several 2-alkyl-7-[3-(substituted amino)-2-hydroxypropoxy]isoflavones 17 (R^1 = Pr, Pr^i, cyclooctyl; R^2 = H, Me, CF$_3$, Pr^i, CH$_2$Ph, fur-2-yl, cyclohexyl) have been prepared in order to study, the importance of the Ph group at the 3-position in promoting antihypertensive activity and the substitution effects at the 2-position of isoflavone (E.S.C. Wu et al., J. Med. Chem., 1992, 35, 3519).

(xviii) Furoisoflavones

The Friedel-Crafts acylation of 6-methoxy-2,3-diphenylbenzofuran with PhCH$_2$COCl in CS$_2$ gives a mixture of the 5-phenylacetyl and 5,7-di(phenyl-acetyl) derivatives, while in C$_6$H$_6$ only the former derivative is obtained. Reaction of either of the acylation products with HC(OEt)$_3$ yields 6,7-di-

phenylfuro[3,2-*g*]isoflavone (1; $R^1 = R^2 = H$) and 9-benzylcarbonyl-6,7-di-
phenylfuro[3,2-*g*]isoflavone (1; $R^1 = COCH_2Ph$, $R^2 = H$), respectively. The
Claisen condensation of 6-methoxy-5,7-di(phenylacetyl)-2,3-diphenylbenzo-
furan with AcOEt affords 9-benzylcarbonyl-2-methyl-6,7-diphenylfuro[3,2-*g*]-
isoflavone (1; $R^1 = COCH_2Ph$, $R^2 = Me$) (O.H. Hishmat *et al.*, Rev. Roum.
Chim., 1988, 33, 741).

(1)

4. Chroman, dihydrochromene, 3,4-dihydro-2*H*-benzo[*b*]pyran and
   derivatives

(a) Chromans

Alkyl- and arylchromans. For studies of a new synthesis of chromans,
see L.A. Shervington, Diss. Abstr. Int. B, 1988, 49, 1184. Treatment of
chroman (1) with Tebbe reagent, obtained by reacting titanocene dichloride
with Me_3Al, in PhMe-THF affords 2-methylenechroman (2) (S.H. Pine, G.
Kim, and V. Lee, Org. Synth., 1990, 69, 72).

(1)   (2)

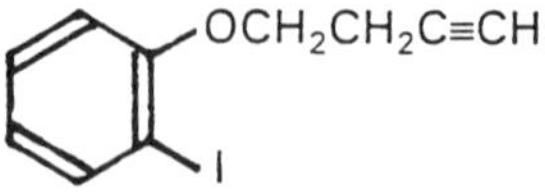

402

(3) 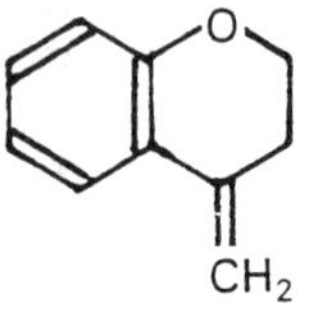   (4)

The Cr(II)-mediated Nickel(II)-catalyzed cyclization of the
iodoaryloxyalkyne 3 gives 4-methylenechroman (4) (D.M. Hodgson and C.
Wells, Tetrahedron Letters, 1994, $\underline{35}$, 1601). The [(η^3-pinene) PdOAc]$_2$
catalyzed cyclization of 2-(pent-3-enyl)phenol yields optically active (+)-2-
-vinylchroman, while (-)-2-(1-hydroxyethyl)chroman (56% e.e.) is formed as
a single diastereomer upon treatment with Me$_3$COOH in the presence of
Ti(OPr^i)$_4$ and L-(+)-diethyltartrate (T. Hosokawa et al., J. Organomet.
Chem., 1989, $\underline{370}$, C13).

The reaction between phenol and isoprene in the presence of aluminium
phenolate in the ratio 3:1:0.1 at 150°C yields 2,2-dimethylchroman (5)
(92%) and three higher-order products [S.A. Butov, Ya.B. Kozlikovskii, and
V.A. Koshchii, Ukr. Khim. Zh. (Russ. Ed.), 1992, $\underline{58}$, 1124]. The
photochemical or AlCl$_3$-catalyzed Fries rearrangement of 4-(4-hydroxy-
phenyl)-2,2,4-trimethylchroman derivatives 6 (R=H, Cl) gives 4-(3-acetyl-4-
-hydroxyphenyl)- and 4-(3-chloroacetyl-4-hydroxyphenyl)-2,2,4-trimethyl-
chroman (7; R=H, Cl), respectively (S.K. Dhande, S.P. Samant, and B.D.
Hosangadi, Indian J. Chem., 1988, $\underline{27B}$, 814). A number of 4-(4-
-hydroxyphenyl)-2,2,4-trimethylchroman derivatives (V. Mark, E. Mark, and
C. Mark, U.S. US 4,749,799, 1988), 7-substituted 2,2-dimethyl-6-(2-alkyl-
phenylethynyl)chromans as retinoic acid like drugs (R.A.S. Chandraratna,
Eur. Pat. Appl. EP 290,130, 1988), and 7-substituted 2,2,4,4-tetramethyl-6-
-[2-(disubstituted phenyl)ethynyl]chromans (idem, PCT Int. Appl. WO 92
14,725, 1992) have been prepared.

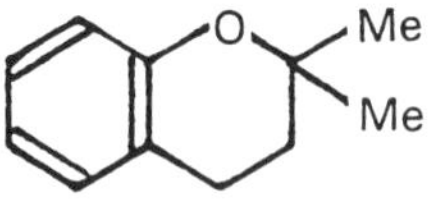

(5) 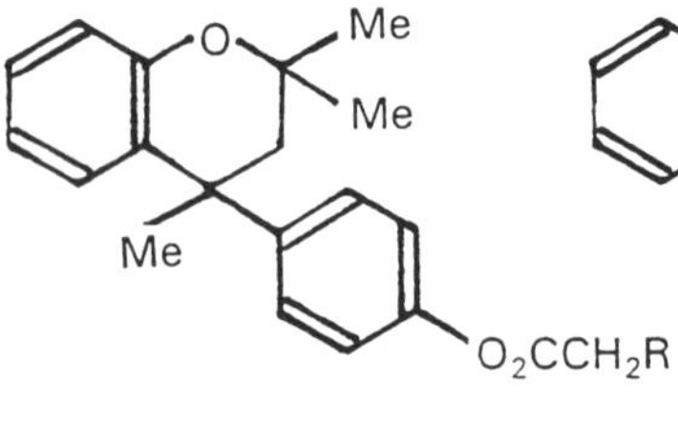   (6) 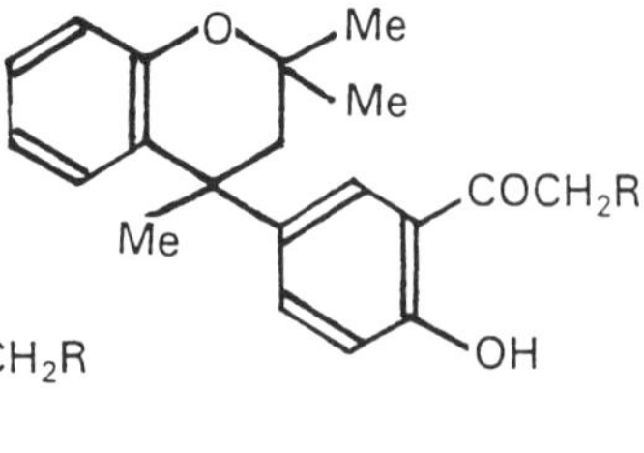   (7)

The catalytic hydrogenation of 4-(4-hydroxyphenyl)-7-methoxy-2,2-
-dimethyl-3-phenylchromene in THF over Pd/C at 1000 psi and 60-70°C
affords a mixture of *cis*-7-methoxy-2,2-dimethyl-4-(4-oxocyclohexyl)-3-
-phenylchroman (8) and *cis*-4-(4-hydroxycyclohexyl)-7-methoxy-2,2-dimethyl-
-3-phenylchroman along with the major product, 4-(4-hydroxyphenyl)-7-
-methoxy-2,2-dimethyl-3-phenylchroman (9) (I. Sharma, S. Ray, and A.H.
Ansari, Indian J. Chem., 1988, 27B, 516).

(8)   (9)

4-[2-(4,4-Dimethylchroman-6-yl)prop-1-enyl]benzoic acid (10) and its
methyl and ethyl esters have been prepared as anticancer agents (K.D. Berlin
et al., U.S. US 4,826,984, 1989). Potential metabolites of ethyl (*E*)-4-[2-
-(4,4-dimethylchroman-6-yl)prop-1-enylbenzoate have been synthesized and
include the 6-substituted 4,4-dimethylchromans 11 [R = (*E*)-CMe = CHCO$_2$Et,
(*E*)-CMe = CHCO$_2$H, CO$_2$H, (*E*)-HOCH$_2$C = CHC$_6$H$_4$CO$_2$Et-4, (*E*)-
OCHC = CHC$_6$H$_4$CO$_2$Et-4, (*E*)-HO$_2$CC = CHC$_6$H$_4$CO$_2$Et-4, 2,5-dihydro-2-oxo-
fur-4-yl] (P.S. Sunthanker *et al.*, J. Pharm. Sci., 1993, 82, 543). The
prepararation of various derivatives of 4-[2-(4,4-dimethylchroman-6-yl)-
ethynyl]benzoic acid (12) and of some of its esters as retinoate analogues
(R.A.S. Chandraratna, U.S. US 4,980,369, 1990; Eur. Pat. Appl. EP
419,132, 1991; PCT Int. Appl. WO 92 06,084, 1992) and of 6-(substituted
buta-1,3-dienyl)-4,4-dimethylchroman useful in cosmetics and medicinal
compositions (G. Lang *et al.*, Brit. UK Pat. Appl. GB 2,188,634, 1987) have
been reported.

(10)

(11)

(12)

(13)

Some chromans 13 (R^1 = H, CH_2OH, Ph, 2-FC_6H_4, CH_2Ph; R^2 = H, CH_2OH, Ph, 4-$O_2NC_6H_4$, 4-$MeOC_6H_4$) substituted at the 3- or/and 4-position have been obtained by [Rh η^5-C_5EtMe_4) (η^6-C_6H_6)] [PF_6]$_2$ catalyzed, intramolecular nucleophilic substitution reactions, from 2-$FC_6H_4CHR^1CHR^2$-CH_2OH in $MeNO_2$-Me_2CO (R.P. Houghton and L.A. Shervington, J. Chem. Res., S, 1989, 239). A number of 4-benzoquinonylchromans have been prepared and investigated for use in the treatment of cerebral disorders (T. Tatsuoka *et al.*, Eur. Pat. Appl. EP 322,248, 1989). 6-Hydroxy-2,2-
-dimethyl-, 6-hydroxy-2,2,5,8-tetramethyl-, 2,2,5,7,8-pentamethyl-, and 6-
-hydroxy-2,5,7,8-tetramethyl-2-phytylchromans and related derivatives have been prepared by reacting the appropriate substituted phenol with the necessary disubstituted allylic alcohol in a trifluoroacetic acid/water mixture under argon at room temperature (F.M.D. Ismail, M.J. Hilton, and M. Stefinovic, Tetrahedron Letters, 1992, **33**, 3795). Several 4-aryl- and 4-(*N*-
-heteroaryl)chroman derivatives have been prepared as potassium channel activators (M.R. Attwood, P.S. Jones, and S. Redshaw, Eur. Pat. Appl. EP 298,452, 1989).

Acylchromans and chromancarboxaldehydes. 2*R*-Benzyl- and 2*R*-
-hydroxymethylchroman (14; R = CH$_2$Ph, CH$_2$OH) and (chroman-2-yl)methyl
triflate (14; R = CH$_2$OSO$_2$CF$_3$) have been prepared and the formylation of
2*R*-benzylchroman by *N*-methylformanilide in the presence of POCl$_3$ afforded
2*R*-benzylchroman-6-carboxaldehyde (15) (F.J. Urban Eur. Pat. Appl. EP
448,254, 1991). The synthesis of 8-hydroxychroman-6-carboxaldehyde
(16) has been reported (J.A. Stafford and N.L. Valvano, J. Org. Chem.,
1994, <u>59</u>, 4346).

(14)   (15)   (16)

2,2,4,6-Tetra substituted 3-benzoylchromans 17 (R^1 = H, Me, Ph;
R^2 = alkyl, vinyl, Ph, anisyl, O$_2$NC$_6$H$_4$, furyl, PhCH = CH; R^3 = Me, Ph; R^4 = H,
NO$_2$) have been prepared by the reaction of appropriate aldehydes and
ketones with 3-R^{4}C$_6$H$_4$C(OH)R^{3}C$\equiv$CPh. The preparation of 3-benzoyl-4-
-phenyl-2.2-spirocyclopentylchroman [17; R^{1}R^2 = (CH$_2$)$_5$, R^3 = Ph, R^4 = H] has
been reported (E.A. Adegoke *et al.*, J. Heterocyclic Chem., 1987, <u>24</u>, 1705).
Friedel-Crafts acylation of 4,4-dimethylchroman with 4-(MeO$_2$C)C$_6$H$_4$COCl in
ClCH$_2$CH$_2$Cl with AlCl$_3$ catalyst, followed by hydrolysis of the resulting ester
with KOH in boiling EtOH, yields 6-(4-carboxybenzoyl)-4,4-dimethylchroman
(18). A number of related derivatives have been prepared and formulated as
antiproliferative agents, especially for dermatological use (J. Maignan *et al.*,
Belg. BE 1,000,195, 1988).

(17)   (18)

406

6-Acetylchroman undergoes a condensation reaction with 4-methoxybenzaldehydes and hydrogenation of the products afford the 6-[3--(substituted 4-methoxyphenyl)propionyl]chroman derivatives **19** (R^1 = Ph, Me; R^2 = H, Me; R^3 = Me, H, OMe; R^4 = OH, H; R^5 = H, OH; R^6 = H, OMe), which showed usefulness as bactericides and fungicides (V.K. Ahluwalia *et al.*, Indian J. Chem., 1987, <u>26B</u>, 384). Dehydration of acetylalkyl-, alkyl-, and alkylhydroxyphenols **20** (R = Ac, H, OH) gives the substituted 2,2-dimethylchromans **21** (M. Tsukayama, M. Kikuchi, and Y. Kawamura, Heterocycles, 1994, <u>38</u>, 1487).

(19)

(20)

(21)

Chromancarboxylic acids. Ethyl chroman-2-carboxylate has been used as a starting material for the synthesis of the hypoglycemic agent, englitazone. An alternative synthesis has also been reported (F.J. Urban and B.S. Moore, J. Heterocyclic Chem., 1992, <u>29</u>, 431). An efficient synthesis of the leukotriene antagonist ablukast sodium **22** has been achieved starting from 2,4-dihydroxyacetophenone (P.S. Manchand, R.A. Micheli, and S.J. Saposnik, Tetrahedron, 1992, <u>48</u>, 9391). The preparations of several related compounds (S.W. Djuric, R.L. Shone, and S.S.T. Yu, Eur. Pat. Appl. EP 292,977, 1988; N. Cohen *et al.*, J. Med. Chem., 1989, <u>32</u>, 1842; A.J. Laurenzano and J.J. Partridge, Eur. Pat. Appl. EP 336,068, 1989; P.S. Manchand and R.A. Micheli, *ibid.*, EP 355,617, 1990) have been reported. A thiophene analogue **23** of leukotriene D_4 antagonist Ro23-3544 has been synthesized but it showed almost no effect on leukotriene D_4-induced bronchostriction in anesthetized guinea pigs [D. Binder *et al.*, Arch. Pharm. (Weinheim, Ger.), 1992, <u>325</u>, 797).

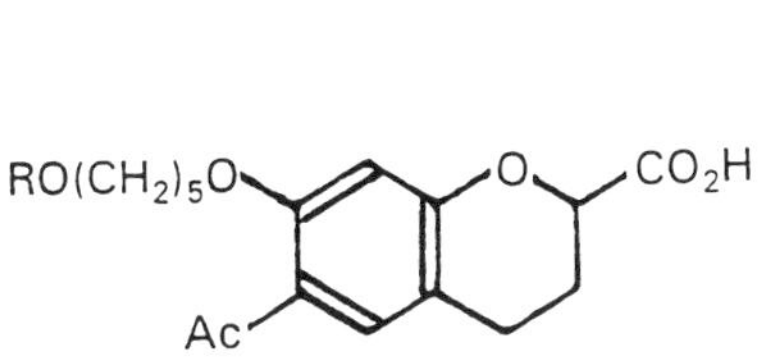

RO(CH₂)₅O— ... —CO₂H

Ac

(22) R 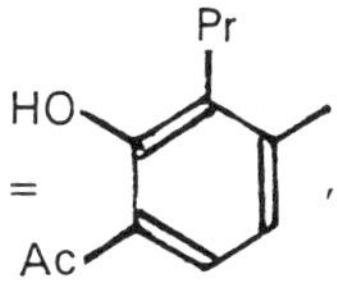 = , H = Na

HO

Pr

Ac

(23) R =

Ac S

HO Pr

RO

Pr

Ac

O O CO₂H

Pr

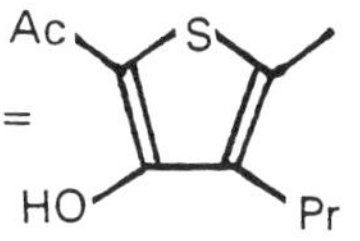

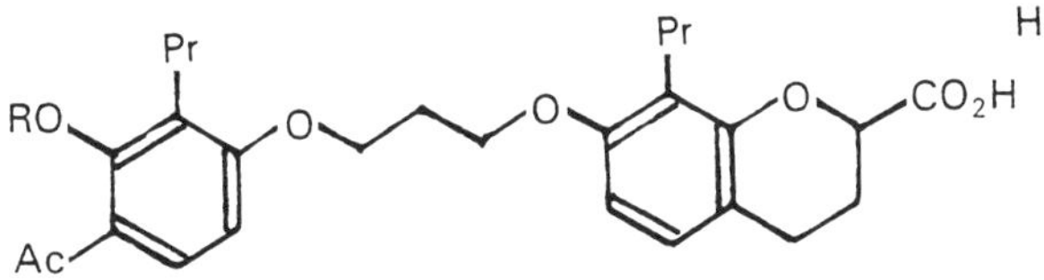

(24)

7-[3-(4-Acetyl-3-methoxy-2-propylphenoxy)propoxy]-8-propylchroman-2-
-carboxylic acid (24; R = Me) has been prepared by the sequential coupling
of the two aromatic groups with $Cl(CH_2)_3Br$ and it was the first orally active,
selective leukotriene B_4 receptor antagonist. The related compound 24
(R = H) was inactive (S.W. Djuric *et al.*, J. Med. Chem., 1989, **36**, 1145). 7-
-{3-[2- -(Cyclopropylmethyl)-4-[(methylamino)carbonyl]phenoxypropoxy}-8-
-propylchroman-2-propanoic acid (25; R = H) a novel high potency
leukotriene B_4 receptor antagonist SC-53228 (S.W. Djuric *et al.*, Bioorg.
Med. Chem. Letters, 1994, **4**, 811), and (+) and (-)-7-{3-[2-(cyclopropylmethyl)-3-methoxy-4-[(methylamino)carbonyl]-phenoxypropoxy}-8-propyl-
chroman-2-propanoic acid (25; R = OMe) leukotriene B_4 antagonists, useful
as antiinflammatory agents and in treating disease conditions mediated by
LTB_4 (S.W. Djuric, S.H. Docter, and S.S. Yu, U.S. US 5,310,951, 1994)
have been synthesized. The effect of a chromancarboxylic acid side chain
on ortho-alkoxyphenyl leukotriene B receptors has been discussed (M.J.
Sofia *et al.*, Bioorg. Med. Chem. Letters, 1992, **2**, 1675). For the
preparation of the individual enantiomers of 6-methoxy-2,5,7,8-tetramethyl-
chroman-2-carboxylic acid, see T. Netscher and I. Gautschi, Ann., 1992,
543; for the preparation of 7,8-dichloro-6-(thien-2-ylcarbonyl)chroman-2-
-carboxylic acid a structural analogue of tienilic acid, G. De Saqui-Sannes *et
al.*, Helv., 1992, **67**, 172; and of chroman-2-carboxylic acid and related
compounds with a complex C-6 side chain, for liquid crystal compositions, K.
Tsucha *et al.*, Jpn. Kokai Tokkyo Koho JP 05 25,158 [93 25,158], 1993.

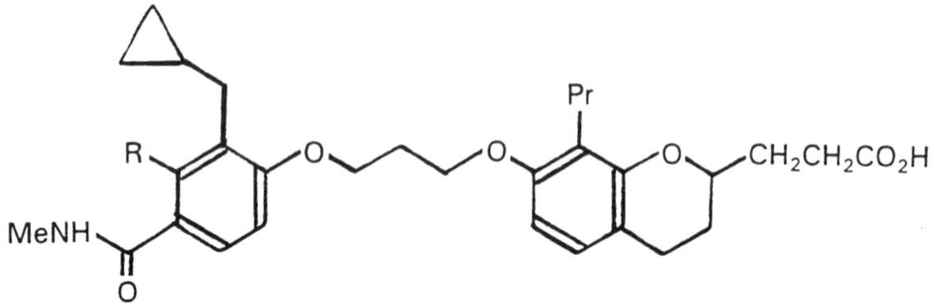

(25)

4-Bromo-4-methyl- and 4-bromo-8-methoxy-4-methyloxepin-5-one (26; R=H, OMe, respectively) on heating with $ZnCl_2$ in MeOH afford methyl 4--methylchroman-4-carboxylate (27; R=H) and methyl 7-methoxy-4-methyl--4-carboxylate (27; R=OMe), respectively. Similarly the related acetals 28 (R=H, OMe) when heated in a protic solvent give the corresponding methyl chroman-4-carboxylates 29 (S.B. Maiti, S.R.R. Chaudhuri, and A. Chatterjee, Synthesis, 1987, 806).

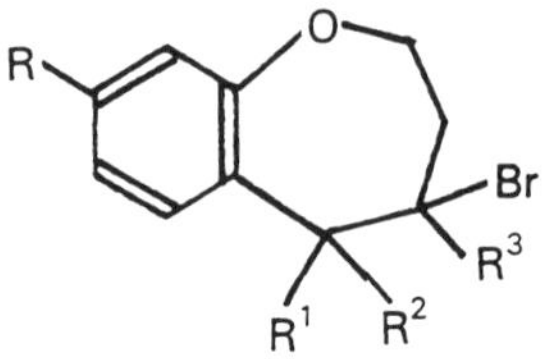

(26) $R^1R^2 = O$, $R^3 = Me$
(28) $R^1 = R^2 = OMe$, $R^3 = H$

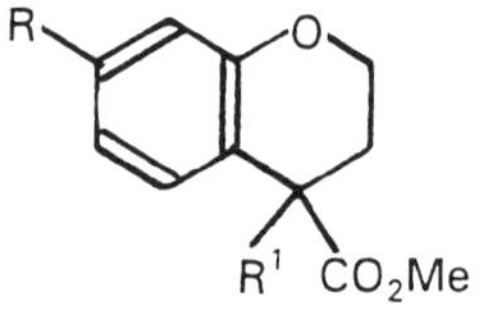

(27) $R^1 = Me$
(29) $R^1 = H$

$4,2\text{-}Me(HO)C_6H_3CMe_2CH_2CMe_2OH$ has been cyclized with H_2SO_4 and the product converted in 2 steps to 2,2,4,4,7-pentamethylchroman-6-car-boxylic acid (30; R=H), which has been esterified with $4\text{-}HOC_6H_4CO_2Et$ to

yield 4-ethoxycarbonylphenyl 2,2,4,4,7-pentamethylchroman-6-carboxylate
(30; $R = C_6H_4CO_2Et$). The preparations of related phenyl chromancarboxy-
lates and analogues having retinoid activity have been reported (R.A.S.
Chandraratna, U.S. US 5,006,550, 1991). For the preparation of (1-butyl-
piperid-4-yl)methyl 6-chlorochroman-8-carboxylate 31 and related derivatives
as 5-HT4 antagonists, see F.D. King, L.M. Gaster, and G.F. Joiner, PCT Int.
Appl. WO 93 16,072, 1993; and 4-ethoxycarbonylphenyl 2,2,4,4-tetra-
methylchroman-7-carboxylate (32), which has retinoic acid-like activity,
R.A.S. Chandraratna, U.S. US 5,134,159, 1992.

(30)

(31)

(32)

Halogenochromans. The reaction of 4-bromo-6-cyano-3,3-(ethylenedi-
oxy)-2,2-dimethylchroman (33) with secondary amide anions furnishes C-5
products, for example, with AcNMe⁻K⁺ 5-acetamido-6-cyano-3,3-(ethylene-
dioxy)-2,2-dimethylchroman (34) is obtained by a S_N2' reaction followed by
a hydride shift (C.S.V. Hoage-Frydrych, A. Marsh, and I.L. Pinto, Chem.
Comm., 1989, 1258). A stereoselective synthesis of 3-bromo-4-(2-bromo-4-
-methylphenoxy)chroman (35) has been reported (K.C. Santosh and K.K.
Balasubramanian, Synth. Comm., 1994, _24_, 1049).

410

(33) R^1 = H, R^2 = Br
(34) R^1 = MeNAc, R^2 = H

(35)

Perfluoroindan-1-one in the electrophilic system H_2O_2-HF-SbF$_5$ undergoes rearrangement to yield perfluorochroman (36) and perfluorocarboxyethyl-phenol (37) (I.P. Chuikov, V.M. Karpov, and V.E. Platonov, Izv. Akad. Nauk SSSR, Ser. Khim., 1990, 2463).

(36)

(37)

Amino-, cyano-, and nitrochromans and chromancarboxamides. (3*S*,4*R*)--6-Cyano-3-hydroxy-2,2-dimethyl-4-(trimethylammonium)chroman iodide on treatment with Bu^{t}OK in THF affords (3*S*,4*S*)-6-cyano-3,4-epoxy-2,2-di-methylchroman (P. Smith and M. Ghavshou, Eur. Pat. Appl. EP 430,621,1991).

The preparation of several 6-cyano-3-hydroxy-2,2-dimethylchromans with a derivative of an amino group in the 4-position and possessing anti-

hypertensive and coronary vasodilating activities have been reported, including the following compounds; 6-cyano-4-(1,2-diacetylhydrazino)-3-
-hydroxy-2,2-dimethylchroman (1) and related derivatives (T. Yamanaka *et al.*, Eur. Pat. Appl. EP 339,562, 1989); *N*''-cyano-*N*-(6-cyano-3-hydroxy-
-2,2-dimethylchroman-4-yl)-*N*'-methylguanidine (2) and analogues (G. Stemp, G. Burrell, and D.G. Smith, *ibid.*, 359,537, 1990; K. Atwal, G.T. Grover, and K.S. Kim, U.S. US 5,140,031, 1992); it has been reported that the electronwithdrawing group at the C-6 position is required for anti-ischemic activity (K.S. Atwal *et al.*, J. Med. Chem., 1993, <u>36</u>, 3971) and cromakalim analogues have been prepared and investigation carried out into whether K channel openers cromakalim and pinacidl share common pharmacophoric features (*idem*, Bioorg. Med. Chem. Letters, 1992, <u>2</u>, 87). The methylthio-amidino derivatives 3 and related derivatives (R. Tsuzuki *et al.*, Jpn. Kokai Tokkyo Koho JP 02 172,984 [90 172,984], 1990); and other related compounds with a 4-amidino group containing substituents (S. Katoh *et al.*, PCT Int. Appl. WO 92, 19611, 1992) have also been prepared. The preparations of 6-cyano-3-hydroxy-2,2-dimethyl-4-(3-oxocyclopent-1-en-1-
-yl)oxy chroman and related compounds as antihypertensives, spasmolytics, and as hair growth stimulators have also been described (S. Blarer, Brit. UK Pat. Appl. GB 2,204,863, 1988).

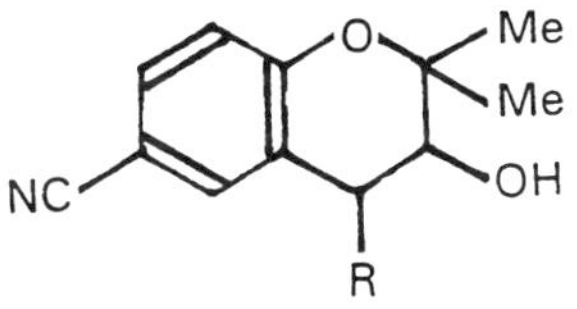

(1) R = NAcNHAc
(2) R = NHC(= NCN)NHMe
(3) R = NHC(SMe) = NCN

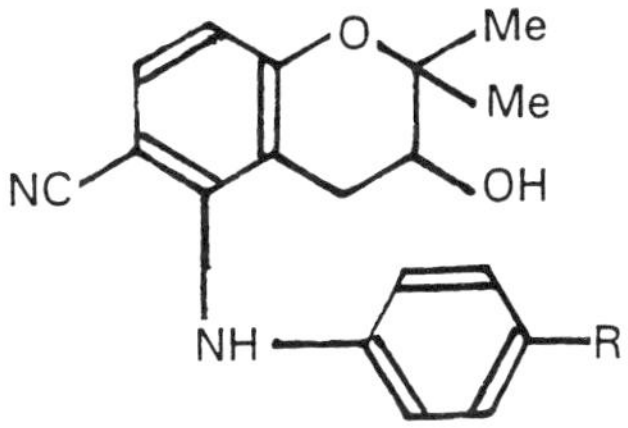

(4)

A number of 4-arylamino-6-cyano (halogeno, MeCO, CF_3CO)-3-hydroxy-
-2,2-dimethylchromans and related chromens and analogues, for example, 4-
-(4-chlorophenyl)-6-cyano-3-hydroxy-2,2-dimethylchroman (4; R = Cl) and 4-
-(4-trifluoromethylphenyl)-6-cyano-3-hydroxy-2,2-dimethylchroman (4; R = CF_3) have been prepared (G. Garcia, A. Di Malta, and P. Soubrie, Eur. Pat. Appl. EP 370,901, 1990). 3-Amino-6-cyano-4-hydroxy-2,2-dimethyl-chroman condenses with Bu^{t}NCO to give *N*-(6-cyano-4-hydroxy-2,2-di-methylchroman-3-yl)-*N*'-tert-butylurea (5) (G. Stemp, *ibid.*, 375, 449, 1990).

The preparations of many 6-cyano-3-hydroxy-2,2-dimethyl-chromans containing a 5 or 6 membered heterocyclic system linked by the ring nitrogen at the 4-position of the chroman, as cardiovascular agents have been reported, for example, 6-cyano-3-hydroxy-4-(3-hydroxy-2-oxopyrolidin--1-yl)-2,2-dimethylchroman (6) (F. Cassidy and E.A. Faruk, *ibid.*, 274,821, 1988). Also, the preparations of related derivatives containing a different group in the C-6 position to the cyano group have been described.

(5)

(6)

The preparations of 4-cyano-4-{2-[2-(3,4-dimethoxyphenyl)ethylmethyl-amino]ethyl}chroman and 4-cyano-4-{2-[2-(3,4-dimethoxyphenyl)ethyl-methylamino]ethyl}-3-methylchroman (7) and the related thiochroman and tetralene derivatives have been reported and tested for acute toxicity and calcium antagonist activity (G. Butora *et al.*, Cesk. Farm., 1990, <u>39</u>, 289; Czech. CS 266,192, 1990).

(7)

The preparation of four optically active isomers 8 and 9 of nipradilol (R^1 =H, R^2 = ONO_2; R^1 = ONO_2, R^2 =H), has been described (M. Shiratsuch *et al.*, Chem. Pharm. Bull., 1987, 35, 3691; K. Kawamura, T. Ohta, and G. Otani, *ibid.*, 1990, 38 2092; K. Kawamura and T. Ohta, Jpn. Kokai Tokkyo Koho JP 05 70,452 [93, 70,452], 1993).

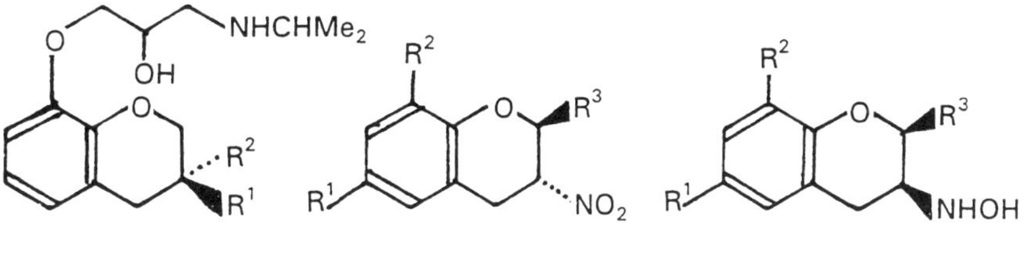

(8) *S*-OH
(9) *R*-OH

(10)

(11)

3-Nitrochromans 10 [R^1 =H, Cl, NO_2; R^2 =H, OMe; R^3 = Ph, $(EtO)_2C_6H_3$, naphthyl] have been obtained by the $NaBH_4$ reduction of the corresponding 3-nitrochromenes in MeOH-THF. The hydroxylamine analogues 11 [R^3 = Ph, $Me_2CHC_6H_4$, $(EtO)_2 C_6H_3$, naphthyl] have been obtained from nitrochromenes and BH_3-THF complex in THF containing $NaBH_4$ (R.S. Varma, Y.Z. Gai, and G.W. Kabalka, J. Heterocyclic Chem., 1987, 24, 767).

The preparation of 4-{2-[4-(3,4-dimethoxyphenyl)piperidin-1-yl]ethyl-chroman *N*-oxide (12) and related derivatives as antiarrhythmics and antifibrillants have been reported (M. Barreau, J.C. Hardy, and C. Renault, Eur. Pat. Appl. EP 379,440, 1990). 2,2-Dimethyl-6-nitrochroman-4-yl carboxylates 13 (R = Me, Ph, $4-O_2NC_6H_4$, $4-MeOC_6H_4$, $ClCH_2$, PhCH = CH, $PhCH_2$) and the dicarboxylates 14 (R = Me, Et, Ph, $4-O_2NC_6H_4$) have been prepared, starting from 3-nitro-6-hydroxyacetophenone (Z. Yan *et al.*, Zhongguo Yaoke Daxue Xuebao, 1993, 24, 129).

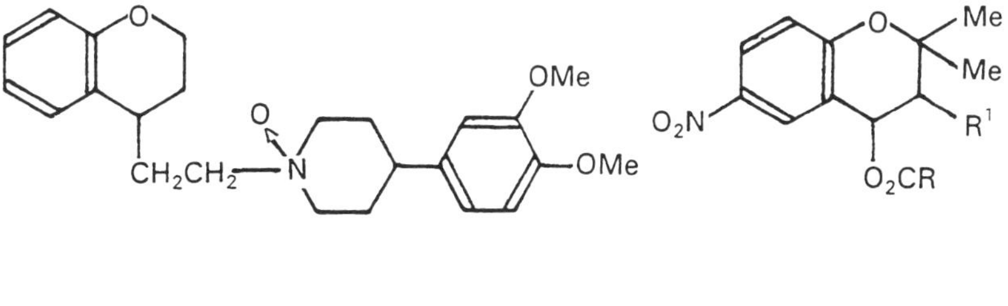

(12)

(13) R^1 =H
(14) R^1 = O_2CR

414

3-Acetamidochroman on treatment with succinic anhydride in
$CICH_2CH_2CI$ containing $AICl_3$ affords a chromanyloxobutyric acid, which on
heating with aqueous HCl, followed by esterification with acidic MeOH gives
methyl 4-(3-aminochroman-6-yl)-4-oxobutyrate (15). The latter compound
has been converted into the related 3-(4-chlorophenyl)sulphonamidochroman-
yloxobutyric acid, a platelet aggregation inhibitor and thromboxane A_2
antagonist (J. Nickl *et al.*, Ger. Offen. DE 3,743,793, 1989). For the
synthesis of tritium labelled 3-dipropylamino-5- and -8-methoxychromans as
D_2 receptors, see L. Pichat, Synth. Appl. Isot. Labelled Cpd., 1988, Proc. Int.
Symp., 1988 (Pub. 1989), 21, Ed. T.A. Baillie and J.R. Jones, Elsevier; 4-
-methyl-3-propylaminochroman, A.J. Davies *et al.*, Bioorg. Med. Chem.
Letters, 1992, 2,481; and 3-dipropylamino-6- and -8-hydroxychromans,
L.D. Wise *et al.*, J. Med. Chem., 1988, <u>31</u>, 688.

(15)

(16) $R^1 = R^2 = Pr$, $R^3 = OH$, $R^4 = H$
(17) $R^3 = H$, $R^4 = OH$; R^1, R^2 (see text)

3-Dipropylamino-8-hydroxychroman (16), an oxygen isostere of the
dopamine receptor agonist of 2-dipropylamino-5-hydroxytetralin, has been
synthesized in 7 steps from 8-methoxychroman-4-one (A.S. Horn *et al.*, Eur.
J. Med. Chem., 1988, <u>23</u>, 325). A number of *N,N*-disubstituted 3-amino-5-
-hydroxychromans 17 [R^1 = (un)saturated alkyl, (un)substituted phenyalkyl;
R^2 = H, alkyl; R^1R^2 = a 5- or 6-membered ring containing 1 or 2 N, O, and/or
S atoms] and their enantiomers and salts, having a therapeutic effect on the
central nervous system, have been synthesized (H.O. Hall, L.G. Johansson ,
and S.O. Thorberg, PCT Int. Appl. WO 88 04,654, 1988).

A method for the synthesis of enantiomers of 3-aminochroman
derivatives 18 [e.g., R^1,R^2 = H, alkyl; n = 1-6; R^3 = CN, (alkyl-, acyl-, alkyl-
sulphonyl-, arylsulphonyl-substituted) amino] and related 3-aminothio-
chromans with 5-HT_{1A} receptor activity (G. Guillaumet *et al.*, Eur. Pat. Appl.
EP 521,766, 1993) and 3-(*N*-isopropyl-*N*-propylamino)-5-(*N*-isopropyl)-

carbamoylchroman (19) as a selective 5-HT (5-hydroxytryptamine) receptor
stimulant (E.M. Hammarberg *et al.*, *ibid.*, 538,222, 1993) have been
reported. For the preparation of 8-bromo-3-(*N*-cyclopropylmethyl-*N*-propyl-
amino)chroman and derivatives, see B.R. Andersson *et al.*, PCT Int. Appl.
WO 90 12,795, 1990; and 3-[(imidoalkyl)amino]chromans and analogues as
5-HT$_{1A}$ receptor agonists, G. Guillaumet and B. Guardiola, Eur. Pat. Appl. EP
452,204, 1991; T. Podona *et al.*, J. Med. Chem., 1994, <u>37</u>, 1779.

(18)

(19)

The sulphur containing 3-dialkylamino-5-hydroxy(methoxy)chromans 20
[R^1 = SMe, R^2 = Me, R^3 = R^4 = Pr; R^1 = H, R^2 = H, Me, R^3 = (CH$_2$)$_2$SMe, R^4 = Pr]
have been prepared from 8-bromo-3-hydroxy-5-methoxychroman, which was
obtained from allyloxybromoanisole 21 by a thallium(III)-mediated ring
closure reaction. The preparation of 8-bromo-3-dipropylamino-5-methoxy-
chroman has been described (B. Andersson, H. Wikstroem, and A. Hallberg,
Acta Chem. Scand. 1990, <u>44</u>, 1024). A number of hydroxy and methoxy
derivatives of 3-dipropylchroman have also been synthesized and their
dopaminergic activity measured (M. Al Neirabeyeh *et al.*, Eur. J. Med.
Chem., 1991, <u>26</u>, 497). The synthesis of racemic monomethoxy- and
monohydroxy-3-[di(³H-propyl)amino]chromans 22 (R^1 = 5-, 6-, 7-, 8-MeO, 5-
-HO; R^2-R^4 = H, T), new radioligands for 5-HT$_{1A}$- and D$_2$-receptor-site
labelling have been reported (J.M. Cossery *et al.*, J. Labelled Compd.
Radiopharm., 1988, <u>25</u>, 833).

416

The reductive addition reaction of $H_2C=CHCO_2Et$ with the appropriate 3-
-nitrochroman derivative in the presence of $NaBH_4$ affords adducts, which
upon reductive cyclization yield the spiro(chroman)pyrrolidinones 23 ($R^1 = H$,
OMe; $R^2 = H$, Me, OMe) (S. Subramanian, U.R. Desai, and G.K. Trivedi,
Synth. Comm., 1990, 20, 1733). For the preparation of 2- and 3-(piperidin-
-1-yl)methylchromans, see G. Mouysset et al., Eur. J. Med. Chem., 1987,
22, 539); cis-4-hydroxy-3-(pyrid-3-yl)methyl-6-(quinol-2-yl)methoxychroman
and related compounds useful as lipoxygenase inhibitors and/or leukotriene
antagonists, J.F. Eggler, A. Marfat, and L.S. Melvin, Jr., Eur. Pat. Appl. EP
313,295, 1989; and 1-(3-chromanylalkyl)piperidine carboxylates, for
example, methyl 1-(chroman-3-yl)methyl-1,2,5,6-tetrahydropyridine-3-
-carboxylated [3-(3-methoxycarbonyl-1,2,5,6-tetrahydropyridin-1-yl)chroman]
(24) and related compounds as nootropics, W. Froestl and A. Zuest, U.S. US
4,957,928, 1990).

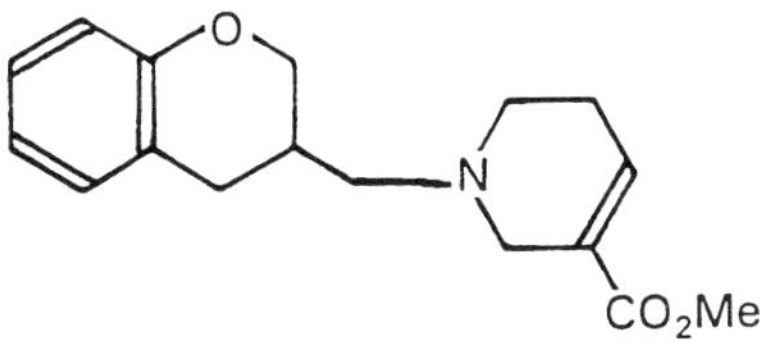

(24)

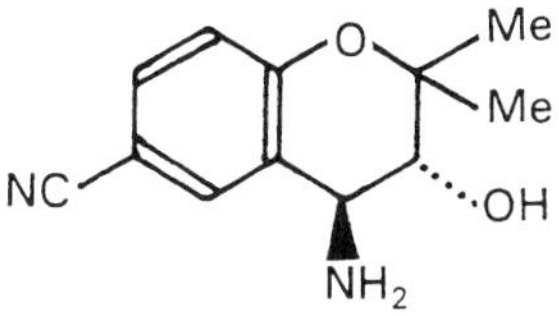

(25)

Chromanylcyanoguanidine derivatives as antiischemic agents (K. Atwal, K.S. Kim, and G.J. Grover, Eur. Pat. Appl. EP 501,797, 1992; K. Atwal, U.S. US 5,276,168, 1994) and 6-cyano-4-[(*N*-cyanoimidoyl)amino]-2,2-dimethylchromans and analogues as antihypertensives and bronchodilators (K. Ohtuka *et al.*, Eur. Pat. Appl. EP 412,531, 1991) have been prepared from *trans*-4-amino-6-cyano-3-hydroxy-2,2-dimethylchroman (25).

The addition of 6-cyano-3,4-epoxy-2,2-dimethylchroman to a solution of 3-amino-1-methyl-1,6-dihydropyridazin-6-one and NaH in DMSO affords 6--cyano-4-(1-methyl-1,6-dihydro-6-oxopyridazin-3-yl)amino-3-hydroxy-2,2--dimethylchroman (26) (R. Gericke *et al.*, *ibid.*, 489,327, 1992).

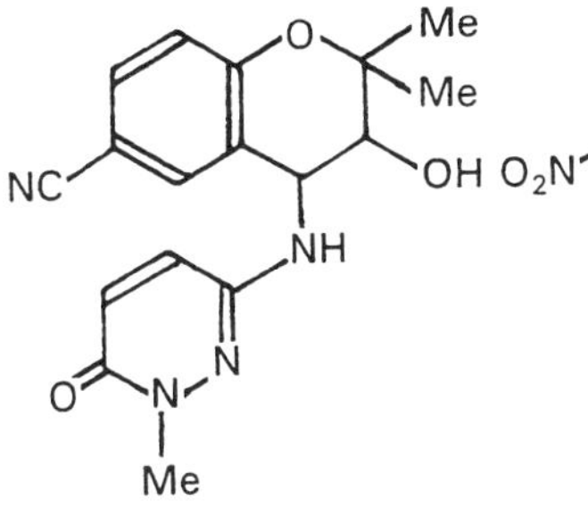

(26)

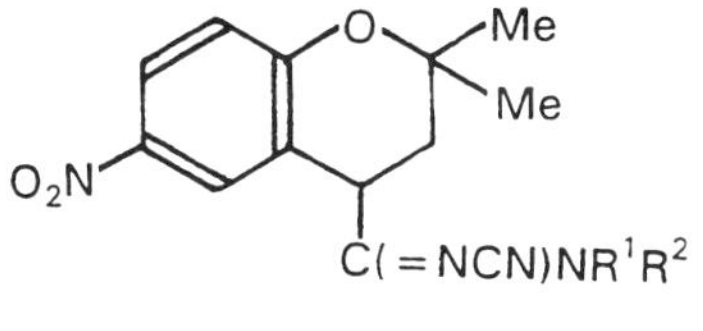

(27)

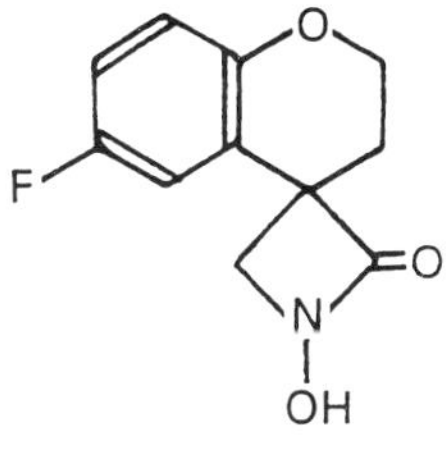

(28)

418

Several 4-(aminomethyl)chroman and -thiochroman derivatives, for
example, 4-[*N*-methyl-*N*-2-(3-chlorophenyl)ethylamino]methyl-7-methoxy-
chroman.HCl, as selective α_2-adrenergic antagonists have been prepared
(J.F. De Bernardis, D.L. Arendsen, and R.E. Zelle, *ibid.*, 325,964, 1989).
For the preparation of 1-[2-(chroman-4-yl)ethyl]-4-(3,4-dimethoxyphenyl)-
piperidine.HCl and related compounds as antiarrhythmics, see J.C. Hardy and
C. Renault, *ibid.*, 300,908, 1989 and M. Barreau *et al.*, *ibid.*, 379,441,
1990; *N*-ethyl-*N*-2-{4-[2-(xanthen-9-yl)ethyl]-1-piperazinyl}ethylchroman-4-
-ylmethylamine trihydrochloride and analogues as antioxidants, T. Nozawa
and S. Kanao Jpn. Kokai Tokkyo Koho JP 01 146,876 [89 146,876], 1989;
3,4-*trans*-7-methoxy-2,2-dimethyl-3-phenyl-4-(4-pyrrolidinoethoxyphenyl)-chroman
(Centchroman) a non-steroidal oral contraceptive for women, A.K.
Srivastava, R.C. Gupta, and P.K. Grover, Indian J. Chem: Org. Chem. Ind.
Med. Chem., 1994, <u>33B</u>, 540; and *N*-cyano-2,2-dimethyl-6-nitrochroman-4-
-carboxamidines 27 (R^1 = H, Me, Et; R^2 = Me, Et), cromakalim analogues
with selective activity for guinea pig trachealis, H. Koga *et al.*, Bioorg. Med.
Chem. Letters, 1993, <u>3</u>, 1111.

Several spiroheterocycle-chromans, for example, 6-fluoro-4,4-spiro(1-
-hydroxy-2-oxoazetidin-3-yl)chroman (28), as aldose reductase inhibitors
have been synthesized (J.F. Eggler and E.R. Larson, U.S. US 5,039,672,
1991).

2-Aminomethylchromans and analogues have been prepared as $5HT_1$
receptor ligands, for example, (+)-29 obtained by condensing (+)-2-amino-
methyl-8-methoxychroman with *N*-(4-bromobutyl)saccharin in DMF
containing Et_3N (B. Junge *et al.*, *ibid.*, 5,137,901, 1992).

(29)

The preparations of 7-hydroxy-6-(methoxymethoxy)-5-(pyrid-3-ylmethyl)-chroman and derivatives (S. Hibi *et al.*, Jpn. Kokai Tokkyo Koho JP 05 310,724 [93 310,724], 1993), and *N*-[2-(6-fluoro-2-methylchroman-8-yl-oxy)ethyl]-4-methylphenethylamine.HCl and derivatives having selective affinity to serotonin (5-HT$_{1A}$) receptors (T. Yasunage *et al.*, *ibid.*, 05 255,302 [93 255,302], 1993) have been reported.

6-Hydroxy-2,5,7,8-tetramethylchroman-2-carboxylic acid has been sequentially *O*-acetylated, converted to the acid chloride, amidated with PhNH$_2$, sulphurated with Lawesson's reagent, and hydrolyzed with aqueous NaOH to afford *N*-phenyl 6-hydroxy-2,5,7,8-tetramethylchroman-2-thiocar-boxamide (30). Related derivatives have been prepared as cell protectors (G. Le Baut *et al.*, Eur. Pat. Appl. EP 587,499, 1994). The preparations of the *N*-(2-carboxyphenyl) (2-methylchroman-2-yl)alkenylcarboxamides 31 (R^1, R^4=H, alkyl; R^2=H, alkoxy; R^3=H, alkyl, alkoxy; n=0,1,2), useful as antiallergics (K. Ejiri *et al.*, Jpn. Kokai Tokkyo Koho JP 62 132,879 [87 132,879], 1987); 2-[(6-hydroxy-2,5,7,8-tetramethylchroman-2-yl)carbonyl]--1,1,1-trimethylhydrazinium 4-methylbenzenesulphonate and derivatives (J.M. Grisar, M.A. Petty, and F. Bokenius, PCT Int. Appl. WO 93 20,059, 1993); *tert*-butyl chroman-3-carbamates 32 (R^1=H, Me, R^2=H, Br, Cl, CN) B.S. Orlek and G. Stemp, Tetrahedron Letters, 1991, <u>32</u>, 4045); and several 3-[(substituted-carbonyl)amino]chromans from 6-substituted 3-amino--4-hydroxy-2,2-dimethylchroman, which were tested for antihypertensive activity (F. Cassidy *et al.*, J. Med. Chem., 1992, <u>35</u>, 1623) have been reported.

(30)

(32)

(31)

420

The synthesis and antihypertensive activity of a series of 6-cyano-3-hydroxy-2,2-dimethylchroman 33 (R=H, Me, Bu^t; X=NCN, CHNO$_2$, NCONH$_2$, NCOMe, NSO$_2$NH$_2$) and corresponding benzopyrans related to cromakalim, in which the pyrolidinone ring at position C-4 has been replaced by a number of potential amide and urea isosteres have been reported (G. Burrell *et al.*, Bioorg. Med. Chem. Letters, 1993, 3, 999).

(33) R^1=NHC(=X)NHR
(35) R^1=NHCONHPh

(34)

The amidation of *trans*-6-cyano-4-[2(dimethylamino)ethyl]amino-3-hydroxy-2,2-dimethylchroman with 3,4,5-trimethoxybenzoyl chloride and K$_2$CO$_3$ in CH$_2$Cl$_2$ yields *trans*-6-cyano-4-[2-(dimethylamino)ethyl-1-(3,4,5-trimethoxybenzoyl]amino-3-hydroxy-2,2-dimethylchroman (34) (C. Hoornaert and J.C. Muller Fr. Demande FR 2,670,780, 1992). For the preparation of several derivatives of chroman-4-carboxamide and related compounds see T. Yamanaka, T. Nakajima, and O. Yaoka, Jpn. Kokai Tokkyo Koho JP 04 264,080 [92 264,080], 1992; T. Yamanaka *et al.*, *ibid.*, 04 36,278 [92 36,278], 1992; 05 32,655 [93 32,655], 1993; and H. Koga, H. Nabata, and H. Nishina, PCT. Int. Appl. WO 92 14,439, 1992; *cis*-(+)-6-acetyl-4-(4-fluorobenzoylamino)-3-hydroxy-2,2-dimethylchroman and *cis*-6-ethyl-4S-(4-fluorobenzoylamino)-3S-hydroxy-2,2-dimethylchroman, W.N. Chan, J.M. Evans and G. Stemp, *ibid.*, 94 13,657, 1994, and related compounds J.M. Evans, M. Thompson, and N. Upton, *ibid.*, 92 22,293, 1992 and I.L. Pinto and G. Stemp, Eur. Pat. Appl. EP 398,665, 1990.

Trans-4-amino-6-cyano-3-hydroxy-2,2-dimethylchroman on boiling with PhNCO in EtOH gives *N*-(6-cyano-3-hydroxy-2,2-dimethylchroman-4-yl)-*N'*-phenylurea (35). The preparations of related compounds as antiischemic agents have been described (K. Atwal, *ibid.*, 462,761, 1991; 587,180, 1994). Also reported are the preparations of related compounds containing a 4-propenamido group at the C-4 positions (K. Atwal, Z.A. Syed, and P.S. Dinos, *ibid*, 587,188, 1994).

For the preparation of some 4-(phenylsulphonamido)chromans and analogues as thrombocyte aggregation inhibitors and thromboxane A_2 antagonists, see U. Niewoehner *et al.*, *ibid.*, 425,946, 1991; and *N*-chroman-4-yl- and *N*-(chroman-4-ylmethyl)benzenesulphonamides, *idem, ibid.* 315,009, 1989.

N-Aralkyl-2-aminomethylchromans and analogues have been prepared as serotoninergic agents, for example, *N*-naphth-1-yl-2-aminomethyl-8-methoxychroman (36) obtained by the reduction of the product formed on condensing ethyl 8-methoxychroman-2-carboxylate with 1-aminomethylnaphthalene (R. Schohe-Loop *et al.*, Ger. Offen. DE 4,135,474, 1993).

(36)

(37)

Treating 8-(2-bromoethoxy)chroman with 4-methoxyphenethylamine in DMF at 50°C affords after salination with HCl, 8-[2-(4-methoxyphenethylamino)ethoxy]chroman.HCl (37) (T. Yasunaga *et al.*, Jpn. Kokai Tokkyo Koho JP 05 125,024 [93 125,024], 1993).

Compounds related to chroman. The preparations of a number of 3,4--benzo-2-oxabicyclo[4.1.0]heptane (1) derivatives as psychotropic agents have been reported (T. Tatsuoka *et al.*, *ibid.*, 62 198,676 [87 198,676], 1987). Cyclocondensation of 2-$HOC_6H_4CH=CHCOMe$ with EtO_2CCH_2COMe in the presence of base affords the tetrahydromethanobenzoxocinone 2. The preparations of some related derivatives have been described (F. Eiden and P. Gmeiner, Arch. Pharm. (Weinheim, Ger.), 1987, <u>320</u>, 213). The Friedel--Crafts inter- and intramolecular alkylation reaction of cyclic allylic alcohols, for example, 3, with resorcinols in the presence of $BF_3.OEt_2$ or $BF_3.OEt_2$/-alumina yields cyclohexenylresorcinol 4 and tetrahydrobenzoxocin 5 (S.H. Baek, Bull. Korean Chem. Soc., 1993, <u>14</u>, 144). For the preparations of (±)-8-chloro-2,6-methano-2*H*-3,4,5,6-tetrahydrobenzoxocin-10-carboxylic acid (6) and related derivatives and their esters and amides, having 5-HT_3 antagonistic properties useful in treatment of gastric disorders and emesis, see J.E. Airey *et al.*, U.S. US 5,288,731, 1994.

(1)

(2)

(3)

(4)

(5)

(6)

The irradiation of 2-allyl- and 2-(2-methylallyl)-5-methyl-1,4-benzoquinones 7 (R^1=H, Me) in the presence of $H_2C = CR^2R^3$ (R^2=H, Me, R^3=Ph, 4-ClC_6H_4, 4-$MeOC_6H_4$, CO_2Me, CN) yields tricyclic compounds 8 and 9 (H. Iwamoto and A. Takuwa, Chem. Letters, 1993, 5).

(7)

(8)

(9)

The high vacuum thermolysis of 2-phenyl-4*H*-benzo[*d*]-1,3,2-dioxaborin affords a mixture of the known trimer 10 and unknown tetramer of 2-methylenequinone 11 (R. Faure *et al.*, Bull. Soc. Chim. Fr., 1991, 378).

(10)   (11)

5,6,7,8-Tetrahydro-2H-benzo[b]pyrans, 5,6,7,8-tetrahydrochromenes. 7,7-Disubstituted 3-(benzoylamino)-5-oxo-5,6,7,8-tetrahydrocoumarins [*N*-(7,7-disubstituted 2,5-dioxo-5,6,7,8-tetrahydro-2*H*-benzo[*b*]pyran-3-yl)-benzamide] 1 ($R^1 = R^2 = H$, Me; $R^1 = H$, $R^2 = Me$) have been obtained from cyclohexane-1,3-diones, hippuric acid, and a one-carbon synthon such as $HC(OEt)_3$ (M. Kocevar *et al.*, synth. Comm., 1989, 19, 1713; Ann., 1990, 501). 3-(Benzoylamino-2,5-dioxo-5,6,7,8-tetrahydro-2*H*-benzo[*b*]pyrans have been prepared by the reaction of hippuric acid in acetic acid with 2-dimethylaminomethylenecyclohexane-1,3-diones (L. Mosti *et al.*, Farmaco, 1994, 49, 45). Reacting 2-aminomethylene-5,5-dimethylcyclohexane-1,3-dione with *N*-methylpiperid-2-one acetals in toluene and then heating the product in aqueous HCl yields 3-(2-methylaminoethyl)-7,7-dimethyl-5-oxo-5,6,7,8-tetrahydrocoumarin (2) (A.K. Shanazarov *et al.*, Khim. Geterotsikl. Soedin., 1987, 1477). Methyl 5,6,7,8-tetrahydrocoumarin-3-carboxylate (3) undergoes a Diels-Alder reaction with $(MeO)_2C = CH_2$ to give methyl 2-methoxy-5,6,7,8-tetrahydronaphthalene-3-carboxylate (4) (D.L. Boger and M.D. Mullican, Org. Synth., 1987, 65, 98).

(1)   (2)

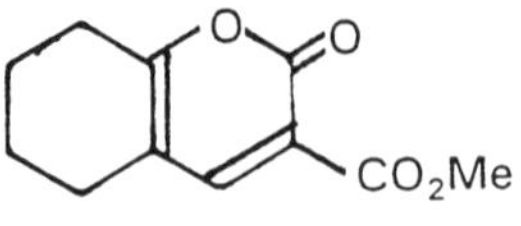

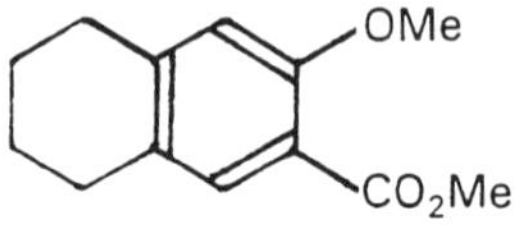

CO_2Me

(3)

OMe

CO_2Me

(4)

The Michael reaction of $MeO_2CC\equiv CCO_2Me$ with cyclopentane-1,3-
-dione, cyclohexane-1,3-dione, and dimedone in the presence of MeONa or
$KF-Al_2O_3$ affords mixtures of products, containing methyl 5-oxo-5,6,7,8-
-tetrahydrocoumarin-4-carboxylate (5; R=H) and methyl 7,7-dimethyl-5-oxo-
-5,6,7,8-tetrahydrocoumarin-4-carboxylate (5; R=Me) (P. Hrnciar, V.
Padychova, and J. Sraga, Chem. Pap., 1989, **43**, 87).

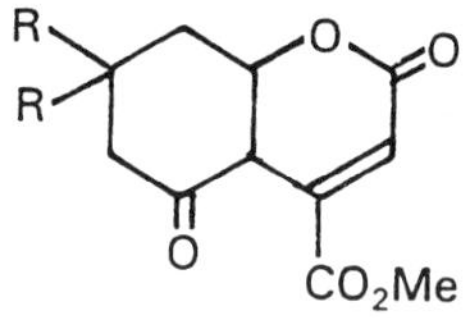

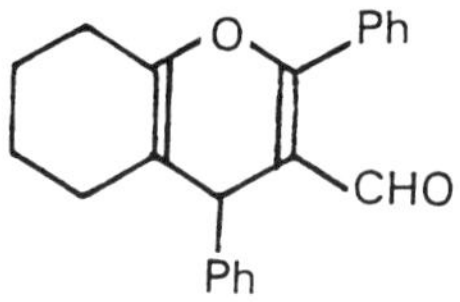

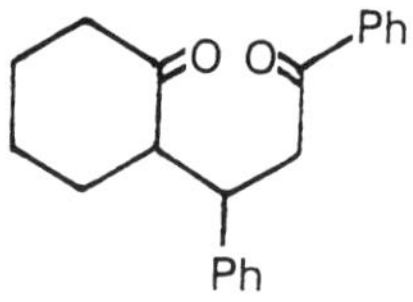

CO_2Me

(5)

Ph

CHO

Ph

(6)

Ph

Ph

(7)

2,4-Diphenyl-5,6,7,8-tetrahydro-4*H*-benzo[*b*]pyran-3-carboxaldehyde (6)
has been synthesized from the 1,5-diketone 7 (M. Venugopal *et al.*,
Tetrahedron Letters, 1991, **32**, 3235). 2-Aryl-7,7-dimethyl-5-oxo-4-phenyl-
-5,6,7,8-tetrahydro-4*H*-benzo[*b*]pyran 8 (R^1 =H, MeO; R^2 =H, MeO, Me, Cl;
R^{1}R^2 =OCH$_2$O) have been obtained by the condensation of 5,5-dimethyl-
cyclohexane-1,3-dione with PhCH = CHCOC$_6$H$_3$R^{1}R^2-3,4 (V.K. Ahluwalia *et
al.*, Indian J. Chem., 1991, **30B**, 1095). 2-Amino-3-cyano-5-oxo-4-phenyl-
-5,6,7,8-tetrahydro-4*H*-benzo[*b*]pyran (9) has been obtained from cyclo-
hexane-1,3-dione (N. Martin-Leon *et al.*, Ann., 1990, 101) and the
preparations of 2-amino-4-aryl-3-cyano-7,7-dimethyl-5-oxo-5,6,7,8-tetra-
hydro-4*H*-benzo[*b*]pyrans 10 (R=Ph, 4-FC$_6$H$_4$) have been reported (Yu.A.
Sharanin and A.M. Shestopalov, Zh. Org. Khim., 1989, **25**, 1331).

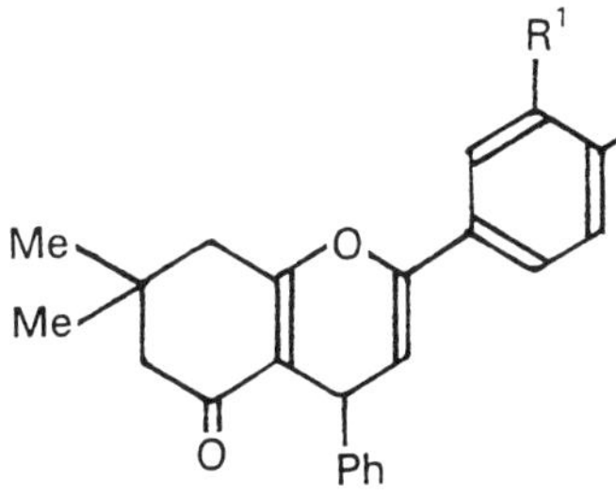

(8) 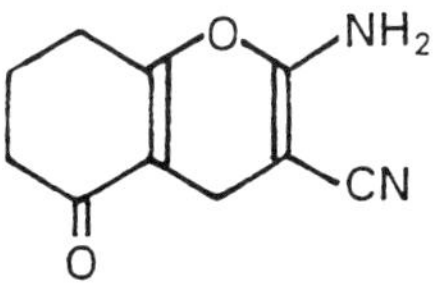 (9) 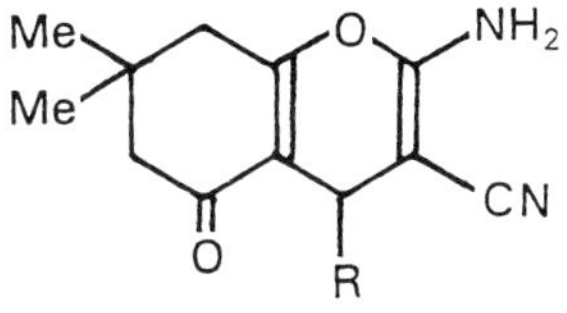 (10)

Photolysis of 7-methyl-4a,5,6,8a-tetrahydrocoumarin 11 ($R^1 = CO_2Et$, $R^2 = Me$; $R^1 = Me$, $R^2 = H$) results in the formation of oxatetracyclodecanones 12 by intramolecular cycloaddition and oxatricyclodecenones 13 by a photochemical ene reaction (Z. Ahmad, U.K. Ray, and R.V. Venkateswaran, Chem. Comm., 1990, 708).

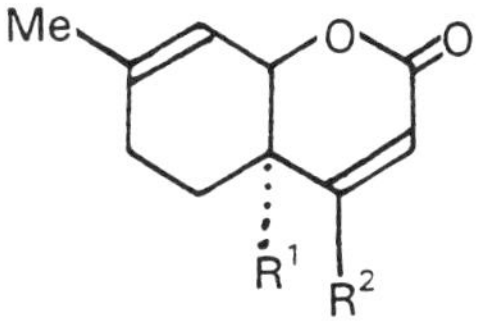

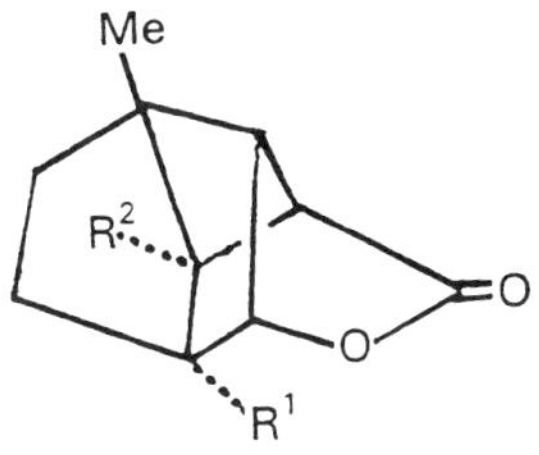

(11) (12)

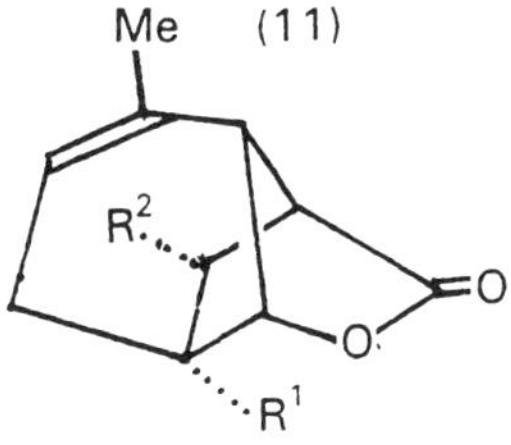

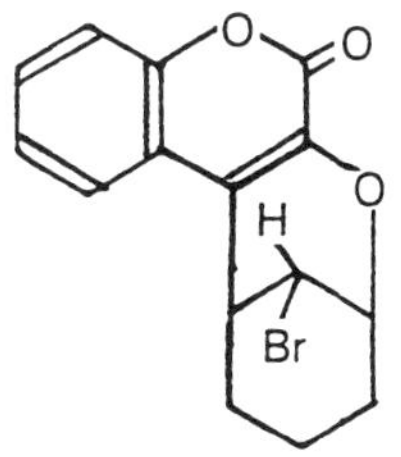

(13) (14)

426

4-(Cyclohex-2-enyl)-3-hydroxycoumarin on acetylation, followed by bromination and treatment with alcoholic KOH affords the fused coumarin derivative 14 (K.C. Majumdar *et al.*, Synth. Comm., 1990, <u>20</u>, 1249).

Perhydrochromans. Photolysis of arylidene-ß-ionones (*E,E*)-1 [R = (un)-substituted phenyl, fur-2-yl) in benzene furnishes oxabicyclodecadienes (*E*)-2 [1,7,7-trimethyl-3-(*E*-2-arylethenyl)-2-oxabicyclo[4.4.0]deca-3,5-dienes] (R.P. Gandhi *et al.*, *ibid.*, 1989, <u>19</u>, 1759).

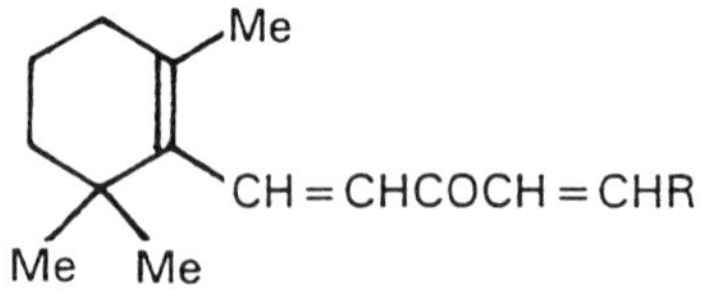

(1)

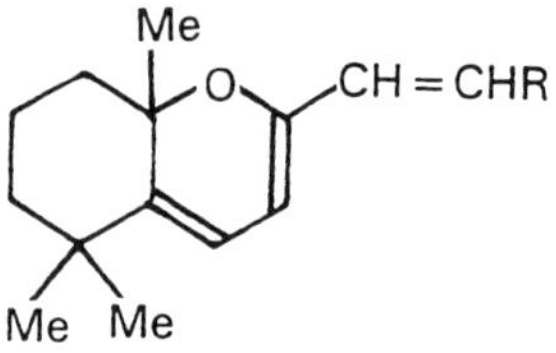

(2)

The cyclization of alkenoic acid 3 using 5% Pd(OAc)$_2$ in the presence of 2 equivalents of NaOAc and 1 atmosphere of oxygen yields 4a,5,6,8a-tetra-hydrochroman-2-one (4) (R.C. Larock and T.R. Hightower, J. Org. Chem., 1993, <u>58</u>, 5298).

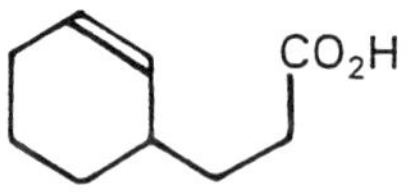

(3)

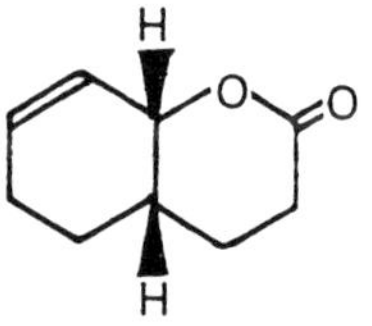

(4)

Isopulegol (5) on addition to boiling EtCHO, followed by distillation gives a mixture of 2-ethyl-4-hydroxy-4,7-dimethylperhydrochroman (6), 2-ethyl-4,7-dimethyl-3,5,6,7,8,8a-hexahydro-2H-benzo[b]pyran (7), 2-ethyl--4,7-dimethyl-4a,5,6,7,8,8a-hexahydro-2H-benzo[b]pyran (8), and 2-ethyl-7--methyl-4-methyleneperhydrochroman (9), having a floral, magnolia, fruity and peach aroma profile (M.A. Sprecker *et al.*, U.S. US 4,999,439, 1991).

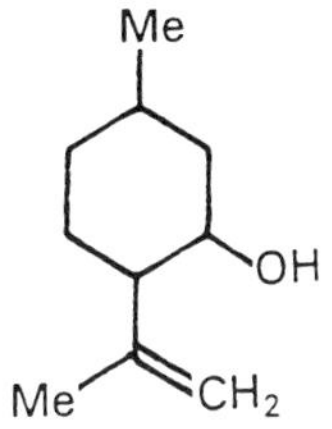

(5) 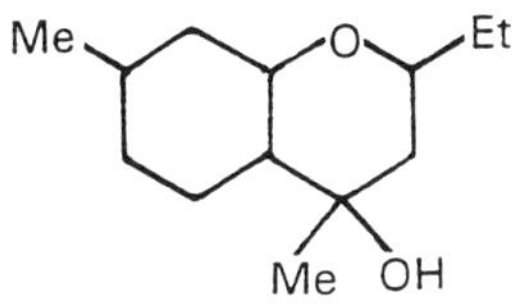 (6) 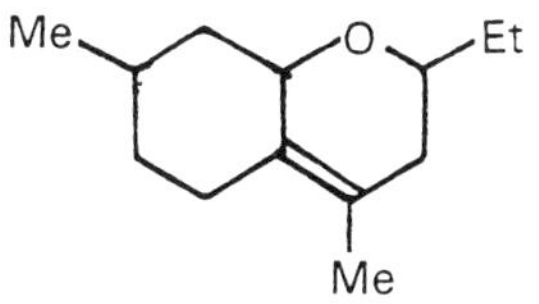 (7)

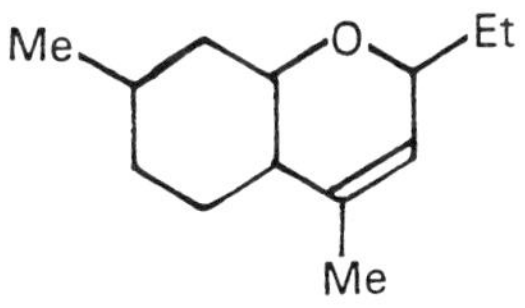

(8) 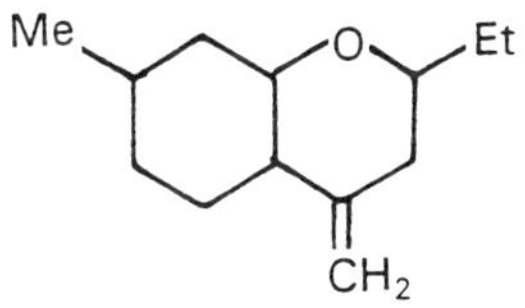 (9)

5,5-Dimethylcyclohexane-1,3-dione reacts with chloromethyloxirane in basic medium to give only one heterocyclic derivative, 3-hydroxy-7,7-dimethyl-5,6,7,8-tetrahydrochroman-5-one (10) along with normal alkylation products. No reaction takes place with other 1,3-dicarbonyl compounds, e.g. cyclohexane-1,3-dione, dibenzoylmethane, and ethyl cyclopentan-2--onoate (N.K. Sangwan and K.S. Dhindsa, Indian J. Chem., 1989, 28B, 316).

428

(10)   (11)   (12)

In an aqueous solution 2-hydroxy-4a,7,8,8a-tetrahydrochroman-6-car-boxaldehyde (glutaraldehyde dimer) (11), useful as a crosslinking agent for the immobilization of biological substances, is in equilibrium with dialdehyde 12 (T. Nakagawa, Y. Kuroda, and T. Tajima, Jpn. Kokai Tokkyo Koho JP 03 294,274 [91 294,274], 1991). 5,6,7,8-Tetrahydrochroman (13) on treatment with pyridinium chlorochromate give the keto lactone 14 (S. Baskaran *et al.*, Chem. Letters., 1987, 1175).

(13)   (14)   (15)

The cis and trans isomers of perhydrochroman (perhydrobenzo[*b*]pyran) (15) have been prepared and their conformations studied by low-temperature ^{13}C and high yield ^{1}H NMR spectroscopy (D.V. Griffiths and G. Wilcox, J.

Chem. Soc., Perkin 2, 1988, 431). Perhydrochroman has also been obtained by the reaction of $(MeO)_2C(CH_2)_3SnMe_3$ with the appropriate cyclic enol ether (T.V. Lee and K.L. Ellis, Tetrahedron Letters, 1989, 30, 3555). Diethyl perhydrochroman-2,2-dicarboxylate (oxabicyclodecanedicarboxylate) (16) on saponification, followed by oxidative bisdecarboxylation with $Ce(NH_4)_2(NO_3)_6$ affords perhydrochroman-2-one (17) (R.G. Salomon, S. Roy, and M.F. Salomon, *ibid.*, 1988, 29, 769).

(16)

(17)

(18)

(19)

Cyclization of cyclohexenyl alcohol 18 with benzenselenyl triflate yields stereoselectively 8-phenylselenoperhydrochroman (19) (S. Murata and T. Suzuki, *ibid.*, 1987, 28, 4415). For the preparation of 3-(4-phenylsulphon-yl)oxyperhydrochroman, see E. Schaumann and A. Kirschning, J. Chem. Soc., Perkin 1, 1990, 1481; perhydrochroman-2-ones with C_{1-5}alkyl substituents in positions-5, -6, -7, and -8 or in some of them, K. Sakai, Y. Nishida, and T. Shirafuji, Eur. Pat. Appl. EP 533,378, 1993; 2,4,4,7,7--pentamethylperhydrochroman-5-one, M.G. Ahmed *et al.*, J. Bangladesh Chem. Soc., 1990, 3, 189); and 3-methylene-perhydrochroman-2-one (α--methylenevalerolactone), E. Ghera, T. Yechezkel, and A. Hassner, J. Org. Chem., 1990, 55, 5977.

The propargylic silane 20 undergoes smooth cyclization in boiling benzene in the presence of AIBN and tributyltin hydride to give 5-cyano-5-
-trimethylsiloxy-4-(2-trimethylsilylethylidene)perhydrochroman (21) (D. Schinzer, P.G. Jones, and K. Obierey, Tetrahedron Letters, 1994, $\underline{35}$, 5853).

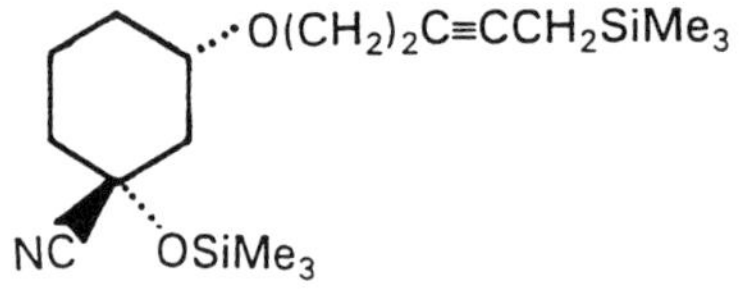

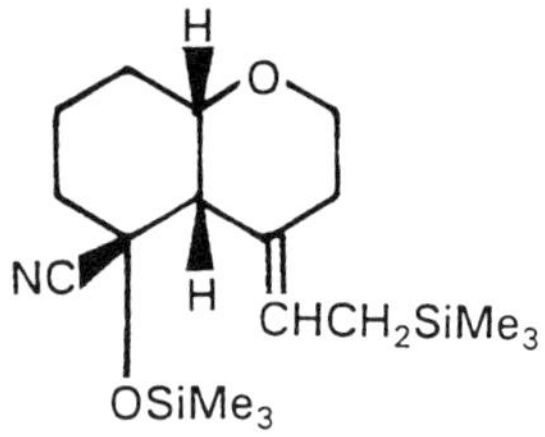

Treating compound 22 with RCHO (R = Ph, 4-MeOC$_6$H$_4$) in EtOH containing aqueous NaOH affords the 3,3-spiro 2-aryl-8a-hydroxyperhydro-chroman 23 (T.I. Akimova, S.V. Kosenko, and M.N. Tilichenko, Zh. Org. Khim., 1993, $\underline{29}$, 644).

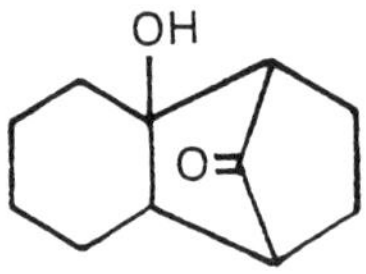

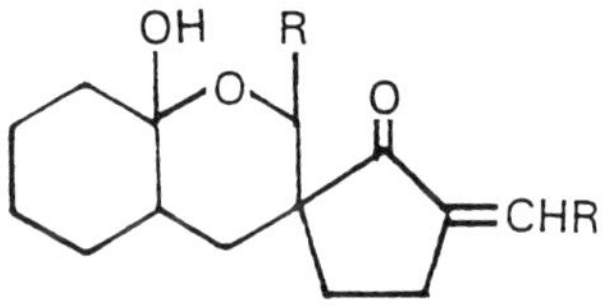

The cycloaddition of cyclohexane-1,3-dione, HCHO, and 2,3-dihydro-
furan furnishes 5-oxo-2,3,3a,5,6,7,8-9a-octahydro-4H-furo[2,3-b]benzo-
pyran, which on aromatization with Davis reagent in the presence of 3,5-
-(MeO)$_2$C$_6$H$_3$NH$_2$ yields 5-hydroxy-2,3,3a,9-tetrahydro-4H-furo[2,3-b]benzo-
pyran as a model compound for alboatrin (M. Krause and H.M.R. Hoffmann,
Synlett, 1990, 485).

(b) Tocopherols

The preparations of a number of intermediates for a-tocopherol and some
which can be regarded structurally as compounds in the a-tocopherol class
have been described. Related compounds useful as antioxidants, 5-lipoxy-
genase inhibitors, allergy inhibitors, antihistamines, and anticholesteremics
have been reported.

The thermostability of the tocopherols has been studied by dynamic
programmed thermogravimetry and has been found to increase in the order
a-tocopherol > γ-tocopherol > δ-tocopherol. Also racemic-a-tocopheryl
acetate was more stable than d-a-tocopheryl succinate (T. Ushikusa, T.
Maruyama, and I. Niiya, Yukagaku, 1991, <u>40</u>, 1073).

The following compounds have been reported as intermediates for the
synthesis of a-tocopherol and related derivatives. 6-Hydroxy-2-hydroxy-
methyl-2,5,7,8-tetramethylchroman (1) (K. Sato et al., Jpn. Kokai Tokkyo
Koho JP 01 68,366 [89 68,366], 1989); methyl and ethyl 6-hydroxy-
-2,5,7,8-tetramethylchroman-2-carboxylates (M.S. Raju and N. Huh, U.S. US
5,348,973, 1994); (2R)-6-hydroxy-2(4-methylpent-3-enyl)-2,5,7,8-tetra-
methylchroman prepared from nerol (K. Sato et al., Jpn. Kokai Tokkyo Koho
JP 63 63,675 [88 63,675], 1988); derivatives of 6-hydroxy-2-methyl-2-(4-
-methylpent-3-enyl)chroman 2 (R^1 = H, protecting group; R^2-R^4 = H, Me) (N.
Hirose, K. Hamamura, and S. Kijima, ibid., 02 11,585 [90 11,585], 1990);
and derivatives of 6-hydroxy-2,5,7,8-tetramethylchroman 3 [R^1 = Me, labile
HO-protecting group; R^2 = halogeno, prop-2-enyl, CH(CO$_2$R^3)$_2$, (CH$_2$)$_3$CHMe-
(CH$_2$)$_3$CHMe(CH$_2$)$_3$CHMe$_2$; R^3 = lower alkyl] (Hoffmann-La Roche, F., und
Co. A.-G. ibid., 62 178,581 [87 178,581], 1987).

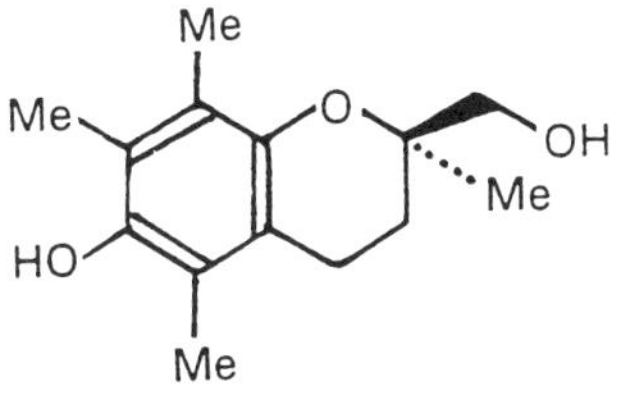

(1) 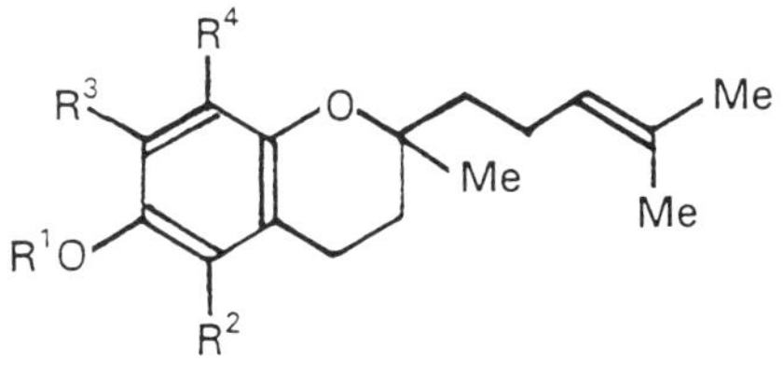   (2)

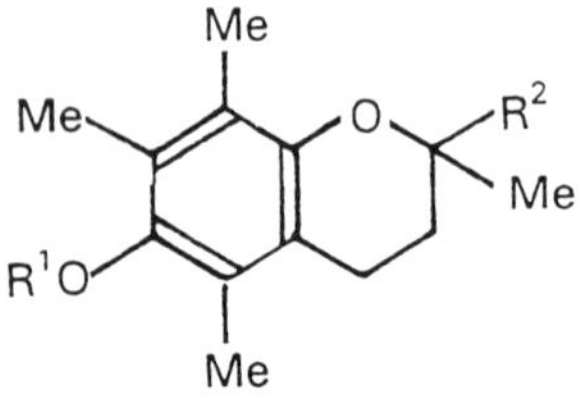

432

(3) (see text)
(4) $R^1 = PhCH_2$, $R^2 = R$

The iodination of ß-tocopherol, followed by reaction with CF_3I in the presence of Cu powder and treatment with NH_3 in MeOH, affords 6-hydroxy--2-phytyl-7-trifluoromethyl-2,5,8-trimethylchroman (T. Nakada, *ibid*., 62 292,777 [87 292,777], 1987). Treatment of 2-methoxychroman derivatives in the *α*-tocopherol structural class, for example, 4 (R = MeO) with Me_3SiCN in the presence of $TiCl_4$ or $BF_3.Et_2O$ generally leads to the 2-cyanochroman, for example, 4 (R = CN) (N. Cohen *et al.*, J. Org. Chem., 1989, <u>54</u>, 3282).

(c) Phenylchromans (flavans and isoflavans) phenyl-2H-3,4-dihydrobenzo[b]-pyrans

(i) Flavan, 2-phenylchroman, 2-phenyl-2H-3,4-dihydrobenzo[b]pyran derivatives

Treatment of $2\text{-MeSCH}_2\text{C}_6\text{H}_4\text{OH}$ with $BF_3.Et_2O$ in C_6H_6, followed by addition of styrene affords 2-phenylchroman (1) (T. Inoue, S. Inoue, and K. Sato, Chem. Letters, 1990, 55). The condensation of phenol with ethyl-

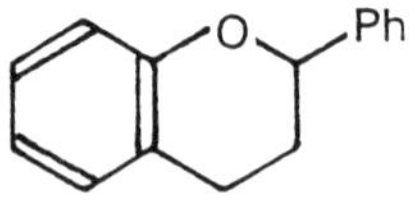

(1)

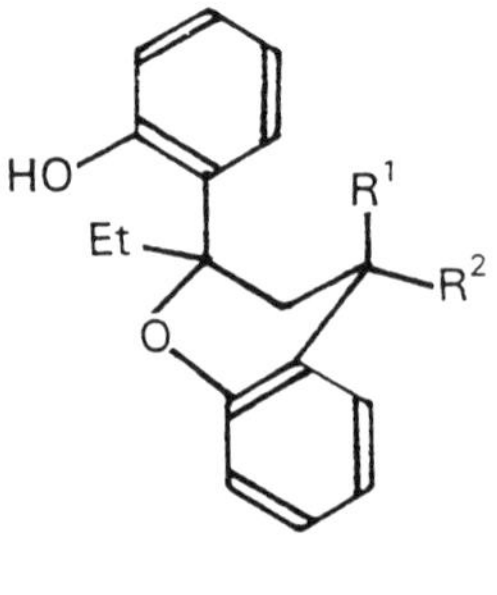

(2)

433

methylketone catalyzed by $(PhO)_3Al$ gives a mixture of products including 2,4-diethyl-2-(2-hydroxyphenyl)-4-methylchromans 2 ($R^1 = Et$, $R^2 = Me$) (13%) and ($R^1 = Me$, $R^2 = Et$) (9%) (Ya.B. Kozlikovskii, B.V. Chernyaev, and V.V. Trachevskii, Zh. Org. Khim., 1987, <u>23</u>, 614).

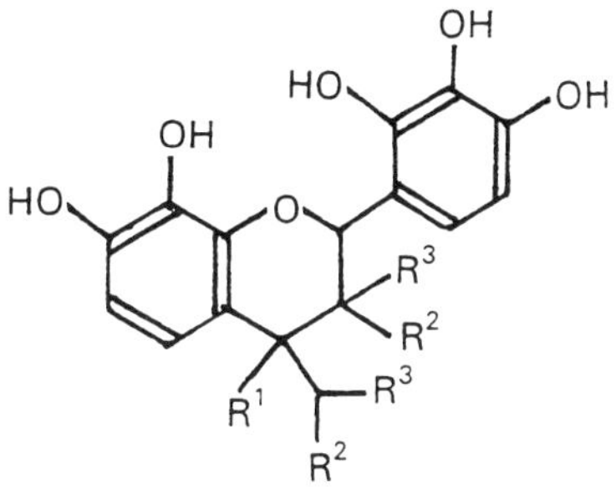

(3)

The cyclocondensation of carbonyl compounds $R^1COCHR^2R^3$ ($R^1, R^2, R^3 = H$, alkyl, alkenyl, cycloalkyl, aralkyl, aryl) with pyrogallol in a organic solvent (e.g., EtOAc) in the presence of an acid catalysis (e.g., 4-$MeC_6H_4SO_3H$) furnishes 7,8-dihydroxy-2-(2,3,4-trihydroxyphenyl)chromans 3 (N. Suzuki, H. Nakanishi, and J. Tomioka, Eur. Pat. Appl. EP 582,309, 1994).

(ii) Isoflavans, 3-phenylchromans, 3-phenyl-2H-3,4-dihydrobenzo[b]pyran derivatives

2,3-Diarylpropanoates serve as precursors to 3-phenylchromans, for example, 3-phenylchroman (1), following consecutive reduction and cyclization steps (M. Versteeg, B.C.B. Bezuidenhoudt, and D. Ferreira, Heterocycles, 1993, <u>36</u>, 1743). Several derivatives of 7-hydroxy-3-aryl-chroman 2 [$R^1 = H$, (un)substituted alkyl; one of R^2 and $R^3 = HO$, alkoxy, alkanoyloxy, alkyl and the other is H, or OR^1R^2; Ar is (un)substituted by alkyl, phenylalkyl, alkanoyloxy, halogeno, NH_2] and their salts useful for treatment of vascular diseases have been prepared (A.. Albert and F.W. Zilliken, Eur. Pat. Appl. EP 267,155, 1988).

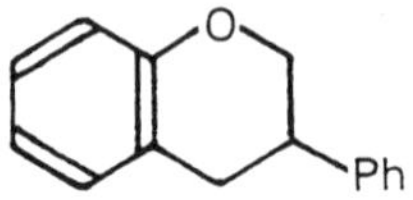

(1) 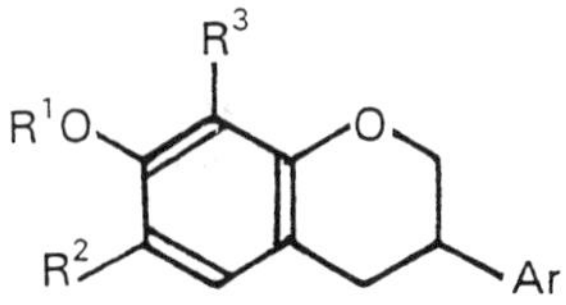   (2)

(d) Chromanols, hydroxychromans

2-Ethoxy-4-phenylchroman (1) is obtained in a *trans:cis* ratio 3:97 by the cycloaddition of 2-[(Me$_2$CHS)CHPh]C$_6$H$_4$OH with EtOCH=CH$_2$ (T. Inoue, S. Inoue, and K. Sato, Bull. Chem. Soc. Jpn., 1990, 63, 1062). Several 6- and 8-mono and -disubstituted 2-ethoxychromans have been prepared (A.R. Kratritzky and X. Lan, Synthesis, 1992, 761). The asymmetric syntheses of some chroman derivatives including 2-ethoxy-6-hydroxy-2,5,7,8-tetramethyl-chroman (2) have been described (S. Yang *et al.*, Youji Huaxue, 1990, 10, 346). 2-Substituted 6-phenylchroman-2-ols 3 (R = Et, anisyl, tolyl, Ph, cyclohexyl) have been obtained from 6-phenylchroman-2-one and the appropriate Grignard reagent (M.A. Sayed *et al.*, Orient. J. Chem., 1987, 3, 174).

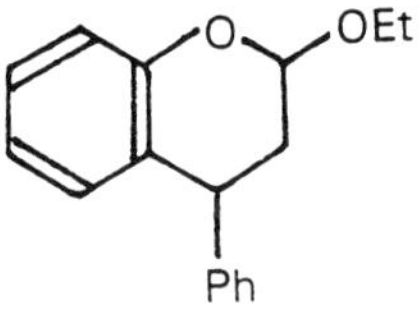

(1) 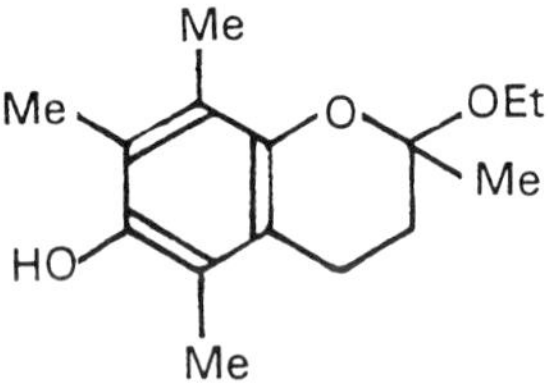   (2) 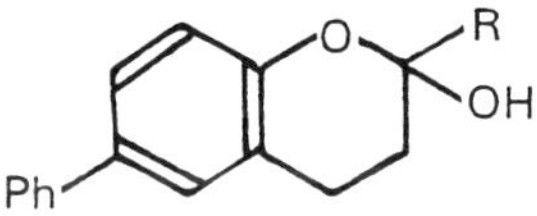   (3)

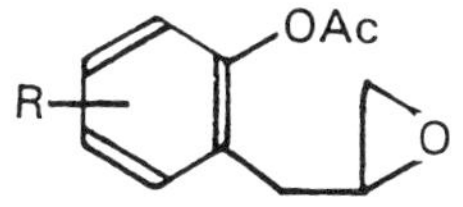

(4)

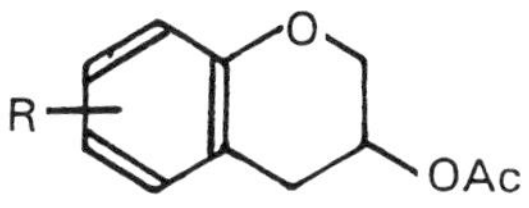

(5)

2-(2,3-Epoxypropyl)phenyl acetates 4 (R=H, 4-Br, 6-OMe, 6-NO$_2$) on treatment with catalytic amounts of NaI in organic solvent give directly the corresponding chroman-3-yl acetates 5 (K. Kawamura, T. Ohta, and G. Otani, Chem. Pharm. Bull., 1990, 38, 2088). For the preparation of 2--hydroxymethyl-2-methylchroman-3-ol and its derivatives and their ^{1}H and ^{13}C NMR spectral data, see M. David, J. Sauleau, and A. Sauleau, Bull. Soc. Chim. Fr., 1993, 130, 527; and the preparation of 7-(2,3-epoxypropoxy)-5--methyl- and -5,8-dimethylchroman-3-ol, E. Raidma et al., Zh. Org. Kihm., 1991, 27, 2629. Treatment of allylphenols 6 (R^1=H, Me; R^2=H, Cl; R^3=H, Cl, Br, Me) with 3-ClC$_6$H$_4$CO$_3$H in CHCl$_3$ gives chroman-3-ols 7 (V. Satyanarayana et al., Synth. Comm., 1991, 21, 1455).

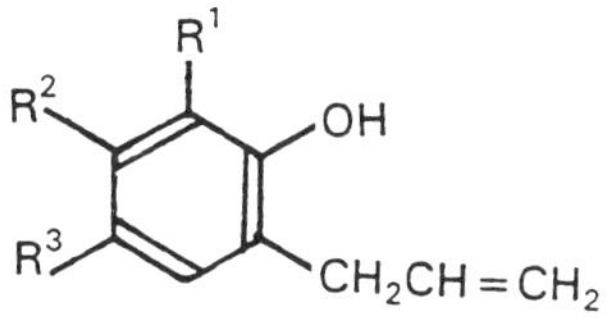

(6)

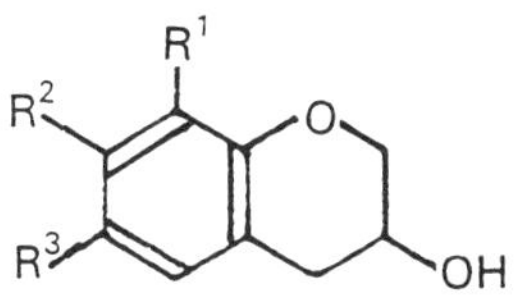

(7)

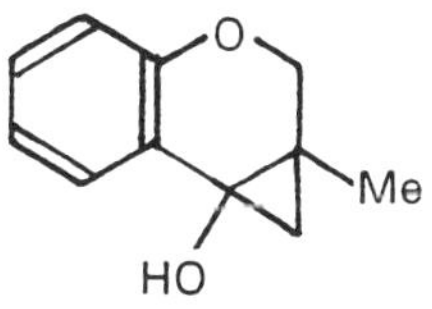

(8)

Reduction of chroman-4-one and thiochroman-4-one by *Mortierella isabellina* ATCC 42613 proceeds to give (*S*)-chroman-4-ol and (*S*)--thiochroman-4-ol in high yield and enanthiomeric excess (H.L. Holland, T.S. Manoharan, and F. Schweizer, Tetrahedron Asymmetry, 1991, <u>2</u>, 335). Directed lithiation of chroman-4-ols occurs predominantly at the C-5 position under kinetic control but at C-8 at higher temperatures. 2,2,5,7-Tetra-methylchroman-4-ol is preferentially lithiated at the 5-Me group (R.J. Bethune *et al.*, J. Chem. Soc., Perkin 1, 1994, 1925). 2-(2-Methylallyloxy)-benzoyl chloride, $2\text{-}(CH_2=CMeCH_2O)C_6H_4COCl$ on treatment with SmI_2 in THF affords 3-methyl-3,4-methanochroman-4-ol (8) (M. Sasaki, J. Collin, and H.B. Kagan, Tetrahedron Letters, 1988, <u>29</u>, 6105).

Mixtures of *cis*- and *trans*-3-benzylchroman-4-ols 9 ($R^1 = R^2 = H$, $R^3 = H$, OMe; $R^1 = H$, $R^2 = R^3 = OMe$) have been obtained by the reduction of 3-aryl-idenechroman-4-ones 10 (M. Gomis *et al.*, Bull. Soc. Chim. Fr., 1988, 585).

(9)

(10)

(11)

(12)

2,3-*trans*-7-Methoxy-2,3-diphenylchroman-4-one on sodium borohydride reduction provides a mixture of isomeric 7-methoxy-2,3-diphenylchroman-4--ols (11) (B.S. Verma, K.S. Dhindsa, and N.K. Sangwan, Indian J. Chem., 1993, <u>32B</u>, 239). 7-(2-Carboxy-5-chlorophenyl)-3-(4-phenylbenzyl)chroman--4-ol and related compounds, useful as LTB_4 (leukotriene B_4) receptor antagonists, and for the treatment of LTB_4-induced illnesses, have been prepared (K. Koch, PCT Int. Appl. WO 93 15,067, 1993). The synthesis of 6-aryloxy-3-aroylmethylchroman-4-ols and -4-ones and analogues, for example, 3-(pyrid-3-ylmethyl)-6-(quinol-2-ylmethyl)chroman-4-ol, as lipoxygenase and LTD_4 inhibitors have been reported (J.F. Eggler, A. Marfat, and L.S. Melvin, Jr., U.S. US 5,059,609, 1991). For the preparation of enantiomeric *cis*-3-(4,6-dihydroxychroman-3-ylmethyl)benzoic acid (12), see C.W. Murtiashaw and B.C. Vanderplas, *ibid.*, 94 08,986, 1994; 6,7-di-chloro-4-hydroxy-2-methylchroman-4-acetic acid (13; $R^1 = R^2 = Cl$) and 7--chloro-6-fluoro-4-hydroxy-2-methylchroman-4-acetic acid (13; $R^1 = F$, $R^2 = Cl$), C.A. Lipinski *et al.*, J. Med. Chem., 1992, <u>35</u>, 2169; and some optically active derivatives of ($\pm$)-*trans*-3-bromo-6-cyanochroman-4-ol, T. Yamanaka, Eur. Pat. Appl. EP 456,266, 1991.

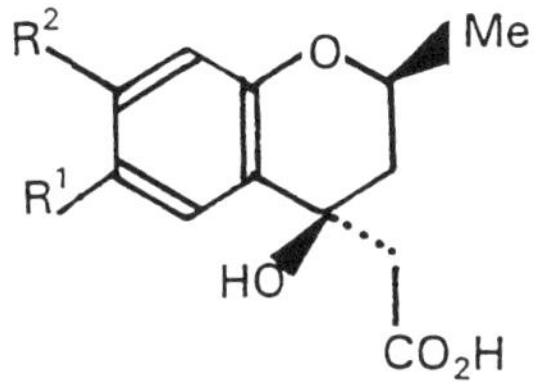

(13)

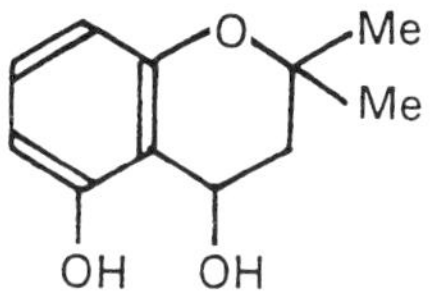

(14)

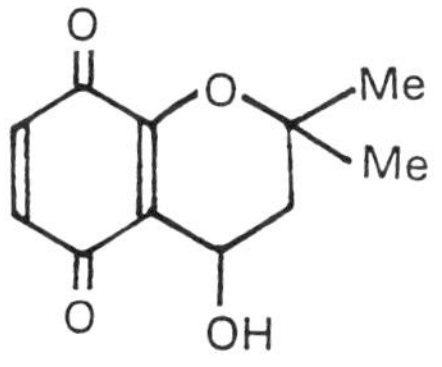

(15)

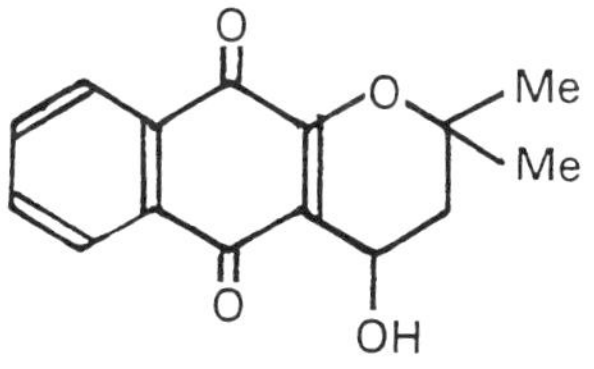

(16)

The oxidation of 4,5-dihydroxy-2,2-dimethylchroman (2,2-dimethyl-chroman-4,5-diol) (14) with $PhI(OAc)_2$ in $MeCN/H_2O$ affords 2,2-dimethyl--5,8-dioxo-5,8-dihydrochroman-4-ol (15), which reacts with $Me_3SiOCH=CH-CH=CH_2$ to yield the pyranonaphthoquinone 16 (B.C. Saitz *et al.*, Synth. Comm., 1992, <u>22</u>, 955).

438

The oxidation of chroman-6-ols by Cu(OAc)$_2$ furnishes 5-acetyloxy-chroman-6-ols 17 (Y. Takizawa *et al.*, Chem. Comm., 1991, 104).
Metallation of 7-hydroxymethyl-5-methoxychroman (18) with butyllithium under kinetically controlled conditions, in benzene at 20°C or ether, TMEDA (2 equivalents) at 20°C results in preferred deprotonation at the C-8 position (L.A. Paquette and M.M. Schulze, Tetrahedron Letter, 1993, 34, 3235).
Derivatives 19 (R^1=H, OH; R^2=H, Me) of 2,2-dimethylchroman-5-ol, -7-ol, and -5,7 diol on treatment with PhCH$_2$CN, 4-MeOC$_6$H$_4$CN, and 3,4-(methylenedioxy)phenylacetonitrile and ZnCl$_2$ and subsequent hydrolysis yield various 6- and 8-phenylacylchromans (K.S.R.M. Rao, C.S.R. Iyer, and P.R. Iyer, Indian J. Chem., 1987, 26B, 676).

(17)

(18)

(19)

(20)

6-Methoxy-2-methylchroman (20) has been obtained from 2,5-dimethoxybenzaldehyde *via* demethylation of 2,5-(MeO)$_2$C$_6$H$_3$CH$_2$CH$_2$CHMeOH with HBr-AcOH (A.R. Tapia *et al.*, Bol. Soc. Chil. Quim., 1992, 37, 305).
Some acyl derivatives of 2-alkylchroman-6-ol have been prepared as liquid crystals (S. Takehara *et al.*, Jpn. Kokai Tokkyo Koho JP 06 256,339 [94 256,339], 1994). Montmorillonite clay catalyzes the rearrangement of 4--MeOC$_6$H$_4$OCH$_2$CH=CMe$_2$ to 2,5-HO(MeO)C$_6$H$_3$CH$_2$CH=CMe$_2$, which with longer reaction time yields 6-methoxy-2,2-dimethylchroman (21) (W.G. Dauben, J.M. Cogen, and V. Behar, Tetrahedron Letters, 1990, 31, 3241).

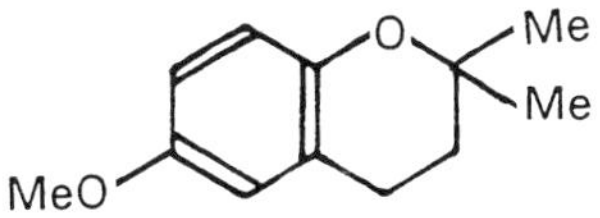

(21)

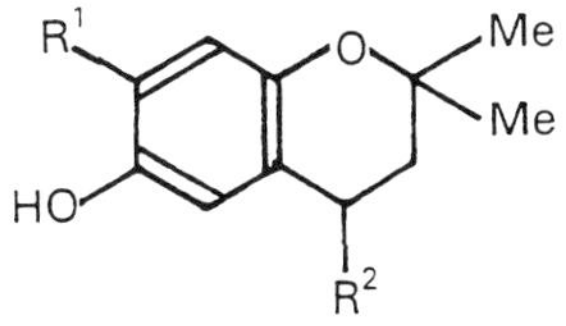

(22)

7-Substituted 4-isopropyl-2,2-dimethyl- and 2,2,4-trimethylchroman-6-ols
22 [R^1 = Bu^t, $CMe_2CH_2CMe_3$, $(CH_2)_7Me$; R^2 = Me, Pr^i] have been prepared by
the cyclocondensation of 2,5-$(HO)_2C_6H_3R^1$ with $HOMe_2CH_2CHR^2OH$ in the
presence of an Amberlyst cation exchange resin (J.R.I. Eubanks and J.G.
Pacifici, Synth. Comm., 1987, 17, 829). The preparations of 2-[(4-amino-
phenoxy)methyl]-2,5,7,8-tetraalkylchroman-6-ols have been reported (J.
Heveling, Eur. Pat. Appl. EP 556,830; 556,831, 1993). (2-R,S)-6-Hydroxy-
-2,5,7,8-tetramethylchroman-2-carboxylic acid has been converted into
methyl (2-R,S)-6-acetyloxy-2,5,7,8-tetramethyl-2H-benzo[b]pyran-2-carboxy-
late (J.M. Grisar, M.A. Petty, and F. Bolkenius, PCT Int. Appl. WO 93
20,058, 1993).

4-(4-Hydroxyphenyl)-7-methoxy-2-phenylchroman (23; R = H) and some
2,4-diaryl-7-methoxychromans have been prepared by the Grignard reaction
between a 7-methoxyflavone and the appropriate aryl halide and subsequent
hydrogenation. Compound 23 (R = H) has been converted into the corresponding ethers 23 (R = pyrrolidinoethyl, glycidyl) and ester 23 (R = Ac), none
of which showed estrogenic, antiestrogenic on antifertility activity V.
Srivastava, S. Ray, and M.M. Singh, Indian J. Chem., 1993, 32B, 569).

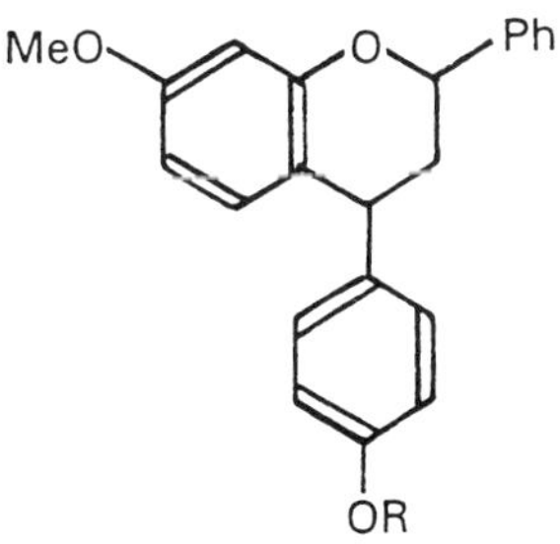

(23)

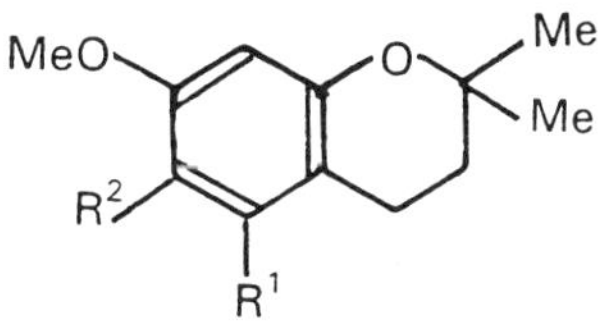

(24)

440

Several 8-alkyl-7-[(carbamoylphenoxy)alkoxy]chroman-2-alkanoates and
analogues have been prepared as LTB$_4$ antagonists (S.W. Djuric, S.H.
Docter, and S.S.T. Yu, PCT Int. Appl. WO 92 00,011, 1992). 5-Acetoxy-7-
-methoxy-2,2-dimethylchroman (24; R^1 = OAc, R^2 = H) on irradiation yields
6-acetyl-5-hydroxy-7-methoxy-2,2-dimethylchroman (24; R^1 = OH, R^2 = Ac)
(25%), also obtained on treating 5-hydroxy-7-methoxy-2,2-dimethylchroman
(24; R^1 = OH, R^2 = H) with AcOH in POCl$_3$ containing AlCl$_3$ (C.S.
Vijayalakshmi, P. Shanmugam, and K.J.R. Prasad, Indian J. Chem., 1989,
<u>28B</u>, 510).

The synthesis of 3,4-*trans*-3-(3-hydroxyphenyl)-2,2-dimethyl-4-[4-(pyrro-
lidinoethoxy)phenyl]-7-methoxychroman-bovine serum albumin (BSA)
conjugate 25, an immunogen for antibody production against centchroman,
has been reported (A.K. Srivastava, R.C. Gupta, and P.K. Grover, Indian J.
Chem., Org. Chem. Incl. Med. Chem., 1994, <u>33B</u>, 671).

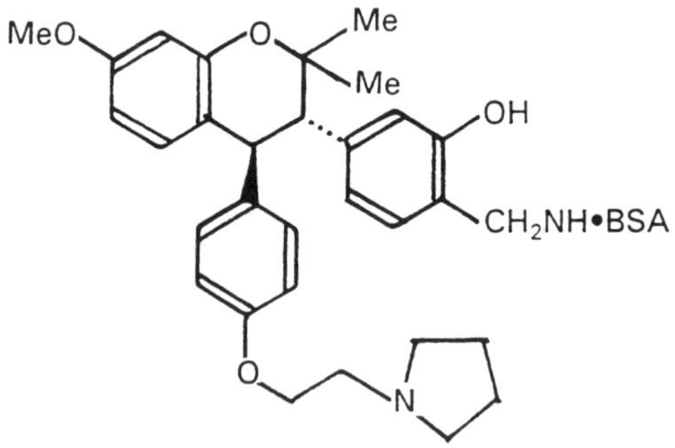

(25)

The substituted 7-hydroxy-2,2-dimethylchromans 26 (R^1 = H, Me, R^2 = H,
OMe, R^3 = CH$_2$CONHC$_6$H$_4$R^5, 4-BrC$_6$H$_4$CH$_2$, CH$_2$Ph, R^4 = H, Me, R^5 = 2-Me-6-
-Et, 2,6-Et$_2$, 2-Cl, 2-Br) have been prepared by the hydrogenation of the
related benzopyran in acetic acid solution in the presence of 10% Pd-C
catalyst (A. Levai, Monatsh., 1992, <u>123</u>, 461). Reversely the following
derivatives of 7-hydroxy-4-methoxy-2,2-dimethylchroman 27 (R^1 = R^2 = H,
R^3 = R^{4}C$_6$H$_4$NHCOCH$_2$, 4-O$_2$NC$_6$H$_4$CH$_2$, PhSO$_2$; R^1 = Me, R^2 = H, R^3 = 4-ClC$_6$-
H$_4$NHCOCH$_2$; R^1 = H, R^2 = R^3 = 4-O$_2$NC$_6$H$_4$CH$_2$; R^4 = H, 2-Me, 4-Me, 2-Cl, 3-
-Cl, 4-Cl, 2-Br, 4-Br) on treatment with 4N HCl in Me$_2$CO or 4-MeC$_6$H$_4$SO$_3$H
in C$_6$H$_6$ afford the corresponding chromens (A. Levai and T. Timar, Synth.
Comm., 1990, <u>20</u>, 641).

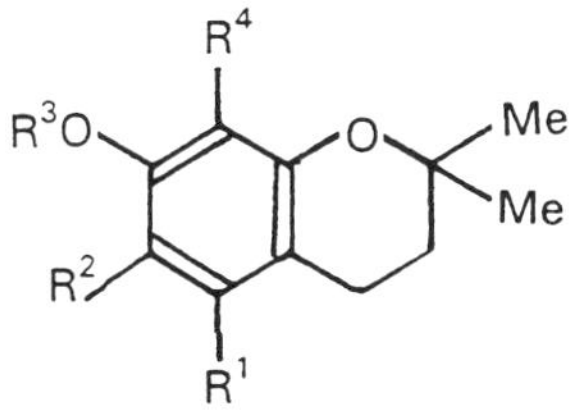

R⁴, R³O, R², R¹, O, Me, Me

(26)

R³O, R², R¹, O, Me, Me, OMe

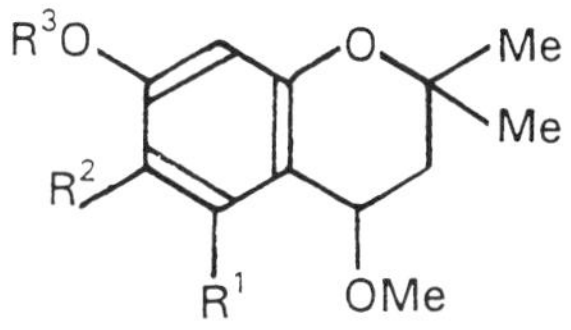

(27)

Aryloxymethylene radicals generated by the decarboxylation of the thio-hydroxamate esters derived from acids 2,6-$R^1R^2C_6H_3OCH_2CO_2H$ (R^1 = alkyl, R^2 = OMe; R^1 = COCH = CHPh, R^2 = H) and 1,2-$(Me_2C = CHCH_2)C_{10}H_6OCH_2$-$CO_2H$ undergo 6-endo cyclization giving 8-methoxy-3-methylchroman (28), 1,2-di(4-oxochroman-3-yl)-1,2-diphenylethane (29), and 2-isopropyl-2,3-di-hydro-1*H*-naphtho[2,1-*b*]pyran (30), respectively, mimicking the unusual biosynthetic reactions involved in the biosynthesis of scabequinone, benzyl-chromanone, and stachyoidin in nature (A.J. Walkington and D.A. Whiting, Tetrahedron Letters, 1989, <u>30</u>, 4731).

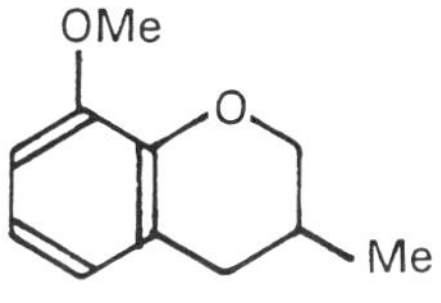

OMe, O, Me

(28)

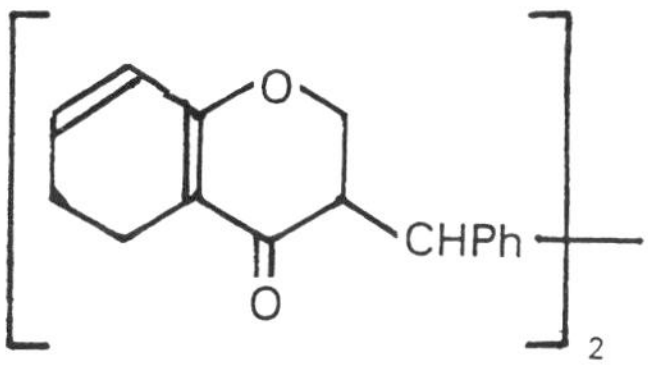

O, O, CHPh, 2

(29)

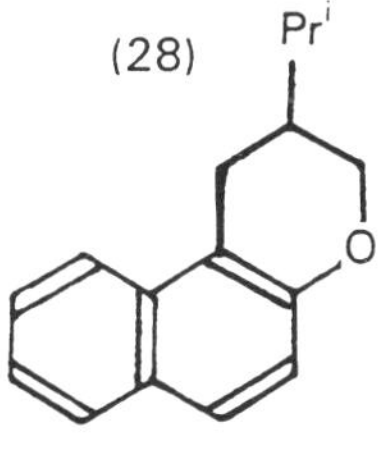

Pr^i, O

(30)

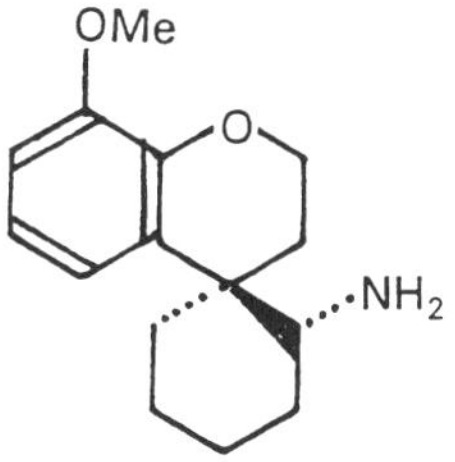

OMe, O, NH₂

(31)

The synthesis of 4,4-spiro(2-aminocyclohexyl)-8-methoxychroman (31) has been reported (A. Bedoui *et al.*, J. Hetercyclic Chem., 1992, <u>29</u>, 547).

(e) Flavanols, 3-hydroxy-2-phenylchromans, 3-hydroxy-2-phenyl-3,4-dihydro-2H-benzo[b]pyran, catechins and related condensed tannins

(i) Flavanols

3,5-$(MeO)_2C_6H_3OH$ cyclocondenses with $PhCOCH = CHPh$ in CH_2Cl_2 in the presence of $TiCl_4$ to afford 5,7-dimethoxy-2,4-diphenyl-4*H*-benzo[*b*]pyran (1), which on hydroboration-oxidation yields 5,7-dimethoxy-4-phenylflavan-3--ol (2) (F. Nakatsubo *et al.*, Mokuzai Gakkaishi, 1987, <u>33</u>, 226).

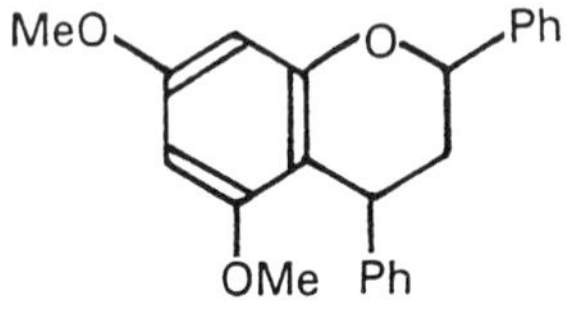

(1)

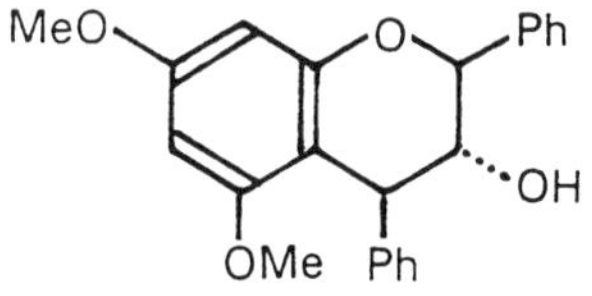

(2)

(ii) Catechins

3-Acylated catechins 1 ($R^1 = C_{1-10}$alkyl, $R^2 = H$, HO) have been prepared as antioxidative agents (M. Sakai *et al.*, Eur. Pat. Appl. EP 618,203, 1994).

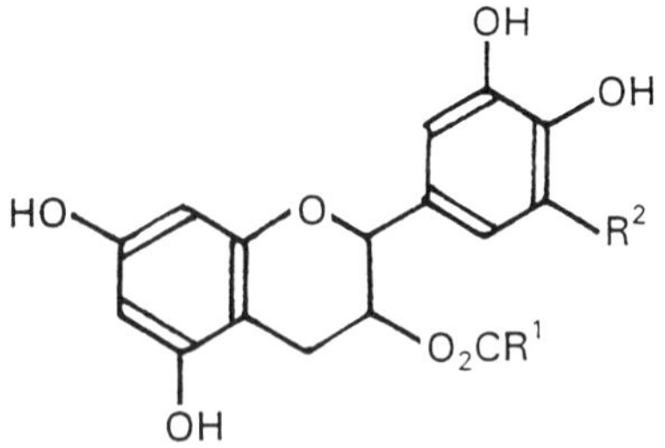

(1)

Flavonoid polymers having d.p. 3-10 and MW 800-5000 have been prepared as potent glucosyltransferase inhibitors, by treating flavonoids [e.g. (+)-catechin] with peroxidase (M. Konya, H. Ono, and T. Tanaka, Jpn. Kokai Tokkyo Koho JP 06 247,959 [94 247,959], 1994).

(iii) Flavan-4-ols

A convenient synthesis of 3-chloro-3-nitroflavan-4-ols (3-chloro-4-
-hydroxy-3-nitro-2-phenyl-3,4-dihydro-2*H*-benzo[*b*]pyrans), for example,
flavanol 1, from (2-chloro-2-nitroethenylbenzenes, $PhCH = C(Cl)NO_2$ has been
reported (D. Dauzonne and P. Demerseman, Synthesis, 1990, 66).

(1)

(f) Chromanones, dihydrobenzo[b]pyranones

(i) Chroman-2-ones, dihydrocoumarin

Dihydrocoumarin has been obtained by heating 3-(2-oxocyclohexyl)prop-
ionic acid lower alkyl esters at 230-260°C in the presence of Pd/Al_2O_3
catalysts or supported Pd catalysts and Al_2O_3 (K. Okumura, A. Imamura, and
H. Mizumoto, Jpn. Kokai Tokkyo Koho JP 03 99,076 [91 99,076], 1991).
This cyclization-dehydrogenation method has been used for the preparation
of dihydrocoumarin derivatives (H. Mizumoto, A. Imamura, and K. Okumura,
ibid., 03 120,268 [91 120,268], 1991). Dihydrocoumarin has been
prepared by the hydrogenation of coumarin in isopropanol in the presence of
Raney Ni at 20-120°C and 30kg/cm²G (K. Onishi *et al.*, *ibid.*, 03 232,877
[91 232,877], 1991). A process for purifying (T. Shirafuji *et al.*, Eur. Pat.
Appl. EP 492,940, 1992) and a method for storage or transport of dihydrocoumarin (H. Tsukass *et al.*, Jpn. Kokai Tokkyo Koho JP 06 92,392 [94
92,392], 1994) have been described.

444

The reactions of coumarin and 4-methylcoumarin with single-electron transfer reagent sodium naphthalenide results in the formation, from reductive dimerization, of two diastereoisomeric products 1 (R=H, Me) in each case (A. Banerji *et al.*, Indian J. Chem: Org. Chem. Incl. Med. Chem., 1994, <u>33B</u>, 576).

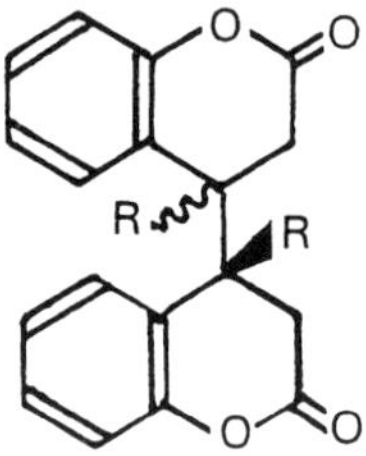

(1)

Chiral ß-carboxylic acids 2 (R^1 = alkyl; R^2 = 2-HOC$_6$H$_4$, 2-HOAr) have been transformed into chiral 3-alkylindan-1-ones 3 (R^1 = alkyl) and 4-alkyldihydrocoumarins 4 (R^1 = alkyl). Ring opening of 4-ethyldihydrocoumarin (4; R^1 = Et) affords (*S*)-2-benzoyl-ß-ethylbenzenepropanoic acid (E. Stephan *et al.*, Tetrahedron: Asymmetry, 1994, <u>5</u>, 41).

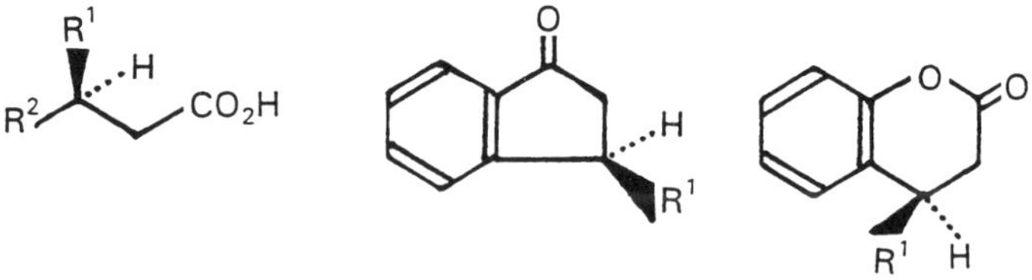

(2)   (3)   (4)

6-Alkenyl-, 6-alkynyl-, and 6-aryldihydrocoumarins have been obtained from 6-bromodihydrocoumarin *via* palladium-catalyzed coupling reactions. Also 6-ketodihydrocoumarins have been prepared *via* palladium-catalyzed carbonylations of 6-iododihydrocoumarins (S.G. Davies, D. Pyatt, and C. Thomson, J. Organometal. Chem., 1990, <u>387</u>, 381). The preparations of 6--chloro-, 4-methoxy-, and 8-methyldihydrocoumarin (J.E. Pickett and P.C. Van Dort, Tetrahedron Letters, 1992, <u>33</u>, 1161), 5-hydroxydihydrocoumarin (2-oxo-5-hydroxychroman) (K. Imaki, H. Wakatsuka, and N. Hamanaka, Jpn. Kokai Tokkyo Koho JP 05 25,159 [93 25,159], 1993), and dihydro-coumarin-4-acetic acid derivatives (A. Bozhilova *et al.*, Synth. Comm., 1992, <u>22</u>, 741) have been reported.

An air-saturated alcoholic solution of 3-acetyldihydrocoumarin (5) on UV irradiation furnishes 3-acetylcoumarin and 3,3'-diacetyl-3,3',4,4'-tetrahydro--4,4'-biscoumarin (I. Petkov, A. Bozhilova, and P. Markov, Monatsh., 1990, <u>121</u>, 85).

(5)

(6)

AcOCHMeC($=CH_2$)CO_2Et on treatment with PhONa gives PhOCHMeC-($=CH_2$)CO_2Et, which rearranges in CF_3CO_2H at 32°C to yield 4-methyl-3--methylenedihydrocoumarin (6) (S.E. Drewes *et al.*, Synth. Comm., 1990, <u>20</u>, 1437). 4-Phenyl-6,8-di(*tert*-butyl)dihydrocoumarin condenses with PhCHO to give 4-phenyl-3-phenylmethylene-6,8-di(*tert*-butyl)dihydrocoumarin [A.Z. Haikal *et al.*, Arch. Pharm. (Weinheim, Ger.), 1990, <u>323</u>, 185). The sequential chlorination and Friedel-Crafts cyclization of $HO_2CCHR^2CHR^1Ar$ (R^1 = Me, Et, H, Pr^i; R^2 = H, Me; Ar = Ph, 3-MeC$_6$H$_4$) yields the indanones 7 (R^3 = H, Me), which undergoes Baeyer-Villiger oxidation to afford the dihydrocoumarins 8 containing up to three substituents (H. Poras *et al.*, Chem. Ind., 1993, 206).

$$(7) \qquad\qquad (8)$$

The cyclocondensation of phenols with cinnamic acid derivatives affords
4-aryldihydrocoumarins, for example, 7-methoxy-4-(4-methoxyphenyl)-
dihydrocoumarin (9), rather than the reported 2,3-dihydrobenzofuran-2-
-ones, for example, 6-methoxy-3-(4-methoxybenzyl)-2,3,-dihydrobenzo[*b*]-
furan-2-one (10) (J.K. Kirtany, Indian J. Chem., 1993, 32B, 993).

$$(9) \qquad\qquad (10)$$

6-Cyclohexylamidomethoxy-4,7-dimethyldihydrocoumarin (11) has been
obtained by the etherification of 6-hydroxy-4,7-dimethyldihydrocoumarin
with $ClCH_2CONH(CH_2)_6$-c in DMF containing K_2CO_3. A number of related
derivatives have also been prepared [K. Vogt, G. Kempter, and R. Schindler,
Ger. (East) DD 270,709, 1989]. The acylation of 7-amino-4-methyldihydro-
coumarin with the acid chlorides of alanine, phenylalanine, and leucine
hydrochlorides gives the corresponding 4-methyldihydrocoumarin-7-ylamides
(Z. Tetere *et al.*, Latv. PSR Zinat. Akad. Vestis, Kim. Ser., 1987, 728).

(11)　　　　　　　　　　(12)

Meldrum's acid on treatment with phloroglucinol, followed by RCHO (R=H, Ph, 4-HOC$_6$H$_4$) affords 6,8-dihydroxydihydrocoumarin and the 4--substituted 6,8-dihydroxydihydrocoumarin 12 (V. Nair, Synth. Comm., 1987, _17_, 723). For the preparation of substituted derivatives 13 (2 of R^1--R^4=H and the others = H, Me, Et) of dihydrocoumarin, see T. Shirafuji _et al._, Eur. Pat. Appl. EP 420,532, 1991; derivatives 13 (R^2=Ph, 4-MeOC$_6$H$_4$, Pr; R^3=R^4=CO$_2$Me; R^3=CO$_2$Me, Ac, R^4=H), E. Wada, S. Kanemasa, and O. Tsuge, Bull. Chem. Soc. Jpn., 1989, _62_, 1198; and 8-carboxyethyl--3,4,5,6,7,8-hexahydrocoumarin (14), H. Huang, Y. Li, and Z. Wang, Gaodeng Xuexiao Huaxue Xuebao, 1987, _8_, 533. 3-Pivaloyl-2_H_-benzo[_b_]-pyran-2-one reacts with (R^{1}R^{2}CHCO)$_2$O (R^1=H, R^2=H, Me, Et; R^1=R^2=Me) in the presence of Et$_3$N to yield 4-(2-oxoalkyl)dihydrocoumarins 15 and 3--pivaloyldihydrocoumarin (A. Boilova _et al._, Ann., 1991, 1279).

(13)　　　　　　　　　　(14)　　　　　　　　　　(15)

448

(ii) Chroman-3-ones

Hydroboration, followed by pyridinium chlorochromate oxidation of 4-
-alkyl- and 4-aryl-2*H*-benzo[*b*]pyrans and 4-alkyl- and 4-aryl-7-methoxy-2*H*-
-benzo[*b*]pyrans (R^1 = H, MeO; R^2 = Me, Et, Ph, 4-MeOC$_6$H$_4$) leads to the
corresponding 4-substituted chroman-3-ones 1 (B.S. Kirkiacharian, A. Danan,
and P.G. Koutsourakis, Synthesis, 1991, 879).

The photochemical behaviour of 3,4-epoxychromans, intermediates in
the synthesis of pterocarpans, has been studied in various solvents and it
has been found that while ring contraction of epoxychromans 2 (R^1 = H,
OMe, Me, Cl, R^2 = H; R^1 = OMe, R^2 = Me), resulting in the formation of 2,3-
dihydro-benzofurans 3 through a photodecarbonylation process occurred in
cyclo-hexane, photoisomerization leading to chroman-3-ones 4 was observed
in acetone (G. Ariamala, T. Sumathi, and K.K. Balasubramanian, Proc.-Indian
Acad. Sci., Chem. Sci., 1992, <u>104</u>, 753). 6-Cyano-2,2-dimethyl-4-(2-oxo-
pyrrolidin-1-yl)chroman-3-one (5) and derivatives with smooth muscle
relaxant activity have been prepared (C.S.V. Frydrych and J.M. Evans, Eur.
Pat. Appl. EP 366,273, 1990).

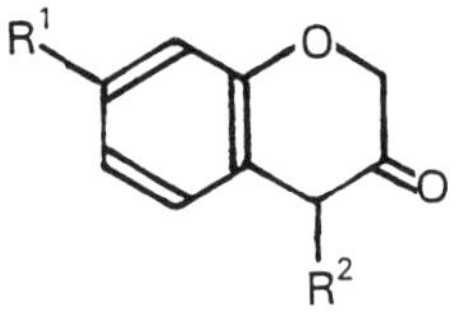

(1)

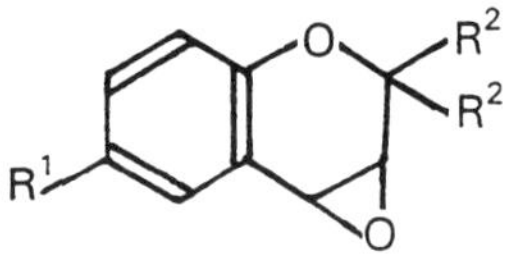

(2)

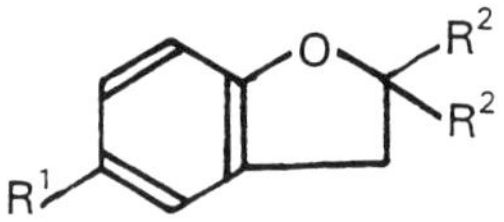

(3)

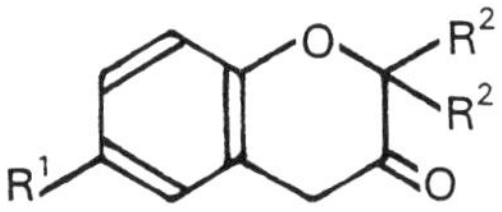

(4)

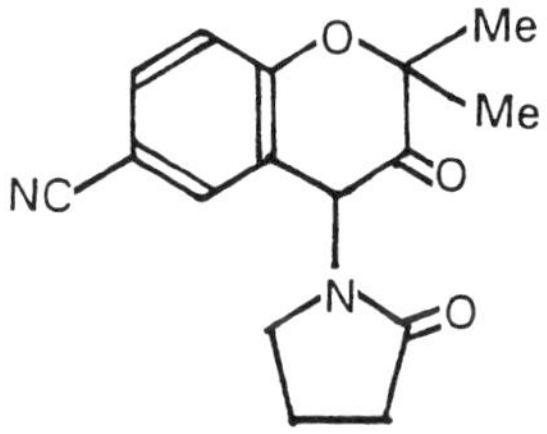

(5)

(iii) Chroman-4-ones

The reaction between monosubstituted phenols and 3-methylbut-2-enoic acid in $POCl_3/ZnCl_2$ and $POCl_3/AlCl_3$ to give substituted 2,2-dimethylchroman-4-ones is strongly influenced by the substituents and their position in the starting phenol (P. Sebok *et al.*, Tetrahedron Letters, 1992, <u>33</u>, 2791; Heterocycles, 1994, <u>38</u>, 2099). (Thio)chroman-4-one 1 (X = S, O) on treatment with DMF-$POCl_3$ yields 4-chloro-2*H*-benzo[*b*](thio)pyran-3-carboxaldehyde 2 (X = S, O). However, with excess DMF-$POCl_3$ at 100°C, chroman-4-one gives 3-chloromethylchroman-4-one (3) (P.R. Giles and C.M. Marson, Tetrahedron, 1991, <u>47</u>, 1303).

(1)   (2)   (3)

Highly enantioselective hydrogenation of chroman-4-ones to the corresponding alcohols has been catalyzed by BINAP-iridium(I)-aminophosphine systems (X. Zhang *et al.*, J. Amer. Chem. Soc., 1993, <u>115</u>, 3318). The reactions and rearrangements of chroman-4-ones have been discussed (C.D. Gabbutt, Diss. Abstr. Int. B, 1990, <u>50</u>, 4521). The reaction of chroman-4-one and 2-phenylchroman-4-one with RCHO [R = (un)substituted Ph, 1-naphthyl, 2-naphthyl, 2-thienyl, 5-chloro-2-thienyl] gives the 3--(substituted methylene)chroman-4-ones 4 and 3-(substituted methylene)-2--phenylchroman-4-ones 5, respectively (A. Levai and Z. Szabo, Pharmazie, 1992, <u>47</u>, 56).

(4)   (5)

6-Methylchroman-4-one has been obtained on heating the aryl γ-halo-genopropargyl ether, $4\text{-MeC}_6\text{H}_4\text{OCH}_2\text{C}\equiv\text{CCl}$ in $\text{HOCH}_2\text{CH}_2\text{OH}$ (G. Ariamala and K.K. Balasubramanian, Tetrahedron, 1989, __45__, 309). Decarboxylation of compound 6 ($R^1 = H$, Me, MeO; $R^2 = $ Me, Et) and the addition of R^3OH ($R^3 = $ Me, Et) affords ketals 7, however the reaction of compound 6 with H_2O yields the 2-(2-substituted 2-oxoethyl)chroman-4-one 8 (S.J. Coutts and T.W. Wallace, Tetrahedron Letters, 1987, __28__, 5195).

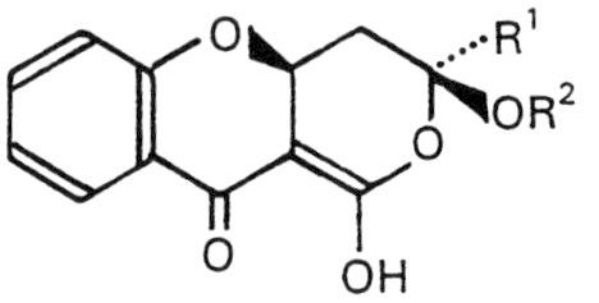

(6)

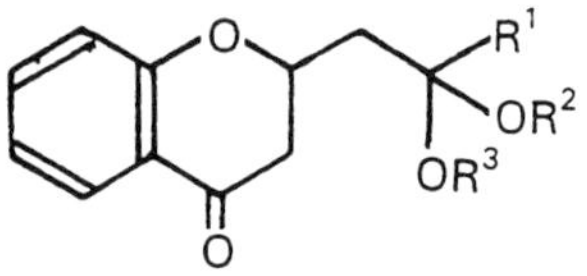

(7)

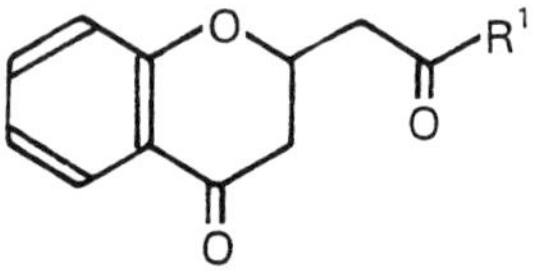

(8)

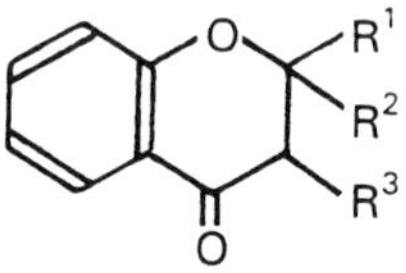

(9) R^1, R^2 (see text), $R^3 = H$
(10) $R^1 = R^2 = R^3 = $ Me

Treating chroman-4-one, 6-chlorochroman-4-one, 3-methylchroman-4--one, and 6-chloro-4-methylchroman-4-one with a pinch, of I_2 and a drop of H_2SO_4 in Me_2SO or with I_2 without H_2SO_4 in Me_2SO affords the corresponding chromones. 4H-Naphtho[1,2-b]pyran-4-one and 1H-naphtho-[2,1-b]pyran-1-one have been prepared in a similar manner (R.P. Kapoor, O.V. Singh, and C.P. Garg, J. Indian Chem. Soc., 1991, __68__, 367).

An alternative to the Kabbe condensation for the synthesis of 2-substi-tuted chroman-4-ones from enolizable aldehydes and ketones, involved a Mukaiyama aldol addition of $2\text{-Me}_3\text{SiOC}_6\text{H}_4\text{C(OSiMe}_3) = \text{CH}_2$ with $R^1\text{COR}^2$ [$R^1 = H$, $R^2 = \text{PhCH}_2$, PhCH_2CH_2; $R^1 = $ Me, $R^2 = $ Ph; $R^1R^2 = (CH_2)_5$] to afford 2--$\text{HOC}_6\text{H}_4\text{COCH}_2\text{CR}^1\text{R}^2\text{OH}$, which on acidic treatment yields chroman-4-ones 9 (S.E. Kelly and B.C. Vandeplas, J. Org. Chem., 1991, __56__, 1325). Treatment of $2\text{-(Me}_2\text{CHO)C}_6\text{H}_4\text{COCN}_2\text{Me}$ with Rh(II) N-benzenesulphonyl-L--prolinate in CH_2Cl_2 at 0°C gives $(+)$-2,2,3-trimethylchroman-4-one (10) in 70% enantiomeric excess (M.A. McKervey and T. Ye, Chem. Comm., 1992, 823).

Treatment of 2-($R^1CH=CR^2CH_2$)$C_6H_4COSePh$ (R^1 =H, R^2 =H, Me, Ph; R^1 = Me, R^2 =H) with Bu_3SnH and AIBN in PhMe affords the 3-substituted and the 3,3-dimethylchroman-4-ones 11, respectively (M.D. Bachi and D. Denenmark, Heterocycles, 1989, <u>28</u>, 583).

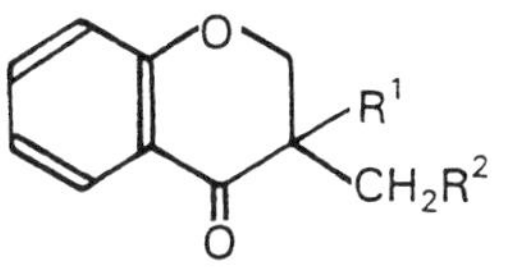

(11) 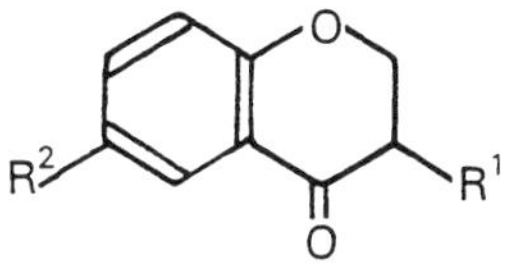   (12)

The preparations of several 3,6-disubstituted chroman-4-ones 12 (R^1 =H, halogeno, alkyl, aryl; R^2 =halogeno, NO_2, NH_2, alkyl) (J.R. Patton and N. Gurusamy, Eur. Pat. Appl. EP 228,172, 1987) as pharmaceutical intermediates have been reported. Chroman-4-one enol ethers 13 (R^1 =H, Me; R^2 = Pr^i, MeO) have been prepared from chromone-3-carboxaldehyde (H. Iwasaki et al., Heterocycles, 1988, <u>27</u>, 1599) and some derivatives of the ethyl isobutyrate 14 of 3-benzylidenechroman-4-one have been synthesized (S. Kirkiacharian, M. Gomis, and P. Koutsourakis, Eur. J. Med. Chem., 1989, <u>24</u>, 309). For the preparation of cyclopenta[b]chromanones 15 (R^1-R^4 =H, Me), see V.C. Waghulde and J.R. Merchant, Indian J. Chem., 1989, <u>28B</u>, 419.

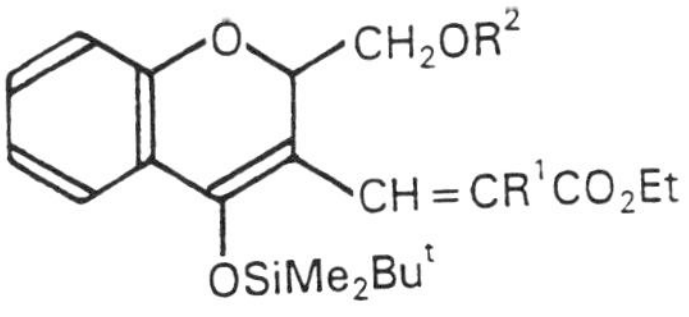

(13) 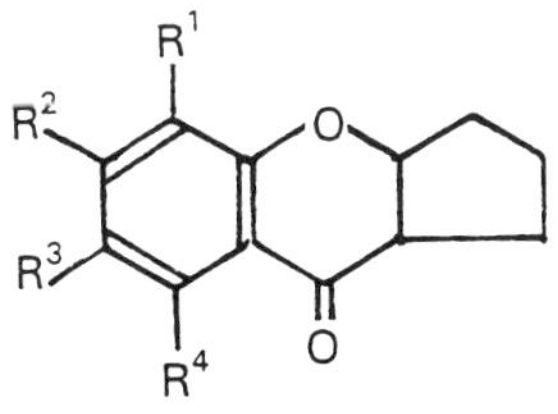   (15)

452

(14)

2-Hydroxychroman-4-ones 16 ($R^1 = R^2 = H$, $R^3 = H$, Me; $R^1 = Me$,
$R^2 = MeO$, $R^3 = H$; $R^1 = H$, $R^2 = R^3 = Me$) on heating above their m.p. afford
(hydroxybenzoyl)(oxobenzopyranyl)ethene derivatives, which are cyclized to
2,3'-bischromones 17 by SeO_2 in boiling isoamyl alcohol (R.R. Soni and K.N.
Trivedi, Indian J. Chem., 1988, <u>27B</u>, 811). The preparation of substituted
6-(benzyloxy)-3-hydroxychroman-4-ones and related compounds for the
treatment of asthma, arthritis, and related diseases, have been described
(J.F. Eggler et al., Eur. Pat. Appl. EP 391,625, 1990). The acid-catalyzed
cyclization of 1-(2,6-dihydroxyphenyl)-3-methylbut-2-en-1-one affords 5-
-hydroxy-2,2-dimethylchroman-4-one (18) (C.S. Vijayalakshmi, M.
Subramanian, and K.J.R. Prasad, Indian J. Chem., 1990, <u>29B</u>, 661). 5,7-
-Diacetoxy-2,2-dimethylchroman-4-one on treatment with porcine pancreas
lipase in THF gives 5-acetoxy-7-hydroxy-2,2-dimethylchroman-4-one (19)
(V.S. Parmar et al., Chem. Comm., 1993, 27).

(16)

(17)

(18) $R^1 = H$, $R^2 = OH$
(19) $R^1 = OH$, $R^2 = OAc$

(20)

Several 3-benzylidenechroman-4-ones and 3-benzylidenethiochroman-4-
-ones have been synthesized and tested for their in vitro antifungal activity.
6-Methoxy-3-(4-methoxybenzylidene)chroman-4-one (20) showed good
activity against *Cryptococcus neoformans* and *Torulopsis glabrata* (T. Al
Nakib *et al.*, Eur. J. Med. Chem., 1990, 25, 455). 3-Diazo-6-(quinol-2-yl)-
methoxychroman-4-one, obtained from 6-hydroxychroman-4-one, and
cyclohexanol when treated with $Rh(OAc)_2$ in PhMe at 70°C give 3-cyclo-
hexyloxy-6-(quinol-2-yl)methxychroman-4-one (J.F. Eggler, L.S. Melvin Jr.,
and A. Marfat, Eur. Pat. Appl. EP 313,296, 1989).

The regioselective synthesis of 7-hydroxychroman-4-one has been
improved by the Friedel-Crafts acylation of resorcinol with 3-chloropropionic
acid in trifluoromethanesulphonic acid, followed by intramolecular base-
mediated cyclization (K. Koch and M.S. Biggers, J. Org. Chem., 1994, 59,
1216).

The 2-hydroxyaryl ß-ketol, $Me_2C = CHCH_2CH_2CMe(OH)CH_2COC_6H_3(OH)$-
OMe-2,4, on cyclodehydration with $(Me_2N)_3P$ gives 7-methoxy-2-methyl-2-
-(4-methylpent-3-en-1-yl)chroman-4-one (21; R = Me) a cannabinoid synthon
(A. Banerji and G.P. Kalena, Synth. Comm., 1989, 19, 159). The
intramolecular 1,2-arene-alkene photocycloaddition of 7-hydroxy-2-methyl-2-
-(4-methylpent-3-en-1-yl)chroman-4-one (21; R = H) and its alkyl ethers 21
(R = Me, Et) to trimethyloxatetracyclotetradecenedione 22 has been
described (G.P. Kalena, P. Pradhan, and A. Banerji, Tetrahedron Letters,
1992, 23, 7775).

(21)

(22)

454

7-Methoxy- and 7,8-dimethoxy-3-methylchroman-4-ones 23 (R = H and
MeO, respectively) have been obtained on treating, the appropriate α-
-hydroxymethyl derivative 24 derived from the related propiophenone, with
2% aqueous ethanolic Na_2CO_3 (A.C. Jain, O.D. Tyagi, and R. Saksena,
Indian J. Chem., 1989, 28B, 15).

(23)

(24)

(25)

(26)

The ethoxymethylation of chroman-4-ones 25 (R^1 = H, Me) with
$EtOCH_2Cl$ in the presence of K_2CO_3 in acetone yields 7-ethoxymethoxy-3,3-
-bis(hydroxymethyl)chroman-4-one (25; R^1 = $EtOCH_2$, R^2 = CH_2OH) and 3,3-
-bis(hydroxymethyl)-7-methoxychroman-4-one (25; R^1 = Me, R^2 = CH_2OH),
respectively (A.C. Jain and R. Saksena, Proc.-Indian Acad. Sci., Chem. Sci.,
1991, 103, 25). The preparations of 3,3-spiro(cyclohexyl)-4-oxochroman-7-
-yloxyacetic acid (26) and related compounds, useful as antihyperuricemics
(H. Harada et al., Eur. Pat. Appl. EP 415,566, 1991) and of 2-[(3S,4R)-4-
-hydroxy-3-benzylchroman-7-yl]cyclopentanecarboxylic acid from 3-benzyl-7-
-(trifluoromethyl)sulphonyloxychroman-4-one (K. Koch et al., PCT Int. Appl.
WO 93 15,066, 1993) have been reported.

The enolate anion of 7-methoxy-2,2-dimethylchroman-4-one obtained
from its enol acetate, undergoes aldol-type condensation with aldehydes,

RCHO (R = H, Me), in the presence of $ZnCl_2$. Dehydration of the resulting (hydroxyalkyl)chroman-4-ones affords the 3-alkylidene-7-methoxy-2,2--dimethylchroman-4-ones 27, which have been reduced to the corresponding 3-alkylchroman-4-ones. Only the reduction of 7-methoxy-2,2-dimethyl-3--methylenechroman-4-one with $NaBH_4$ in the presence of $CeCl_3$ yields 7--methoxy-2,2-dimethyl-3-methylenechroman-4-ol (28) (P. Anastasis et al., J. Chem. Res., S, 1989, 36).

(27)

(28)

7-Hydroxy-, 7-methoxy-, and 7-ethoxychroman-4-ones 29 ($R^1 = OH$, OMe, OEt; $R^2 = H$, OH, OMe) with or without a OH or a OMe group in position-6 have been obtained by treating the appropriate substituted phenol 30 with dimethylacrylic acid in the presence of trifluoroacetic acid (R. Chaturvedi and N.B. Mulchandani, Indian J. Chem., 1992, 31B, 338). The tandem Claisen Diels-Alder reaction of 2,2,8-trimethyl-7-(1,1-dimethylprop-2--ynyloxy)chroman-4-one (31) affords the novel 2,2,7,7,9-pentamethylcyclopenta[h]-chroman-4,8-dione (32) (R.R. Soni and K.N. Trivedi, ibid., 1993, 32B, 991).

(29)

(30)

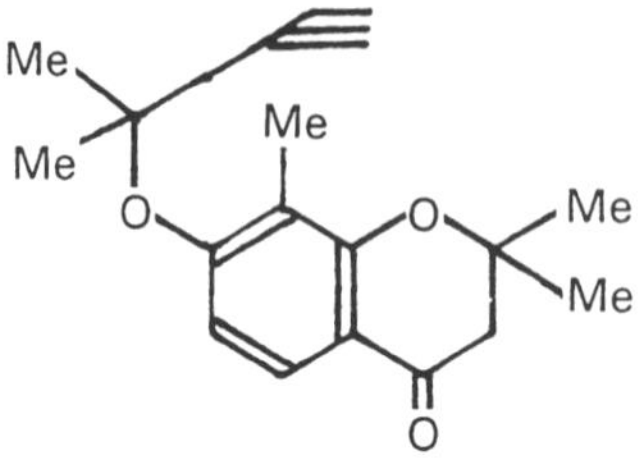

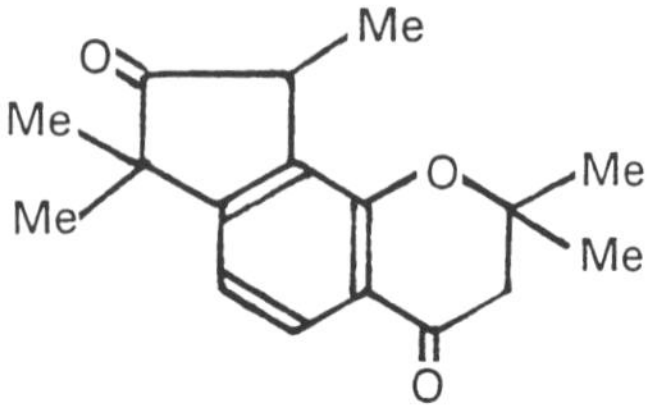

(31) (32)

5,7-Dihydroxy-8-(γ,γ-dimethylallyl)-2-(4-hydroxy-2-methoxyphenyl)-chroman-4-one (33) has been prepared from chalcone 34 in 3 steps involving prenylation, cyclization, and demethoxymethylation (R. Chopra and M. Krishnamurti, J. Indian Chem. Soc., 1987, $\underline{64}$, 444). 2,2-Dimethylchroman-4-ones 35 (R^1-R^3 = H, H, Me, Cl, CHO) with substituents in the 5-, 6-, and 8-positions have been prepared from the corresponding 7-hydroxy-2,2-dimethylchroman-4-one (P. Sebok et al., Hung. Teljes HU 59,121, 1992).

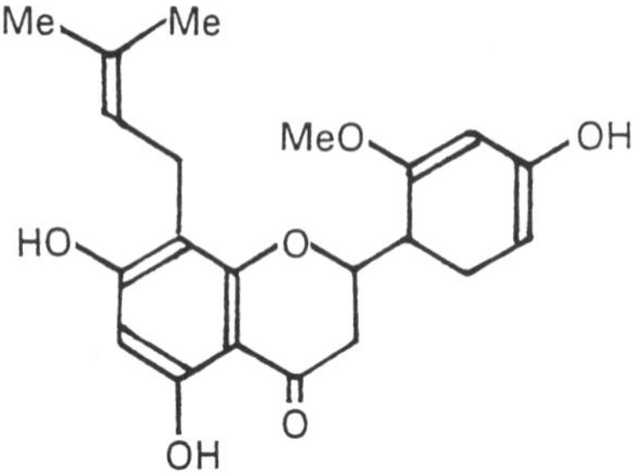

(33) 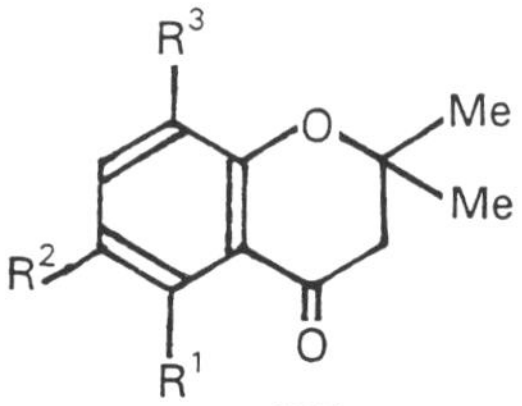 (35)

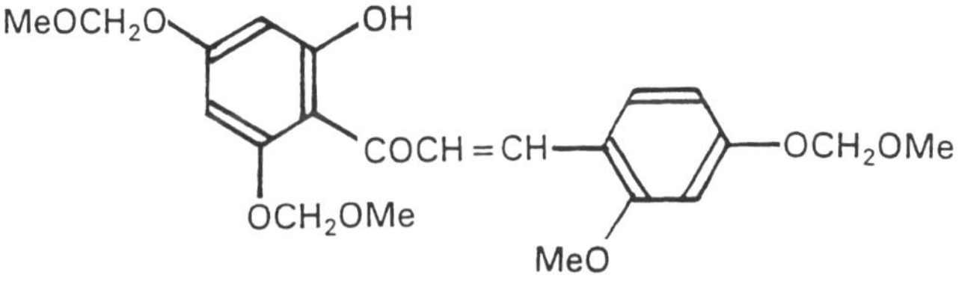

(34)

Methyl chromone-2-carboxylate undergoes radical alkylation with RCO_2H ($R = Bu^t$, Pr^i, Et) to furnish dimeric products 36 (M. Tada and N. Mori, Heterocycles, 1991, _32_, 749). For the preparation of optically enriched methyl 4-oxochroman-2-acetate (37), see T.W. Wallace *et al.*, Chem. Cmm., 1991, 1707).

(36)

(37)

(38) $R^1 = H$
(39) $R^1 = CO_2Me$
(40) $R^1 = COR$

6-Fluorochroman-4-one (38), useful as an intermediate for aldolase reductase inhibitor sorbinil has been obtained by reacting $ClCH_2CH_2COCl$ with $4\text{-}FC_6H_4OMe$ in $(ClCH_2)_2$ containing $AlCl_3$ (D. Wirth and D. Gibert, Fr. Demande FR 2,588,860, 1987). It has also been prepared from byproducts obtained during the synthesis of sorbinil (C.W. Berkeley, Jr., *et al.*, Rom. RO 91,017, 1987) and in 4 steps from $4\text{-}HOC_6H_4NHAc$ (S. Hu, J. Chen, and J. Jian, Zhongguo Yiyao Gongye Zazhi, 1990, _21_, 220).

The ester $(Z)\text{-}4\text{-}FC_6H_4OC(CO_2Me) = CHCO_2Me$ in concentrated H_2SO_4 at 25-30°C yields methyl 6-fluoro-4-oxochroman-carboxylate (39) (M. Kurono *et al.*, Eur. Pat. Appl. EP 331,078, 1989). Some 6-fluoro-4-oxochroman-2-carboxylic acid esters and amides 40 [$R = OR^1$, NR^2R^3; $R^1 = $ (un)substituted alkyl; $R^2,R^3 = H$, (un)substituted alkyl, aryl] have been prepared with high selectivity by catalytic hydrogenation of the corresponding chromones in the presence of metal catalysts, for example, Pt/C (T. Koizumi, Jpn. Kokai Tokkyo Koho JP 03 17,076 [91 17,076], 1991). A number of optically

active 6-fluoro-4-oxochroman-2-carboxylic acid derivatives, for example, 40
[R = (*S*)-NHCHMePh] with (+)-:(-)-isomer ratio 80:20, have been prepared
(*idem, ibid.*, 03 44,384 [91 44,384], 1991). Also, racemic 40 (R = OH) has
been treated with *R*-(+)-PhCHMeNHCH$_2$Ph and the salt decomposed to yield
(*S*)-40 (R = OH) (M. Kurono *et al.*, Eur. Pat. Appl. EP 488,047, 1992). The
preparation of a number of optically active (*S*)-6-halogeno-4-oxochroman-2-
-carboxylic acid derivatives, useful as intermediates for treating agents for
diabetes complications, have been reported (*idem*, Jpn. Kokai Tokkyo Koho
JP 06 92,956 [94 92,956], 1994). For the preparation of 2-hydroxy-2-
-trifluoromethylchroman-4-one, see V.Ya. Sosnovskikh and I.S. Ovsyannikov,
Zh. Org. Khim., 1993, <u>29</u>, 89).

Cyclization of Mannich base salts 41 (R^1 = H, Cl; R^2 = H, Cl) with
aqueous KOH affords 6-chloro-, 7-chloro-, and 6,7-dichlorochroman-4-ones
42 (B. Cox and R.D. Waigh, Synthesis, 1989, 709). The Ag ion assisted
solvolysis of the trichlorobenzocyclopropadihydropyran 43 yields 3-dichloro-
methylene-7-methoxy-2,2-dimethylchroman-4-one (44) (P.E. Brown and Q.
Islam, Tetrahedron Letters, 1987, <u>28</u>, 3047).

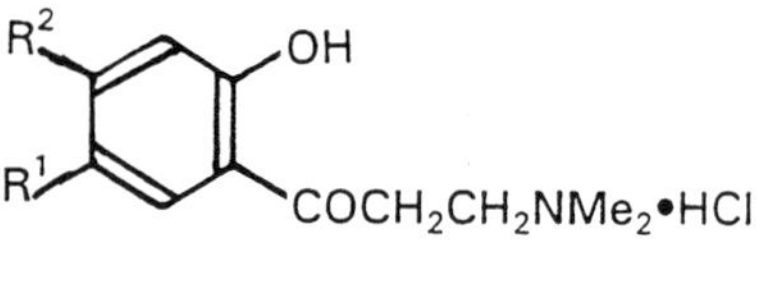

(41)

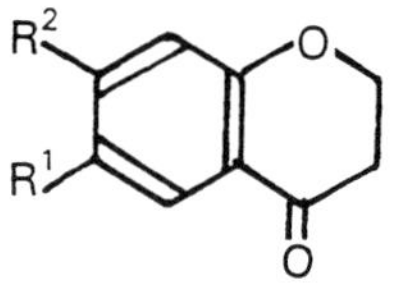

(42)

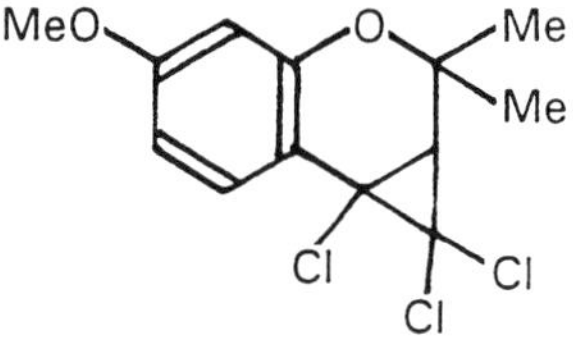

(43)

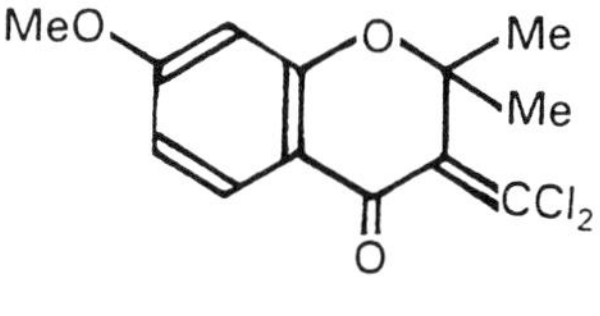

(44)

The oxidative coupling of 2,4-dimethylphenol with 3,5-dichloro-4-(benzyl-oxy)phenylacetic acid, followed by Fries rearrangement of the product in the presence of $AlCl_3$ and intramolecular oxidative cyclization yields the 2,2-spiro derivative 45, an analogue of the natural product thelepin, isolated from the marine annelid, *Thelepus setosus* (B.S. Ko and T. Oritani, Han'guk Nonghwa Hakhoechi, 1992, <u>35</u>, 470).

(45)

2,2-Dimethyl- and 2,2,6-trimethylchroman-4-ones and 2,2-dimethyl- and 2,2,6-trimethylthiochroman-4-ones react with an excess of thionyl chloride to give 3-chloro-3-chlorosulphenylchroman- and thiochroman-4-ones 46 (R^1 = H, Me; X = O, S), which form sulphenamides 47 (R^2 = Et_2N, morpholino, pyrrolidino) when treated with secondary amines. Sulphenamide 47 (R^2 = Et_2N, X = S) on hydrolysis undergoes a ring contraction to afford the benzo[*b*]thiophen-3-ones 48 (C.D. Gabbutt *et al.*, Tetrahedron, 1994, <u>50</u>, 827). The preparation of a number of derivatives of 3-chloro-3-chlorosulphenylchroman-4-one have been reported, direct oxidation of which provides 3,3-dichlorochroman-4-ones, while conversion to the sulphenamides before oxidation furnishes a facile route to 3-chlorochroman-4-ones (C.D. Gabbutt, J.D. Hepworth, and B.M. Heron, *ibid.*, p.5245).

(46) Y = CO
(47) Y = CH_2, SCl = SR^2

(48)

The addition of bromine to the exocyclic α,β-unsaturated ketones 49
(X = O, S, CH_2, CH_2CH_2; R^1 = H, Ph; R^2 = H, Me) gives the dibromides 50,
for example, 3-bromo-3-phenylbromomethylchroman-4-one 50 (X = O,
$R^1 = R^2 = H$) (G. Toth, F. Janke, and A. Levai, Ann., 1989, 651).

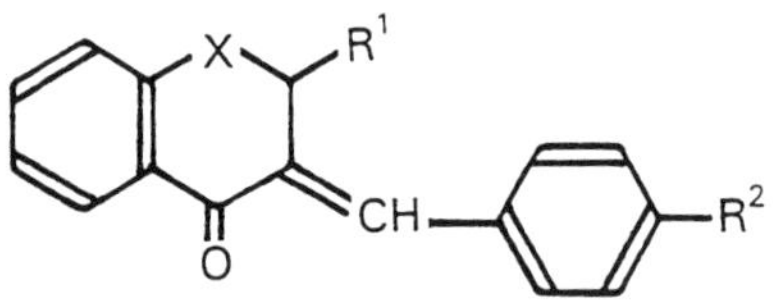

(49)

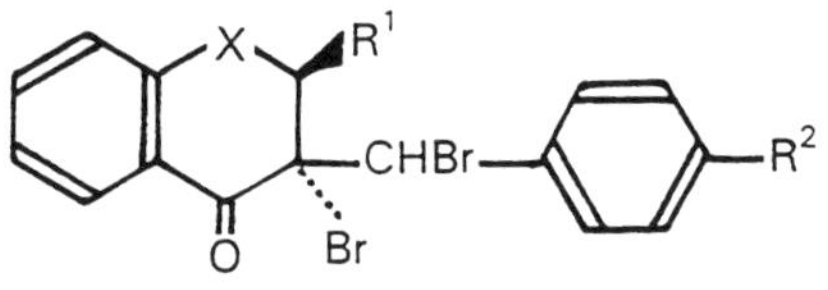

(50)

2,2-Dimethyl-6-nitrochroman-4-one (51) reacts with tetrahydrothiophene
S-oxide to give the 4-hydroxy-2,2-dimethyl-6-nitro-4-thienylchroman 52,
which on dehydration yields the related chromene. The latter two
compounds are effective potassium channel activators, but are less active
than the corresponding pyridine N-oxide (S.E. Yoo et al., Bioorg. Med. Chem.
Letters, 1993, 3, 553).

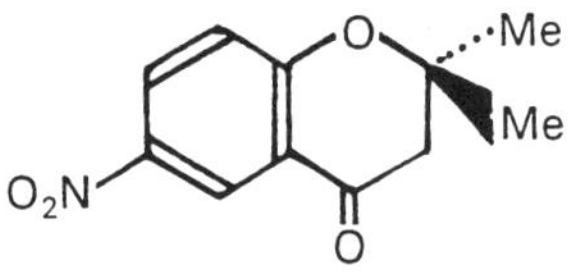

(51)

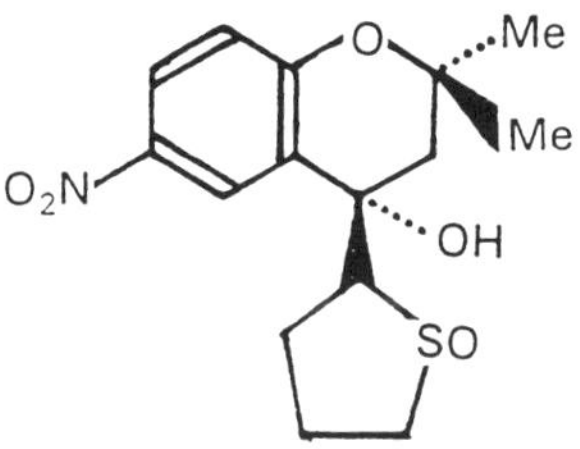

(52)

The reaction of 8-amino-7-decyloxychroman-4-one with *tert*-BuCOCl in the presence of Et_3N affords 8-(tert-butyloxyamino)-7-decyloxychroman-4--one (53). A number of related derivatives, (T. Shiota *et al.*, PCT Int. Appl. WO 93 15,065, 1993) and derivatives of (7-methyl-4-oxochroman-8-yl)urea for example, 54 (*idem, ibid.*, 92 01,681, 1992), useful as anticholesteremics, have been prepared. For the preparation of 2,2-dimethylchroman-4--ones and 2,2-dimethylchromenes with carboxamide containing side chain, see A. Levai and T. Timar, Pharmazie, 1990, <u>45</u>, 728.

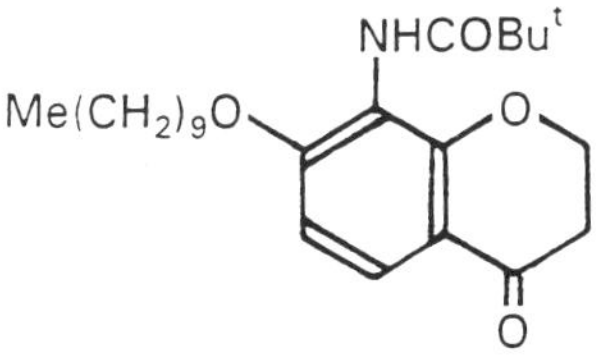

(53)

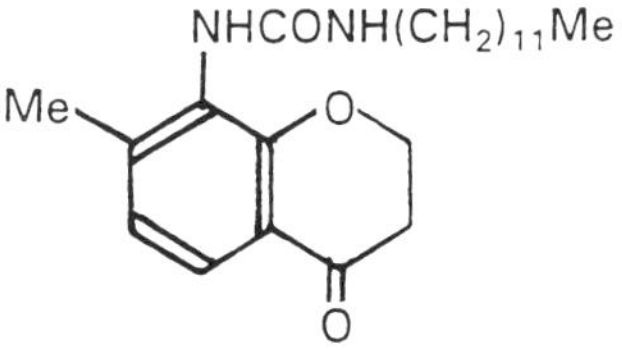

(54)

(iv) Chromandiones

Chroman-2,4-dione (1) reacts with $PhI(OAc)_2$ to give ylide 2, which reacts with $MeSO_2Cl$ in the presence of Et_3N to yield 3-chloro-3-methyl-sulphonylchroman-2,4-dione (3) [W. Hanefeld and B. Spangenberg, Arch. Pharm. (Weinheim, Ger.), 1987, <u>320</u>, 666).

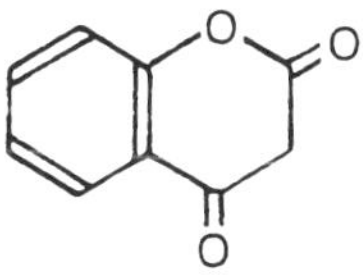

(1)

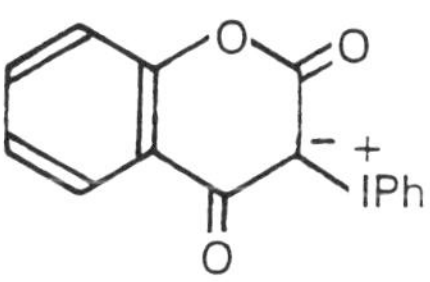

(2)

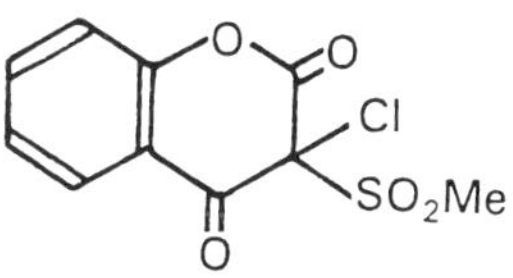

(3)

The mechanism for the photochemical conversion of 4,4,7,7-tetramethyl-
-4,6,7,8-tetrahydro-2*H*-benzopyran-2,5(3*H*)-dione (4) into 3,3,6,6-tetra-
methyl-3,5,6,7-tetrahydrobenzofuran-4(2*H*)-one (5) (K. Hobel and P.
Margaretha, Ber., 1990, <u>123</u>, 101) and the equilibrium between *Z* and *E*
isomers of Schiff bases 6 (X = S, Se; R = alkyl, 4-MeC$_6$H$_5$) derived from 3-
-formyl-4-thio(seleno)coumarin (7) (A.L. Nivorozhkin *et al.*, J. Chem. Soc.,
Perkin 2, 1993, 2423) have been discussed.

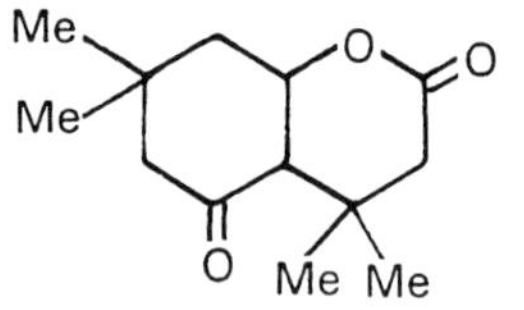

(4)

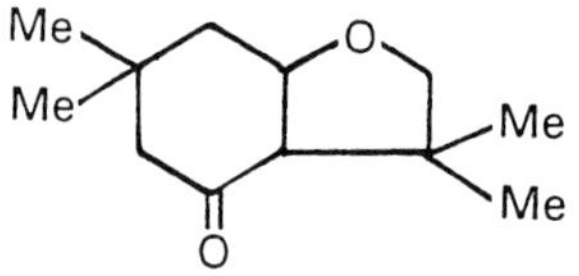

(5)

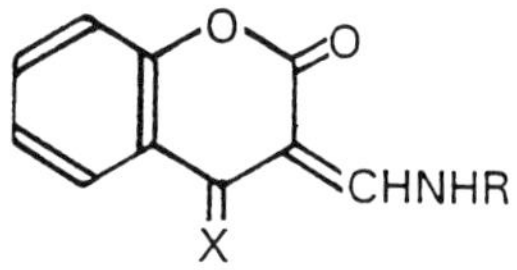

(6)

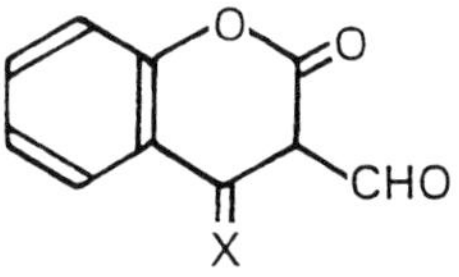

(7)

(g) *Flavanones, 2-phenylchroman-4-ones, 2-phenyl-3,4-dihydro-2H-*
benzo[b]pyran-4-one

The abnormal base-catalyzed reaction of benzaldehydes, 4-RC$_6$H$_4$CHO
(R = H, MeO, Cl, O$_2$N) with 2-HOC$_6$H$_4$COCH$_2$COPh yields the 4'-substituted
flavanones [2-(4-substituted phenyl)chroman-4-ones] 1 (S.J. Joglekar and
S.D. Samant, Tetrahedron Letters, 1988, <u>29</u>, 241).

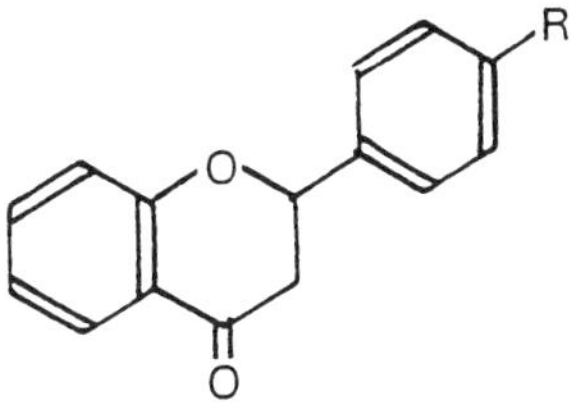

Silyl enol ethers of flavanones 2 (R = H, 4'-OMe, 4'-Cl, 3'-Br) undergo oxido-rearrangement on treatment with PhIO-BF$_3$ to give, primarily the corresponding isoflavanones in ~70% yield (L. Li and Y. Rui, Gaodeng Xuexiao Huaxue Xuebao, 1991, 12, 777; L. Li et al., Chin. Chem. Letters, 1994, 5, 285).

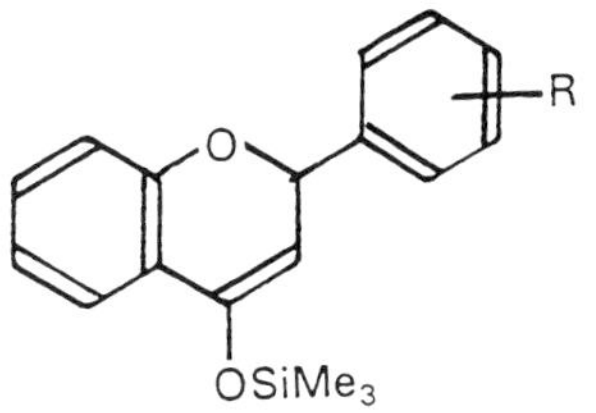

A tentative mechanism has been suggested for the methylene blue sensitized photooxygenation of 3-arylideneflavanones 3 (R^1-R^3 = H; R^1 = H, R^2 = R^3 = OMe; R^1 = OMe, R^2 = R^3 = H, Cl) to give the 3-aroylflavones 4 (H.M. Chawla and S.K. Sharma, Synth. Comm., 1990, 20, 301).

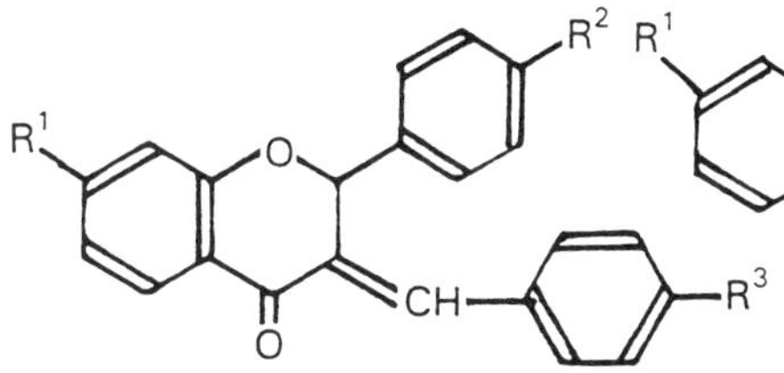

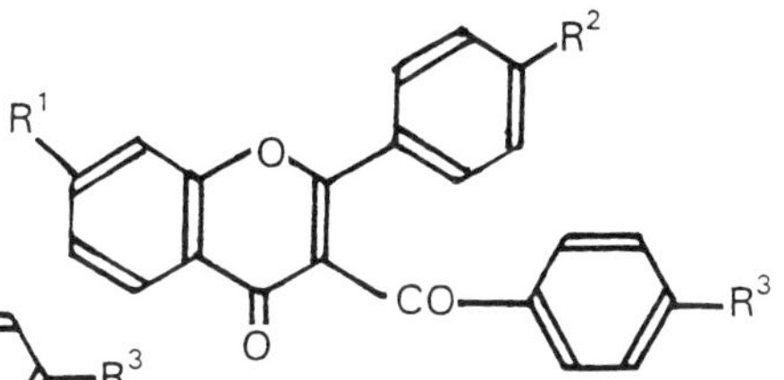

464

N-Salicyloylflavanone hydrazones 5 (R^1=H, Me; R^2,R^3=H, MeO) have been oxidized with SeO$_2$ in AcOH to afford flavones 6, and flavanone azines 7, and 2-HOC$_6$H$_4$CO$_2$H (A.Y. Kale, A. Shivhare, and D.D. Berge, J. Indian Chem. Soc., 1987, <u>64</u>, 642).

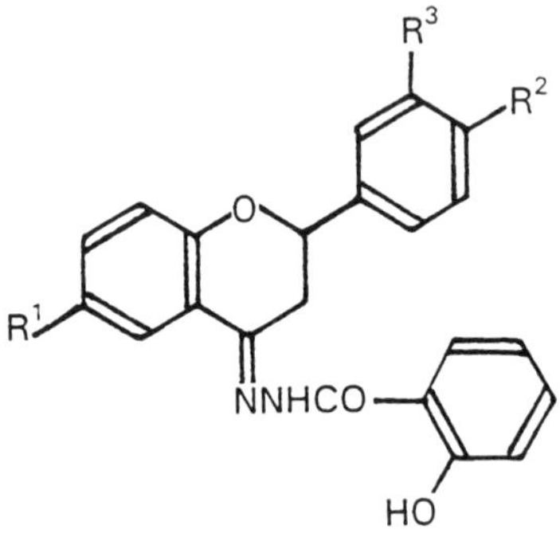

(5)

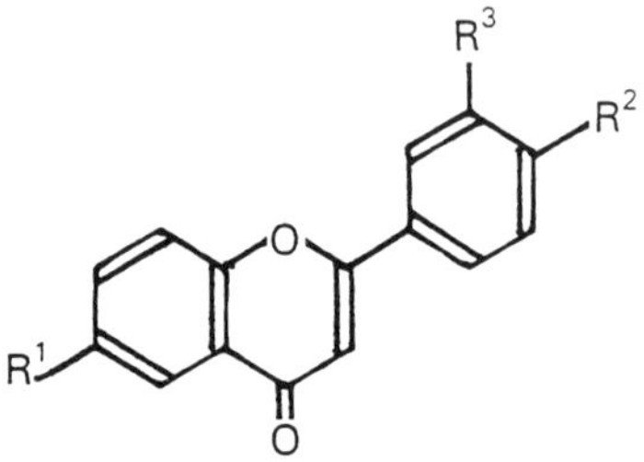

(6)

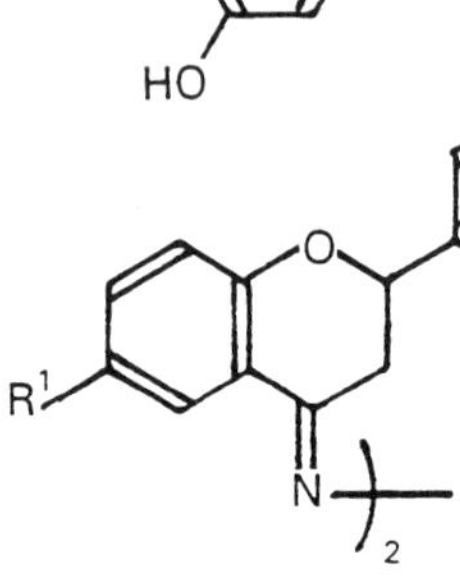

(7)

(h) Hydroxyflavanones

Heating chalcones 1 [R=H, 2-Me, 4-Me, 4-F, 3-O$_2$N, 2,5-HO(MeO), 3,4--methylenedioxy] with concentrated HCl in AcOH at 80-85 °C affords 5,7--dihydroxyflavanones 2 (Y. Yao, F. Yang, and F. Gao, Zhonggu Yiyao Gongye Zazhi, 1992, <u>23</u>, 211). RCH=C(CN)$_2$ [R=(un)substituted Ph] on treatment with phloroglucinol gives 5,7-dihydroxyflavanones 2 (J. Chen and Y. Li, Xiamen Daxue Xuebao, Ziran Kexueban, 1992, <u>31</u>, 651). The mass spectra of 5,7-dihydroxyflavanones (R=H, HO, Cl, NO$_2$MeO) have been reported (*idem, ibid.*, 1994, <u>33</u>, 123).

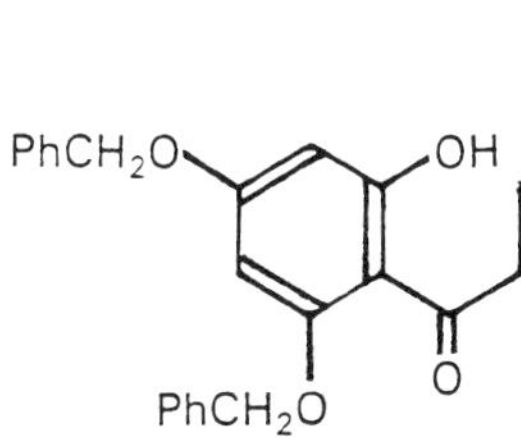

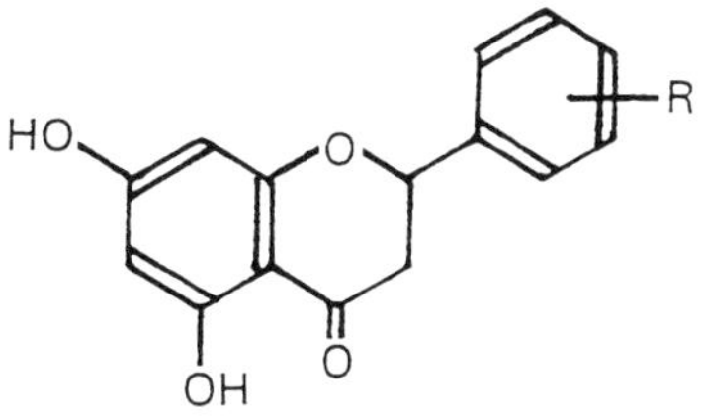

(1) (2)

5,7-Dihydroxy-4'-methoxy-8,3'-di(3-methylbut-2-enyl)flavanone (3) has been obtained in 3 steps from chalcone 4 (R. Chopra and M. Krishnamurti, Indian J. Chem., 1989, 28B, 507).

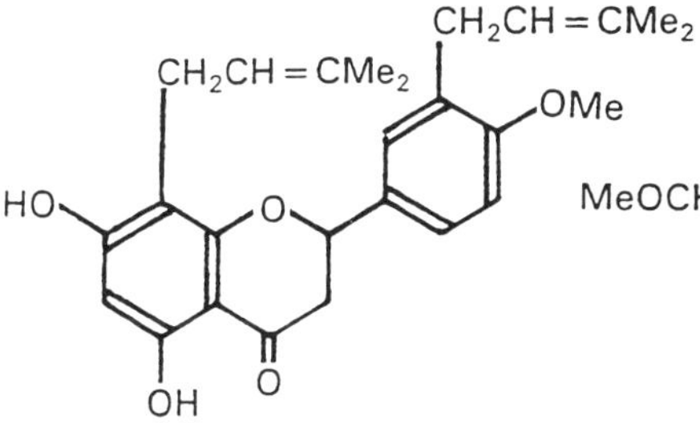

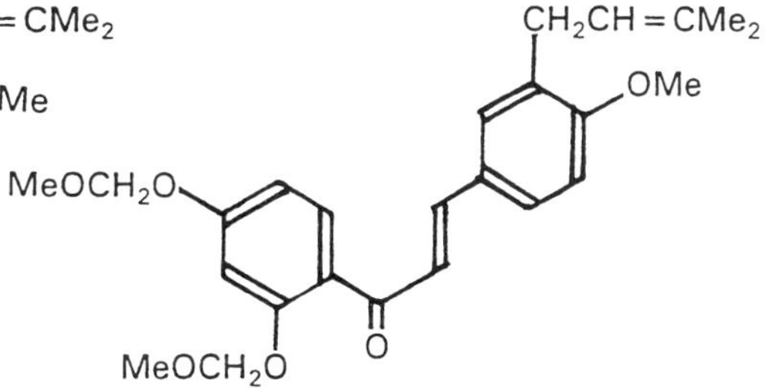

(3) (4)

Pure (+)-aromadendrin trimethyl ether (3-hydroxy-4',5,7-trimethoxy-flavanone) (5) and its enantiomer have been prepared and on reduction with NaBH$_4$ they afford four pure flavan-3,4-diol trimethyl ethers (3,4-dihydroxy--4',5,7-trimethoxyflavan) 6 (H. Takahashi et al., Chem. Pharm. Bull., 1988, 36, 1877).

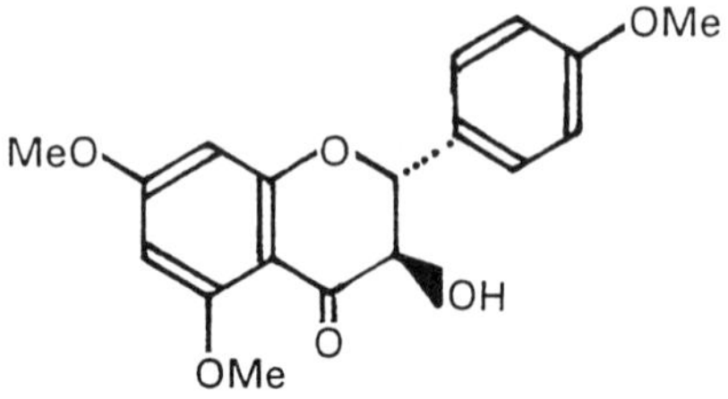

466

(5) 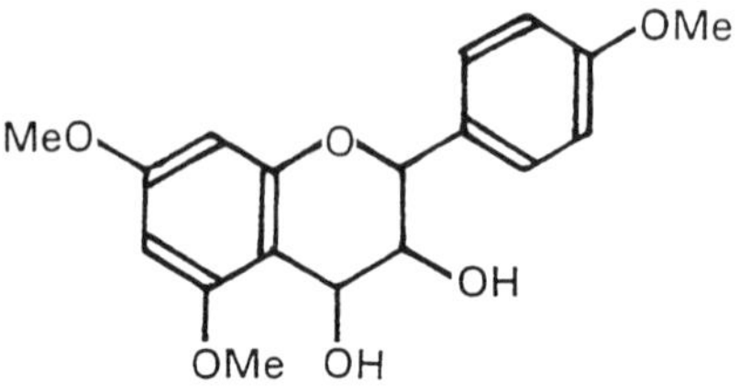   (6)

It has been reported that phloroglucinol reacted with 4-hydroxyphenyl-methylenepropanedinitrile and phenylmethylenepropanedinitrile, respectively, to afford 5,7-dihydroxy-4-(4-hydroxyphenyl)-3,4-dihydrocoumarin and 5,7--dihydroxy-4-phenyl-3,4-dihydrocoumarin, but not naringenin and pinocembrin (K.-K. Lu, Z. Tan, and Yu-L. Li, Gaodeng Xuexiao Huaxue Xuebao, 1994, <u>15</u>, 864). $2,4,5\text{-}(HO)R^1R^2C_6H_2COCH=CHC_6HCO_2H\text{-}2$ $(R^1=R^2=H;\ R^1=OMe,\ R^2=H;\ R^1=H,\ R^2=OMe)$ rapidly isomerizes to compound 7 in acid or on heating, but both compounds on treatment with 5% aqueous $NaHCO_3$ or 10% NaOH/MeOH give 2'-carboxyflavanone, 2'--carboxy-6-methoxy- and 2'-carboxy-7-methoxyflavanones (8), respectively (N. Abe, Akita Daigaku Kyoikugakubu Kenkyu Kiyo Shizen Kagaku, 1988, 107).

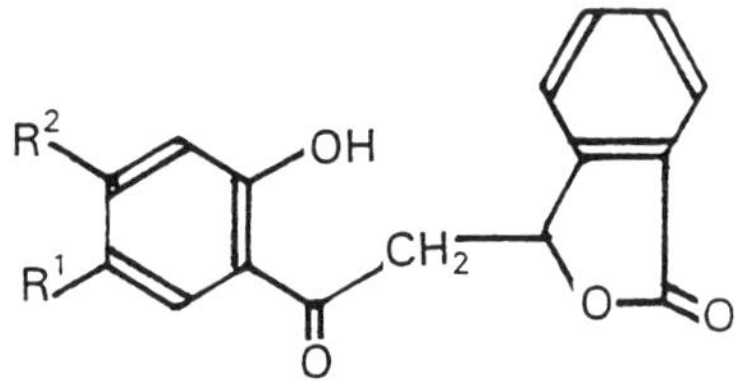

(7) 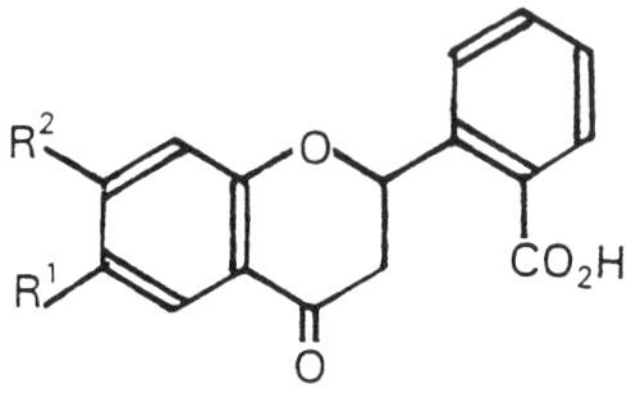   (8)

Treatment of appropriate 2-hydroxyacetophenones with 4-$R^1C_6H_4$CHO (R^1 = H, OMe, Cl, NO_2) and NaOH yields the substituted 3-benzylideneflavanones 9 (R^2 = H, OH; R^3 = H, OMe) (H.M. Chawla and S.K. Sharma, Indian J. Chem., 1987, <u>26B</u>, 1075; J. Chem. Res., S, 1988, 96). Preparations of 3--benzylidene-6-hydroxyflavanone and 6-hydroxy-4'-methoxy-3-(4-methoxy-benzylidene)flavanone have been reported (U.K. Mallik, M.M. Saha, and A.K. Mallik, J. Indian Chem. Soc., 1990, <u>67</u>, 478).

(9)

(10)

(11)

Reaction of chalcones 10 (R^1 = aryl, R^2 = H, NO_2) with H_2SO_4 gives 7--butoxy- and 7-butoxy-6-nitroflavanones 11, respectively (M.D. Ankhiwala and H.B. Naik, *ibid.*, 1989, <u>66</u>, 482). A number of B-ring substituted 7--hydroxyflavanones have been synthesized (X. He *et al.*, Zhongguo Yiyao Gongye Zazhi, 1989, <u>20</u>, 392).

2,4-(HO)RC_6H_3COCH = $CHC_6H_4CO_2$H-4 (R = H, MeO) on boiling in 5% aqueous $NaHCO_3$ cyclizes to give 4'-carboxyflavanone and 4'-carboxy-7--methoxyflavanone (12). Oxidation of the starting compound (R = MeO) with H_2O_2 in MeOH containing 15% aqueous NaOH affords the 7-methoxy-flavonol 13 (R = Me) which on boiling with HI in AcOH gives 7-hydroxy-flavonol 13 (R = H) (K. Sato, G. Saitoh, and N. Abe, Akita Daigaku Kyoikugakubu Kenkyu Kiyo, Shizen Kagaku, 1988, 117). For the synthesis of flavanoximinotosylates 14 (R^1 = H, OMe; R^2 = H, Me, Cl, OMe; R^3 = H, OMe) and their antimicrobial activity see V. Joshi and P.N. Patil, J. Indian Chem. Soc., 1991, <u>68</u>, 295; and for the preparation of 8-iodo-7-phenoxy-flavanone, U.K. Mallik and A.K. Mallik, Indian J. Chem., 1992, <u>31B</u>, 696.

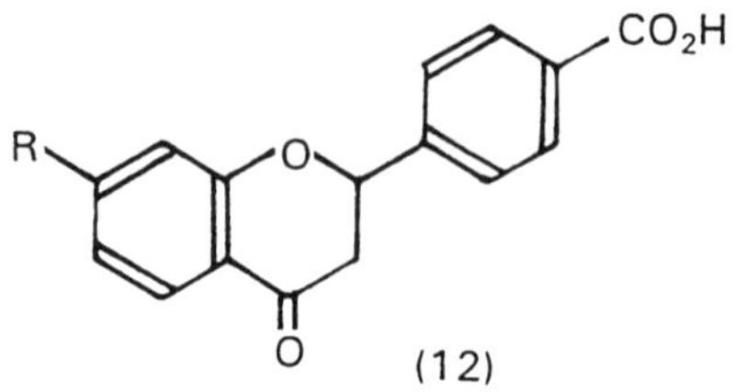

468

(12)

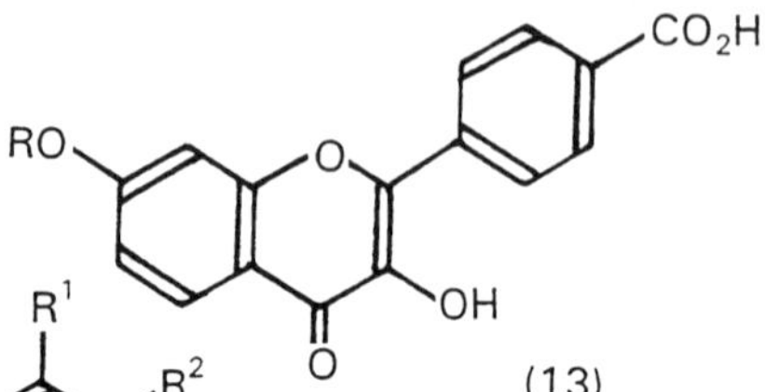

(13)

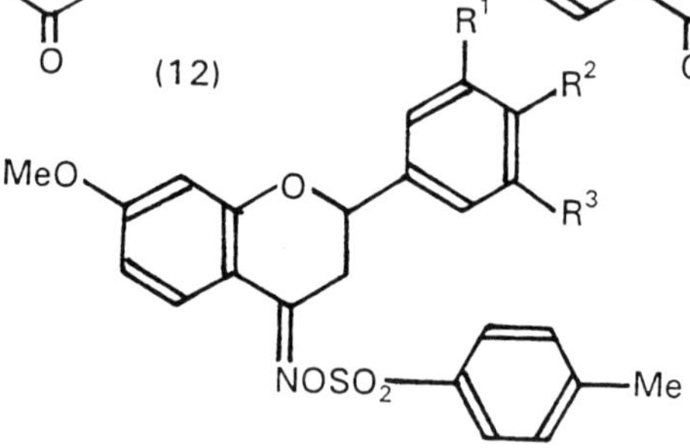

(14)

(i) Biflavanones

Flavones 1 (R = H, MeO) undergo reductive dimerization on treatment with BuMgBr in the presence of manganous chloride in THF to give mixtures of racemic and meso 2,2'-biflavanones 2. The configuration of meso 2--(flavanon-2-yl)flavanone (2; R = H) has been determined by x-ray crystal analysis (Y. Li, F. Zhang, and Q. Wang, Chin. J. Chem., 1992, 10, 359).

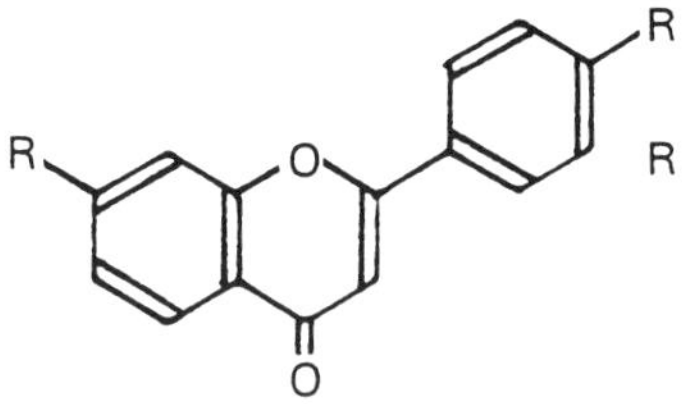

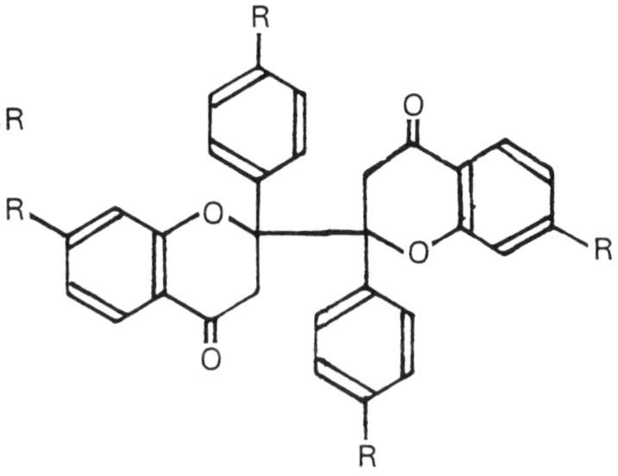

(1)

(2)

*(j) Isoflavanones, 3-phenylchroman-4-ones, 3-phenyl-3,4-dihydro-2H-benzo-
[b]pyran-4-ones*

Arylation of the A-ring substituted and unsubstituted 3-allyloxycarbonyl-
chroman-4-ones 1 (R^1 =H, R^2 = CO_2CH = CH_2, R^3 = R^4 = OMe; R^1 =H,
R^2 = CO_2CH_2CH = CH_2, R^3 = OMe, R^4 =H) with appropriate aryllead tri-
acetates, followed by selective catalytic deallyloxycarbonylation yields isoflavanones 1 [R^1 = Ph, 4-MeC_6H_4, 4-$MeOC_6H_4$, 2,4-$(MeO)_2C_6H_3$, 2,4,6-
-$(MeO)_3C_6H_2$, R^2 =H] or isoflavones 2 (D. M.X. Donnelly, J.P. Finet, and B.A.
Rattigan, J. Chem. Soc., Perkin 1, 1993, 1729).

(1)

(2)

5. 1*H*-Benzo[c]pyran, 1*H*-2-benzopyran, isobenzopyran, 3,4-benzoypran,
isochromene, and its derivatives

(a) 1H-Benzo[c]pyrans

Silylketenes easily undergo a [4 + 2] cycloaddition reaction with *o*-quino-
methanes to give the corresponding 1*H*-benzo[c]pyrans, for example, 1-
-methoxy-3-*tert*-butyldimethylsilylmethyl-1*H*-benzo[c]pyran (1) (T. Ito, T.
Aoyama, and T. Shioiri, Tetrahedron Letters, 1993, 34, 6583).

(1)

Nafion-H has been shown to be an effective catalyst for the transformation of the epoxidized benzene/furan cycloadduct 2 to the indan-2-one 3. However, when the rearrangement was carried out using acidic alumina as catalyst, 1*H*-benzo[*c*]pyran-1-carboxaldehyde (4) was obtained in addition to the expected product 3 (L.G. French and T.P. Charlton, Heterocycles, 1993, <u>35</u>, 305).

(2)   (3)   (4)

(b) Benzo[c]pyrylium salts

1-Substituted 3- aryl-6,7-dimethoxybenzo[*c*]pyrylium perchlorates 1 (R^1 = Me, Ph, R^2 = Ac; R^1 = PhCH$_2$, Ph, R^2 = H; R^1 = PhCH$_2$, R^2 = COCH$_2$Ph) are prepared by the acylation of 3,5-di-*tert*-butyl-4-hydroxy-3',4'-dimethoxy-benzoin in the presence of 70% HClO$_4$ (I.V. Shcherbakova *et al.*, Khim. Geterotsikl. Soedin., 1987, 1032). The preparations of 4-acetyl-6,7-dimeth-oxy-1,3-dimethylbenzo[*c*]pyrylium perchlorate and 4-acetyl-1-alkyl(Me,Et,Pr)--3-aminobenzo[*c*]pyrylium perchlorates have been reported (Yu.A. Nikolyukin, S.L. Bogza, and V.I. Dulenko, *ibid.*, 1990, 465).

(1)

(2) X = F$_2$
(3) X = O

Treatment of perfluoro-1-methylindane at 130°C with antimony pentafluoride system affords the perfluoro-4-methylbenzo[c]pyrylium salt. If the reaction is carried out for 49h at 130°C perfluoro-4-methyl-1H-benzo[c]-pyran (2) and perfluoro-4-methyl-1H-benzo[c]pyran-1-one (3) are obtained (V.M. Karpov et al., Izv. Akad. Nauk SSSR, Ser. Khim., 1991, 745).

The reaction of benzophenone derivative 4 with $(MeO)_2CMeNMe_2$ in the presence of $HClO_4$ affords condensation product 5, which has been cyclized to 1-(4-methoxyphenyl)-3-methyl-5,7-dinitrobenzo[c]pyrylium perchlorate (6) (S.L. Bogza, N.Yu. Zubritskii, and V.I. Dulenko, Zh. Org. Khim., 1993, 29, 1640).

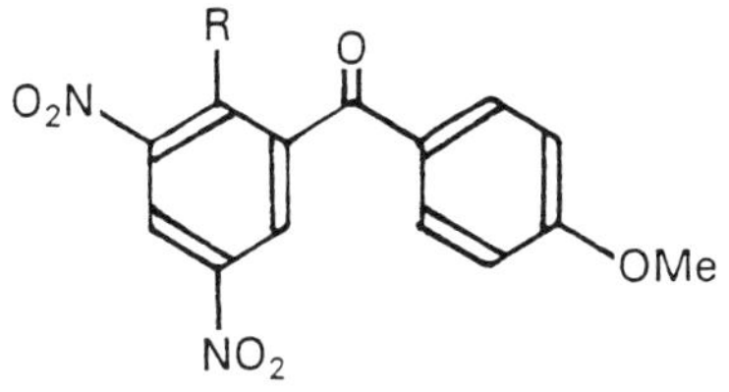

(4) R = Me
(5) R = CH = CMeNMe₂

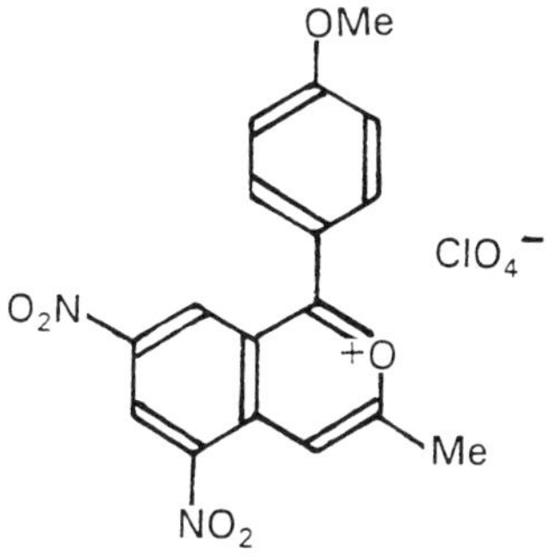

(6)

3-Substituted 6,7-dimethoxy-1-methylbenzo[c]pyrylium perchlorates 7 [R = Me, Ph, 3,4-(MeO)₂C₆H₃] on treatment with weak bases yield anhydro bases, which rapidly react with the original cations to form the spiro dimers 8. The molecular structure of 8 (R = Me) has been determined by x-ray analysis (S.V. Verin et al., Mendeleev Comm., 1991, 104).

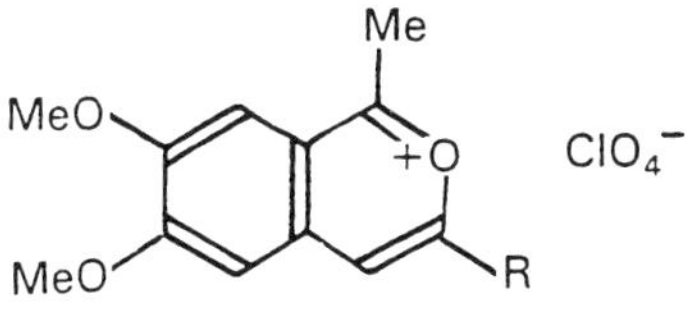

(7)

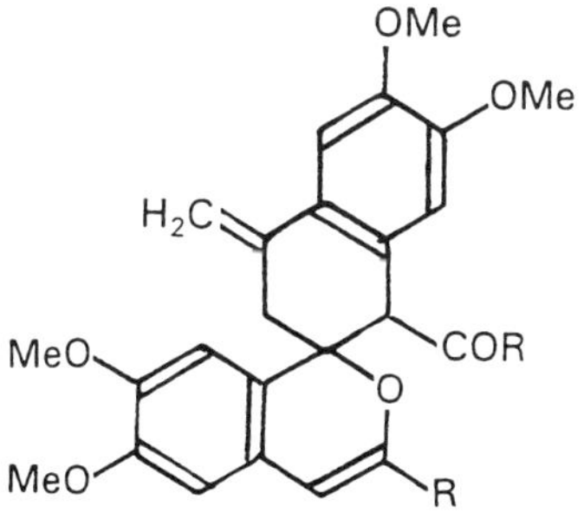

(8)

Dimers 9 [R = Ph, 3,4-(MeO)$_2$C$_6$H$_3$], of 3-aryl-6,7-dimethoxypyrylium perchlorates 10 on cyclization give naphthobenzo[c]pyrans 11, which upon hydrolysis undergo ring cleavage to furnish naphthalenes 12 (S.V. Verin, D.E. Tosunyan, and E.V. Kuznetsov, Khim. Geterotsikl. Soedin., 1991, 175).

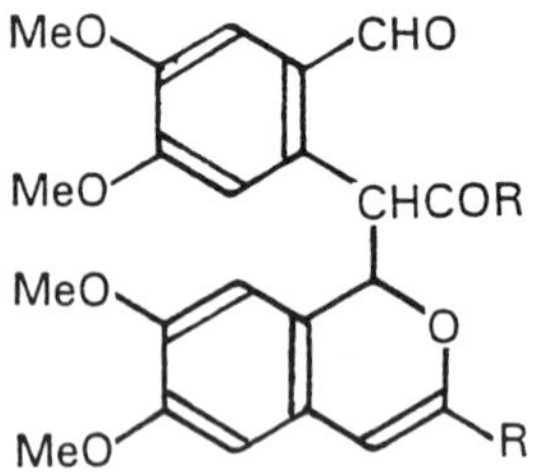

(9)

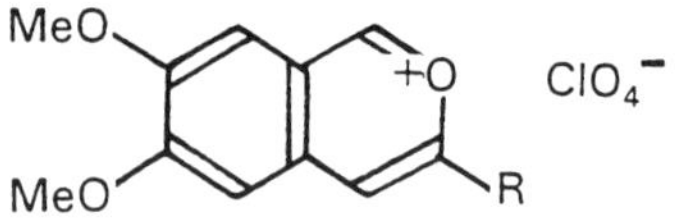

(10)

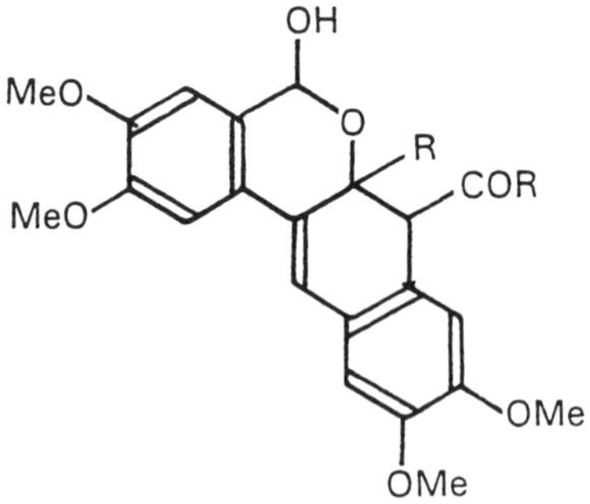

(11)

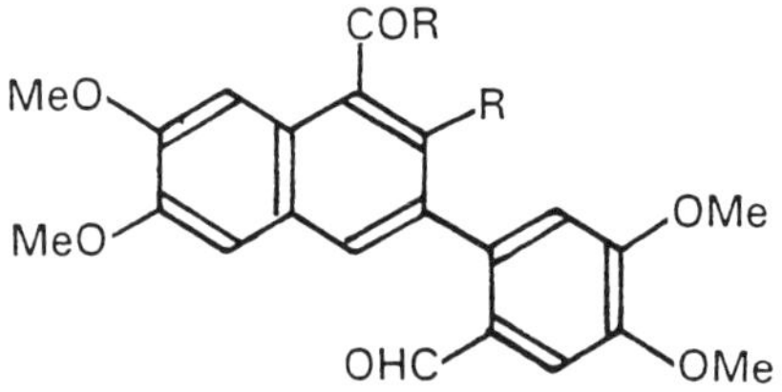

(12)

1,4-Bis(1-substituted 6,7-dimethoxybenzo[c]pyrylium-3-yl)benzene diperchlorates 13 (R = H, Me, Et, PhCH$_2$, Ph, 4-MeOC$_6$H$_4$) have been prepared by the acylation of 1,4-bis(3,4-dimethoxyphenylacetyl)benzene by anhydrides in the presence of HClO$_4$ or by carboxylic acids in polyphosphoric acid (V.I. Dulenko and Yu.A. Nikolyukin, *ibid.*, 1987, 600).

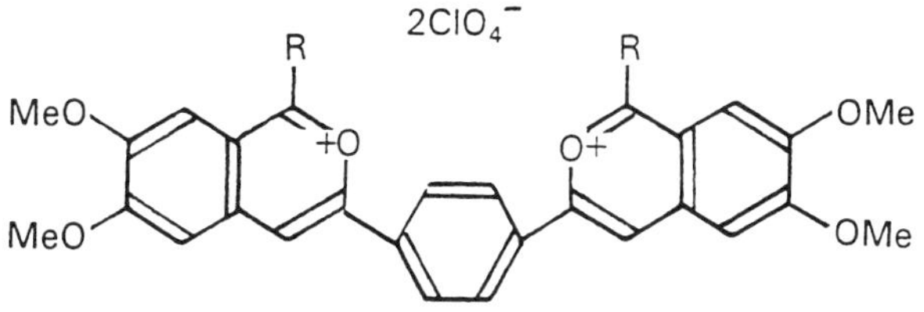

2ClO$_4^-$

(13)

(c) Isocoumarin, 1H-benzo[c]pyran-1-one

The Pd-catalyzed reaction of 2-IC$_6$H$_4$CH$_2$COMe with CO gives 3-methyl-isocoumarin (3-methyl-1*H*-benzo[c]pyran-1-one) (1) (I. Shimoyama *et al.*, Tetrahedron Letters, 1990, 31, 2841). The reaction of 2-cyanophenyl-phenylacetylene with acetic acid in the presence of the catalyst Ru$_3$(CO)$_{12}$ affords 3-phenylisocoumarin (2) (N. Menashe and Y. Shvo, Heterocycles, 1993, 35, 611). Also, for the preparation of 3-ethyl- and 3-phenyliso-coumarins, see N.K. Sinha *et al.*, Indian J. Heterocyclic Chem., 1992, 1, 235). Phthalide-3-phosphonates 3 (R = Me, Et, Ph, AcOC$_6$H$_4$) on treatment with Zn in AcOH afford 3-substituted isocoumarins 4.

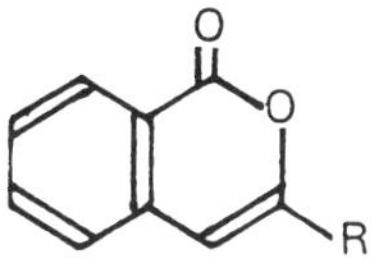

(1) R = Me
(2) R = Ph
(4) (see text)

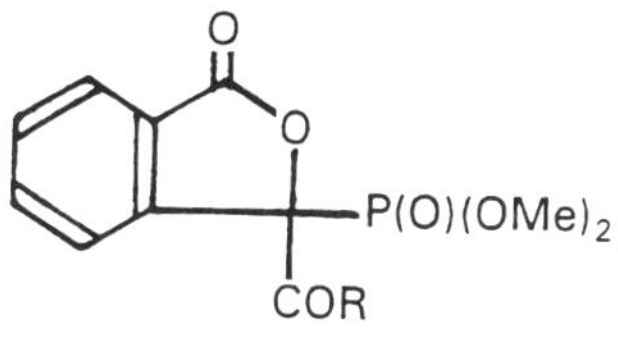

(3)

474

2,3-Diphenylindenone is electrolyzed in MeOH containing AcONa to yield the cyclic ortho ester 5, which on boiling in toluene containing 4-MeC$_6$H$_4$-SO$_3$H yields 3,4-diphenylisocoumarin (6) (J. Simonet and J. Delaunay, Fr. Demande FR 2,597,099, 1987). Similar preparations of 4-arylisocoumarins 7 (R^1 = H, MeO; R^2 = Ph, 4-MeOC$_6$H$_4$; R^3 = Me, Ph) have been reported (J. Delaunay and J. Simonet, Tetrahedron Letters, 1988, <u>29</u>, 543). For the preparations of a number of substituted 3-methylisocoumarins 8 (R^1 = R^3 = R^4 = H, R^2 = OMe; R^1 = R^2 = OMe, R^3 = R^4 = H; R^{1}R^2 = benzo, R^3 = R^4 = H; R^1 = R^2 = H, R^{3}R^4 = benzo) see B.H. Bhide, V.D. Akolkar, and D.I. Brahmbhatt, Proc. Indian Acad. Sci., Chem. Sci., 1989, <u>101</u>, 301; and 5,6,8-trimethoxy-3,4,7-trimethylisocoumarin, M.E. Botha *et al.*, J. Chem. Soc., Perkin 1, 1991, 89.

(5)

(6)

(7)

(8)

The reaction of 4,5-dimethoxyhomophthalic anhydride with phenol derivatives, such as anisole, phenetole, cresol, etc., in the presence of AlCl$_3$ or SnCl$_4$, affords 3-aryl-6,7-dimethoxyisocoumarins, for example, 3-(4--methoxyphenyl)-6,7-dimethoxyisocoumarin (9). These isocoumarins on hydrolysis yield acids, for example, 2-(4-methoxyphenylcarbonyl)methyl-4,5--dimethoxybenzoic acid (10) (S.U. Kulkarni and R.N. Usgaonkar, J. Indian Chem. Soc., 1991, <u>68</u>, 525).

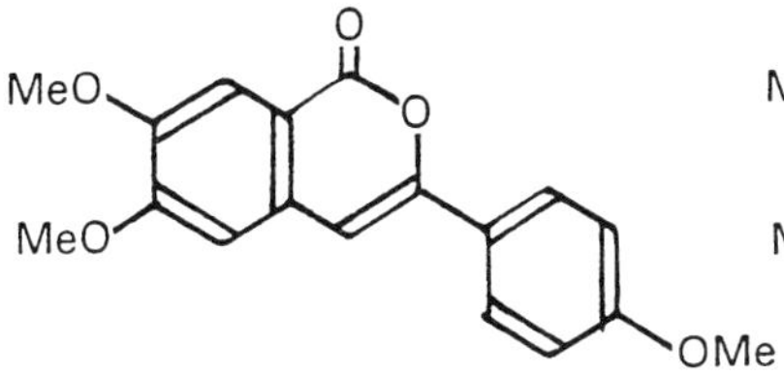

(9)

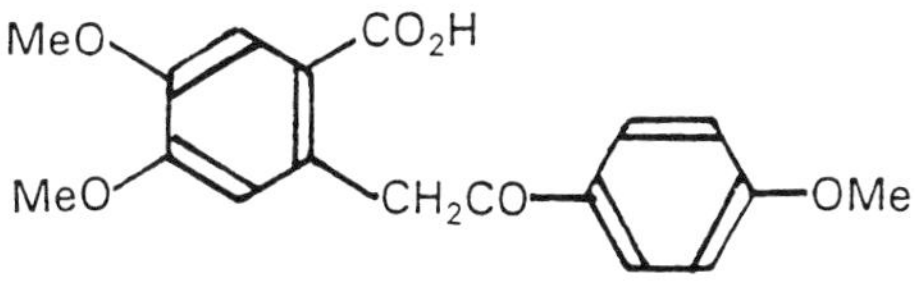

(10)

The synthesis of 3-aryl-4-halogenoisocoumarins 11 (R^1 = H, Cl, Br, Me; R^2 = H, Br, Cl, iodo, Ac) involves reaction of 2-$MeO_2CC_6H_4C{\equiv}CC_6H_4R^1$-4 with $Hg(OAc)_2$ to give isocoumarin mercurials 11 (R^2 = HgCl), which undergo substitution reactions to yield the required 4-halogenoisocoumarins 11 (A. Nagarajan and T.R. Balasubramanian, Indian J. Chem., 1987, 26B, 917). The 3-substituted 4-halogenoisocoumarins 12 (R^1 = Ph, BrC_6H_4, tolyl, Bu; R^2 = Br, Cl, iodo) have been obtained by the treatment of the appropriate ethynylbenzoic acid 13 with the respective N-halogenosuccinimides, $NaHCO_3$, and Triton B (A. Nagarajan and T.R. Balasubramanian, ibid., 1988, 27B, 380).

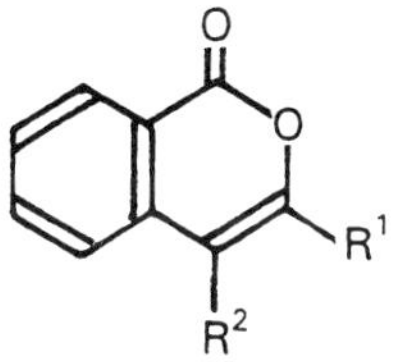

(11) R^1 = 4-$R^1C_6H_4$
(12) (see text)

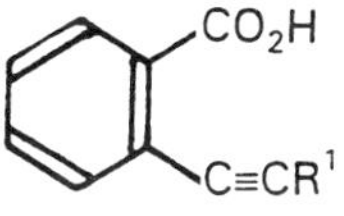

(13)

476

Treatment of 2-methoxycarbonylstyrene with propane-1,3-diol, palladium chloride, and CuCl in 1,2-dimethoxyethane under oxygen at 50-60°C affords derivative 14 (R^1-R^4 =H), which on treatment with 5% HCl in EtOH gives isocoumarin. This method has been used to prepare isocoumarin derivatives 15 (R^1-R^4 =H, F, Cl, C$_{1-3}$alkyl, C$_{1-3}$alkoxy, NO$_2$), which are useful as materials for isoquinolines and are widely used as perfumes, pharmaceuticals, and agrochemicals (A. Kasahara and T. Izumi, Jpn. Kokai Tokkyo Koho JP 03 178,973 [91 178,973], 1991).

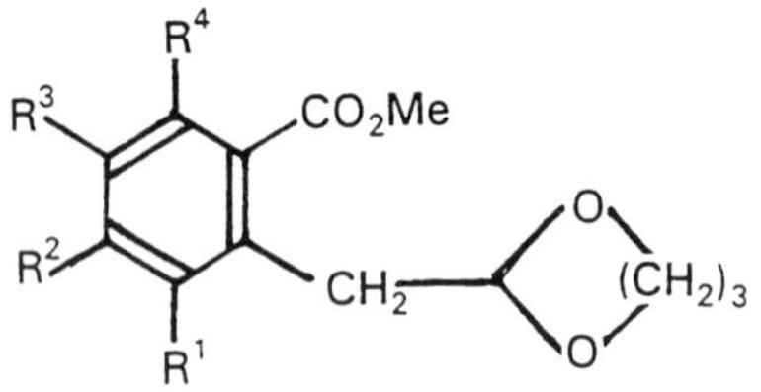

(14) 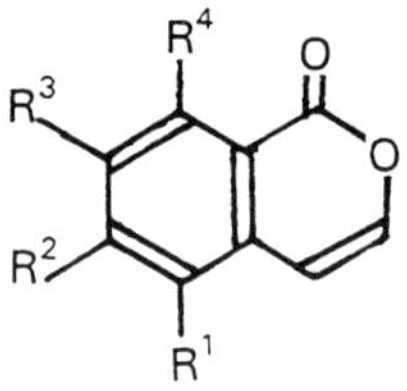 (15)

The reaction of ethyl propynoate with 3,4-bis(trifluoromethyl)furan at 150°C gives a mixture of ethyl 6,7-bis(trifluoromethyl)isocoumarin-3- and -4-carboxylate (16 and 17) in the ratio 52:9 (A.B. Abubakar, B.L. Booth, and A.E. Tipping, J. Fluorine Chem., 1990, 47, 355; A.B. Abubakar *et al.*, *ibid.*, 1992, 56, 359).

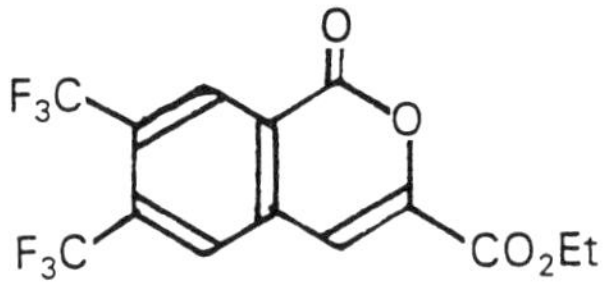

(16) 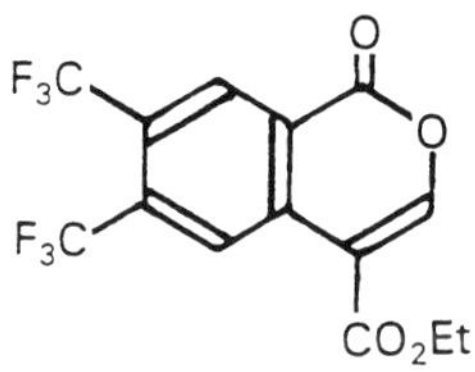 (17)

The absorption, fluorescence, and phosphorescence spectra of a number of isocoumarins have been reported [S. Bakalova *et al.*, God. Sofii. Univ. "Kliment Okhridski", Khim. Fak., 1985 (Pub. 1990), <u>79</u>, 367], some derivatives of isocoumarin, which inhibit human leukocyte elastase (J.C. Powers *et al.*, U.S. US 5,089,633, 1992) have been investigated, and the preparation of isocoumarins with basic substituents as serine proteases inhibitors, anticoagulants, and antiinflammatory agents have been discussed (J.C. Powers and C.M. Kam, *ibid.*, 4,845,242, 1989; J.C. Powers *et al.*, *ibid.*, 4,954,519, 1990; C.M. Kam *et al.*, J. Med. Chem., 1994, <u>37</u>, 1298).

(d) 3H-*Benzo[c]pyran-3-ones*

3*H*-Benzo[*c*]pyran-3-ones 1 (R^1 =H, Me, R^2, R^3,R^4 =H, MeO) have been prepared by treating the corresponding 2-acylphenylacetic acid 2 with DCC/ 2-hydroxypyridine. Some of the benzopyran-3-ones have been trapped with dienophiles to yield Diels-Alder adducts (P.I. Van Broeck *et al.*, J. Chem. Soc., Perkin 1, 1991, 639).

(1)　　　　　　　(2)　　　　　　　(3)

3*H*-Benzo[*c*]pyran-3-ones 3 (R^1 = Ph, R^2 = OMe, R^3 = OMe, H, R^2R^3 = OCH$_2$O; R^1 = Me, R^2R^3 = OCH$_2$O) are stable and easily isolated, but 7- -methoxy-1-phenyl-3*H*-benzo[*c*]pyran-3-one (3; R^1 = Ph, R^2 = H, R^3 = OMe) cannot be isolated. The stabilizing effect of the 6,7-methylenedioxy group is therefore due to the alkoxy group at the C-6 position. This is consistent with a donor-acceptor interaction involving the C-6 alkoxy group and the pyrone carbonyl, which decreases reactivity towards nucleophilic attack by water (D.P. Bradshaw, D.W. Jones, and J. Tideswell, *ibid.*, p.169).

6. Isochroman, 3,4-dihydro-1*H*-benzo[*c*]pyran, 3,4-dihydro-1*H*-2-benzopyran and derivatives

(a) Isochroman

Isochromans have been obtained from appropriate carbenes generated by photolysis of tosylhdrazone or diazo precursors in protic solvents (W. Kirmse and K. Kund, J. Org. Chem., 1990, $\underline{55}$, 2325) and by the intramolecular Friedel-Crafts cycloalkylation of oxonium species generated by reaction of MEM ethers with Lewis acids (M. Ahmar and R. Bloch, Synth. Comm., 1992, $\underline{22}$, 1417).

(2-Bromophenyl)acetaldehydes acetals 1 ($R^1 = Me$, $R^1R^1 = CH_2CH_2$) are transformed with *n*-BuLi and aldehydes R^2CHO ($R^2 = H$, Me, Ph) to the hydroxyacetals 2. Derivatives 2 ($R^1 = Me$) are cyclized with 4-toluene-sulphonic acid to the 3-methoxyisochromans 3, which on hydrolysis with dilute HCl afford the corresponding 3-hydroxyisochromans 3 ($R^1 = H$) [B. Wuensch, Arch. Pharm. (Weinheim, Ger.), 1990, $\underline{323}$, 493]. The regioselective introduction of a methoxy group at the benzylic position of isochroman derivatives by cerium (IV) oxidation in MeOH has been discussed (K. Isobe *et al.*, Chem. Pharm. Bull., 1989, $\underline{37}$, 3390).

(1)   (2)   (3)

ω-Hydroxy-2-quinodimethane 4, generated by photoenolization of 2--methylbenzaldehyde in benzene, is quantitatively trapped with trifluoroacetone to give 1-hydroxy-3-methyl-3-trifluoromethylisochroman (5) and hydroxyaldehyde, $2\text{-OHCC}_6\text{H}_4\text{CH}_2\text{CMe(CF}_3)\text{OH}$ (A.G. Griesbeck and S. Stadtmueller, Ber., 1993, $\underline{126}$, 2149).

(4)

(5)

5,8-Dimethoxyisochroman on treatment with DDQ and then Ph_3Sn-CH_2=CH_2 in CH_2Cl_2 affords the allylated derivative, 1-allyl-5,8-dimethoxy-isochroman (6) stereoselectively (Y.C. Xu, C. Roy, and E. Lebeau, Tetra-hedron Letters, 1993, 34, 8189). The preparations of the racemates of 6,8--dimethoxy-1,3-dimethylisochroman (7) and 1-methyl-, 1-ethyl-, 1-propyl-, and 1-isopropyl-6,8-dimethoxyisochroman (N.H. Rama and A. Saeed, J. Chem. Soc. Pak., 1992, 14, 286), 1-ethyl-6,7-dimethoxyisochroman-1-yl-alkanamines as calcium antagonists (M. Combourieu and J.C. Laigle, Eur. Pat. Appl. EP 450,689, 1991) have been reported.

(6)

(7)

The isomerization of 1-(3-chlorophenyl)-6,7-dimethoxy-3-methyliso-chroman (8; $R^1 = 3\text{-}ClC_6H_4$, $R^2 = OMe$) and 1-(4-aminophenyl)-3-methyliso-chroman (8; $R^1 = 4\text{-}H_2NC_6H_4$, $R^2 = H$) with CF_3CO_2H in $CHCl_3$ yields a mixture of *cis*- and *trans*-isomers 8 and 9. 1-(4-Nitrophenyl)-3-methyliso-chroman (8; $R^1 = 4\text{-}O_2NC_6H_4$, $R^2 = H$) isomerizes only in neat CF_3CO_2H (E. Koltai *et al.*, Acta Chim. Hung., 1990, <u>127</u>, 3).

(8)   (9)

The oxidation of 2-[2-(3,4-dimethoxybenzylhydroxymethyl)-4,5-di-methoxyphenyl]ethanol by chromium (VI) reagents yields 6,7-dimethoxy-1--(3,4-dimethoxybenzyl)isochroman (10) and 6,7-dimethoxy-1-(3,4-dimethoxy)-benzyl)isochroman-3-one (11) (E. Dominguez *et al.*, Bull. Soc. Chim. Belg., 1989, <u>98</u>, 133).

(10) $X = H_2$
(11) $X = O$

Ozonolysis of 1-substituted 2,3-diphenylindenes 12 (R = Me, Ph) and 2-
-(1-substituted-3-methoxy-2-phenylprop-2-enyl)benzophenones 13 (R = Me,
Ph) in $MeOH-CH_2Cl_2$ at -70°C gives the stereoisomeric 4-substituted 1-
-methoxy-1,3-diphenylisochroman-3-hydroperoxides (14 and 15),
respectively (N. Nakamura *et al.*, J. Amer. Chem. Soc., 1989, <u>111</u>, 1799).

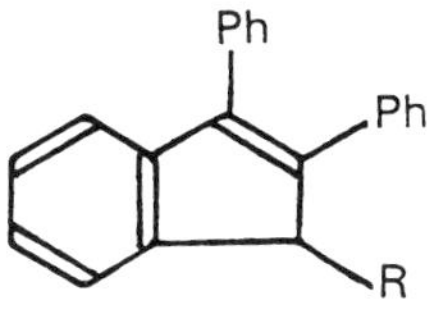

(12)

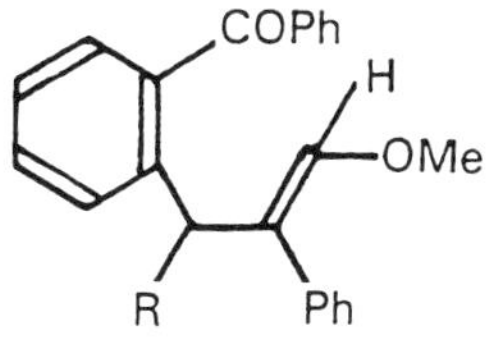

(13)

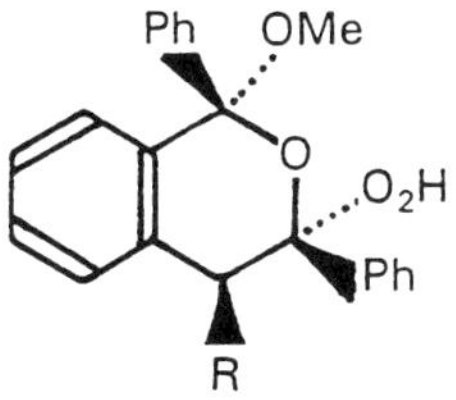

(14)

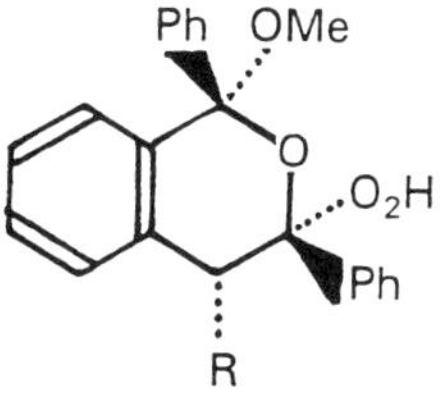

(15)

For the preparations of 9-methylcyclopenta[*f*]isochroman (9-methylcyclo-
penta[*e*]chroman) (16), see F.E. Meyer and A. De Meijere, Synlett, 1991,
777; the fused isochroman derivatives 17 and 18, R. Grigg *et al.*,
Tetrahedron, 1991, <u>47</u>, 9703; and 4,6,6,7,8,8-hexamethylcyclopenta[*g*]-
isochroman (19; $R^1 = R^2 = H$), S. Yan and N. Liu, Faming Zhuanli Shenging
Gongkai Shuomingshu, CN 1,041,362, 1990, and derivatives 19 ($R^1 = H$,
Me; $R^2 = C_{1-3}$alkyl), B. Mi, K. Yan, and C. Wang, *ibid.*, CN 1,072,179, 1993.

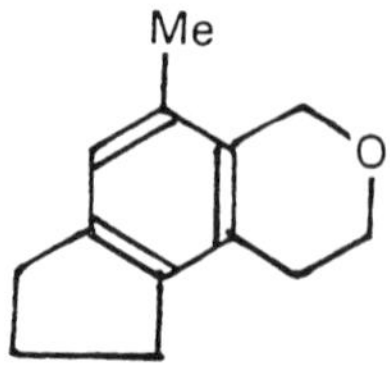

(16)

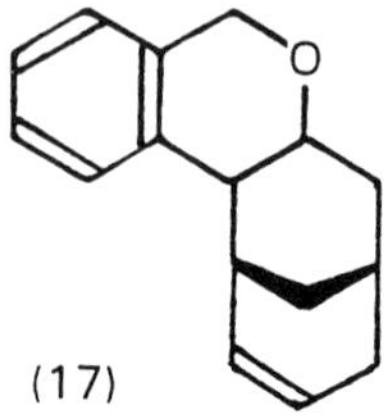

(17)

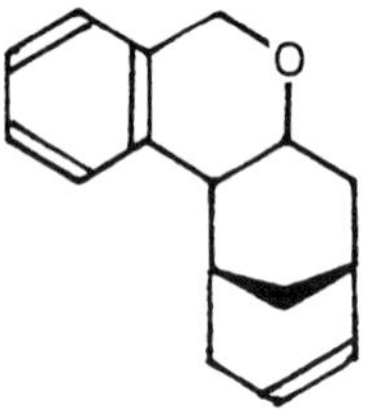

(18)

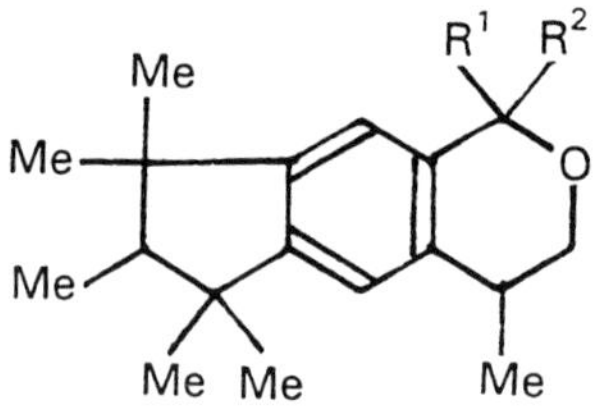

(19)

The DDQ induced oxidative coupling of isochromans, for example, 3-
-acetyl-5,8-dimethoxyisochroman (20), and alcohols proceeds with good
regio- and stereochemical control to give *trans* acetals 21 [R = Me, Pr^i, Bu^t,
$(CH_2)_2NHCO_2Bu^t$, $(CH_2)_3NHCO_2CH = CH_2$] as exclusive or predominant
products (Y.C. Xu *et al.*, Tetrahedron Letters, 1993, <u>34</u>, 3841).

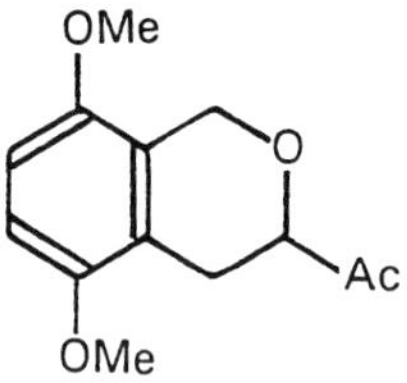

(20)

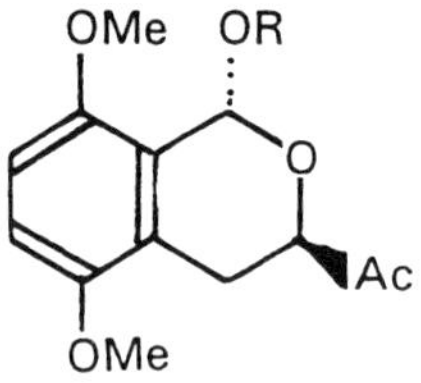

(21)

Methyl (*S*)-(3,4-dihydroxyphenyl)lactate undergoes the oxa-Pictet-
-Spengler reaction with aromatic and aliphatic aldehydes and ketones to give
the 1-substituted alkyl (3*S*)-6,7-dihydroxyisochroman-3-carboxylate 22
[R^1 = Pr, Ph, R^2 = H, R^3 = Me; R^1R^2 = $(CH_2)_5$, $(CH_2CH_2)_2NAc$, $(CH_2CH_2)NMe$,
R^3 = Me, Et]. However, the reaction of unsubstituted (*S*)-3-phenylacetic acid
succeeds only with aromatic aldehydes (B. Wuensch and M. Zott, Ann.,
1992, 39). The reaction between methyl (*S*)-3-(3,4-dihydroxyphenyl)lactate
and methyl levulinate to give a methyl isochroman-3-carboxylate derivative
and its conversion to other derivatives has been discussed [B. Wuensch, M.
Zott, and G. Hoefner, Arch. Pharm. (Weinheim, Ger.), 1992, <u>325</u>, 733].

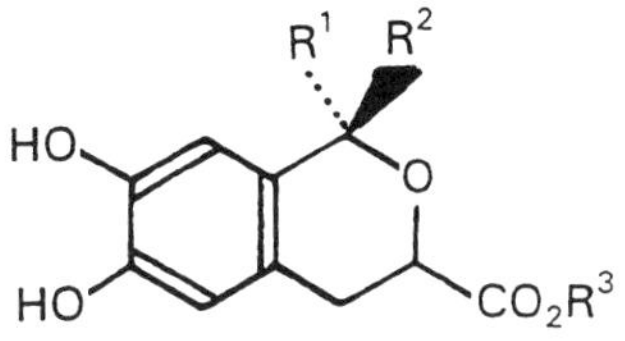

(22) 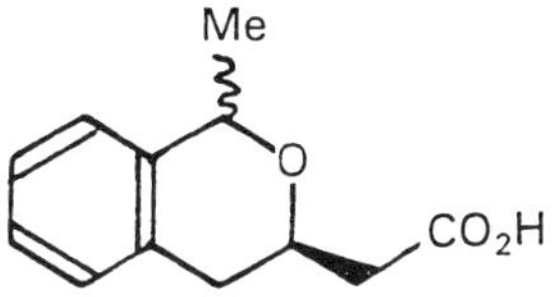   (23)

Methyl 4-(2*a*-acetoxyphenyl)but-3-enoate on heating with KOH in MeOH
results in double bond migration and intramolecular addition to afford after
acidification a mixture of *cis*- and *trans*-1-methylisochroman-3-acetic acid
(23) (U.P. Dhokte and A.S. Rao, Indian J. Chem., 1991, <u>30B</u>, 68).

1-Bromoisochroman on treatment with $RNaC(CO_2Et)_2$ (R = H, Pr)
furnishes diethyl 2-(isochroman-1-yl)malonate (24; R = H) and diethyl 2-(iso-
chroman-1-yl)-2-propylmalonate (24; R = Pr), respectively (A.G. Samodurova
and E.A. Markaryan, Arm. Khim. Zh., 1990, <u>43</u>, 332). For the preparation
of 10-hydroxy-1-methyl-6,9-dioxo-6,9-dihydrobenzo[*g*]isochroman-3-acetic
acid, see G. Bach *et al.*, Ger. Offen. DE 4,121,468, 1993). The preparations
of derivatives of 4-dichloromethylenechroman-3-ol (P.E. Brown *et al.*, J.
Chem. Soc., Perkin 1, 1990, 139) and 1-bromomethyl-5,6-(spirocyclohexyl-
methylenedioxy)3-*tert*-butylisochroman (25) (R.W. Schoenleber *et al.*, U.S.
US 4,963,568, 1990) have been reported.

CR(CO$_2$Et)$_2$

(24)

CH$_2$Br

Bu^t

(25)

The cyclization of *N*-(2-phenylethyl)-2-(2-hydroxyethyl)benzamide with SOCl$_2$ gives *N*-(2-phenylethyl)isochroman-1-ylimine (26; R^1-R^4 = H). Related derivatives 28 (R^1-R^4 = MeO; R^1 = H, MeO, R^2 = MeO, R^3 = R^4 = H) [W. Meise and H.J. Mika, Arch. Pharm. (Weinheim, Ger.), 1989, 322, 245) and 27 (R = H, OMe) (W. Meise and G. Schlueter, *ibid.*, p.581) have also been prepared.

R^2

R^1

N(CH$_2$)$_2$

R^3

R^4

(26)

R

R

N(CH$_2$)$_3$

(27)

The synthesis of derivatives of the tetracyclic compound 28 containing an isochroman system (M.O. Lursmanashvili *et al.*, Soobshch. Akad. Nauk Gruz. SSR, 1990, 137, 525) and tricyclic compound 29 and derivatives [B. Wuensch, M. Zott, and G. Hoefner, Arch. Pharm. (Weinheim, Ger.), 1993, 326, 823] have been described.

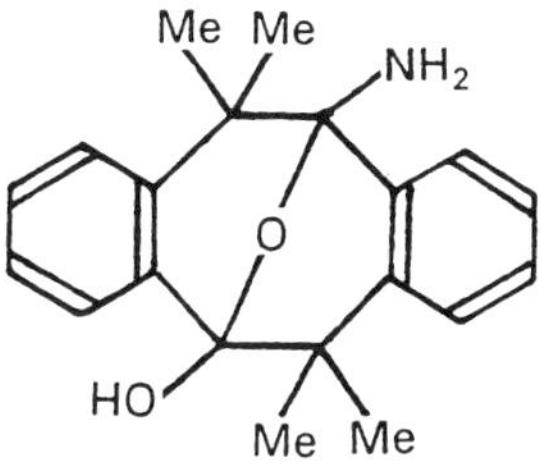

(28)

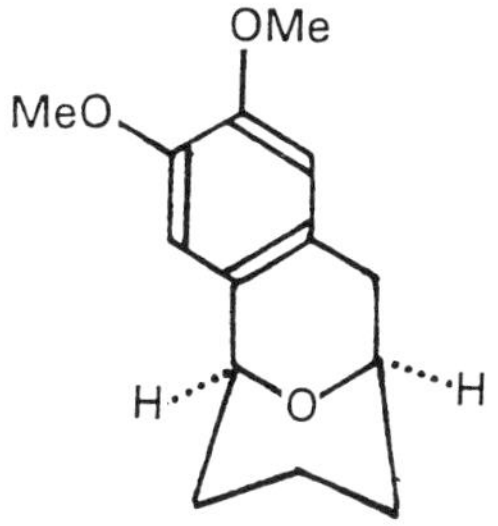

(29)

(b) Isochromanones, 3,4-dihydroisocoumarins, and isochromandiones

(i) Isochroman-1-ones. 2-Hydroxyindan-1-ones have been transformed
to isochroman-1-ones by oxidative cleavage followed by reduction and acid
treatment [K. Shishido *et al.*, Heterocycles, 1990, <u>30</u> (1, Spec. Issue), 253].
3-Substituted isochroman-1-ones have been obtained *via* radical and
photochemical fragmentations, for example, radical cyclization resulting from
tert-butylhydroxyisochroman 1 and its ring-opened tautomer 2 yields 3-
-methylisochroman-1-one (3) (K. Kobayashi *et al.*, J. Chem. Soc., Perkin 1,
1993, 111).

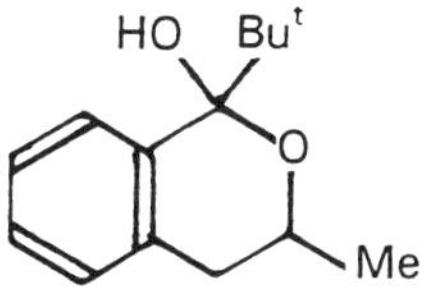

(1)

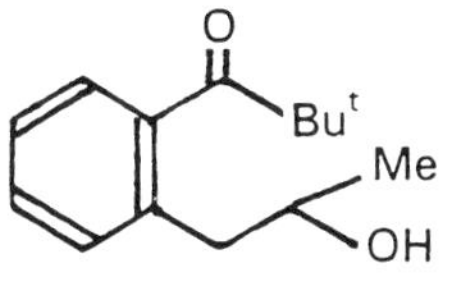

(2)

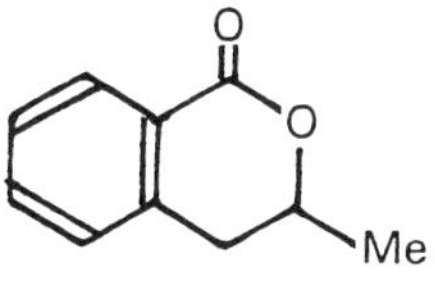

(3)

Isochroman-1-ones are readily prepared by the palladium-catalyzed coupling of 2-vinylbenzoic acids with 1-alkenyl halides or triflates, for example, 2-vinylbenzoic acid and hex-1-enyl bromide or iodide affords 3--(hex-1-enyl)isochroman-1-one as the predominant product (R. Larock and H. Yang, Synlett, 1994, 748).

Substituted 3-ethylisochroman-1-ones 4 (R = 5-, 6-, 8-MeO, 6-, 7-Me) have been obtained by ortho-lithiation of the corresponding N-methylbenzamide $RC_6H_4CONHMe$, followed by alkylation with 1,2-butylene oxide (D.I. Brahmbhatt, J. Indian Chem. Soc., 1989, <u>66</u>, 481). 3,8-Disubstituted isochroman-1-ones 5 (R^1 = H, OMe, R^2 = H, Me, Et, Ph, 4-MeOC$_6$H$_4$) have been prepared by the Nef oxidation of 3,2-R^1(HO$_2$C)C$_6$H$_3$CH$_2$CHR^{2}NO$_2$, followed by cyclization and dehydration of the resulting product (F.M. Hauser and V.M. Baghdanov, J. Org. Chem., 1988, <u>53</u>, 4676). A number of 3,3,4--trisubstituted isochroman-1-ones have been obtained from the dianions of o--toluic acid and α-phenyl-o-toluic acid, which with lithium diisopropylamide were condensed with appropriate aldehydes or ketones and the resulting products acid-cyclized (K.C. Hildebran *et al.*, Synth. Comm., 1994, <u>24</u>, 779).

(4) (5)

A stereocontrolled synthesis of *cis*-3,4-diarylisochroman-1-ones through the diastereoselective reaction of benzaldehydes and α-lithio-2-cyanodiaryl-methane intermediates has been described, for example, *cis*-8-methoxy-5--methyl-3,4-diphenylisochroman-1-one (6) from benzonitrile 7 (L. Crenshaw *et al.*, Tetrahedron Letters, 1988, <u>29</u>, 3777).

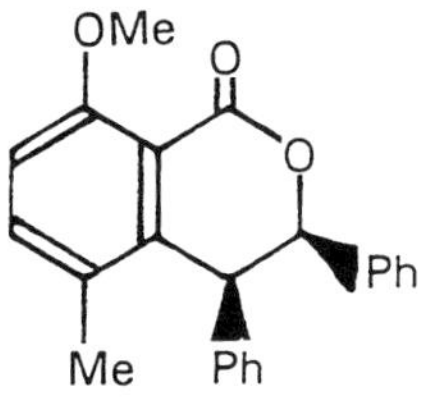

(6) 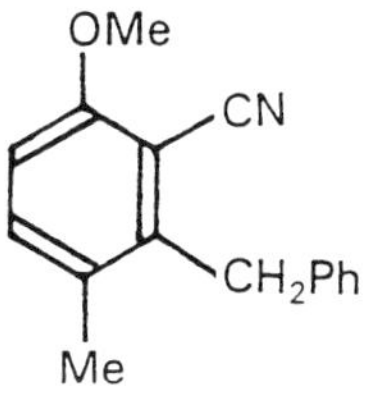   (7)

Triisopropoxytitanium derivatives of homophthalic acid, for example, 8, react with aldehydes or ketones to yield, for example, methyl 3-substituted 1-oxoisochroman-4-carboxylates 9 [R = Pr^i, CH$_2$CHMe$_2$, Ph, fur-2-yl, (*E*)--MeCH = CH] (A.N. Kasatkin *et al.*, Zh. Org. Khim., 1991, 27, 700).

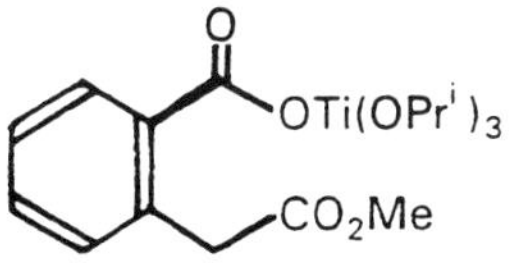

(8) 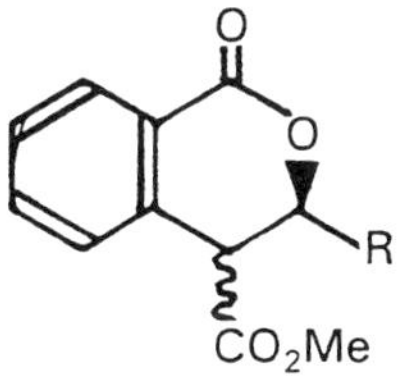   (9) 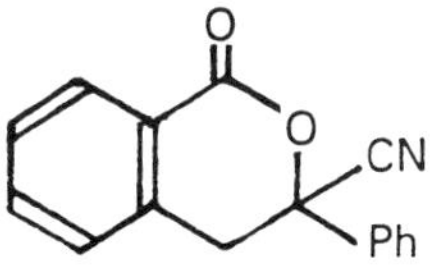   (10)

The irradiation of 2-MeC$_6$HCOCN in presence of 2 equivalents of benzoyl cyanide in benzene or acetonitrile affords 3-cyano-3-phenylisochroman-1-one (10) (R. Connors and T. Durst, Tetrahedron Letters, 1992, 33, 7277). For the preparation of 7-methyl-5-phenylthioisochroman-1-one, see G.H. Posner *et al.*, J. Org. Chem., 1990, 55, 2132.

488

The lithiation of diamide 11 with BuLi and subsequent condensation with propylene oxide furnished 3,3'-dimethyl-8,8'-biisochroman-1,1'-dione (12) (D.I. Brahmbhatt, J. Indian Counc. Chem., 1987, *3*, 53).

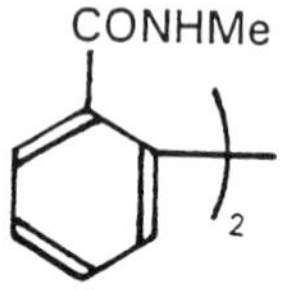

(11)

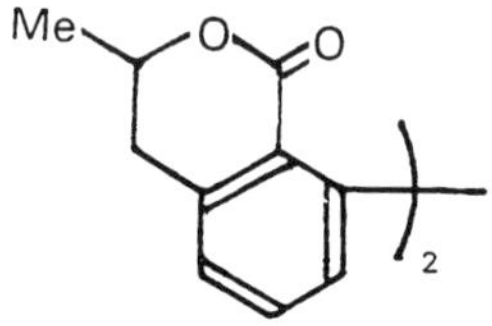

(12)

(ii) Isochroman-3-one. The ethoxycarbonylation of 2-acylbenzyllithiums with ethyl chloroformate and the subsequent $NaBH_4$ reduction and cyclization of the resulting 2-acylphenylacetic acid derivatives results in the formation of isochroman-3-ones, for example, 1-*tert*-butyl-4-methyliso-chroman-3-one (1) (K. Kobayashi *et al.*, Bull. Chem. Soc. Jpn., 1994, *67*, 582).

The thermolysis of alkenylbenzocyclobutenylcarboxylic acids 2 $[R^1,R^2 = H, OMe, R^3 = Me, 4\text{-}MeOC_6H_4, R^4 = H;\ R^3R^4 = (CH_2)_4, (CH_2)_5, R^5 = H]$ yields 4-alkylideneisochroman-3-ones 3, *via* an electrocyclic [1,5]sigmatropic process of *o*-quinodimethane, whereas esters 2 (R^1-R^4 same, $R^5 = Me$) afford dihydronaphthalenes 4 (K. Shishido *et al.*, Chem. Letters, 1987, 2117; Heterocycles, 1987, *26*, 2361). The reduction of 1-(3,4-dimethoxybenzyl-idene)-6,7-dimethoxyisochroman-3-one (5) with $LiAlH_4$ affords 1-(3,4--dimethoxybenzylidene)-6,7-dimethoxyisochroman together with small amounts of naphthalene derivatives (E. Dominguez *et al.*, J. Heterocyclic Chem., 1989, *26*, 549).

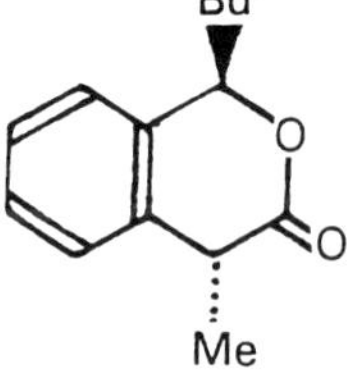

(1)

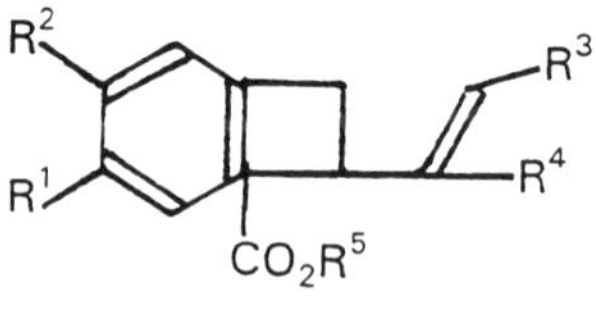

(2)

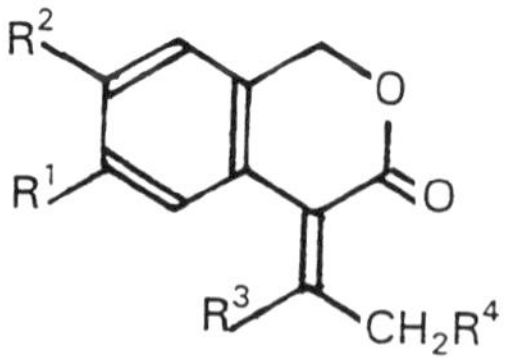

(3)

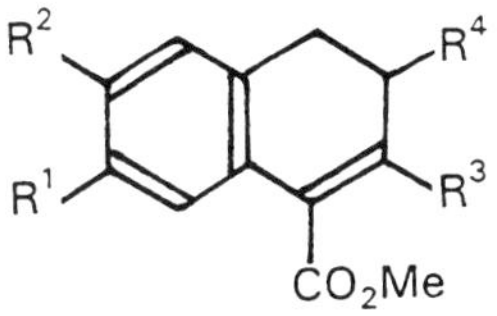

(4)

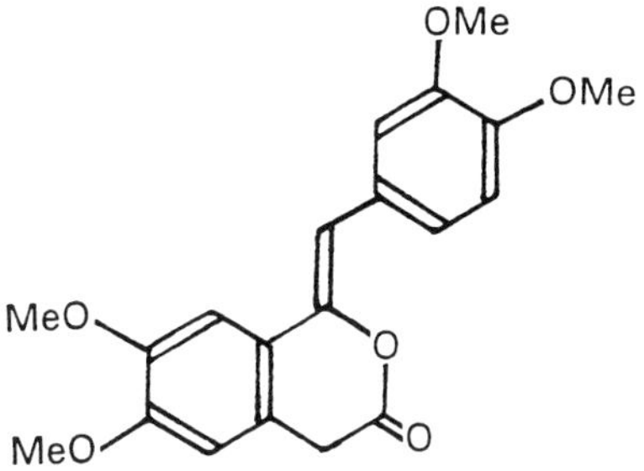

(5)

5-(Hydroxymethyl)-2-isopropylphenol (6) has been converted in a number of steps to 4-(2-iodoethyl)-7-isopropyl-8-methoxy-4-methylisochroman-3-one (7) (M. Mallaiah and B.M. Bhawal, Indian J. Chem., 1987, **26B**, 768).

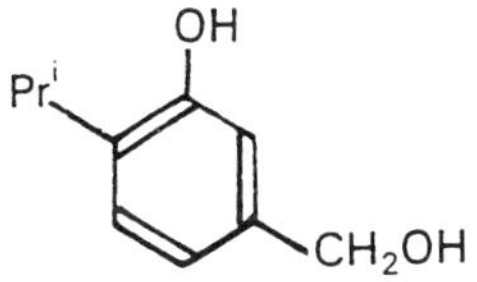

(6)

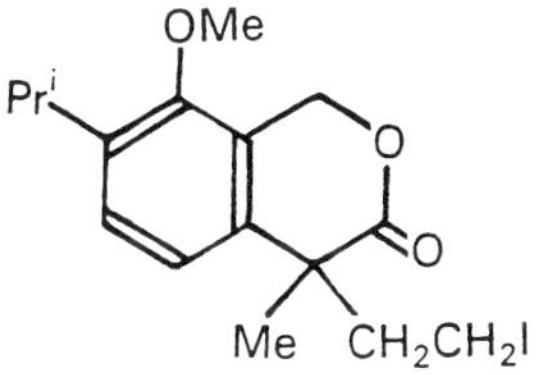

(7)

(iii) Isochromandiones. The alkoxide-induced ring-opening of 4-methoxy-methyleneisochroman-1,3-dione (1) leads to methyl isocoumarin-4-carboxylate (2), methyl *cis-* and *trans*-3-methoxy-1-oxoisochroman-4-carboxylates (3) and (E)-MeOCH=C(CO$_2$H)C$_6$H$_4$CO$_2$Me-2. A X-ray structure of *trans*-3 reveals a trans-diaxial geometry (M.G. Hutchings *et al.*, Tetrahedron, 1991, **47**, 7869).

(1)

(2)

(3)

Isochroman-1,3-dione (homophthalic anhydride) (4) on treatment with MeOH, followed by MeI and K_2CO_3 in acetone yields 12-methoxybenzo[d]-naphtho[2,3-b]pyran-5-one (5) along with dimethyl homophthalate (W.V. Murray and S.K. Hadden, J. Chem. Res., S, 1991, 279). The preparation of 3-methyl-5,8-dihydroisochroman-5,8-dione (6) has been reported (A.J. Esterhuyse *et al.*, S. Afr. J. Chem., 1993, <u>46</u>, 34).

(4)

(5)

(6)

(c) Tetrahydro- and perhydroisochroman derivatives and related compounds

The preparations of optically active δ-lactone derivatives related to 8-
-hydroxy-3-methyl-4a,5,6,7-tetrahydroisochroman-1-one (1), useful as
intermediates in the synthesis of antibiotics have been reported (S. Takano
et al., Jpn. Kokai Tokkyo Koho JP 03 112,975 [91 112,975]; 03 112,976
[91 112,976]; 03 112,980 [91 112,980], 1991).

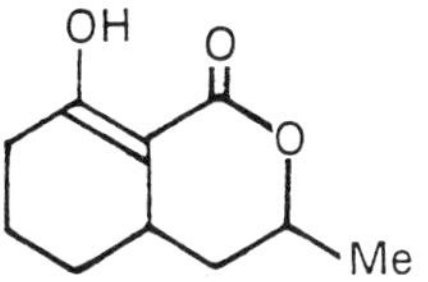

(1)

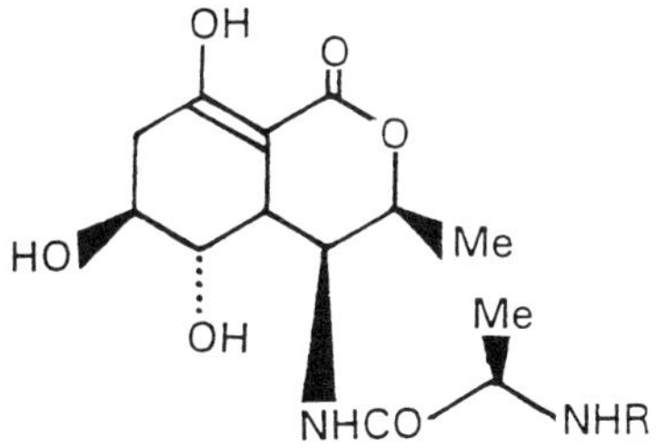

(2) R = H
(3) R = COCH$_2$Ph

Amines and hydroxyamines react selectively with benzyl 4-nitrophenyl
carbonate, PhCH$_2$OCO$_2$C$_6$H$_4$NO$_2$-4, to afford the benzyloxycarbonyl
protected amines, for example, actinobolin salt (2) on treatment with benzyl
4-nitrophenyl carbonate gives the carbamate 3 (D.R. Kelly and M. Gingell,
Chem. Ind., 1991, 888). The preparations of some 3,3,8-trisubstituted
4a,7,8,8a-tetrahydroisochromans have been described (D. Craig and J.C.
Reader, Synlett, 1992, 757).

Four stereoisomers of 3-methylperhydroisochroman-1-one (3-methylocta-
hydroisocoumarin) (4, 5, 6, and 7) and two 1-methylperhydroisochroman-3-
-ones (8 and 9) have been synthesized and their stereochemistry determined
(Y. Fujiwara and M. Okamoto, Chem. Pharm. Bull., 1989, 37, 1458).

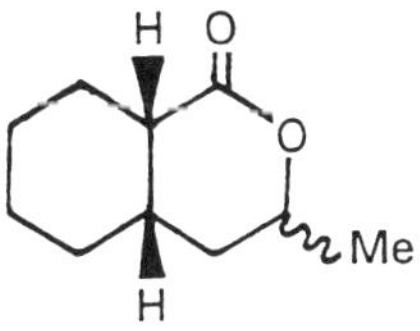

(4) β-Me
(5) α-Me

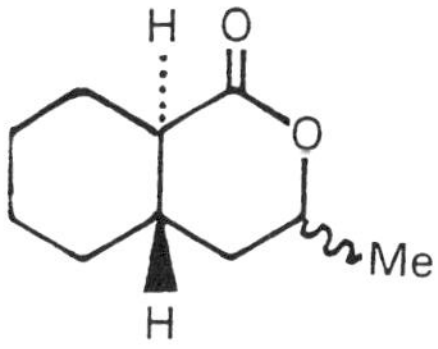

(6) β-Me
(7) α-Me

(8)   (9)

The structures and configurations of 3-substituted *trans*-perhydroiso-chromans 10 (R=H, Me, Et, Pr, Ph) have been confirmed by spectral data (N.P. Volynskii and M.B Smirnov, Izv. Akad. Nauk, Ser. Khim., 1992, 861). The ring cleavage of *trans*-perhydroisochromans 10 with PBr_3/HBr has been investigated (N.P. Volynskii and O.L. Alikhanova, Izv. Akad. Nauk SSSR, Ser. Khim., 1991, 1877). The cyclocondensation of 2,2-dimethyltetrahydro-pyran-4-one with MeCOCH=CHR (R=fur-2-yl, Ph) affords, 8-(fur-2-yl)- and 8-phenyl-4a-hydroxyperhydroisochroman-6-ones 11, respectively (T.M. Mukhametkaliev *et al.*, Izv. Akad. Nauk Kaz. SSR, Ser Khim., 1989, 74). The preparation of methyl 1,1,6-trimethyl-3-oxoperhydroisochroman-4--carboxylate (12) has been reported (L.F. Tietze *et al.*, Org. Synth., 1990, <u>69</u>, 31).

(10)   (11)

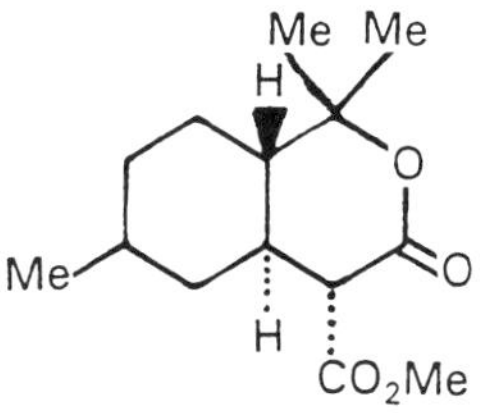

Me Me

Me

H

CO$_2$Me

(12)

R

R

N

HCl

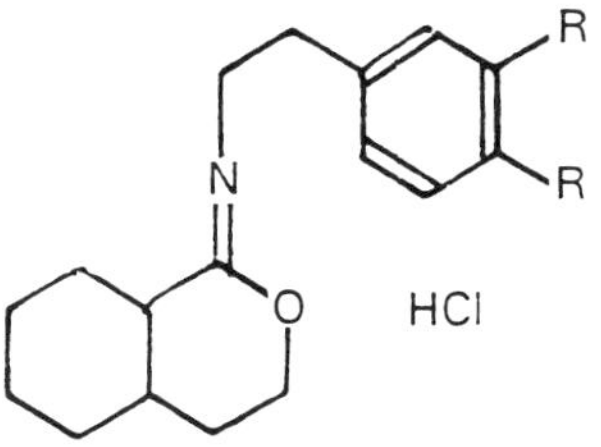

(13)

The aminolysis of *cis*-perhydroisochroman-1-one with 3,4-R$_2$C$_6$H$_3$CH$_2$-CH$_2$NH$_2$ (R = H, OMe), followed by treatment of the product with SOCl$_2$ yields *N*-(2-phenylethyl)- and *N*-[2-(3,4-dimethoxyphenyl)ethyl]perhydroisochroman-1-imines (13), respectively [W. Meise and U. Mika, Arch. Pharm. (Weinheim, Ger.), 1989, 322, 573].

3-Substituted 5,5-dimethyl-4,4a,5,8-tetrahydro-3*H*-isochromen-8-ones 14 (R = OEt, OAc, 4-MeOC$_6$H$_4$, 3,4-(MeO)$_2$C$_6$H$_3$, 3,4,5-(MeO)$_3$C$_6$H$_2$) have been prepared by the stereoselective hetero-Diels-Alder reaction of 2-formylcyclohexa-2,5-dien-1-one (15) with electron-rich olefins, CH$_2$ = CHR (S.R. Desai, V.K. Gore, and S.V. Bhat, Synth. Comm., 1992, 22, 97).

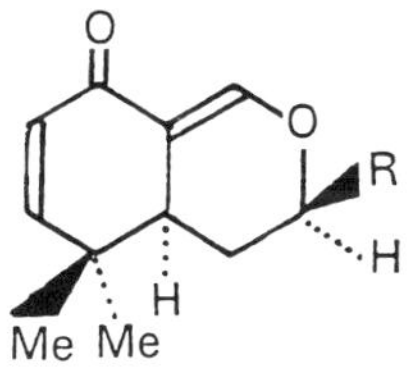

O

O

R

H

H

Me Me

(14)

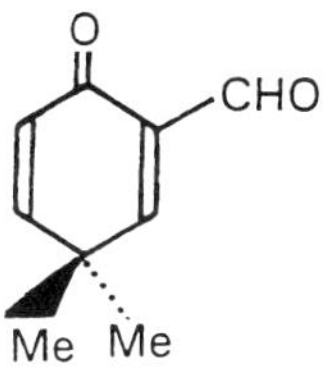

O

CHO

Me Me

(15)

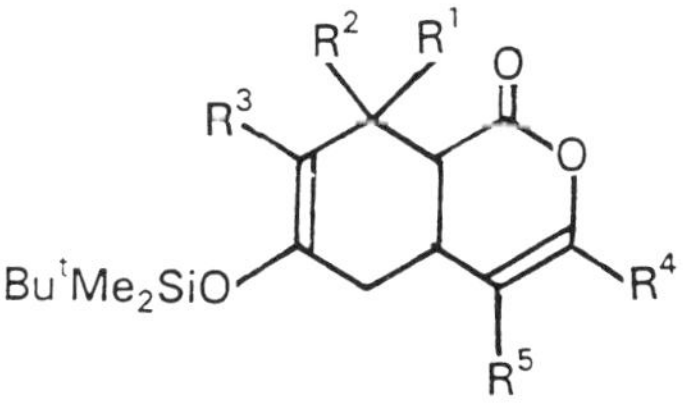

R^2 R^1

R^3

O

O

Bu^{t}Me$_2$SiO

R^4

R^5

(16)

O

Me

Me

CpCo

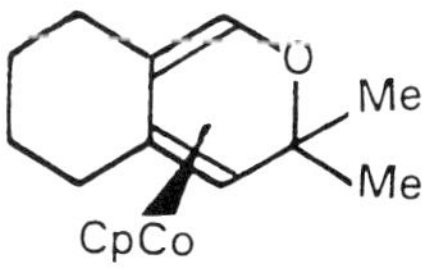

(17)

The preparations of the substituted 6-*tert*-butyldimethylsiloxy-4a,5,8,8a-
-1H-isochromen-1-one 16 [R^1,R^2 = H, Me, Ph, R^3 = H; R^1 = H, R^2R^3 = $(CH_2)_n$,
n = 3,4; R^4 = Me, R^5 = H; R^4R^5 = $(CH=CH)_2$] (K. Ohkata *et al.*, J. Org.
Chem., 1991, <u>56</u>, 5052) and the η^5-cyclopentadienylcobalt complex of 3,3-
-dimethyl-5,6,7,8-tetrahydro-3H-isochromen (17) (D.F. Harvey *et al.*, Synlett,
1989, 15) have been reported.

7. Xanthene, dibenzo[*b,e*]pyran, 2,3:5,6-dibenzopyran and derivatives

(a) Xanthene derivatives

An electron diffraction study of gaseous xanthene (1) indicated that the
molecule is not planar and that the bond angle at the O atom is 120.9°. The
C-O distance of 1.352Å is shorter than usual, implying double-bond
character (K. Iijma, T. Misu, and S. Onuma, J. Mol. Struct., 1990, <u>221</u>,
315). The photoisomerization of xanthene (1) to 6H-dibenzo[*b,d*]pyran (2)
has been reported and a reaction mechanism proposed (C.G. Huang,
D. Shukla, and P. Wan, J. Org. Chem., 1991, <u>56</u>, 5437).

(1)   (2)

The oxidation of xanthenes with $Mn(OAc)_2$ in the presence of active
methylene compounds, such as 1,3-dicarbonyl compounds, $CH_2(CN)_2$
derivatives, Me_2CO and $MeNO_2$ selectively affords 9-substituted xanthene
derivatives (H. Nishino *et al.*, *ibid.*, 1992, <u>57</u>, 3551). The synthesis and
chemical behaviour of the 9-(adamant-2-yl)xanthene systems have been
discussed (I.T. Badejo, Diss. Abstr. Int. B, 1990, <u>50</u>, 4531).

Condensation of PhOH, Ph_2CO, and $Al(OPh)_3$ at 190°C yields 9,9-di-
phenylxanthene (3) (1.8%) fuchsone (4) (21.3%) along with other products
(Ya.B. Kozlikovskii, B.V. Chernyaev, and N.I. Kuz'mina, Izv. Vyssh. Uchebn.
Zaved., Khim. Khim. Tekhnol. 1987, <u>30</u>, 31). The Wittig reaction of tributyl-
phosphoniumxanthen-9-ylide with $4-O_2NC_6H_4CHO$ yields 9-(4-nitrophenyl)-

methylenexanthene (5) (K.L. Handoo, L. Kishan, and A. Kaul, Indian J. Chem., 1990, 29B, 274). The treatment of (9*H*-xanthyl)tropylium tetrafluoroborate with either methanol, *N,N*-diisopropylethylamine, or DMSO furnishes 9-[1-(cyclohepta-2,4,6-trienylidene)]xanthene (6) (I.T. Badejo *et al.*, J. Org. Chem., 1990, 55, 4327).

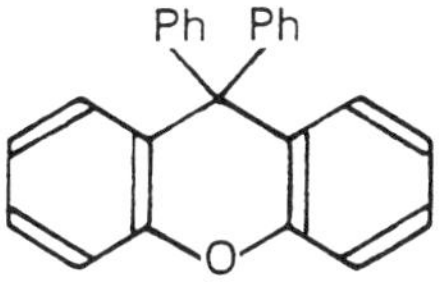

Ph   Ph

(3)

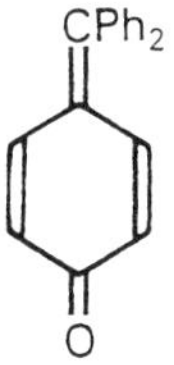

CPh$_2$

O

(4)

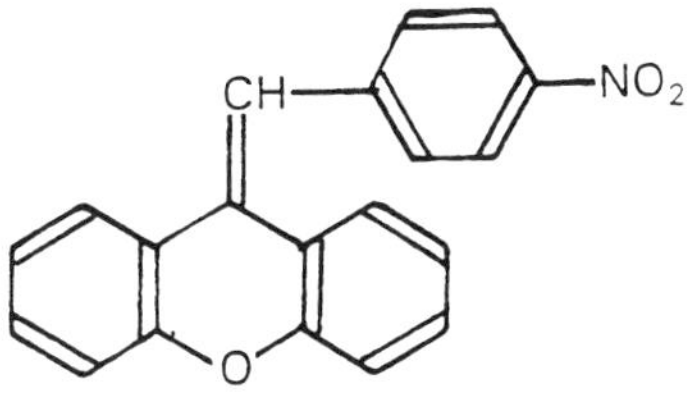

CH—〈 〉—NO$_2$

(5)

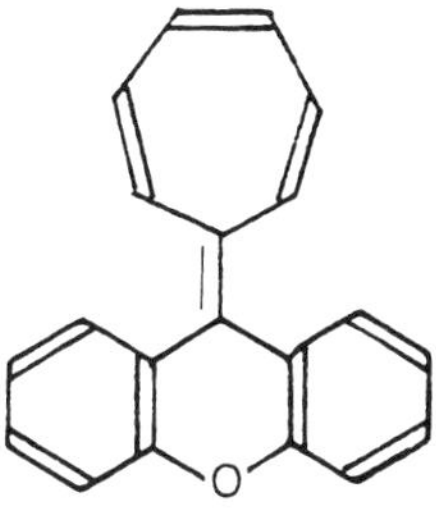

OR

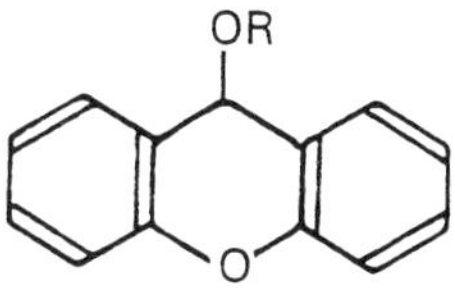

(7)

(6)

The alkylation of ROH [R = CH$_2$=CHC≡CC(Me$_2$)O, CH$_2$=CMeC≡CC-(Me,Et)O, (CH$_2$=CHC≡C)$_2$C(Me)O] with xanthydrol (xanthen-9-ol) in AcOH affords xanthydrol enyne peroxides 7 (G.A. Petrovskaya, Yu.V. Panchenko, and V.A. Puchin, Zh. Org. Khim., 1987, 23, 2474). Several 9-alkyl derivatives of xanthen-9-ol have been obtained by reactions of the dianion of xanthen-9-one with alkyl halides, for example, the lithiation of xanthen-9-one in the presence of 4,4'-di-*tert*-butylbiphenyl with sonication, followed by alkylation with RX (R = Me, Pr, tropyl, PhCH$_2$, X = halogeno) affords 9-alkyl-xanthen-9-ols 8 (I.T. Badejo *et al.*, Chem. Comm., 1989, 566).

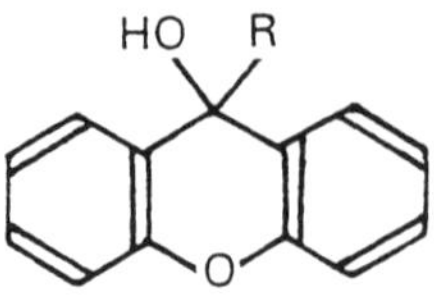

496

(8)

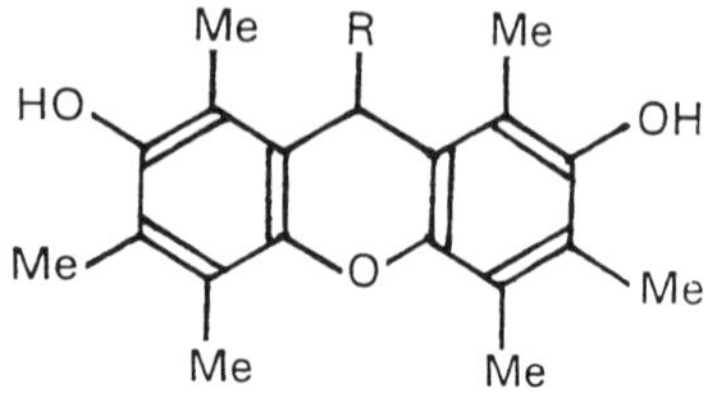

(9)

The antioxidant activities of xanthene-2,7-diols 9 (R=H, Me, Et, Pr^i, Ph), prepared by reacting 2,3,5-trimethylhydroquinone with RCHO, have been evaluated by means of an oxygen absorption method at 60°C for tetralin. Very good activities were observed for 1,3,4,5,6,8-hexa- and 1,3,4,5,6,8,9-heptamethylxanthen-2,7-diols (9; R=H, Me) (T. Nishiyama et al., Bull. Chem. Soc. Jpn., 1993, 66, 2430).

Xanthene-9-carboxaldehyde (10; R=H) and xanthen-9-yl ketones (10; R=Me, tert-Bu, mesityl, 2,4,6-Pr$^i_3$C$_6$H$_2$) are found to be ketonic in CDCl$_3$, but in DMSO-d_6, compounds 10 (R=H, mesityl, 2,4,6-Pr$^i_3$C$_6$H$_4$) are in rapid equilibrium with the xanthenylidene enols 11 (E. Rochlin and Z. Rappoport, J. Amer. Chem. Soc., 1992, 114, 230).

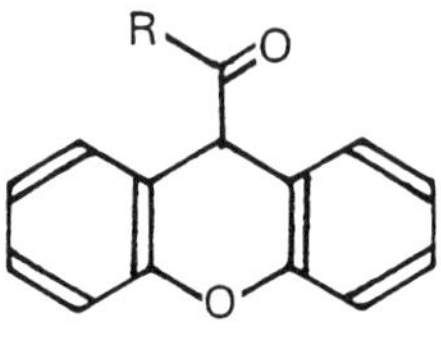

(10)

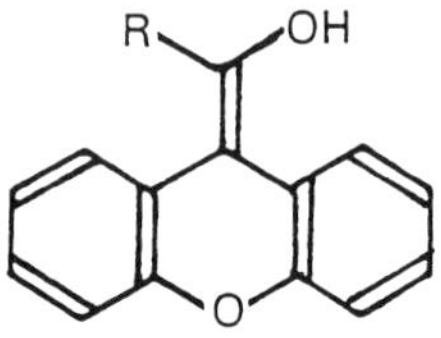

(11)

Propantheline bromide a known drug with anticholinergic, para-sympatholytic, and spasmolytic action, has been prepared by esterification of xanthene-9-carboxylic acid (12) with $Me(Me_2CH)_2N^+CH_2CH_2OH$ Br^- (D. Breazu *et al.*, Rom. RO 91,931, 1987).

(12)

(13)

2,7-Di-*tert*-alkyl-9,9-dimethylxanthene-4,5-dicarboxylic acids 13 (R = Bu^t, EtCMe$_2$) and their derivatives have been prepared and shown to contain intramolecular hydrogen bonds that organize the binding sites and modify their chemical properties (J.S. Nowick *et al.*, J. Amer. Chem. Soc., 1990, 112, 8902). The preparations of several 2,4,7-trisubstituted 9-chloromethyl-enexanthenes 14 (R^1 = Me, Cl, Et; R^2 = H, Me; R^3 = Me, Et, F, Cl) and their ^{1}H and ^{13}C NMR spectral data have been reported (N.H. Martin *et al.*, J. Elisha Mitchell Sci. Soc., 1992, 108, 102).

(14)

(15)

498

The synthesis of a series of *N*-(2-chloroethyl)-*N'*-(xanthen-9-yl)-*N*-nitro-
ureas and related thioxanthenes, for example, the 2-methoxy derivative 15,
have been described and their antitumor activity investigated [E. Filippatos *et
al.*, Arch. Pharm. (Weinheim, Ger.), 1993, <u>326</u>, 451). 2,7-Diamidino-
xanthene (16) (P.M.S. Chauhan *et al.*, Indian IN 167,210, 1990) and *N*-(2,6-
-diisopropylphenyl)-2-(xanthen-9-yl)acetamide (A. Yoshida *et al.*, PCT Int.
Appl. WO 93 06,096, 1993) have been prepared.

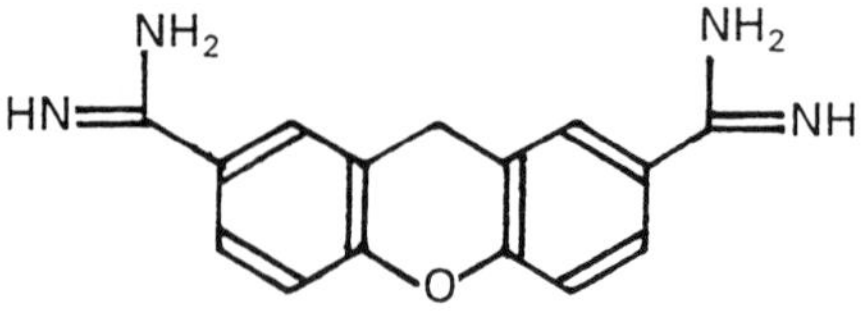

(16)

9-(1-Methylpiperidin-4-yl)xanthen-9-ol (17) on treatment with HCO_2H
gives a mixture of dehydrated product, 9-(1-methylpiperidin-4-ylidene)-
xanthene (18) and reduced product, 9-(1-methylpiperidin-4-yl)xanthene (19).
Derivative 18 on reaction with HCO_2H yields 19. The corresponding
products are obtained by the same reaction with 9-(1-methylpiperidin-4-yl)-
thioxanthen-9-ol (D.G. Loughhead, Tetrahedron Letters, 1988, <u>29</u>, 5701).

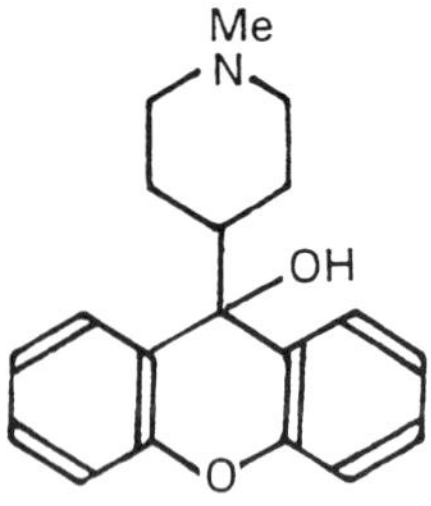

(17) 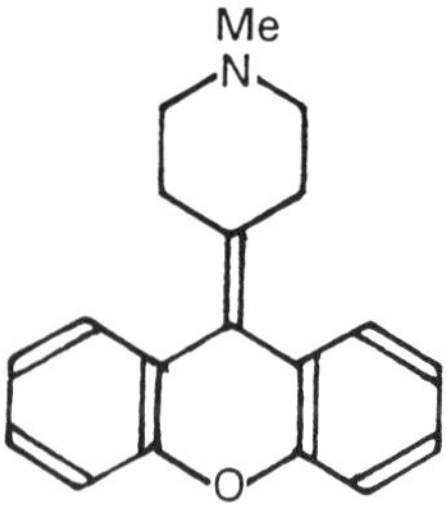   (18) 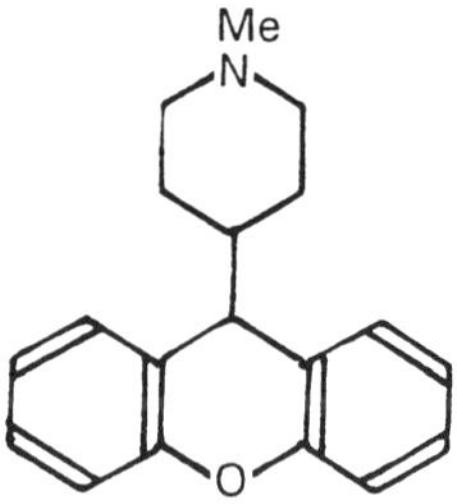   (19)

Both photolysis and thermolysis of 9-arylxanthen-9-yl azides 20 (R = Ph,
4-anisyl, 4-F$_3$CC$_6$H$_4$, 4-ClC$_6$H$_4$) give 9-(arylimino)xanthenes 21 and 11-aryldi-
benz[*b,f*]oxazepines 22 (P. Coombes, A. Goosen, and B. Taljaard,
Heterocycles, 1989, 28, 559).

(20)
(21)
(22)

(b) Xanthylium salts and xanthene colouring matters

Xanthylium salts have been obtained by the thermolysis of diphenylcyclo-
propanes fused with bromo-substituted 1,4-benzoquinones, for example,
derivative 1 resulted in the formation of 3-bromo-2-hydroxy-9-phenylxanthyl-
ium bromide (2), *via* ring opening accompanied by 6π electrocyclization
(T. Oshima and T. Nagai, Tetrahedron Letters, 1993, 34, 649).

(1)
(2)

500

Irradiation of the bromonaphthoquinone-fused diphenylcyclopropane 3 in the presence of naphthalene, dimethoxybenzene, or triphenylamine affords the benzo[c]xanthylium bromide 4. The corresponding chloro and methyl derivatives 3 (Br = Cl, Me) do not undergo a photochemical reaction (*idem*, Chem. Comm., 1994, 1681).

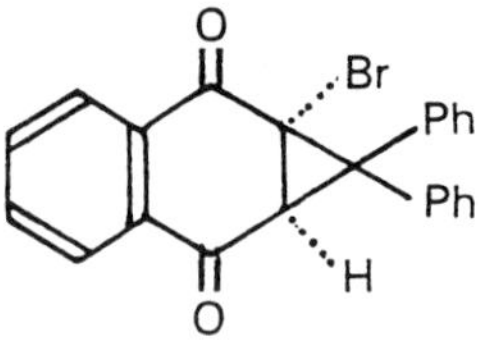

(3)

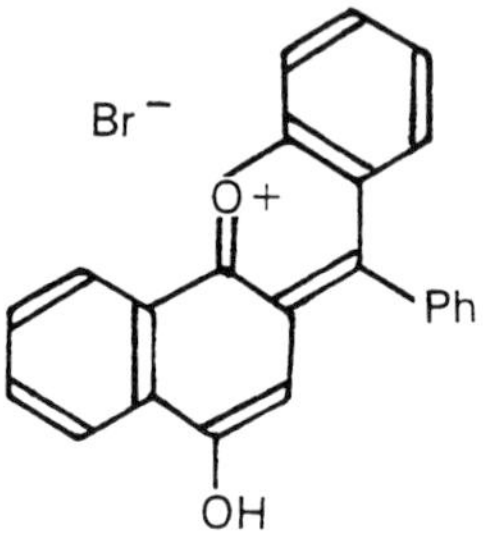

(4)

4,5-Bis{[aryl(hetaryl)amino]methylene}-1,2,3,4,5,6,7,8-octahydroxanthylium perchlorates 5 [R = (un)substituted Ph, pyrid-2-yl, pyrimidin-2-yl] have been prepared by the reaction of 4,5-bis(ethoxymethylene)octahydroxanthylium perchlorate with the appropriate amine (E.P. Olekhnovich *et al.*, Zh. Org. Khim., 1990, 26, 664). 4-Arylmethylene-1,2,3,4,5,6,7,8-octahydroxanthylium perchlorates 6 ($R^1 = R^2 = H$, $R^3 = NO_2$, NMe_2; $R^1 = OH$, $R^2 = R^3 = H$; $R^1 = H$, $R^2 = R^3 = OMe$) show moderate antimicrobial and antiphagic activities (V.G. Kharchenko, N.I. Kozhevnikova, and L.K. Kulikova, Khim-Farm. Zh., 1990, 24, 36).

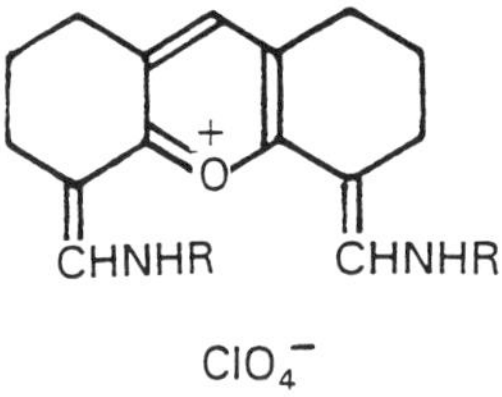

(5)

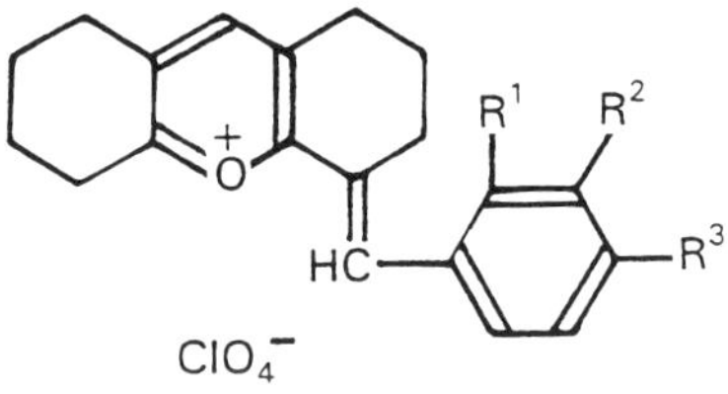

(6)

The cyclocondensation of $PhC{\equiv}CCH(OEt)_2$ with cyclic ketones tetralin-1-one and cyclohexanone affords the xanthylium perchlorates 7 and 8, respectively, containing a phenylacetylenic substituent (E.P. Olekhnovich *et al.*, Zh. Org. Khim., 1990, **26**, 2615).

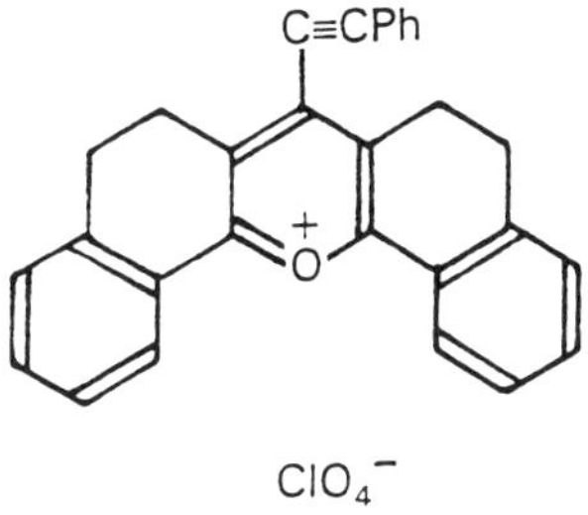

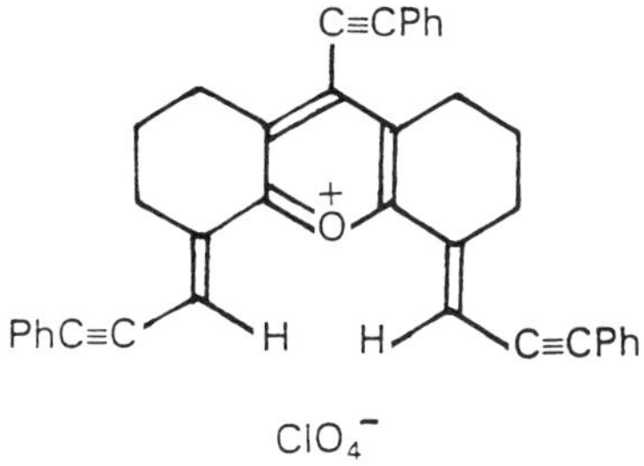

Fluorescein (9; R = OH) on boiling with $PhSO_2Cl$ in benzene and $AcNMe_2$ gives 3,6-dichlorofluorane (9; R = Cl). Its fluorescence, IR, NMR, and mass spectra have been reported (L. Wang *et al.*, Huaxue Shiji, 1990, **12**, 341). The bromination of fluorescein (9; R = OH) in aqueous EtOH at 40-80°C in the presence of 25-40% H_2O_2 furnishes free-acid eosine (10) (J. Prachar and M. Bednarova, Czech. CS 268,474, 1990).

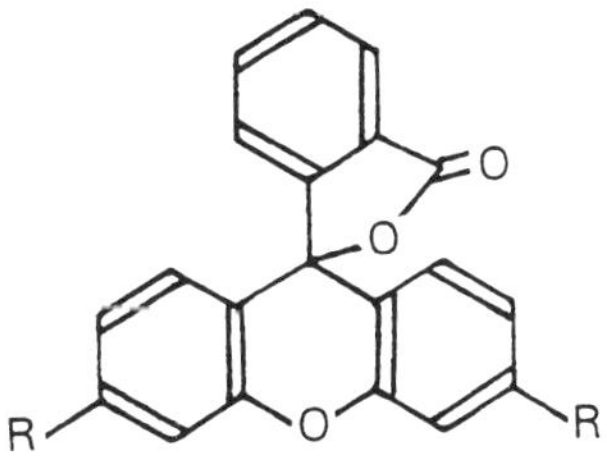

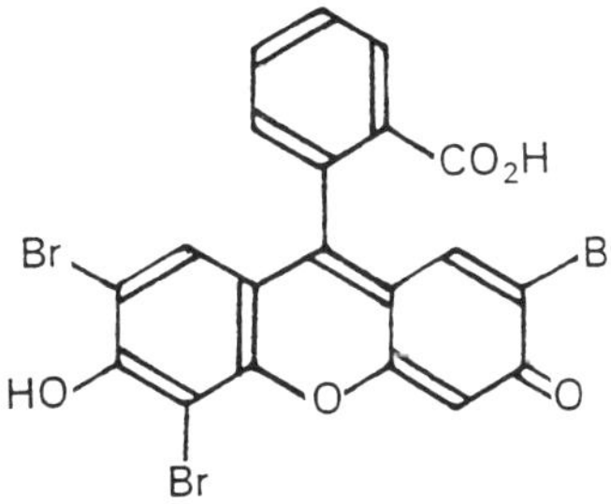

502

The preparations of some (hydroxyalkyl)fluoresceins for the labelling of nucleosides and nucleotides, including 6-(*O*-hydroxybutyl)fluorescein methyl ester, which when stirred with diisopropylammonium tetrazolide and bis(di-isopropylamino)methoxyphosphine in CH_2Cl_2 affords phosphoramidite [F. Schubert, K. Ahlert, and D. Cech, Ger. (East) DD 288,604, 1991]. Phosphoramidiate 11 used to label oligonucleotides with fluorescein during automated DNA synthesis, has been prepared by alkylation of fluorescein methyl ester with 4-chloro-(4,4-dimethoxytrityl)-butan-1-ol, detritylation, and phosphitylation with $(Pr_2^iN)_2POMe$ (*idem*, Nucleic Acids Res., 1990, <u>18</u>, 3427).

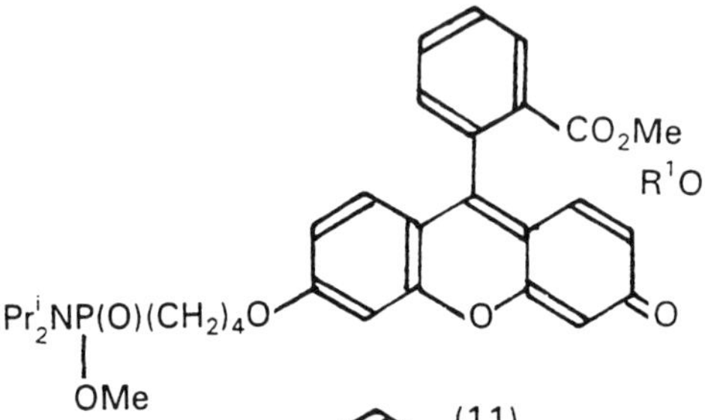

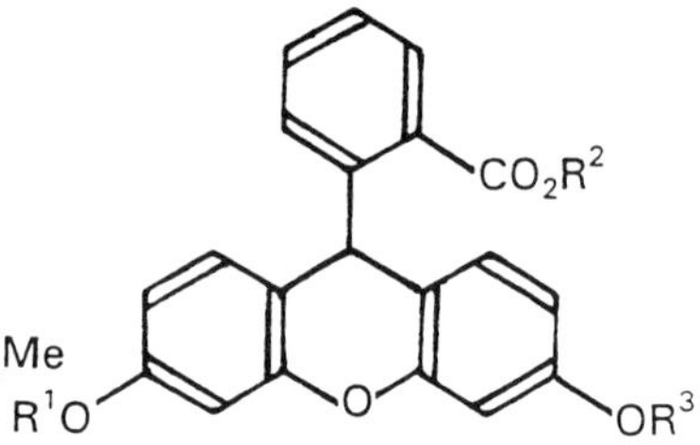

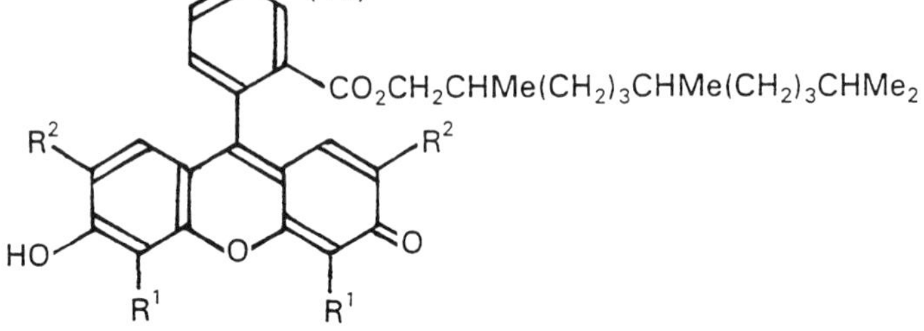

A series of water insoluble indicators, including trimethyldodecyl fluorescein esters 12 ($R^1 = Br$, $R^2 = H$; $R^1 = R^2 = Br$) have been synthesized and used to investigate pH changes accompanying enzyme-catalyzed reactions (L. Brown *et al.*, Tetrahedron Letters, 1990, <u>31</u>, 5799) and the preparations of some dihydrofluorescein ethers and esters 13 ($R^1 = H$, Ac, PhCH_2; $R^2 = H$, PhCH_2, Me; $R^3 = H$, Ac, Me, PhCH_2, MeOCH_2CH_2OCH_2$) have been reported (D. Tadic and A. Brossi, Heterocycles, 1990, <u>31</u>, 1975).

(c) Xanthones

A direct method for selective deuteration of xanthone (1) by the lithium salt of 1,3-diaminopropane-*N,N,N',N'-d$_4$* (S.R. Abrams, J. Labelled Compd. Radiopharm., 1987, <u>24</u>, 941) and the formation of xanthones by the KF/Al$_2$O$_3$ mediated *O*-alkylation of 2-hydroxybenzophenone (M. Patek, M. Lebl, and M. Budesinsky, Tetrahedron Letters, 1992, <u>33</u>, 4507) have been described.

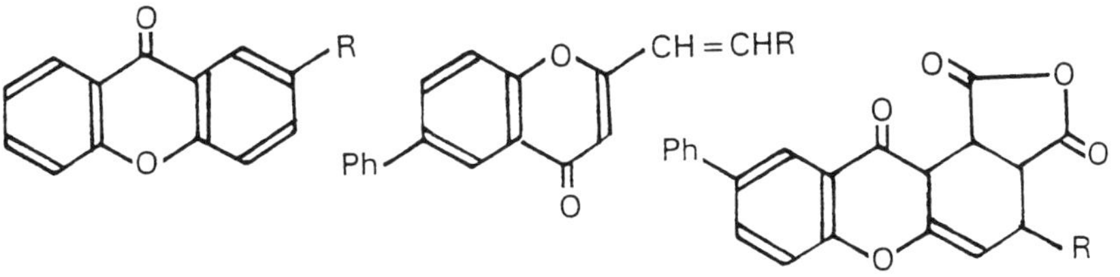

(1) R = H

(2) (see text)

(3)

(4)

2-Substituted xanthones 2 (R = H, Me, Cl, MeO) have been obtained from diPh ethers *via* a metalation-CO$_2$ reaction sequence (A.A. Vitale *et al.*, J. Chem. Res., S, 1994, 82).

Diels-Alder reactions of 6-phenyl-2-(2-substituted vinyl)chromones 3 (R = Ph, 4-Me$_2$NC$_6$H$_4$, PhCH = CH) with maleic anhydride give santhone derivatives 4 (M.R. Mahmoud, A.Y. Soliman, and H.M. Bakeer, Phosphorus, Sulphur, Silicon Relat. Elem., 1990, <u>53</u>, 135).

(i) Halogenoxanthones. 2-Mercaptomethylxanthone (1; R = H) and its 7-chloro derivative 1 (R = Cl) on treatment with ClCH$_2$CO$_2$Na in aqueous NaOH give the xanthon-2-ylmethylthioacetic acids 2, respectively. Other xanthon-2-ylmethylthioalkanoic acids and their 7-chloro derivatives and their methyl and ethyl esters have been prepared (H. Marona, Acta Pol. Pharm., 1988, <u>45</u>, 31).

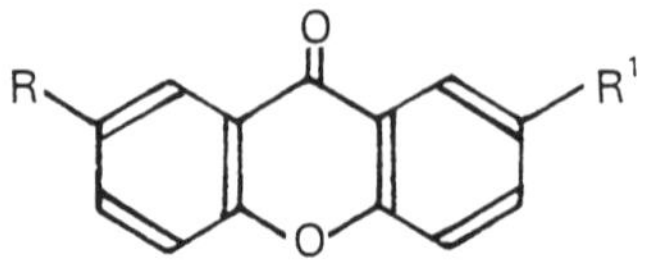

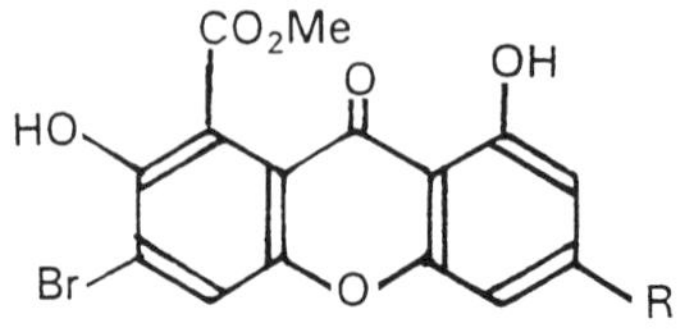

504

(1) R^1 = CH$_2$SH

(2) R^1 = CH$_2$SCH$_2$CO$_2$H

(3) R = H

(4) R = Me

The bromination of hydroxyxanthones with pyridine hydrobromide perbromide affords mono bromination products, for example, the bromination of 1-hydroxyxanthone gives 2-bromo-1-hydroxyxanthone. The bromination of xanthones with bromine in AcOH yields dibromo derivatives, thus 1--hydroxyxanthone gives 2,3-dibromo-1-hydroxyxanthone (G.N. Patel *et al.*, Pol. J. Chem., 1988, 62, 409).

1-Hydroxy-8*a*-methoxycarbonyl-7-oxo-5,6,7,8-tetrahydroxanthone and 1--hydroxy-3-methyl-8*a*-methoxycarbonyl-7-oxo-5,6,7,8-tetrahydroxanthone on treatment with phenyltrimethylammonium perbromide in THF give methyl 6--bromo-1,7-dihydroxyxanthone-8-carboxylate (3) and 6-bromopinselin (methyl 6-bromo-1,7-dihydroxy-3-methylxanthone-8-carboxylate) (4), respectively V.O.T. Omuaru and W.B. Whalley, Indian J. Chem., 1988, 21B, 1041).

Xanthone sulphonamides 5 (R^1 = alkyl, alkoxy, halogeno; R^2 = halogeno, electron-attracting group, SO$_2$NMeCH$_2$CO$_2$H), useful as aldose reductase inhibitors, have been prepared, for example, boiling 2-chloroxanthone in ClSO$_3$H affords 2-chloro-7-(chlorosulphonyl)xanothone, which on treatment with *N*-methylglycine in 1N KOH gave sulphonamide 5 (R^1 = H, R^2 = Cl) (G. Caccia and M. Baldacci, Eur. Pat. Appl. EP 307,879, 1989).

A mixture of 4-allyl-2-chloro-3-hydroxyxanthone, 3-ClC$_6$H$_5$C(O)OOH, and CHCl$_3$ on stirring at room temperature affords 4-chloro-2-hydroxymethyl-1,2--dihydrofuro[2,3-*c*]xanthone (6), which on oxidation with CrO$_3$ in aqueous H$_2$SO$_4$ gives 4-chloro-1,2-dihydrofuro[2,3-*c*]xanthone-2-carboxylic acid, useful along with related derivatives as diuretics and uric acid-excreting agengs (H. Koga *et al.*, Jpn. Kokai Tokkyo Koho JP 63 10,722 [88 10,722], 1988). A series of xanthon-3-yloxyacetic acids 7 (R^1 = H, F, Cl; R^2 = H, Cl, Me; R^3 = H, Cl) and 1,2-dihydrofuro[2,3-*c*]xanthone-2-carboxylic acids 8 (R^1 = H, F, Cl; R^2 = H, Cl, Me; R^3 = H, Br, Cl, Me) have been prepared and tested for their diuretic and uricosuric activity (H. Sato *et al.*, Chem. Pharm. Bull., 1990, 38, 1266).

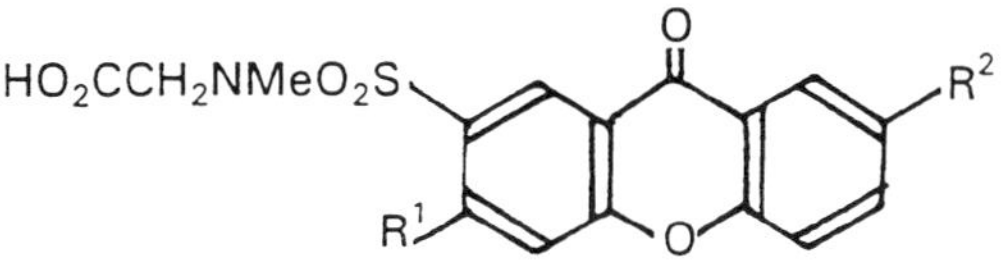

(5)

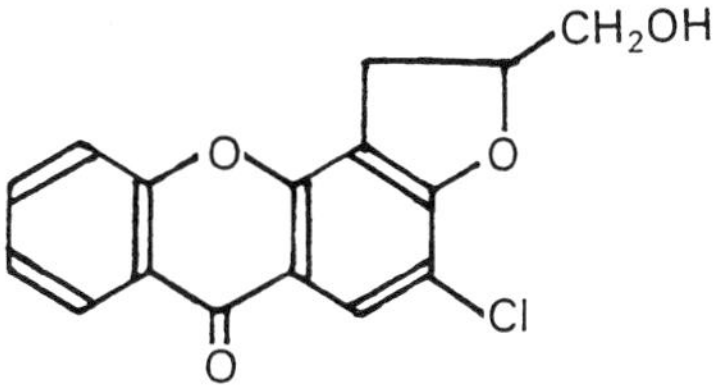

(6)

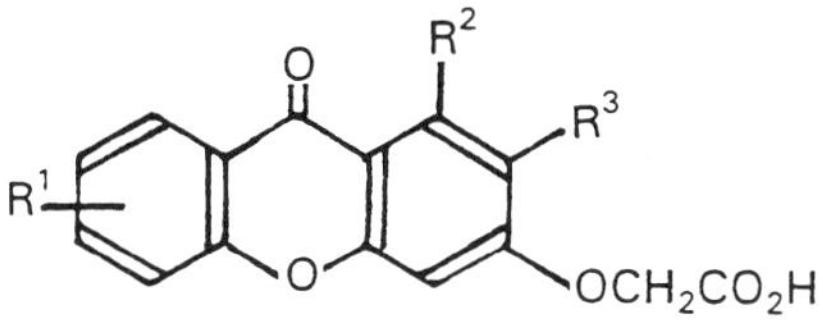

(7)

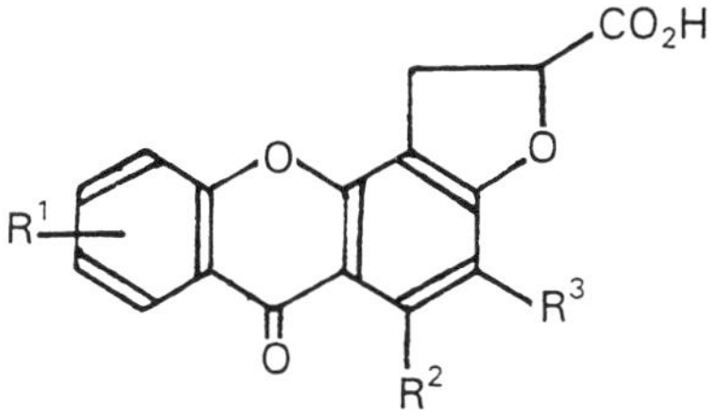

(8)

(ii) Nitro- and aminoxanthones. 3-Nitroxanthone (1) reacts with CN⁻ to
yield 4-cyano-3-hydroxyxanthone (2) (60-70%) and a resin which contains 3-
-amino-4-cyanoxanthone (3) and 4-cyano-3-(xanthon-3-yl)aminoxanthone (4)
(J.H. Gorvin, J. Chem. Res., S, 1991, 88). 3,6-Dinitroxanthone has been
obtained on boiling bis(2-iodo-4-nitrophenyl)ketone with Cu powder in DMF
(P.G. Farrell, D. Moskowitz, and F. Terrier, Synth. Comm., 1993, $\underline{23}$, 231).

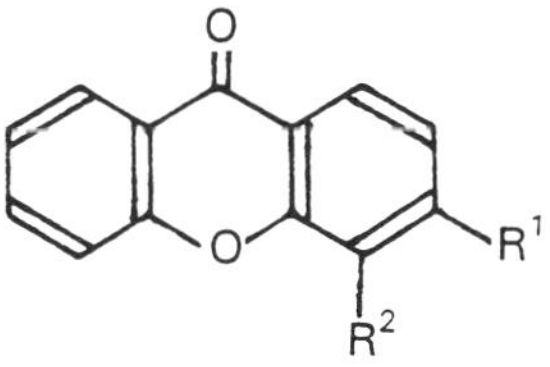

(1) $R^1 = NO_2$, $R^2 = H$
(2) $R^1 = OH$, $R^2 = CN$
(3) $R^1 = NH_2$, $R^2 = CN$

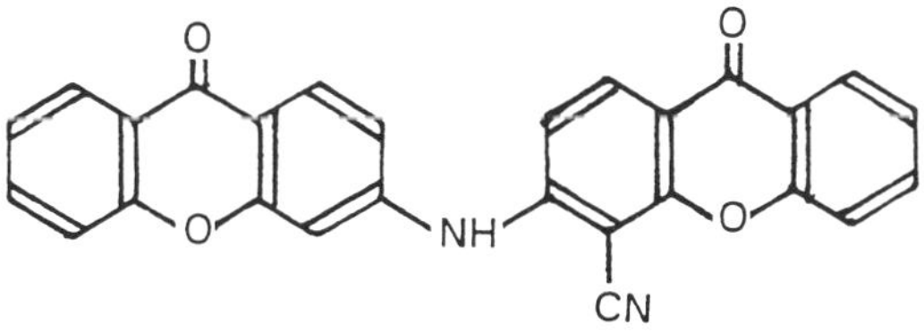

(4)

1-Fluoro-4-methoxyxanthone has been demethylated, protected, and treated with $H_2NCH_2CH_2NHCH_2CH_2OH$, and deprotected to afford the 1-(substituted)amino-4-hydroxyxanthone 5 (M.L. Mancini and J.F. Honek, *ibid.*, 1989, 19, 2001).

The synthesis of some alkanol derivatives 6 [R = NHCHEtCH$_2$OH, NHCH$_2$-CHMeOH, NHCMe$_2$CH$_2$OH, 3-(morpholin-1-yl)propylamino] of 2-aminomethylxanthone have been prepared by reaction of 2-bromomethylxanthone (6; R = Br) with the necessary amine, RH, in toluene in the presence of K_2CO_3. They have been tested for biological properties (H. Marona *et al.*, Acta Pol. Pharm., 1989, 46, 227). 2,7-Diamidinoxanthone (7) has been obtained from 2,7-dibromoxanthone *via* the dicarbonitrile (P.M.S. Chauhan *et al.*, Indian J. Chem., 1987, 26B, 248; Indian IN 167,932, 1991).

(iii) Hydroxyxanthones and related compounds. 4-(Alkylaminoalkoxy)-xanthones 1 (R^1 = H, OH; R^2 = H, Me; R^3 = C_{1-4}alkyl, cyclohexyl; n = 0-2) have been prepared by etherification of the Na salt of 4-hydroxyxanthone with dialkylaminoalkyl chloride hydrochlorides in EtOH in the presence of NaOEt (M. Protiva, I. Cervena, and V. Hola, Czech. CS 262,293, 1989). 4-(Methylaminobutoxy)xanthone (1; R^1 = R^2 = H, R^3 = Me, n = 2) and 4-(methylaminobutoxy)thioxanthone (1; R^1 = R^2 = H, R^3 = Me, n = 2, O = S) have been prepared and are cyclic analogues of the antidepressant and cerebral activator bifemelane (I. Cervena *et al.*, Coll. Czech. Chem. Comm., 1988, 53, 1307).

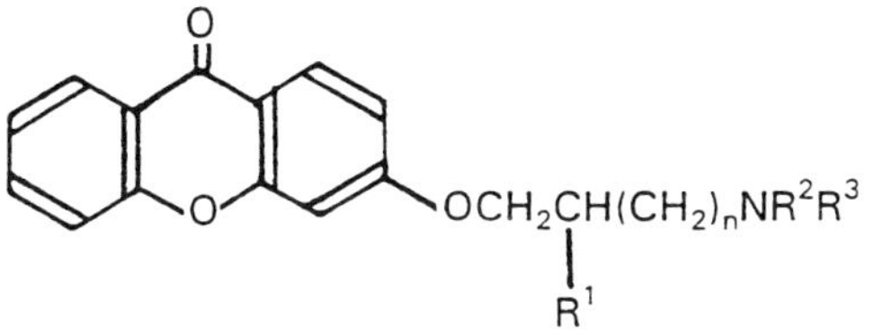

(1)

Benzophenones 2 (R^1 =H, OH; R^2 =OMe, Cl) have been cyclized
intramolecularly by NaH in DMSO to 3-hydroxyxanthone (3; R^1 =H) and 3,4-
-dihyroxyxanthone (3; R^1 =OH), respectively (P. Ravi, P. Vathani, and
G.C. Reddy, Indian J. Heterocyclic Chem., 1994, _3_, 209). The preparations
of 3-alkoxy-6-alkylxanthones 4 [R^1 = C_{3-8}alkyl (linear, branches, cyclic),
R^2 = C_{1-8}alkyl (linear, branched, cyclic)] have been reported and their use for
treatment of osteoporosis investigated (P. Da Re *et al.*, Eur. Pat. Appl. EP
586,960, 1994). A number of 3-alkoxy-1-hydroxyxanthones and analogues
as light stabilizers have been prepared (G. Berner and A. Valet, *ibid.*,
434,618, 1991).

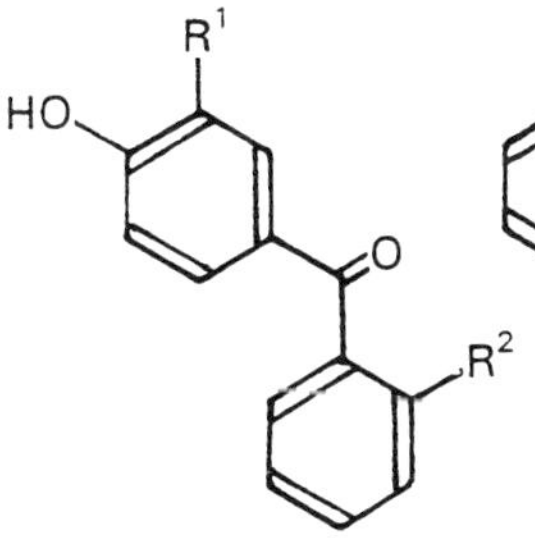

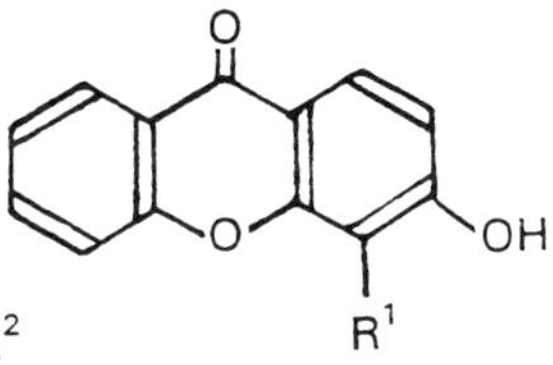

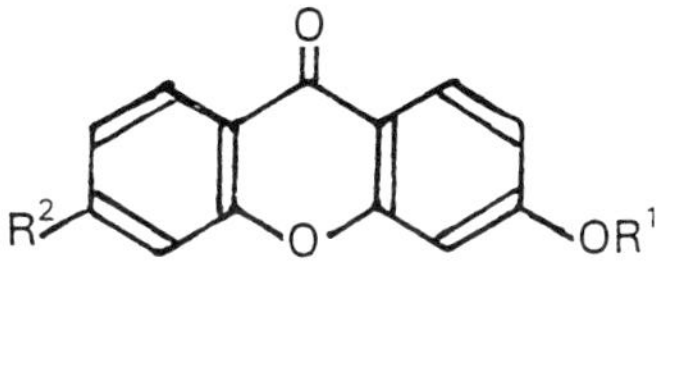

(3)

(4)

(2)

508

4-Prenyl-1-hydroxy-3-methoxyxanthone (5) undergoes an unusual Cope rearrangement in the presence of base to give regioisomer, 2-(1,1-dimethyl-allyl)-1,3-dimethoxyxanthone (6) besides 4-prenyl-1,3-dimethoxyxanthone (7) (S. Raghavan and G.S.R.S. Rao, Tetrahedron Letters, 1992, _33_, 119), 3--Hydroxyxanthone, 2,3-dihydroxyxanthone diacetate and 3,4-dihydroxy-xanthone and its acetate have been prepared and showed potent antiplatelet effects on arachidonate- and collagen-induced aggression, while 3,5-dihydr-oxyxanthone and its acetate, 1,6-dimethoxyxanthone, and 3,6-dihydroxyxan-thone and its diacetate showed potent antiplatelet effects on arachidonate--induced aggregation. The preparation of 3-[2-hydroxy-3-(propylamino)prop-oxy]xanthone has also been reported (C.N. Lin _et al._, J. Pharm. Sci., 1993, _82_, 11).

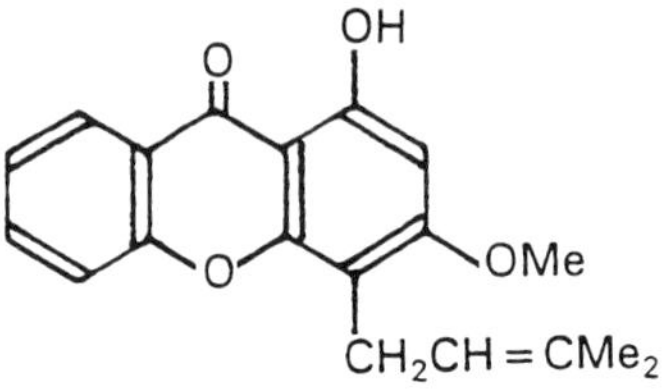

(5)

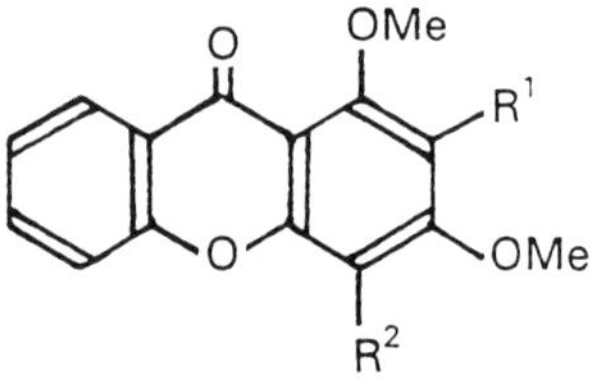

(6) $R^1 = CMe_2CH = CH_2$, $R^2 = H$
(7) $R^1 = H$, $R^2 = CH_2CH = CMe_2$

Demethylation of 1-hydroxy-2-prenyl-3,5,8-trimethoxyxanthene (8) with hot aqueous morpholine results in rearrangement and formation of a mixture containing 2-prenyl-1,3,5,8-tetrahydroxyxanthone, 5-methoxy-2-prenyl--1,3,8-trihydroxyxanthone, 2-prenyl-1,3,7,8-tetrahydroxyxanthone, and 7--methoxy-2-prenyl-1,3,8-trihydroxyxanthone (G.J. Bennett and H.H. Lee, Tetrahedron Letters, 1989, _30_, 7265).

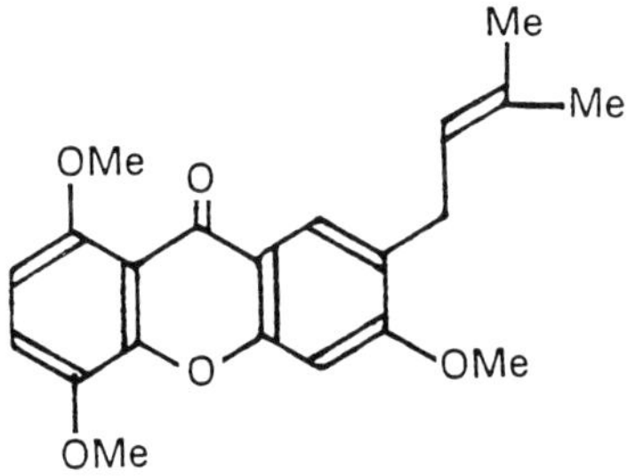

(8)

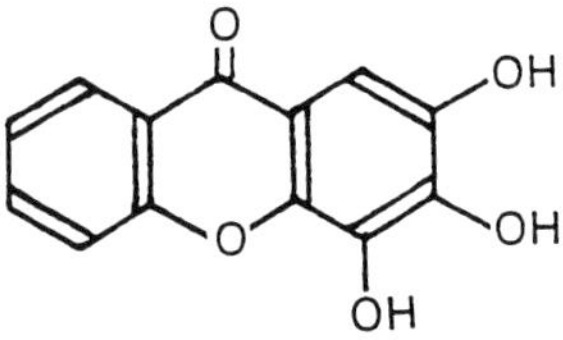

(9)

The benzophenone derivative, obtained by the acylation of 1,2,3,4-tetra-methoxybenzene with $2\text{-MeOC}_6\text{H}_4\text{CO}_2\text{H}$, on heating with pyridine hydrochloride yields 2,3,4-trihydroxyxanthone (9) (M.M.M. Pinto, Rev. Port. Farm., 1987, __37__, 29). Norathyriol (1,3,6,7-tetrahydroxyxanthone) and its 1,3,5,6-, 3,4,5,6-, 3,4,6,7-, and 2,3,6,7-tetrahydroxy analogues have been synthesized from benzophenone precursors by Friedel-Crafts acylation and base-catalysed cyclization. Both 3,4,6,7- and 2,3,6,7-tetrahydroxyxanthone tetraacetates showed potent inhibition of arachidonic acid-induced platelet aggregation and 3,4,6,7-tetrahydroxyxanthene tetraacetate and 1,3,5,6--tetrahydroxyxanthone showed potent and significant inhibition of collagen--induced platelet aggregation (C.N. Lin *et al.*, J. Pharm. Sci., 1992, __81__, 1109). Also for the preparations of polyoxygenated 2-hydroxyxanthones, see N.M.A. Mahfouz *et al.*, Arch. Pharm. (Weinheim, Ger.), 1990, __323__, 163.

9-(4-Formylphenyl)-2,3,7-trihydroxyfluor-6-one (10) has been prepared by cyclocondensation of terephthaldehyde with 1,2,4-triacetoxybenzene (R. Zhou and P. Zheng, Huaxue Shiji, 1989, __11__, 251).

(10)   (11)   (12)

1-, 2-, 3-, and 4-Mercaptoxanthones have been prepared from the corresponding hydroxyxanthones (P. Montanari, P. Valenti, and P. Da Re, Indian J. Chem., 1989, __28B__, 354). The fluoride ion induced thiophilic addition of allylsilanes to thio ketones has been applied to xanththione (11), thus, it reacted with $\text{CH}_2 = \text{CHCH}_2\text{SiMe}_2$ in DMF in the presence of $\text{Bu}_4\text{N}^+\text{F}^-$ to afford 9-allylthioxanthene (12) (A. Capperucci *et al.*, Synlett, 1992, 880).

(iv) Acylxanthones, xanthonecarboxylic acids and derivatives. Treatment of 3-acetyl-2-methylchromone (1) with sodium or DMF-POCl_3 gives 2-(2--hydroxybenzoyl)-1,3-dimethylxanthone (2) or 1-chloroxanthone-2-carbox-aldehyde (3), respectively. A number of related conversions have been

reported (C.K. Ghosh, S. Sahana, and A. Patra, Tetrahedron, 1993, 49, 4127). The preparations of 7-chloro-2-(2-hydroxybenzoyl)-4-phenylxanthone and 7-chloro-2-(5-chloro-2-hydroxybenzoyl)-4-phenylxanthone have been described (C.K. Ghosh and S. Sahana, Indian J. Chem., 1992, 31B, 346).

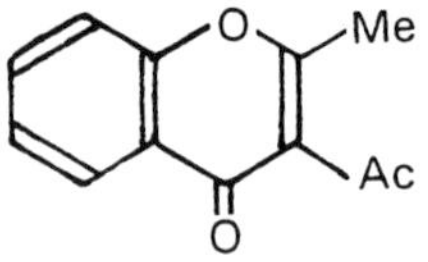

(1) 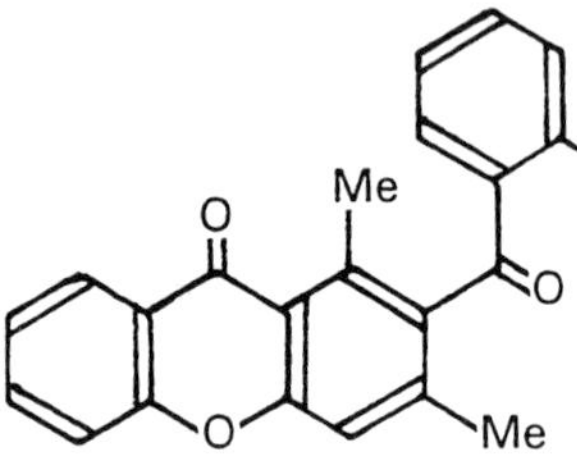 (2) 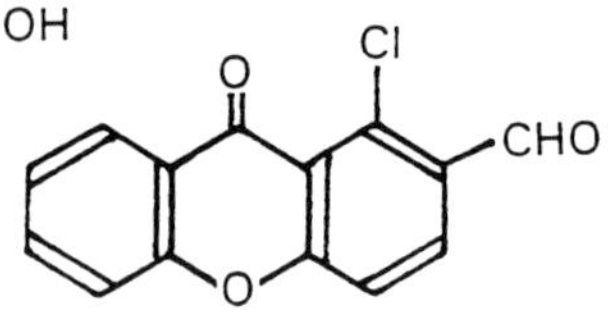 (3)

The preparations of 7-substituted xanthone-2-carboxylate 4 (X = NNH$_2$, NNHCONH$_2$, NOH) (D. Ge et al., Yaoxue Xuebao, 1987, 22, 822), 5-substituted xanthone-4-acetic acids 5 (R = CF$_3$, halogeno, Ph, alkyl, alkoxy, CH$_2$CO$_2$H) (G.J. Atwell et al., J. Med. Chem., 1990, 33, 1375) and (R = CO$_2$H, CH$_2$CH$_2$CO$_2$H, OCH$_2$CO$_2$H, CH$_2$SO$_3$H, CH$_2$CO$_2$CH$_2$CH$_2$NMe$_2$) (G.W. Rewcastle et al., ibid., 1991, 34, 2864), and other derivatives of xanthone-4-acetic acid (W.A. Denny et al., Eur. Pat. Appl. EP 278,176, 1988) have been reported.

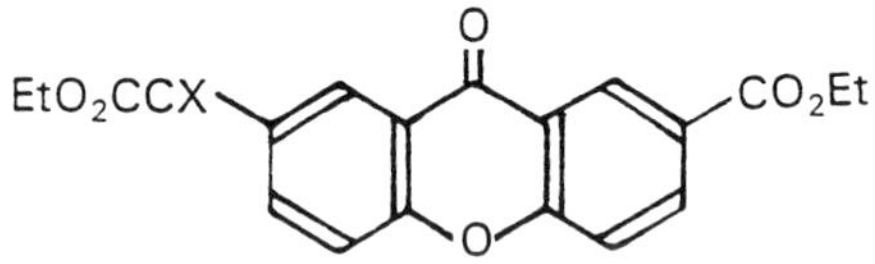

(4) 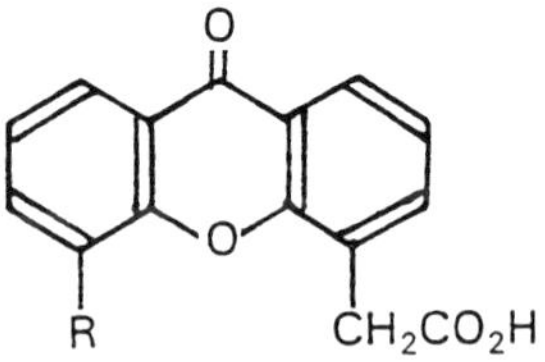 (5)

Xanthone-2-acetic acid on boiling with $HOCH_2CHEtNH_2$ in xylene affords racemic N-[2-(1-hydroxybutyl)xanthone-2-acetamide (6). (R)- And (S)-6 have been obtained analogously (H. Marona, Acta Pol. Pharm., 1990, **47**, 65). In an effort to develop increasingly potent and specific leukotriene B_4 (LTB_4) receptor antagonists, several xanthone dicarboxylic acids, for example 7, have been synthesized and evaluated (W.T. Jackson et al., J. Med. Chem., 1993, **36**, 1726).

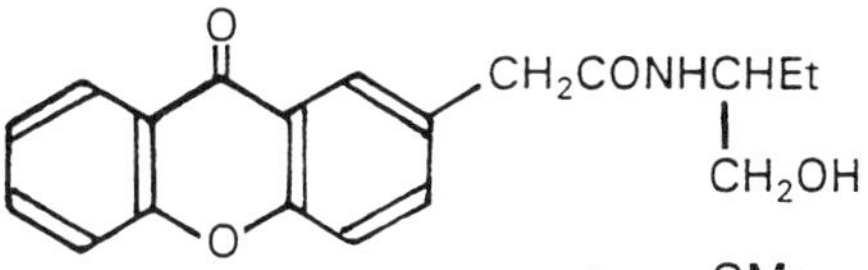

(6)

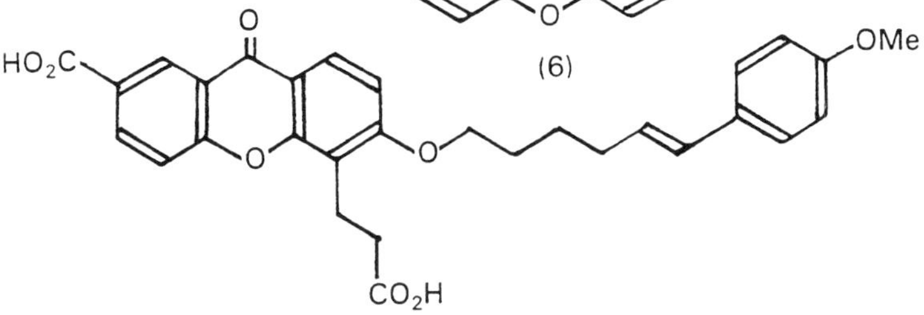

(7)

(v) Tetra-, hexa-, octa-, deca-, and perhydroxanthones and related compounds. 3-Formylchromone and methyl chromone-3-carboxylate react with $H_2C = CMeCMe = CH_2$ in the presence of $TiCl_4$ to yield the corresponding [4 + 2] cycloadducts 1 ($R^1 = CHO$, $R^2 = R^3 = Me$; $R^1 = CO_2Me$, $R^2 = R^3 = Me$), the former undergoing facile deformylation to afford 2,3--dimethyl-1,4,4a,9a-tetrahydroxanthone (1; $R^1 = H$, $R^2 = R^3 = Me$). 3-Aryl-sulphinylchromones [e.g., 3-(4-chlorophenylsulphinyl)chromone] on heating with the diene, $MeOCH = CHC(OSiMe_3) - CH_2$ furnish 3-hydroxyxanthone (P.J. Cremins, S.T. Saengchantara, and T.W. Wallace, Tetrahedron, 1987, **43**, 3075). Substituted 1,4,4a,9a-tetrahydroxanthones have been obtained by the cycloaddition between 4-*tert*-butyldimethylsiloxybenzopyrylium salts and α,β-unsaturated ketones (Y.G. Lee et al., Heterocycles, 1989, **29**, 35).

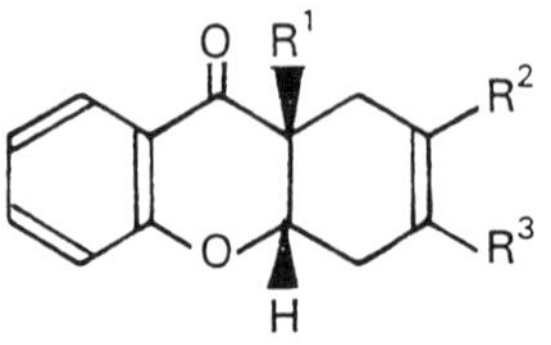

512

(1) 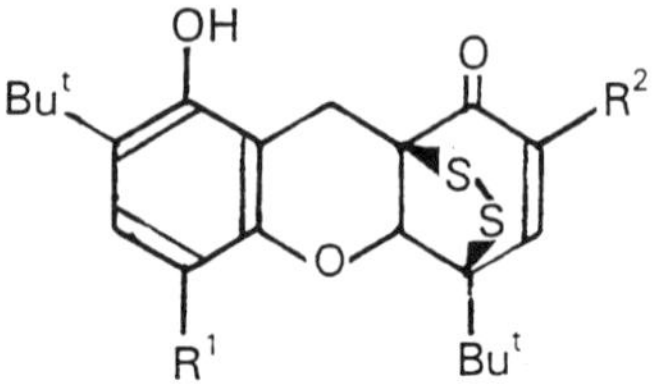   (2)

2,2'-Methylenebisphenols on treatment with S_2Cl_2 in PhMe containing pyridine give 4,9a-(epidithio)-1,4,4a,9a-tetrahydroxanthen-1-ones [4,10a-(epidithio)-4,4a,10,10a-tetrahydro-1H-5-oxaanthracen-1-ones] 2 (R^1 = Bu^t, Me; R^2 = H, Me, Et, Pr^i) (S. Kolly *et al.*, Helv., 1988, **71**, 1101). 4a,6-Bis-(*N*-methylanilino)-7-hydroxy-2,3,8-trimethyl-1,4,4a,9a-tetrahydroxanthene-1,4-dione (3) has been obtained as a thermolysis product of 6-(*N*-methyl-anilino)-7a- -methyl-3a,7a-dihydroindazole-4,7-dione (M. Schubert-Zsilavecz, Monatsh., 1991, **122**, 545).

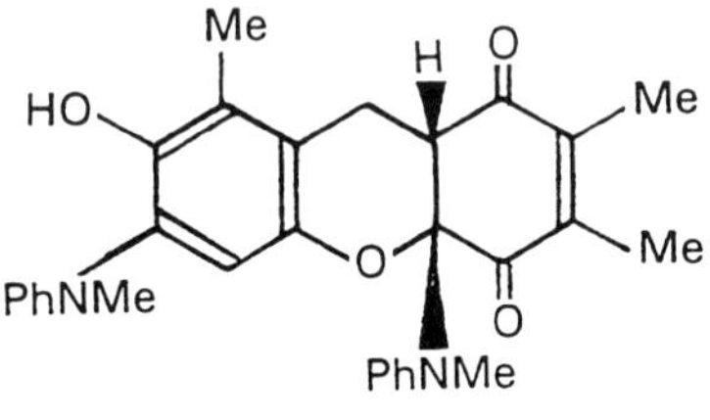

(3)

Diastereoisomeric oxabicycloheptanes 4 (R = H, Me) have been demethylated with excess BCl_3 to yield, methyl 9a-bromo-2,8-dihydroxy-9-
-oxo-1,2,3,4,4a,9a-hexahydroxanthene-1-carboxylate (5; R = H) and methyl 9a-bromo-2,8-dihydroxy-6-methyl-9-oxo-1,2,3,4,4a,9a-hexahydroxanthone-
-1-carboxylate (5; R = Me) (V.O.T. Omuaru and W.B. Whalley, Indian J. Chem., 1989, 28B, 1009).

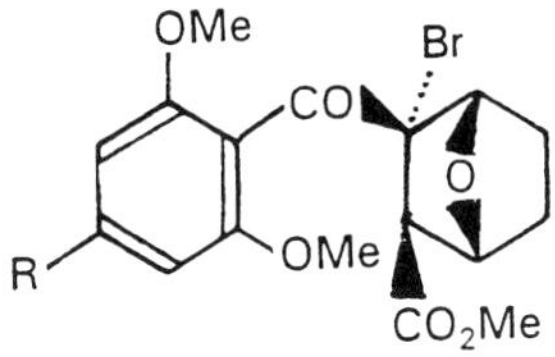

(4) 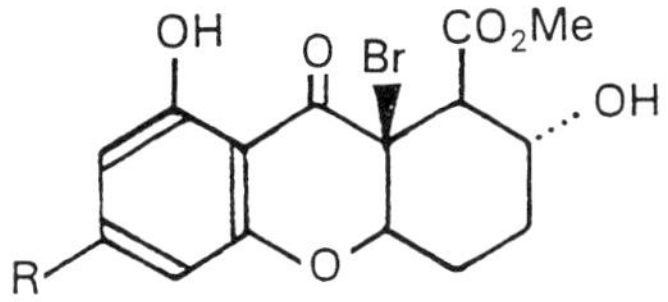   (5)

2,6-Diarylidenecyclohexanones 6 (R = Ph, 2-HOC_6H_4, 4-$MeOC_6H_4$, 2-ClC_6H_4, fur-2-yl) undergo Diels-Alder reaction with 1,4-benzoquinone to yield 9-aryl-5-arylidene-1,4,4a,5,6,7,8,9a-octahydroxanthene-1,4-diones 7 (S.A. Mahgoub and A.S. Yanni, Bull. Fac. Sci., Assiut Univ., 1991, 20, 9).

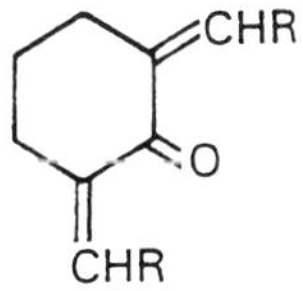

(6) 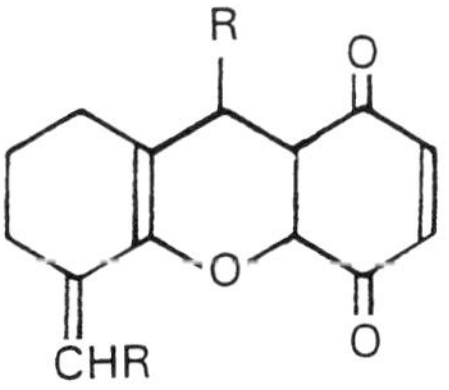   (7)

514

9-(3-Methoxyphenyl)-3,3,6,6-tetramethyl-1,2,3,4,5,6,7,8-octahydro-xanthene-1,8-dione (8) reacts with chlorosulphonic acid to give 9-[2,4-di-(chlorosulphonyl)-5-methoxyphenyl]-3,3,6,6-tetramethyl-1,2,3,4,5,6,7,8-
-octahydroxanthene-1,8-dione (9).

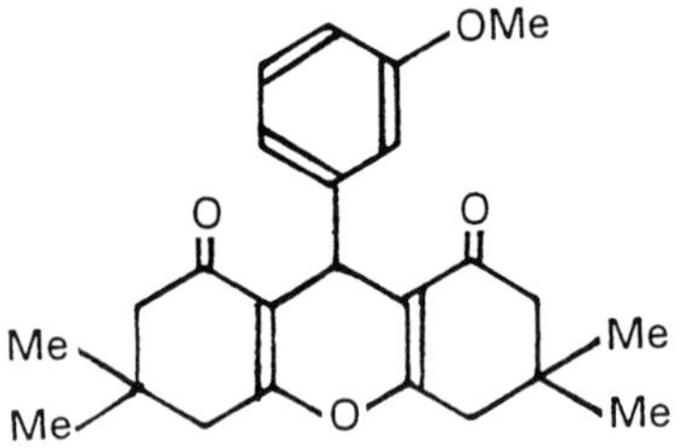

(8)

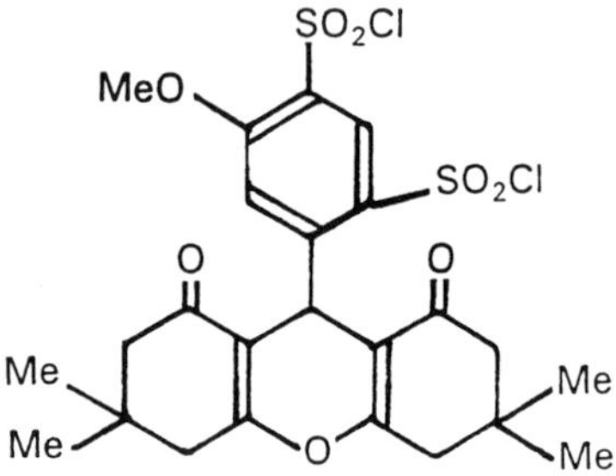

(9)

Condensation of $RCCl = CHCHO$ ($R = Me_3Si$, Ph) with dimedone affords dimedone derivatives 10, which on cyclization with H_2SO_4 gives 9-(2-chloro-
-2-trimethylsilyl)vinyl-3,3,6,6-tetramethyl-1,2,3,4,5,6,7,8-octahydroxanth-one-1,8-dione (11; $R = Me_3Si$) and 9-(2-chloro-2-phenyl)vinyl-3,3,6,6-tetra-methyl-1,2,3,4,5,6,7,8-octahydroxanthone-1,8-dione (11; $R = Ph$),
respectively (O.I. Yurchenko *et al.*, Zh. Obshch. Khim., 1988, 58, 1182).

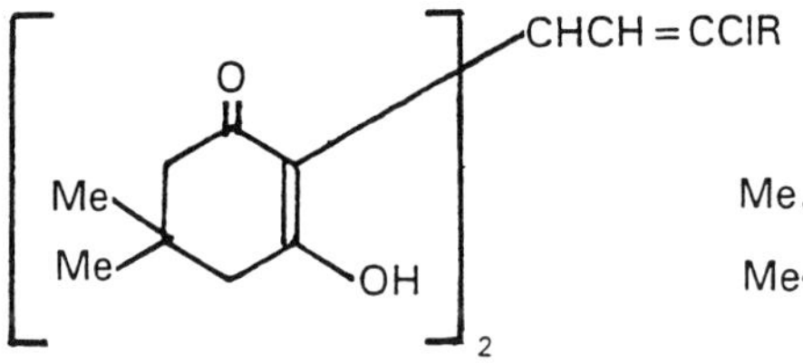

(10)

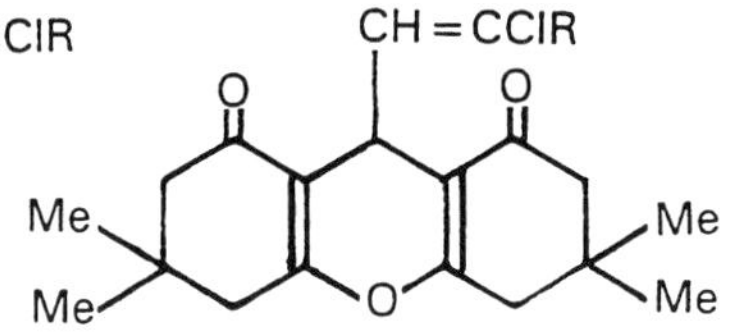

(11)

Pyrolysis of a mixture of dimedone and 1-methoxycyclohexene results in
an ene reaction and the formation of an intermediate enedione, which undergoes both a Diels-Alder reaction with 1-methoxycyclohexene to form 9-spiro-
cyclohexyl-3,3-dimethyl-10a-methoxy-1,2,3,4,5,6,7,8,8a,10a-decahydroxan-
then-1-one 12 and a reaction with oxygen to yield a cyclic peroxide 13
(J.T. Pinhey and Phan Thanh Xuan, Austral. J. Chem., 1988, <u>41</u>, 331).

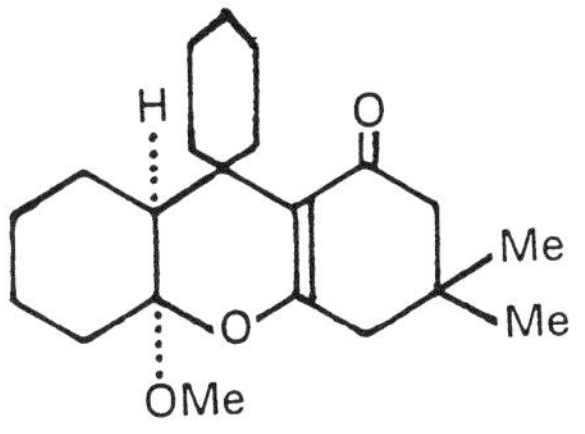

(12)

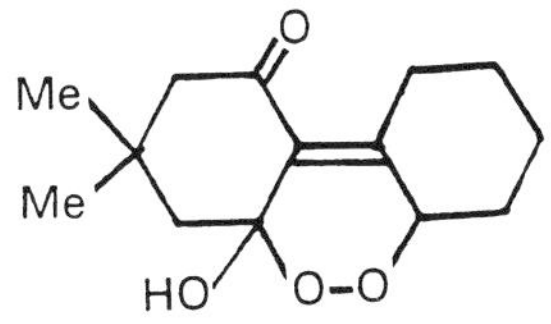

(13)

Reductive cyclization of 1,5-diketones 14 (R = H, Me) by Raney Ni/H$_2$
yields 1,2,3,4,5,6,7,8,8a,10a-decahydroxanthen-1-ones 15, which on
further hydrogenation over Rh/C afford perhydroxanthenes 16. Treatment of
diketones 14 with NaBH$_4$ give decahydroxanthen-1-ones 17, which on
hydrogenation over Rh/C furnish perhydroxanthenes 18 (V.G. Kharchenko *et
al.*, Zh. Org. Khim., 1987, <u>23</u>, 576).

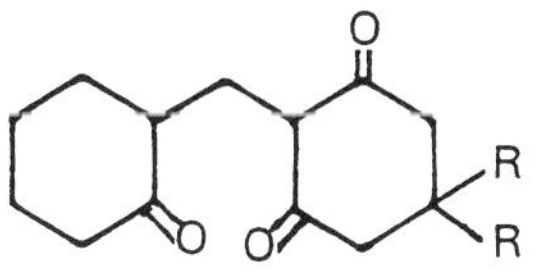

(14)

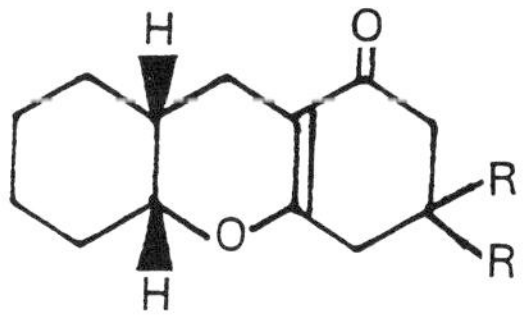

(15)

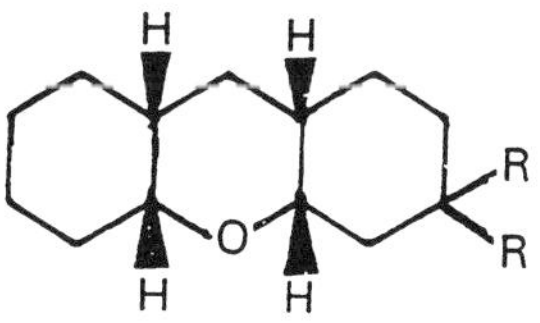

(16)

516

(17)   (18)

For the regio- and diastereoselective synthesis of hexahydro-4a,9-prop-
anoxanthenones, see F. Eiden, P. Gmeiner, and H. Lotter, Ann., 1988, 125;
the synthesis of cyclopenta- and cyclohexa[*a*]xanthones, R.M. Letcher and
T.Y. Yue, Chem. Comm., 1992, 1310; R.M. Letcher, T.Y. Yue, and
K.K. Cheung, *ibid.*, 1993, 159; and indeno[1,2-*b*]pyrans, J. Bloxham and
C.P. Dell, Tetrahedron Letters, 1991, $\underline{32}$, 4051.

(vi) Furoxanthones. 6-Phenyl-2-styrylchromones 1 (R = Ph, 4-Me$_2$N,
PhCH = CH) undergo Diels-Alder reactions with maleic anhydride to give the
furo[3,4-*a*]xanthones 2. Various reactions of 2-methyl-6-phenylchromone, 2-
-methyl-6-phenyl-4*H*-benzo[*b*]pyran-4-thione, and xanthone 2
(R = CH = CHPh) with hydrazine, phenylhydrazine, primary aromatic amines,
and Grignard reagents have been reported (M.R. Mahmoud, A.Y. Soliman,
and H.M. Bakeer, Phosphorus, Sulphur and Silicon Relat. Elem., 1990, $\underline{53}$,
135).

(1)   (2)

NMR studies have been carried out on the furo[3,4-*a*]xanthone 3 obtained from (*E*)-2-styrylchromone 4 and maleic anhydride in order to determine its stereochemistry. The resulting data suggested that all proposed 1,2,3,9a-tetrahydroxanthone structure of the products from similar reactions *i.e.* Diels-Alder adducts should be revised to 1,2,3,4-tetrahydroxanthone structures (R.M. Letcher and T.Y. Yue, J. Chem. Res., S, 1992, 248).

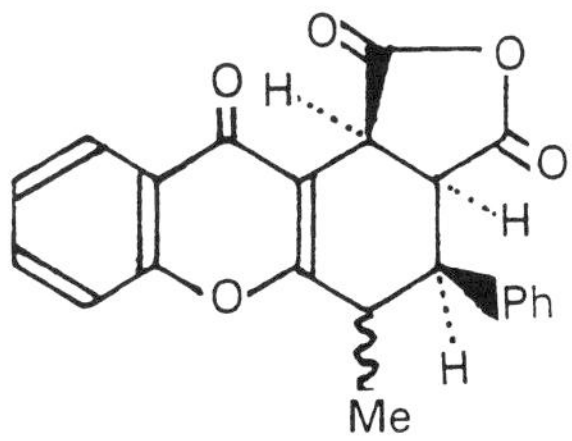

(3) 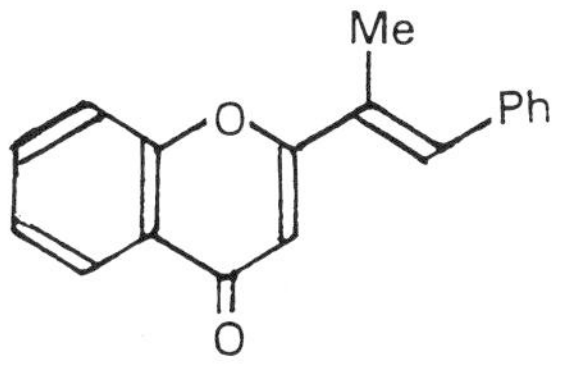 (4)

Some derivatives of 8*H*-benzofuro[2,3-*c*]xanthen-8-one and 12*H*-benzofuro[3,2-*b*]xanthen-12-one have been prepared (V.K. Ahluwalia *et al.*, Indian J. Chem., 1989, **28B**, 1021).

8. 6*H*-Dibenzo[*b,d*]pyran, 3,4-benzochromone and its derivatives

2-PhOC$_6$H$_4$CH$_2$OH undergoes photochemical cyclization in aqueous solution to give 6*H*-dibenzo[*b,d*]pyran (1) (P. Wan and C.G. Huang, Chem. Comm., 1988, 1193). 4-(2-Methoxyphenyl)-10-methoxy-6*H*-dibenzo[*b,d*]pyran (2) has been obtained by treating 2-IC$_6$H$_4$OMe with K$_2$CO$_3$, Bu$_4$NBr in DMF containing Pd(OAc)$_2$. Other derivatives have been prepared by this method (G. Dyker, Angew. Chem., 1992, **104**, 1079). 4,9-Dimethyl-10-methoxy-6*H*-dibenzo[*b,d*]pyran has been obtained by the same method (*idem*, Ber., 1994, **127**, 739).

OMe

OMe

R

(1)　　　　　　　　(2)　　　　　　　　(3) R = NH₂

(4) R = Nphthyl

A suspension of 2-amino-6*H*-dibenzo[*b,d*]pyran (3) and 3,4,5,6-tetra-hydrophthalic anhydride in boiling AcOH affords dibenzo[*b,d*]pyran derivative 4, which shows complete control of *Indian mallow* (M. Enomoto *et al.*, Jpn. Kokai Tokkyo Koho JP 63 264,582 [88 264,582], 1988). The synthesis and mesomorphic properties of 3-(4-*n*-alkoxybenzylideneamino)dibenzo[*b,d*]-pyrans (S. Subramanyam, Diss. Abstr. Int. B, 1991, <u>51</u>, 5878) and the preparation of 1-aminoalkyl-4-hydroxy(or ether groups)-6*H*-dibenzo[*b,d*]pyran derivatives for the treatment of dopaminergic hyperkinesia (A. Nakazato *et al.*, Jpn. Kokai Tokkyo Koho JP 05 155,875 [93 155,875], 1993) have been discussed.

The reaction of 4-substituted bis-(3-alkoxybenzoyl) peroxides in neat phenols yields mainly 8-alkoxy-6*H*-dibenzo[*b,d*]pyran-6-ones and 2-benzoyl-oxylation products of the phenol, for example, 3,5-(MeO)₂C₆H₃CO₂O₂CC₆H₃-(OMe)-3,5 and 4-MeC₆H₄OH afford 2-methyl-8,10-dimethoxy-6*H*-dibenzo-[*b,d*]pyran-6-one (5) (41%) and 2-hydroxy-5-methylphenyl 3,5-dimethoxy-benzoate (6) (28%). Diaroyl peroxides without electron-releasing meta substituents furnish only analogues of hydroxyphenylbenzoate 6 (S. Auricchio, A. Citterio, and R. Sebastiano, J. Org. Chem., 1990, <u>55</u>, 6312).

Me

OMe

MeO

O

O

(5)

MeO

CO₂

OH

OMe

Me

(6)

A number of dibenzo[*b,d*]pyran-6-ones 7 (R^1 =H, R^2 =H, Me, OMe, Br, NO_2, CF_3; R^1 = R^2 = Me; R^3 =H, Me, OMe) with up to three substituents in the 2-, 4-, and 9-positions have been obtained by *ortho*-arylation of appropriate phenols by (2-cyanoaryl)azo sulphides 8 (R^3 =H, Me, OMe, R^4 = Bu^t; R^3 =H, R^4 =Ph) in $S_{RN}1$ conditions, followed by a silica-gel--catalyzed lactonization of the resulting 2-cyano-2'-hydroxybiphenyls 9 (G. Petrillo *et al.*, Tetrahedron, 1991, 47, 9297).

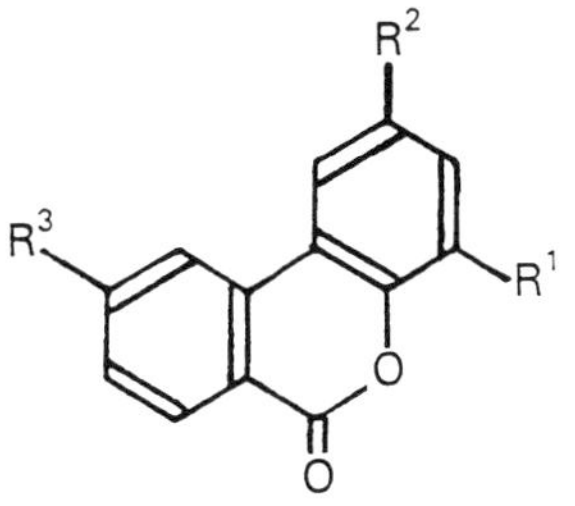

(7)

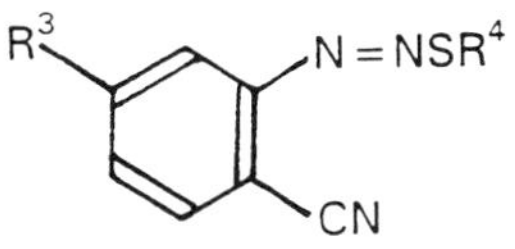

(8)

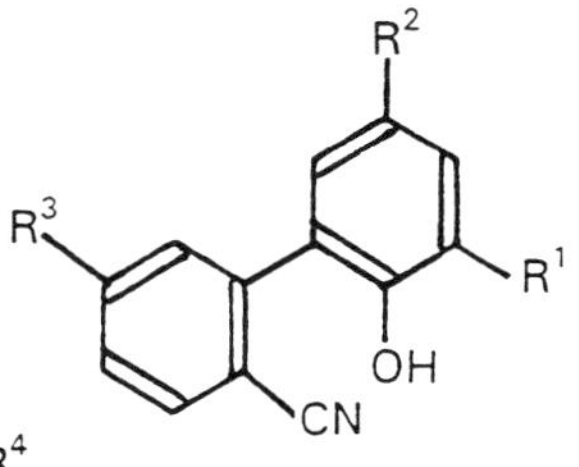

(9)

The decarboxylation of 10-carboxy-3,8-dinitro-6*H*-dibenzo[*b,d*]pyran-6--one (10) in DMSO, DMF, or HMPT yields 10-hydroxy-3,8-dinitro-6*H*--dibenzo[*b,d*]pyran-6-one and 3,8-dinitro-6*H*-dibenzo[*b,d*]pyran-6-one (A.M. Andrievskii *et al.*, Khim. Geterotsikl. Soedin., 1989, 164). X-ray analysis of ketone 10 in a crystal containing toluene and water reveals that the pyran ring exists in a boat form and H_2O molecules participate in 4H bonds; with the CO_2H and CO groups and with both O atoms of one NO_2 group. Toluene molecules lie between molecules of ketone 10 (L.A. Chetkina *et al.*, Doklady Akad. Nauk SSSR, 1988, 301, 350).

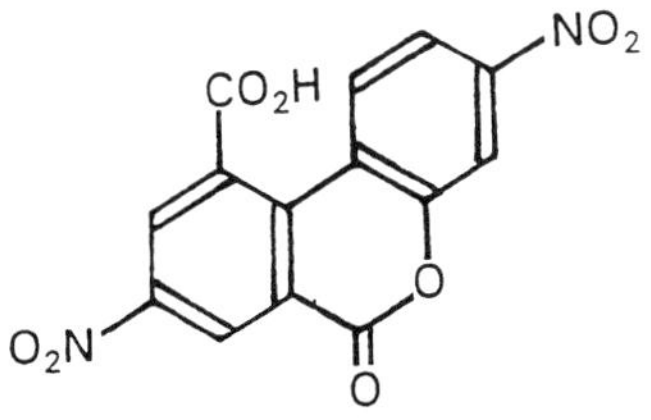

(10)

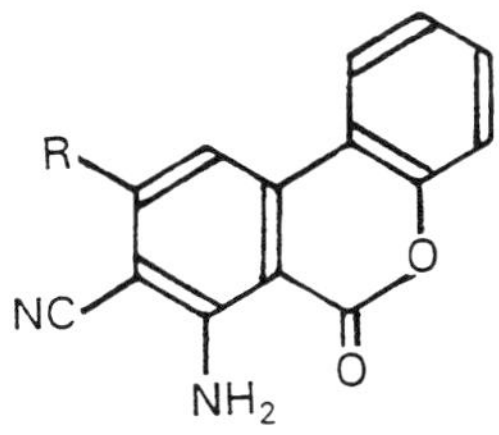

(11)

520

9-Substituted 7-amino-8-cyano-6*H*-dibenzo[*b,d*]pyran-6-one 11 [R = pyrid-
-3-yl, thien-2-yl, (un)substituted Ph] are conveniently prepared in a one-step
synthesis by the ternary condensation of RCHO, $H_2C(CN)_2$ and 3-cyano-4-
-methylcoumarin (F.F. Abdel-Latif, Gazz., 1991, 121, 9). A number of 6*H*-
-dibenzo[*b,d*]pyran-6-ones with up to 8 substituents have been synthesized
and their use as aldose reductase inhibitors investigated (H. Nakayama *et al.*,
Jpn. Kokai Tokkyo Koho JP 01 250,373 [89 250,373], 1989; 02
304,080 [90 304,080], 1990). Several 6*H*-dibenzo[*b,d*]pyran-6-ones with
up to five substituents in the 2-, 3-, 8-, 9-, and 10-positions have been
synthesized *via* dienone-phenol rearrangements of 1-spiro(cyclohexa-2,5-
-diene)-1,3-dihydroisobenzofuran-3-ones (D.J. Hart *et al.*, Tetrahedron,
1992, 48, 8179).

3,4,4a,10a-Tetrahydro-6*H*-dibenzo[*b,d*]pyran (12; R = H) and 2-methyl-
-3,4,4a,10a-tetrahydro-6*H*-dibenzo[*b,d*]pyran (12; R = Me) have been
obtained by the cyclic carbopalladation of appropriate alkenyl ethers
(E. Negishi *et al.*, Heterocycles, 1989, 28, 55). For the synthesis of 6,6-
-dimethyl-3-methoxy-6a,7,8,9-tetrahydro-6*H*-dibenzo[*b,d*]pyran-9-one (13),
which on Clemmensen reduction affords 6,6-dimethyl-3-methoxy-6a,7,8,9-
-tetrahydro-6*H*-dibenzo[*b,d*]pyran, see M.B. Hogale, D.R. Kharade, and
S.D. Shirke, Indian J. Chem., 1993, 32B, 1156; 6,6-dimethyl-1-methoxy-
-6a,7,10,10a-tetrahydro-6*H*-dibenzo[*b,d*]pyran-9-carboxaldehyde (14),
J.W. Huffman, M.J. Wu, and H.H. Joyner, J. Org. Chem., 1991, 56, 5224;
3-(5-bromopentyl)-1-hydroxy-6,6,9-trimethyl-6,7,10,10a-tetrahydro-6*H*-
-dibenzo[*b,d*]pyran (15), A. Charalambous *et al.*, J. Med. Chem., 1992, 35,
3076; and 1-(aminoalkoxy)-6,6,9-trimethyl-3-pentyl-6a,7,8,10a-tetrahydro-
-6*H*-dibenzo[*b,d*]pyrans 16 (n = 2-20) and related tetrahydrocannabinol
derivatives and their conjugates with proteins, Y. Uemachi *et al.*, Jpn. Kokai
Tokkyo Koho JP 02 101,072 [90 101,072], 1990.

(12)   (13)   (14)

(15)

(16)

The preparation and chemistry of 4a-hydroxy-6-methyl-7,10-dioxo-
-1,2,3,4,4a,7,10,10a-octahydrodibenzo[*b,d*]pyran (17) [U. Kucklaender *et
al.*, Arch. Pharm. (Weinheim, Ger.), 1991, <u>324</u>, 11) and 9-methyl-1,2,3,4,-
4a,6a,7,8,10a,10b-decahydro-6*H*-dibenzo[*b,d*]pyran-6-one (18) (A. Alexakis,
D. Jachiet, and L. Toupet, Tetrahedron, 1989, <u>45</u>, 6203) have been
discussed.

R = CHO

(17)

(18)

522

9. Naphthopyrans, benzochromenes and derivatives

(a) Naphtho[1,2-b]-, [2,1-b]-, and [2,3-b]pyrans

Since the preparation and properties of some naphtho[1,2-b]- and [2,1--b]pyrans are reported together in the literature they are, therefore, conveniently discussed in the same section (iii).

(i) Naphtho[1,2-b]pyrans. The reactions of 4-methyl-2*H*-naphtho[1,2-*b*]-pyran-2-one (1) with esters, amines, hydrazines, Grignard reagents, bromine, P_2S_5, and CH_2=CHCN and its conversion to 4-(2-oxo-2*H*-naphtho[1,2-*b*]-pyran-4-yl)butyramide (2) have been described (R.M. Saleh *et al.*, Phosphorus, Sulphur Silicon Relat. Elem., 1990, 48, 285).

(1) R = Me
(2) R = $(CH_2)_3CONH_2$

(3)

(4)

The cycloaddition of 4-[2-(4-methoxyphenyl)vinyl]-2*H*-naphtho[1,2-*b*]-pyran-2-one (3) with maleic anydride yields the adduct 4 (S.A. Essawy *et al.*, Indian J. Chem., 1992, 31B, 39). The ^{13}C NMR, UV, and fluorescence spectra and fungicidal activities of 7-hydroxy-4-methyl-2*H*-naphtho[1,2-*b*]-pyran and related compounds have been reported (X. Qian, Y. Zhang, and C. Ni, Huadong Ligong Daxue Xuebao, 1994, 20, 336).

Condensation of 2-methyl-4*H*-naphtho[1,2-*b*]pyran-4-one (5) with aromatic aldehydes affords 2-styryl derivatives 6 (R = 4-ClC_6H_4, 3-$O_2NC_6H_4$, 2-HOC_6H_4, Ph), which react with maleic anhydride to afford the

tetrahydroxanthone derivatives 7 (A.Y. Soliman, M.R. Mahmoud and
H.M. Bakeer, Acta Chim. Hung., 1991, _128_, 345). 6-Bromo-2-methyl-4H-
-naphtho[1,2-b]pyran-4-one has also been converted to the 2-styryl
analogues by reaction with aromatic aldehydes (I.S. Al Naimi and
B.A. Hussain, Qatar Univ. Sci. J., 1992, _12_, 73).

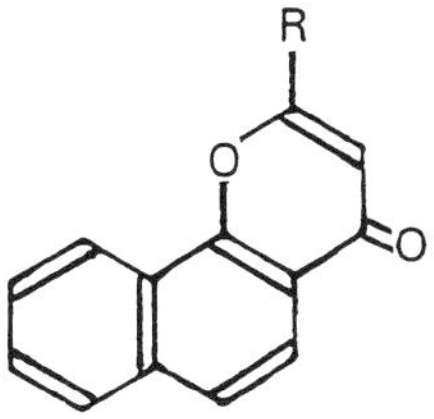

(5) R = Me
(6) R = CH = CHR

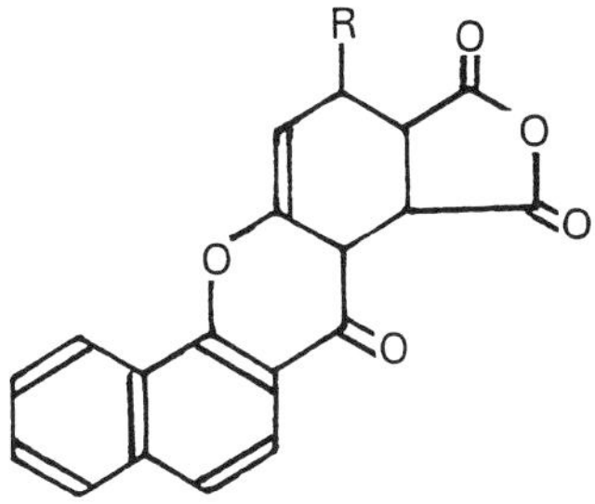

(7)

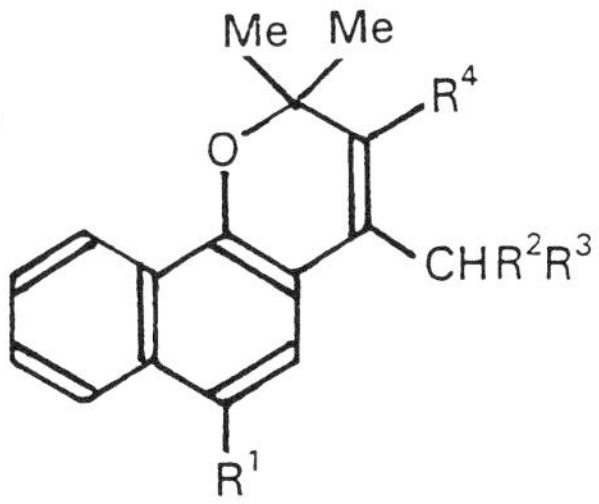

(8)

The reactions of 2,2,4-trimethyl-2H-naphtho[1,2-b]pyran (8;
$R^1 = R^2 = R^3 = R^4 = H$) with 3 and with 4 or 5 mol give 3,6-dibromo-4-bromo-
methyl-2,2-dimethyl-2H-naphtho[1,2-b]pyran (8; $R^1 = R^2 = R^4 = Br$, $R^3 = H$)
and 3,6-dibromo-4-dibromomethyl-2,2-dimethyl-2H-naphtho[1,2-b]pyran (8;
$R^1 = R^2 = R^3 = R^4 = Br$), respectively. The latter derivative is also obtained by
the addition of 3 mol of bromine to 6-bromo-2,2,4-trimethyl-2H-naphtho[1,2-
-b]pyran (8; $R^1 = Br$, $R^2 = R^3 = R^4 = H$) (E. Bradley _et al._, J. Chem. Res., S,
1993, 314).

Cyclocondensation of 4-chloro-1-hydroxynaphthalene-2-carboxaldehyde
with ethyl acetoacetate or cyanoacetate yields 3-acetyl-6-chloro-2H-naphtho-
[1,2-b]pyran-2-one (9; R = Ac) and 6-chloro-3-cyano-2H-naphtho[1,2-b]-
pyran-2-one (9; R = CN), respectively (A.M. El-Agrody _et al._, Proc. Indian
Natl. Sci. Acad., Part A, 1991, _57_, 579). Cyclocondensation of 2-amino-4-
-(2-chlorophenyl)-3-cyano-2H-naphtho[1,2-b]pyran (10) with malonitrile and
acetaldehyde yields naphthopyranopyridine 11 (A.E.A. Harb, A.M. El-
Maghraby, and S.A. Metwally, Coll. Czech. Chem. Comm., 1992, _57_,
1570). For the preparations of 2-amino-3-cyano-4-(4-substituted 3-nitro-
phenyl)-4H-naphtho[1,2-b]pyrans 12 (R = amino, pyrrolidinyl, morpholinyl),
see. J. Bloxham, C.P. Dell and C.W. Smith, Heterocycles, 1994, _38_, 399;
2-amino-4-aryl-6-chloro-3-cyano(ethoxycarbonyl)-4H-naphtho[1,2-b]pyrans
13 [$R^1 = CN$, EtO_2C; $R^2 = $(un)substituted Ph], A.M. El-Agrody, J. Chem.
Res., S, 1994, 280; and antiproliferative derivatives of 4H-naphtho[1,2-b]-
pyran with up to four substituents; C.P. Dell and C.W. Smith, Eur. Pat. Appl.
EP 537,949, 1993.

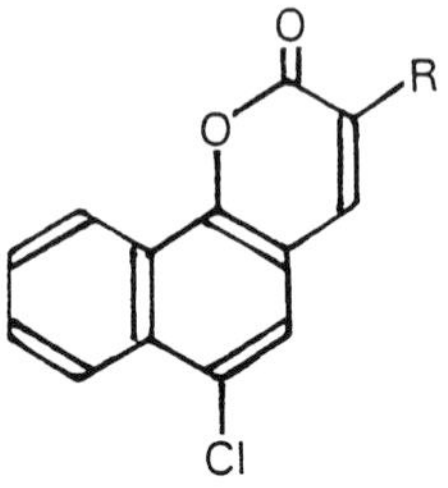

524

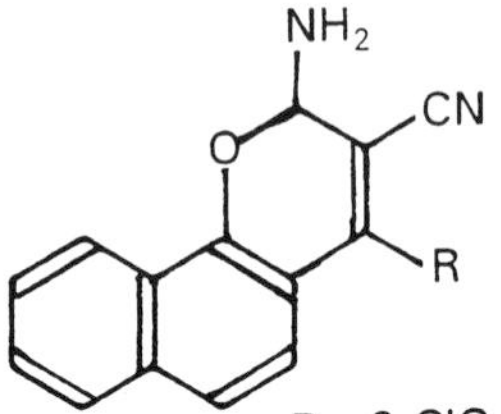

R = 2-ClC$_6$H$_4$

(9)　　(10)　　(11)

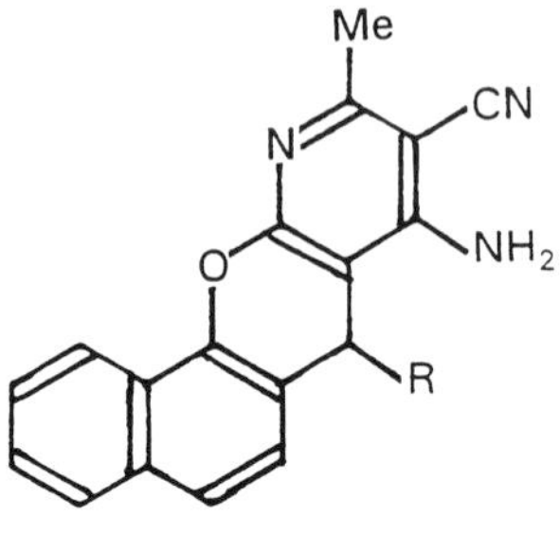

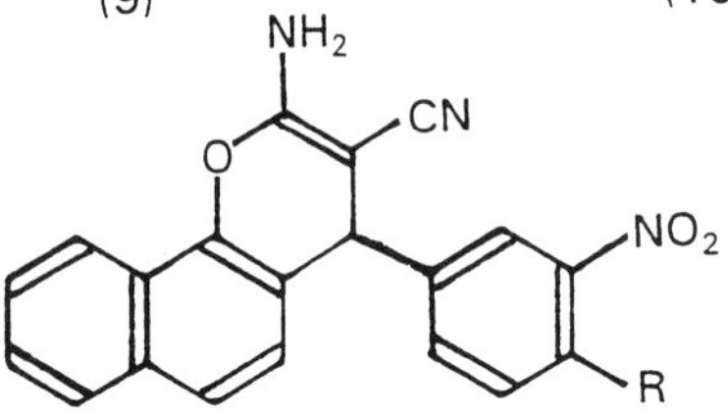

(12) 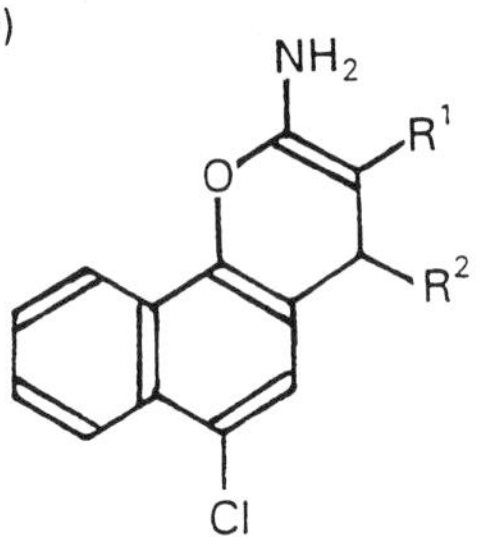　　(13) 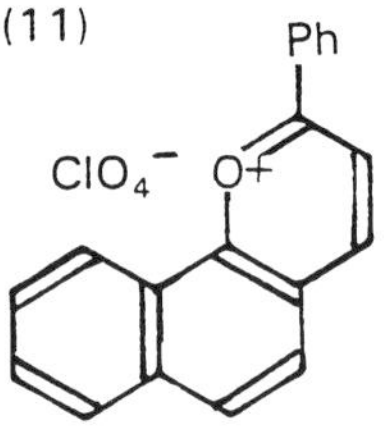　　(14)

2-(3-Phenyl-3-oxopropyl)naphth-1-ol on treatment with Ph$_3$CClO$_4$ affords 2-phenylnaphtho[1,2-*b*]pyrylium perchlorate (14) [J.N. Chatterjea and N. Ojha, Natl. Acad. Sci. Letter (India), 1988, 11, 311).

The cyclization of *a*-(3-methyl-2-butenoyloxy)naphthalene with ZnCl$_2$ gives 2,2-dimethyl-3,4-dihydro-2*H*-naphtho[1,2-*b*]pyran-4-one (15) (V.K. Tandon *et al.*, Indian J. Pharm. Sci., 1991, 53, 22). The preparations of methyl 4-(4-nitrophenyl)-2-(pyrrolidin-1-yl)-5,6-dihydro-4*H*-naphtho[1,2-*b*]-pyran-3-carboxylate (J. Bloxham and C.P. Dell, J. Chem. Soc., Perkin 1, 1993, 3055), 5,6-dihydro-4*H*-naphtho[1,2-*b*]pyrans with up to four substituents in the 2-, 3-, and 4-positions and one of the 7-, 8-, 9-, or 10--positions, which have been tested as cell antiproliferation agents (M. Brunavs *et al.*, Eur. Pat. Appl. EP 557,075, 1993), and a number of related derivatives (A.C. Williams, *ibid.*, 618,206, 1994) have been reported.

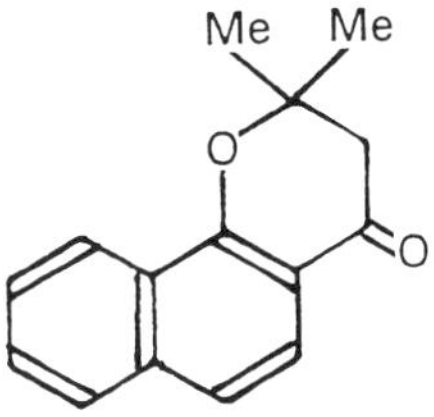

(15)

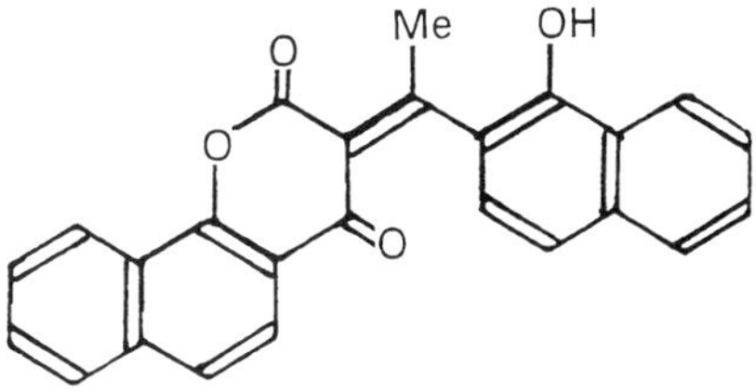

(16)

The K_2CO_3-catalyzed cyclization of 2-acetylnaphth-1-yl methyl carbonate
yields, besides the expected 4-hydroxybenzo[*h*]coumarin, 3-[1-(1-hydroxy-
naphth-2-yl)ethylidene]-3,4-dihydro-2*H*-naphtho[1,2-*b*]pyran-2,4-dione (16)
as a minor product, identified on the basis of NMR and mass spectral data
(C.B. Pandey *et al.*, Indian J. Chem., 1987, <u>26B</u>, 1067).

3,4,4a,5,6,10b-Hexahydro-2*H*-naphtho[1,2-*b*]pyran-2-one (17) has been
synthesized starting from tetralone (D. Ramesh, M.I. Sami and J.K. Ray,
ibid., 1989, <u>28B</u>, 76). 2,4,4a,5,6,10b-Hexahydro-2*H*-naphtho[1,2-*b*]pyran-
-2-one has been converted into some 1-benzoquinonyl-1,2,3,4-tetrahydro-
naphthalene derivatives, useful for treatment of cerebral disorders
(T. Tatsuoka *et al.*, Eur. Pat. Appl. EP 322,248, 1989). (±)-, (*R*)- And (*S*)-9-
-dipropylamino-3,4,7,8,9,10-hexahydro-2*H*-naphtho[1,2-*b*]pyran (18) and
7,8,9,10-tetrahydro-2*H*-naphtho[1,2-*b*]pyran have been prepared and
evaluated for central 5-hydroxytryptamine and dopamine receptor activity
(B.B. Hoeoek *et al.*, Eur. J. Med. Chem., 1991, <u>26</u>, 215).

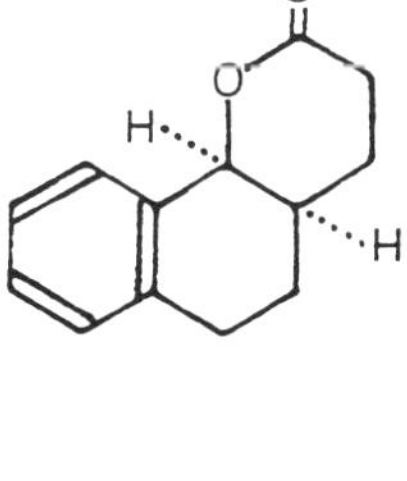

(17)

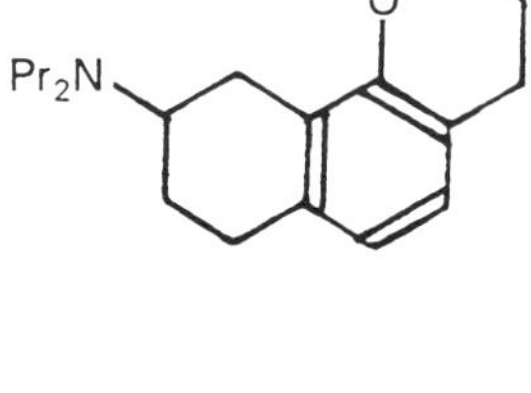

(18)

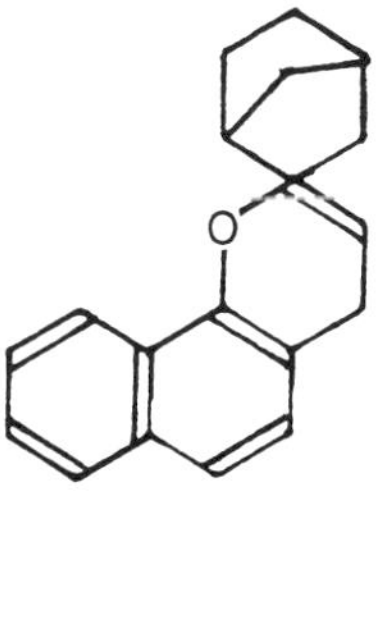

(19)

The preparation of 2-spiro 2*H*-naphtho[1,2-*b*]pyran 19 and related derivatives as photochromic compounds has been reported (T. Tanaka *et al.*, Eur. Pat. Appl. EP 401,958, 1990; T. Kobayakawa and K. Kuramoto, Jpn. Kokai Tokkyo Koho JP 05 306,280 [93 306,280], 1993).

(ii) Naphtho[2,1-b]pyrans. 3,3-Dimethyl-3*H*-naphtho[2,1-*b*]pyran (1), on boiling either with formic acid or acetic acid containing a very small amount of sulphuric acid or with methanolic hydrogen chloride, gives dimer 2, mp 207°C, containing four methyl groups. Treatment of 4-(2-hydroxynaphth-1--yl)-2-methyl-3-en-2-ol (3) with either formic acid or acetic acid containing a very small amount of sulphuric acid also yields dimer 2, but with methanolic hydrogen chloride a second dimer 4, mp 286-287°C, of the naphthopyran 1, containing three methyl groups is obtained along with 2.

The proposed structured for dimer 2 was supported by its ^{1}H NMR and mass spectral data and the structure of dimer 4 was elucidated by comparing its ^{1}H NMR spectrum with that of dimer 5 obtained from 4-(2--hydroxynaphth-1-yl)-2-([^{2}H$_3$]methyl) [1,1,1-^{2}H$_3$]but-2-en-2-ol (6) (J.R. Beswick *et al.*, J. Chem. Res., S, 1994, 8).

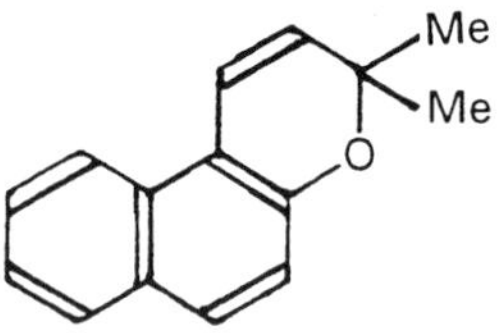

(1)

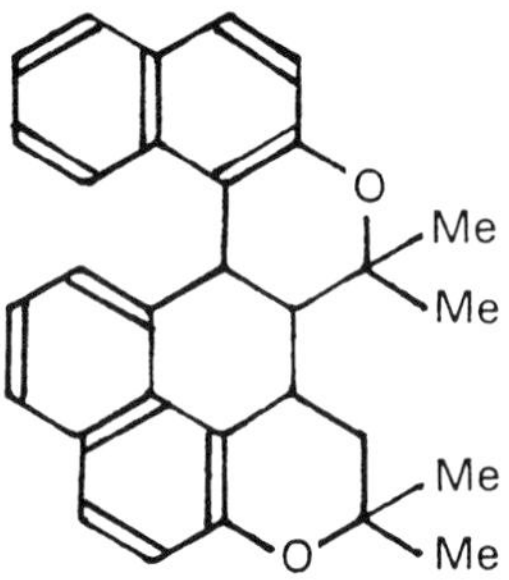

(2)

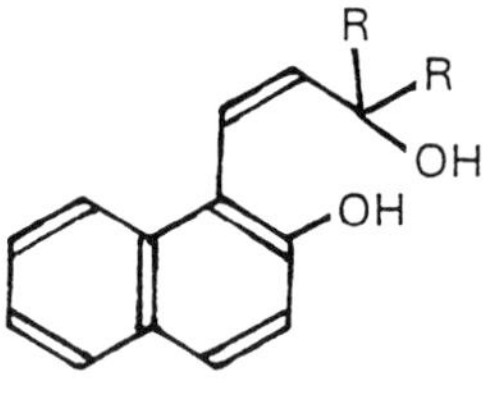

(3) R = Me
(6) R = CD$_3$

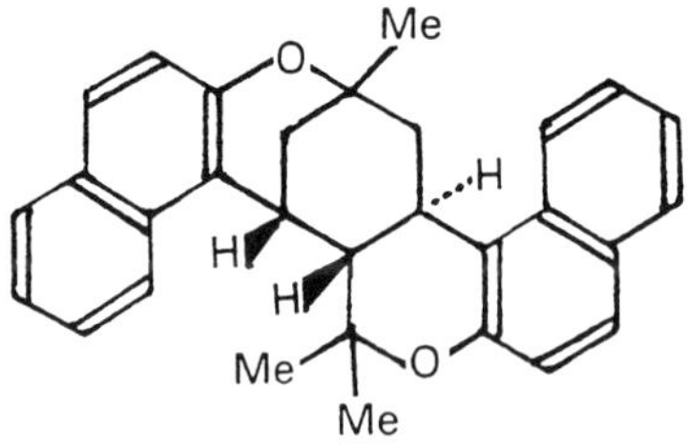

(4)

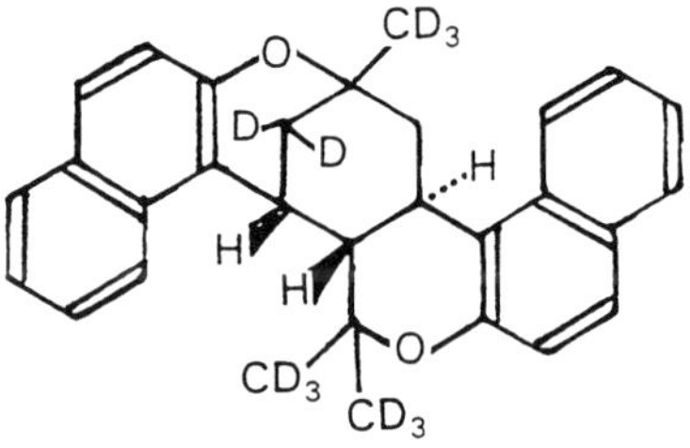

(5)

3*H*-Naphtho[2,1-*b*]pyrans react with 1,1-diphenylethene in acetic acid containing a small amount of sulphuric acid to afford adducts, 5,5-diphenyl--3,3a,4,5-tetrahydro-2*H*-phenaleno[1,9-*bc*]pyrans, for example, 3,3-dimethyl--3*H*-naphtho[2,1-*b*]pyran (1) with 1,1-diphenylethene yields 2,2-dimethyl--5,5-diphenyl-3,3a,4,5-tetrahydro-2*H*-phenaleno[1,9-*bc*]pyran (7). This like dimer 2 involves a ring closure at the C-10 position of the naphthopyran molecule (*idem*, J. Heterocyclic Chem., 1993, 30, 623), but as expected, 3,3,10-trimethyl-3*H*-naphtho[2,1-*b*]pyran (8; R = Me) did not react under similar conditions with 1,1-diphenylethene, because of the presence of the naphthopyran C-10 methyl group. However, 10-methyl-3,3-diphenyl-3*H*--naphtho[2,1-*b*]pyran (8; R = Ph) did react with 1,1-diphenylethene to yield the expected adduct, 10-methyl-3,3-diphenyl-1-(2,2-diphenylvinyl)-2,3--dihydro-1*H*-naphtho[2,1-*b*]pyran (9). Treatment of the 3,3,10-trimethyl-naphthopyran 8 (R = Me) or the diol 10 in acetic acid with a small amount of sulphuric acid gives the dimer 11 (W.D. Cotterill *et al.*, J. Chem. Res., S, 1994, 254).

(7)

(8)

(9)

(10)

(11)

The preparations of 3*H*-naphtho[2,1-*b*]pyran with a number of various substituents have been reported, including, 3-aryl-3-aralkenyl- and 3,3-diaryl--3*H*-naphtho[2,1-*b*]pyrans and related compounds as photochromic substances (B. Van Gemert and M.P. Bergomi, U.S. US 5,066,818, 1991; B. Van Gemert, PCT Int. Appl. WO 92 09,593, 1992; B. Van Gemert and M.P. Bergomi, *ibid.*, 93 10,112, 1993; B. Van Gemert, U.S. US 5,274,132, 1993; M. Rickwood *et al.*, PCT Int. Appl. WO 94 22,850, 1994) and 1-aryl-1*H*-naphtho[2,1-*b*]pyrans for pharmaceutical use (S.J. Amber *et al.*, Eur. Pat. Appl. EP 619,314, 1994).

1-Methyl-3*H*-naphtho[2,1-*b*]pyran-3-one (4-methyl-5,6-benzocoumarin (12) is prepared by the reaction of naphth-2-ol with ethyl chloroacetate. It condenses with aromatic aldehydes to give the corresponding 1-vinyl-naphthopyrans [M. El-Mobayed, B.E. Bayoumy and A.F. Haikel, An. Quim., 1990, <u>86</u>, 59; A.F. El-Farargy *et al.*, Egypt. J. Chem., 1987 (Pub. 1989), <u>30</u>, 497].

(12)   (13)

The reaction of rhodanine with 2-hydroxy-1-naphthaldehyde, followed by reduction of the resulting 2,2-dithionaphthopyranone affords 2-mercapto-3*H*--naphtho[2,1-*b*]pyran-3-one (13), which undergoes Michael addition reactions with acrylonitrile and methyl acrylate to give 2-(2-cyanoethyl)thio- and 2-(2-carboxyethyl)thio-3*H*-naphtho[2,1-*b*]pyran-3-ones (13; R = CH$_2$CH$_2$CN, CH$_2$CH$_2$CO$_2$H, respectively) (W. Wang and G. Lin, Beijing Daxue Xuebao, Ziran Kexueban, 1987, 120). The condensation of ethyl 8--bromo-3-oxo-3*H*-naphtho[2,1-*b*]pyran-2-carboxylate with ketones under Michael conditions has been investigated (A.H. Bedair *et al.*, Acta Pharm. Jugosl., 1986, <u>36</u>, 363).

2-Hydroxy-1-naphthaldehyde on treatment with furandiones 14 (R = H, Me, Cl, Br, MeO) affords 2-benzoyl-3*H*-naphtho[2,1-*b*]pyran-3-ones 15 (Yu.S. Andreichikov and T.N. Tokmakova, Zh. Org. Khim., 1987, <u>23</u>, 880).

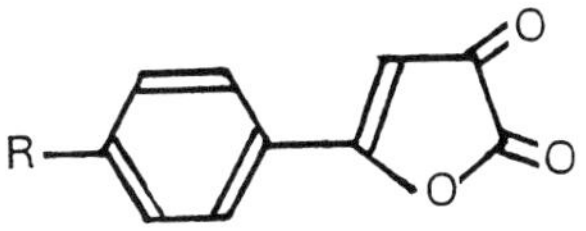

(14)

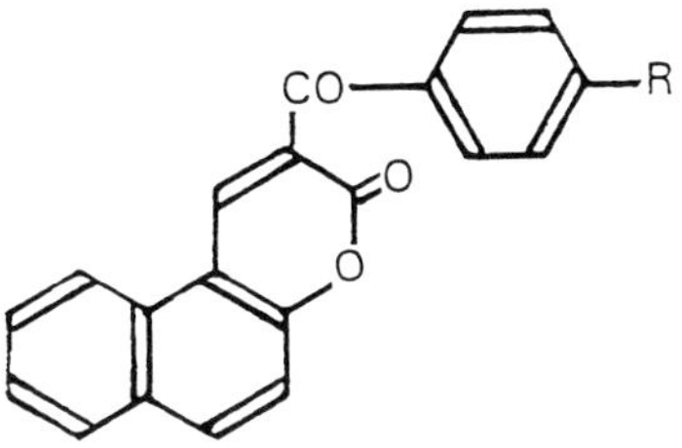

(15)

2-Acetyl-3*H*-naphtho[2,1-*b*]pyran-3-one on bromination gives 2-bromo-acetyl-3*H*-naphtho[2,1-*b*]pyran-3-one, which reacts with Ph_3P to afford the phosphorus ylide 16, whose reactions with aldehydes and alkylating agents have been described. The bromoacetyl derivative also reacts with NH_2-containing compounds, pyridines, and quinolines to yield nitrogen hetero-cycles (O.V. Skripskaya *et al.*, Zh. Obshch. Khim., 1992, <u>62</u>, 661). 2-
-Acetyl-3*H*-naphtho[2,1-*b*]pyran-3-one reacts with activated nitriles to give a range of naphthopyran-3-one derivatives, for example, the 2-[(arylamino)-cyanoalkyl]-3*H*-naphtho[2,1-*b*]pyran-3-ones 17 (R = Ph, heteroaryl) and 2-[(3-
-amino-2,4-dicyano-5-phenyl)phenyl]-3*H*-naphtho[2,1-*b*]pyran-3-one (18) (A.M. El-Agrody, J. Chem. Res., S, 1994, 50).

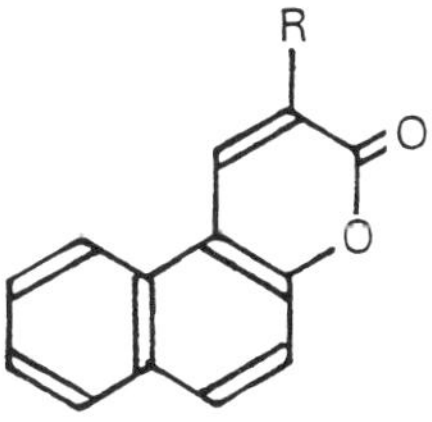

(16) R = COCH = PPh_3
(17) R = MeC(NHR)CH(CN)$_2$

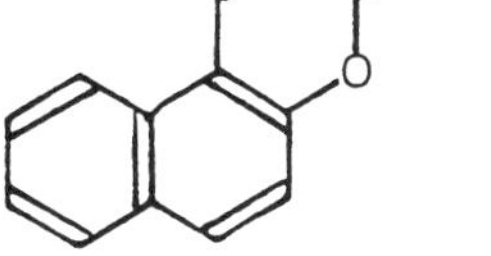

(18)

530

The condensation of 2-hydroxy-1-naphthaldehyde with diethyl malonate and ethyl acetoacetate in pyridine containing a catalytic amount of piperidine affords ethyl 3-oxo-3*H*-naphtho[2,1-*b*]pyran-2-carboxylate and 2-acetyl-3*H*--naphtho[2,1-*b*]pyran-3-one, respectively. Hydrolysis of the former affords 3-oxo-3*H*-naphtho[2,1-*b*]pyran-2-carboxylic acid (G.S.S. Murthi and M. Basak, J. Indian Chem. Soc., 1993, <u>70</u>, 170) and its treatment with an amine in boiling ethanol or xylene or neat at 150-170°C affords carbox-amides (A.M. El-Agrody *et al.*, Afinidad, 1993, <u>50</u>, 253).

2-Substituted 8-bromo-3*H*-naphtho[2,1-*b*]pyran-3-ones 19 (R^1 = CN, CO_2H, CO_2Et, CONHPh) have been treated with $R^2CH_2COCH_2R^3$ (R^2,R^3 = H, Me) in the presence of NH_4OAc or $MeNH_2$ to furnish naphthoxazocines 20 (R^4 = H, Me) (F.M. Aly *et al.*, *ibid.*, 1987, <u>44</u>, 390). The reactions between ethyl 8-bromo-3-oxo-2*H*-naphtho[2,1-*b*]pyran-2-carboxylate and its 2-phenyl-amide and alkylmagnesium bromides have been discussed (A.H. Bedair *et al.*, J. Indian Chem. Soc., 1988, <u>65</u>, 519).

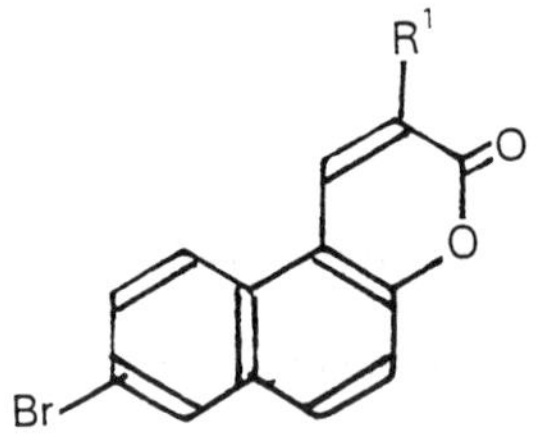

(19) 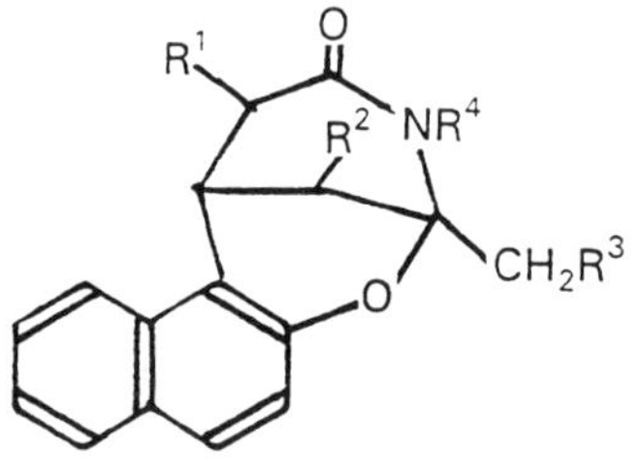   (20)

Cyclocondensation of 2-hydroxy-6-nitro-1-naphthaldehyde with diethyl malonate yields ethyl 8-nitro-3-oxo-3*H*-naphtho[2,1-*b*]pyran-2-carboxylate (21) (M.R. Selim *et al.*, Sci. Phys. Sci., 1990, <u>2</u>, 100). For the preparation of some 8-nitronaphtho[2,1-*b*]pyran-3-ones and their chemical reactions see *idem, ibid.*, p.94. Nitration of methyl 9-hydroxy- and 9-methoxy-3-oxo-3*H*--naphtho[2,1-*b*]pyran-2-carboxylates by HNO_3-AcOH affords methyl 9--hydroxy- and 9-methoxy-10-nitro-3-oxo-3*H*-naphtho[2,1-*b*]pyran--carboxylates, respectively. Also nitration of 3-hydroxy-6*H*-dibenzo[*b,d*]-pyran-6-one yields 3-hydroxy-2-nitro- and 3-hydroxy-2,4-dinitro-6*H*-dibenzo-[*b,d*]pyran-6-ones. Cyclocondensation of 2,7-dihydroxy-1-naphthaldehyde with $O_2NCH_2CO_2Et$ yields 9-hydroxy-2-nitro-2*H*-naphtho[2,1-*b*]pyran-3-one (22) (J.R. Yang *et al.*, Org. Prep. Proced. Int., 1991, <u>23</u>, 621).

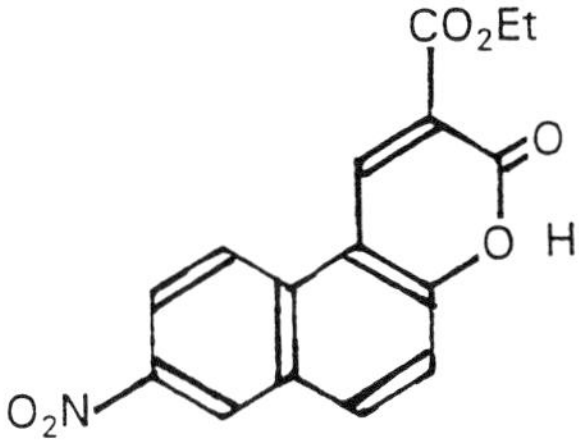

(21) 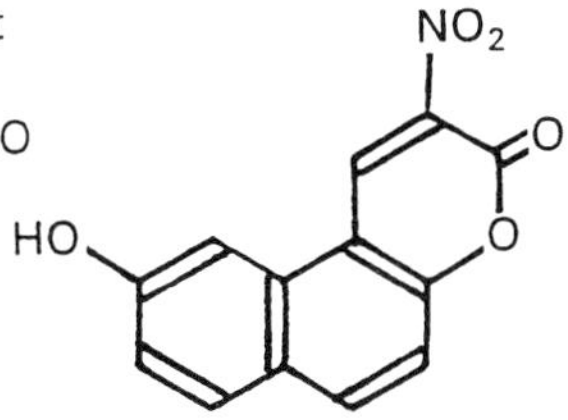   (22) 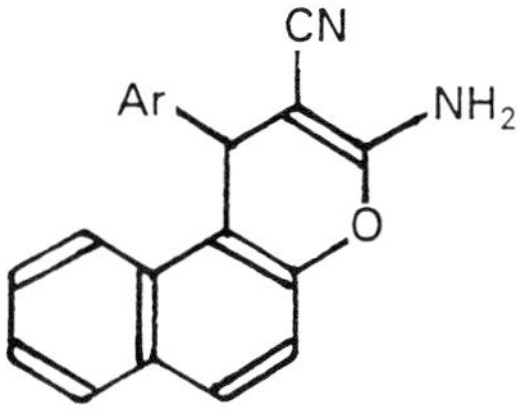   (23)

Condensation of ArCHO [Ar = (un)substituted Ph] with $CH_2(CN)_2$ and naphth-2-ol in ethanolic piperidine affords 3-amino-1-aryl-2-cyano-1*H*--naphtho[2,1-*b*]pyrans 23 (F.F. Abdel-Latif, Indian J. Chem., 1990, 29B, 664). Also for the preparations of other 3-amino-1-aryl-2-cyano(ethoxy-carbonyl)-1*H*-naphtho[2,1-*b*]pyrans, see V.E. Shklover, Yu.A. Sharanin and Yu.T. Struchkov, Zh. Org. Khim., 1987, 23, 412.

A number of 1-(arylsulphonylaminomethyl)-3*H*-naphtho[2,1-*b*]pyran-3--ones 24 (R = H, Me, NH_2, NHAc, Br, 2,5-Br_2) have been prepared (C.D. Lakkannavar, M.V. Kulkarni and V.D. Patil, J. Indian Chem. Soc., 1992, 69, 399).

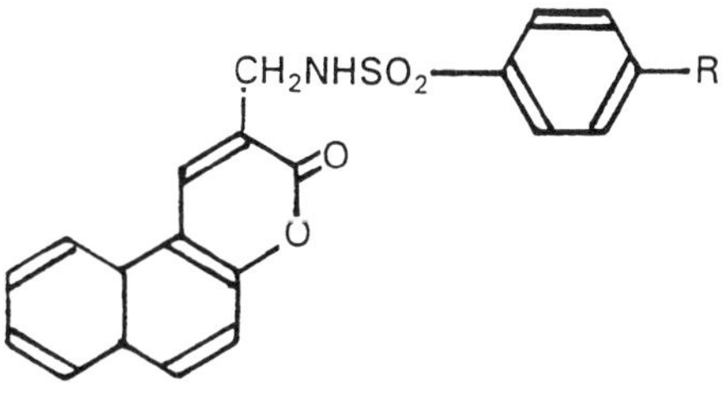

(24) 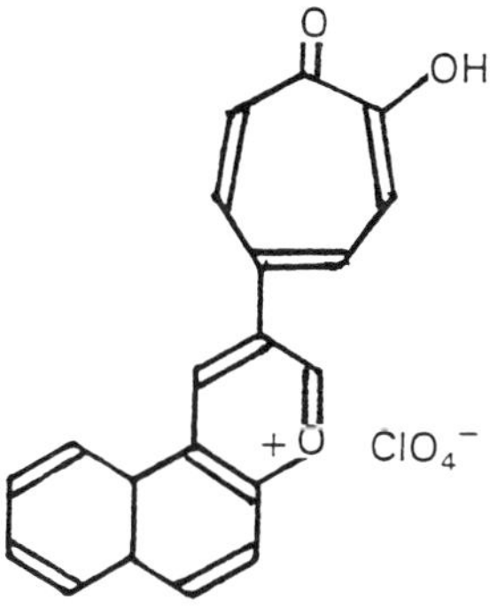   (25)

532

3-Acetyltropolone reacts with 2-hydroxy-1-naphthaldehyde in the presence of ethyl formate and perchloric acid to yield the naphtho[2,1-*b*]pyrylium tropolone perchlorate 25 (D.L. Wang and Z.T. Jin, Chin. Chem. Letters, 1993, <u>4</u>, 877).

2-Isopropyl-2,3-dihydro-1*H*-naphtho[2,1-*b*]pyran (26) has been synthesized by methods mimicking biosynthetic reactions (A.J. Walkington and D.A. Whiting, Tetrahedron Letters, 1989, <u>30</u>, 4731). The preparations of 2,3-dihydro-1*H*-naphtho[2,1-*b*]pyrans 27 (R^1 = H, Ph, 4-$Me_2NC_6H_4$; R^2 = OEt, pyrrolidin-2-on-1-yl) (A.R. Katritzky and X. Lan, Synthesis, 1992, 761) have been reported.

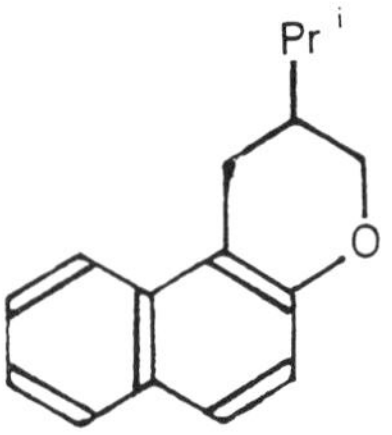

(26) 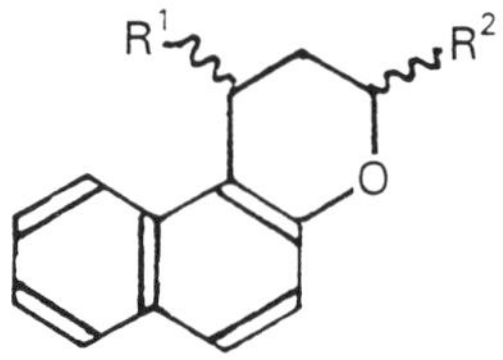   (27) 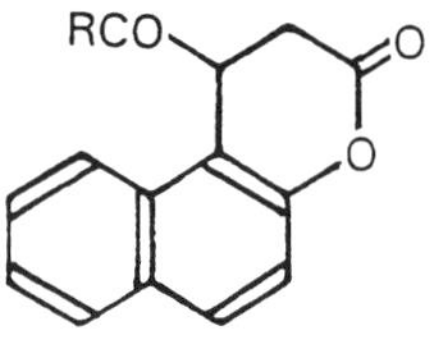   (28)

1-Aroyl-2,3-dihydro-1*H*-naphtho[2,1-*b*]pyran-3-ones 28 (R = 4-$MeOC_6H_4$, 3,4-$Me_2C_6H_3$) undergo condensation with aromatic aldehydes to yield 2-arylidene derivatives and bromination afford 1,2-dibromo or 5-bromo derivatives depending on the reaction conditions (M.A. El-Kasaby *et al.*, Indian J. Chem., 1990, <u>29B</u>, 928; Chin. J. Chem., 1990, 446). Ethyl 3-oxo-3*H*-naphtho[2,1-*b*]pyran-2-carboxylate on treatment with anthranilic acid in boiling C_6H_6 yields ethyl 3-oxo-2,3-dihydro-1*H*-naphtho[2,1-*b*]pyran-2-carboxylate and ethyl 1-(2-carboxyphenylamino)-3-oxo-2,3-dihydro-1*H*-naphtho[2,1-*b*]pyran-2-carboxylate (M.A.I. Salem and M.A. El-Kasaby, J. Chem. Soc. Pak., 1987, <u>9</u>, 177). The preparations of a number of derivatives of forskolin (derivatives of 5,6,10,10b-tetrahydroxy-3,4a,7,7,10a-pentamethylperhydronaphtho[2,1-*b*]pyran-1-one) have been reported (S. Ikegami, S. Hashimoto and S. Sakata, Jpn. Kokai Tokkyo Koho JP 01 186,833 [89 186,833], 1989).

2-Methyl-3*H*-naphtho[2,1-*b*]pyran with a spiro-fused norbornane at the 3-position and analogues have been synthesized as photochromic agents (T. Tanaka *et al.*, *ibid.*, 02 69,471 [90 69,471], 1990). The 2,3-dispiro-

(1,2-dihydronaphthalen-2-one)-2,3-dihydro-1*H*-naphtho[2,1-*b*]pyrans 29
(R = H, Br, Bu^t, 1-methylcyclohex-1-yl) on irradiation yield a wide range of
products resulting from ß-cleavages and intramoledular cycloadditions
(T.R. Kasturi *et al.*, Indian J. Chem., 1988, <u>27B</u>, 5).

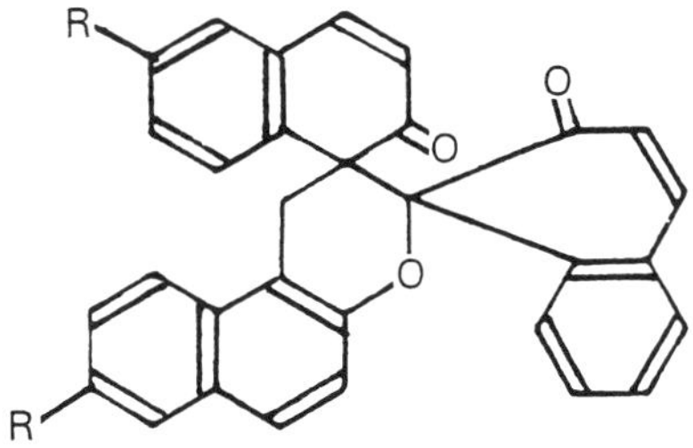

(29)

(iii) Naphtho[1,2-b]- and [2,1-b]pyrans. 2-Phenyl-4*H*-naphtho[1,2-*b*]-
pyran (30) and 3-phenyl-1*H*-naphtho[2,1-*b*]pyran (31) are prepared by
reduction of their respective pyrylium perchlorates [N. Ojha, J. Inst. Chem.
(India), 1989, <u>61</u>, 111].

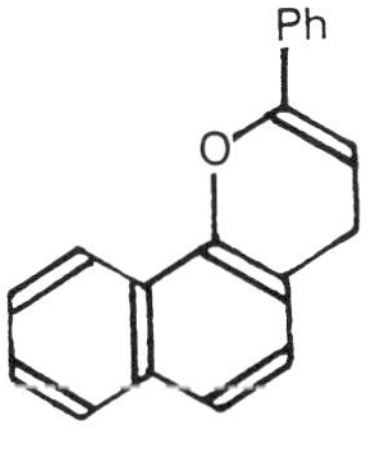

(30)

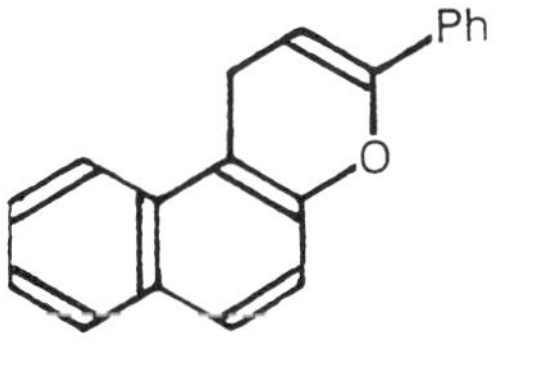

(31)

Cyclocondensation of 1- and 2-naphthols with $RCOCH_2CO_2Et$ (R = Me, Ph, 4-FC$_6$H$_4$) furnishes 2-methyl-4H-naphtho[1,2-b]pyran-4-one (32; R = Me) and 2-aryl-4H-naphtho[1,2-b]pyran-4-one (32; R = Ph, 4-FC$_6$H$_4$) and 3-methyl-1H-naphtho[2,1-b]pyran-1-one (33; R = Me) and 3-aryl-1H-naphtho-[2,1-b]pyran-1-one (33; R = Ph, 4-FC$_6$H$_4$), respectively (M. Zhang, J. Yang and Z. Chen, Beijing Daxue Xuebao, Ziran Kexueban, 1990, 26, 435). Lawesson reagent converts 2-phenyl-4H-naphtho[1,2-b]pyran-4-one (32; R = Ph) and 3-phenyl-1H-naphtho[2,1-b]pyran-1-one (33; R = Ph) into their corresponding 4- and 1-thiones (T.S. Hafez, Phosphorus, Sulphur Silicon Relat. Elem., 1991, 63, 249). Compounds 32 (R = Ph) and 33 (R = Ph) are sometimes referred to as naphthoflavones.

(32)　　　　　　　　　　(33)

4-Methyl-2H-naphtho[1,2-b]pyran-2-one and 2-oxo-2H-naphtho[1,2-b]-pyran-4-acetic acid and 1-methyl-3H-naphtho[2,1-b]pyran-3-one and 3-oxo-3H-naphtho[2,1-b]pyran-1-acetic acid have been prepared (S. El-Nagdy et al., J. Chem. Soc. Pak., 1990, 12, 230). Esterification of the two acetic acids with n-BuOH affords the corresponding butyl esters, the reactions of which with hydrazine hydrate, hydroxylamine hydrochloride, acrylonitrile, maleic anhydride, amines, and aromatic aldehydes have been investigated (idem, Acta Chim. Hung., 1991, 128, 35).

2-Amino-4-aryl-3-cyano(ethoxycarbonyl)-4H-naphtho[1,2-b]pyrans and 3-amino-1-aryl-2-cyano(ethoxycarbonyl)-1H-naphtho[2,1-b]pyrans 34 and 35 (R^1 = aryl, R^2 = CN, CO$_2$Et) have been prepared by the reaction of appropriate cinnamonitriles with 1- and 2-naphthol, respectively (A.G.A. Elagamey and F.M.A.A. El-Taweel, Indian J. Chem., 1990, 29B, 885).

(34)

(35)

The opening of the pyranone ring in 3-benzoylamino-2*H*-naphtho[1,2-*b*]-
pyran-2-one (36) and 2-benzoylamino-3*H*-naphtho[2,1-*b*]pyran-3-one (37;
R = H) and its 9-methoxy derivative (37; R = OMe) with nucleophiles gives 3-
-naphth-2-yl)- andd 3-(naphthyl-1-yl) propenoates, respectively (B. Ornik,
B. Stanovnik and M. Tisler, J. Heterocyclic Chem., 1992, 29, 1241).

(36)

(37)

The addition reaction of naphth-2-ol with acrylaldehyde, followed by ring
closure of the product in the presence of NaOH in benzene affords 2,3-
-dihydro-1*H*-naphtho[2,1-*b*]pyran-3-ol (38). The reaction with crotonaldehyde and methacrylaldehyde yields 1- and 2-methyl-2,3-dihydro-1*H*-naphtho-
[2,1-*b*]pyran-3-ols. However, the reaction of *trans*-cinnamaldehyde with

naphth-2-ol yields 1-phenyl-1*H*-naphtho[2,1-*b*]pyran as the main product, rather than 1-phenyl-2,3-dihydro-1*H*-naphtho[2,1-*b*]pyran-3-ol. The reaction of naphth-1-ol with acrylaldehyde gives 3,4-dihydro-2*H*-naphtho[1,2-*b*]pyran--2-ol (39) along with resinous products (T. Nakazawa *et al.*, Nippon Kagaku Kaishi, 1989, 244). The ^{13}C NMR of 1,2- and 1,4-naphthoquinones having a pyran ring attached have been recorded (F.W. Ribeiro *et al.*, J. Braz. Chem. Soc., 1991, <u>1</u>, 55).

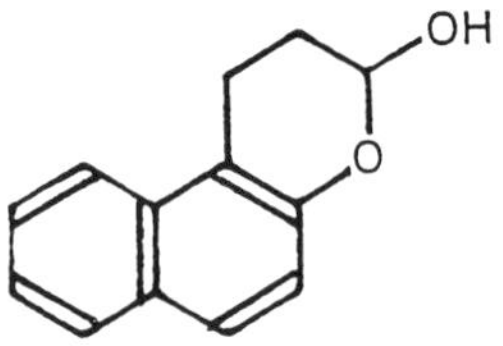

(38) 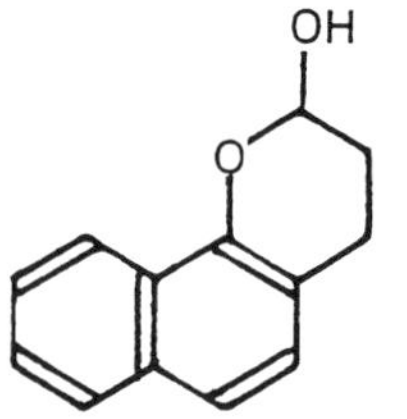 (39)

(iv) Naphtho[2,3-b]pyrans. 6,9-Dimethoxy-2-methyl-3,4,5,10-tetrahydro--2*H*-naphtho[2,3-*b*]pyran-5,10-dione (40) has been synthesized from 2--hydroxy-5,8-dimethoxy-1,4-naphthoquinone *via* Michael adduct 41 (B.C. Saitz, J.A. Valderrama and R. Tapia, Synth. Comm., 1992, <u>22</u>, 2411).

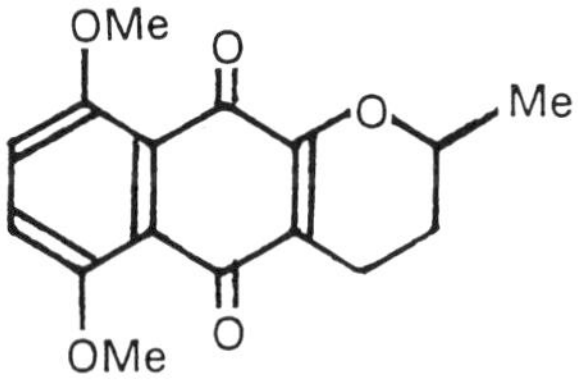

(40) 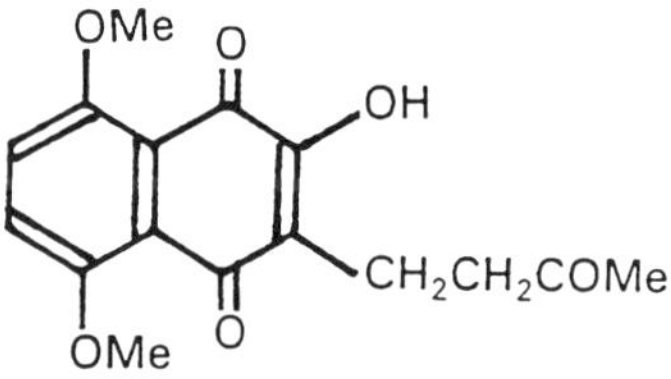 (41)

(b) Naphtho[1,2-c]- and [2,3-c]pyrans, isonaphthopyrans

The stereoselective isomerization of the pair of epimers rel-(2R, 4S, 5S)- and rel-(2S, 4S, 5S)-4-(4-isopropoxy-5,7-dimethoxynaphthalen-2-yl)-2,5--dimethyldioxolane (1) with an excess of a mixture TiCl$_4$ and titanium tetraisopropoxide gives two products rel-(1R, 3S, 4S)-4-hydroxy-6-isopropoxy-7,9-dimethoxy-1,3-dimethyl-3,4-dihydronaphtho[1,2-c]pyran (2), and the C-1 epimer. Isomerization of rel-(2S, 4S, 5S)-4-(8-bromo-4-isopropoxy-5,7--dimethoxynaphthalen-2-yl)-2,5-dimethyldioxolane (3) with an excess of TiCl$_4$ alone yields rel-(1S, 3S, 4R)-4-hydroxy-6-isopropoxy-7,9-dimethoxy--1,3-dimethyl-3,4-dihydronaphtho[1,2-c]pyran (4) and its 5-bromo derivative, formed by bromine migration on the aromatic nucleus (R.G.F. Giles et al., J. Chem. Soc., Perkin 1, 1994, 865; Chem. Comm., 1991, 287).

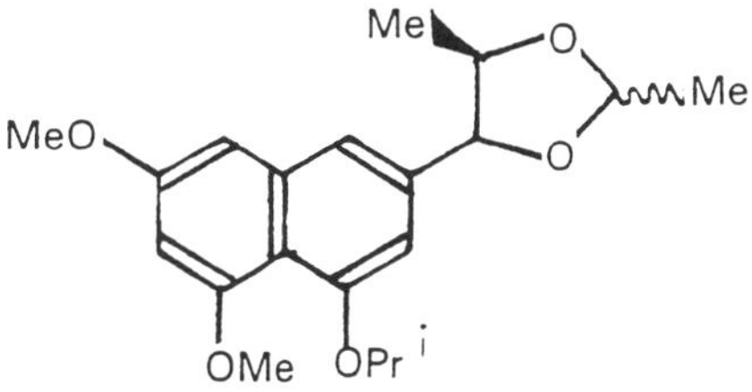

(1)

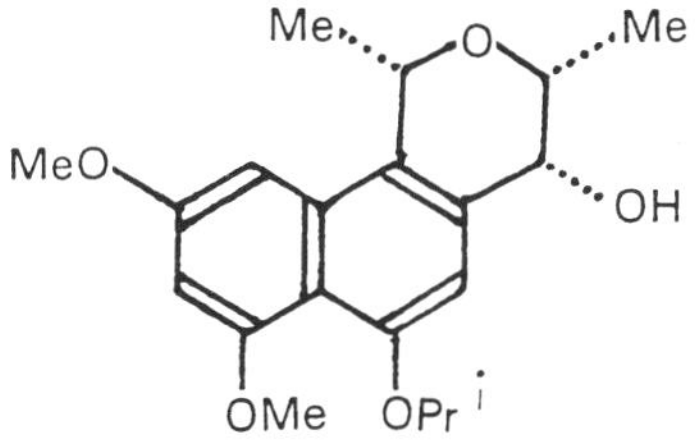

(2)

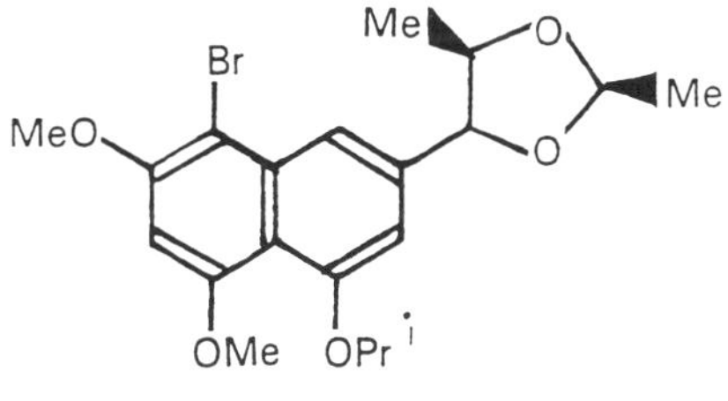

(3)

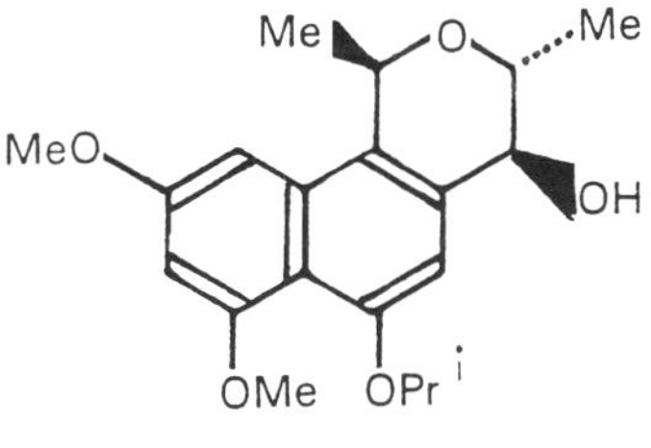

(4)

2-Methyl-1,4-naphthoquinone has been transformed into 3-(fur-2-yl)- and 3-(thien-2-yl) – 5,10-dihydro-1H-naphtho[2,3-c]pyran-5,10-dione (5; X = O,S) by use of acylated pyridinium ylides (M.F. Aldersley et al., J. Chem. Soc., Perkin 1, 1990, 2163). The synthesis of 1-hydroxy-7,9-dimethoxy-1-methyl--3-propyl-3,4,5,10-tetrahydro-1H-naphtho[2,3-c]pyran-5,10-dione

538

(V.I. Hugo, Synth. Comm., **1994**, _24_, 2563), _trans_-10-methoxy-1,3-
-dimethyl-3,4-dihydro-1_H_-naphtho[2,3-_c_]pyran (6) and related derivatives
(R.G.F. Giles _et al._, _ibid._, p.859), and several derivatives of 3,4-dihydro-1_H_-
-naphtho[2,3-_c_]pyran-1-one with a wide range of substituents as inhibitors of
leukotriene biosynthesis (R.N. Young _et al._, Eur. Pat. Appl. EP 501,579,
1992) have been reported.

(5)   (6)

(c) Naphtho[1,8-b,c]- and [1,8-c,d]pyrans

Naphtho[1,8-_b,c_]pyran (oxaphenalene) (1) has been obtained from
acenaphthenone by Baeyer-Villiger oxidation, followed by reduction of the
resulting lactone, and dehydration of the ensuing lactol (Y. Miki _et al._,
Synthesis, 1990, 312). Several 2-substituted naphtho- and 6-methoxy-
naphtho[1,8-_b,c_]pyrans 2 (R^1 = H, MeO; R^2 = NO_2, Ac, PhCO, $MeOC_6H_4CO$,
CO_2H, CO_2Me, CO_2H, CO_2Me, CO_2Et, $CONH_2$) (J.P. Buisson and R. Royer, J.
Heterocyclic Chem., 1988, _25_, 539), and 2-acyl-3-hydroxy-6-methoxy-
naphtho[1,8-_b,c_]pyrans 3 (R = Me, Et, Ph, $4\text{-}MeOC_6H_4$, $4\text{-}O_2NC_6H_4$) and
related derivatives (V.V. Mezheritskii, L.G. Minyaeva and O.M. Golyanskaya,
Zh. Org. Khim., 1993, _29_, 1187) have been prepared.

(1) $R^1 = R^2 = R^3 = H$
(2) $R^3 = H$, R^1, R^2 (see text)   (4)   (5)
(3) $R^1 = OMe$, $R^2 = RCO$, $R^3 = OH$

539

2-Acetoxy- and 2-hydroxy-2,3-dihydronaptho[1,8-*b,c*]pyrans (4; R = Ac,
H, respectively) have been prepared by the decomposition of 1-hydroperoxy-
-1,2-dihydroacenaphthene (5) in acetic acid in the presence of perchloric acid
(Z. Stec, Monatsh., 1993, <u>124</u>, 331).

1*H*,3*H*-Naphtho[1,8-*c,d*]pyran-1,3-dione (naphthalic anhydride) (6) has
been obtained by heating acenaphthene with air under pressure in AcOH/-
PhCl containing Co(OAc)$_2$·4H$_2$O, Mn(OAc)$_2$·4H$_2$O, and NaBr at 200°C
(T. Yamada and T. Kageyama, Jpn. Kokai Tokkyo Koho JP 62 195,375
[87, 195,375], 1987). Modifications to this process (I.I. Kon'kov *et al.*,
U.S.S.R. SU 1,518,335, 1989) and the preparations and properties of
related thioanhydrides (1*H*,3*H*-naphtho[1,8-*c,d*]pyran- and -thiopyran-1,3-
-diones and -dithiones) 7 (X^1 = X^3 = O, X^2 = S; X^1 = X^3 = S, X^2 = O;
X^1 = X^2 = S, X^3 = O; X^1 = X^2 = X^3 = S) (M.P. Cava and M.V. Lakshmikantham,
Phosphorus, Sulphur Silicon Relat. Elem., 1989, <u>43</u>, 95;
M.V. Lakshmikantham, W. Chen and M.P. Cava, J. Org. Chem., 1989, <u>54</u>,
4746) have been reported.

(6)

(7)

Sulphuration of 6-bromo-1*H*,3*H*-naphtho[1,8-*c,d*]pyran-1,3-dione (8) with
Na$_2$S-DMF affords sulphide 9 (V.V. Plakhtinskii *et al.*, Izv. Vyssh. Uchebn.
Zaved., Khim. Khim. Tekhnol., 1989, <u>32</u>, 24). For the preparation of 6-(2-
-aminophenylthio)-1*H*,3*H*-naphtho[1,8-*c,d*]pyran-1,3-diones, see W. Bauer
and W. Koumbouris, Eur. Pat. Appl. EP 570,778, 1993; and aniline
derivatives of 6-chloro-5-nitro-1*H*,3*H*-naphtho[1,8-*c,d*]pyran-1,3-dione
J. Pytlarz and J. Taborska, Pol. PL 138,108, 1987.

540

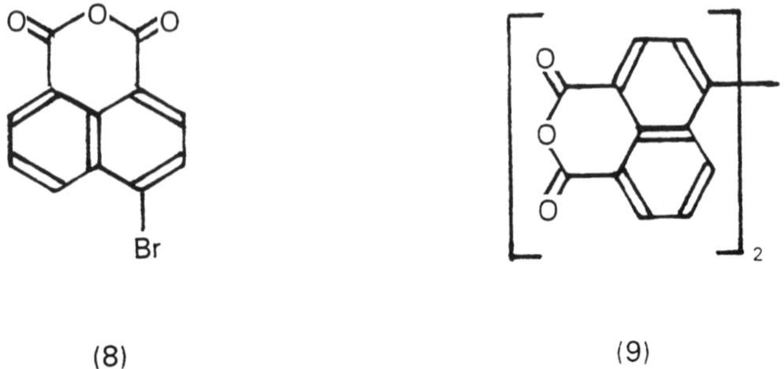

(8)   (9)

10. Benzo- and dibenzoxanthone derivatives and miscellaneous pyran ring systems

8,10-Dimethoxy-6-phenyl-5,6-dihydro-7*H*-benzo[*c*]xanthen-7-one (1) has been obtained by the photochemical cyclization of 5,7-dimethoxy-2'-styryl-flavone (M.R. Parthasarathy and N. Grover, Indian J. Chem., 1991, <u>30B</u>, 440) and dibenzo[*a,c*]xanthen-9-one (2) by photolysis of PhCOCOPh in MeOH (S.S. Kim *et al.*, Bull. Korean Chem. Soc., 1993, <u>14</u>, 661). Preparations of 9-acetoxy-7-hydroxy-5-trifluoroacetyl-7-trifluoromethylbenzo-[*c*]xanthene (3) and its related benzo[*c*]thioxanthene have been described (E. Okada *et al.*, Synthesis, 1992, 536).

(1)   (2)

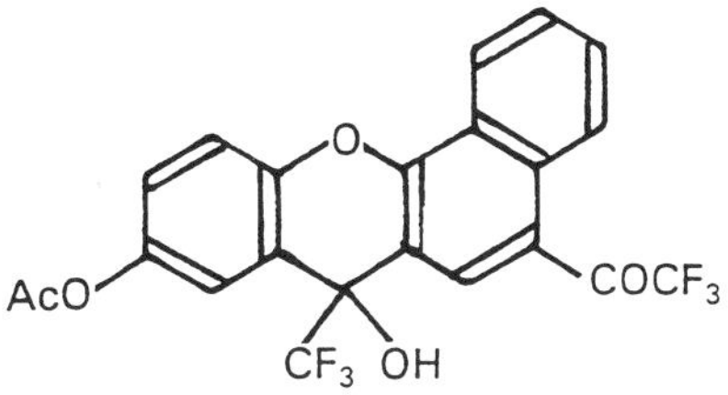

(3)

9-Hydroxy-3-methoxy-5*H*-phenanthro[4,5-*b,c,d*]pyran (4) and 9-hydroxy-
-3-methoxy-5*H*-phenanthro[4,5-*b,c,d*]pyran-5-one (5) have been obtained *via*
dibenzo[*b,d*]pyran 6 (R^1 = R^2 = H, R^{1}R^2 = O, respectively (P.L. Majumder and
A.K. Sarkar, J. Indian Chem. Soc., 1989, <u>66</u>, 673).

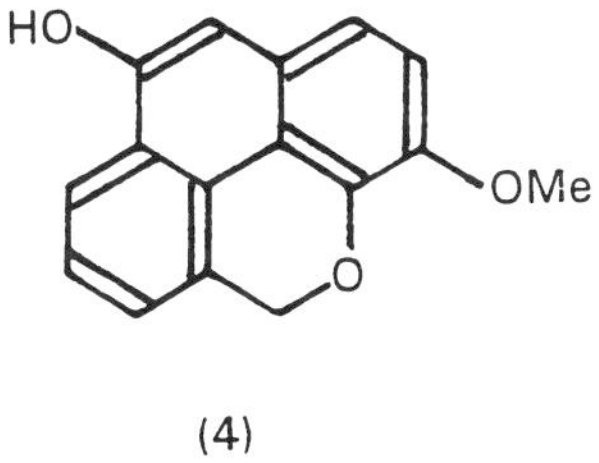

(4)

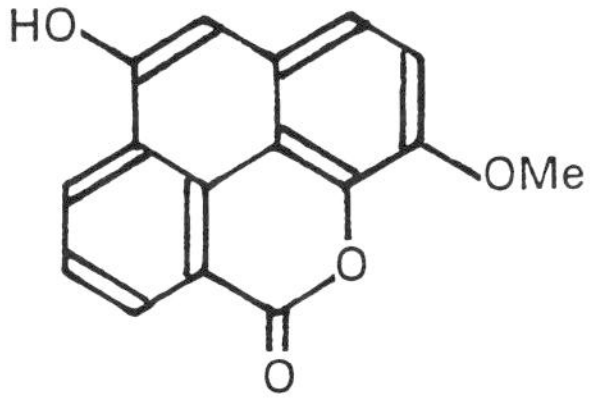

(5)

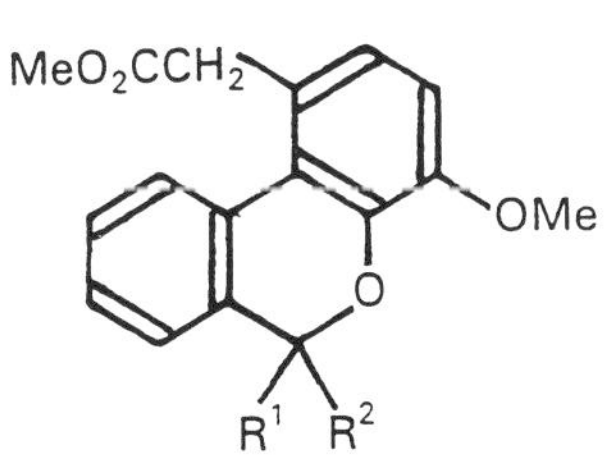

(6)

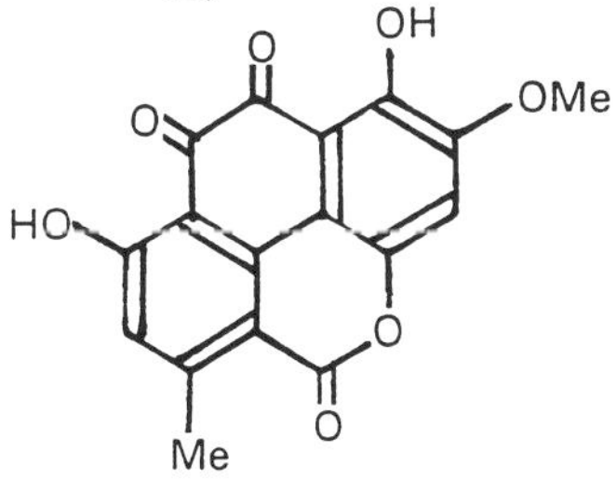

(7)

542

•   The structure of biruloquinone, a 9,10-phenanthrenequinone isolated from *Mycosphaerella rubella* has been revised to 1,8-dihydroxy-2-methoxy-6-methyl-9,10-dihydro-5*H*-phenanthro[4,5-*b,c,d*]pyran-5,9,10-trione (7) on the evidence of ^1H and ^{13}C NMR spectral data (A. Arnone, G. Nasini and O. Vajna de Pava, Phytochem., 1991, <u>30</u>, 2729). 9,10-Dihydro-5*H*-phenanthro[4,5-*b,c,d*]pyrans 8 (X = H$_2$, R^1,R^3 = OAc, OMe; R^2 = H, OAc, OMe; R^4 = H, OMe) have been converted to their respective dihydrophenanthropyran-5-ones 8 (X = O) or phenanthropyran-5-ones by using varying amounts of DDQ in boiling benzene (M.U. Bhaskar, L.J.M. Rao and N.S.P. Rao, Indian J. Chem., 1993, <u>32B</u>, 975).

(8)

(9)

Cis-4,5-dihydroxy-3-nitro-4,5-dihydropyrene on KO$_2$ oxidation affords 3-nitro-5*H*-phenanthro[4,5-*b,c,d*]pyran-5-one (9). Similarly 1-nitro-5*H*-phenanthro[4,5-*b,c,d*]pyran-5-one is obtained from the analogous dihydroxynitrodihydropyrene (S. Abdel-Baky, C. Sotiriou-Leventis and R.W. Giese, Tetrahedron, 1991, <u>47</u>, 5667).

The Wittig reaction of phenanthrene-9,10-quinone with Ph$_3$P = CHCOPh affords 2-benzoyl-3-phenyl-4*H*-phenanthro[9,10-*b*]pyran along with two phenanthrofurans and a butene-1,4-dione (D.N. Nicolaides *et al.*, J. Chem. Soc., Perkin, 1989, 2329). The reactions of 9-dicyanomethylene-9,10-dihydrophenanthren-10-one with Ph$_3$P = CHCOR (R = OMe, OEt, Ph) in benzene at 25°C yields the adducts 10, which on heating to 200°C undergo intramolecular cyclisations to afford 2-methoxy-, 2-ethoxy-, and 2-phenyl-4,4-dicyano-4*H*-phenanthro[9,10-*b*]pyrans (11), respectively (W.M. Abdou and N.A.F. Ganoub, Chem. Ind., 1991, 217).

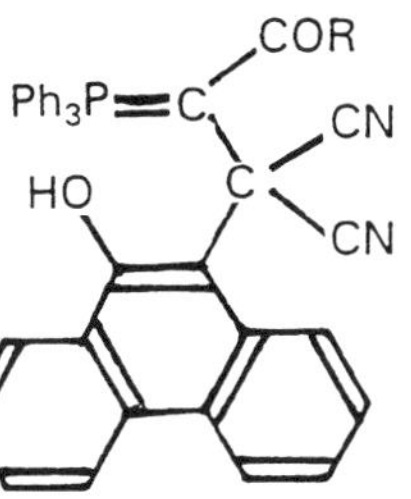

(10)

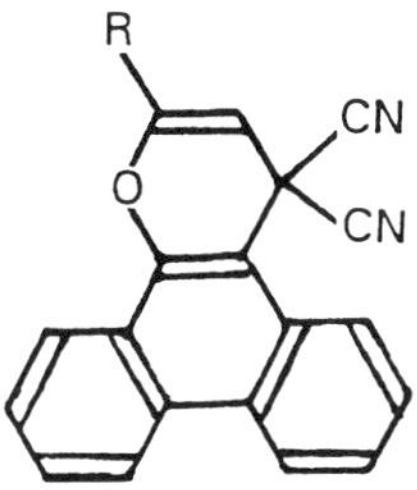

(11)

The preparations of some 1,3-dialkyl- and 1,3-dialkoxy-6*H*-benzo[*b*]-naphtho[2,1-*d*]pyran-6-ones (G. Bringmann *et al.*, Ann., 1992, 225), 1,3--dialkyl- and 1,3-dimethoxy-6*H*-benzo[*b*]naphtho[2,1-*b*]pyrans (*idem, ibid.*, p.769), 6-methoxy-6*H*-benzo[*b*[naphtho[2,1-*b*]pyran (*idem*, Ann., 1994, 439), 8-ethyl-1,10,12-trimethoxy-6*H*-benzo[*d*]naphtho[1,2-*b*]pyran-6-one (12), structurally related to the aglycon of gilvocarins, and its 1,4,12-tri-methoxy isomer (D.H. Hua *et al.*, J. Org. Chem., 1992, <u>57</u>, 399), 1,10,12--trialkoxy-8-methyl-6*H*-benzo[*d*]naphtho[1,2-*b*]pyran-6-ones (H. Suda *et al.*, Jpn. Kokai Tokkyo Koho JP 04 26,685 [92 26,685], 1992), and 4*H*--dinaphtho[2,1-*b*:1',2'-*d*]pyran-4-one (13) (G. Bringman *et al.*, Ann., 1994, 91) have all been described.

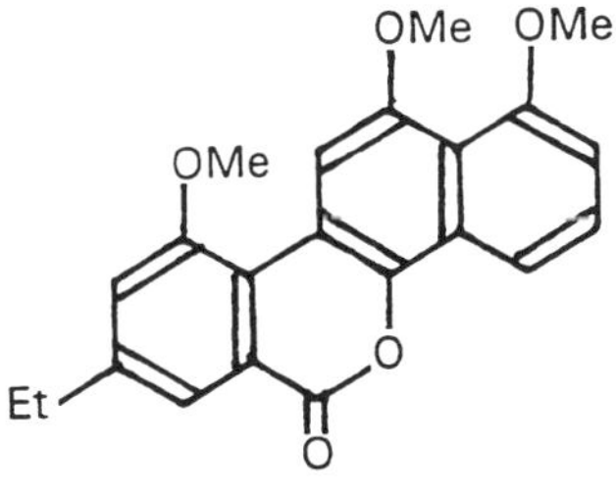

(12)

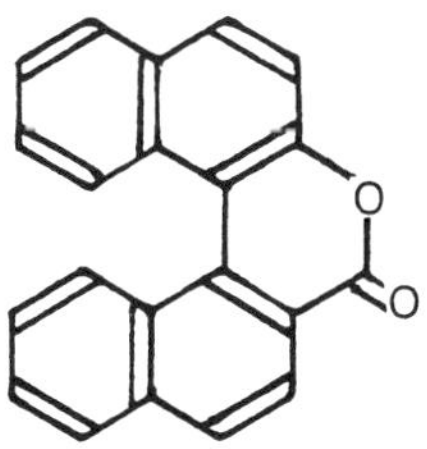

(13)

The addition of piperidine to the $C \equiv C$ bond of the alkynoylanthraquin-ones 14 [$R^1R^2 = CH = CH\text{-}CH = CH$, $(CH_2)_4$, $(CH_2)_3$; $R^1 = R^2 = Me$], followed by cyclization of the adducts on silica gel furnishes the 2-alkenyl-4*H*-anthra[1,2--*b*]pyran-4,7,12-triones 15 (M.A. Mzhel'skaya, A.A. Moroz and M.S. Shvartsberg, Izv. Akad. Nauk SSSR, Ser. Khim., 1991, 1656). 2-(3--Hydroxyhex-4-en-2-yl)- and 2-(2,3-epoxyhex-4-en-2-yl-11-hydroxy-5-methyl--4*H*-anthra[1,2-*b*]pyran-4,7,12-triones 16 ($R^1 = OH$, $R^2 = H$; $R^1R^2 = $epoxy, respectively) and their salts have been manufactured microbially by *Streptomyces* sp. HP530 (FERM BP-2786) as neoplasm inhibitors (N. Enoki *et al.*, Jpn. Kokai Tokkyo Koho JP 04 117,379 [92 117,379], 1992).

For the synthesis of benzo[6,7]cyclohepta[1,2-*b*]pyran derivatives, for example, 4-dimethylamino-3-phenyl-2*H*-benzo[6,7]cyclohepta[1,2-*b*]pyran-2--one (17), with platelet antiaggregating and other activities, see A. Bargagna *et al.*, Farmaco, 1991, <u>46</u>, 461; and benzotropolonepyran 18, C. Plueg and W. Friedrichsen, Tetrahedron Letters, 1992, <u>33</u>, 7509.

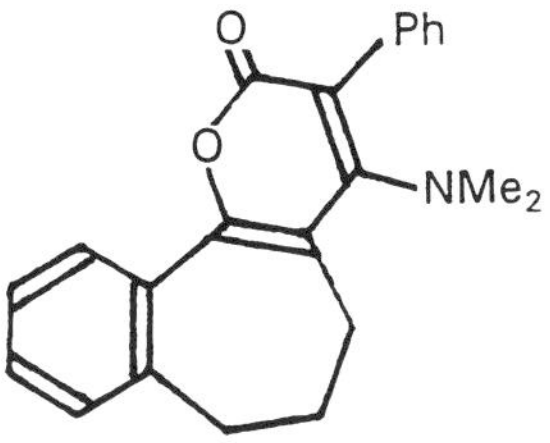

(17)

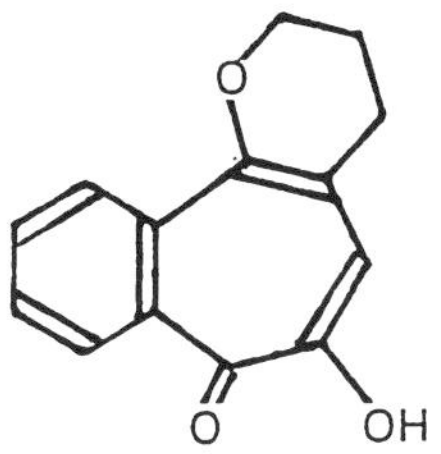

(18)

Unsymmetrical analogues of 5,6,11,12-tetrahydro-5,11-epoxydibenzo-[a,e]cyclooctene (Kagan's ether) (19) have been synthesized from methyl 1,3-dihydro-2H-indenone-1-carboxylate (M. Harmata and T. Murray, J. Org. Chem., 1989, 54, 3761). The structure of the product of the transannular reaction of 1,2,5,6-dibenzo-3,3,7,7-tetramethylcyclooctane-4,8-dione with methylamine has been shown to be 11-hydroxy-5-methylamino-6,6-12,12--tetramethyl-5,6,11,12-tetrahydro-5,11-epoxydibenzo[a,e]cyclooctene (20) from ^{13}C NMR and mass spectral data and x-ray analysis (R.M. Lagidze, Yu.V. Gatilov and Yu.A. Soobshch. Akad. Nauk Gruz. SSR, 1989, 135, 345).

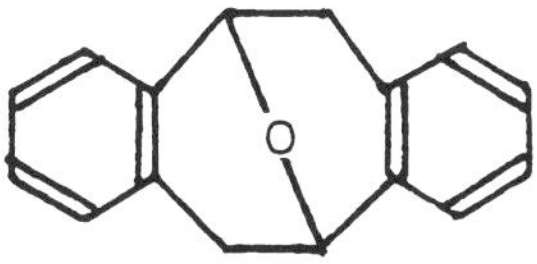

(19)

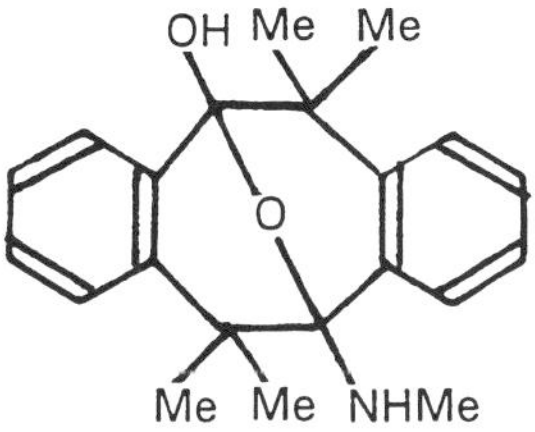

(20)

546

3-(4-Methoxyphenyl)methylene-1*H*-phenanthro[1,10-*c,d*]pyran (21) has
been obtained along with other products on the irradiation of 10-hydroxy-
methylphenanthrene-9-carboxylic lactone with (*E*)-4-MeOC$_6$H$_4$CH = CHMe in
benzene (H. Itoh *et al.*, Chem. Letters, 1991, 1683). The synthesis of 3-
-methyl-2*H*-pyreno[1,2-*b*]pyran-2-one (22) (R.G. Harvey *et al.*, J. Org.
Chem., 1988, 53, 3936), and 1-methyl 1*H*-3,4-dihydroacenaphtho[1,2-*c*]-
pyran 23 (A.C. Razus, M.D. Gheorghiu and E. Bartha, Rev. Roum. Chim.,
1991, 36, 215) have been reported.

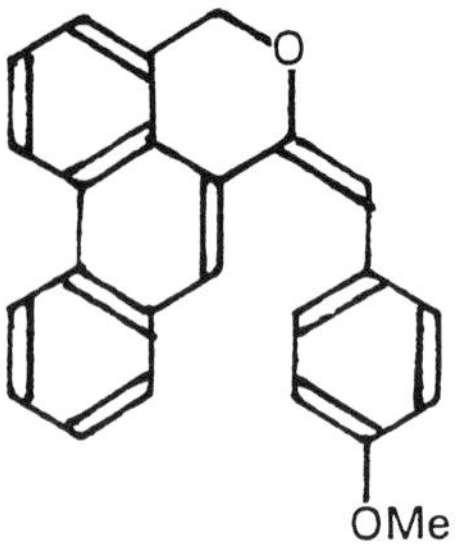

(21) 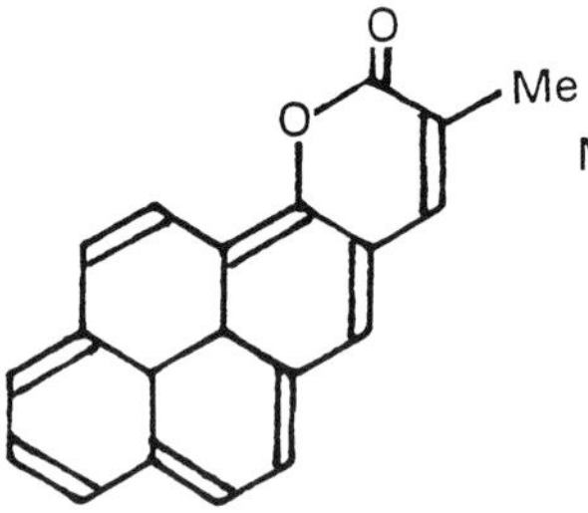   (22) 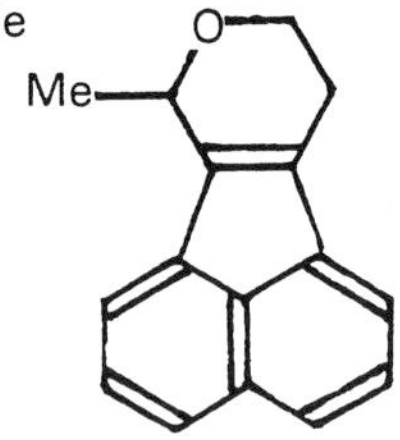   (23)

6-Substituted 1-bromomethyl-2-tetrahydropyran-2-oxynaphthalenes 24
(R^1 = H, Br, Bu^t) on reaction with tetrachloro- or tetrabromocatechol in
EtCOEt or MeCOEt afford spiro(benzoxanthennaphthofuran)ones 25 (R^2 = Cl,
Br; R^3 = H, *a*-Me, ß-Me; R^4 = H, Me) (T.R. Kasturi *et al.*, Tetrahedron, 1991,
47, 5245).

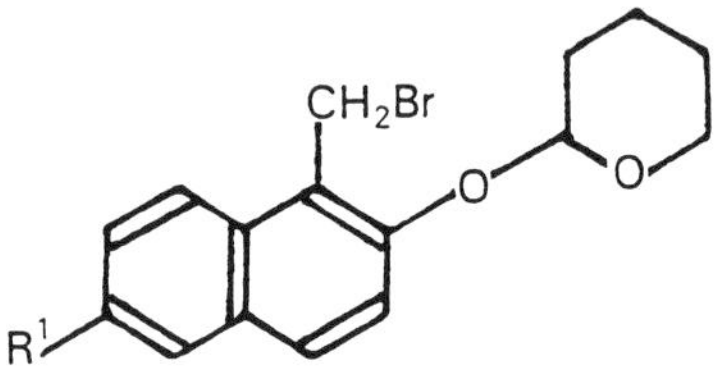

(24)

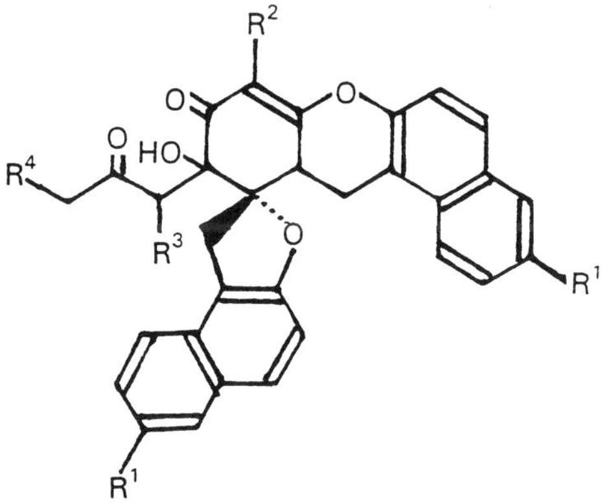

(25)

*Second Supplements to the 2nd Edition of Rodd's Chemistry
of Carbon Compounds, Vol.IV E*, edited by M. Sainsbury
© 1997 Elsevier Science B.V. All rights reserved.

Chapter 21

SIX-MEMBERED RING COMPOUNDS WITH ONE HETERO ATOM: SULPHUR, SELENIUM, TELLURIUM, SILICON, GERMANIUM, AND TIN

R. LIVINGSTONE

1. Thiopyran derivatives

(a) Thiopyrans, thiopyrylium salts and thiopyrones

Recent advances in the chemistry of thiopyrans, thiins continues to be mainly concerned with the synthesis, structure and properties of their derivatives. Some of the derivatives are used as pharmaceuticals and others as intermediates for the synthesis of pharmaceuticals, electron donors, sensitizers, and dyes used for research on organic conductors and photoconductors.

(i) Thiopyrans

1H-Thiopyrans. 1-Alkyl-2-aroyl- and 1-alkyl-2-cyano-thiabenzenes 1 have been obtained by the deprotonation of the corresponding thiopyrylium salts with triethylamine in ethanol. The ylide structures of 1 were supported by spectral and chemical evidence. Thermal reactions of 1 in several solvents resulted in S-alkyl rearranged products or ring-contracted products, depending upon the solvent, but in every case a mixture of products is obtained, in some just rearranged products in others a mixture of both. (H. Shimizu *et al.*, Tetrahedron Letters, 1990, <u>31</u>, 115).

R^1 = H,Me
R^2 = 4-BrC$_6$H$_4$CO,PhCO,CN
R^3 = Me, Et

(1)

(2)

(3)

2H-*Thiopyrans*. $R^1C_6H_4CSCH=CHNMe_2$ (R^1 = 3-NO_2, 4-NO_2, 3-CF_3, 4-CF_3) reacts with activated alkenes $CH_2=CHR^2$ (R^2 = CHO, Ac) to give the 3-substituted 2*H*-thiopyrans 2 after elimination of dimethylamine (U. Dietrich *et al.*, J. Prakt. Chem./Chem. Ztg., 1994, <u>336</u>, 434).

In related chemistry 3-acyl-2*H*-thiopyrans are obtained from the cyclo-addition of the enamines of monothio-ß-dicarbonyl compounds and unsaturated aldehydes and ketones. Thus cycloaddition of c-$C_6H_{11}NCH=CPhCSPh$ with $CH_2=CHCHO$ furnishes 86% of 5,6-diphenyl-2*H*-thiopyran-3-carboxaldehyde 3, whereas a yield of 57% is obtained from $Me_2NCH=CPhCSPh$ and $CH_2=CHCHO$ (D. Greif, M. Pulst, and M. Weissenfels, Synthesis, 1987, 456). For similar chemistry leading to 4 see, R. Hoffmann *et al.*, Z. Chem., 1990, <u>30</u>, 247.

R¹ = H,Me,Ph
R² = H,Me
R¹R² = (CH₂)₃

(4)

R¹ = Ar
R² = H,Ph

(5)

R¹ = MeO,Me,Prⁱ,H,R² = H
R¹ = H,R² = Ph

(6)

2-(Methoxycarbonylmethylene)-2*H*-thiopyran-3-carboxylic esters 5 are obtained by reacting $R^1C(S)CR^2=CHNMe_2$ with $MeO_2CCH=C=CCHCO_2Me$ in benzene, and also by treating the same allene with $HSCR^1=CR^2CHO$ [M. Pulst *et al.*, J. Chem. Res. (S), 1989, 300; Ger. (East) DD 272,848, 1989].

For the preparation of methyl 2-(2-methoxycarbonyl-2*H*-thiopyran-3-yl)-glyoxylates 6, see T. Blitzke *et al.*, J. Prakt. Chem./Chem.-Ztg, 1994, <u>336</u> 163; and 2-alkoxycarbonylmethylene-2*H*-thiopyran-3-carboxylates as polymethine dye intermediates, D. Greif, M. Pulst, and M. Weissenfels, Ger. (East) DD 258,008, 1988.

2-(2-Hydroxy-3-oxocycloalk-1-enyl)-2*H*-thiopyrans 9 are prepared by condensations of thiopyrylium salts 7 with cycloenols 8 in ethanol containing piperidine acetate [T. Zimmermann *et al.*, Ger. (East) DD 280,324, 1990].

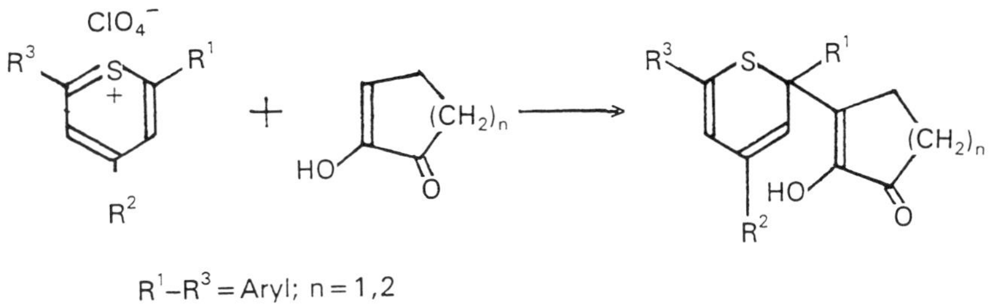

$R^1-R^3 = Aryl; n = 1,2$

(7)   (8)   (9)

Treatment of dimethylamino(thienyl)fulvene 10 with lithium naphthalene, followed by quenching with water, results in a novel reductive rearrangement to give 2-cyclopentadienylidene-2*H*-thiopyran 11 (T. Kawase, S. Fujino, and M. Oda, Tetrahedron Letters, 1991, *32*, 3499).

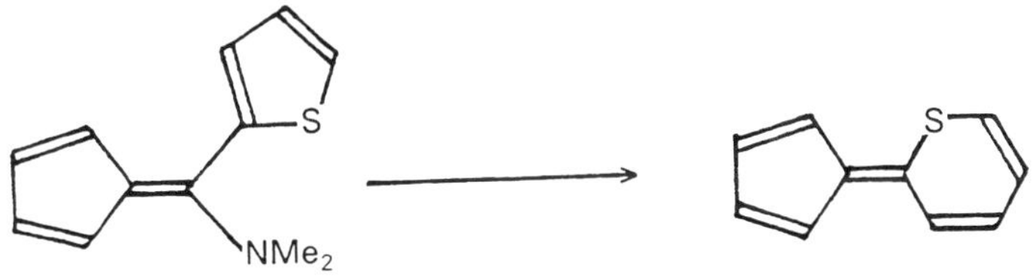

(10)   (11)

The Diels-Alder reactivity of some 2*H*-thiopyrans, containing electron-donating substituents has been investigated (D.E. Ward, Y. Gai, and W.M. Zoghaib, Canad. J. Chem., 1991, *69*, 1487), and the preparation of 2-*tert*--butyl-2-trimethylsilyl-2*H*-thiopyran has been reported (K.-T. Kang, C.H. Park, and U.C. Yoon, Bull. Korean Chem. Soc., 1992, *13*, 41). For a report on the chemistry of stabilized 2*H*-thiopyrans see M. Weissenfels, M. Pulst, and D. Greif, Phosphorus, Sulphur, Silicon Relat. Elem., 1991, 59, 311; for the formation of 6-methylthio-2*H*-thiopyran by a pericyclic cyclization from an unstable 2,4-pentadienedithioate, Y. Vallee, M. Khalid, and J.L. Ripoll, Tetrahedron Letters, 1993, 34, 2605; and for the preparation of perfluoroalkyl-2*H*-thiopyrans, K.I. Pashkevich, M.B. Bobrov, and V.I. Saloutin, Sulphur Letters, 1992, 14, 195.

4H-Thiopyrans. 2,4,4,6-Tetraaryl-4*H*-thiopyrans are prepared by reacting 1,3,3,5-tetraarylpentane-1,5-diones with tetraphosphodecasulphide in xylene at elevated temperature. The tetraarylthiopyrans on exposure to UV undergo a reversible photochemical colour change (S. Nespurek *et al.*, J. Chem. Soc., Perkin II, 1992, 1310). Also, the crystal and molecular structure, and photochromic properties of 2,4,4,6-tetraphenyl-4H-thiopyran 12 (J. Vojtechovsky *et al.*, Coll. Czech. Chem. Comm., 1992, 57, 1326), and the preparation of substituted 4*H*-thiopyrans 13 (R^1, R^3, R^5 = Ph, R^4 = Me, R^2 = H; R^1, R^5 = Ph, R^2, R^4 = Me, R^3 = H) (S.N. Petrakov *et al.*, U.S.S.R. SU 1,583,421, 1990) and 3,5-dibromo- and -dichloro-2,4,4,6-tetraphenyl-4*H*--thiopyran (M. Schwarz, P. Sebek, and J. Kuthan, Coll. Czech. Chem. Comm., 1992, 57, 546) have been reported.

(12)   (13)

2,6-Diamino-3,5-dicyano-4-phenyl(aryl)-4*H*-thiopyrans are prepared by stirring an ethanolic solution of equimolar amounts of RCHO (R = Ph, substituted Ph), $CH_2(CN)_2$ and $NCCH_2CSNH_2$ in the presence of piperidine at room temperature (F.F. Abdel-Latif, Bull. Chem. Soc. Jpn., 1989, 62, 3768). The 3-aryl-2-cyanothioacrylamides 14 react with electron-deficient alkynes, methyl propiolate and dimethyl acetylenedicarboxylate to give derivatives of methyl and ethyl 2-amino-3-cyano-4-aryl-4*H*-thiopyran-5-carboxylates 15 (J. Bloxham and C.P. Dell, J. Chem. Soc., Perkin I, 1994, 989).

R^1 R^2 = H,alkyl; R^3 = NO$_2$,OMe,Cl; R^4 = H,OMe,Cl;
R^5 = Me,Et; R^6 = H,CO$_2$Me

(14)

(15)

2,4,6-Triphenyl-4H-thiopyran (16; R = H) and 4-(4-methoxyphenyl)-2,6-
-diphenyl-4H-thiopyran (16; R = Me) on treatment with selenium dioxide in
pyridine afford the related thiophenes 17 (B.L. Drevko, L.A. Fomenko, and
V.G. Kharchenko, Khim. Geterotsikl. Soedin., 1989, 767).

(16)

(17)

552

Bis(thiopyranylidene. 2,2'-Diaryl-6,6'-diphenyl-4*H*-bithiopyranylidenes
(2,2'-diaryl-6,6'-diphenyl-$\Delta^{4,4'}$-bi-4*H*-thiopyrans) 19 have been obtained in
excellent yields by the reaction of 1,5-diarylpent-4-yne-1,3-diones 18 with
phosphorus pentasulphide in boiling pyridine [M.G. Marei, Afinidad, 1994, <u>51</u>
<u>(449)</u>, 43]. They have also been prepared by reacting 2-aryl-3-halogeno-6-
-phenylpyrid-4-ones with phosphorus pentasulphide (*idem*, Sulphur Letters,
1993, <u>16</u>, 131). 2,2',6,6'-Tetraphenyl-4*H*-bithiopyranylidene has been
prepared by coupling two molecules of diphenylthiopyrylium perchlorate
using Zn in MeCN (E.M. Gluzman, L.V. Gavriko, and V.A. Starodub, Elektron.
Org. Mater., 1985, 40).

PhC≡CCOCHRCOAr

Ar = Ph,4-MeC$_6$H$_4$,4-ClC$_6$H$_4$
R = H,Cl

(18)

(19)

The catalytic and ionic reduction of the double bonds in some 4*H*-cyclo-
penta[*b*]thiopyrans 20 to furnish 2-thiabicyclo[4.3.0]nonanes 21 (S.K.
Klimenko, T.I. Tyrina, and N.N. Sorokin, Zh. Org. Khim., 1990, <u>26</u>, 401)
have been reported.

R^1 = Ph,4-MeOC$_6$H$_4$,R^2 = H; R^1 = R^2 = Ph

(20)

(21)

(ii) Thiopyrylium salts

2,4,6-Tris(dialkylamino)thiopyrylium salts 22 (R = alkyl) have been obtained by a multi-step route from 1,1,5,5-tetrachloropenta-1,4-dien-3-one. They undergo electrophilic substitution and are better represented as polymethine-like systems (W. Schroth, R. Spitzner, and E. Kleinpeter, Tetrahedron Letters, 1988, **29**, 4695).

ClO_4^-

X^-

(22) $R^2 = R^4 = R^5 = NR_2, R_3 = H$
(23) $R^5 = CH_2R^1, R^1 = H, Me, Ph$
$\quad\ R^2 = R^3 = Ph, aryl, R^4 = H$

$R = H, Me, MeOC_6H_4$
$X = ClO_4, BF_4, Cl$

(24)

2-Alkyl-5,6-diarylthiopyrylium salts 23 are prepared in good yields *via* the aldol condensation of chloroacroleins, $ClCR^2 = CR^3CHO$ with methyl ketone, R^1CH_2COMe, followed by reaction with Na_2S and cyclization using $POCl_3$ (D. Greif *et al.*, Synthesis, 1989, 515). A number of 2,6-diphenyl- and -diarylthiopyrylium salts 24 with substituents at the 4-position (B.I. Drevko *et al.*, U.S.S.R. SU 1,703,649, 1992) and 3-substituted 2-(4-aminobutadienyl)--4,6-diphenylthiopyrylium salts and 2-(3-formylallylidene)-2*H*-thiopyrans have been prepared (M. Pulst, F. Kropfgans, M. Weissenfels, Z. Chem., 1987, **27**, 443).

2,4,6-Triarylthiopyrylium salts have been converted into 2-(2-hydroxy-3--oxocycloalk-1-enyl)-2*H*-thiopyrans by reaction with cycloalkane-1,2-diones in the presence of an appropriate acid binding reagent, for example, sodium acetate or triethylamine (T. Zimmermann, G.W. Fischer, and B. Olk, J. Prakt. Chem., 1989, **331**, 853).

2,4,6-Triarylthiopyrylium salts 25 on treatment with acetic anhydride in the presence of sodium acetate undergo ring transformation to afford aryl-benzenes 26 and thiobenzophenones 27. Under similar reaction conditions 3,5-dimethyl-2,4,6-triphenyl thiopyrylium perchlorate yields 2-acetoxy-3,5--dimethyl-4,6-diphenylthiobenzophenone (T. Zimmermann and G.W. Fischer, *ibid.*, 1988, **330**, 35).

554

(25)   (26)   (27)

(iii) Thiopyranones

2H-Thiopyran-2-ones, 2H-thiin-2-ones. The reaction between the
cyanothioacetamide, $PhCH = C(CN)CSNH_2$ and $NCCH_2CONHNH_2$ in ethanol in
the presence of triethylamine furnishes 6-amino-3,5-dicyano-4-phenyl-*2H*-
-thiopyran-2-one 28 (B.Y. Riad and S.M. Hassan, Sulphur Letters, 1989, <u>10</u>,
1). For the preparation of methyl 6-methoxy-3,4-diphenyl-2-oxo-2*H*-thio-
pyran-5-carboxylate, see Y. Ikemi, J. Heterocyclic Chem., 1990, <u>27</u>, 1597.

(28)

4H-Thiopyran-4-ones, 4H-thiin-4-ones. 2,6-Diaryl-4*H*-thiopyran-4-ones
are key building blocks for the synthesis of numerous electron donors,
sensitizers, and dyes used for research on organic conductors and

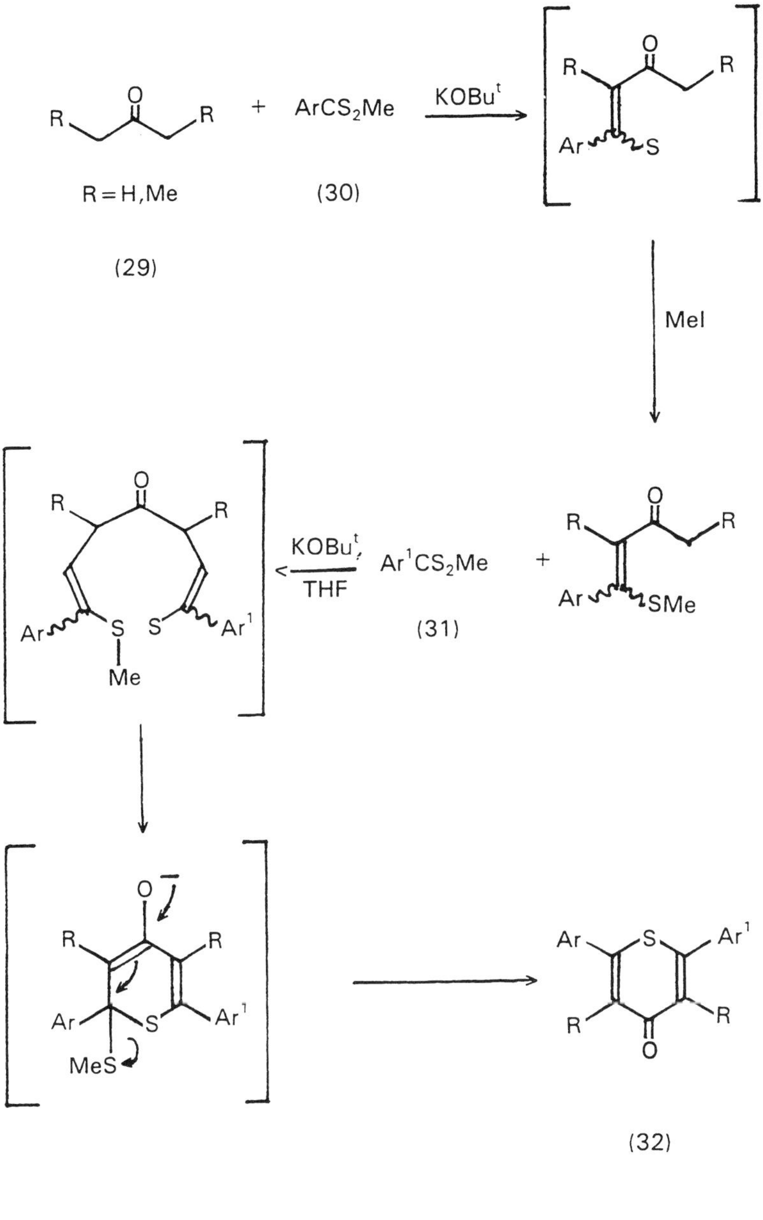

R
O
R
+
ArCS$_2$Me
KOBu^t
R = H,Me
(30)
(29)
R
O
R
Ar
S
MeI
R
O
R
Ar
SMe
KOBu^t,
Ar^{1}CS$_2$Me
THF
+
(31)
R
O
R
Ar
S
S
Ar^1
Me
O$^-$
R
R
Ar
S
Ar^1
MeS
Ar
S
Ar^1
R
R
O
(32)

556

photoconductors. The main intermediate for most syntheses of 2,6-diaryl-
thiopyran-4-ones is the corresponding 2,6-diaryltetrahydrothiopyran-4-one (4-
-thiacyclohexanone). The dehydrogenation conditions used to oxidize the
2,6-diaryltetrahydrothiopyran-4-ones are usually harsh and intolerant of
substitution. One exception involves the mild initial oxidation using *N*-
-chlorosuccinimide in pyridine to give the corresponding dihydrothiopyran-4-
-one, followed by one of several oxidation methods to yield the 4*H*-thio-
pyran-4-one.

Now a novel synthesis of 2,6-diaryl-4*H*-thiopyran-4-ones 32 has been
reported and involves two sequential thio-Claisen condensations of dialkyl
ketone 29 and two dithioesters 30 and 31. The intermediate ß-thioxo
ketone from the first condensation is converted to the corresponding ß-
-(methylthio)enone for both protection and reactivity purposes. Facile
addition-elimination of the methylthio moiety by a ß-thio ketone enolate
generated in the second thio-Claisen condensation gives the required diaryl-
-4*H*-thiopyran-4-one 32. This method is rapid and simple and the only
requirement being moderate substituent alkaline stability (N.W. Boaz and
K.M. Fox, J. Org. Chem., 1993, <u>58</u>, 3042).

For the preparation of some 2,6-diaryl- and 2,6-diaryl-2-chloro-4*H*-thio-
pyran-4-ones, see M. El-Ghanam, Alfinidad, 1994, <u>51</u>, 299.

The reaction of 1-phenylbut-1-yn-3-one and 1-phenylpent-1-yn-3-one,
$PhC\equiv CCOCH_2R$ (R=H, Me respectively) with carbon disulphide in the
presence of sodium hydride in dimethylformamide, followed by alkylation
gives a mixture of derivatives of 6-phenyl-4*H*-thiopyran-4-one 33 and thiophenes 34 (W.D. Rudorf and R. Schwarz, Tetrahedron Letters, 1987, <u>28</u>,
4267). The condensation of 2-(4-butoxyphenyl)-6-phenyl-4*H*-thiopyran-4-
-one 1,1-dioxide (35; R=BuO) with malononitrile in ethanol containing
piperidine gives 2-(4-butoxyphenyl)-4-dicyanomethylene-6-phenyl-4*H*-thio-
pyran 1,1-dioxide (36; R=BuO) (N.G. Rule and T.M. Kung, U.S. US
4,968,813, 1990).

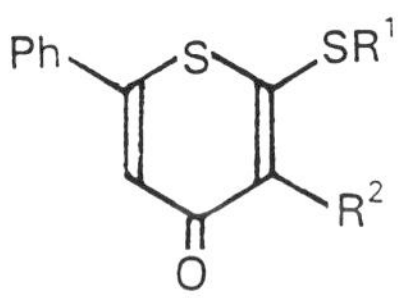

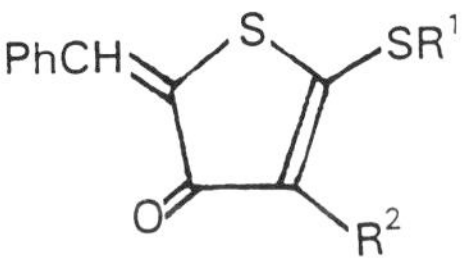

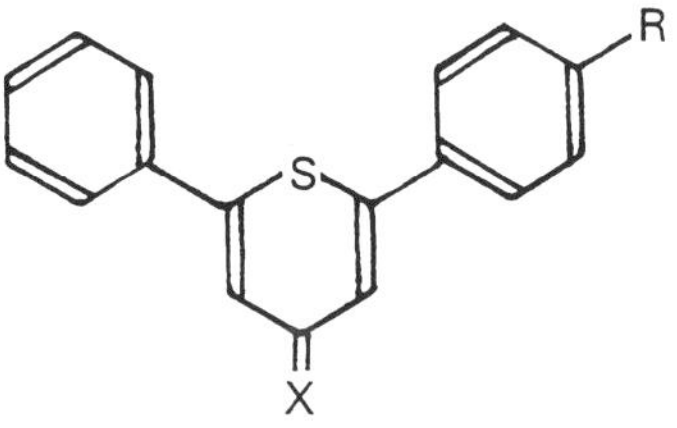

$R^1 = Me, PhCH_2, 4\text{-}BrC_6H_4$
$R^2 = H, Me$

$R = C_{4\text{-}8}alkoxy$

(33)   (34)   (35) X = O
(36) X = C(CN)$_2$

Treatment of 1-aryl-5-phenylpent-4-yne-1,3-diones, $PhC \equiv CCOCHRCOAr$ (R = H, Cl) with phosphorus pentasulphide in pyridine at room temperature affords 2-aryl-6-phenyl-4H-thiopyran-4-thiones 37 (M.G. Marei, Phosphorus, Sulphur, Silicon Relat. Elem., 1993, <u>81</u>, 101).

R = H,Cl

(37)   (38)   (39)

2,6-Bis(dialkylamino)-4*H*-thiopyran-4-ones 38 (R = alkyl) have been prepared by a multi-step route starting from 1,1,5,5-tetrachloropenta-1,4--dien-3-one, they show a significantly increased donor-reactivity at the carbonyl-oxygen compared to the non-amino analogues (W.Schroth, R. Spitzner, and E. Kleinpeter, Tetrahedron Letters, 1988, <u>29</u>, 4695).

4*H*-Benzo[*b*]thieno[3,2-*b*]thiopyran-4-one on oxidation with either Ph_3CClO_4 or SeO_2 affords 4*H*-benzo[*b*]thieno[3,2-*b*]thiopyran-4-one 5,5--dioxide 39 (K.Goerlitzer and R. Vogt, Arch. Pharm. (Weinheim, Ger.), 1990, <u>323</u>, 837).

(b) Di- and tetra-hydro-thiopyrans and -thiopyranones

(i) Dihydrothiopyrans, dihydrothiins, dihydrothiapyrans and related 2-, 3-, and 4-ones

3,4-Dihydro-2H-thiopyrans. α,ß-Unsaturated thioketones 1 (1-thiabuta--1,3-dienes, thiochalcones) are regenerated from their precursors the dithiin--type dimers 2, and readily undergo the Lewis acid-promoted Diels-Alder reactions with a variety of carbonyl-activated dienophiles 3 to give the appropriate [4 + 2] cycloadducts 4. $AlCl_3$ and $EtAlCl_2$, diethyl ether and dichloromethane accelerate the reaction even at lower temperatures. The reaction is very slow in the absence of the catalyst. When the dimer 2 is reacted with methyl acrylate, methyl 4,6-diphenyl-3,4-dihydro-2*H*-thiopyran--3-carboxylate is obtained. The configuration and relative conformations of the adducts have been determined (S. Motoki *et al.*, J. Chem. Soc., Perkin 1, 1992, 2943; 1991, 2281). For related chemistry see T. Karakasa, S. Satsumabayashi, and S. Moriyama, Nippon Shika Daigaku Kiyo, Ippan Kyoiku-kei, 1989, <u>18</u>, 81.

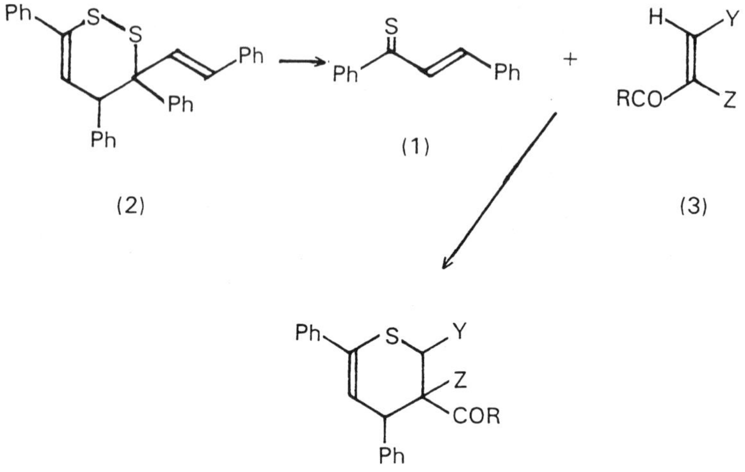

(1)

(2)

(3)

(4)

The thermal isomerization of thiochalcone dimers 5, 6, and 7 at 61.5°C in chloroform, ethylene glycol-dimethyl ether, or benzene has been investigated by HPLC (T. Karakasa, S. Satsumabayashi, and H. Kurihara, *ibid.*, 1991, <u>20</u>, 65).

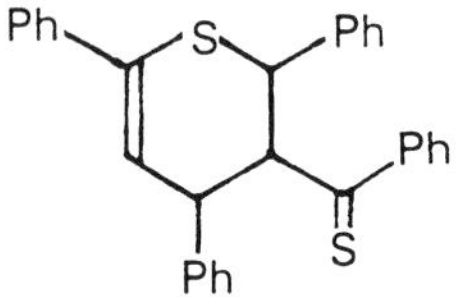

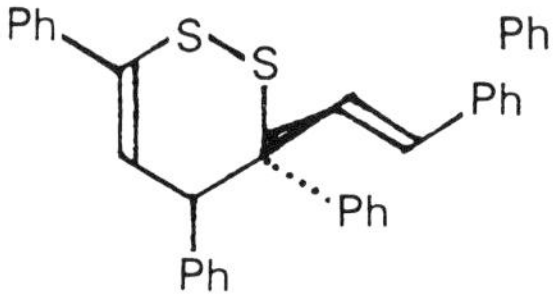

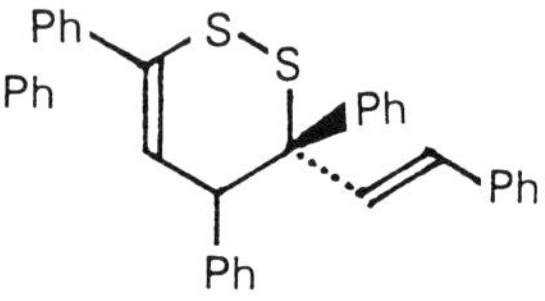

(5)

(6)

(7)

When diallyl sulphoxide 8 is heated with dienophiles at 90-110°C the thioacrolein Diels-Alder adducts are formed. Thus the reaction of sulphoxide 8 with dimethyl fumarate 9 affords dimethyl 3,4-dihydro-2*H*-thiopyran- -2,3-dicarboxylate 10 (E. Block and S.H. Zhao, Tetrahedron Letters, 1990, <u>31</u>, 5003).

A number of derivatives of 3,4-dihydro-2*H*-thiopyran have been obtained by making use of the Diels-Alder reaction between cyanothioacetamide derivatives, $R^1CH = C(CN)CSNH_2$ ($R^1 = $ 2-furyl, 2-thienyl) and $CH_2 = CHCN$, $CH_2 = CHCO_2Et$, and $PhCH = CHNO_2$ to afford dihydrothiopyrans 11 (B.Y. Riad *et al.*, Sulphur Letters, 1987, <u>6</u>, 105); 3-aryl-2-cyanothioacetamides, $R^1CH = C(CN)CSNH_2$ and $R^2CH = CHR^3$ to yield dihydrothiopyrans 12 (S.A. El- -Sharabasy *et al.*, Indian J. Chem., 1988, <u>27B</u>, 472); and enaminothiones and $RCH = CHNO_2$ ($R = Ph$, $4-MeOC_6H_4$, 2-furyl) to give in very good yields (78-90%) (*2R, 3S, 4R*)-6-aryl-2-aryl/furyl-4-dimethylamino-3-nitro-3,4- -dihydro-2*H*-thiopyrans (P.D. Baruah, S. Mukherjee, and M.P. Mahajan, Tetrahedron, 1990, <u>46</u>, 1951).

$(H_2C = CHCH_2)_2SO$

(8)

+

$(E)\text{-}MeO_2CCH = CHCO_2Me$

(9)

(10)

(11) $R^1 = $ 2-furyl, 2-thienyl
$R^2 = CN$, CO_2Et, $R^3 = H$
$R^2 = NO_2$, $R^3 = Ph$

(12) $R^1 = Ar$
$R^2 = CN$, CO_2Et, R^3H;
$R^2 = NO_2$ $R^3 = Ph$

2-(*N*-acylamino)-1-thiaalka-1,3-dienes, which are generated by acylation of the corresponding thioamides, undergo irreversible regio- and stereo-specific Diels-Alder reaction with the starting thioamides to furnish 6-(*N*--acylamino)-3,4-dihydro-2*H*-thiopyrans, for example, treatment of *(E)*-H$_2$NCS-CH = CHPh with acetyl chloride in acetone and pyridine affords the thiopyran-thiocarboxamide 13 (R = Ac). NMR techniques have been used to investigate the regiochemistry, relative configurations, and major solution conformations of 13 (R = Ac, Et) and some related 3,4-dihydro-2*H*-thiopyrans. It has also been reported that these systems exhibit a number of unique physical and chemical properties (I.T. Barnish *et al.*, Tetrahedron, 1989, 45, 6771). Acylation of R^{1}HNCSCH = CHPh (R^1 = H, Et) with acetyl chloride and subsequent reaction with electron deficient, non-activated, and electron-rich alkenes, R^{2}CH = CHR^3 (R^2 = R^3 = CO$_2$Me; R^2 = H, R^3 = CO$_2$Me, CN, OEt) affords the 3,4-dihydro-2*H*-thiopyrans 14 as mixtures of stereoisomers (*idem, ibid.*, p.4449).

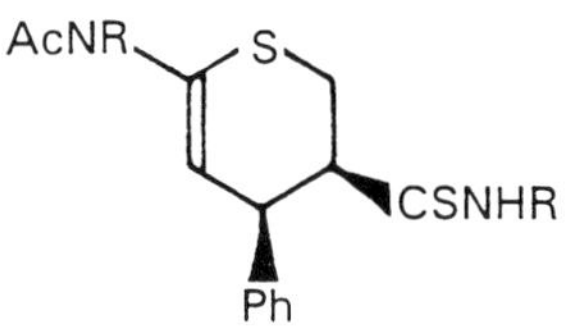

(13)

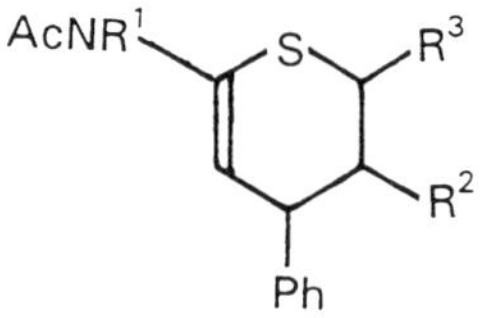

(14)

1,5-Diarylpenta-1,4-dien-3-ones, RCH = CHCOCH = CHR (R = Ph, tolyl) react with P$_4$S$_{10}$ to yield the thiaphosphothiaphosphorin sulphides 15. Compounds 15 when heated with norbornene afford 1,5-diarylpenta-1,4--diene-3-thiones, which undergo a Diels-Alder reaction with the norbornene to give the fused 3,4-dihydro-2*H*-thiopyrans 16 (S. Motoki, Y. Terauchi, and S. Kametani, Chem. Letters, 1988, 717).

(15)

(16)

3,4-Dihydro-2H-thiopyran-2- and -4-ones. Treatment of 2-oxocyclo-pentanepropionic acid with Lawesson's reagent in benzene gives 3,4--dihydrocyclopenta[b]thiopyran-2-one 17 (51%) along with 3,4-dihydro-cyclopenta[*b*]pyran-2-one, 3,4-dihydrocyclopenta[*b*]thiopyran-2-thione 18 and cyclopenta[*b*]thiopyran-2-thione 19 (E.V. Trushina *et al.*, Doklady Akad. Nauk, 1992, <u>327</u>, 354).

The reaction of 1,4-diarylbut-1-en-3-ones, $R^1CH = CHCOCH_2R^2$ (R^1 = Ph, ClC_6H_4, R^2 = Ph, anisyl) with arylisothiocyanates R^3NCS (R^3 = Ph, tolyl,

(17)

(18)

(19)

BrC_6H_4) and sodium in methanol affords the 2,5-diaryl-6-arylamino-3,4--dihydro-2*H*-thiopyran-4-ones 20. Similarly, ethyl 2,2-disubstituted-4-oxo-6--phenylamino-3,4-dihydro-2*H*-thiopyran-5-carboxylates 21 are obtained from $R^4_2 C = CHCOCH_2CO_2Et$ [R^4 = Me, R^4R^4 = $(CH_2)_4$, $(CH_2)_5$] and PhNCS (J. Becher *et al.*, J. Heterocyclic Chem., 1988, <u>25</u>, 795).

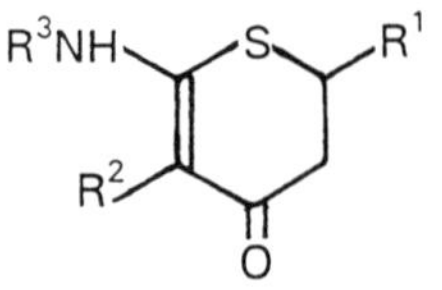

(20)

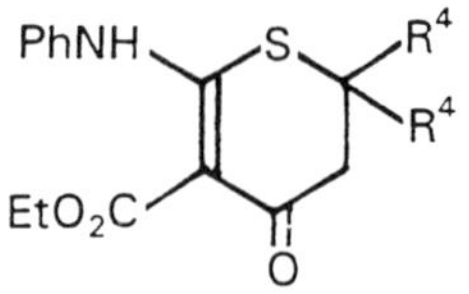

(21)

Treatment of 5,6-diphenyl-3,4-dihydro-2*H*-thiopyran-4-one 22 with phosphoryl chloride-dimethylformamide yields 4-chloro-5,6-diphenyl-2*H*-thio-pyran-3-carboxaldehyde 23 (N. Vasumathi, D.V. Ramana, and S.R. Ramadas, Synth. Comm., 1990, 20, 2749).

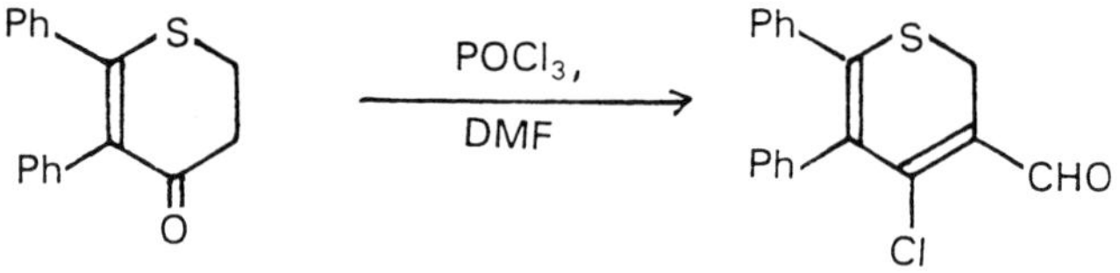

POCl₃, DMF

(22)

(23)

A number of fused thiopyrans, for example, 2,3,4,5-tetrahydro-4-oxo-indeno[1,2-*b*]thiopyran -2-carboxylic acid, 2,3-dihydro-4-oxo-4*H*-benzo[*b*]-furo[3,2-*b*]thiopyran-2-carboxylic acid, and 2,3-dihydro-4-oxo-4*H*-benzo[*b*]-thieno[3,2-*b*]thiopyran-2-carboxylic acid 24 have been prepared and converted to the respective thiopyranonecarboxaldehyde 25 [K. Goerlitzer and R. Vogt, Arch. Pharm. (Weinheim, Ger.), 1990, 323, 841; 847]. 1,4--Dihydropyridine derivatives 26 have also been prepared along with the 1,2--dihydropyridine derivatives (*idem, ibid.*, p.853).

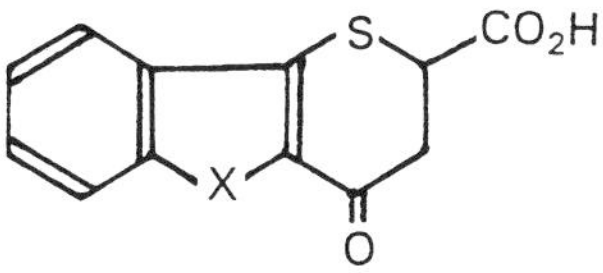

$X = CH_2,O,S$

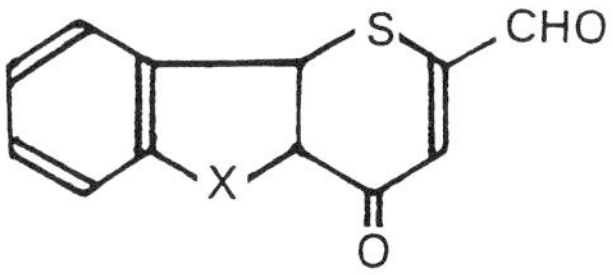

(24)

(25)

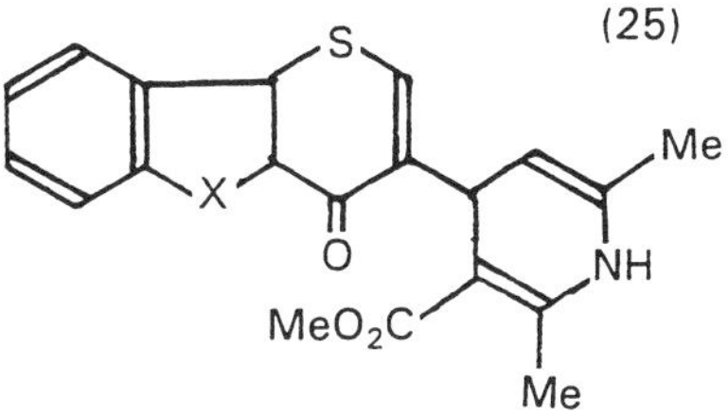

(26)

$X = CO,NH,O,S$

5,6-Dihydro-2H-thiopyrans (3,6-dihydro-2H-thiopyrans). In recent literature both nomenclature systems are used.

The stereochemical course of the 1,4-cycloaddition of monosubstituted sulphines (thioaldehyde *S*-oxides) $R^1CH = S(O)$ ($R^1 = Ph$, substituted Ph, Bu^t) to 2,3-dimethylbuta-1,3-diene, buta-1,3-diene, and *cis*- and *trans*-penta-1,3--diene has been investigated. It is reported that the reactions of buta-1,3--diene and 2,3-dimethylbuta-1,3-diene with (*Z*)-monoaryl sulphines give unexpectedly cis/trans mixtures of 6-aryl-5,6-dihydro-2*H*-thiopyran 1-oxide 27 and 6-aryl-3,4-dimethyl-5,6-dihydro-2*H*-thiopyran 1-oxide 28, respectively, in which the relative amounts of the two isomers depends upon the initial diene/sulphine ratio. In fact a *Z* to *E* isomerization of the dienophiles during the cycloaddition is responsible. However, in the case of *Z/E* mixture of *tert*-butyl sulphine with 2,3-dimethylbuta-1,3-diene, only the *trans*-cycloadduct 29 is formed. Catalysis of the reaction by Lewis acids has also been investigated (G. Barbaro *et al.*, J. Org. Chem., 1991, <u>56</u>, 2512).

$R^1 = Ph, 4\text{-}MeC_6H_4, Bu^t$

$R^2 = H, Me$

(27) $R^2 = H$
(28) $R^2 = Me$

(29)

The cycloadduct 30 (R = H) of anthracene and the labile thioaldehyde
$HCSCO_2Et$ is converted into the α-lithio derivative with lithium diisopropylamide. Subsequent treatment with methyl iodide, ethyl iodide, allyl bromide,
or benzyl bromide yields the corresponding 12-alkyl derivative 30 (R = Me, Et,
allyl, $PhCH_2$). Heating 30 in toluene at 111 °C liberates anthracene and a
thioketone, which when trapped in situ with 2,3-dimethylbuta-1,3-diene
affords 2-substituted ethyl 4,5-dimethyl-3,6-dihydro-2H-thiopyran-2-
-carboxylate 31. Similar treatment of the adduct 30 (R = Me) with either
cyclopentadiene or cyclohexadiene furnishes the appropriate cycloadducts
32 (n = 1,2) of the cyclic dienes and ethyl thioxopropionate (G.W. Kirby,
M.P. Mahajan, and M.S. Rahman, J. Chem. Soc., Perkin I, 1991, 2033;
3225). Diethyl 4,5-dimethyl-3,6-dihydro-2H-thiopyran-2,2-dicarboxylate has
been obtained in a similar manner (G.W. Kirby and W.M. McGregor, *ibid.*,
1990, 3175).

Related derivatives 31 (R = H, Me, Ac; OEt = Me, Ph, OEt) and 32
(Me = H, Me, Ac; OEt = Me, Ph, OEt) have been prepared (G. Capozzi *et al.*,
Tetrahedron, 1992, __48__, 9023) and the synthesis of some 5,6-dihydro-2H-
-thiopyrans from dialkyl sulphoxides and 1,3-dienes has been catalyzed by
certain metal complexes, ML_n (M = Ni, Al, Cu, Pd, Co, L = acetylacetonato)
(U.M. Dzhemilev *et al.*, Izv. Akad. Nauk SSSR, Ser. Khim., 1992, 160;
U.S.S.R. SU 1,680, 699, 1991).

(30)

(31)

(32)

Heating di-*n*-octyl sulphoxide in an autoclave at 150°C in the presence
of Pd(acac)$_2$ or Ni(acac)$_2$-(Ph$_2$PCH$_2$)$_2$-AlEt$_3$ produces 1-octene, water and the
thioaldehyde C$_7$H$_{15}$CHS, which is trapped by buta-1,3-diene to furnish 6-*n*-
-heptyl-5,6-dihydro-2*H*-thiopyran (*idem*, Izv. Akad. Nauk SSSR, Ser. Khim.,
1988, 2646). The reaction of equimolar amounts of PrCHO or Me(CH$_2$)$_4$-
CHO with CS$_2$ and isoprene in DMF containing Co(acac)$_2$ and Ph$_3$P under
carbon dioxide [(Co):(CO$_2$) = 1:50] at 50 atmospheres gives 45:55 mixtures
of 6-ethyl-3- and -4-methyl-5,6-dihydro-2*H*-thiopyran and 6-butyl-3- and
-4-methyl-5,6-dihydro-2*H*-thiopyran in 80 and 90% overall yields,
respectively (*idem, ibid.*, 1992, 2813). For an account of catalysts used in
the preparation of 5,6-dihydro-2*H*-thiopyrans from dialkyl sulphoxides and
dienes, see *idem*, U.S.S.R. SU 1,710,560, 1992.

Pichia farinosa IAM 4682, effects kinetic hydrolysis of enol esters 33 of
sulphur-containing cyclic ketones. The resulting 3-benzyl-tetahydrothio-
pyran-4-one 34, can be easily desulphurized to yield the corresponding
acyclic ketone 35, without loss of optical purity (Y. Kume and H. Ohta,
Tetrahedron Letters, 1992, <u>33</u>, 6367).

(33)

(34)

(35)

566

R-(-)-Methyl-*α*-naphthyl-phenylsilyl phenyl thioketone, prepared by acid catalyzed reaction of the corresponding ketone with hydrogen sulphide, reacts with methyllithium to give two diastereomers of MeSCHPhSiMePh-naphth-1yl in a 70:30 ratio, which undergo a Diels-Alder addition with buta--1,3-diene to yield diastereomeric dihydrothiopyrans 36 in a 75:25 ratio. After protodesilylation with tetrabutylammonium fluoride in THF/H_2O, the mixture 36 yields 2-phenyl-3,6-dihydro-2*H*-thiopyran 37, m.p. 44-45°C, $[\alpha]_D^{20}$ + 12.9°, in 51% enantiomeric excess (B.F. Bonini *et al.*, Chem. Comm., 1988, 365).

(36) R = SiMe(Ph)naphth-l-yl   (38)
(37) R = H

$X = O, \ Y = H_2$
$X = H_2, Y = O$

(39)

The preparations of 2-trimethylsilyl-2-phenyl-3,6-dihydro-2*H*-thiopyran (*idem*, Tetrahedron Letters, 1991, <u>32</u>, 815) and some 4,5-dimethyl-3,6--dihydro-2*H*-thiopyran-2-carboxamide-2-yl alkyl and aralkyl sulphides [H.W. Modrow *et al.*, Ger. (East) DD 250,318, 1987] have been reported. Thermal cycloaddition of *tert*-butyl trimethylsilyl thioketone with isoprene and 2--trimethylsilyloxybuta-1,3-diene occurs smoothly at 80°C to give regio-isomeric mixtures of 2-*tert*-butyl-5-methyl-2-trimethylsilyl-3,6-dihydro-2*H*--thiopyran (38; R^1 = Me, R^2 = H) and 2-*tert*-butyl-4-methyl-2-trimethylsilyl--3,6-dihydro-2*H*-thiopyran (38; R^1 = H, R^2 = Me) and the tetrahydrothio-pyrans 39, respectively. Protodesilylation of the silylated adducts can be performed with ease. The preparation of 2-*tert*-butyltrimethylsilyl-2*H*-thio-pyran is also reported (K.-T. Kang, C.H. Park, and U.C. Yoon, Bull. Korean Chem. Soc., 1992, <u>13</u>, 41).

The reaction of Bunte salt, 2-$F_3CC_6H_4C$ (SSO$_3$Na)CO$_2$Et with NEt$_3$/CaCl$_2$/EtOH, generates *α*-thioketo ester, 3-$F_3CC_6H_4C(S)CO_2Et$, which reacts with buta-1,3-diene to yield ethyl 2-(3-trifluoromethylphenyl)-3,6-dihydro-2*H*--thiopyran-2-carboxylate 40. Ester 40 is converted in 4 steps to the analogue 41 of aprikalim, a potassium channel activator (I.L. Pinto *et al.*, Tetrahedron Letters, 1992, <u>33</u>, 7597.

(40)   (41)

The lithio derivatives of sulphoxides 42 (R = Me, Ph) give, after
protonation, 2ß,4-dimethyl-3,6-dihydro-2*H*-thiopyran 1α-oxide (43; R = Me)
and 4-methyl-2ß-phenyl-3,6-dihydro-2*H*-thiopyran 1α-oxide (43; R = Ph),
respectively. They have been converted into their corresponding 1ß-oxides
and sulphones (M. Reglier and S.A. Julia, Bull. Soc. Chim. Fr., 1990, 226).

R = Me,Ph

(42)   (43)

Sulphuration reagents such as 5*H*-1,2,3,4-benzotetrathiepin and 6*H*-
-1,2,3,4,5-benzopentathiocin react with relatively stable phosphorus ylides,
$Ph_3P = CR^1R^2$ (R^1 = H; R^2 = Ac, CO_2Me, CO_2Et, COPh; R^1R^2 = 9*H*-fluoren-9-
-ylidene) to give efficiently thiocarbonyl compounds, such as thioaldehydes

568

and thioketones, which further react with 2,3-dimethylbuta-1,3-diene or
cyclopentadiene in situ to afford dihydrothiopyrans, for example, 2-acetyl-
-4,5-dimethyl-3,6-dihydro-2*H*-thiopyran (44; R^1 = H, R^2 = Ac), and thiabi-
cyclohept-5-enes 45 (R^1 = H, R^2 = Ac), respectively (R. Sato and S. Satoh,
Synthesis, 1991, 785).

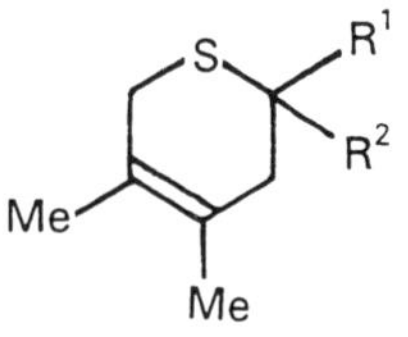

(44) 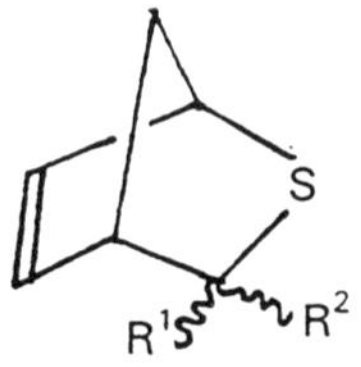   (45)

The Diels-Alder reaction between thiobenzophenone and *trans*-1-
-phenylbuta-1,3-diene, methyl *trans*-penta-2,4-dien-oate, and *trans*-penta-
-1,3-diene gives 2,2,6-triphenyl-3,6-dihydro-2*H*-thiopyran (46; R = Ph),
methyl 2,2-diphenyl-3,6-dihydro-2*H*-thiopyran-6-carboxylate (46;
R = CO_2Me), and 6-methyl-2,2-diphenyl-3,6-dihydro-2*H*-thiopyran (46;
R = Me), respectively (J. Huang and Y. Jin, Huaxue Xuebao, 1989, 47,
1124).

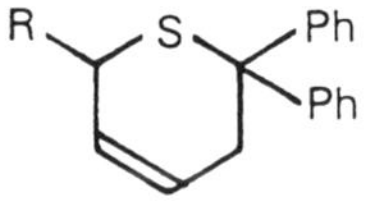

(46)   (48)

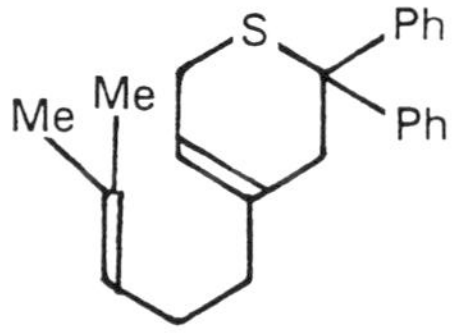

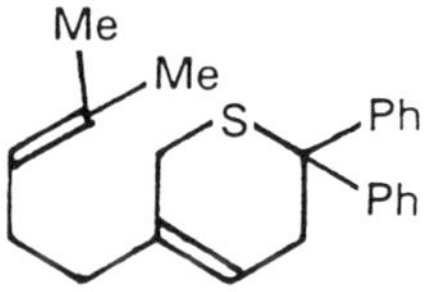

(47)

Thiobenzophenone reacts with myrcene at room temperature to give a mixture 4- and 5-(4-methylpent-3-ene-1-yl)-2,2-diphenyl-3,6-dihydro-2*H*-thiopyran (47 and 48) (Z. Xie, L. Hu, and N. Ying, Hecheng Huaxue, 1994, 2, 151). 4,5-Dimethyl-2-perfluorobutyl-2-(propylthio)-3,6-dihydro-2*H*-thiopyran 49 has been prepared by the reaction of 2,3-dimethylbuta-1,3-diene with H(CF$_2$)$_4$C(S)SPr and has been shown to possess cardiotonic activity equal to that of amrinone [N.M. Reonido *et al.*, Jpn. Kokai Tokkyo Koho JP 06,100,555 (94,100,555), 1994].

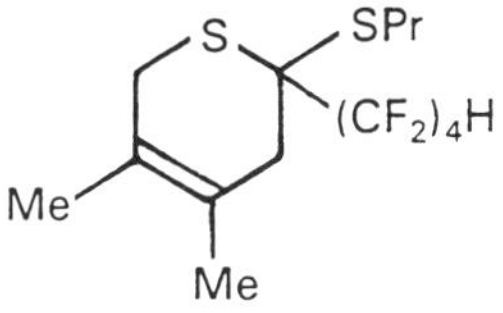

(49)

3,6-Dihydro-2H-thiopyran-3-ones. 3-Bromo-3-methylbutan-2-one reacts with mercapto esters, R$_2$C(SH)CO$_2$Et (R = H, Me) to give 5-oxo-3-thia-hexanoates, which cyclize to tetrahydrothiopyran-3,5-diones 50 (thiane-3,5--diones). Conversion of the diones to ethyl enol ethers, followed by reduction with LiAlH$_4$ yields 3,6-dihydro-2*H*-thiopyran-3-ones 51 (E. Er and P. Margaretha, Helv., 1992, 75, 2265).

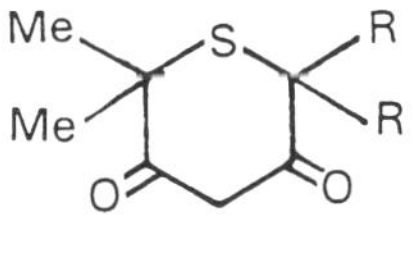

R = H,Me

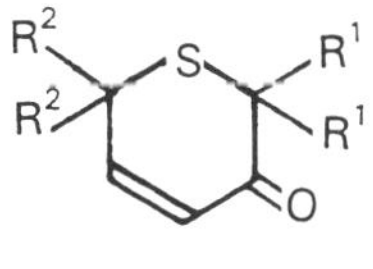

R^1,R^2 = H,Me

(50)

(51)

570

Miscellaneous dihydrothiopyrans. Bicyclic adducts are obtained by the
intramolecular Diels-Alder reaction of thioaldehydes generated from dienals.
Thus, treating $CH_2 = CHCH = CH(CH_2)_3CHO$ with $(Me_3Si)_2S$ and BuLi in
boiling THF affords *cis-* and *trans-*4a,7a-dihydro-2*H*-cyclopenta[b]thiopyrans
(2-thiabicyclo[4.3.0]non-4-enes) (52) in 67:33 ratio (M. Segi *et al.*, Synth.
Comm., 1989, <u>19</u>, 2431).

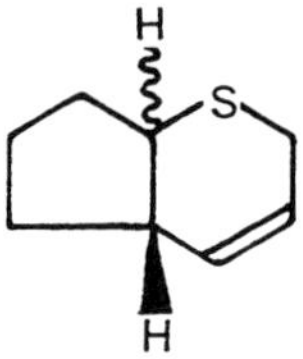

(52)

a-Chlorovinyl-3-nitrophenyl disulphides 53 (R = CF$_3$, Cl, Cl$_2$C = CCl) readily
afford *a*-chlorovinylthiolates that lose HCl to yield thioketenes, RCIC = C = S,
which react with cyclopentadiene to give 3-(substituted chloromethylene)-
-2-thiabicyclo[2.2.1]hept-5-ens 54. In the absence of cyclopentadiene, 2,4-
-disubstituted-1,3-dithietanes 55 are formed. It has been indicated that
thioketenes are potential biological reactive intermediates (W. Dekant *et al.*,
J. Amer. Chem. Soc., 1991, <u>113</u>, 5120).

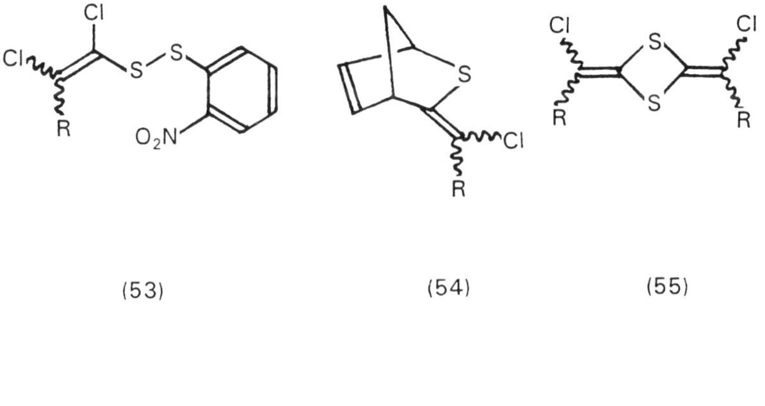

(53)   (54)   (55)

[4 + 2] Cycloaddition reactions occur between $4\text{-RC}_6\text{H}_4\text{CSCH}=\text{CHNMe}_2$ (R = H, Me, Cl) and norborn-2-ene and dicyclopentadiene resulting in the formation of *exo*-methanobenzothiopyrans 56 and methanoindenothiopyrans 57, respectively (P.D. Barush and M.P. Mahajan, J. Chem. Res. (S), 1993, 420).

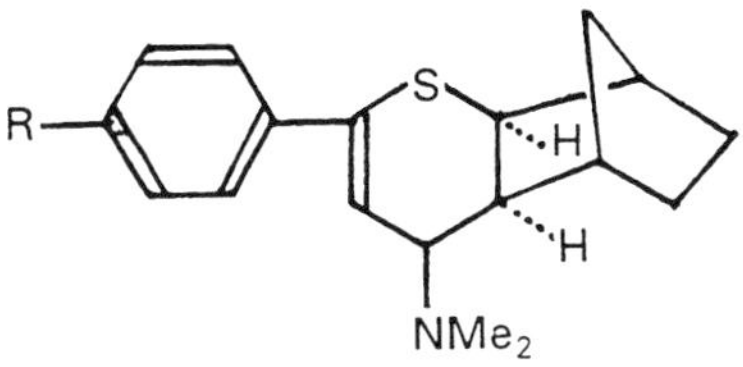

(56)

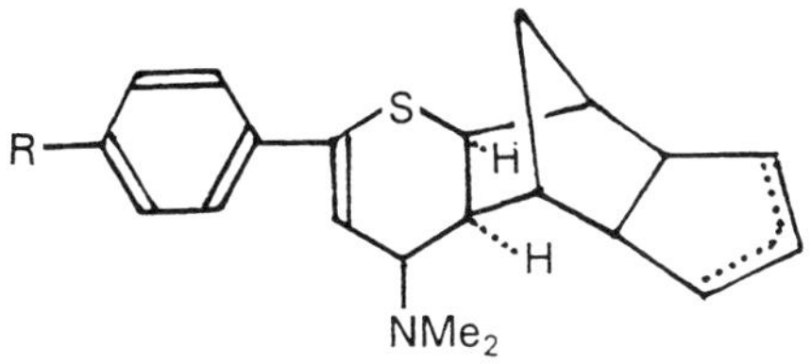

(57)

Pentacarbonyl (thiobenzaldehyde) complexes $(\text{CO})_5\text{M}[S=\text{C(Ph)}]$ (M = W, Cr) react with cyclopentadiene and cyclohexa-1,3-diene *via* a [4 + 2] cycloaddition to afford 2-metal-coordinated 2-thiabicyclo[2.2.1]hept-5-ene 58 and 2-thiabicyclo[2.2.1]oct-5-ene 59, respectively (H. Fischer *et al.*, Ber., 1990, 123, 725). A number of 2-thiabicyclo[2.2.1]hept-5-ene-3-carbox-amides (2-thianorborn-5-ene-3-carboxamides) 60 (R^1 = alkyl, aralkyl, aryl; $\text{R}^2 = \text{NH}_2$, NHR^1, NR^1_2) have been prepared [W.Thiel, R. Mayer, and H.W. Modrow, Ger. (East) DD 250,317, 1987].

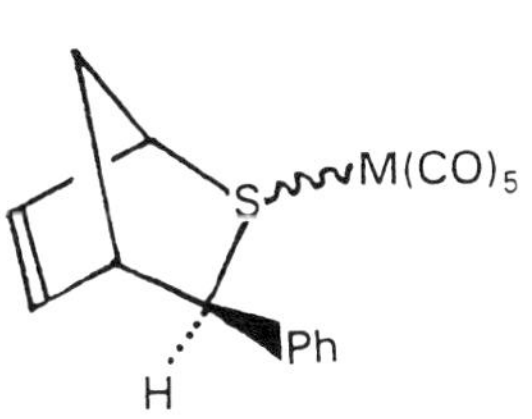

(58)

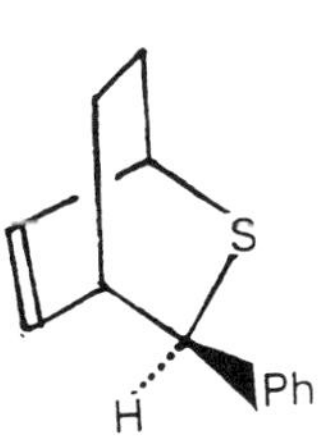

(59)

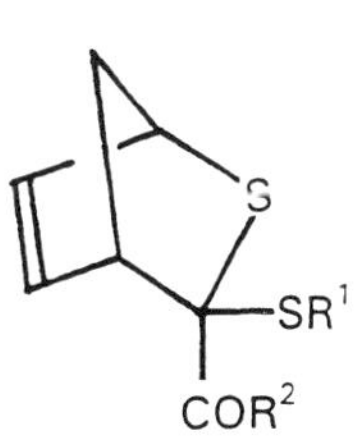

(60)

572

*(ii) Tetrahydrothiopyrans, thianes, tetrahydrothiapyrans and related 3-
    and 4-ones*

Tetrahydrothiopyrans. The preparation of optically active tetrahydrothio-
pyran-2- and -3-carboxylic acids (1 and 2) has been facilitated by amidation
of the racemic 1 and 2 by α-aminoesters and hydrolysis of the individual
diasteromeric amides (E. Fritz-Langhals, Eur. Pat. Appl. EP 521,496, 1993;
Angew. Chem., 1993, *105*, 785).

$R = Me, Pr^i, Bu^t$

(1)   (2)   (3)

2-Tetrahydrothiopyranyl aryl ethers 3 and thioethers are obtained by the
reaction of phenols and arenethiols with 1-(4-toluenesulphonylimino)tetra-
hydrothiopyran in the presence of NaH in DMF, respectively, or by acid-
-catalyzed transacetalization of 2-methoxytetrahydrothiopyran with
secondary alcohols and arenethiols (E. Giovani and E. Napolitano, Gazz.,
1993, *123*, 257).

2-(Methylthio)tetrahydrothiopyran [2-(methylthio)thiane] is deprotonated
with butyllithium in the 2-position and the resulting carbanion reacts with
electrophiles to yield the 2-substituted derivatives 4 (R = Et, Pr, Bu, Pr^i,
CH_2Ph, Bz) (A. Boege, G. Schwaar, and J. Voss, Phosphorus, Sulphur,
Silicon Relat. Elem., 1993, *83*, 175).

$R^1 = aryl$
$R^2 = alkyl$
$n = 0,1$

(4)   (5)

2-Aryltetrahydrothiopyran-2-carbothioamides and their 1-oxides 5 have been prepared as vasodilators. Thus, *cis*- and *trans*-($\pm$)-2-phenyltetrahydrothiopyran 1-oxide in THF on treatment with LDA and MeNCS affords ($\pm$)--*trans*-2-phenyltetrahydrothiopyran-2-carbothiomethylamide 1-oxide (5; R^1 = Ph, R^2 = Me, n = 1) (D.C. Cook *et al.*, Eur. Pat. Appl. EP 326,297, 1989). 6-Substituted 2-methyltetrahydrothiopyrans 7 have been obtained, as a mixture of stereoisomers, by the free radical mediated cyclization of substituted vinylcyclopropanes 6 possessing a 4-mercaptopentyl moiety (K.S. Feldman and H.M. Berven, Synlett, 1993, 827).

$$R^1 = R^2 = Cl, \; Br; \; R^1 = Ph, \; R^2 = H$$

(6)   (7)

Cis- and *trans*-tetrahydrothiopyran-4-carboxylic-4-acetic diacid 1-oxides 8 and *cis*- and *trans*-3,4-dihydro-2*H*-thiopyran-4-carboxylic-4-acetic diacid 1--oxide 9 (sulphoxide analogues of tetrahydro- and dihydroprephenate) have been prepared starting from α-ketoglutaric acid and have been shown to be modest, reversible inhibitors of prephenate dehydratase, the enzyme responsible for the Grob-type fragmentation of prephenic acid to phenyl-pyruvic acid (J.H. Bushweller and P.A. Bartlett, J. Org. Chem., 1989, 54, 2404). For the preparation of 3,5-di(1,4-dithiapentyl)tetrahydropyran 10 see J.M. Desper and S.H. Gellman, Angew. Chem., 1994, 106, 335) and 1--(tetrahydrothiopyran-4-yl)prop-1-en-3-ones, S. Kozima and M. Hatano, Eur. Pat. Appl. EP 406,839, 1991).

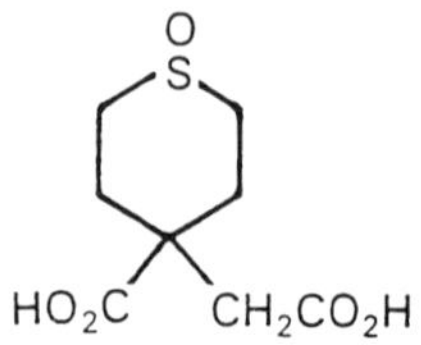

574

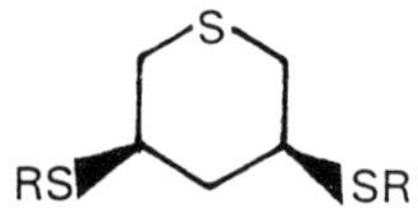

(8) 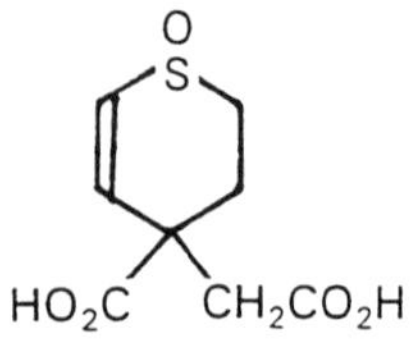 (9) (10)

R = (CH₂)₂SMe

4-Diazo-3,3,5,5-tetramethyltetrahydrothiopyran 11 reacts with freshly prepared carbon monosulphide in toluene at -78°C to yield 3,3,5,5-tetramethyltetrahydrothiopyran-4-thioketene 12 (E.K. Moltzen *et al.*, J. Org. Chem., 1991, <u>56</u>, 1317). Appropriate thioketenes react with tetracyanoethylene oxide to yield dicyanomethylenetetrahydrothiopyrans, for example, 1,4-dicyanomethylene-3,3,5,5-tetramethyltetrahydrothiopyran (13) (G. Adiwidjaja *et al.*, J. Chem. Soc., Perkin 1, 1993, 2385). 2,2-Dimethyl-4--(propyn-2-yl)tetrahydrothiopyran-4a-ol (14), 2,2-dimethyl-4-(4-dialkylaminobutyn-2-yl)tetrahydrothiopyran-4a-ols 15 (R¹ = alkyl), and their sulphones have been synthesized (A.A. Akhrem *et al.*, Vestsi Akad. Navuk BSSR, Ser. Khim. Navuk, 1990, 82).

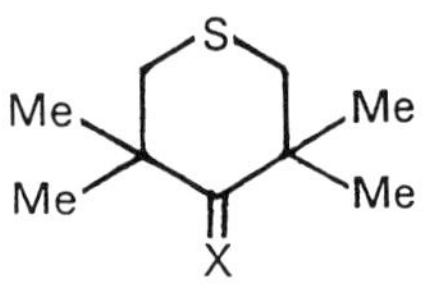

(11) X = N₂
(12) X = CS

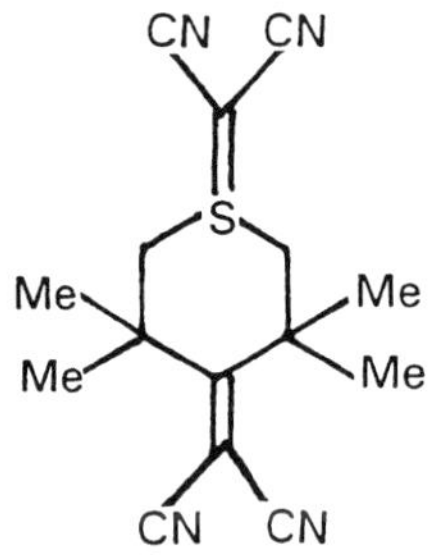

(13)

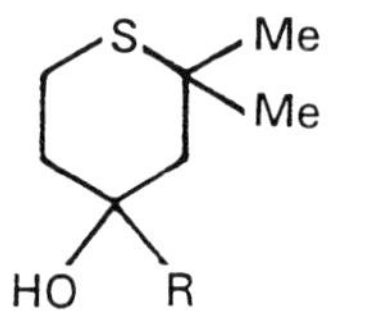

(14) R = CH₂C≡CH
(15) R = CH₂C≡CCH₂NR¹₂

Reaction of the *α*-sulphinyl carbanion derived from 4-(*tert*-butyldiphenyl-siloxy)tetrahydrothiopyran 1-oxide 16 with either methyl iodide, benzyl bromide or trimethylsilyl chloride results in substitution in the 2-position trans to the sulphoxide oxygen to afford derivatives 17. Acylation occurs in a stereocomplementary fashion to yield 2-keto product 18, with the new substituent cis to the sulphoxide oxygen. The change in stereochemistry is accompanied by an inversion of the tetrahydrothiopyran ring (R. Armer *et al.*, J. Chem. Soc., Perkin 1, 1993, 3105).

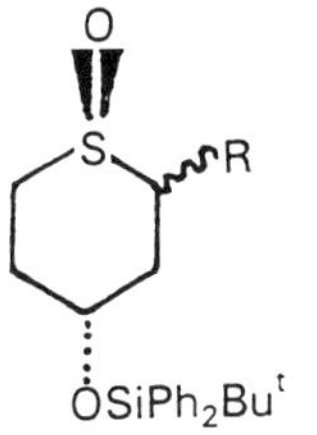

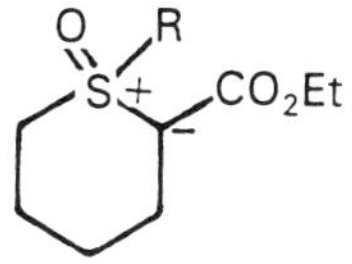

(16) R = H

(17) R = α-Me,-PhCH$_2$, -TMS

(18) R = β-COEt

(19)

The Rh$_2$(OAc)$_4$-catalyzed decomposition of RS(O)(CH$_2$)$_3$COC(:N$_2$)CO$_2$Et (R = Et, PhCH$_2$, allyl, PhCH = CHCH$_2$) gives the tetrahydrothiopyran sulphoxonium ylides 19. The structure of 19 (R = allyl) has been determined by X-ray crystallography (C.J. Moody *et al.*, Tetrahedron Letters, 1988, <u>29</u>, 6009).

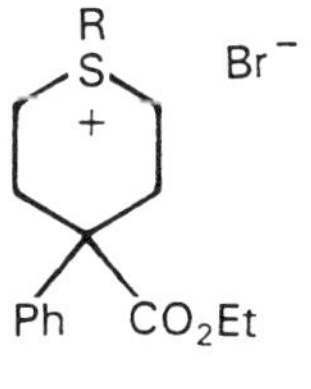

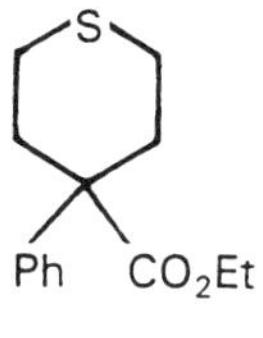

(20)

(21)

Hexahydrothiopyrylium salts (thianium salts) 20 (R = alkyl) analogues of the piperidine analgesics have been prepared by the reaction of ethyl 4--phenyltetrahydrothiopyran-4-carboxylate 21 with RBr; they are insufficiently active to warrant further investigation (G.W. Hardy, P.M. Doyle, and T.W. Smith, Eur. J. Med. Chem., 1987, <u>22</u> 331).

2-Alkyltetrahydrothiopyrans 22 (R = Me, Et, Pr^i) on treatment with *tert*--butyl hypochlorite give *trans*-1-*tert*-butoxyhexahydrothiopyrylium salts 23, which are hydrolyzed to 2-alkyltetrahydrothiopyran 1-oxides 24 and 25 (I. Jalsovszky *et al.*, Synth., 1990, 1037).

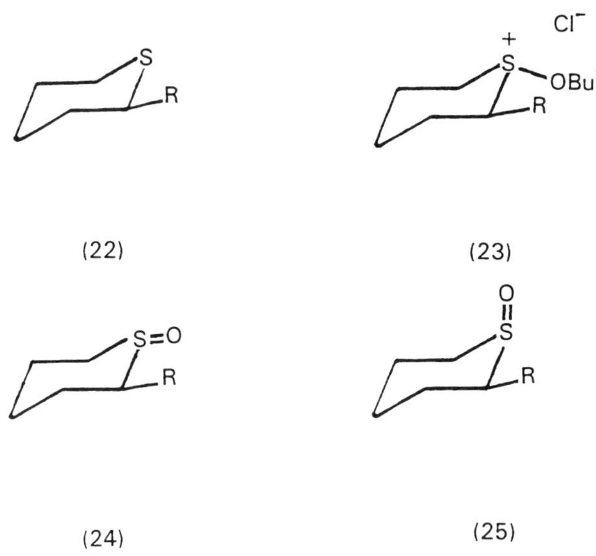

(22)   (23)

(24)   (25)

Deprotonation of 4-(*tert*-butyldiphenylsiloxy)tetrahydrothiopyran-1-oxide (26) with the asymmetric base 27 and then reaction with TMSCl gives (28) in modest e.e. (R. Armer *et al.*, J. Chem. Soc., Perkin 1, 1993, 3099; P.J. Cox, A. Persad, and N.S. Simpkins, Synlett, 1992, 194, 197).

(26) R = H
(28) R = TMS

(27)

(29)

Tetrahydrothiopyran-3-ones. Treatment of tetrahydrothiopyran-3-one
with pyrrolidine in benzene in the presence of 4-toluenesulphonic acid gives
the corresponding enamine, which on treatment with 2,4-dichlorobenzoyl
chloride in methylene dichloride in the presence of triethylamine yields 2-
-(2,4-dichlorobenzoyl)tetrahydrothiopyran-3-one 29 (J. Watanabe *et al.*, Jpn.
Kokai Tokkyo Koho JP 01 93,583 [89 93,583], 1987; U.S. US 5,132,434,
1992).

Treatment of $RS(CH_2)_3COC(:N_2)CO_2Et$ (R = Et, $PhCH_2$) with Rh_2 $(OAc)_4$ in
benzene gives the cyclic ylides 30, which rearrange to the ethyl 2-
-substituted 3-oxotetrahydrothiopyran-2-carboxylates 31 upon heating.
However when R in the starting material is an allyl group, ethyl 2-allyl-3-oxo-
tetrahydrothiopyran-2-carboxylate (31; R = allyl) is obtained directly (C.J.
Moody and R.J. Taylor, Tetrahedron Letters, 1988, <u>29</u>, 600). For the
preparation of related cyclic sulphonium ylides and derivatives by the same
procedures see *idem*, Tetrahedron, 1990, <u>46</u>, 6501. Also, the $Rh_2(OAc)_4$-
-catalyzed cyclization of $RS(O)(CH_2)_3COC(:N_2)CO_2Et$ (R = Et, allyl, $PhCH_2$,
$PhCH = CHCH_2$) affords the sulphoxonium ylides 32 (*idem, ibid.*, p.6525).

(30)

(31)

(32)

A cycloadduct, formed from 1-ethoxy-3-(trimethylsiloxy)buta-1,3-diene and sulphine, 1-oxo-2-thioxoindane *S*-oxide, can either undergo mild hydrolysis to give the 6,6-disubstituted spiro derivative of 5-ethoxytetra-hydrothiopyran-3-one 1-oxide 33 or elimination to give the related 3,6--dihydro-2*H*-thiopyran-3-one 1-oxide 34 (J.B.M. Rewinkel and B. Zwanenburg, Compt. Rend., 1990, <u>109</u>, 190).

(33)   (34)

The ylides 35 have been synthesized in a number of stages from D- and L-methionine (G.A. Tolstikov, F.Z. Galin, and S.N. Lakeev, Izv. Akad. Nauk SSSR, Ser. Khim., 1989, 974).

(35)   (36) X = O
                        (37) X = $NOCH_2CH = CH_2$

2-Methyltetrahydrothiopyran-3,5-dione on boiling with butyryl chloride in pyridine containing zinc chloride gives 2-methyl-4-butyryltetrahydrothiopyran--3,5-dione 36, which on treatment with $CH_2=CHCH_2ONH_2$. HCl in ethanol containing triethylamine affords 2-methyl-4-(1-oximinoalkyl)tetrahydrothiopyran 37 a herbicide (T. Gilkerson, R.W. Shaw, and D.C. Jennens, Brit. UK Pat. Appl. GB 2,202,529, 1988).

Tetrahydrothiopyrans-4-ones. Recent literature is mainly concerned with the conversion of various tetrahydrothiopyran-4-ones into related derivatives. Treatment of tetrahydrothiopyran-4-one with DMF-POCl$_3$ affords 4-chloro--5,6-dihydro-2*H*-thiopyran-3-carboxaldehyde (P.R. Giles and C.M. Marson, Tetrahedron, 1991, <u>47</u>, 1303). Tetrahydrothiopyran-4-one {[(dimethylethyl)-amino]hydroxypropyl}oxime monohydrochloride 38 has been obtained from tetrahydrothiopyran-4-one and $Me_3CNHCH_2CH(OH)CH_2ONH_2$.2HCl and along with related compounds have been tested for use as ß-blockers (J.J. Baldwin *et al.*, U.S. US 4,803,286, 1989).

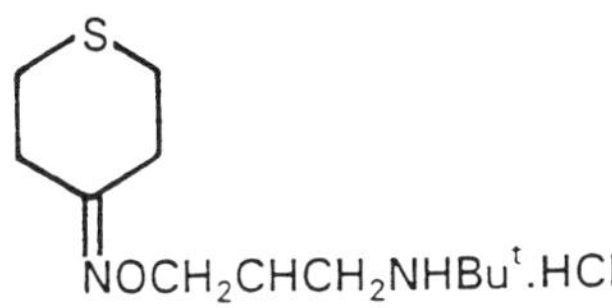

(38)

3-Alkyltetrathiopyran-4-ones (3-alkylthian-4-ones) 39 (R = Me, PhCH$_2$, allyl) are conveniently prepared by the alkylation of 3-allyloxycarbonyltetra-hydrothiopyran-4-one followed by deallyloxycarbonylation facilitated by the use of tetrakis(triphenylphosphin)palladium in the presence of morpholine. Also prepared by analogous procedures are the corresponding sulphones and 2,3-dialkyltetrahydrothiopyran-4-ones (G. Casy *et al.*, Synthesis, 1989, 767). The lithium enolate of tetrahydrothiopyran-4-one undergoes an aldol reaction with aldehydes to furnish threo isomers, for example, isomer 40. These highly diastereoselective aldol reactions have been investigated (T. Hayashi, Tetrahedron Letters, 1991, <u>32</u>, 5369).

(39)

(40)

The olefination of tetrahydrothiopyran-4-one with $(EtO)_2P(O)CH_2CO_2Et$ yields the alkylidene 41, which following the addition of $4\text{-}MeOC_6H_4CH_2SH$ and subsequent saponification affords 4-[(4-methoxyphenyl)methylthio]tetra-hydrothiopyran-4-acetic acid 42 (B. Lammek, I. Derdowska, and P. Rekowski, Pol. J. Chem., 1990, 64, 351).

(41)

(42)

Tetrahydrothiopyran-4-one undergoes a Mannich reaction with dimethyl-amine to give 3-(dimethylaminomethyl)tetrahydrothiopyran-4-one, which reacts with arylmagnesium bromides to yield mixtures of *trans*- and *cis*-4--aryl-3-(dimethylaminomethyl)tetrahydrothiopyran-4-ols 43. The predominating trans-isomers are obtained by crystallization of the hydrochlorides or chromatography of the bases (J. Urban *et al.*, Coll. Czech. Chem. Comm., 1987, <u>52</u>, 1340; Czech. CS 255,617, 1988). 4-Substituted 2,6-diphenyltetrahydrothiopyran-4-ols 44 [R = C≡CH, CH = CH$_2$, P(O)(OMe)$_2$, P(O)(OEt)$_2$] have also been prepared (R.D. Dil'barkhanov *et al.*, Khim.-Farm. Zh., 1988, <u>22</u>, 1375).

(43)

(44)

Methyl 3-alkyl-4-oxotetrahydrothiopyran-3-carboxylates (3-alkyl-3--methoxycarbonylthian-4-ones) have been monofluorinated at the 2-position by controlled-potential electrolysis in acetonitrile containing Et$_3$N.3HF on a Pt anode (S. Narizuka *et al.*, Denki Kagaku oyobi Kogyo Butsuri Kagaku, 1993, <u>61</u>, 868). 2,5-Dimethyltetrahydrothiopyran-4-one reacts with the appropriate primary amine to give the Schiff base 45 [R = MeOCH$_2$CH$_2$, 3,4--(MeO)$_2$C$_6$H$_3$CH$_2$CH$_2$] and spirooxazolidine 46 (Yu. G. Bosyakov, G.T. Maishinova, and N. Yu. Kuz'mina, Izv. Akad. Nauk Resp. Kaz., Ser. Khim., 1992, 77).

(45)

(46)

582

Both 2- and 3-methyl-4-sulphoaminotetrahydrothiopyran monosodium
salts (47 and 48), thia analogues of cyclamates have been prepared from the
methyltetrahydrothiopyran-4-ones *via* their oximes or dimethoxybenzhydryl-
imines. *Cis*- 47 and *trans*- 58 are sweet but *trans*- 47 and *cis*- 48 have no
taste [B. Unterhalt and M. Moellers, Arch. Pharm. (Weinheim, Ger.), 1991,
<u>324</u>, 525].

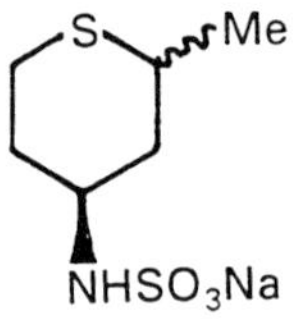

(47) 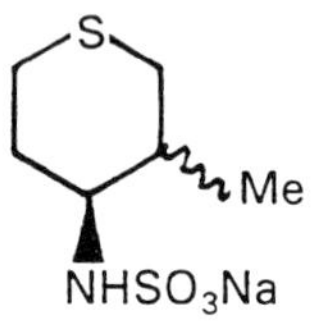   (48)

The Michael addition of methyl acrylate and chiral imines 49 (R = Me,
alkyl) yields tetrahydrothiopyran-4-ones 50 with 93% (e.e.). NaBH$_4$
reduction of ketone 50 affords the lactones 51 and 52 ($\sim$ 80:20) and
desulphurization of 51 (R = allyl) with nickel boride gives lactone 53 (H.
Matsuyama, S. Fujii, and N. Kamigata, Heterocycles, 1991, <u>32</u>, 1875). For
the synthesis of optically active lactones and related compounds from
optically active 3,3-dialkyltetrahydrothiopyran-4-ones see H. Matsuyama *et
al.*, Bull. Chem. Soc. Jpn., 1993, <u>66</u>, 1743.

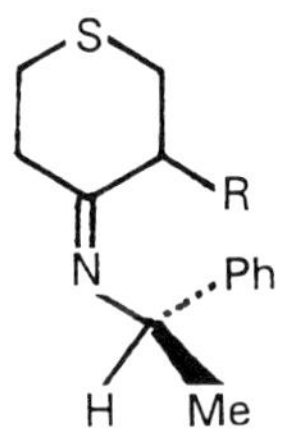

(49) 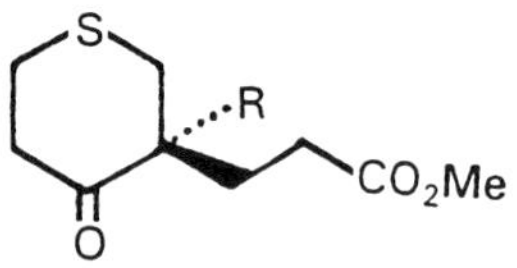   (50) 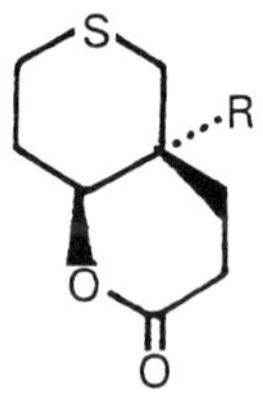   (51)

(52)   (53)

Treatment of 2,2,6,6-tetramethyltetrahydrothiopyran-4-one 54 or its 1-oxide with sulphuric acid gives a species the ESR of which suggest it is the cation radical of 54 (T. Endo, Kyushu Joshi Daigaku Kiyo, 1987, 23, 205). The reactions of 1,2,2,6,6-pentamethyl-4-oxotetrahydrothiopyrylium tetrafluoroborate with alkali metals halides in the solid state have been investigated (T. Ohashi, K. Tokuno, and Y. Otoda, Nippon Kagaku Kaishi, 1987, 1359).

(54)   (55)   (56)

Compounds related to tetrahydrothiopyran. 2-Substituted 3,3-dialkoxy-6-thiabicyclo[3.1.0]hexane 6,6-dioxides 55 (R^1 = H, R$_2^2$ = CH$_2$CH$_2$, CH$_2$CMe$_2$CH$_2$; R^1 = H, R^2 = Bu; R^1 = PhCH$_2$, allyl, R^2 = CH$_2$CH$_2$) have been fully characterized and the x-ray crystal structures have been obtained for some of the compounds (S.M. Jeffery *et al.*, J. Chem. Soc., Perkin 1, 1993, 2317).

The reaction of cycloocta-1,5-diene with SCl_2 in CH_2Cl_2 yields 2,6-
-dichloro-9-thiabicyclo[3.3.1]nonane 56. Treatment of 56 with Na_2CO_3 in
acetone-water yields the corresponding dihydroxy derivative, which upon
treatment with ClCOCOCl and DMSO in CH_2Cl_2 at <-60°C affords 9-thia-
bicyclo[3.3.1]nonane-2,6-dione (R. Bishop, Org. Synth., 1992, 70, 120).
For the preparation and properties of methyl 5-oxo-7-thiabicyclo[2.2.2]-
octane-2-carboxylate and related compounds see D.E. Ward and Y. Gai,
Canad. J. Chem., 1992, 70, 2627 and of some 2,6-substituted derivatives
of 9-thiabicyclo[3.3.1]nonane, N.I. Andreeva et al., Izv. Akad. Nauk SSSR,
Ser. Khim., 1989, 1869).

Phenyl bicyclo[3.3.1]non-6-ene-3-(selenothioperoxy)carboxylate 57 on
heating with AIBN yields (phenylseleno)thiahomoprotoadamantanone 58 (T.
Toru et al., Bull. Chem. Soc. Jpn., 1988, 61, 2675).

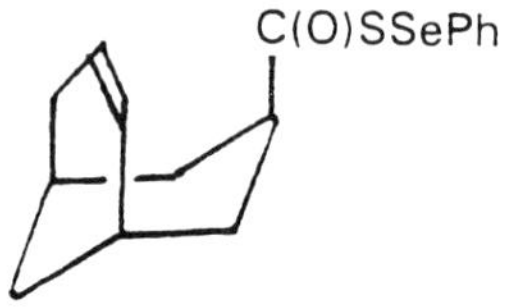

(57)

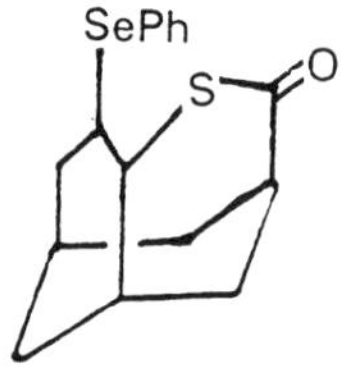

(58)

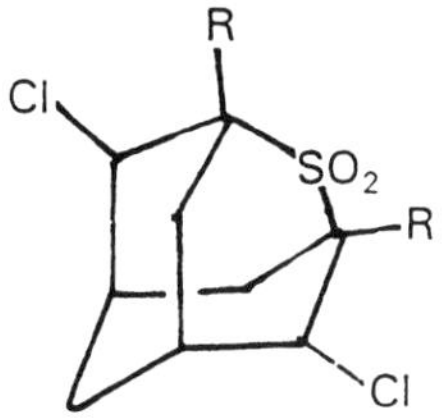

(59)

4,8-Dichloro-2-thiaadamantane 2,2-dioxide (59; R=H), obtained by the
oxidation of 4,8-dichloro-2-thiaadamantane with H_2O_2 in acetic acid,
undergoes chlorination by CCl_4 in Bu^{t}OH containing KOH to afford 3,4,8,9-
-tetrachloro-2,2-thiaadamantane 2,2-dioxide (59; R=Cl) (N.S. Zefirov et al.,
Doklady Akad. Nauk, 1992, 327 223).

2. Benzo[*b*]thiopyrans, 1-benzothiopyrans, benzo[*b*]thiins, thiochromens, 5,6-benzothiapyrans and derivatives

(a) Benzo[b]thiopyrans and benzo[b]thiopyranones and related naphtho derivatives

(i) Benzo[b]thiopyrans, 1-benzothiopyrans, benzo[b]thiins, thiochromenes

2*H*-Benzo[*b*]thiopyran (1; $R^1 = R^2 = H$), 3-phenyl-2*H*-benzo[*b*]thiopyran (1; $R^1 = H$, $R^2 = Ph$), and 3-(4-anisyl)-2-isopropyl-2*H*-benzo[*b*]thiopyran (1; $R^1 = Pr^i$, $R^2 = 4$-anisyl) have been obtained by the cyclization of 2-HSC$_6$H$_4$-CH$_2$P$^+$Ph$_3$Br$^-$ with R^2COCH(R^3)R^1 ($R^1 = R^2 = H$, $R^3 = Cl$; $R^1 = H$, $R^2 = Ph$, $R^3 = Br$; $R^1 = Pr^i$, $R^2 = 4$-anisyl, $R^3 = Br$, respectively) (A. Arnoldi and M. Carughi, Synthesis, 1988, 155. Related 2,3-disubstituted 6-nitro-2*H*-benzo-[*b*]thiopyrans 2 and 3 have been prepared (M. Weissenfels, M. Pulst, and D. Greif, J. Prakt. Chem./ Chem.-Ztg., 1992, <u>334</u>,147). Also see Y. Abe *et al.*, Chem. Express, 1992, <u>7</u>, 769; R. Asakawa and I. Arai, Jpn. Kokai Tokkyo Koho JP 05, 186,457 [93, 186,457], 1993; and M. Hirano, A. Miyashita, and H. Nohira, Chem. Letters, 1991, 209; for the preparation of photochromic spiroindolinobenzo[*b*]thiopyran 4 and related compounds.

(1) R^3 = H
(2) R^3 = NO$_2$

(3)

(4)

3,4-Epoxy-2,2,6-trimethyl-3,4-dihydro-2*H*-benzo[*b*]pyran 1,1-dioxide 5 reacts with pyrrolid-2-one in Me_2SO in the presence of NaH at room temperature under nitrogen to afford 2,2,6-trimethyl-4-(pyrrolid-2-one-1-yl)- -2*H*-benzo[*b*]thiopyran 6, which reduces systolic blood pressure in rats and shows bronchodilating activity in vitro in guinea pigs (D.G. Smith and G. Stemp, *ibid.*, 322,251, 1989).

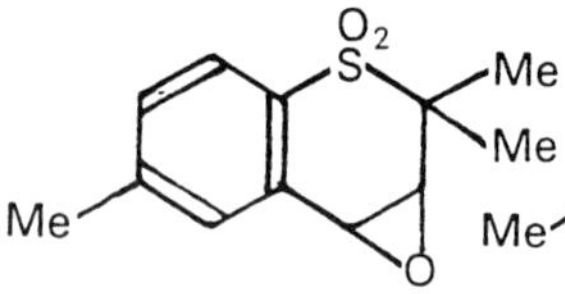

(5)

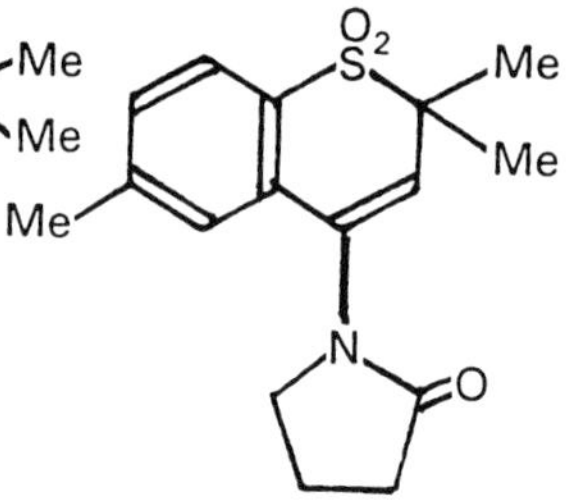

(6) 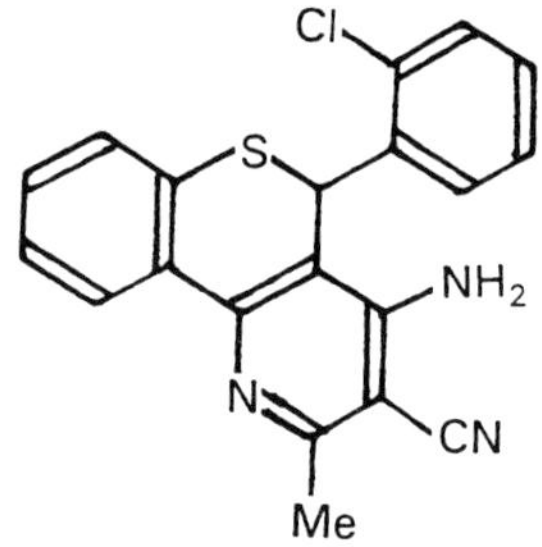   (7)

The cyclocondensation of acetaldehyde, malononitrile, and 4-amino-2- -(2-chlorophenyl)-3-cyano-2*H*-benzo[*b*]thiopyran yields the benzothiopyrano- pyridine 7 (A.E.A. Harb, A.M. El-Maghraby, and S.A. Metwally, Coll. Czech. Chem. Comm., 1992, <u>57</u>, 1570).

(ii) Benzothiopyrylium salts, 1-benzothiopyranylium salts, thiochromylium salts

4-Chloromethylene-2-methyl- and 4-chloromethylene-2,6-dimethylbenzo- thiopyrlium perchlorates 8 and 9 are prepared by the cyclization of $4\text{-}RC_6H_4\text{-}SCHMeCH_2COCH_2Cl$ with $HClO_4$ (S.V. Tolkunov and V.I. Dulenko, Khim. Geterotsikl. Soedin., 1987, 766).

(8) R = H
(9) R = Me

(10)

(11)

A number of benzothiopyrylium salts have been investigated for use as dyes, colouring materials for polymers, dichroic dyes for liquid crystals, electrophotographic photoreceptors, recording materials for optical disks, and near-IR filters. For instance, 2-phenyl-4-(4-*N*,*N*-dimethylaminophenyl)vinylbenzothiopyrylium perchlorate 10, is obtained from 4-methyl-2-phenylbenzothiopyrylium perchlorate 11 and 4-Me$_2$NC$_6$H$_4$CHO (T. Kitao and H. Nakasumi, Jpn. Kokai Tokkyo Koho JP 01, 242,579 [89, 242,579], 1989). A number of derivatives related to compound 10 have also been synthesized (*idem, ibid.*, 02 62,874 [90 62,874], 1990; H. Nakasumi, S. Watanabe, and T. Kitao, *ibid.*, 03 74,382 [91 74,382], 1991).

The hydrogenation of 5,6,7,8-tetrahydrobenzothiopyrylium salts 12 over Pd/C under pressure affords *cis*-1-thiadecalins with substituents R^1 and R^2 in the equatorial orientation. The same products are formed by the hydrogenation of the 5,6,7,8-tetrahydro-4*H*-benzo[*b*]thiopyrans 13 and 5,6,7,8,8a--pentahydrobenzo[*b*]thiopyrans 14. Oxidation of the above products furnishes the corresponding sulphoxides and sulphones (S.K. Klimenko, T.I. Tyrina, and N.N. Sorokin, Khim. Geterotsikl. Soedin., 1987, 614).

X = BF$_4$, CF$_3$CO$_2$
R^1, R^2 = H,aryl

(12)   (13)   (14)

The reaction between the 1,5-diketone 15 and H$_2$S in AcOH containing BF$_3$.Et$_2$O results in the formations of 2-(4-methoxyphenyl)-5,6,7,8-tetrahydrobenzothiopyrylium tetrafluoroborate 16 along with a number of other products (*idem.*, Zh. Org. Khim., 1988, <u>24</u>, 2008).

(15)   (16)

When 2-phenyl- and 2,4-diaryl-5,6,7,8-tetrahydrobenzothiopyrylium salts 17 are treated with bases, products 18 of oxidative dimerization are isolated. Dehydrogenated intermediates have also been identified. The reaction of 2,4-diphenylcyclopenta[*b*]thiopyrylium perchlorate with pyridine yields the cyclopentathiopyran dimer 19 (S.K. Klimenko, N.N. Ivanova, and I.M. Yudovich, Khim. Geterotsikl. Soedin., 1992, 26). Treatment of benzothio-

pyrylium tetrafluoroborates 17 (R^1 = Ph, 4-MeOC$_6$H$_4$; R^2 = Ph) with
K$_3$Fe(CN)$_6$ in alcoholic NaOH gives dimers 18 (R^1 = Ph, 4-MeOC$_6$H$_4$; R^2 = Ph)
in yields of 90 and 94%, respectively. Reaction of these dimers with 70%
HClO$_4$ affords the diperchlorates 20 (R = Ph, 4-MeOC$_6$H$_4$) in yields of 95 and
99% (S.K. Klimenko, N.N. Ivanova, and N.N. Sorokin, Zh. Org. Khim., 1987,
<u>23</u>, 2019).

R = H, Ar
X = BF$_4$, ClO$_4$

(17)

(18)

(19)

(20)

Benzothiopyrylium tetrafluoroborates 21 (R^1 = R^2 = H; R^1 = Ph, R^2 = H;
R^1 = 4-MeC$_6$H$_4$, R^2 = H; R^1 = CN, R^2 = H; R^1 = H, R^2 = Me; R^1 = Ph, R^2 = Me)
undergo polar cycloadditions with conjugated dienes, CH$_2$ = CR^{3}CR^4 = CH$_2$
(R^3 = R^4 = Me, H; R^3 = Me, Ph, R^4 = H) to afford benzo-fused bicyclic
sulphonium salts 22 with a bridgehead sulphur. Compound 22 (R^1 = CN,
R^2 = H, R^3 = R^4 = Me) on treatment with base is readily transformed into the
spiro compound 23 and also the dioxolane 24. The latter may arise through
the oxygenation of a 2-alkdienylbenzothiopyran intermediate (H. Shimizu *et
al.*, Chem. Comm., 1992, 1586).

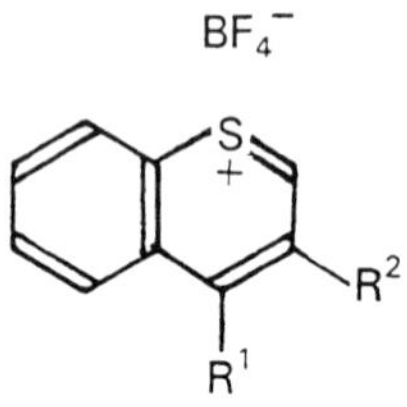

BF$_4^-$

(21)

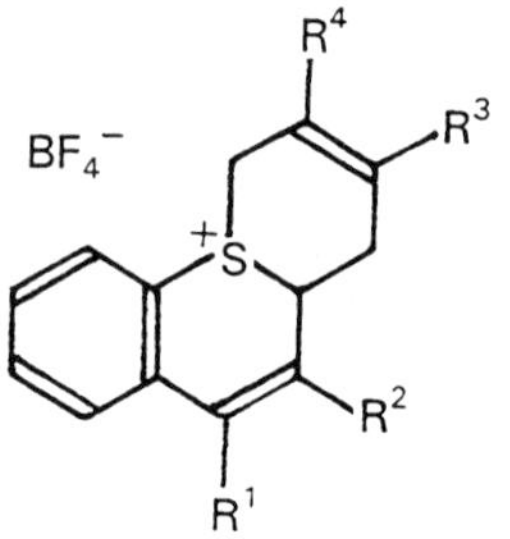

R^1
R^2

R^4
R^3
BF$_4^-$
R^2
R^1

(22)

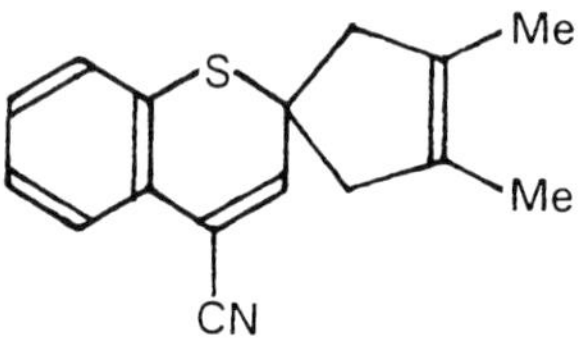

Me
Me
CN

(23)

Me
Me
CH$_2$
CN

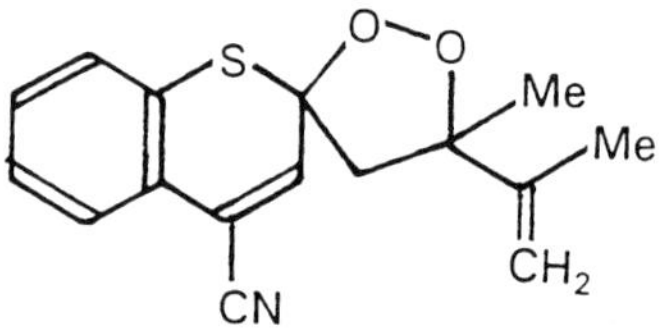

(24)

Preparations of thiochromenium salts 25 and ylide 26 have been reported. The ylide 26 on treatment with MeO$_2$CC≡CCO$_2$Me yields the benzothioin 27 (M. Hori *et al.*, Chem. Pharm. Bull., 1988, <u>36</u>, 3816).

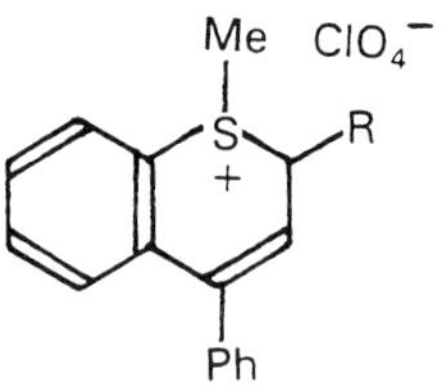

Me ClO$_4^-$
R
Ph

(25)

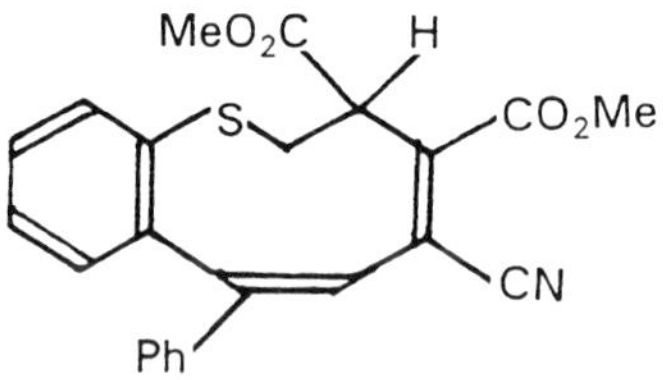

MeO$_2$C H
CO$_2$Me
CN
Ph

(27)

Me
CN

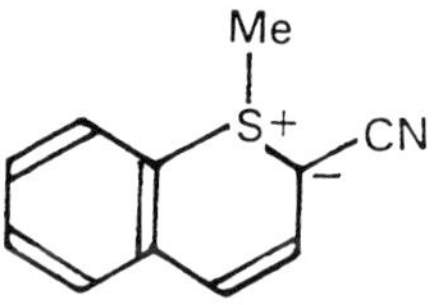

(26)

(iii) Benzothiopyranones

2H-Benzo[b]thiopyran-2-ones, 2H-1-benzothiopyran-2-ones, 2H-benzo-thiin-2-ones, thiocoumarins. Although in this section the parent name *2H-*-benzo[*b*]thiopyran-2-one is used, thiocoumarin still appears often in the literature. 2-Bu^{t}SC$_6$H$_4$CH = C(CN)CONHCH$_2$CH$_2$R (R = NMe$_2$, NEt$_2$, morpholino) on heating in polyphosphoric gives the carbamoyliminobenzo-thiopyrans 1, which on hydrolysis afford 3-carbamoylbenzothiopyran-2-ones 2 (Y. El-Ahmad and P. Reynaud, J. Heterocyclic Chem., 1988, 25, 711).

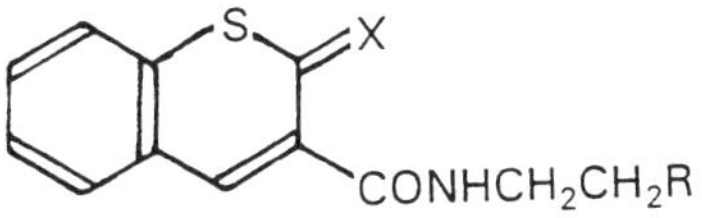

(1) X = NH
(2) X = O

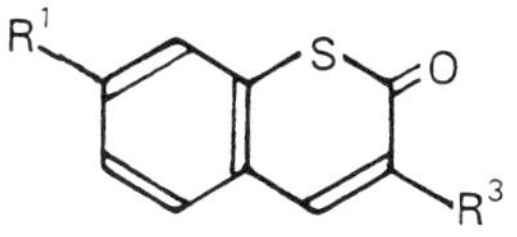

(3) R^3 = C(= NH)CH$_2$CH$_2$NR$^2_2$
(4) R^3 = C(SMe) = NH

A new method for the preparation of *N*-(2-dialkylaminoalkyl)-substituted 2-oxo-2*H*-benzothiopyran-3-carboxamidines 3 (R^1 = H, OMe, OCH$_2$Ph, OCH$_2$CO$_2$Et; NR$^2_2$ = NMe$_2$, NEt$_2$, morpholine) from the corresponding methyl 2-oxo-2*H*-benzothiopyran-3-carboximidothioates 4 has been reported (Y. El-Ahmad, J.D. Brion, and P. Reynaud, Heterocycles, 1993, 36, 1979).

The phase-transfer catalyzed alkylation of 4-hydroxy-2*H*-benzothiopyran--2-thione (5) with dimethyl sulphate or alkyl halides affords 2-alkylthio-4*H*--benzothiopyran-4-ones 6 [R = Me, CH$_2$Ph, (*E*)-CH$_2$CH = CHR^1, CH$_2$C(Cl) = CH$_2$, CH$_2$CH = CMe$_2$; R^1 = Me, Ph] (K.C. Majumdar, A.T. Khan, and S. Saha, Indian J. Chem., 1991, 30B, 643).

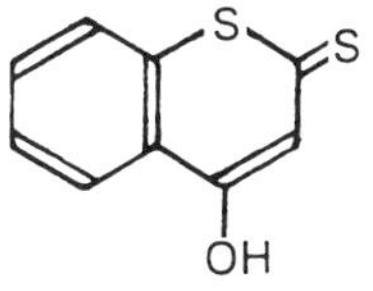

(5)

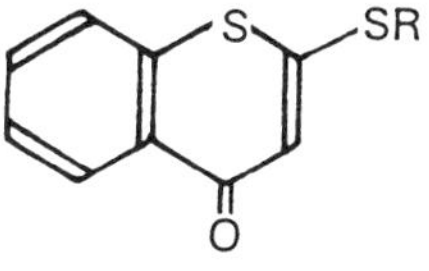

(6)

4H-Benzo[b]thiopyran-4-ones, 4H-1-benzothiopyran-4-ones, 4-benzothiin--4-one, thiochromones. N.B. both the first and last names are equally used in the literature. The reaction of 2-RC(O)SC$_6$H$_4$CO$_2$H (R = Me, Et, Ph, 4--ClC$_6$H$_4$, 4-MeOC$_6$H$_4$) with Ph$_3$P$^+$C$^-$ = C = NPh leads to acylphosphoranes, which subsequently undergo an intramolecular Wittig reaction to yield the respective 2-substituted 4*H*-benzo[*b*]thiopyran-4-one 7 (P. Kumar, A.T. Rao, and B. Pandey, Chem. Comm., 1992, 1580). The synthesis of a number of 3-methylthiochromone derivatives has been reported (L. Fang, G. Zhang, and B. Wei, Shenyang Yaoxueyuan Xuebao, 1989, <u>6</u>, 51). The SeO$_2$ oxidation of the appropriate 2-methyl-4*H*-benzo[*b*]thiopyran-4-one yields the corresponding 4-oxo-4*H*-benzo[*b*]thiopyran-2-carboxaldehyde 8 (R = H, Me) (K. Ito and K. Nakajima, J. Heterocyclic Chem., 1988, <u>25</u>, 511).

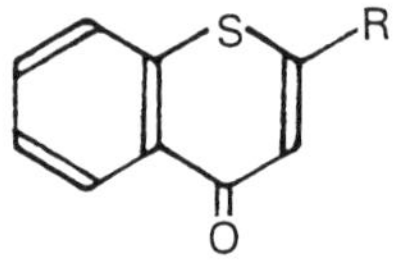

(7) 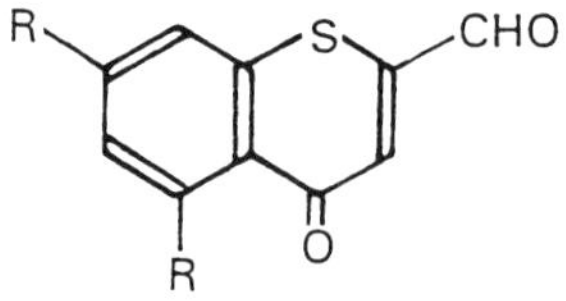 (8)

4-(Trimethylsiloxy)thiochrom-3-ene reacts with SOCl$_2$ and olefins and alkynes to yield 3-(substituted thio)thiochromones 9 (R = CH$_2$CMe$_2$Cl, *trans*--2-chlorocyclopentyl, CMe = CMeCl, CH = CBuCl) (I.W.J. Still *et al.*, Canad. J. Chem, 1989, <u>67</u>, 369).

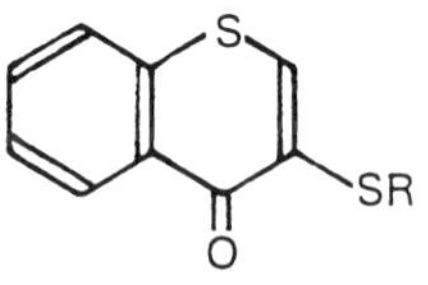

(9) 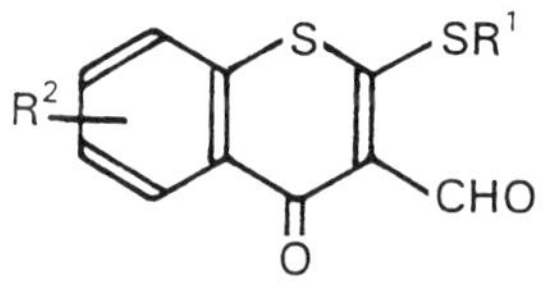 (10)

2-Methylthio-4-oxo-4*H*-benzo[*b*]thiopyran-3-carboxaldehyde (10; R^1 = Me, R^2 = H) is prepared by heating the sodium salt of 2-chlorobenzoyl-acetaldehyde with carbon disulphide and methyl iodide in the presence of sodium hydride. A number of related compounds have also been reported [W.D. Rudorf and J. Koeditz, Ger. (East) DD 292, 454, 1991; Synthesis, 1992, 667].

The piperidine-catalyzed reaction of the thiochroman-4-ones 11 (R^1 = H, Ph) with aromatic aldehydes affords the 3-benzyl-4*H*-benzo[*b*]thiopyran-4--ones 12 (R^1 = H, Ph; R^2 = 2-Br, 3-NO_2, 4-CN, 4-Cl, 2,4-Cl_2, 3,4-Cl_2). The reaction of 12 with Lawesson's reagent furnishes the corresponding 3-benzyl-4H-benzo[b]thiopyran-4-thiones (A. Levai and Z. Szabo, Bull. Soc. Chim. Fr., 1991, 976).

(11)

(12)

Thiochromone 1,1-dioxides when treated with diphenylsulphilimine, Ph_2S = NH, afford the azirinobenzothiopyrans 13 (R^1 = H, Me, Ph; R^2 = H, Me, Ph) (K. Buggle and B. Fallon, J. Chem. Res., S, 1988, 49).

(13)

(14)

594

The reaction of 2,3-dibromothiochromone 1,1-dioxide with alkyl
amines and primary aromatic amines yields 2-amino derivatives 14
(R^1 = NHMe, NMe_2, NHPh, $HNC_6H_4R^3$-4; R^2 = Br; R^3 = Me, MeO, MeCO,
NO_2). The reaction with the secondary aromatic amine, PhNHMe affords the
2-amino derivative 14 (R^1 = NMePh, R^2 = Br) along with the 2-aryl derivative
14 (R^1 = 4-C_6H_4-NHMe, R^2 = Br). Tertiary and some secondary aromatic
amines yield only 2-aryl derivatives 14 (R^1 = 4-$C_6H_4NR^4R^5$, R^2 = Br;
R^4 = R^5 = Me, Et; R^4 = H, R^5 = Et, Ph) (S. Watanabe, H. Nakazumi, and T.
Kitao, J. Chem. Soc., Perkin 1, 1988, 1829). 2,3-Dibromo-5,8-dimethoxy-
-4H-benzo[b]thiopyran-4-one 1,1-dioxide has been used as a starting material
to prepare sulphone analogues of 1,4-naphthoquinone dyes and its reactions
with aliphatic and aromatic amines has been investigated (S. Watanabe *et
al.*, J. Heterocyclic Chem., 1990, <u>27</u>, 1241).

3-Hydroxymethylthiochromone on treatment with $SOCl_2$ gives 3-chloro-
methylthiochromone, which reacts with 2-ethylimidazole in DMF in the
presence of NaH with ice cooling to afford 3-(2-ethyl-1-imidazolyl)methylthio-
chromone. The preparations of a number of related derivatives have also
been reported (T. Sano *et al.*, Jpn. Kokai Tokkyo Koho JP 03 291,280 [91
291,280], 1991), along with compounds derived from 3-chloromethyl-6,8-
-dimethylthiochromone as 5-HT_3(serotonin 3) receptor antagonists (*idem.*,
ibid., 04 82,887 [92 82,887], 1992). Some thiochromone derivatives 15
[R^1 = $CH_2CO_2R^3$, $CH_2CH_2CO_2R^3$, CH = $CHCO_2R^3$, R^3 = H, lower alkyl; R^2 = H,
halogeno, lower alkyl or alkoxy, (halogeno-substituted)phenoxy] have been
prepared and investigated for use as hypolipemics (K. Tomizawa *et al.*, *ibid.*,
02 255,676 [90 255,676], 1990).

(15)

2-Phenyl-4H-benzo[b]thiopyran-4-one, 2-phenyl-4H-1-benzothiopyran-4-one, 2-phenyl-4H-benzo[b]thiin, thioflavones. Treatment of thioflavones with the iodine-cerium(IV) ammonium nitrate system furnishes the 3-iodo derivatives (F. Zhang and Y. Li, Synthesis, 1993, 565). Thioflavone oxime diethylaminoethyl ether hydrochloride is prepared by alkylation of thioflavone oxime with diethylaminoethyl chloride in the presence of an alkali metal alcoholate on alkali metal (especially Na) in boiling EtOH and/or MeOH. In vitro, it acts as a fungicide and prevents growth of *Candida albicans* and *Geotrichum candidum* strains (W. Basinski, Pol. PL 147,858, 1989).

(b) Thiochromans, thiochromanones and related compounds

A variety of other names for these compounds are known, but since the term thiochroman is well accepted it will be used here.

(i) Thiochromans, 3,4-dihydro-2H-benzo[b]thiopyrans

The synthesis of a number of 2,2- and 4,4-dimethyl and -dialkyl derivatives of thiochroman have been described and are reported below. 7-Alkyl-4-(alkylimino)-5-hydroxy-2,2-dimethylthiochromans, for example, 1 (R = alkyl) have been obtained from the parent benzothiopyran-4-one and the appropriate amine (A. Arnoldi *et al.*, J. Med. Chem., 1990, **33**, 2865). Nitrate esters of the general type 2 of 2,2-dialkyl-3-hydroxythiochromans have been prepared and investigated as antihypertensives and bronchodilators (J.M. Evans and C.S.V. Frydrych, PCT Int. Appl. WO 89 05,808, 1989). 4-Aminothiochromans have also been prepared for a variety of purposes (K. Atwal, Eur. Pat. Appl. EP 462,761, 1991; K. Atwal, Z.A. Syed, and P.S. Dinos, *ibid.*, 587,188, 1994; K. Atwal and G.C. Rovnyak, *ibid.*, 587,180, 1994; T. Yamanaka *et al.*, Jpn. Kokai tokkyo Koho JP 05 32,655 [93 32,655], 1993).

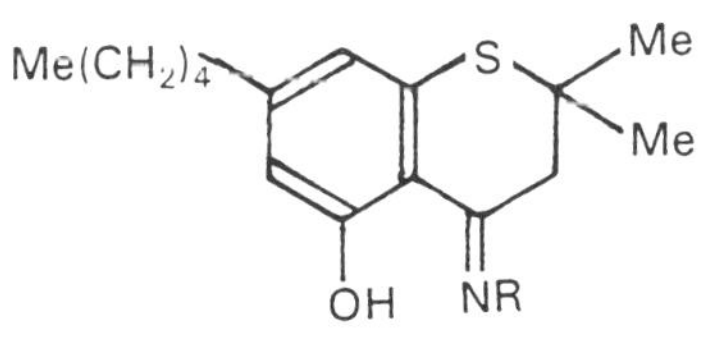

(1)

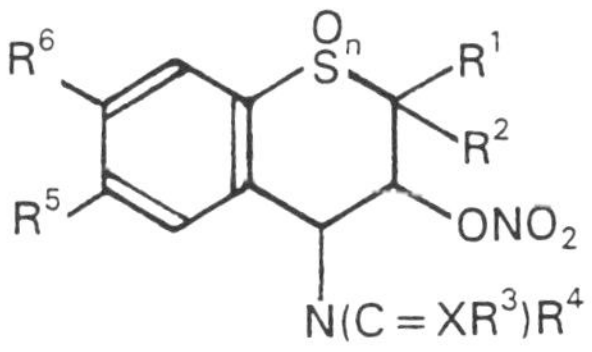

(2)

2-Phenyl-3-phenylthiothiochroman 3 has been prepared by the reaction between PhSH and $PhC \equiv CCH_2OH$ in $ClCH_2CH_2Cl$ in the presence of 4--$MeC_6H_4SO_3H$ at 80°C. Other analogues have been obtained by similar methods (Y. Ishino, T. Masaoka, and T. Hirashima, Chem. Letters, 1990, 1185; Synlett, 1991, 633).

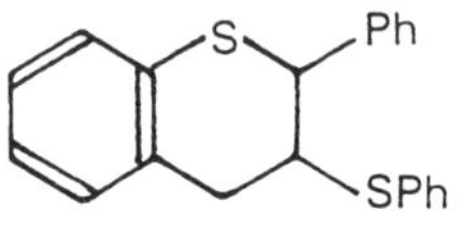

(3) 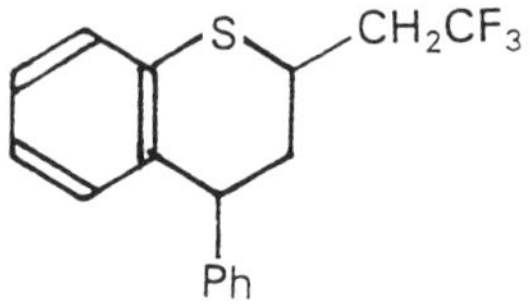   (4)

The O,S-acetal, $PhSCH(OAc)CH_2CF_3$ reacts with styrene to furnish 2--(2,2,2-trifluoroethyl)-4-phenylthiochroman (4) in excellent yield (T. Fuchigami, K. Yamamoto, and H. Yano, J. Org. Chem., 1992, *57*, 2946).

Thiochroman-4-carbonitriles 6 (R = H, CN, CO_2Me) have been obtained by the thermal decomposition of 2-azidobenzo[b]thiophene 5 in the presence of various (*E*)- and (*Z*)-alkenes. The *o*-quinoidal enethione 7 may be an intermediate (M. Funicello, P. Spangnolo, and P. Zanirato, J. Chem. Soc., Perkin 1, 1990, 2971). For the preparation of 4-cyanothiochromans 8, see M. Smid and J. Taimr, Cesk. Farm., 1990, *39*, 289; G. Butora *et al.*, Czech. CS 266,192, 1990; for methods for the synthesis of 3-aminothiochroman derivatives with 5-HT_{1A} receptor activity, G. Guillaumet *et al.*, Eur. Pat. Appl. EP 521,766, 1993; and 4-(aminomethyl)thiochroman derivatives as selective *σ*-adrenergic antagonists, J.F. De Bernardis, D.L. Arendsen, and R.E. Zelle, *ibid.*, 325,964, 1989.

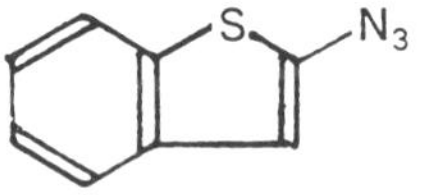

(5) 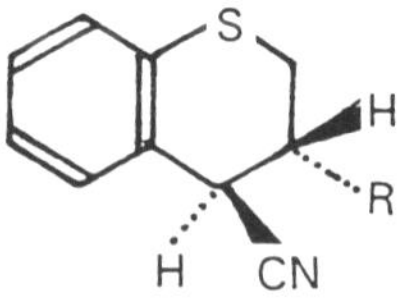   (6) 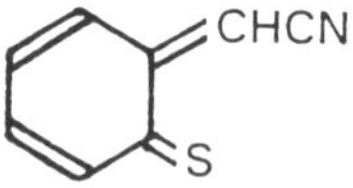   (7)

R = H,Me

(8)

5-Methoxythiochroman-3-carboxylic acid is obtained by heating 3-
-(MeO)C_6H_4SCH$_2$C(CO$_2$H)=CH$_2$ at 200°C in 2-ClC$_6$H$_4$Cl containing Et$_3$N.
On boiling with (PhO)$_2$P(O)N$_3$ in Bu^{t}OH containing Et$_3$N it gives 9 which can
be converted in four steps to 5-ethoxy-3-dipropylaminothiochroman 10. A
number of related derivatives have been synthesized and tested as neuro-
transmitter agonists (A.J. Hutchinson, *ibid.*, 280,269, 1988). A number of
derivatives of 6-hydroxy-5,7,8-trimethylthiochroman-2-carboxylic acid 11
(R^1=H, lower alkyl; R^2=H, protective group) and its 4-one have been used
as intermediates for the synthesis of thiovitamin E (*a*-thiotocopherol) (K.
Matsuo, S. Sakane, and M. Shiono, Jpn. Kokai Tokkyo Koho JP   02,
212,483 [90,   212,483], 1990).

(9)   (10)   (11)

Thiochroman-8-ol has been condensed with epichlorohydrin and the
product condensed with 1-(2-methoxyphenyl)piperazine to yield the 8-
-(aminoalkoxy)thiochroman 12 (G. Guillaumet *et al.*, Eur. Pat. Appl. EP
571,243, 1993).

(12)

The reaction of 3-phenylthiopropionaldehyde 13 and its methyl derivatives with a variety of thiols in acetonitrile at room temperature in the presence of sulphuric acid gives the corresponding 4-arylthio- and alkylthio-thiochromans 14 (R^1 = H, Me, Cl; R^2,R^3 = H, Me; R^4 = Ph, 4-MeC$_6$H$_4$, Bu) (T. Nakazawa, K. Sato, and K. Itabashi, Nippon Kagaku Kaishi, 1990, 115).

(13)   (14)

(ii) Thiochromanones, 3,4-dihydro-2H-benzo[b]thiopyranones

A number of derivatives 1 [R^1,R^2-H, alkyl; $R^1R^2 = C_{2-7}$ polymethylene; R^3 = H, (substituted) alkyl, aryl; R^4 = H, alkyl; or R^3R^4 = linking chain; R^5,R^6 = H, (substituted) alkyl, alkoxy, fluoroalkoxy; X = O,S] of thiochroman--3-one possessing smooth muscle relaxant activity have been prepared (C.S. Frydrych and J.M. Evans, Eur. Pat. Appl. EP 366,273, 1990).

(1)

The alkylation of thiochroman-4-one oxime with *N,N*-diethylaminoethyl chloride in boiling EtOH and/or MeOH in the presence of an alkali metal alcoholate affords 4-[2-(*N,N*-diethylamino)ethoxyimino]thiochroman 2, the hydrochloride of which acts as a fungicide (W. Basinski, Pol. PL 147,859, 1989). The preparations of a number of derivatives of (dioxocyclohexanecarbonyl)thiochroman 1,1-dioxide useful as herbicides have been reported, for example, 4-methoxyimino-5,8-dimethyl-6-(2,6-dioxocyclohexylcarbonyl)-thiochroman 1,1-dioxide (3) (I. Nasuno *et al.*, PCT Int. Appl. WO 94 08,988, 1994).

(2)

(3)

The action of DMF-POCl$_3$ on thiochroman-4-one yields 4-chloro-3-formyl-
-2*H*-thiochromen 4, but with excess DMF-POCl$_3$ at 100°C 3-formylthio-
chromone 5 is obtained (P.R. Giles and C.M. Marson, Tetrahedron, 1991,
<u>47</u>, 1303).

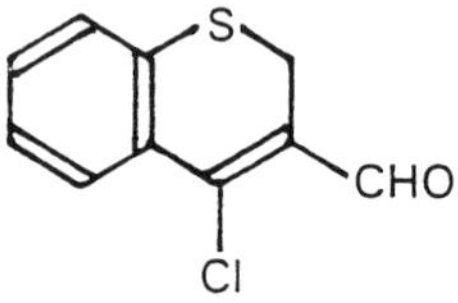

(4) 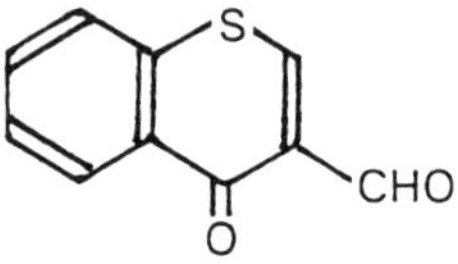   (5)

Treatment of 3-(hydroxymethylene)thiochroman-4-one (6) with *N*-
-chlorosuccinimide provides efficient routes to 3-chlorothiochroman-4-one
(7), 3-chlorothiochromone (8), and 2,3-dichlorothiochromone (9). The
products formed are rationalized in terms of *C*-chlorination, in contrast to *S*-
-chlorination which has been postulated for the action of *N*-chlorosuccinimide
on thiochroman-4-one (*idem*, Austral. J. Chem., 1992, <u>45</u>, 439).

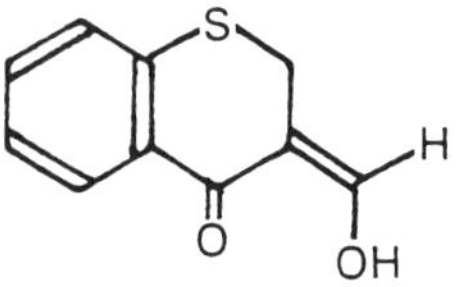

(6) 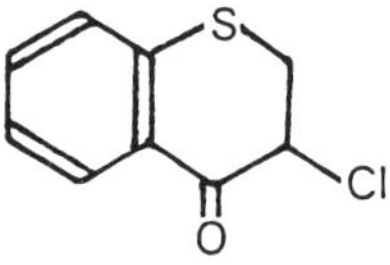   (7) 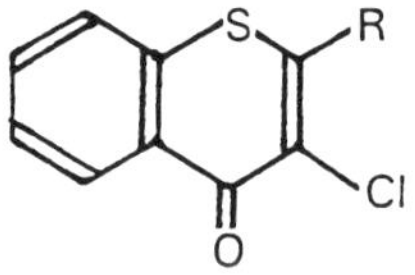   (8) R = H
                                                      (9) R = Cl

The addition of bromine to the exocyclic α,β-unsaturated ketone 10 affords 2,3-dibromo-3-(4-methylphenylbromomethyl)-2-phenylthiochroman-4--one (11) (G. Toth, F. Janke, and A. Levai, Liebigs Ann., 1989, 651).

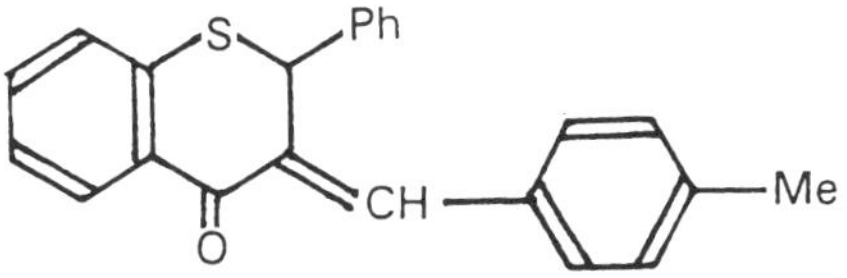

(10) 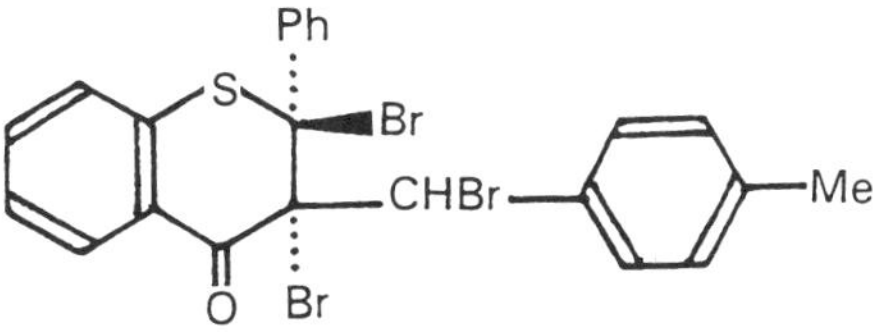 (11)

2,2-Dimethyl- and 2,2,6-trimethylthiochroman-4-one react with an excess of thionyl chloride to yield the α-chlorosulphenyl chlorides 12 (R^1 =H, Me), which when treated with secondary amines form the sulphenamides 13 (R^2 = Et$_2$N, morpholino, pyrrolidino). On hydrolysis the sulphenamides 13 undergo ring contraction to give benzo[b]thiophen-3-ones 14 (C.D. Gabbutt et al., Tetrahedron, 1994, 50, 827).

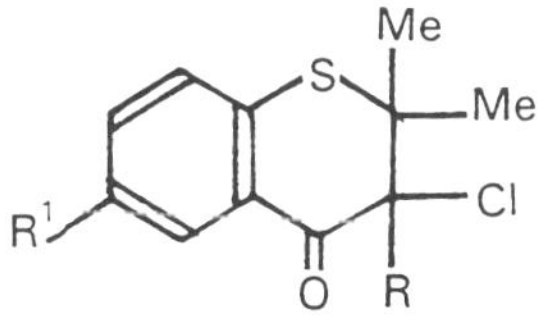

(12) R = SCl
(13) R = SR^2

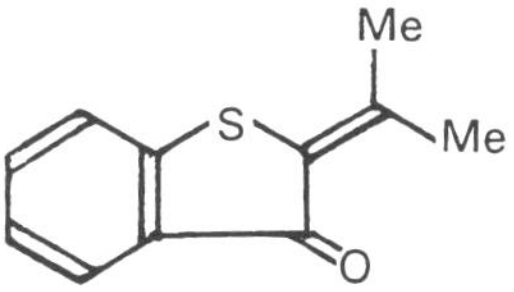

(14)

The preparations of 3-chloro- and 3,3-dichlorothiochroman-4-ones have been described (C.D. Gabbutt, J.D. Hepworth, and B.M. Heron, *ibid.*, p5245).

3,4-Dihydro-2*H*-benzo[*b*]thiopyran-4-ols 16 (NR_2^1 = Me_2N, Et_2N, R^2 = MeO, R^3 = Ph; R_2^1 N = pyrrolidino, R^2 = R^3 = H) have been prepared as potential anti-estrogenic agents, for instance, cyclization of $3\text{-MeOC}_6\text{H}_4\text{SCH}_2\text{CHPhCO}_2\text{H}$ gives 7-methoxy-3-phenylthiochroman-4-one 15, which undergoes a Grignard reaction with $4\text{-Me}_2\text{NCH}_2\text{CH}_2\text{OC}_6\text{H}_4\text{MgBr}$ to yield 7-methoxy-4-(2--dimethylaminoethoxyphenyl)-3-phenylthiochroman-4-ol (16; NR_2^1 = NMe_2, R^2 = MeO, R^3 = Ph) (V.L. Pruitt, K.D. Berlin, and K.S. Hirsch, Org. Prep. Proced. Int., 1990, 22, 235).

(15)

(16)

Heating $4\text{-MeCOC}_6\text{H}_4\text{S(CH}_2)_3\text{CO}_2\text{H}$ in polyphosphoric acid at 120°C gives 7-acetyl-3,4-dihydro-2*H*-benzothiepin-5-one, and the product obtained by reacting it with SO_2Cl_2 under reflux, gives on treatment with Et_3N in CH_2Cl_2 the acylcyclopropylbenzothiopyranone 17 (R = Me). Related derivatives 17 (R = C_{1-5} linear or branched alkyl, C_{3-8} cyclo alkyl, C_{6-10} aryl) have been synthesized as ulcer inhibitors (M. Okitsu, N. Taniguchi, and K. Yoshida, Jpn. Kokai Tokkyo Koho JP 02 142,784 [90 142,784], 1990).

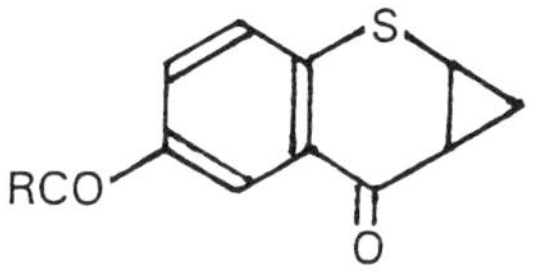

(17)

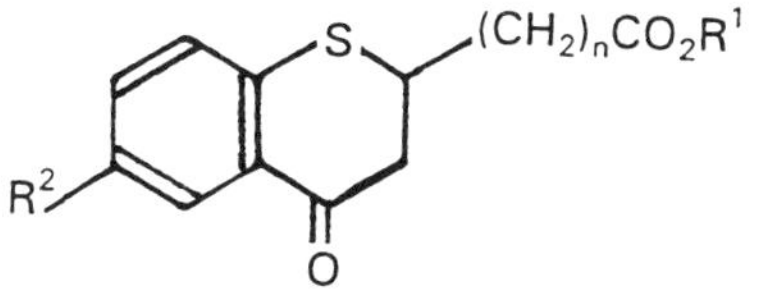

(18)

4-Phenoxyphenylthio)succinic anhydride on treatment with $AlCl_3$ at room temperature gives 6-phenoxythiochroman-4-one-2-carboxylic acid (18; $R^1 = H$, $R^2 = PhO$, $n = 0$). A number of related derivatives have been prepared as hypolipemics (K. Tomizawa, K. Kameo, and K. Hatayama, Jpn. Kokai Tokkyo Koho JP 02 255,677 [90 255,677], 1990).

The preparation of thiochroman-4-ones 19 ($R^1 = H$, 5-, 6-, 7-, 8-Me, 8-Me-6-Bu^t, 5,6-benzo, 7,8-benzo, 6-Cl, 5,7-Me$_2$, $R^2 = H$, Me) from the corresponding thiophenols 20 and 3-methylbut-2-enoic acid in the presence of methane sulphonic acid is accompanied by *tert*-butylation of the aromatic ring (S.E. Clayton *et al.*, Tetrahedron, 1993, <u>49</u>, 939).

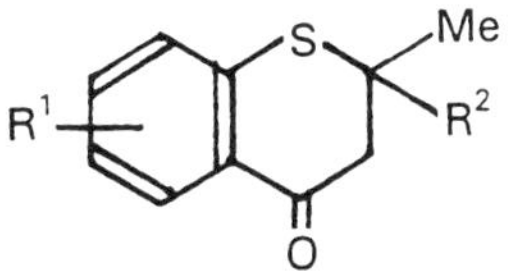

(19)

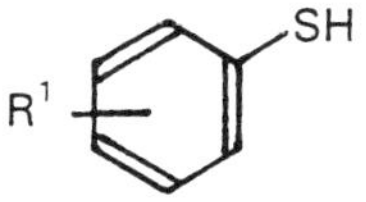

(20)

604

For the preparation of heterocyclylmethoxy substituted thiochroman-4-ones
as drugs see J.F. Eggler, L.S. Melvin Jr., and A. Marfat, Eur. Pat. Appl. EP
313,296, 1989; and of *N*-amidino-*N'*-benzothiopyranylhydrazones as
cardiovascular agents, A.-G. Bayer, Chem. Abst. 1988, 109, 92711d; for
the hydrogenation of a number of thiochroman-4-one derivatives to their
corresponding 4-alcohol catalyzed by BINAP-iridium(I)-aminophosphine
systems see X. Zhang *et al.*, J. Amer. Chem. Soc., 1993, 115, 3318); and
the formation of thiochroman-3,4-diones by the reaction of thiochroman-4-
-ones with isoamyl nitrile, C.D. Gabbutt, J.D. Hepworth, and B.M. Heron,
Tetrahedron, 1994, 50, 7865.

Thioflavanone (2-phenylthiochroman-4-one, 2-phenyl-3,4-dihydro-2*H*-
-benzo[*b*]thiopyran-4-one) 21 reacts with aromatic aldehydes, for example,
4-RC$_6$H$_4$CHO (R-H, OH, OMe, Pr^i, Cl, F) in the presence of piperidine as a
catalyst to yield 3-arylidenethioflavanones 22, but with 4-RC$_6$H$_4$CHO (R = Br,
CN, NO$_2$) under the same conditions 3-benzylthioflavones 23 are obtained
(A. Levai, A. Szollosy, and G. Toth, Acta Chim. Hung., 1991, 128, 359).

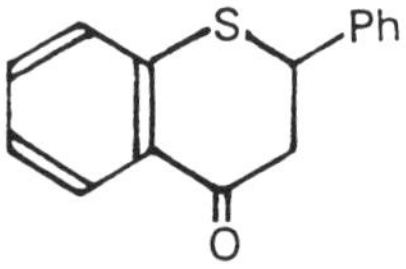

(21)

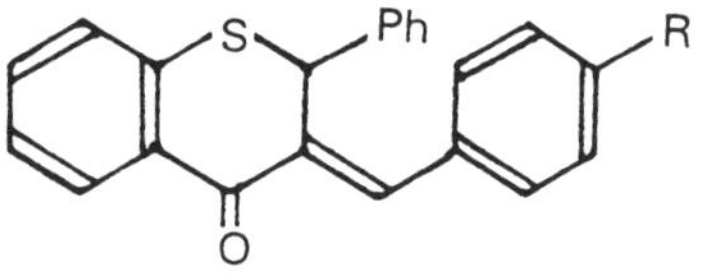

(22)

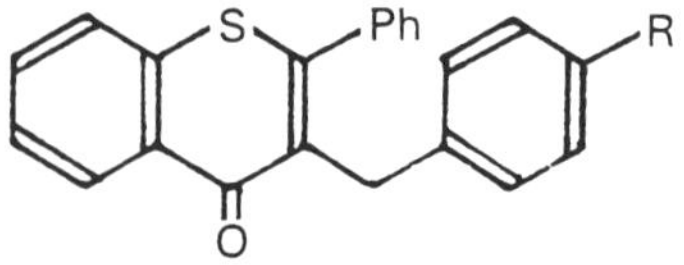

(23)

(iii) Thiochromanols, 3,4-dihydro-2H-benzo[b]thiopyranols
Oxidation/epoxidation of 6-cyano-2,2-dimethylthiochromen 1 with 70%
3-ClC$_6$H$_4$CO$_2$OH and 4,4'-thiobis(6-*tert*-butyl-3-methylphenol) in CH$_2$Cl$_2$
affords the epoxide-dioxide 2, which on treatment with Me$_3$SiN$_3$ and
(Me$_2$CHO)$_4$Ti in THF yields the azide 3. Conversion of azide 3 to the amino
derivative 4 is accomplished by dissolving it along with Ph$_3$P in THF and
treating with NH$_4$OH. The reaction of amino derivative 4 with Cl(CH$_2$)$_3$COCl
and Et$_3$N in CH$_2$Cl$_2$ furnishes 4(4ß-chloro-1-oxobutylamino)-6-cyano-2,2-
-dimethylthiochroman-3α-ol 1,1-dioxide [5; R^1 = OH, R^2 = R^4 = H,
R^3 = Cl(CH$_2$)$_3$, R^5 = CN]. A number of derivatives 5 (R^1 = OH; R^2 = H;
R^{1}R^2 = bond; R^3 = alkyl; R^4 = H; R^{3}R^4 = alkylene, alkenylene; R^5 = H,
halogeno, CN, alkoxy, arylsulphonyloxy) have also been prepared and found
to be effective as coronary vasodilators (R. Tsuzuki *et al.*, Jpn. Kokai Tokkyo
Koho JP 03 112,984 [91 112,984], 1991).

(1)

(2)

(4)

(3)

(5)

Thiochroman-4-one has been reduced by *Mortierella isabellina* ATCC 42613 to give (S)-thiochroman-4-ol in high yield and enantiomeric excess and the (R) enantiomer is obtained by bioconversion of thiochroman-4-one or (±)-thiochroman-4-ol by *Helminthosporium* species NRRL 4671 (H.L. Holland, T.S. Manoharan, and F. Schweizer, Tetrahedron Asymmetry, 1991, 2, 335).

1-*tert*-Butylamino-3-(4-hydroxy-8-thiochromanyloxy)propan-2-ol (6) has been prepared as a ß-blocking agent, by reducing 8-methoxythiochroman-4--one with $NaBH_4$ and demethylating the product with Me_2CHSNa. The

606

resulting diol was boiled with 1-chloro-2,3-epoxypropane in Me_2CO containing K_2CO_3 to give an epoxy ether, which reacts with Bu^tNH_2 in hot Pr^iOH to afford the propan-2-ol 6 (B. Marchand and Y.M. Gargouil, Fr. Demande FR 2,588,260, 1987).

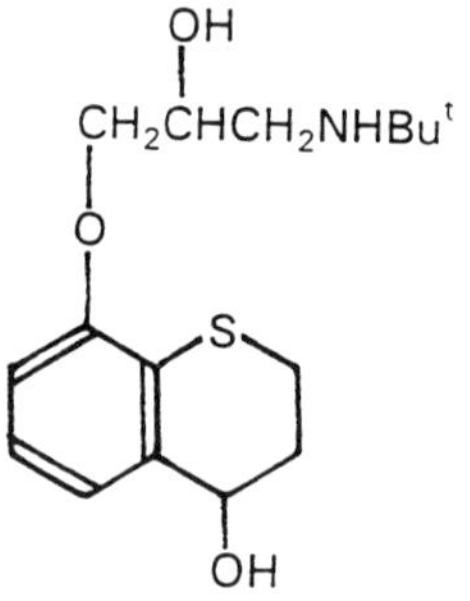

(6)

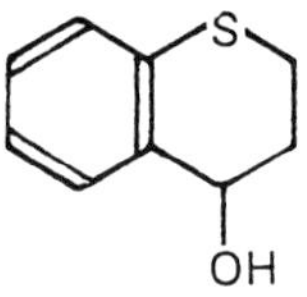

(7)

    Thiochroman-4-ol 7 with a number of substituents have been prepared (K. Koch, PCT Int. Appl. WO 93 15,067, 1993; K. Koch and L.S. Melvin, *ibid.*, 93 15066, 1993).

    Some 6-[dioxocyclohexyl)carbonyl] thiopyran-4-ol derivatives 8 ($R^1 = C_{1-6}$ alkyl; $R^2 = C_{1-4}$ alkyl; R^3, R^4, $R^5 = C_{1-4}$ alkyl; n = 0-2) have been prepared as herbicides, for example, 8 ($R^1 = Et$, $R^2 = Me$, $R^3 = 8$-Me, $R^4 = R^5 = H$, n = 2) has been shown to kill *Setaria viridis* and *anthium strumarium* (M. Sakamoto *et al.*, *ibid.*, 94 04,524, 1994).

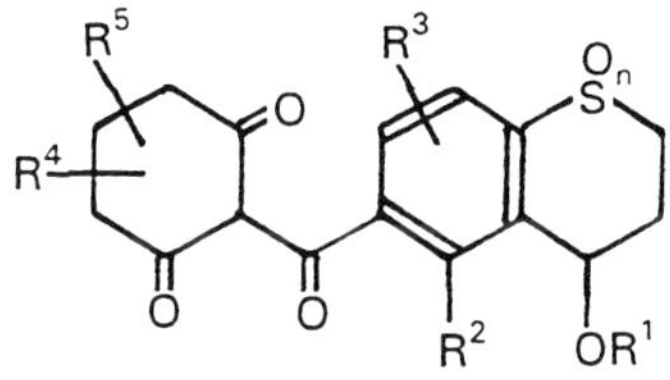

(8)

(iv) Hexahydrobenzo[b]thiopyrans, hexahydro-1-benzothiopyrans, hexa-hydrothiochromens and octahydrobenzo[b]thiopyrans, octahydro-1--benzothiopyrans, octahydrothiochromens, hexahydrothiochromans
Hexahydrobenzo[b]thiopyrans. 4-Aryl-2-phenyl-4a,5,6,7,8,8a-hexa-hydro-4*H*-benzo[*b*]thiopyran-8a-thiols (4-aryl-2-phenyl-1-thiabicyclo[4.4.0]-dec-2-ene-10-thiols) 2 (R = H, Ph, 4-MeOC$_6$H$_4$) are prepared simply by the reaction of diketones 1 with H$_2$S in formic acid in the presence of concentrated hydrochloric acid (B.L. Drevko *et al.*, U.S.S.R. SU 1,705,295, 1992).

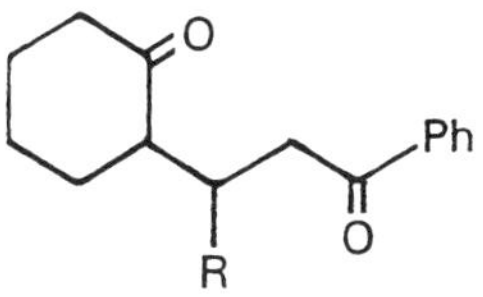

(1) 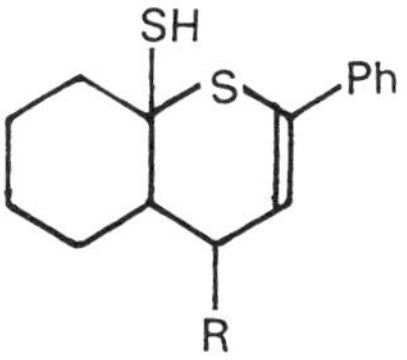 (2)

Treatment of ketones 3 (R = H, Me, Cl, MeO) with Lawesson's reagent affords dimers of the corresponding thioketones, which react with norborn-ene to give 2-cyclopropyl-4-aryl-5,8-methano-4a,5,6,7,8,8a-hexahydro-4*H*--benzo[*b*]thiopyrans 4 (S. Satsumabayashi and T. Yamaguchi, Nippon Shika Daigaku Kiyo, Ippan Kyoiku-kei, 1992, 21, 83).

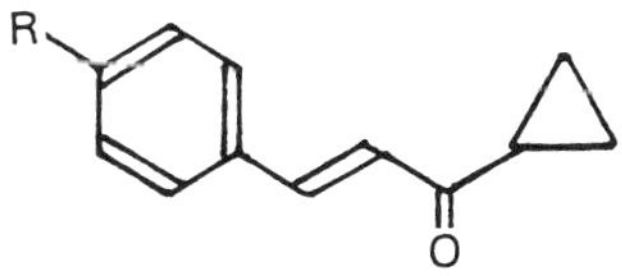

(3)

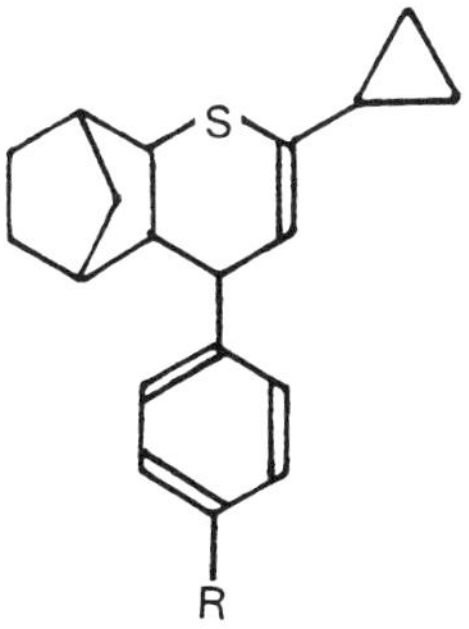

(4)

The intramolecular Diels-Alder reaction of (E)-CH_2=CHCH=CH$(CH_2)_4$-C(S)H yields 4a,5,6,7,8,8a-hexahydro-2H-benzo[b]thiopyran (1-thiabicyclo-dec-3-ene) 5 in a 2:1 *cis-trans* ratio (E. Vedejs, T.H. Eberlein, R.G. Wilde, J. Org. Chem., 1988, $\underline{53}$, 2220).

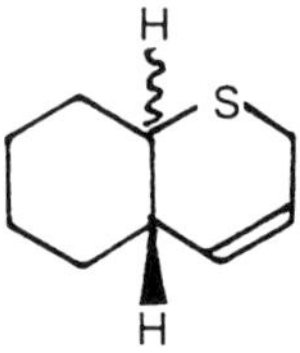

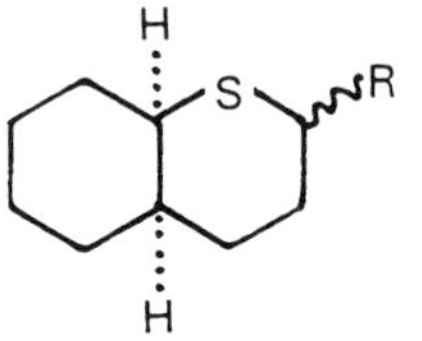

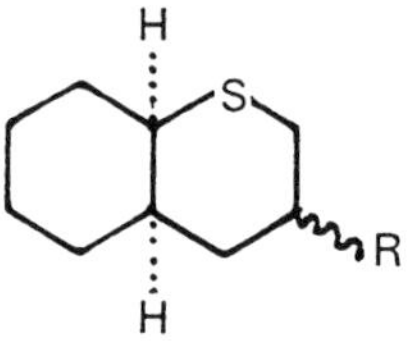

(5) (6) (7)

Perhydrobenzo[b]thiopyrans, 1-thiadecalins. *Cis*- and *trans*-2- and 3-
-alkyl-*cis*-1-thiadecalins 6 (R=H, Me) and 7 (R=Me, Pr) have been prepared
and subjected to conformational analysis (N.P. Volynskii, L.A. Zegelman, and
M.B. Smirnov, Izv. Akad. Nauk SSSR, Ser. Khim., 1989, 393).

Dialkylaminoacetates of *trans*-2e-methylhexahydrothiochroman-4-ols 8
(NR$_2$=Me$_2$N, Et$_2$N, piperidino, morpholino, n=0) and their sulphones 8
(n=2) have been prepared from the corresponding hexahydrothiochroman-4-
-ols and sulphones by esterification with ClCH$_2$COCl, followed by amination
with R$_2$NH (V.V. Kokhomskaya *et al.*, Vestsi Akad Navuk BSSR, Ser. Khim.
Navuk, 1987, 59).

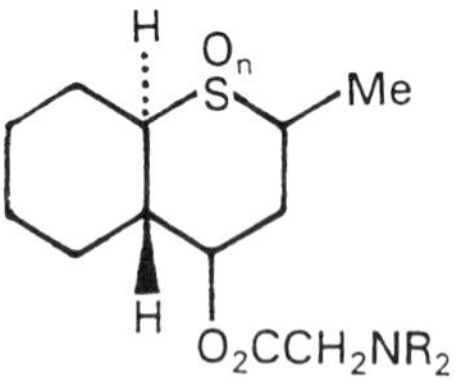

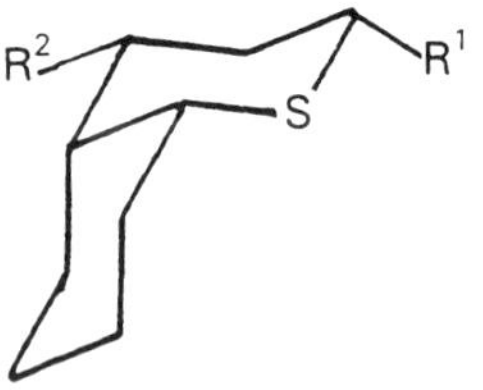

(8) (9)

Oxidation of *cis*-1-thiadecalins 9 (R^1 = Ph, R^2 = H, Ph, 4-MeOC$_6$H$_4$, 3,4--(MeO)$_2$C$_6$H$_3$; R^1 = 4-MeOC$_6$H$_4$, R^2 = H, Ph, 4-MeOC$_6$H$_4$) by H$_2$O$_2$-AcOH for 24h at 20°C affords first the corresponding sulphoxides then the sulphones (S.K. Klimenko *et al.*, Khim. Geterotsikl. Soedin., 1987, 463).

The synthesis of *trans*-2e-methyl-4-[3(dialkylamino)propyn-1-yl]hexahydrothiochroman-4-ols, their sulphoxides and sulphones (A.A. Akhrem *et al.*, Vestsi Akad. Navuk BSSR, Ser. Khim. Navuk, 1990, 59), *trans*-2e--methyl-4e-(4-dialkylaminobutyn-2-yl)hexahydrothiochroman-4a-ols, their sulphoxides and sulphones (*idem, ibid.*, p.82), and *trans*-2-methyl-4-(2--propynyl- and 2-propenyl)hexahydrothiochroman-4-ols and *trans*-2-methyl--4e-(2-propynyl)hexahydrothiochroman-4a-ol sulphoxides have been reported (*idem, ibid.*, p.61). Some of these compounds showed radioprotective activity.

The *trans*-1-thiadecalin-3,8-dione 12 and the *trans*- and *cis*-thiaindandiones 13 and 14 have been prepared from the corresponding a-thioenones 10 and 11 by a proline-catalyzed intramolecular Michael process. The e.e. of the reaction is, however, only 19-28%. Reduction of thiadecalindione 12 by actively fermenting baker's yeast provided exclusively the product 15. Similarly *trans*-thioindandione 13 afforded product 16 (A.P. Kozikowski and B.B. Mugrage, J. Org. Chem., 1989, <u>54</u>, 2274).

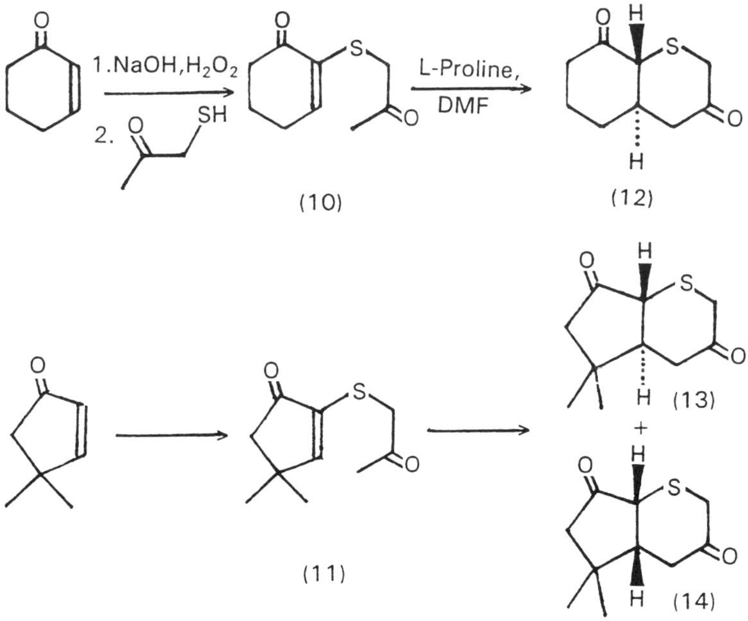

610

(12) → [Baker's Yeast] → (15)   (13) → [Baker's Yeast] → (16)

3. 1*H*-Benzo[*c*]thiopyran, 1*H*-2-benzothiopyran, isothiochromene, 3,4--benzothiopyran and its derivatives

(a) 1H-Benzo[*c*]thiopyran, isothiochromene and derivatives

The stable and crystalline benzo[*c*]thiopyran derivatives 1 (2-thianaphthalenes) (R = CN, Bz) are prepared in high yields by proton abstraction from the corresponding benzo[*c*]thiopyrylium salts on reaction with Et_3N in EtOH. Their ylide nature is supported by spectral and chemical evidence. The reaction of 1 (R = CN) with $Bu^tC \equiv CCO_2H$ in EtOH affords dimethyl 1-cyano--2,3-naphthalene-2,3-dicarboxylate, and in benzene the dihydrocyclopropa-[*a*]naphthalene derivative 2 and 5*H*-benzocycloheptene derivative. Other reactions of benzo[*c*]thiopyran derivatives 1 have been reported (M. Hori *et al.*, J. Chem. Soc., Perkin 1, 1988, 1885).

(1)   (2)

The polar cycloaddition of benzo[c]thiopyrylium ions 3 (2-thianaphthylium ions) (R^1 =H, COPh) with $H_2C = CR^2CR^3 = CH_2$ ($R^2 = R^3 = Me$, H; $R^2 = Me$, $R^3 = H$) yields benzo-fused bicyclic sulphonium salts 4 in excellent yields. The sulphonium salt 4 ($R^1 = COPh$, $R^2 = R^3 = H$) undergoes retro-addition to regenerate the cation, which is easily trapped with buta-1,3-diene. Treatment of 4 ($R^1 = COPh$, $R^2 = R^3 = Me$) with several nucleophiles causes ring opening to give 1-allyl and 1-homoallyl substituted isothiochromenes 5 ($R^4 = CH_2CMe = CMeCH_2R^5$, $R^5 = OH$, OMe, SePh, CN; $R^4 = CH_2CR^5MeCMe = CH_2$, $R^5 = OH$, OMe) (H. Shimizu et al., Tetrahedron Letters, 1991, <u>32</u>, 5571).

(3) (4) (5)

2-$LiCH_2C_6H_4CONEt_2$ on treatment with RC(S)OEt (R = Ph, 4-MeC_6H_4, 4-$MeOC_6H_4$, 4-ClC_6H_4, 1,3-benzodioxol-5-yl, 2-thienyl, Me, cyclohexyl) followed by acid workup affords the 3-aryl- and 3-alkyl-1H-benzo[c]thio-pyran-1-ones 6 (isothiocoumarins) (A. Coutore, H. Cornet, and P. Grandelaudon, Synthesis, 1990, 1133).

(6)

612

(b) 3,4-Dihydro-1H-benzo[c]thiopyran, 3,4-dihydro-1H-2-benzothiopyran, isothiochroman and its derivatives

(S)-(+)-4-Ethyl-3,4-dihydro-1H-benzo[c]thiopyran 1 has been prepared starting with (S)-(+)-PhCHEtCO$_2$H and it has been suggested from ^{1}H NMR and CD spectral data that its conformation is predominantly *cis-ax*-half chair. It has been found that on treatment with HgCl$_2$ it gave predominantly a (+)-[(R)(S)(M)-1, HgCl$_2$]$_2$ complex with the ligand in the (M)-trans-eq-half chair conformation (M. Bavia and Paolo Biscarini, J. Chem. Res. S, 1987, 44).

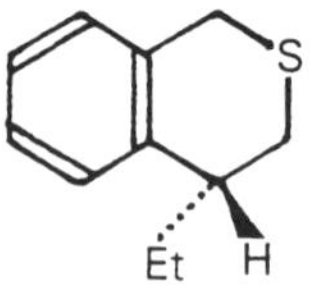

(1) 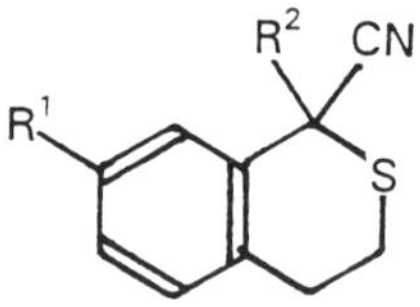   (2) 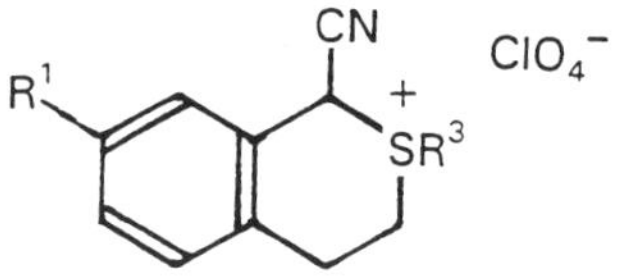   (3)

1-Cyanoisochromans 2 (R^1 = H, Me; R^2 = H) are prepared by the intra-molecular Pummerer reaction of 4-R^{1}C$_6$H$_4$CH$_2$CH$_2$S(O)CH$_2$CN with (CF$_3$CO)$_2$O and on alkylation they give the *S*-alkyl derivatives 3 (R^1 = H, R^3 = Me, Et; R^1 = R^3 = Me). Deprotonation of 3 with Et$_3$N or NaH yields ylides 4. Thermal reaction of ylide 4 (R^1 = H, R^3 = Me) in C$_6$H$_6$ or DMF affords the 1,2-rearranged product 2 (R^1 = H, R^2 = Me) accompanied by dimers 5, whereas reaction in MeCN furnished (E)-5 (31.6%) and (Z)-5 (22.6%) as major products. Some 1-benzylated isothiochromans have been prepared (M. Hori *et al.*, Chem. Pharm. Bull., 1988, <u>36</u>, 1698).

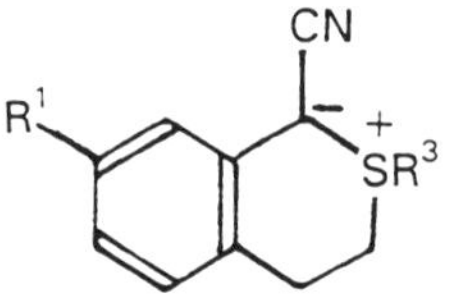

(4) 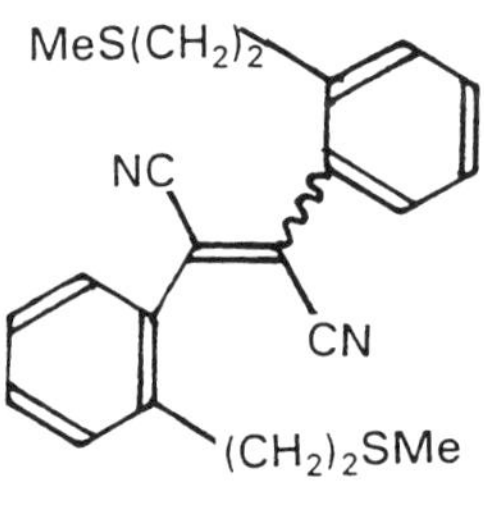   (5)

The novel 10-membered ring 1-thiadec-3,8-diyn-5-ene (6) on treatment with alcoholic potassium hydroxide gives 4-methoxy and ethoxy-isothiochroman (7; R-Me, Et) and with DBU in CCl_4, 8-chloroisothiochromans 8 ($R^1 = H$, $R^2 = CCl_3$; $R^1R^2 = $ bond), 8-chloroisothiochroman-4-one 9 and 8- -chloro-4-dichloromethyleneisothiochroman 10 through a free radical pathway (K. Toshima *et al.*, Tetrahedron Letters, 1991, <u>32</u>, 291).

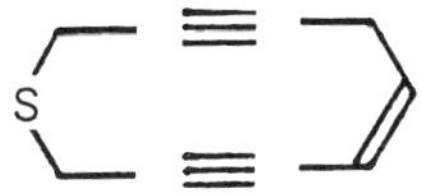

(6)

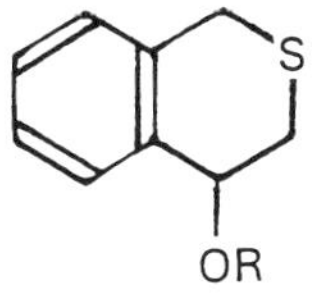

(7)

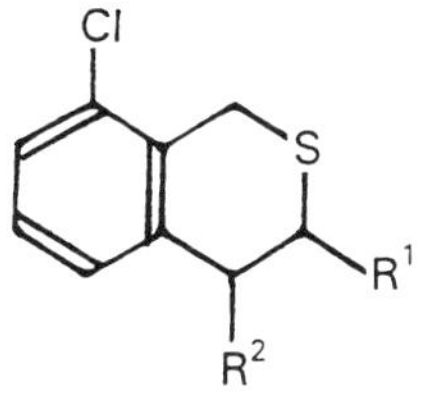

(8)

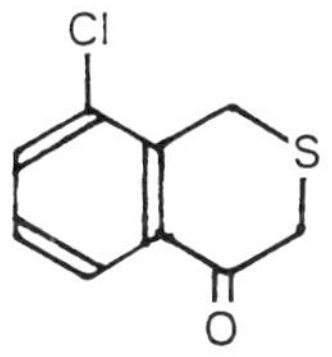

(9)

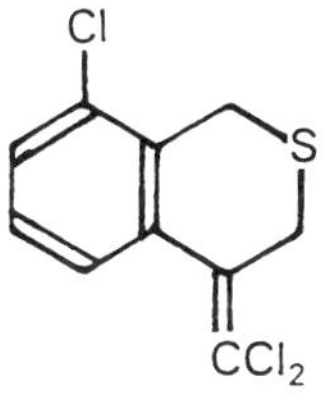

(10)

Methyl and ethyl 5,8-dimethoxyisothiochroman-3-carboxylates (11; R = OMe, OEt) and 3-acetyl-5,8-dimethoxyisothiochroman (11; R = Me) have been prepared in a one-pot process from 3,6-dimethoxy-*o*-xylylene and methyl and ethyl thioglycolate and $BzSCH_2COMe$, respectively. It has been suggested that this method could be extended to the synthesis of 1*H*- -anthraceno[2,3-*c*]thiopyran 12 (Y.C. Xu *et al.*, Synthesis, 1994, 363).

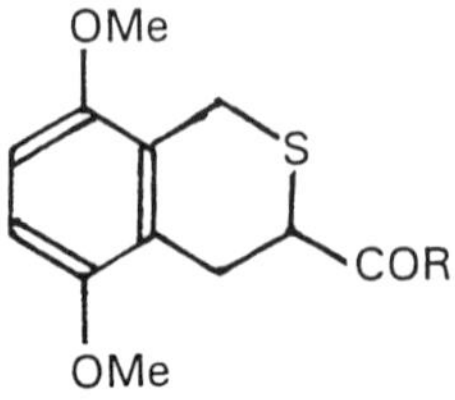

The DDQ induced oxidative coupling of 3-acetyl-5,8-dimethoxyisothio-
chroman (11; R = Me) and alcohols proceeds in high yield with good regio-
and stereochemical control to give trans acetals 13 [R = Me, Pr^i, Bu^t,
(CH$_2$)$_2$NHCO$_2$Bu^t, (CH$_2$)$_3$NHCO$_2$CH = CH$_2$] as exclusive or predominant
products (*idem*, Tetrahedron Letters, 1993, 34, 3841).

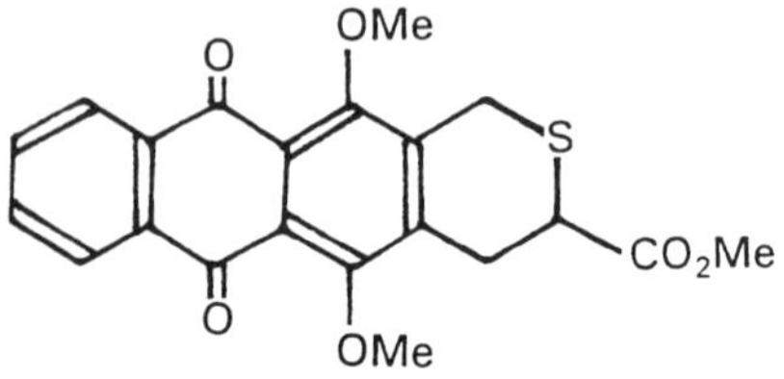

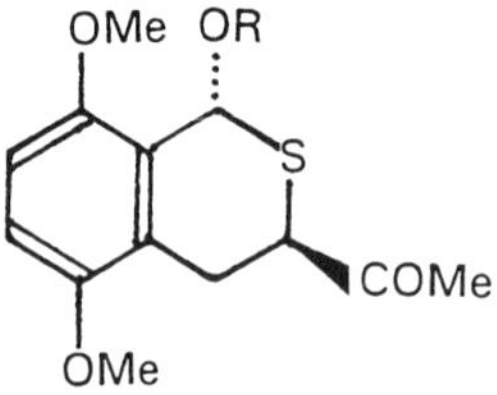

The reaction of methoxycarbonyl(4-chlorophenyl)methylthioacetic acid
with oxalyl chloride in DMF, followed by treatment with AlCl$_3$ in CH$_2$Cl$_2$
yields methyl 6-chloro-4-oxo-3,4-dihydro-1*H*-benzo[*c*]thiopyran-1-carboxy-
late, which on saponification affords 6-chloro-4-oxo-3,4-dihydro-1*H*-benzo-
[*c*]thiopyran-1-carboxylic acid (14). Acid 14 has been tested for the
treatment of osteroporosis (T. Sohda, M. Tsuda, and I. Yamazaki, Eur. Pat.
Appl. EP 376,197, 1990).

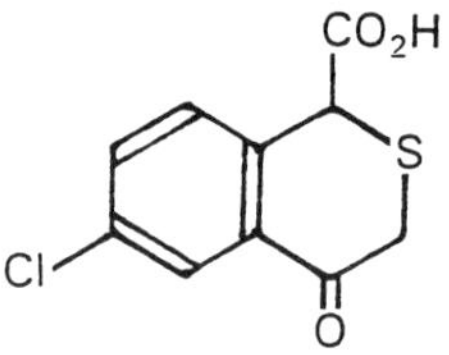

The cyclocondensation of furfurylidenecyclohexylidenemalononitrile 15 with dithietanes 16 (R = alkyl) affords the 5-furfurylidene-5,6,7,8-tetrahydro--1*H*-benzo[*c*]thiopyran-1-ylidenecyanoacetates 17 [K. Peseke, J. Quincoces Suarez, and O. Zayas, Ger. (East) DD 277,461, 1990).

(15)

(16)

(17)

The lithio derivative of sulphide 18 after protonation yields *trans*- (45%) and *cis*-1-phenyl-3,5,6,7,8,8a-octahydro-1*H*-benzo[*c*]thiopyran (15%) 19 and 20. The sulphoxide and sulphone of 18 react similarly and give the bicyclic analogues (M. Reglier and S.A. Julia, Bull. Soc. Chem. Fr., 1990, 236). A number of 3-substituted *trans*-octahydrobenzo[*c*]thiopyrans (*trans*-2-thiadecalins) have been synthesized (N.P. Volynskii and O.L. Alikhanova, Izv. Akad. Nauk SSSR, Ser. Khim., 1991, 1877).

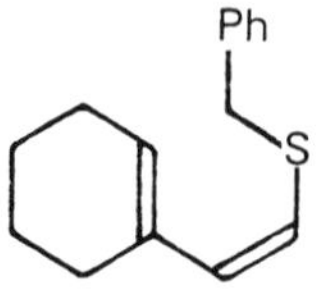

616

(18) 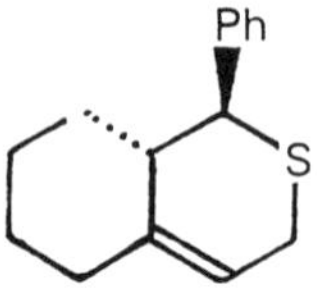 (19) 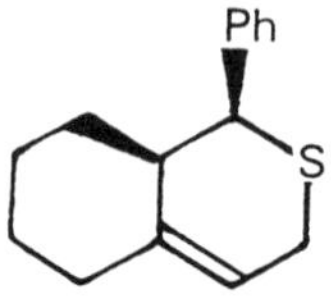 (20)

The reaction of oxathiaspirononanone and oxathiaspirodecanone (21; n = 1 and 2, respectively) with RPh (R = H, Me, Br, Cl, OMe, Ph, 1,4-Me$_2$) under the catalytic action of AlCl$_3$ gives the spiro(cycloalkane-1,1'--isochroman)-4'-ones 22 (M.F. El-Zohry, Phosphorus, Sulphur, Silicon Relat. Elem., 1992, $\underline{66}$, 311).

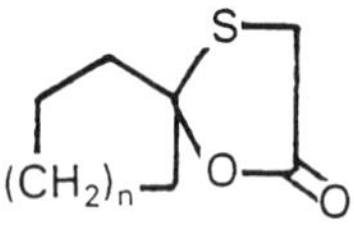

(21) 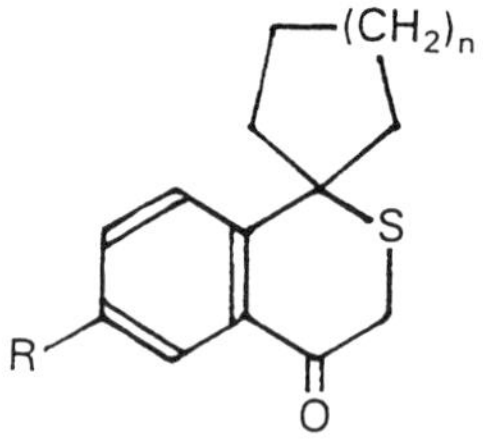 (22)

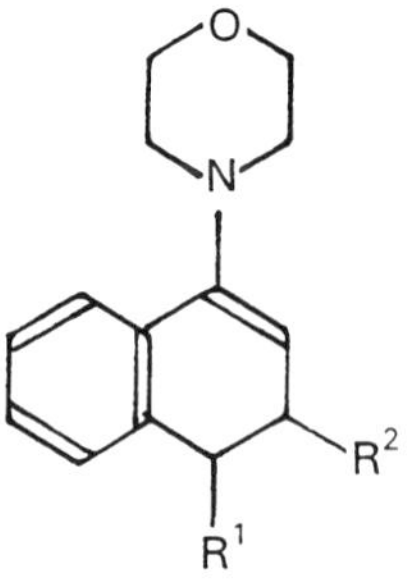

(23) 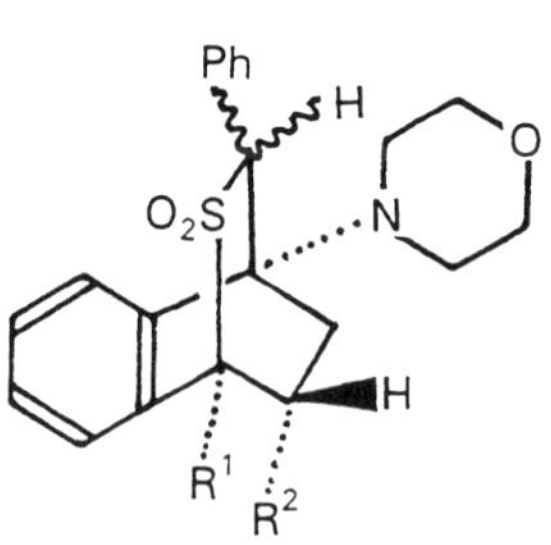 (24)

The [4 + 2] adducts 24 are obtained by the reaction of morpholino enamines 23 (R^1 = H, R^2 = H, Me; R^1 = Me, R^2 = H) of 1-tetralones with PhCH = SO_2 prepared in situ from $PhCH_2SO_2Cl$ and Et_3N, along with [2 + 2] adducts (S. Kitane *et al.*, Bull. Soc. Chim. Belg., 1989, <u>98</u>, 109).

4. Dibenzothiopyran, thioxanthene and derivatives

(a) Thioxanthenes, thioxanthylium salts, and thioxanthenium salts

The reaction between 9-phenythioxanthene 10-oxide 1 and Grignard or organolithium reagents yields 9-substituted 9-phenylthioxanthenes 2, possibly by a mechanism involving a 9-phenylthioxanthylium ion intermediate. The structure of 9-aryl-9-phenylthioxanthenes 2 (R = aryl) have been confirmed by an alternative synthesis *via* acid-catalyzed cyclization of triarylmethanol derivatives, 2-PhSC$_6$H$_4$CRPhOH (H. Shimizu, T. Kataoka, and M. Hori, Chem. Pharm. Bull., 1993, <u>41</u>, 842).

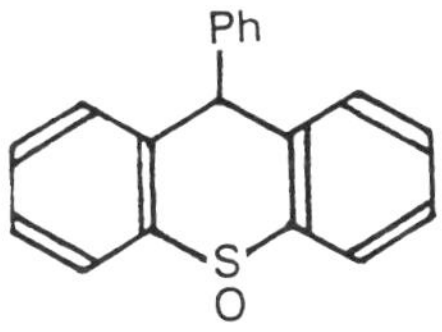

(1)

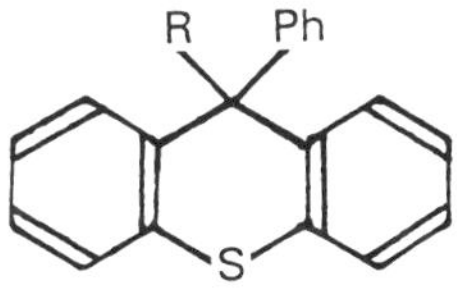

(2)

2,2'-Dimethoxy- and -dimethylbithioxanthylidenes 3 (R^1, R^2 = Me, MeO) have been synthesized and resolved by chiral HPLC and the x-ray molecular structure of 9-(2'-methyl-9'*H*-thioxanthene-9'-ylidene)-9*H*-xanthene confirms the folded form of these molecules. The UV and CD studies of several bithioxanthylidenes have been reported (W.F. Jager *et al.*, Tetrahedron Asymmetry, 1993, <u>4</u>, 1481).

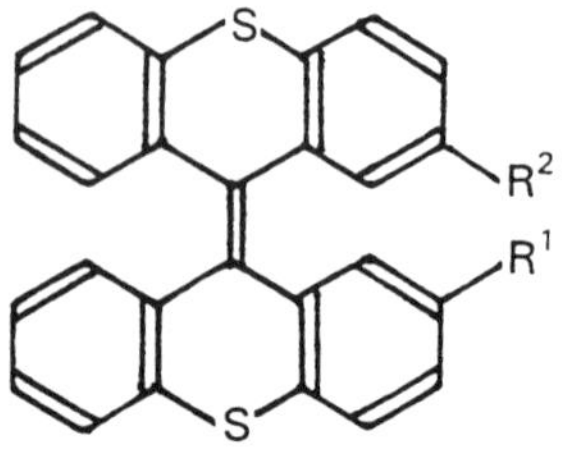

(3)

2,3-Dihydro-7-methylhydrazone-4(1*H*)-phenanthrene has been converted to its diazo compound and cyclocondensed with 2-methoxy-9*H*-thioxanthene-9-thione to afford the thioepoxides 4 (R^1 = MeO, R^2 = H; R^1 = H, R^2 = MeO), which have been desulphurized to yield the methyl(methoxythioxanthylidene)phenanthrenes 5 (R^1 = MeO, R^2 = H; R^1 = H, R^2 = MeO) (B.L. Feringa *et al.*, J. Amer. Chem. Soc., 1991, <u>113</u>, 5468).

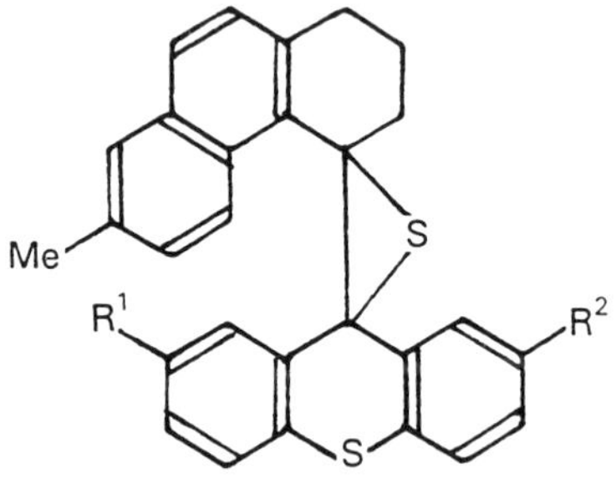

(4)

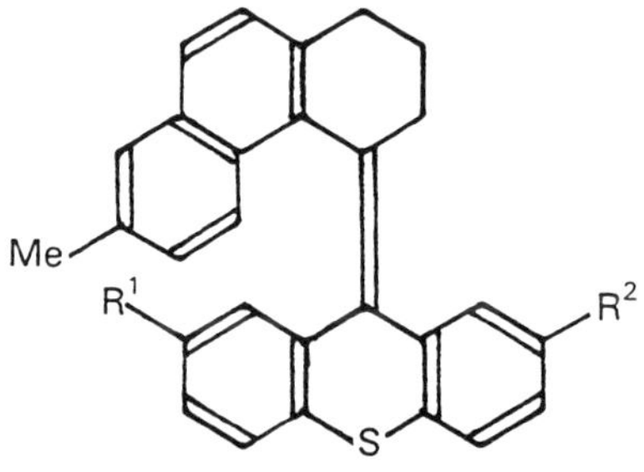

(5)

Treatment of thioxanthene with BuLi in THF affords a dark red solution
of lithium thioxanthenide, which on condensation with hexachlorocyclo-
propane, followed by treatment with BuLi yields tris(thioxanthen-9-ylidene)-
cyclopropane 6. The radical cation of 6 has been obtained by oxidation with
TI $(O_2CCF_3)_3$ and its ESR spectrum has been reported (T. Sugimoto *et al.*,
Angew. Chem., 1988 <u>100</u>, 1129).

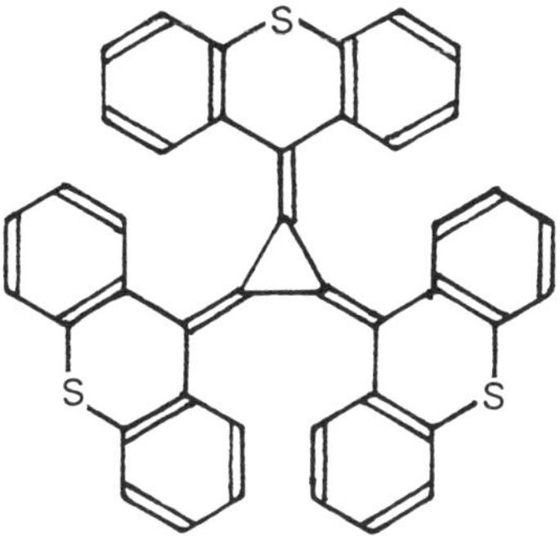

(6)

2,7-Dibromothioxanthene on heating with CuCN and pyridine at 200°C
for 48h gives 2,7-dicyanothioxanthene, which on treatment with HCl-EtOH-
-dioxane at 0°C yields an imino ether. Reaction of the imino ether with
EtOH-NH_3 affords 2,7-diamidinothioxanthene 7, useful in the treatment of
human Kala-azar (P.M.S. Chauhan *et al.*, Indian J. Chem., 1987, <u>26B</u>, 248;
Indian IN 167,210, 1990).

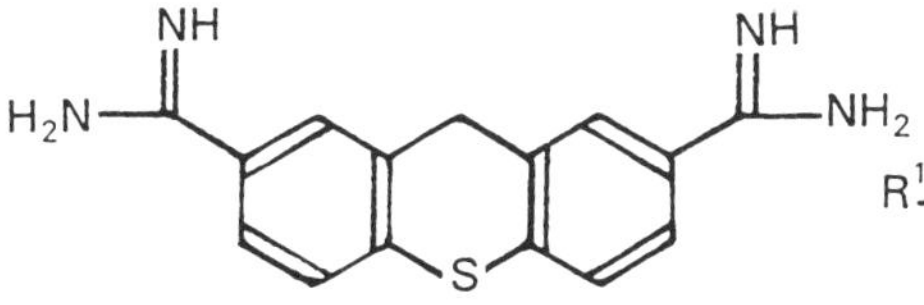

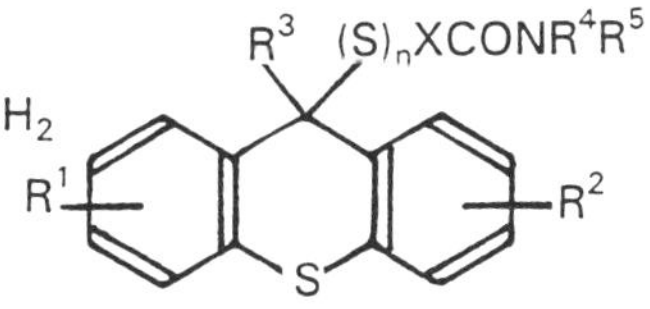

(7)   (8)

620

Several substituted thioxanthenes 8 (R^1, R^2 = H, alkyl, alkoxy, halogeno, halogenoalkyl; R^3,R^5 = H, alkyl; R^4 = substituted heterocyclyl,, naphthyl; X = alkylene; n = 0,1) with a side-chain containing an amido group at the 9--position have been synthesized (A. Yoshida *et al.*, PCT Int. Appl. WO 93 06,096, 1993).

A number of thioxanthene derivatives have been prepared from thio-xanthen-9-ol 9, for example, it condenses with nitriles, R^1CH_2CN (R^1 = CN, CO_2Et) to give thioxanthen-9-ylacetonitriles 10 (R = CHR^1CN) and with thiols R^2SH (R^2 = CH_2CO_2H, Ph, CH_2CO_2Me, $CH_2CONHNH_2$ to afford thioethers 10 [I.A. El-Sakka *et al.*, Arch. Pharm. (Weinheim, Ger.), 1994, <u>327</u>, 133].

(9)　R = OH
(10) R = SR^2

(11) R^1 = OH, R^2 = $(CH_2)_3NMe_2$
(12) R^1 R^2 = $CH(CH_2)_2NMe_2$

A number of 2-(disubstituted amino)-9-(3-dimethylaminopropylidene)thio-xanthenes 12 (R = NMe_2, pyrrolidino, piperidino, morpholino, 4-methylpiper-azino) and their salts, have been prepared by dehydration of the correspond-ing derivative 11 of 9-hydroxy-9-(dimethylaminopropyl)thioxanthene (M. Protiva *et al.*, Czech. CS 247,595, 1988). The preparation of 9-[3--(dimethylamino)propylidene]-2-[(methylthio)methyl]-thioxanthene and its oxalate as tranquilizers has been described (M. Protiva, V. Kmonicek, and S. Wildt, *ibid.*, 236,339).

The reaction of thioxanthen-9-ol 9 with nucleophiles such as pyrazoles, succinimide, phthalimide, and aminotriazoles has been investigated and it has been found that the thioxanthen-9-yl group is introduced in the 4-position of pyrazoles (e.g. 13) on the nitrogen in the carboximides and on the amino group of aminotriazoles (e.g. 14). Compounds 13 and 14 have been tested for molluscicidal and nematocidal activity (I. El-Sakka, I.F. Zeid, and A.E.S. Ahmed, Coll. Czech. Chem. Comm., 1993, <u>58</u>, 2197).

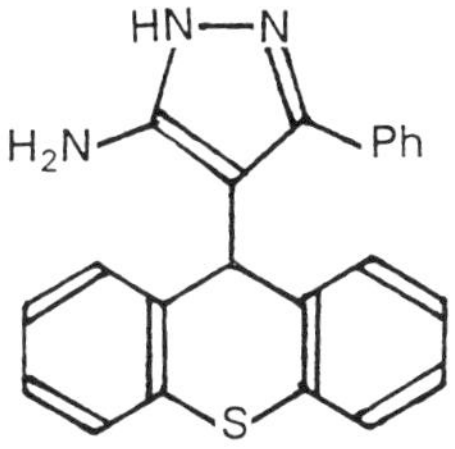

(13)

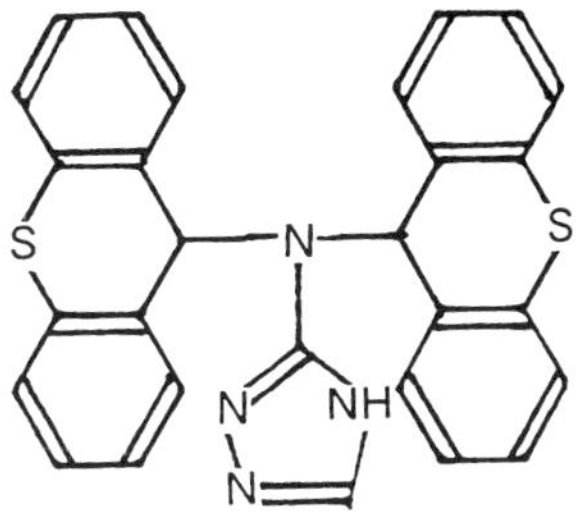

(14)

The preparation of 3,6-diamino-9-arylthioxanthylium salts (thiorhod-amines) 15 ($R^2 = CO_2R^3$; R^1, $R^3 = H$, C_{1-5} alkyl; Z = anion) as antitumor agents and fluorescent dyes has been reported. The starting material for these salts is the corresponding 3,6-diaminothioxanthone 16 (C.H. Chen and J.L. Fox, *ibid.*, 330,444, 1989).

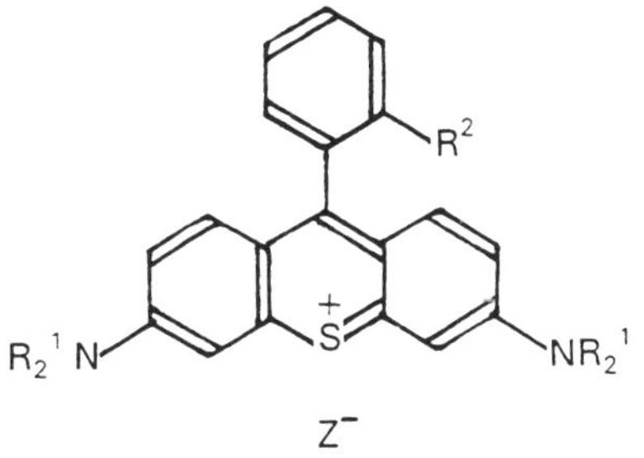

(15)

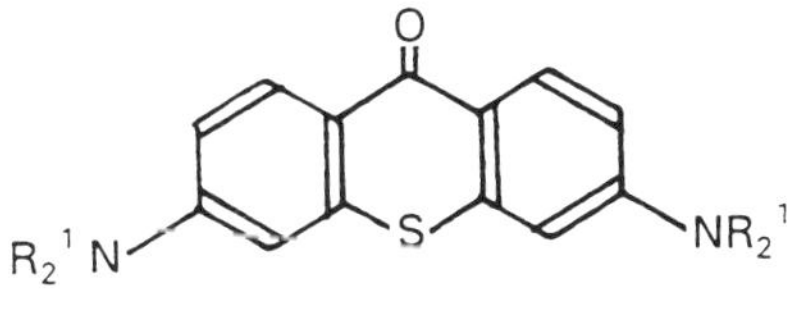

(16)

The hydrogenation of *sym*-octahydrothioxanthylium salts 17 (R=H, Me, Z=BF$_4$; R=Ph, Z=SbCl$_6$) over Pd/C is stereoselective and affords 65-86% *cis*, *syn*, *cis*-perhydrothioxanthenes 18, which have been converted to 10,10-dioxides by treatment with H$_2$O$_2$-AcOH (V.G. Kharchenko *et al.*, Khim. Geterotsikl. Soedin., 1987, 1187).

R

Z$^-$

(17)   (18)   (19)

1,2,3,4,5,6,7,8-Octahydrothioxanthenes 19 (R=H, Ph, 4-MeOC$_6$H$_4$) are conveniently dehydrogenated by 3,5-di-*tert*-butyl-*o*-benzoquinone in HClO$_4$ to give the octahydrothioxanthylium salts 17 (R=H, Ph, 4-MeOC$_6$H$_4$; Z=ClO$_4$). The synthesis of compound 19 (R=Ph) has also been reported (V.T. Abaev *et al.*, *ibid.*, 1991, 51).

The reaction of *sym*-octahydrothioxanthylium salts 20 with bases, for instance, pyridine or an aqueous solution of NaHCO$_3$ in EtOH affords dimer 21 the structure of which has been confirmed by x-ray analysis (V.G. Kharchenko, A.A. Shcherbakov, and Yu. T. Struchkov, *ibid.*, 1989, 479).

Cl$^-$

(20)

(21)

10-Phenylthioxanthenium salts 22 react with aryllithiums to afford 9-
-phenylthioxanthenes 23 in good yields. However 22 ($R^1 = R^2 = R^3 = H$),
when treated with PhLi at -15 to -20°C yields 9-phenylthioxanthen-9-ol (23;
$R^4 = OH$) (17%) together with 9-phenylthioxanthene (23; $R^4 = H$), because of
the lability of 10-phenylthioxanthene (10-phenyl-10-thiaanthracene) to air.
10-Phenyl-9-(4-tolyl)thioxanthene (24; $R^5 = 4\text{-MeC}_6H_4$) generated in situ
from the sulphonium salt 22 ($R^1 = R^3 = H$, $R^2 = 4\text{-MeC}_6H_4$) fails to react with
4-tolyllithium. An isolable ylide, 9-benzoyl-10-phenylthioxanthene (24;
$R^5 = COPh$) has been treated with 4-tolyllithium at 0°C to yield 9-benzoyl-
thioxanthene (82%). On the other hand, treatment of 9,9,10-triphenylthio-
xanthenium salt (22; $R^1 = R^2 = Ph$, $R^3 = H$) with PhLi affords a ring-opened
product, a ring-contracted product, diPh sulphide, and 9,9-diphenylthio-
xanthene (23; $R^4 = Ph$). A number of 9-aryl-10-phenylthioxanthenium salts
22 ($R^1 = R^2 = R^3 = H$; $R^1 = R^3 = H$, $R^2 = COPh$, Ph, substituted Ph; $R^1 = H$,
$R^2 = Ph$, $R^3 = Me$; $R^1 = R^2 = Ph$, $R^3 = H$) have been prepared and their stereochemistry determined by ^1H NMR spectroscopy (M. Hori *et al.*, J. Chem.
Soc., Perkin 1, 1988, 1209).

(22) (23) (24)

10-(4-Methoxyphenyl)-9-phenyl-9*H*-thioxanthene [10-(4-methoxyphenyl)-
-9-phenyl-10-thiaanthracene] generated by proton abstraction of 10-(4-
-methoxyphenyl)-9-phenylthioxanthenium perchlorate (25) with base undergoes 1,4-rearrangement to give 9-(4-methoxyphenyl)-9-phenylthioxanthene
(26) similar to some of the above transformations. A number of related
compounds generated from the corresponding thioxanthenium salts by
treatment with base also decompose thermally to yield the corresponding
1,4-rearranged products. The rearrangement has been shown to be an
intramolecular 1,4-sigmatropic rearrangement (H. Shimizu, T. Kataoka, and
M. Hori, Gifu Yakka Daigaku Kiyo, 1992, 41, 37).

The potassium salts of benzoic acid, 4-nitrobenzoic acid, phthalimide, 1,2-2*H*-benzisothiazol-3-one 1,1-dioxide, 2-acetyl-2,3-dihydro-1*H*-inden-1- -one, and 2-acetyl-1,2,3,4-tetrahydronaphthalen-1-one have been methylated in the solid state with 10-methylthioxanthenium tetrafluoroborate 27. The temporal change of each mixture has been followed by IR spectroscopy and x-ray diffraction analysis at room temperature, 40°C, and 70°C. Peaks attributable to a methylation product, thioxanthene (the demethylated product), and the ion-exchanged product have been indicated (T. Ohashi, K. Tokuno, and Y. Otoda, Yakugaku Zasshi, 1988, <u>108</u>, 733).

(b) Thioxanthones, thioxanthen-9-ones

The molecular structure of thioxanthone has been determined from gas electron-diffaction data and it has been found that the dihedral angle between the two benzene rings is much larger than that of related molecules (K. Iijima *et al.*, Bull. Chem. Soc., Jpn., 1987, <u>60</u>, 3887). 2-Carboxamido- and 2-*O*-carbamoyl-substituted diaryl sulphones, for example, 2-Et₂NCOC₆-H₄SO₂Ph undergo LDA-mediated amide alternative ring migration and cyclization to furnish thioxanthone 10,10-dioxide 1 derivatives (F. Beaulieu and V. Snieckus, J. Org. Chem., 1994, <u>59</u>, 6508).

(1)   (2)

2-Methylthioxanthone (2; R = 2-Me) has been prepared by heating 2-
-mercaptobenzoic acid with toluene in concentrated sulphuric acid at 50-
-100°C. Other mono-substituted derivatives 2 (R = alkyl, OH, alkoxy, NO_2)
of thioxanthone have been synthesized and their IR, UV, and mass spectral
data reported (M. Xue, J. Wang, and G. Qi, Huadong Huagong Xueyuan
Xuebao, 1993, 19, 310). The synthesis, stereochemistry and molecular
spectra of 2-substituted thioxanthones, their oxides, and related ylides have
been discussed (C.P. Yu, Diss. Abstr. Int. B, 1989, 50, 1425).

The Na salt of 4-hydroxythioxanthone reacts with (dimethylamino)alkyl
chlorides to afford ethers 3 [R = $CH_2CH_2NMe_2$, $(CH_2)_3NMe_2$]. Partial
demethylation of 3 [R = $(CH_2)_3NMe_2$] via the carbamate yields 3 [R = $(CH_2)_3$-
NHMe].

Reaction of thioxanthone 3 [R = $(CH_2)_4Br$] with amines yields 3
[R = $(CH_2)_4NHR^1$; R^1 = Me, Bu, cyclohexyl]. These compounds, especially 3
[R = $(CH_2)_4NHMe$] are cyclic analogues of the antidepressant and cerebral
activator bifemelane, but they do not exhibit the pharmacological profile of
antidepressants (I. Cervena et al., Coll. Czech. Chem. Comm., 1988, 53,
1307). A number of related compounds 4 (R^1 = H, OH; R^2 = H, Me;
R^3 = C_{1-4}alkyl, cyclohexyl; n = 0-2) have been synthesized as bactericides
and fungicides (M. Protiva, I. Cervena, and V. Hola, Czech. CS 262,293,
1989).

(3)

(4) $R = CH_2CHR^1 (CH_2)_nNR^2 R^3$

(5)

A number of 1-aryloxythioxanthones 5 [R^1-R^5 = H, alkyl, alkoxy, halogeno, CF_3, CN, NO_2, OH, (substituted) amino, CO_2H, $CONH_2$, trazin-2-yl; R^3R^4, R^4R^5 = CH = CHCH = CH; R^6-R^{12} = H, alkoxy, acyloxy, halogeno, CF_3, CN, (substituted) alkyl, aryl, aralkyl, aroyl, aryloxy, amino; or adjacent R^6-R^{12} = COOCO, $CONR^{13}CO$; R^{13} = H, (substituted) alkyl, acyl, Ph, CH_2Ph; n = 0-2] have been prepared as photosensitizers, colour indicators and photo--switchable elements (E. Fischer et al., Eur. Pat. Appl. EP 430,881, 1991).

The photo-Fries rearrangement of 4-ethoxycarbonylphenyl thiosalicylate; (6; R^1 = H, R^2 = CO_2Et) gives benzophenone 7, which undergoes thermal cyclization to ethyl thioxanthone-2-carboxylate 8 (ethyl 9-oxothioxanthene--2-carboxylate). A number of thiosalicylates 6 (R^1 = H, Ac; R^2 = MeO, Me) have been prepared and their photoreactions studied (C. Belled, M.A. Miranda, and A. Simon-Fuentes, An. Quim., Ser. C, 1989, 85, 39).

(6)

(7)

(8)

1,5-Dichloro-9H-thioxanthen-9-one 12 has been prepared by the cyclization of 2-chloro-6-[(2-chlorophenyl)thio]benzoic acid 11 obtained from 2-chloro-6-iodobenzoic acid 9 and 2-chlorobenzenethiol 10. Similarly, 1,7--dichloro-9H-thioxanthen-9-one 15 is prepared from 9 and 4-chlorobenzenethiol 13 *via* 2-chloro-6-[(4-chlorophenyl)thio]benzoic acid 14 and also by the condensation of 2-chloro-6-mercaptobenzoic acid 16 with chlorobenzene 17 in the presence of sulphuric acid (I. Okabayashi, N. Murakami, and K. Sekiya, J. Heterocyclic Chem., 1989, 26, 635). 1,6-, 1,8-, and 3,6-dichloro-9*H*--thioxanthen-9-ones (I. Okabayashi, H. Fujiwara, and C. Tanaka, *ibid.*, 1991, 28, 1977) and various methoxy-substituted thioxanthen-4-ones (M. Watanabe *et al.*, Chem. Pharm. Bull., 1989, 37, 36) have been prepared.

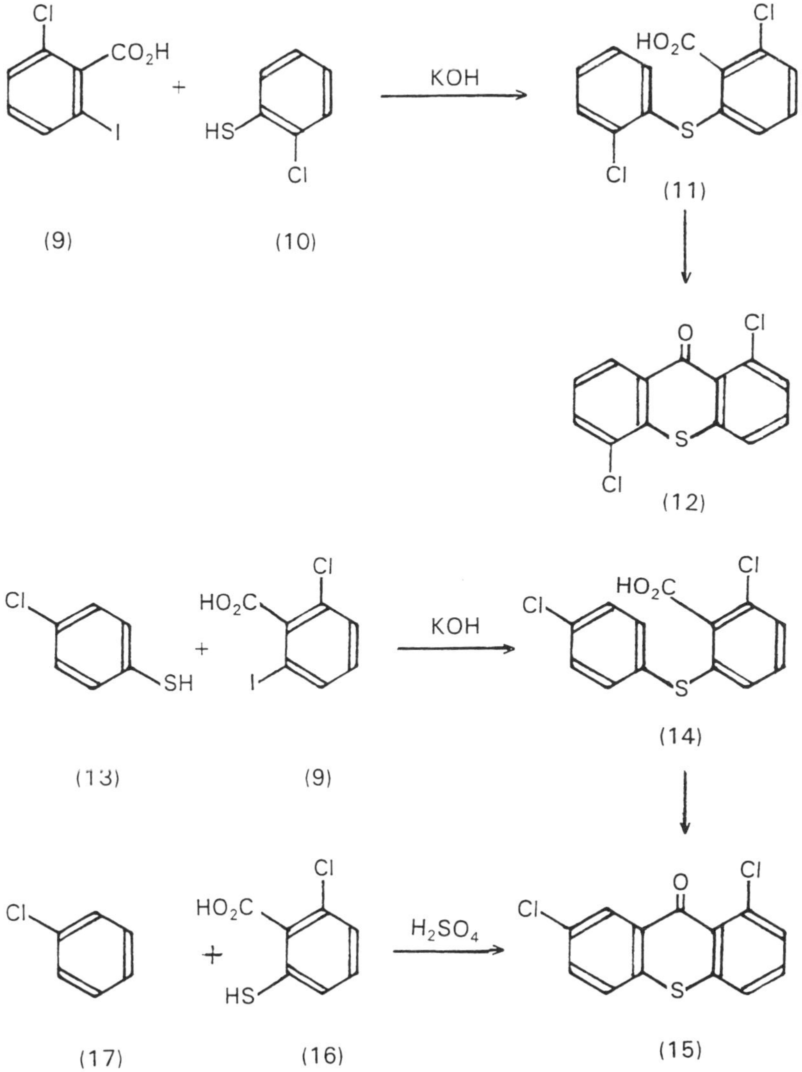

[14]C Labelled and ^{13}C, ^{15}N labelled WIN 33377, 1-(2-diethylaminoethyl-amino)-4-(methanesulphonamidomethyl)thioxanthen-9-one, compounds (18 and 19, respectively), have been prepared by related routes from the known 1-(2-diethylaminoethylamino)thioxanthen-9-one 20 and the labelling has been introduced with cyanide, followed by further elaboration. Compounds 18 and 19 are anticancer agents related to Hycanthone (G.R. Duffin *et al.*, J. Labelled Compd. Radiopharm., 1994, <u>34</u>, 463). The following related derivatives 21 [R^1,R^2 = lower alkyl; R^3 = H, lower alkyl, (un)substituted Ph; R^4 = lower alkyl (un)substituted Ph; R^5 = H, lower alkyl or alkoxy, OH; n = 2,3] have been prepared (T.C. Miller *et al.*, U.S. US 5,346,917, 1994).

(18)

(19)

(20)

(21)

Other 4-substituted 1-(dialkylaminoalkylamino)thioxanthen-9-ones 22 (R^1 = CH_2NH_2, CH_2NHCHO, CH = NPh, $CONH_2$; R^2,R^3 = alkyl; R^4 = H, halogeno, alkyl, alkoxy; n = 2 or 3) have been prepared as antitumor agents (*idem*, Eur. Pat. Appl. EP 518,414, 1992). For the preparation of the di-hydrochloride salt of 2,7-diamidinothioxanthen-9-one 23 as a leishmanicide, see P.M.S. Chauhan *et al.*, Indian IN 167,932, 1991).

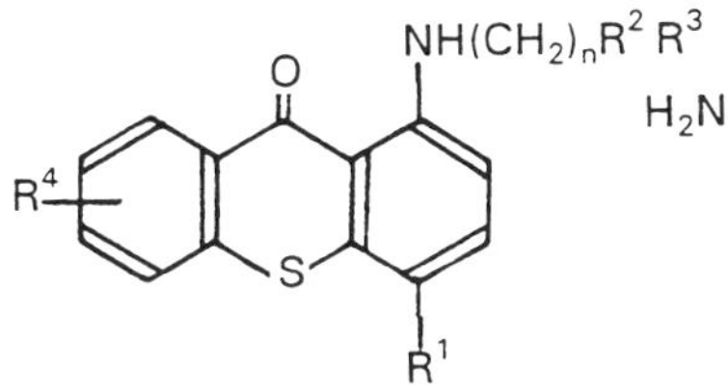

(22)

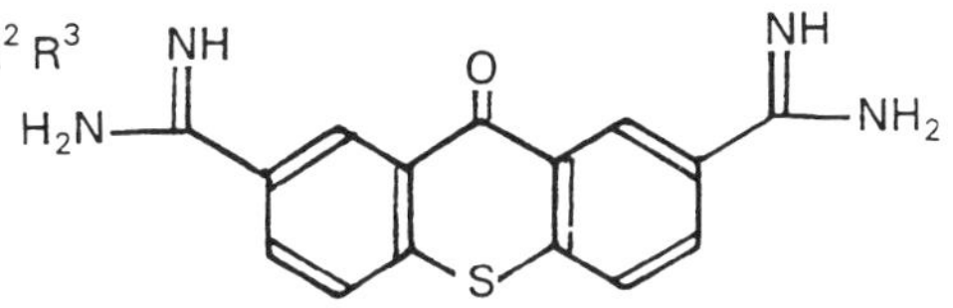

(23)

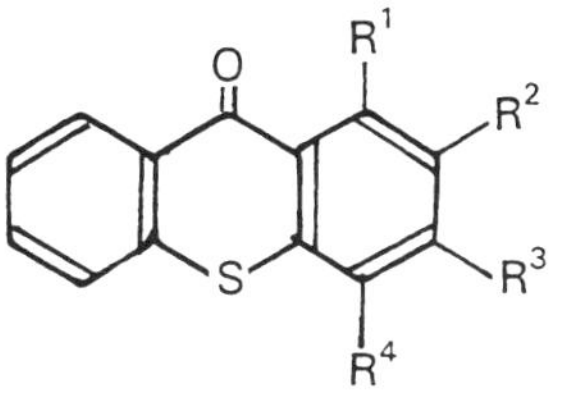

(24)

Derivatives 24 [one of R^2, R^3, $R^4 = OCH_2CH(OH)CH_2N^+R^5R^6R^7A^-$ (one of $R^5,R^6,R^7 =$ alkyl or CH_2Ph, the others are alkyl; $A^- =$ an anion); R^1 and the others of $R^2,R^3,R^4 = H$, alkyl, alkoxy] of thioxanthone useful as photoinitiators for photopolymerizable systems have been prepared (G. Gwane, P.N. Green, and W.A. Green, Eur. Pat. Appl. E.P. 224,967, 1987).

(c) Dibenzo[b,d]thiopyrans

4,4,8-Trimethyl-1,3-diphenyl-2*H*,4*H*-cyclopenta[*c*]benzo[*b*]thiopyran-2-
-one 1 and dimethyl acetylenedicarboxylate in bromobenzene have been
boiled together until TLC indicated that none of the starting material
remained. After removal of the bromobenzene the residue was eluted from a
short silica column with 20% EtAc in hexane to give dimethyl 2,6,6-
-trimethyl-7,10-diphenyl-6*H*-dibenzo[*b,d*]thiopyran-8,9-dicarboxylate 2 (C.D.
Gabbutt, J.D. Hepworth, and B.M. Heron, Tetrahedron, 1994, 50, 7865.

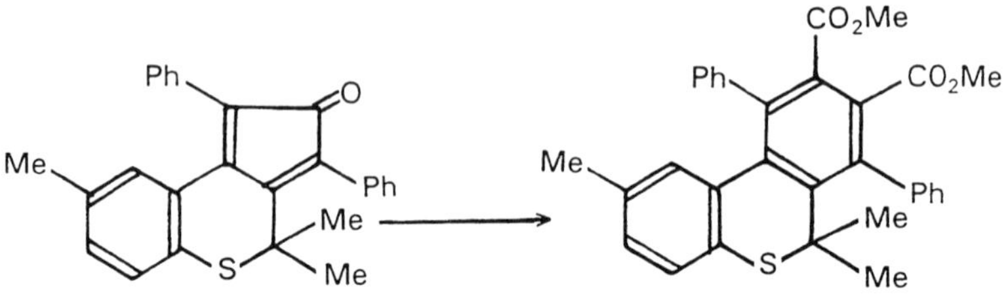

5. Naphthothiopyrans and related compounds

(a) Naphthothiopyrans

2,3-Dihydro-1*H*-naphtho[2,1-*b*]thiopyran-1-ol 1, 3*H*-naphtho[2,1-b]thio-
pyran 2 and 2,3-dihydro-1*H*-naphtho[2,1-*b*]thiopyran 3 have been obtained
by the protic acid-catalyzed ring-closure reaction of 3-(2-naphthylthio)-
propionaldehyde. The ratio of products has been varied by changing the
reaction conditions. In the presence of nucleophiles, for example, thiols,
thiocarboxylic acids and acetonitriles, the ring-closure reaction of the
naphthylthiopropionaldehyde proceeds *via* the addition of nucleophiles to the
formyl group to yield the corresponding 1-substituted 2,3-dihydro-1*H*-
-naphtho[2,1-*b*]thiopyran (T. Nakazawa, M. Shibazaki, and K. Itabashi,
Nippon Kagaku Kaishi, 1987, 45).

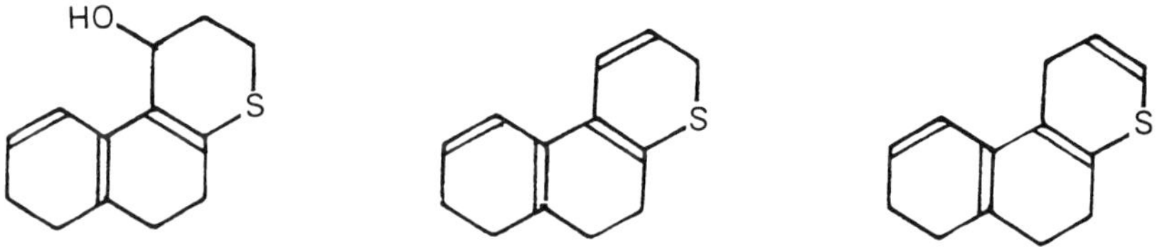

The reaction of 2,3-dihydro-1*H*-naphtho[2,1-*b*]thiopyran-1-ol 1 in dry Et$_2$O containing H$_2$SO$_4$ affords a mixture of 3*H*-naphtho[2,1-*b*]thiopyran 2 and 2,3-dihydro-1*H*-naphtho[2,1-*b*]thiopyran 3, but only the naphthothio-pyran 2 is formed quantitatively in CHCl$_3$ containing HCl. Heating alcohol 1 in polyphosphoric acid furnishes 70% of the dihydronaphthothiopyran 3 and heating a solution of 1 and its methyl derivatives with alcohols, thiols, and thiocarboxylic acids in the presence of H$_2$SO$_4$ at 30° gives the corresponding ethers, sulphides, and thio esters, respectively. Treating 2,3-dihydro-1*H*--naphtho[2,1-*b*]thiopyran-1-ol 1 with AcBr in the presence of Et$_3$N gives the corresponding acetoxy derivative, while in the absence of Et$_3$N brominated compounds are obtained. Similarly, 3,4-dihydro-2*H*-naphtho[1,2-*b*]thiopyran--4-ol 4 reacts with AcBr to give brominated compounds (T. Nakazawa *et al.*, *ibid.*, 1989, 1103). The preparations of some 4-alkyl(aryl)thio and acylthio--3,4-dihydro-2*H*-naphtho[1,2-*b*]thiopyrans 5 (R = Bu, Ph, PhCH$_2$, 1-naphthyl, 2-naphthyl, Ac, Bz) and 1-alkyl(aryl)thio and acylthio-2,3-dihydro-1*H*--naphtho[2,1-*b*]thiopyrans 6 (R same as for 5) have been reported (T. Nakazawa and K. Itabashi, J. Heterocyclic Chem., 1988, <u>25</u>, 301).

(4)	(5)	(6)

The reaction of naphthalene-1- and -2-thiol with malonic acid provides 2--oxonaphtho[1,2-*b*]thiopyran-4-ol 7 and 3-oxonaphtho[2,1-*b*]thiopyran-1-ol 8, respectively. However, the reactions of 1- or 2-methylthionaphthalene with malonic acid or methylmalonic acid in polyphosphoric acid yields 3-hydroxy--1*H*-phenalen-1-ones, also obtained by the reaction of the naphthalenethiols with methylmalonic acid (H. Kamijo *et al.*, Nippon Kagaku Kaishi, 1993, 1257).

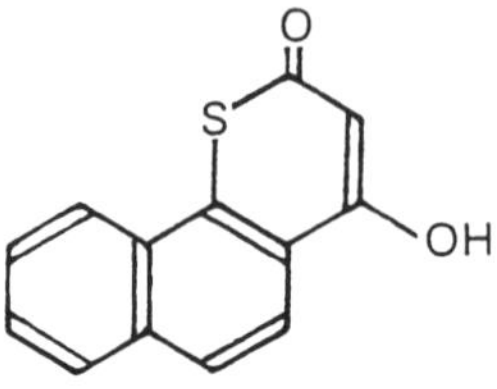

632

(7)

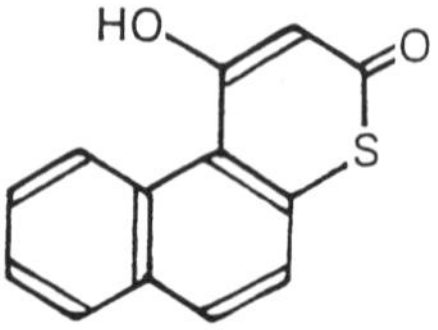

(8)

The tetralinthione obtained on treating 2-[(phenylthio)methylene]tetralin-
-1-one with Lawesson's reagent, undergoes a Diels-Alder reaction with
chalcone to give 3-benzoyl-2-phenyl-4-thiophenyl-3,4,5,6-tetrahydro-2H-
-naphtho[1,2-b]thiopyran 9, which following elimination of PhSH by
treatment with sodium alkoxide and reaction with Lawesson's reagent
affords 2-phenyl-3-(thiobenzoyl)-5,6-dihydro-2H-naphtho[1,2-b]thiopyran 10
(T. Karakasa, S. Moriyama, and S. Motoki, Chem. Letters, 1988, 1029).
The preparations of several derivatives 11 (R^1 =H, Me, Ph, CO_2Me; R^2 =H,
Me; R^3 = CO_2Me, COMe, CN, CHO, Ph; R^2R^3 = CONPhCO, 2-$C_6H_4CH_2$)
related to compound 10 have been described (S. Moriyama and S. Motoki,
Bull. Chem. Soc. Jpn., 1992, 65, 2056).

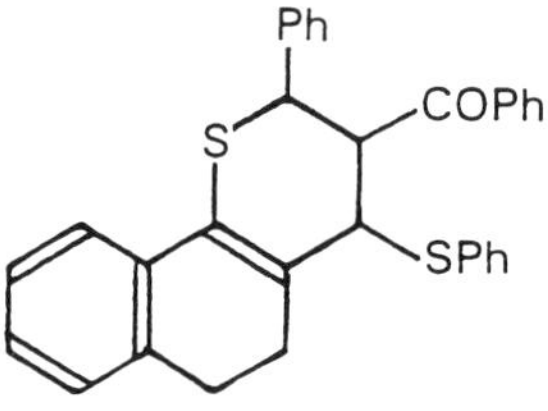

(9)

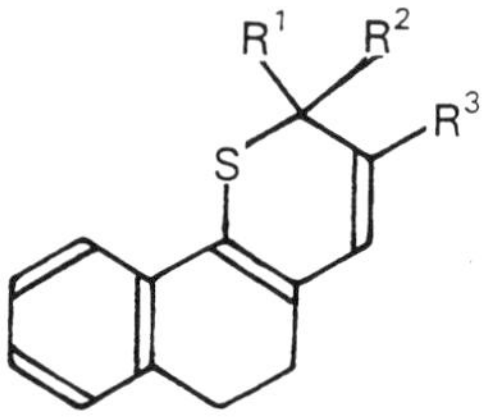

(10) R^1 =H, R^2 =Ph, R^3 =CSPh
(11) (see text)

2-Phenyl-3-(thiobenzoyl)-5,6-dihydro-2*H*-naphtho[1,2-*b*]thiopyrans 12
(R^1 = Ph, 4-ClC$_6$H$_4$, 4-MeOC$_6$H$_4$, 2-thienyl, 2-furyl, R^2 = Ph; R^1 = Ph, R^2 = H)
have been synthesized and 12 (R^1 = R^2 = Ph) reacted as an α,ß-unsaturated,
thioketone with norborn-2-ene (to give adduct 13), norborna-2,5-diene,
diethyl azodicarboxylate, diphenyl fumarate, *N*-phenyl- and *N*-(4-tolyl)-
maleimide, 2-chloroacrylonitrile, methyl methacrylate, acrylonitrile and
styrene to give the corresponding [4 + 2] cycloadducts (S. Moriyama, T.
Karakasa, and S. Motoki, *ibid.*, 1990, <u>63</u>, 2540).

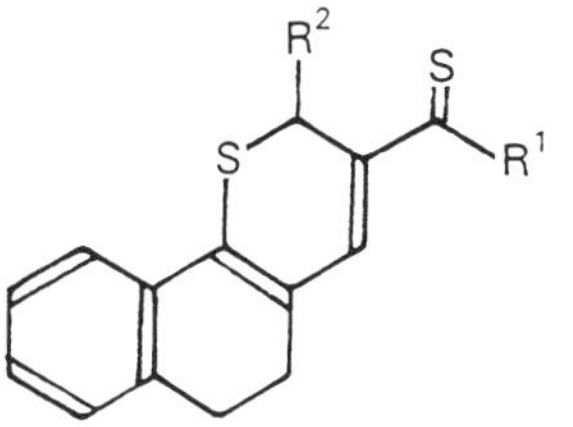

(12) 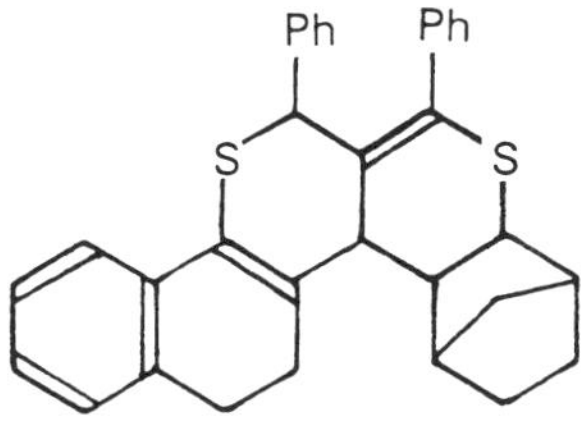   (13)

The α,ß-unsaturated thioketones 14 (R^1 = Ph, 4-MeC$_6$H$_4$, 4-MeOC$_6$H$_4$;
R^2 = H, Me; n = 1,2) and 15 (R = H, Me) undergo intramolecular Diels-Alder
reactions with high regio and diastereoselectivities (trans/cis) to yield the
dihydrothiopyran and dihydronaphthothiopyran-fused cycloadducts 16 and
17, respectively (T. Saito *et al.*, Chem. Co., 1990, 1665).

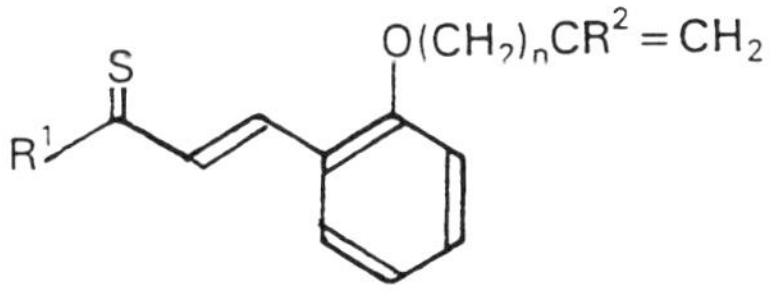

(14) 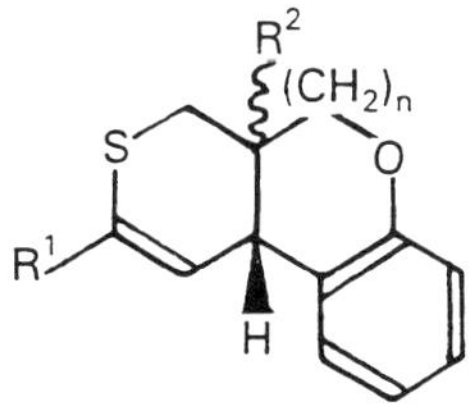   (16)

(15)   (17)

(b) Perinaphthothiopyrans, thiaphenalenes

The naphthalene-1-thiyl radical, formed in thermolysis of naphthalene-1-
-thiol, bis(naphth-1-yl)disulphide, and allyl naphth-1-yl sulphide or in the
reaction of 1-chloro- and 1-bromonaphthalene with H_2S, reacts with
acetylene at 410-700°C to yield naphtho[1,8-*b,c*]thiopyran 1 and
naphtho[1,2-*b*]thiophene 2 (N.V. Russavskaya *et al.*, Zh. Org. Khim., 1991,
<u>27</u>, 1743).

(1)   (2)   (3)

The preparations and properties of 1*H*,3*H*-naphtho[1,8-*c,d*]thiopyran-1,3-
-dione, -1,3-dithione, and -1-one-3-thione ($X^1 = X^3 = 0$, $X^2 = S$;
$X^1 = X^2 = X^3 = S$; $X^1 = 0$, $X^2 = X^3 = S$, respectively) 3 have been reported
(M.P. Cava and M.V. Lakshmikantham, Phosphorus, Sulphur silicon Relat.
Elem., 1989, <u>43</u>, 95).

2,7-Dithia-1,2,3,6,7,8-hexahydropyrene 6 has been synthesized, starting from naphthalene-1,4,5,8-tetracarboxylic acid 4, which on treatment with dimethyl sulphate afforded tetramethyl naphthalene-1,4,5,8-tetracarboxylate, reduced by $(Me_2CHCH_2)_2AlH$ to yield 1,4,5,8-tetra(hydroxymethyl)naphthalene (5; R = OH), which reacts with PBr_3 in dioxane to give 1,4,5,8-tetra-(bromomethyl)naphthalene (5; R = Br). The tetrabromo derivative 5 (R = Br) was then added to an aqueous solution of $Na_2S.9H_2O$ and EtOH and boiled to give 2,7-dithia-1,2,3,6,7,8-hexahydropyrene 6 (T. Kamata, Y. Gama and N. Wasada, Jpn. Kokai Tokkyo Koho, JP 02 184,688 [90 184,688], 1990).

(4) $R^1-R^4 = CO_2H$
(5) $R^1-R^4 = CH_2R$

(6)

(c) Benzothioxanthene derivatives

Benzo[b]thioxanthone oxides 1 [R^1 = (substituted) aryl; R^2 = H, (ar)alkyl, acyl; n = 1,2] have been prepared as photochromic substances, for example, 3-(2-carboxyphenylthio)-1,4-dimethoxynaphthalene has been boiled with $POCl_3$ in 1,2-dichlorobenzene and the product etherified with phenol to yield, after oxidation with H_2O_2, the 10-methoxy-5-phenoxybenzo[b]xanthone 12-oxide 1 (R^1 = Me, R^2 = Ph, n = 1) (W. Fischer, J. Finter, and H. Spahni, Can. Pat. Appl. CA 2,058,393, 1992).

(1)

636

1,4-Dimethoxythioxanthen-9-one on demethylation with borontribromide
yields the corresponding hydroquinone, which on in situ oxidation with silver
oxide followed by addition of various dienes affords 6a,7,10,10a-tetrahydro-
benzothioxanthen-6,11,12-triones 2 (R^1-R^3 = H, Me, MeO) and the regio-
isomers 3 (R^1 = R^3 = H, Me, MeO) (P. Singh *et al.*, Phosphorus, Sulphur
Silicon Relat. Elem., 1994, _86_, 21).

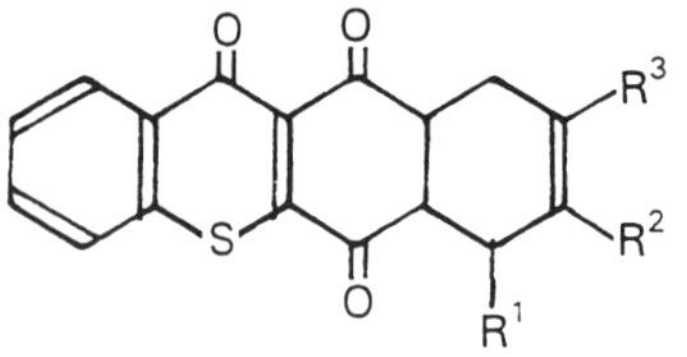

(2) 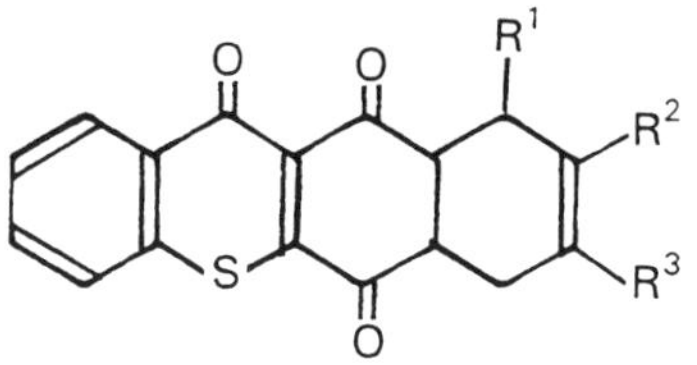   (3)

An improved synthesis of cyclohexanothioxanthenones resulting from the
treatment of thiosalicylic acid or 2,2'-dithiosalicylic acid with 1,2,3,4-tetra-
hydronaphthalene in the presence of concentrated sulphuric acid or a mixture
of 95% sulphuric acid with 27-30% fuming nitric acid (in 5:1 to 2:1 vol./vol.
ratio) has been reported. The crude product consisted of three isomers, 1,2-
-cyclohexanothioxanthenone 4, 2,3- and 3,4-cyclohexanothioxanthenones
(R. Ran and C.U. Pittman, Jr., J. Heterocyclic Chem., 1993, _30_, 1673).

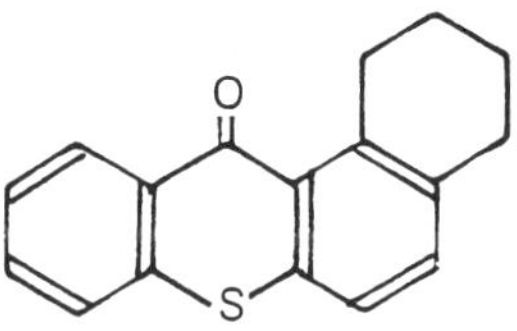

(4)

6. Selenopyrans and related compounds

(a) Selenopyrans and derivatives

For a review of the developments in the chemistry of thiopyrans, selenopyrans, and telluropyrans, see J. Kuthan, P. Sebek, and S. Bohm, Adv. Heterocyclic Chem., 1994, <u>59</u>, 179. 6-Cyano-3,4-dimethyl-2*H*-selenopyran (1) is obtained along with 2-cyano-4,5-dimethyl-3,6-dihydro-2*H*-selenopyran- -2-yl 3-chlorobenzoate (2) following the oxidation of 2-cyano-4,5-dimethyl- -3,6-dihydro-2*H*-selenopyran (3) with 1.5 equivalents of 3-chloroperbenzoic acid. Derivative 3 and analogues with an electron-withdrawing group at the 2-position are prepared by the Diels-Alder reaction of butadienes with selenoaldehydes generated in situ from selenocyanates and NEt_3 (T. Kataoka *et al.*, Chem. Pharm. Bull., 1994, <u>42</u>, 811).

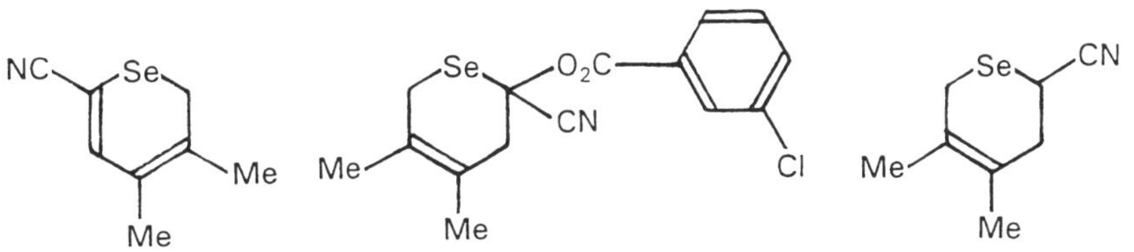

(1)   (2)   (3)

The cyclocondensation of $PhCH = C(CH)_2$ with $NCCH_2CSeNH_2$ affords 2,6-diamino-3,5-dicyano-4-phenyl-4*H*-selenopyran (4; R = Ph), which has been converted into 6-amino-3,5-dicyano-4-phenyl-1,2-dihydropyridine-2- -selenone (5; R = Ph) (V.D. Dyachenko *et al.*, Zh. Obshch. Khim., 1989, <u>59</u>, 881). The cycloaddition of (pyrid-3-ylmethylene)malononitrile with $NCCH_2C$-$SeNH_2$ in ethanol containing 4-methylmorpholine yields 2,6-diamino-3,5- -dicyano-4-(pyrid-3-yl)-4*H*-selenopyran (4; R = pyrid-3-yl), which on treat-ment with HCl rearranges to afford 6-amino-3,5-dicyano-4-(pyrid-3-yl)-1,2- -dihydropyridine-2-selenone (5; R = pyrid-3-yl) (Yu.A. Sharanin and V.D. Dyachenko, Zh. Obshch. Khim., 1987, <u>57</u>, 1662). For the preparation of tetramethyl 4-dimethylamino-4*H*-selenopyran-2,3,5,6-tetracarboxylate and tetramethyl 4-dimethylamino-4-methyl-4*H*-selenopyran-2,3,5,6-tetracarboxy-late, see D. Dubreuil *et al.*, Tetrahedron Letters, 1995, <u>36</u>, 237.

638

(4)   (5)

The oxidation of 2,6-diphenyl- and 2,6-diphenyl-4-methoxy-4*H*-seleno-
pyrans 6 (R-H, OMe) with H_2O_2 in Me_2CO affords 2,6-diphenyl-4*H*-seleno-
pyran-4-one (7). Treatment of 2,4,6-triphenyl-4*H*-selenopyran (6; R = Ph)
with H_2O_2 in C_6H_6 or H_2O_2 in Me_2CO or selenopyrans 6 (R = H, Ph) with
SeO_2 in pyridine furnishes the corresponding selenophenes 8 (B.I. Drevko *et
al.*, Khim. Geterotsikl. Soedin., 1995, 24). Also treatment of 4-(4-methoxy-
phenyl)-2,6-diphenyl- and 2,6-(4-methoxyphenyl)-4-phenyl-4*H*-selenopyrans
with SeO_2 in pyridine affords the corresponding 2,3,5-triarylselenophenes.
Related thiopyrans behave in a similar manner (B.I. Drevko, L.A. Fomenko,
and V.G. Kharchenko, *ibid.*, 1989, 767). The oxidation of 2,6-diphenyl- and
2,6-diphenyl-3,5-dimethyl-4*H*-selenopyrans with $KMnO_4$ in Me_2CO or MeCN
affords 2,6-diphenyl- and 2,6-diphenyl-3,5-dimethyl-4*H*-selenopyran-4-ones,
respectively (S.N. Petrakov *et al.*, *ibid.*, 1991, 996).

(6)   (7)   (8)

The synthesis of electron pi donors including 4,4'-bis(selenopyranyl-idenes) and 4,4'-(telluropyranylidenes) has been reviewed and the optical, magnetic and electronic characteristics of the charge transfer complexes and radical ion salts derived from these electron donors have been discussed (G. Le Coustumer and Y. Mollier, Sulphur Rep., 1993, 15, 67; C. Regnault du Mottier *et al*., Mol. Cryst. Liq. Cryst., 1988, 164, 197). The systematic replacement of the sulphur atoms of 2'-(thiopyran-4-ylidene)-1,3-dithiole (9; $X = Y = S$) by selenium, for example, 9 ($X = S$, $Y = Se$; $X = Se$, $Y = S$; $X = Y = Se$) has been investigated in order to develop novel unsymmetrical electron donors. Other substituted derivatives have also been prepared (T. Otsubo *et al*., J. Chem. Soc., Perkin 2, 1993, 1815; Y. Shiomi *et al*., Chem. Comm., 1988, 822). The synthesis of bis(2,6-diaryl-4,4'-seleno-pyranylidenes) 10 ($R = C_9H_{19}$, $C_{12}H_{25}$, OC_9H_{19}, $OC_{12}H_{25}$, SCC_9H_{19}) has been reported. Some of these compounds (e.g. 10) showed mesomorphic properties (C. Regnault du Mottier *et al*., Synth. Met., 1993, 55., 856).

(9)   (10)

The reactions of selenopyrylium salts 11 ($R^1 = Ph$, 4-MeOC$_6$H$_4$, Bu^t, $R^2 = H$, 4-MeOC$_6$H$_4$, Ph, Me; $Z = BF_4$, Cl, 0.5 ZnCl$_2$ 0.5 ZnBr$_2$) and salts 11 ($Se = S$, O, $R^1 = R^2 = Ph$; $Z =$ as above) with MeONa in C_6D_6 have been followed by NMR spectroscopy. Depending on the number and type of substituents on the heteroaromatic cation, different products form; 4*H*--selenopyrans, mixtures of 2*H*- and 4*H*-selenopyrans with varying isomeric ratios or 4-methyleneselenopyrans (B.I. Drevko *et al*., Zh. Org. Khim., 1994, 30, 115). The preparations of 2,6-di(4-methoxyphenyl)-4-phenyl- and 2,4,6--tri(4-methoxyphenyl)selenopyrylium formates, benzoates, and hydroxy-benzoates have been reported (B.I. Drevko, V.G. Kharchenko, and

640

L.M. Yudovich, U.S.S.R. SU 1,447,824, 1988). A mechanism has been
proposed to explain the heteroatom scrambling observed in the synthesis of
chalcogenopyrylium trimethine dyes 12 (X^1, X^2 = Te, Se) from 4-methyl-
chalogenopyrylium salts and the appropriate aldehydes (M.R. Detty,
D.N. Young, and A.J. Williams, J. Org. Chem., 1995, _60_, 6631).

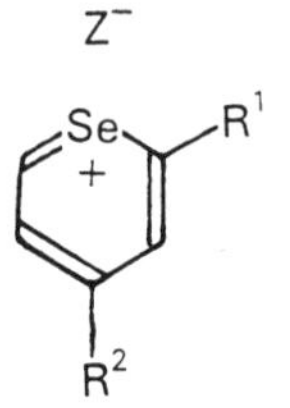

(11)

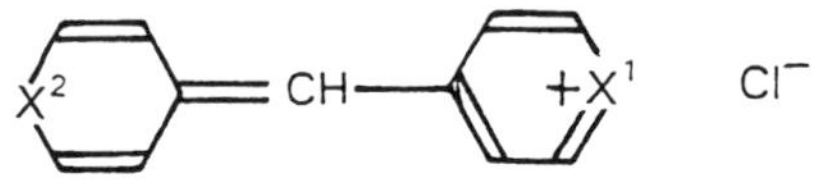

(12)

The reaction of aldehydes with $(Me_3Si)_2Se$ in the presence of $BF_3.Et_2O$
gives 1,3,5-triselenanes *via* seleno aldehydes. The generation of seleno
aldehydes is also achieved by thermal or Lewis acid-induced fragmentation
of 1,3,5-triselenanes, for example, triselenane 13, prepared from PhCHO,
upon treatment with CH_2 = CMeCMe = CH_2 and $SnCl_4$ in CH_2Cl_2 affords 2-
-phenyl-4,5-dimethyl-3,6-dihydro-2*H*-selenopyran (14) (Y. Takikawa *et al.*,
Tetrahedron Letters, 1989, _30_, 6047).

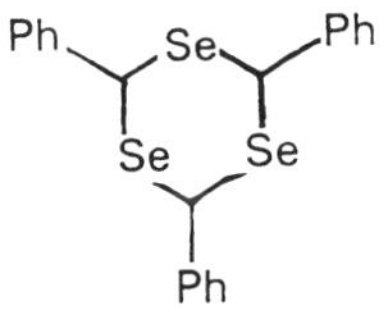

(13)

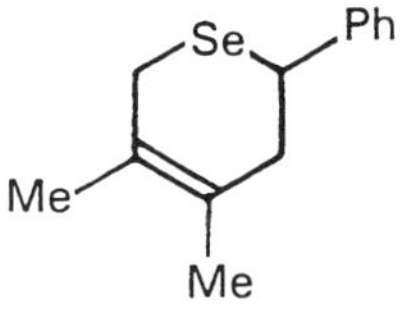

(14)

Selenobenzophenone dimer 15 in solution forms mainly Ph_2CSe, which undergoes cycloaddition with dienes, for example, Ph_2CSe with $H_2C=CMe-CMe=CH_2$ yields 2,2-diphenyl-4,5-dimethyl-3,6-dihydro-2*H*-selenopyran (16) (G. Erker *et al.*, Angew. Chem., 1990, **102**, 1082).

(15)

(16)

Diethyl chloromalonate reacts with Cs_2CO_3 in the presence of elemental S_8 or Se_n to yield the corresponding diethyl thioxo- or selenoxomalonates, which are subsequently trapped in situ with various 1,3-dienes to give thio- or selenopyrans, for example, diethyl 3,6-diphenyl-3,6-dihydro-2*H*-thio- or -selenopyran-2,2-dicarboxylates 17 (X = S, Se) and dihydroselenopyrans 18 (M.M. Abelman, Tetrahedron letters, 1991, **32**, 7389). The pentacarbonyl-(selenobenzaldehyde) and -(diphenylselenoketone) complexes $(CO)_5M$-[Se:C(Ph)R] (R = H, M = W; R = H, M = Cr; R = Ph, M = W) react with cyclohexa-1,3-diene by [4 + 2] cycloaddition to give the metal-coordinated 3,6--dihydro-2*H*-selenopyrans (3-phenyl- and 3,3-diphenyl-2-selenabicyclo[2.2.2]-oct-5-ens) 19 (H. Fischer *et al.*, Ber., 1990, **123**, 725).

(17)

(18)

(19)

The oxidation of 3,6-dihydro-2*H*-selenopyrans with an electron-
-withdrawing group at the 2-position proceeds *via* an unprecedented ring-
-contraction to furnish selenophenes, for example, 2-cyano-4,5-dimethyl-3,6-
-dihydro-2*H*-selenopyran (20) affords 2-cyano-4,5-dimethylselenophene (21)
(T. Kataoka *et al.*, Chem. Comm., 1993, 577).

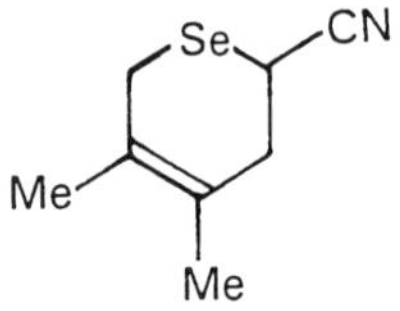

(20) 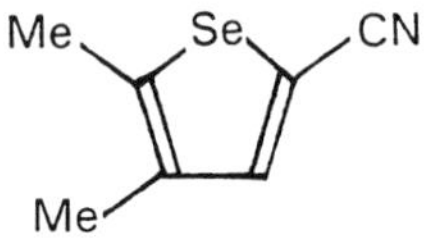   (21)

The cycloaddition reactions of electron-deficient or conjugated
selenoaldehydes with electron-rich dienes, for example, $CH_2 = C(OEt)$-
$CH = CH_2$ proceed efficiently yielding 2-substituted 4-ethoxy-3,6-dihydro-2*H*-
-selenopyrans 22 (R = H, CN, Ph) (P.T. Meinke and G.A. Krafft, J. Amer.
Chem. Soc., 1988, <u>110</u>, 8671). For the preparation of 2,2-diphenyl-3,5-
-dimethyl-3,6-dihydro-2*H*-selenopyran (23) and other highly functionalized
selenopyrans, see *idem, ibid.*, p.8679; and for 6-isopropyl-2,2-dimethyltetra-
hydroselenopyran-3-ol (24), S.B. Kurbanov *et al.*, Zh. Org. Khim., 1991, <u>27</u>,
942.

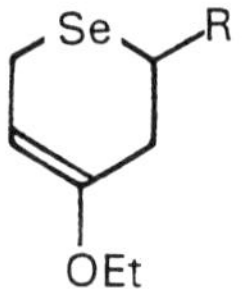

(22) 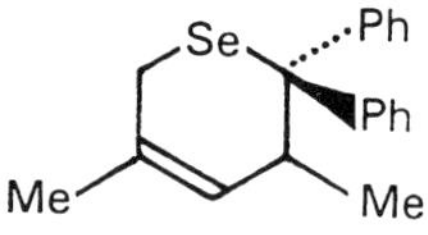   (23) 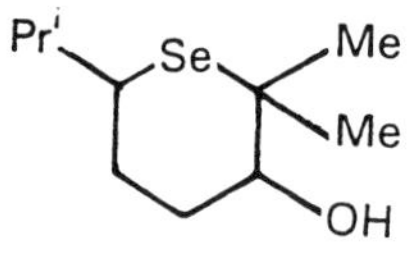   (24)

The reduction of the cis isomers of 2,5-dimethyltetrahydroselenopyran-
-4-one (2,5-dimethylselenan-4-one) with aluminum isopropoxide, $NaBH_4$, or
$LiAlH_4$ furnishes mixtures of 3 isomeric 2,5-dimethyltetrahydroselenopyran-
-4-ols (2,5-dimethylselenan-4-ols) (B.M. Butin *et al.*, Zh. Obshch. Khim.,
1994, <u>64</u>, 771).

(b) Selenopyrans containing fused ring

(i) Benzoselenopyrans and derivatives
The thin-layer chromatographic behaviour on silica gel and consequent
rapid separation of some selenopyans, including 4*H*-benzo[*b*]selenopyran-4-
-one (selenochromone) (1; $R^1R^2 = O$), 4*H*-benzo[*b*]selenopyran-4-ol (1;
$R^1 = H$, $R^2 = OH$) and 6*H*-dibenzo[*b,d*]selenopyran (2) have been described
(G. Bottura and M.A. Pevesi, Microchem. J., 1987, <u>35</u>, 223).

(1)

(2)

Treatment of 3,4-dihydro-2*H*-benzo[*b*]selenopyran-4-one oxime (seleno-
chroman-4-one oxime) (3) with *N,N*-diethylaminoethyl chloride in boiling
EtOH and/or MeOH in the presence of an alkali metal alcoholate affords *N,N*-
-diethylaminoethyl selenochroman-4-one oxime ether (4) (W. Basinski, Pol.
PL 147,859, 1989).

644

(3) R = H
(4) R = CH₂CH₂NEt₂

(5)

2*H*-Selenofuro[3,2-*g*]benzo[*b*]selenopyran-2-one (5) and related sulphur and selenium analogues of psoralen have been synthesized starting from 4,6--dibromoisophthalaldehyde (E.A. Jakobs, L.E. Christiaens, and M.J. Renson, Heterocycles, 1992, $\underline{34}$, 1119).

(6)

(7)

Some 2-methylisoselenochromanium salts have been prepared and their spectroscopic and reactions investigated. The ^{1}H NMR spectra in $CDCl_3$ and electron impact mass spectra show that selenonium salts 6 (X = tosylate, mesylate) are selenuranes and that the tetrafluoroborate and triflate analogues are selenonium salts. Selenonium salts and selenuranes react with some nucleophiles to give styrene derivatives 2-CH_2 = $CHC_6H_4CH_2SeMe$ and methyl phenethyl selenides, for example, 7[R = Me, Et, Ph, $CHAc_2$, $CH(CO_2Me)_2$] (M. Hori *et al.*, Chem. Pharm. Bull., 1990, $\underline{38}$, 779).

(ii) Selenoxanthene and derivatives

The electrochemical properties of selenoxanthene (1), selenoxanthen-9-ol (2), and selenoxanthone (3) in non-aqueous and mixed media have been investigated by voltammetry at rotating platinum and vitreous carbon disk electrodes, cyclic voltammetry, chronopotentiometry and constant potential coulometry. In "dry" acetonitrile selenoxanthene leads to the selenoxanthylium cation. In the presence of water selenoxanthen-9-ol is obtained (B. Dakova *et al.*, Electrochim. Acta, 1990, 35, 1855).

(1) $R^1 = R^2 = H$
(2) $R^1 = H$, $R^2 = OH$
(3) $R^1\ R^2 = O$

(4)

(5)

The preparations of various methoxy-substituted selenoxanthones have been reported (M. Watanabe *et al.*, Chem. Pharm. Bull., 1989, 37, 36). 3,5--Di(*tert*-butyl)-2-benzoquinone is a convenient reagent for the dehydrogenation of 1,2,3,4,5,6,7,8-octahydroselenoxanthenes 4 ($R = H$, Ph, 4-MeOC$_6$H$_4$) in HClO$_4$ to yield the 1,2,3,4,5,6,7,8-octahydroselenoxanthylium perchlorates 5 (V.T. Abaev *et al.*, Khim. Geterotsikl. Soedin., 1991, 53).

The reaction of 4-nitronaphthalene-1,8-dicarboxylic anhydride with the zincate of 2-aminobenzoselenol and subsequent treatment with amyl nitrile furnishes benzo[*k,l*]selenoxanthene-3,4-dicarboxylic acid anhydride (6), which on amination with RNH$_2$ ($R = Bu$, hexyl, octadecyl, CH$_2$CH$_2$OH) affords the selenoxanthenoquinolinediones 7 (A.V. Gutzeit, K.A. Balodis, and I.A. Meirovits, *ibid.*, 1993, 1426).

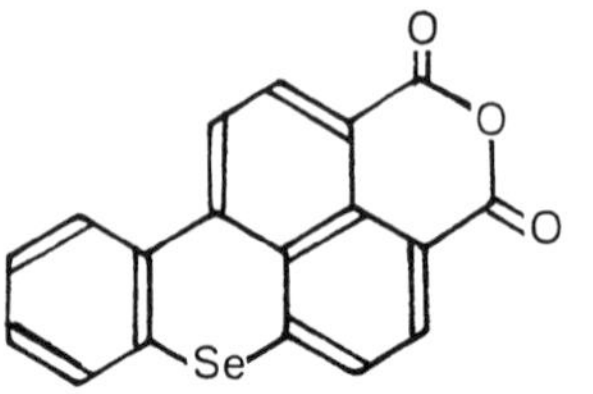

(6)

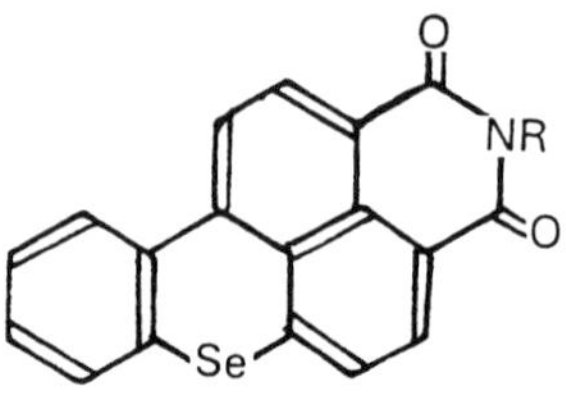

(7)

7. Telluropyrans and related compounds

(a) Telluropyrylium salts and tellurane, tetrahydrotelluropyran derivatives

Telluropyran-4-ones and benzo[*b*]telluropyrans are not oxidised to the corresponding tellurones or telluroxides wth common oxidizing agents, but oxidative addition products are obtained with peracetic acid to yield diacetates, for example, 2,6-di-tert-butyltelluropyran-4-one 1,1-diacetate (1). 2,6--Disubstituted telluropyrylium salts in pyridine with triphenylphosphine under aerobic conditions undergo oxidative dimerization to produce 1,1-dioxo-(telluropyranylidene)telluropyrans, which in the case of 1,1-dioxo-2,2',6,6'--tetraphenyl(telluropyranylidene)telluropyran (2) displays reversible reduction by cyclic voltammetry. The reaction of 2,6-disubstituted telluropyrylim salts and triphenylphosphine oxide in pyridine under argon also furnishes the dioxo(telluropyranylidene)telluropyrans (M.R. Detty and B.J. Murray, J. Org. Chem., 1987, 52, 2123).

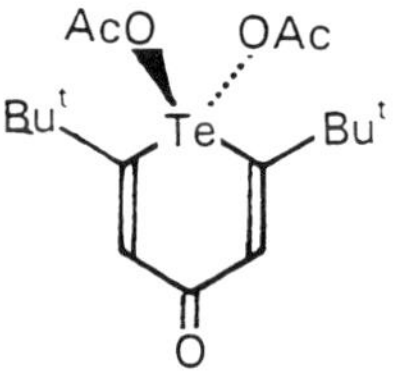

(1)

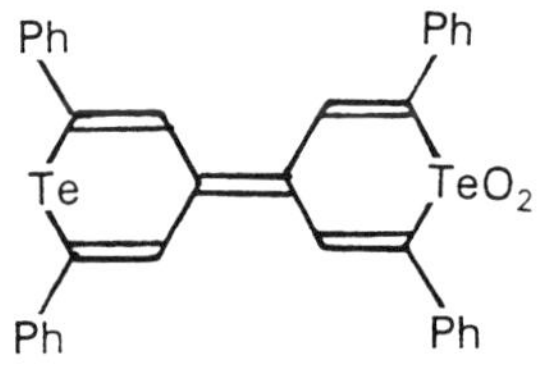

(2)

The dihydroxytelluranes derived from telluropyrylium dyes *via* oxidative addition of hydrogen peroxide or scavenging of singlet oxygen in the presence of water are useful as mild oxidants. The dihydroxytellurane 3 oxidizes leuco dyes 4 (R = OMe, NMe_2) to 5, based on the pseudo-first-order appearance of reduced telluropyrylium dye 6 (M.R. Detty and S.L. Gibson, Organometallics, 1992, **11**, 2147; M.R. Detty, *ibid.*, 1991, **10**, 702; Phosphorus, Sulphur Silicon Relat. Elem., 1992, **67**, 383).

Tetrahydrotelluropyran (telluracyclohexane) reacts with methyl-, ethyl-, and propylmethane and methyl-, ethyl-, and propylbenzene sulphonates to yield white telluronium salts of the corresponding telluride with methyl or benzene sulphonate as ionic species (H.A. Al-Shiray da, A.M. Ali, and I.S. Alnaimy, Int. J. Chem., 1994, **5**, 1). For reviews of the chemistry of telluranes, see M.R. Detty and M.B. O'Ragan, Chem. Heterocyclic Compd., 1994, **53**, 147; and V.I. Naddaka *et al.*, Sulphur Rep., 1988, **8**, 61.

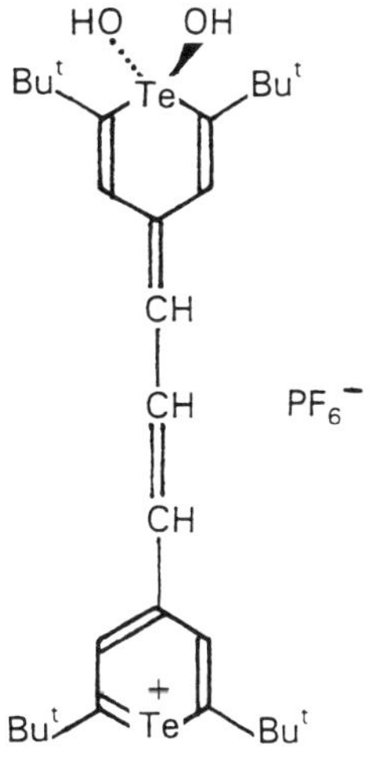

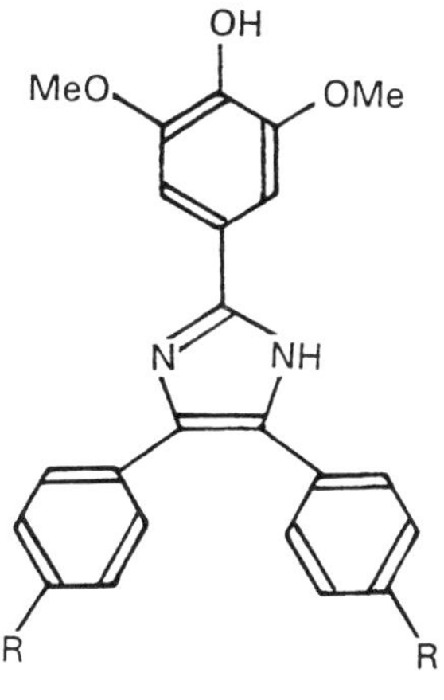

(3)

(4)

(5)

(6)

(b) Telluropyrans containing fused rings

(i) Benzotelluropyrans

(Z)-3-MeOC$_6$H$_4$TeCPh=CHCO$_2$H on treatment with P$_2$O$_5$ in MeSO$_3$H affords 7-methoxy-2-phenyltellurochromone (4H-tellurobenzo[b]pyran-4-one) (1). The preparations of a number of derivatives of tellurochromone have been reported (M.P. Detty, B.J. Murray, and J.H. Perlstein, U.S. US 4,434,098, 1984). Tellurochromylium perchlorate (benzo[b]telluropyrylium perchlorate) (2) has been obtained by treating 2H-benzo[b]telluropyran (tellurochrom-3-ene) (3) with Ph$_3$C$^+$ClO$_4^-$ in boiling CF$_3$CO$_2$H (A.A. Ladatko, V.L. Nivorozhkin, and I.D. Sadekov, Khim. Geterotsikl. Soedin., 1986, 1571).

(1)

(2)

(3)

(ii) Telluroxanthenes

Heating 9-azido-9-(4-methylphenyl)telluroxanthene (1) in boiling xylene affords a mixture of dibenzotelluroazepine 2, 9-(4-methylphenylimino)telluroxanthene, and phenanthridine 3 (A.A. Ladatko, I.D. Sadekov, and V.I. Minkin, *ibid.*, 1987, 279).

(1)   (2)   (3)

The electrochemical behaviour of dihalogenotelluroxanthenes, dihalogenoxanthones, and their selenium analogues (A.A. Bumber, A.A. Ladatko, and I.D. Sadekov, Zh. Obshch. Khim., 1990, $\underline{60}$, 847), the formation of radicals and anion radicals in telluroxanthenes (A.A. Bumber *et al.*, Khim. Geterotsikl. Soedin., 1988, 1196), and the effect of n,.pi.*-configurations on the nature of the lowest electronic states of heteroaromatic compounds (M.B. Ryzhikov *et al.*, Metalloorg. Khim., 1989, $\underline{2}$, 898) have been investigated and the carbonyl complex of rhodium (I) with telluroxanthene has been prepared from $Rh_2(CO)_4Cl_2$ and telluroxanthene in toluene (A.D. Garnovskii *et al.*, Koord. Khim., 1984, $\underline{10}$, 234).

8. Compounds containing an element from Group 4

(a) Silicon compounds

(i) Silabenzene, silacyclohexenes and their derivatives. High-vacuum pyrolysis of 1-silacyclohexa-2,5-dienes 1 (R = SiM_3, $SiPr^iMe_2$) affords 2,4,6--tri(trimethylsilyl)-1-*tert*-butyl-1-silabenzene (2; R = $SiMe_3$) and 2,6-di(isopropyldimethylsilyl)-4-trimethylsilyl-1-silabenzene (2; R = $SiPr^iMe_2$), respectively, which are trapped in a low-temperature Ar or 2-methyltetrahydrofuran matrix and are stable up to 90°K (P. Jutzi *et al.*, Ber., 1989, $\underline{122}$, 1227). Pyrolysis of silacyclohexadienes 3 (R = H, D) yields an

650

intermediate identified by mass spectroscopy as silabenzene (4) (A.K.
Mal'tsev *et al.*, Izv. Akad. Nauk SSSR, Ser. Khim., 1989, 1051). The
conformation of 1-phenylsilabenzene and 1,1'-bisilabenzene have been
examined by the ab initio STO-3G method involving partial geometry
optimization (H.J. Hofmann, R. Cimiraglia, and M. Persico, Chem. Phys.
Letters, 1988, 146, 249). For methods of obtaining heat of formation of
silabenzene, see M.J.S. Dewar and A.J. Holder, Heterocycles, 1989, 28,
1135; physical properties of some monosubstituted silabenzenes, K.K.
Baldridge and M.S. Gordon, Organometallics, 198, 1, 144; and a review on
the gas-phase generation of silabenzene, H. Back, Pure Appl. Chem., 1990,
62, 383.

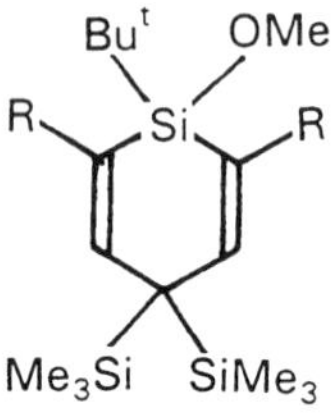

(1) 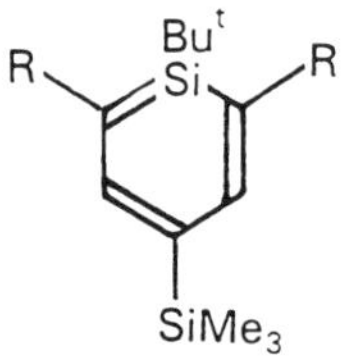　　　　　　　　　　(2) 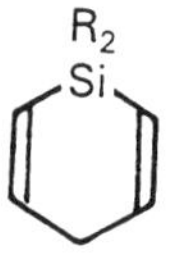　　　　　　　　　　(3)

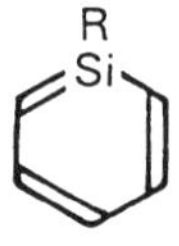

(4) 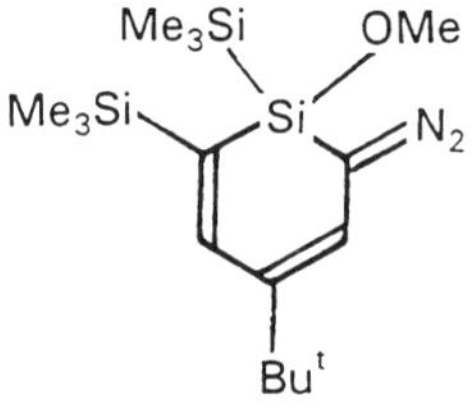　　　　　　　　　　(5)

The photochemical and thermal decomposition of 2-azo-1-methoxy-4-
-*tert*-butyl-1,6-di(trimethylsilyl)-1-silacyclohexa-3,5-diene (5) has been
investigated. Depending on the reaction conditions the carbene formed by
N_2-elimination rearranges by silyl migration to give a silabenzene and by
carbene migration to afford a silafulvene as reactive intermediates
(G. Maerkl, W. Schlosser, and W.S. Sheldrick, Tetrahedron Letters, 1988,
29, 467). The preparation of 1,4-di-*tert*-butyl-2,6-di(trimethylsilyl)sila-
benzene, stable in solution at -100°C has been reported (G. Maerkl and
W. Schlosser, Angew. Chem., 1988, 100, 1009).

1-Methoxy-1-*tert*-butyl-2,4,6-tri(trimethylsilyl)-1-silacyclohexa-3,5-diene (6) on reduction with $LiAlH_4$ affords 1-*tert*-butyl-2,4,6-tri(trimethylsilyl)-1--silacyclohexa-3,5-diene (7), which on bromination with NBS affords 1,4--dibromo-1-*tert*-butyl-2,4,6-tri(trimethylsilyl)-1-silacyclohexa-3,5-diene (8). Bromination of 6 with NBS furnishes 4-bromo-1-methoxy-1-*tert*-butyl-2,4,6--tri(trimethylsilyl)-1-silacyclohexa-2,4-diene (9). 1-Methoxy-1-*tert*-butyl--2,4,4,6-tetra(trimethylsilyl)-1-silacyclohexa-2,4-diene (10), 1-methoxy-1,4--di(*tert*-butyl)-1-silacyclohexa-2,5-diene and related derivatives have been prepared (P. Jutzi and M. Meyer, Ber., 1988, 121, 1393). The metallation of 1-*tert*-butyl- and 1,1-dimethyl-1-silacyclohexa-3,5-dienes with BuLi has been discussed (P. Jutzi *et al.*, J. Amer. Chem. Soc., 1990, 112, 4841).

(6) $R^1 = OMe$, $R^2 = H$
(7) $R^1 = R^2 = H$
(8) $R^1 = R^2 = Br$

(9) $R^1 = OMe$, $R^2 = Br$
(10) $R^1 = OMe$, $R^2 = Me_3Si$

Pyrolysis of 1,1-dichloro-1-silacyclohexa-3,5- and -2,4-dienes (11 and 12) yields cyclopentadiene and $:SiCl_2$ (A.K. Mal'tsev *et al.*, Izv. Akad. Nauk SSSR Ser. Khim., 1989, 1051). For the preparation of 4,5-dimethyl-2,2--diphenyl-1,1-di(trimethylsilyl)-1-silacyclohex-4-ene (13), see J. Ohshita *et al.*, Organometallics, 1993, 12, 876; and the synthesis of sila-vitamin A1 (14) in five steps from sila-ß-ionone, R. Muenstedt and U. Wannagat, J. Organomet. Chem., 1987, 322, 11.

(11)

(12)

(13)

(14)

(ii) Silacyclohexane and derivatives. Cyclization of $ClCHRCH_2SiMe_2CH_2$-
$CH=CH_2$ (R=H, Me) at 80°C in benzene containing Bu_3SnH and AIBN
affords 1,1,3-tri- and 1,1,3,5-tetramethyl-1-silacyclohexanes (1; R=H and
R=Me, respectively) in 4-7% yields (K. Saigo *et al.*, J. Org. Chem., 1988,
53, 1572). Treatment of 1,5-dibromopentane with magnesium and then
with methyldichlorovinylsilane in boiling ether affords 1-methyl-1-vinyl-1-
-silacyclohexane (2) (R. Sugimoto, J. Takeda, and T. Asanuma, Jpn. Kokai
Tokkyo Koho JP 01 110,692 [89 110,692], 1989). Polymers have been
prepared from 1-vinyl-1-silacyclohexane or a derivative containing an alkyl
group on Si (*idem, ibid.*, 63 248,808 [88 248,808, 1988). The ring
cleavage of 1-silacyclohexanes 3 (R=H, D) with metal ions or iron, cobalt,
and nickel in gas-phase reactions has been investigated (A. Bjarnason and
I. Arnason, Org. MassSpectrom., 1993, **28**, 989).

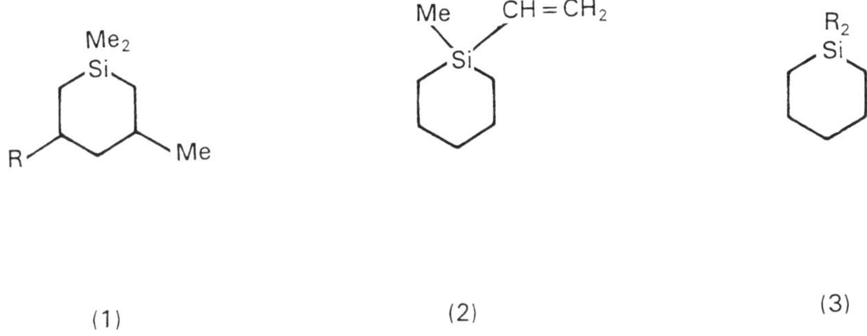

rac-1,1-Dimethyl-1-silacyclohexan-2-ol on esterification with acetic anhydride affords *rac*-2-acetoxy-1,1-dimethyl-1-silacyclohexane, which on enantioselective hydrolysis in acqueous solution catalyzed by a crude lipase preparation of *Candida cylindracea* lead to the formation of (*S*)-1,1-dimethyl--1-silacyclohexan-2-ol (95% ee) (K. Fritsche *et al.*, Appl. Microbiol. Biotechnol. 1989, 31, 107). 1,1-Dichloro-1-silacyclohexane reacts with $(NH_2NH)_2P(S)OPh$ in THF in the presence of NEt_3 to give a spirocyclic derivative (U. Engelhardt and M. Rosefid, Z. Anorg. Allg. Chem., 1994, 620, 620). For the ESR evidence for an asymmetrically distorted structure of the 1-methyl-1-silacyclohexane radical cation see, M. Shiotani *et al.*, Chem. Phys. Letters, 1992, 188, 93 and K. Komaguchi, M. Shiotani, and M. Ishikawa, Proceedings of the 6th Japan-China Bilateral Symposium on Radiation Chemistry, 1994, 527; the molecular structure of conformationally stable *trans*-1-(4-bromophenyl)-4-*tert*-butyl-1-methyl-1-silacyclohexane, H. Sakurai *et al.*, J. Organomet. Chem., 1988, 341, 133; and the ring enlargement of *cis*- and *trans*-1-silacyclopentanes containing a 1-chloromethyl substituent to furnish 1-silacyclohexanes, S.G. Cho, Bull. Korean Chem. Soc., 1994, 15, 183.

4-Dicyanomethylene-1-methyl-1-trimethylsilyl-1-silacyclohexane (4) and the related analogue 5 show intramolecular charge transfer bands in their absorption spectra (M. Kira *et al.*, Chem. Letters, 1988, 353).

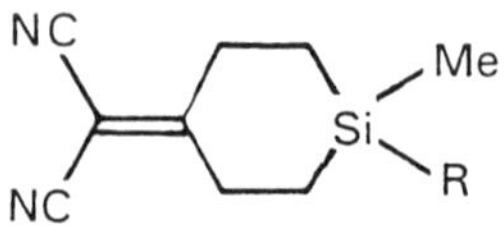

(4) R = SiMe$_3$
(5) R = SiMe$_2$SiMe$_3$

(iii) Polycyclic silanes with six-membered rings. The stereochemistry of
the reaction of some optically active organosilanes, including (-)-2-fluoro-2-
-(naphth-1-yl)-2-sila-1,2,3,4-tetrahydronaphthalene (1) with 4-RC$_6$H$_4$OM
(R = H, MeO, NO$_2$; M = Na, NBu$_4$) and allyllithium emphasizes the relative
influence of the attacking anion. For a given leaving group, the stereochemistry essentially depends upon the electronic character of the *para*
substituents and on the ion-pair dissociation of the phenoxides. This
dependence is also supported by the stereochemical behaviour of allyllithium.
The mechanism requires nucleophilic substitution at Si (R.J.P. Corriu and C.
Guerin, J. Organomet. Chem., 1982, <u>225</u>, 141).

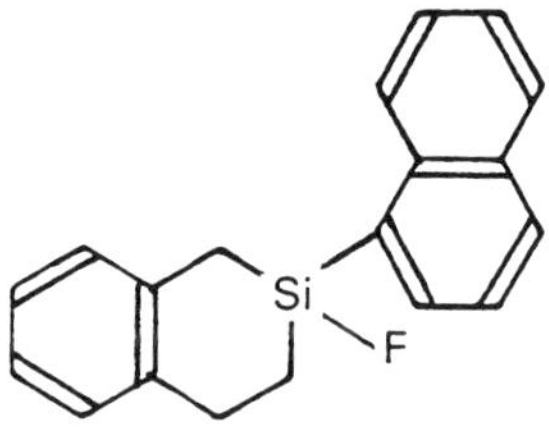

(1) 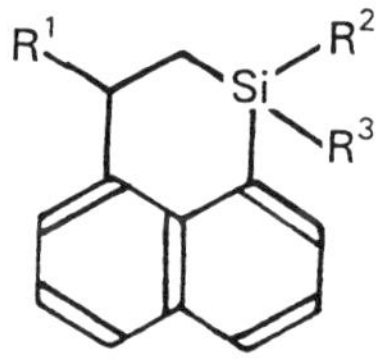   (2)

Heating naphth-1-ylmethylchlorosilane with $CH \equiv CH$ at 600°C affords 1-
-chloro-1-methyl-1-silaphenalene (2; $R^1 = H$, $R^2 = Cl$, $R^3 = Me$). Related
derivatives 2 ($R^1 = H$, Ph; $R^2 = Me$, Cl; $R^3 = H$, Cl, Me) have been prepared
and the crystal structure of 2 ($R^1 = Ph$; $R^2 = R^3 = Me$) has been determined
(E.A. Chernyshev *et al.*, Zh. Obshch. Khim., 1987, <u>57</u>, 1730).

(b) Germanium and tin compounds

(i) Germanium compounds. Although there is a sharp drop in aromati-
city from benzene to silabenzene, those of sila- and germabenzene are
almost the same. Another sharp drop is predicted from germa- to stanna-
benzene (K.K. Baldridge and M.S. Gordon, J. Amer. Chem. Soc., 1988, <u>110</u>,
4204). NMR spectral data and molecular mechanics calculations of germa-
cyclohexane and methyl substituted germacyclohexanes 1 ($R^1 = R^2 = H$,
$R^3 = 3$-Me, 4-Me; $R^1 = Me$, $R^2 = H$, $R^3 = H$, 3-Me, 4-Me; $R^1 = R^2 = Me$, $R^3 = H$)
reveal that the ring is flattened around the Ge and that a Me group on Ge
has a slight preference for the axial orientation rather than the equatorial,
which is in line with the results for the corresponding silacyclohexanes. The
^{13}C and ^{73}Ge chemical shifts of these germacyclohexanes are consistent
with the calculations. Thus, both ^{13}C and ^{73}Ge data indicate that 1-methyl-
-1-germacyclohexane is a 60:40 mixture of axial and equatorial isomers
(Y. Takeuchi *et al.*, J. Chem. Soc., Perkin 2, 1988, 7).

(1)

(ii) Tin compounds. 3-Allyl-1-stannacyclohexanes and 1-methyl-1-
-stannabicyclo[3.3.1]nonane have been prepared starting from (2-allylpent-4-
-enyl)tin hydrides. 1R, and ^{119}Sn and ^{13}C NMR spectral data have been
reported (F. Krech and K. Issleib, Z. Anorg. Allg. Chem., 1988, <u>557</u>, 143).

8,8-Dimethyl-8-stanna-4,7a,8,8a-tetrahydro-*sym*-indacenes 1 (R = H, Me) have been obtained by the reaction of Me_2SnCl_2 with the dilithium salts of $(C_5H_4)_2CR_2$. The crystal structure of 8,8-dimethyl-8-stanna-4,7a,8,8a-tetrahydro-*sym*-indacene (1; R = H) has been determined. The central stannacyclohexane moiety has a flattened chair conformation with $C(sp^3)$ atoms in the 2 and 6 positions (meso-form of the molecule in the crystal structure) and a tetrahedrally coordinated Sn atom (I.E. Nifant'ev *et al.*, Metalloorg. Khim., 1991, **4**, 292).

(1)

Guide to the Index

This index is constructed in a similar manner to the volume indexes of the first edition of the Chemistry of Carbon Compounds. However, to make the index easier to use, more descriptive entries have been made for the commonly occurring individual, and groups of chemicals.

The indexes cover primarily the chemical compounds mentioned in the text, and also include reactions and techniques, where named, and some sources of chemical compounds such as plant and animal species, oils, etc.

Chemical compounds have been indexed alphabetically under the names used by authors, editing being restricted to ensuring uniformity of entries under the same heading. In view of the alternative nomenclature that can often be used, a limited amount of cross-referencing has been done where it is considered to be helpful, but attention is particularly drawn to Convention 2 below.

For this and the succeeding volumes, the indexing conventions listed below have been adopted.

1. Alphabetisation

(a) A letter by letter alphabetical sequence is followed for entries, firstly for the main entry, followed by the descriptive entry.

(b) The following prefixes have not been counted for alphabetising:

n-	*o-*	*as-*	*meso-*	*C-*	*E-*
	m-	*sym-*	*cis-*	*O-*	*Z-*
	p-	*gem-*	*trans-*	*N-*	
	vic-			*S-*	
		lin-		*Bz-*	
				Py-	

Some prefixes and numbering have been omitted in the index, where they do not usefully contribute to the reference.

(c) The following prefixes have been alphabetised:

Allo	Epi	Neo
Anti	Hetero	Nor
Bis	Homo	Pseudo
Cyclo	Iso	

2. Cross references

In view of the many alternative trivial and systematic names for chemical compounds, the indexes should be searched under any alternative names which may be indicated in the main body of the text. Only a limited amount of cross-referencing has been carried out, where it is considered that it would be helpful to the user.

3. Derivatives

Simple derivatives are not normally indexed if they follow in the same short section of the text.

4. Collective and plural entries

In place of "– derivatives" the plural entry has normally been used. Plural entries have occasionally been used where compounds of the same name but differing numbering appear in the same section of the text.

5. Main entries

The main entry of the more common individual compounds is indicated by heavy type. Multiple entries, such as headings and sub-headings over several pages are shown by "–", e.g., 67–74, 137–139, etc.

Index